# Certified Information Systems Auditor

Invent Your Future. Get Certified!

# CISA Review Manual 2009

CISA®
CERTIFIED INFORMATION SYSTEMS AUDITOR™

## ISACA®

With more than 86,000 constituents in more than 160 countries, ISACA (*www.isaca.org*) is a recognized worldwide leader in IT governance, control, security and assurance. Founded in 1969, ISACA sponsors international conferences, publishes the *Information Systems Control Journal*®, and develops international information systems auditing and control standards. It also administers the globally respected Certified Information Systems Auditor™ (CISA®) designation, earned by more than 60,000 professionals since 1978; the Certified Information Security Manager® (CISM®) designation, earned by more than 9,000 professionals since 2002; and the new Certified in the Governance of Enterprise IT™ (CGEIT™) designation.

## IT Governance Institute®

The IT Governance Institute (ITGI™) (*www.itgi.org*) is a nonprofit, independent research entity that provides guidance for the global business community on issues related to the governance of IT assets. ITGI was established by the nonprofit membership association ISACA in 1998 to help ensure that IT delivers value and its risks are mitigated through alignment with enterprise objectives, IT resources are properly managed, and IT performance is measured. ITGI developed *Control Objectives for Information and related Technology* (COBIT®) and Val IT™, and offers original research and case studies to help enterprise leaders and boards of directors fulfill their IT governance responsibilities and help IT professionals deliver value-adding services.

## Disclaimer

ISACA has produced this publication as an educational resource to assist individuals preparing to take the CISA certification exam. It was produced independently from the CISA Certification Board, which has had no responsibility for its content. Copies of past examinations are not released to the public and were not made available to ISACA for preparation of this publication. ISACA makes no representations or warranties whatsoever with regard to these or other ISACA/ITGI publications assuring candidates' passage of the CISA examination.

## Reservation of Rights

## ISACA

3701 Algonquin Road, Suite 1010
Rolling Meadows, Illinois 60008 USA
Phone: +1.847.253.1545
Fax: +1.847.253.1443
Web site: *www.isaca.org*

ISBN 978-1-60420-046-1
*CISA Review Manual 2009*
Printed in the United States of America

# CISA REVIEW MANUAL 2009

ISACA is pleased to offer the 2009 (19th) edition of the *CISA Review Manual*. The purpose of this manual is to provide CISA candidates with updated technical information and references to assist in preparation and study for the Certified Information Systems Auditor examination.

The content in this manual has been based on the current CISA job practice which can be viewed at: *www.isaca.org/cisajobpractice*. As with previous manuals, the 2009 edition is the result of contributions from many qualified authorities who have generously volunteered their time and expertise. We respect and appreciate their contributions, and hope their efforts provide extensive educational value to CISA manual readers.

The *CISA Review Manual* is updated annually to keep pace with the rapid changes in the IT audit, control and security profession. As such, your comments and suggestions regarding this manual are welcomed. A feedback questionnaire is included at the back of the manual. After the examination is over, please take a moment to complete and return the questionnaire to ISACA. Your observations will be extremely valuable as future editions of the manual are prepared.

Members of ISACA's conference and education board and other subject matter experts reviewed the content of the *CISA Review Manual* for appropriateness. Sample questions contained in this manual are designed to assist the understanding of the material in this manual and to depict the type of questions typically found on the CISA examination.

Certification has provided a positive effect on many careers, and the CISA designation is respected and acknowledged by organizations around the world. We wish you success with the CISA examination. Your commitment to pursue the leading certification of IS audit, assurance, control and security skills is exemplary.

Lynn C. Lawton, CISA, FBCS CITP, FCA, FIIA, PIIA
International President
2008-2009

Henny J. Claessens, CISA, CISM
Chairman, Conference and Education Board
2008-2009

# ACKNOWLEDGMENTS

The 2009 edition of the *CISA Review Manual* is the result of the collective efforts of many volunteers. ISACA members from throughout the global IS audit, control and security profession participated, generously offering their talents and expertise. This international team exhibited a spirit and selflessness that has become the hallmark of contributors to this manual. Their participation and insight are truly appreciated. Special thanks go to Stephen Head, CISA, CISM, Manager of Technology Risk Management Services and IT Auditing Subject Matter Expert for Jefferson Wells, who served as technical content project leader and editor.

Listed below are the ISACA members who participated in the review of the *CISA Review Manual 2009*. All are deserving of our thanks and gratitude.

Sunil Bhaskar Bakshi, CISA, CISM
*Wipro Technologies*
*Pune Chapter, India*

David M. Cannon, CISA
*Grand Valley State University*
*Western Michigan Chapter, USA*

Salvador Juan Cárdenas Avendaño, CISA CISM
*Comisión Nacional del Sistema de Ahorro para el Retiro*
*Mexico City Chapter, Mexico*

David Cramm, CISA
*Public Works and Government Services Canada*
*Ottawa, Valley Chapter, Canada*

Faisal R. Danka
*London Chapter, UK*

Luis Díaz Villanueva
*Endesa*
*Madrid Chapter, Spain*

Gemma Deler-Castro, CISA
*Applus Servicios Tenológicos S.L.*
*Barcelona Chapter, Spain*

María Isabel Escobar Gavito
*Deloitte*
*Monterrey Chapter, Mexico*

Jacqueline Galeano, CISA
*Alta Tecnología, Consultrores Asociados*
*North Texas Chapter*

Fernando Hervada Vidal, CISA
*Endesa*
*Madrid Chapter, Spain*

John C. Heaton, CISA
*Deloitte & Touche LLP*
*Toronto Chapter, Canada*

Marcelo Héctor González, CISA
*Banco Central de la República Argentina*
*Buenos Aires Chapter, Argentina*

Alexander Hilgert, CISA
*Deloitte & Touche*
*German Chapter, Germany*

Thomas Hughes
*Canberra Chapter, Australia*

Florin Mihai Iliescu, CISA
*Info-Logica Siverline SRL*
*Romania Chapter, Romania*

Mohamed Iqbal Keeka
*RMB Private Bank*
*South Africa Chapter, South Africa*

Prof. Venugopal R. Iyengar, CISA, CISM
*IOTM*
*Mumbai Chapter, India*

Michael William Krasny, CISA
*Commonwealth Bank of Australia*
*Sydney Chapter, Australia*

Arun Laxminarayanan
*Cochin Chapter, India*

Villy Grove Madsen, CISA
*ATCO I-Tek*
*Edmonton Chapter, Alberta, Canada*

Dr. Derek Oliver, CISA, CISM
*London Chapter, UK*

Manuel Palao, CISA, CISM
*People & Technology: Solutions, SL*
*Madrid Chapter, Spain*

Corrado Pomodoro
*HSPI SPA*
*Milano Chapter, Italy*

María Patricia Prandini, CISA
*Oficina Nacional de Tecnologías de Información*
*Buenos Aires Chapter, Argentina*

Abdul Rafeq, CISA
*A. Rafeq & Associates*
*Bangalore Chapter, India*

Sree Krishna Rao, CISA
*Lovelock & Lewes*
*Bangalore Chapter, India*

Angelo Rodaro, CISA
*Edison S.p.A.*
*Milan Chapter, Italy*

María del Pilar Rodríguez Meriño
*PWC*
*Bogota Chapter, Colombia*

Markus Schiemer, CISA
*UNISYS Österreich GMBH*
*Austria Chapter, Austria*

Yuene C. Shure
*CALSTRS*
*Sacramento Chapter, California, USA*

Dr. Katalin Szenes, CISA, CISM
*Budapest Chapter, Hungary*

Johann Tello Meryk, CISA, CISM
*Primer Banco del Istmo*
*Panama Chapter, Panama*

Dr. Smita Dilip Totade, CISA, CISM
*National Insurance Academy*
*Pune Chapter, India*

Marina Touriño Troitiño, CISA, CISM
*Marina Touriño & Asociados, S.L.*
*Madrid Chapter, Spain*

Carlos E. Villamizar Rodriguez, CISA
*Protela S.A.*
*Bogota Chapter, Colombia*

Jeung-Mo Wang
*Toronto Chapter, Canada*

Dr. Edgar R. Weippl
*Security Research*
*Austria Chapter, Austria*

Kurt Weirich
*University of Arizona*
*Phoenix Chapter, USA*

Hui Zhu, CISA
*PricewaterhouseCoopers LLP*
*Toronto Chapter, Canada*

ISACA has begun planning for the 2010 edition of the *CISA Review Manual*. Volunteer participation drives the success of the manual. If you are interested in becoming a member of the select group of professionals involved in this global project, we want to hear from you.

Susan M. Caldwell
Chief Executive Officer

Terence J. Trsar
Chief Professional Development Officer

## Chapter 4:

# IT Service Delivery and Support

## Chapter 5:

# Protection of Information Assets

## Chapter 6:

# Business Continuity and Disaster Recovery

# ABOUT THIS MANUAL

## OVERVIEW

The *CISA Review Manual 2009* is intended to assist candidates in preparing for the CISA exam. **The manual is one source of preparation for the exam but should not be thought of as the only source nor should it be viewed as a comprehensive collection of all the information and experience that is required to pass the exam.** No single publication offers such coverage and detail.

As candidates read through the manual and encounter a topic that is new to them or one in which they feel their knowledge and experience is limited, additional references should be sought. The exam will be a combination of questions testing candidates' technical and practical knowledge, and their ability to apply the knowledge (based on experience) in given situations.

## FORMAT OF THIS MANUAL

The *CISA Review Manual 2009* provides coverage of the knowledge and activities related to the various functions associated with the content areas as detailed in the CISA job practice as set out in the *Candidates's Guide to the CISA Exam and Certification*.

Each of these areas is tested in the exam in the percentages listed below.

| Area 1 | The IS Audit Process | 10 percent |
|--------|----------------------|------------|
| Area 2 | IT Governance | 15 percent |
| Area 3 | Systems and Infrastructure Life Cycle Management | 16 percent |
| Area 4 | IT Service Delivery and Support | 14 percent |
| Area 5 | Protection of Information Assets | 31 percent |
| Area 6 | Business Continuity and Disaster Recovery | 14 percent |

The objective of each of these areas appears at the beginning of each chapter with the corresponding tasks and knowledge statements which are tested in the exam.

The manual has been developed and organized to assist in the study of these areas. Exam candidates should evaluate their strengths, based on knowledge and experience, in each of these areas. Each chapter consists of four parts: text, case study, practice questions and reference material.

The structure of the content includes numbering to identify first the chapter where a topic is located and then headings of the subsequent levels of topics addressed in the chapter (i.e., 2.6.3 Risk Analysis Methods, is a subtopic of Risk Management in chapter 2). Relevant content in a subtopic is bolded for specific attention.

One's understanding of the textual material is a barometer of one's knowledge, strengths and weaknesses, and is an indication of areas in which one needs to seek reference sources over and above this manual. However, written material is not a substitute for experience. **Actual exam questions will test the candidate's practical application of this knowledge.** Each chapter's case studies present a situation within the profession in the area of study. The case studies include a scenario with topics addressed in the chapter and also include practice questions which assist in understanding how a CISA question could be presented at the CISA exam. The practice questions at the end of each chapter also serve this purpose and should not be used independently as a source of knowledge. Practice questions should not be considered a measurement of one's ability to answer questions correctly in the exam for that area. The questions are intended to familiarize candidates with question structure and general content, and may or may not be similar to questions that will appear on the actual exam. The reference material includes other publications that could be used to further acquire and better understand detailed information on the topics addressed in the manual.

At the end of each chapter is a glossary which contains terms that apply to the material included in that chapter. The glossary includes both terms that are discussed in the text and terms that apply to the different areas but may not have been specifically discussed. The glossary is an extension of the text in the manual and can, therefore, be another indication of areas in which candidates may need to seek additional references.

Throughout the manual, the word "association" refers to ISACA, formerly known as Information Systems Audit and Control Association, and the "institute" or the acronym "ITGI" refers to the IT Governance Institute. Also, please note that the manual has been written using standard American English.

> **Note:** The *CISA Review Manual* is a living document. As technology advances, the manual will be updated to reflect such advances. Further updates to this document before the date of the exam may be viewed at *www.isaca.org/studyaidupdates*.

Suggestions to enhance the materials covered herein, or suggested reference materials, should be directed to:

Mail: ISACA
3701 Algonquin Road, Suite 1010
Rolling Meadows, Illinois 60008 USA
Attention: Manager—Certification Study Program and Educational Development

Phone: +1.847.253.1545
Fax: +1.847.253.1443
E-mail: *education@isaca.org*

# ABOUT THE CISA REVIEW QUESTIONS, ANSWERS AND EXPLANATION MANUAL

Candidates may also wish to enhance their study and preparation for the exam by using the *CISA Review Questions, Answers & Explanations Manual 2008*, and the *2008 and 2009 Supplements*.

The *CISA Review Questions, Answers Explanations Manual 2008* consists of 600 multiple-choice study questions, answers and explanations arranged in the areas of the current CISA job practice. Many of these items appeared in previous editions of the *CISA Review Questions, Answers & Explanations Manual*, but have been rewritten to be in line with current practice, more representative of actual exam items and to provide further clarity.

The *2008 and 2009 Supplements* are the result of ISACA's dedication each year to create 100 new sample questions, answers and explanations for candidates to use in preparation for the CISA exam. Each year, ISACA develops 100 new review questions, using a strict process of review similar to that performed for the selection of questions for the CISA exam by the CISA Certification Board. In the *2008 and 2009 Supplements*, the questions have been arranged in the proportions of the most recent CISA job practice analysis.

Another study aid that is available is the CISA Review Questions, Answers & Explanations CD-ROM 2009. It consists of the 800 questions, answers and explanations included in the *CISA Review Questions, Answers & Explanations Manual 2008*, and the *2008 and 2009 Supplements*. With this product, CISA candidates can identify strengths and weaknesses by taking various length random sample exams and breaking the results down by area. Sample exams also can be chosen by area, allowing for concentrated study, one area at a time, and other sort features such as the omission of previous correctly answered questions.

Questions in these publications are representative of the types of questions that have appeared on the exam and include an explanation of the correct and incorrect answers. Questions are sorted by the CISA content areas and as a sample test. These publications are ideal for use in conjunction with the *CISA Review Manual 2009*. These manuals can be used as study sources throughout the study process or as part of a final review to determine where a candidate may need additional study. It should be noted that these questions and suggested answers are provided as examples; they are not actual questions from the exam and may differ in content from those that actually appear on the exam.

# CISA ONLINE REVIEW COURSE

The CISA Online Review Course is a completely web-based, self-paced study tool. There are no hard copy materials (books, study manuals, etc.) provided with the course. While it is significantly different in terms of how the information is delivered, the course is based off of the *CISA Review Manual 2009*. The course does include over 160 practice questions as well as interactive activities and exercises and an online glossary to reinforce content comprehension.

If you purchase the complete set of 6 modules, the subscription length is 150 days. The subscription length for individual modules is 90 days. You may view the course and take the CPE quizzes as often as you wish during your subscription period. Once your subscription period has expired, you will no longer have access to the course materials. Because the course is completely web-based, you may access the course from multiple locations without difficulty.

Upon purchase of any module, you will receive an e-mail confirming your purchase. Please save this e-mail as it will serve as your proof of purchase. No further invoices will be provided.

To better evaluate whether this is an appropriate study tool for you, please view the course demonstration at *http://company.vcampus.com/isaca/cisa_demo/*. To register for the course, please go to *www.isaca.org/elearningcampus*.

*Chapter 1:*

# The IS Audit Process

Page intentionally left blank

# The IS Audit Process

# 1.1 INTRODUCTION

The IS Audit Process is a chapter that encompasses the entire practice of IS auditing, including procedures and a thorough methodology which allows an IS auditor to perform an audit on any given IT area in a professional manner.

## 1.1.1 OBJECTIVE

The objective of this area is to ensure that the CISA candidate has the knowledge necessary to provide information systems (IS) audit services in accordance with IS audit standards, guidelines and best practices to ensure that an organization's information technology and business systems are protected and controlled.

> **This area represents 10 percent of the CISA examination (approximately 20 questions).**

## 1.1.2 TASKS

There are five (5) tasks within the IS audit process area:

T1.1    Develop and implement a risk-based IS audit strategy for the organization in compliance with IS audit standards, guidelines and best practices.

T1.2    Plan specific audits to ensure that IT and business systems are protected and controlled.

T1.3    Conduct audits in accordance with IS audit standards, guidelines and best practices to meet planned audit objectives.

T1.4    Communicate emerging issues, potential risks and audit results to key stakeholders.

T1.5    Advise on the implementation of risk management and control practices within the organization while maintaining independence.

## 1.1.3 KNOWLEDGE STATEMENTS

There are ten (10) knowledge statements within the IS audit process area:

KS1.1    Knowledge of ISACA IS Auditing Standards, Guidelines and Procedures and Code of Professional Ethics

KS1.2    Knowledge of IS auditing practices and techniques

KS1.3    Knowledge of techniques to gather information and preserve evidence (e.g., observation, inquiry, interview, computer-assisted audit techniques [CAATs], electronic media)

KS1.4    Knowledge of the evidence life cycle (e.g., collection, protection, chain of custody)

KS1.5    Knowledge of control objectives and controls related to IS (e.g., COBIT)

KS1.6    Knowledge of risk assessment in an audit context

KS1.7    Knowledge of audit planning and management techniques

KS1.8    Knowledge of reporting and communication techniques (e.g., facilitation, negotiation, conflict resolution)

KS1.9    Knowledge of control self-assessment (CSA)

KS1.10  Knowledge of continuous audit techniques

## 1.1.4 RELATIONSHIP OF TASK TO KNOWLEDGE STATEMENTS

The task statements are what the CISA candidate is expected to know how to perform. The knowledge statements delineate what the CISA candidate is expected to know in order to perform the tasks. The task and knowledge statements are approximately mapped in **exhibit 1.1** insofar as it is possible to do so. Note that although there is often overlap, each task statement will generally map to several knowledge statements.

| Exhibit 1.1—Tasks and Knowledge Statements Mapping | |
|---|---|
| **Task Statements** | **Knowledge Statements** |
| T1.1 Develop and implement a risk-based IS audit strategy for the organization in compliance with IS audit standards, guidelines and best practices. | KS1.1 Knowledge of ISACA IS Auditing Standards, Guidelines and Procedures and Code of Professional Ethics<br>KS1.5 Knowledge of control objectives and controls related to IS (e.g., CoBiT)<br>KS1.6 Knowledge of risk assessment in an audit context<br>KS1.7 Knowledge of audit planning and management techniques<br>KS1.9 Knowledge of control self-assessment (CSA)<br>KS1.10 Knowledge of continuous audit techniques |
| T1.2 Plan specific audits to ensure that IT and business systems are protected and controlled. | KS1.1 Knowledge of ISACA IS Auditing Standards, Guidelines and Procedures and Code of Professional Ethics<br>KS1.2 Knowledge of IS auditing practices and techniques<br>KS1.3 Knowledge of techniques to gather information and preserve evidence (e.g., observation, inquiry, interview, computer-assisted audit techniques [CAATs], electronic media)<br>KS1.7 Knowledge of audit planning and management techniques<br>KS1.10 Knowledge of continuous audit techniques |
| T1.3 Conduct audits in accordance with IS audit standards, guidelines and best practices to meet planned audit objectives. | KS1.1 Knowledge of ISACA IS Auditing Standards, Guidelines and Procedures and Code of Professional Ethics<br>KS1.2 Knowledge of IS auditing practices and techniques<br>KS1.3 Knowledge of techniques to gather information and preserve evidence (e.g., observation, inquiry, interview, computer-assisted audit techniques [CAATs], electronic media)<br>KS1.4 Knowledge of the evidence life cycle (e.g., collection, protection, chain of custody)<br>KS1.5 Knowledge of control objectives and controls related to IS (e.g., CoBiT)<br>KS1.7 Knowledge of audit planning and management techniques<br>KS1.10 Knowledge of continuous audit techniques |
| T1.4 Communicate emerging issues, potential risks and audit results to key stakeholders. | KS1.1 Knowledge of ISACA IS Auditing Standards, Guidelines and Procedures and Code of Professional Ethics<br>KS1.5 Knowledge of control objectives and controls related to IS (e.g., CoBiT)<br>KS1.8 Knowledge of reporting and communication techniques (e.g., facilitation, negotiation, conflict resolution)<br>KS1.10 Knowledge of continuous audit techniques |
| T1.5 Advise on the implementation of risk management and control practices within the organization while maintaining independence. | KS1.1 Knowledge of ISACA IS Auditing Standards, Guidelines and Procedures and Code of Professional Ethics<br>KS1.5 Knowledge of control objectives and controls related to IS (e.g., CoBiT)<br>KS1.6 Knowledge of risk assessment in an audit context<br>KS1.9 Knowledge of control self-assessment (CSA)<br>KS1.10 Knowledge of continuous audit techniques |

# 1.2 MANAGEMENT OF THE IS AUDIT FUNCTION

The audit function should be managed and led in a manner that ensures that the diverse tasks performed and achieved by the audit team will fulfill audit function objectives, while preserving audit independence and competence. Furthermore, managing the audit function should ensure value added contributions to senior management regarding the efficient management of IT and achievement of business objectives.

## 1.2.1 ORGANIZATION OF THE IS AUDIT FUNCTION

IS audit services can be provided externally or internally.

The role of the IS internal audit function should be established by an audit charter. IS audit can be a part of internal audit, function as an independent group, or integrated within a financial and operational audit (see **exhibit 1.7**) to provide IT-related control assurance to the financial or management auditors. Therefore, the audit charter may include IS audit as an audit support function. The charter should clearly state management's responsibility and objectives for, and delegation of authority to, the IS audit function. This document should outline the overall authority, scope and responsibilities of the audit function. The highest level of management and the audit committee, if available, should approve this charter. Once established, this charter should be changed only if the change can be and is thoroughly justified. ISACA IS Auditing Standards require that the responsibility, authority and accountability of the IS audit function are appropriately documented in an audit charter or engagement letter (S1 Audit Charter). It should be noted that an audit charter is an overarching document that covers the entire scope of audit activities in an entity while an engagement letter is more focused on a particular audit exercise that is sought to be initiated in an organization with a specific objective in mind.

If IS audit services are provided by an external firm, the scope and objectives of these services should be documented in a formal contract or statement of work between the contracting organization and the service provider.

In either case, the internal audit function should be independent and report to an audit committee, if available, or to the highest management level such as the board of directors.

## 1.2.2 IS AUDIT RESOURCE MANAGEMENT

IS technology is constantly changing. Therefore, it is important that IS auditors maintain their competency through updates of existing skills and obtain training directed toward new audit techniques and technological areas. ISACA IS Auditing Standards require that the IS auditor be technically competent (S4 Professional Competence), having the skills and knowledge necessary to perform the auditor's work. Further, the IS auditor is to maintain technical competence through appropriate continuing professional education. Skills and knowledge should be taken into consideration when planning audits and assigning staff to specific audit assignments.

Preferably, a detailed staff training plan should be drawn for the year based on the organization's direction in terms of technology and related risk issues that need to be addressed. This should be reviewed periodically to ensure that the training needs are aligned to the direction that the audit organization is taking. Additionally, IS audit management should also provide the necessary IT resources to properly perform IS audits of a highly specialized nature (e.g., software, scanners for network vulnerability tests, penetration testing).

## 1.2.3 AUDIT PLANNING

### *Annual Planning*
Audit planning consists of both short- and long-term planning. Short-term planning takes into account audit issues that will be covered during the year, whereas long-term planning relates to audit plans that will take into account risk-related issues regarding changes in the organization's IT strategic direction that will affect the organization's IT environment.

# The IS Audit Process

Analysis of short- and long-term issues should occur at least annually. This is necessary to take into account new control issues; changes in the risk environment, technologies and business processes; and enhanced evaluation techniques. The results of this analysis for planning future audit activities should be reviewed by senior audit management and approved by the audit committee, if available, or alternatively by the board of directors and communicated to relevant levels of management.

## *Individual Audit Assignments*

In addition to overall annual planning, each individual audit assignment must be adequately planned. The IS auditor should understand that other considerations, such as the results of periodic risk assessments, changes in the application of technology, and evolving privacy issues and regulatory requirements, may impact the overall approach to the audit. The IS auditor should also take into consideration system implementation/upgrade deadlines, current and future technologies, requirements from business process owners, and IS resource limitations.

When planning an audit, the IS auditor must have an understanding of the overall environment under review. This should include a general understanding of the various business practices and functions relating to the audit subject, as well as the types of information systems and technology supporting the activity. For example, the IS auditor should be familiar with the regulatory environment in which the business operates.

To perform audit planning, the IS auditor should perform the following steps:
• Gain an understanding of the business's mission, objectives, purpose and processes, which include information and processing requirements such as availability, integrity, security and business technology, and information confidentiality.
• Identify stated contents such as policies, standards and required guidelines, procedures, and organization structure.
• Perform a risk analysis to help in designing the audit plan.
• Conduct a review of internal controls related to IT.
• Set the audit scope and audit objectives.
• Develop the audit approach or audit strategy.
• Assign personnel resources to the audit.
• Address engagement logistics.

ISACA IS Auditing Standards require the IS auditor to plan the IS audit work to address the audit objectives and comply with applicable professional auditing standards (S5 Planning). The IS auditor should develop an audit plan that takes into consideration the objectives of the auditee relevant to the audit area and its technology infrastructure. Where appropriate, the IS auditor should also consider the area under review and its relationship to the organization (strategically, financially and/or operationally) and obtain information on the strategic plan, including the IS strategic plan. The IS auditor should have an understanding of the auditee's information technology architecture and technological direction to design a plan appropriate for the present and, where appropriate, future technology of the auditee.

Steps an IS auditor could take to gain an understanding of the business include:
• Touring key organization facilities
• Reading background material including industry publications, annual reports and independent financial analysis reports
• Reviewing business and IT long-term strategic plans
• Interviewing key managers to understand business issues
• Reviewing prior audit reports or IT-related reports (from external or internal audits, or specific reviews such as regulatory reviews)
• Identifying specific regulations applicable to IT
• Identifying IT functions or related activities that have been outsourced

Another basic component of planning is the matching of available audit resources to the tasks as defined in the audit plan. The IS auditor who prepares the plan should consider the requirements of the audit project, staffing resources and other constraints. This matching exercise should consider the needs of individual audit projects as well as the overall needs of the audit department.

# 1.2.4 EFFECT OF LAWS AND REGULATIONS ON IS AUDIT PLANNING

Each organization, regardless of its size or the industry within which it operates, will need to comply with a number of governmental and external requirements related to computer system practices and controls and to the manner in which computers, programs and data are stored and used. Additionally, business regulations can impact the way data are processed, transmitted and stored (stock exchange, central banks, etc.)

Special attention should be given to these issues in those industries that, historically, have been closely regulated. For example, the banking industry worldwide has severe penalties for companies and their officers should a company be unable to provide an adequate level of service because of substandard backup and recovery procedures. Also, Internet service providers (ISPs) are subject, in several countries, to specific laws regarding confidentiality and service availability.

IS auditors should review management's privacy policy to ascertain whether it takes into account the requirements of applicable privacy laws and regulations, including transborder data flow requirements such as Safe Harbor and the Organization for Economic Cooperation and Development (OECD) guidelines governing the protection of privacy and transborder flows of personal data.

Several countries, because of growing dependencies on information systems and related technology, are making efforts to establish added layers of regulatory requirements concerning IS audit. The contents of these legal regulations regard:
• Establishment of the regulatory requirements
• Organization of the regulatory requirements
• Responsibilities assigned to the corresponding entities
• Correlation to financial, operational and IT audit functions

Management personnel as well as audit management, at all levels, should be aware of the external requirements relevant to the goals and plans of the organization, and to the responsibilities and activities of the information services department/function/activity.

There are two major areas of concern: legal requirements (laws, regulatory and contractual agreements) placed on audit or IS audit, and legal requirements placed on the auditee and its systems, data management, reporting, etc. These areas would impact audit scope and audit objectives. The latter is important to internal and external auditors. Legal issues also impact the organizations' business operations in terms of compliance with ergonomic regulations, the US Health Insurance Portability and Accountability Act (HIPAA), Protection of Personal Data Directives and Electronic Commerce within the European Community, fraud prevention within banking organizations, etc.

An example of strong control practices is the US Sarbanes-Oxley Act of 2002, which requires evaluating an organization's IT controls. Sarbanes-Oxley provides for new corporate governance rules, regulations and standards for specified public companies including US Securities and Exchange Commission (SEC) registrants. The SEC has mandated the use of a recognized internal control framework. Sarbanes-Oxley requires organizations to select and implement a suitable internal control framework. The *Internal Control—Integrated Framework* from the Committee of Sponsoring Organizations of the Treadway Commission (COSO) has become the most commonly adopted framework by public companies seeking to comply. Since the US Sarbanes-Oxley Act has as its objective increasing the level of control of business processes and the information systems supporting them, the IS auditors have to consider the impact of Sarbanes-Oxley as part of audit planning.

A similar example is the Basel II Accord which regulates the minimum amount of capital for financial organizations based on the level of risk faced by these organizations. The Basel II Committee on Banking Supervision recommends conditions that should be fulfilled, in addition to capital requirements, to manage credit exposures. These conditions will ideally result in an improvement of:
• Credit risk management
• Operational risk management
• The management of information systems through clearly defined requirements

The following are steps an IS auditor would perform to determine an organization's level of compliance with external requirements:
- Identify those government or other relevant external requirements dealing with:
  – Electronic data, personal data, copyrights, e-commerce, e-signatures, etc.
  – Computer system practices and controls
  – The manner in which computers, programs and data are stored
  – The organization or the activities of information technology services
  – IS audits
- Document pertinent laws and regulations.
- Assess whether the management of the organization and the IS function have considered the relevant external requirements in making plans and in setting policies, standards and procedures, as well as business application features.
- Review internal IS department/function/activity documents that address adherence to laws applicable to the industry.
- Determine adherence to established procedures that address these requirements.
- Determine if there are procedures in place to ensure contracts or agreements with external IT services providers reflect any legal requirements related to responsibilities.

It is expected that the organization would have a legal compliance function on which the IS control practitioner could rely.

> **Note:** A CISA candidate will not be asked about any specific laws or regulations, but may be questioned about how one would audit for compliance with laws and regulations. The examination will only test knowledge of accepted global practices.

# 1.3 ISACA IS AUDITING STANDARDS AND GUIDELINES

## 1.3.1 ISACA CODE OF PROFESSIONAL ETHICS

ISACA sets forth this Code of Professional Ethics to guide the professional and personal conduct of members of the association and/or its certification holders.

Members and ISACA certification holders shall:
1. Support the implementation of, and encourage compliance with, appropriate standards, procedures and controls for information systems.
2. Perform their duties with objectivity, due diligence and professional care, in accordance with professional standards and best practices.
3. Serve in the interest of stakeholders in a lawful and honest manner, while maintaining high standards of conduct and character, and not engage in acts discreditable to the profession.
4. Maintain the privacy and confidentiality of information obtained in the course of their duties unless disclosure is required by legal authority. Such information shall not be used for personal benefit or released to inappropriate parties.
5. Maintain competency in their respective fields and agree to undertake only those activities that they can reasonably expect to complete with professional competence.
6. Inform appropriate parties of the results of work performed, revealing all significant facts known to them.
7. Support the professional education of stakeholders in enhancing their understanding of IS security and control.

Failure to comply with this Code of Professional Ethics can result in an investigation into a member's and/or certification holder's conduct and, ultimately, in disciplinary measures.

> **Note:** A CISA candidate is not expected to have memorized the ISACA IS Auditing Standards, Guidelines and Procedures and the ISACA Code of Professional Ethics, word for word. Rather, the candidates will be tested on their understanding of the standard, guideline or code, its objectives and how it applies in a given situation.

## 1.3.2 ISACA IS AUDITING STANDARDS FRAMEWORK

The specialized nature of IS auditing and the skills and knowledge necessary to perform such audits require globally applicable standards that pertain specifically to IS auditing. One of the most important functions of ISACA is providing information (common body of knowledge) to support knowledge requirements. (See standard S4 Professional Competence.)

One of ISACA's goals is to advance standards to meet this need. The development and dissemination of the IS Auditing Standards is a cornerstone of the association's professional contribution to the audit community. The IS auditor needs to be aware that there may be additional standards, or even legal requirements through legislation, placed on the auditor.

The objectives of the ISACA IS Auditing Standards are to inform:
• IS auditors of the minimum level of acceptable performance required to meet the professional responsibilities set out in the Code of Professional Ethics for IS auditors
• Management and other interested parties of the profession's expectations concerning the work of audit practitioners
• Holders of the Certified Information Systems Auditor™ (CISA®) designation of requirements that failure to comply with these standards may result in an investigation into the CISA holder's conduct by the ISACA Board of Directors or appropriate ISACA committee and, ultimately, in disciplinary action.

The framework for the ISACA IS Auditing Standards provides for multiple levels as follows:
• Standards define mandatory requirements for IS auditing and reporting.
• Guidelines provide guidance in applying IS Auditing Standards. The IS auditor should consider them in determining how to achieve implementation of the above standards, use professional judgment in their application and be prepared to justify any difference.
• Procedures provide examples of processes an IS auditor might follow in an audit engagement. The procedure documents provide information on how to meet the standards when completing IS auditing work, but do not set requirements.

The ISACA Code of Professional Ethics requires members of ISACA and holders of the CISA designation to comply with the IS Auditing Standards adopted by ISACA. Apparent failure to comply with these may result in an investigation into the member's or CISA holder's conduct by ISACA or an appropriate ISACA board or committee. Disciplinary action may ensue.

> **Note:** See Appendix B on IS Auditing Standards, Guidelines and Procedures. The complete text of these guidelines and procedures is available at *www.isaca.org/standards*.

### *Auditing Standards*
The IS Auditing Standards applicable to IS auditing are:
• **S1 Audit Charter:**
  – The purpose, responsibility, authority and accountability of the IS audit function or IS audit assignments should be appropriately documented in an audit charter or engagement letter.
  – The audit charter or engagement letter should be agreed on and approved at an appropriate level within the organization(s).
• **S2 Independence:**
  – *Professional independence*—In all matters related to the audit, the IS auditor should be independent of the auditee in both attitude and appearance.
  – *Organizational independence*—The IS audit function should be independent of the area or activity being reviewed to permit objective completion of the audit assignment.
• **S3 Professional Ethics and Standards:**
  – The IS auditor should adhere to the ISACA Code of Professional Ethics.
  – The IS auditor should exercise due professional care, including observance of applicable professional auditing.
• **S4 Professional Competence:**
  – The IS auditor should be professionally competent, having the skills and knowledge to conduct the audit assignment.
  – The IS auditor should maintain professional competence through appropriate continuing professional education and training.

- **S5 Planning:**
  – The IS auditor should plan the information systems audit coverage to address the audit objectives and comply with applicable laws and professional auditing standards.
  – The IS auditor should develop and document a risk-based audit approach.
  – The IS auditor should develop and document an audit plan detailing the nature and objectives, timing, extent and resources required.
  – The IS auditor should develop an audit program and procedures.
- **S6 Performance of Audit Work:**
  – Supervision—IS audit staff should be supervised to provide reasonable assurance that audit objectives are accomplished and applicable professional auditing standards are met.
  – Evidence—During the course of the audit, the IS auditor should obtain sufficient, reliable and relevant evidence to achieve the audit objectives. The audit findings and conclusions are to be supported by appropriate analysis and interpretation of this evidence.
  – Documentation—The audit process should be documented, describing the audit work and the audit evidence that supports the IS auditor's findings and conclusions.
- **S7 Reporting:**
  – The IS auditor should provide a report, in an appropriate form, upon completion of the audit. The report should identify the organization, the intended recipients and any restrictions on circulation.
  – The audit report should state the scope, objectives, period of coverage and the nature, timing and extent of the audit work performed.
  – The report should state the findings, conclusions and recommendations, and any reservations, qualifications or limitations in scope that the IS auditor has with respect to the audit.
  – The IS auditor should have sufficient and appropriate audit evidence to support the results reported.
  – When issued, the IS auditor's report should be signed, dated and distributed according to the terms of the audit charter or engagement letter.
- **S8 Follow-up Activities:**
  – After the reporting of findings and recommendations, the IS auditor should request and evaluate relevant information to conclude whether appropriate action has been taken by management in a timely manner.
- **S9 Irregularities and Illegal Acts:**
  – In planning and performing the audit to reduce audit risk to a low level, the IS auditor should consider the risk of irregularities and illegal acts.
  – The IS auditor should maintain an attitude of professional skepticism during the audit, recognizing the possibility that material misstatements due to irregularities and illegal acts could exist, irrespective of his/her evaluation of the risk of irregularities and illegal acts.
  – The IS auditor should obtain an understanding of the organization and its environment, including internal controls.
  – The IS auditor should obtain sufficient and appropriate audit evidence to determine whether management or others within the organization have knowledge of any actual, suspected or alleged irregularities and illegal acts.
  – When performing audit procedures to obtain an understanding of the organisation and its environment, the IS auditor should consider unusual or unexpected relationships that may indicate a risk of material misstatements due to irregularities and illegal acts.
  – The IS auditor should design and perform procedures to test the appropriateness of internal control and the risk of management overriding controls.
  – When the IS auditor identifies a misstatement, the IS auditor should assess whether such a misstatement may be indicative of an irregularity or illegal act. If there is such an indication, the IS auditor should consider the implications in relation to other aspects of the audit and, in particular, the representations of management.
  – The IS auditor should obtain written representations from management at least annually, or more frequently, depending on the audit engagement. They should:
    - Acknowledge its responsibility for the design and implementation of internal controls to prevent and detect irregularities or illegal acts.
    - Disclose to the IS auditor the results of the risk assessment that a material misstatement may exist as a result of an irregularity or illegal act.

- Disclose to the IS auditor its knowledge of irregularities or illegal acts affecting the organization in relation to:
  - Management
  - Employees who have significant roles in internal control.
  - Disclose to the IS auditor its knowledge of any allegations of irregularities or illegal acts, or suspected irregularities or illegal acts, affecting the organization as communicated by employees, former employees, regulators and others.
- If the IS auditor has identified a material irregularity or illegal act, or obtains information that a material irregularity or illegal act may exist, the IS auditor should communicate these matters to the appropriate level of management in a timely manner.
- If the IS auditor has identified a material irregularity or illegal act involving management or employees who have significant roles in internal control, the IS auditor should communicate these matters in a timely manner to those charged with governance.
- The IS auditor should advise the appropriate level of management and those charged with governance of material weaknesses in the design and implementation of internal control to prevent and detect irregularities and illegal acts that may have come to the IS auditor's attention during the audit.
- If the IS auditor encounters exceptional circumstances, such as a material misstatement or illegal act, that affect the IS auditor's ability to continue performing the audit, the IS auditor should consider the legal and professional responsibilities applicable in the circumstances, including whether there is a requirement for the IS auditor to report to those who entered into the engagement or, in some cases, those charged with governance or regulatory authorities, or consider withdrawing from the engagement.
- The IS auditor should document all communications, planning, results, evaluations and conclusions relating to material irregularities and illegal acts that have been reported to management, those charged with governance, regulators and others.

- **S10 IT Governance:**
  - The IS auditor should review and assess whether the IS function aligns with the organization's mission, vision, values, objectives and strategies.
  - The IS auditor should review whether the IS function has a clear statement about the performance expected by the business (effectiveness and efficiency) and assess its achievement.
  - The IS auditor should review and assess the effectiveness of IS resource and performance management processes.
  - The IS auditor should review and assess compliance with legal, environmental and information quality, and fiduciary and security requirements.
  - A risk-based approach should be used by the IS auditor to evaluate the IS function.
  - The IS auditor should review and assess the control environment of the organization.
  - The IS auditor should review and assess the risks that may adversely affect the IS environment.
- **S11 Use of Risk Assessment in Audit Planning:**
  - The IS auditor should use an appropriate risk assessment technique or approach in developing the overall IS audit plan and determining priorities for the effective allocation of IS audit resources.
  - When planning individual reviews, the IS auditor should identify and assess risks relevant to the area under review.
- **S12 Audit Materiality**
  - The IS auditor should consider audit materiality and its relationship to audit risk while determining the nature, timing and extent of audit procedures.
  - While planning for audit, the IS auditor should consider potential weakness or absence of controls, and whether such weakness or absence of controls could result in a significant deficiency or a material weakness in the information system.
  - The IS auditor should consider the cumulative effect of minor control deficiencies or weaknesses and absence of controls to translate into significant deficiency or material weakness in the information system.
  - The report of the IS auditor should disclose ineffective controls or absence of controls, and the significance of the control deficiencies and possibility of these weaknesses resulting in a significant deficiency or material weakness.
- **S13 Using the Work of Other Experts**
  - The IS auditor should, where appropriate, consider using the work of other experts for the audit.
  - The IS auditor should assess and be satisfied with the professional qualifications, competencies, relevant experience, resources, independence and quality control processes of other experts, prior to engagement.
  - The IS auditor should assess, review and evaluate the work of other experts as part of the audit, and conclude the extent of use and reliance on the experts' work.
  - The IS auditor should determine and conclude whether the work of other experts is adequate and complete to enable the IS auditor to conclude on the current audit objectives. Such conclusion should be clearly documented.

- The IS auditor should apply additional test procedures to gain sufficient and appropriate audit evidence in circumstances where the work of other experts does not provide sufficient and appropriate audit evidence.
- The IS auditor should provide appropriate audit opinion and include scope limitation where required evidence is not obtained through additional test procedures.
- **S14 Audit Evidence**
  - The IS auditor should obtain sufficient and appropriate audit evidence to draw reasonable conclusions on which to base the audit results.
  - The IS auditor should evaluate the sufficiency of audit evidence obtained during the audit.
- **S15 IT Controls**
  - The IS auditor should evaluate and monitor IT controls that are an integral part of the internal control environment of the organizsation.
  - The IS auditor should assist management by providing advice regarding the design, implementation, operation and improvement of IT controls.
- **S16 E-commerce**
  - The IS auditor should evaluate applicable controls and assess risk when reviewing e-commerce environments to ensure that e-commerce transactions are properly controlled

---

**Note:** The CISA exam does not test whether a candidate knows the specific number of an IS auditing standard. The CISA exam tests how guidelines are applied within the audit process.

---

## 1.3.3 ISACA IS AUDITING GUIDELINES

The objective of the ISACA IS Auditing Guidelines is to provide further information on how to comply with the ISACA IS Auditing Standards. The IS auditor should:
- Consider them in determining how to implement the above standards.
- Use professional judgment in applying them.
- Be able to justify any difference.

---

**Note:** The CISA candidate is not expected to know the specific number of an IS auditing guideline. The CISA exam tests how guidelines are applied within the audit process. The IS auditor should review the IS Audit Guidelines thoroughly to identify the subject matter that is truly needed in the job.

---

### *Index of Guidelines*

G1   Using the Work of Other Auditors, effective 1 June 1998
G2   Audit Evidence Requirement, effective 1 December 1998
G3   Use of Computer Assisted Audit Techniques (CAATs), effective 1 December 1998
G4   Outsourcing of IS Activities to Other Organisations, effective 1 September 1999
G5   Audit Charter, effective 1 September 1999
G6   Materiality Concepts for Auditing Information Systems, effective 1 September 1999
G7   Due Professional Care, effective 1 September 1999
G8   Audit Documentation, effective 1 September 1999
G9   Audit Considerations for Irregularities, effective 1 March 2000
G10  Audit Sampling, effective 1 March 2000
G11  Effect of Pervasive IS Controls, effective 1 March 2000
G12  Organisational Relationship and Independence, effective 1 September 2000
G13  Use of Risk Assessment in Audit Planning, effective 1 September 2000
G14  Application Systems Review, effective 1 November 2001
G15  Planning Revised, effective 1 March 2002
G16  Effect of Third Parties on an Organisation's IT Controls, effective 1 March 2002
G17  Effect of Nonaudit Role on the IS Auditor's Independence, effective 1 July 2002
G18  IT Governance, effective 1 July 2002
G19  Irregularities and Illegal Acts, effective 1 July 2002

G20 Reporting, effective 1 January 2003
G21 Enterprise Resource Planning (ERP) Systems Review, effective 1 August 2003
G22 Business-to-consumer (B2C) E-commerce Review, effective 1 August 2003
G23 System Development Life Cycle (SDLC) Review, effective 1 August 2003
G24 Internet Banking, effective 1 August 2003
G25 Review of Virtual Private Networks, effective 1 July 2004
G26 Business Process Reengineering (BPR) Project Reviews, effective 1 July 2004
G27 Mobile Computing, effective 1 September 2004
G28 Computer Forensics, effective 1 September 2004
G29 Post-implementation Review, effective 1 January 2005
G30 Competence, effective 1 June 2005
G31 Privacy, effective 1 June 2005
G32 Business Continuity Plan Review From IT Perspective, effective 1 September 2005
G33 General Considerations on the Use of the Internet, effective 1 March 2006
G34 Responsibility, Authority and Accountability, effective 1 March 2006
G35 Follow-up Activities, effective 1 March 2006
G36 Biometric Controls, effective 1 March 2007
G37 Configuration Management, effective 1 November 2007
G38 Access Control, effective 1 February 2008
G39 IT Organizations, effective 1 May 2008

---

**Note:** The CISA candidate should be familiar with the section in the IS Audit Guidelines related to the "Contents of the Audit Charter" (G5).

- Also important are the "Audit Considerations for Irregularities" (G9) in relation to the standard "Irregularities and Illegal Acts" (S9) for the purpose of reporting irregularities such as fraud.
- The IS Auditor should be familiar with the IS Audit Guideline related to Maintaining Independence "Effect of Nonaudit Role on the IS Auditor's Independence" (G17) and the related standard "Independence" (S2).
- Knowledge on "Follow-up Activities (G35) should be further identified by the IS auditor in the IS Audit Guidelines.

---

## 1.3.4 ISACA IS AUDITING PROCEDURES

Procedures developed by the ISACA Standards Board provide examples of possible processes an IS auditor might follow in an audit engagement. In determining the appropriateness of any specific procedure, IS auditors should apply their own professional judgment to the specific circumstances. The procedure documents provide information on how to meet the standards when performing IS auditing work, but do not set requirements.

It is not mandatory for the IS auditor to follow these procedures; however, following these procedures will provide assurance that the standards are being followed by the auditor.

### *Index of Procedures*
P1 IS Risk Assessment, effective 1 July 2002
P2 Digital Signatures, effective 1 July 2002
P3 Intrusion Detection, effective 1 August 2003
P4 Viruses and Other Malicious Code, effective 1 August 2003
P5 Control Risk Self-assessment, effective 1 August 2003
P6 Firewalls, effective 1 August 2003
P7 Irregularities and Illegal Acts, effective 1 November 2003
P8 Security Assessment—Penetration Testing and Vulnerability Analysis, effective 1 September 2004
P9 Evaluation of Management Controls Over Encryption Methodologies, effective 1 January 2005
P10 Business Application Change Control, effective 1 October 2006
P11 Electronic Funds Transfer (EFT), effective 1 May 2007

## 1.3.5 RELATIONSHIP AMONG STANDARDS, GUIDELINES AND PROCEDURES

Standards defined by ISACA are to be followed by the IS auditor. Guidelines provide assistance on how the auditor can implement standards in various audit assignments. Procedures provide examples of steps the auditor may follow in specific audit assignments so as to implement the standards; however, the IS auditor should use professional judgment when using guidelines and procedures.

## 1.3.6 INFORMATION TECHNOLOGY ASSURANCE FRAMEWORK (ITAF™)

The Information Technology Assurance Framework (ITAF™) is a comprehensive and good-practice-setting model that:
• Provides guidance on the design, conduct and reporting of IT audit and assurance assignments
• Defines terms and concepts specific to IT assurance
• Establishes standards that address IT audit and assurance professional roles and responsibilities, knowledge and skills, and diligence, conduct and reporting requirements

ITAF is focused on ISACA material as well as content and guidance developed by the IT Governance Institute® (ITGI™) and other organizations, and, as such, provides a single source through which IT audit and assurance professionals can seek guidance, research policies and procedures, obtain audit and assurance programs, and develop effective reports. ITAF includes three categories of standards—general, performance and reporting—as well as guidelines and tools and techniques:
• **General Standards**—The guiding principles under which the IT assurance profession operates. They apply to the conduct of all assignments, and deal with the IT audit and assurance professional's ethics, independence, objectivity and due care as well as knowledge, competency and skill.
• **Performance Standards**—Deal with the conduct of the assignment, such as planning and supervision, scoping, risk and materiality, resource mobilization, supervision and assignment management, audit and assurance evidence, and the exercising of professional judgment and due care.
• **Reporting Standards**—Address the types of reports, means of communication and the information communicated
• **Guidelines**—Provide the IT audit and assurance professional with information and direction about an audit or assurance area. In line with the three categories of standards outlined above, guidelines focus on the various audit approaches, methodologies, tools and techniques, and related material to assist in planning, executing, assessing, testing and reporting on IT processes, controls and related audit or assurance initiatives. Guidelines also help clarify the relationship between enterprise activities and initiatives, and those undertaken by IT.
• **Tools and Techniques**—Provide specific information on various methodologies, tools and templates—and provide direction in their application and use to operationalize the information provided in the guidance. Note that the tools and techniques are directly linked to specific guidelines. They take a variety of forms, such as discussion documents, technical direction, white papers, audit programs or books—e.g., the ISACA publication on SAP, which supports the guideline on enterprise resource planning (ERP) systems.

### Section 2200—General Standards
General standards are the guiding principles under which the IT assurance profession operates. They apply to the conduct of all assignments and deal with the IT audit and assurance professional's ethics, independence, objectivity and due care, as well as knowledge, competency and skill.

General standards include:
• **Independence and Objectivity**—The IT audit and assurance professional should maintain an independent and objective state of mind in all matters related to the conduct of the IT assurance assignment. The IT audit and assurance professional must conduct the IT assurance assignment with an impartial and unbiased frame of mind in addressing assurance issues and reaching conclusions. It is important that the IT audit and assurance professional not only be independent, but also appear to be independent at all times.
• **Reasonable Expectation**—The IT audit and assurance professional should have a reasonable expectation that the IT assurance assignment can be completed in accordance with these IT assurance standards or other appropriate professional, regulatory or industry standards, and result in a professional opinion. The scope of the audit or assurance engagement should be sufficient to permit a conclusion to be drawn on the subject matter and the ability to address any restrictions.

- **Management's Acknowledgement**—The IT audit and assurance professional should be satisfied that management understands his/her obligations and responsibilities with respect to the provision of appropriate, relevant and timely information that may be required in the performance of the assignment and his/her responsibility to ensure the co-operation of personnel during the audit or assurance activity.
- **Training and Proficiency**—The IT audit and assurance professional and others assisting with the assignment should collectively possess adequate skills and proficiency in conducting IT audit and assurance assignments to enable the professionals to perform the work required.
- **Knowledge of the Subject Matter**—The IT audit and assurance professional and others engaged in performing the IT assurance assignment should collectively possess adequate knowledge of the subject matter.
- **Due Professional Care**—The IT audit and assurance professional should exercise due care in planning, performing and reporting on the results of the IT assurance assignment.
- **Suitable Criteria**—IT audit subject matter should be evaluated against suitable and appropriate criteria. The characteristics of suitable criteria include:
- **Objectivity**—Criteria should be free from bias that may adversely impact the IT audit and assurance professional's findings and conclusions, and, accordingly, may mislead the user of the IT assurance report.
- **Measurability**—Criteria should permit consistent measurement of the subject matter and the development of consistent conclusions when applied by different IT audit and assurance professionals in similar circumstances.
- **Understandability**—Criteria should be communicated clearly and not be subject to significantly different interpretations by intended users.
- **Completeness**—Criteria should be sufficiently complete so that all criteria that could affect the IT audit and assurance professional's conclusions about the subject matter are identified and used in the conduct of the IT assurance assignment.
- **Relevance**—Criteria should be relevant to the subject matter and contribute to findings and conclusions that meet the objectives of the IT assurance assignment.

Current ISACA IS Auditing Standards include the following general standards:
- S2 Independence
- S3 Professional Ethics and Standards
- S4 Competence
- S6 Performance of Audit Work

## Section 2400—Performance Standards

Performance standards establish baseline expectations in the conduct of IT assurance engagements. While these standards apply to assurance professionals performing any assurance assignment, compliance is particularly important when the IT audit and assurance professional is acting in an audit capacity. Accordingly, the performance standards focus on the IT audit and assurance professional's attention to the design of the assurance work, the conduct of the assurance, the evidence required, and the development of assurance and audit findings and conclusions.

Performance standards include:
- **Planning and Supervision**—IT assurance work should be adequately planned and the IT audit and assurance professional should ensure that other persons performing the IT assurance assignment are properly supervised. Planning of the IT assignment should address the:
  - Objective of the IT audit or assurance assignment
  - Criteria to be used in conducting the IT assurance assignment
  - Level of assurance required. This includes whether the engagement is to be conducted at the examination or review level, or as an advisory or consulting assignment; what type of findings and conclusions will be required; and what format reporting will take.
  - Nature of the subject matter and the likely items within the assertion
  - Possible sources of information and evidence, including the tools, techniques and skills required to obtain the evidence. Considerations may include the use of computer-assisted audit techniques (CAATs), audit software and unique analyses.
  - Availability of appropriate and skilled IT audit and assurance resources
  - Availability and access to records and other information
  - Preliminary conclusions on assignment and audit risks, and the means by which these risks will be mitigated
  - Resource and expertise requirements—as well as their source, critical skills required and the timing of their participation in the IT assurance activity

&ndash; Nature, extent and timing of the various IT assurance tasks and, if an audit is being performed, audit tests
&ndash; Conditions that may require extension of modification of assurance work and audit tests
&ndash; Anticipation of time requirements and the establishment of time and cost budgets
&ndash; Nature of the expected report

Planning and supervision work should be documented and form part of the IT assurance work papers. This documentation should clearly indicate the nature, extent and timing of IT assurance work performed; the information and documents obtained; and the conclusions reached regarding the subject matter.

• **Obtaining Sufficient Evidence**—When an audit is being performed, the IT audit and assurance professional should obtain sufficient evidence to provide a reasonable basis for the conclusions drawn and expressed in the IT audit report:
&ndash; IT audit procedures should be applied to obtain and accumulate sufficient and appropriate audit evidence to provide a reasonable basis for conclusions to be drawn and expressed in the IT auditor's report. Sufficiency addresses the concept of quantity of evidence, and appropriateness addresses the quality of evidence in support of measuring achievement of the audit objective. In determining the sufficiency and appropriateness of IT audit evidence, the IT audit and assurance professional should consider the level of assurance being provided and the assessment of risk.
&ndash; Evidence is normally obtained through inspection, observation, enquiry, confirmation, re-performance analysis and discussion. The IT audit and assurance professional may seek corroborating evidence from different sources when forming a conclusion on the results of an IT audit procedure.
&ndash; The IT audit and assurance professional should ensure that the source of evidence is considered when assessing its appropriateness in supporting the audit procedure.
&ndash; The IT audit and assurance professional should document the test performed and the results obtained in sufficient detail to support the conclusions reached.

For other assurance work where an audit is not being performed, the IT audit and assurance professional should look to the requirements as a basis for establishing the sufficiency of information to be obtained and the documentation required. Usually, assurance documentation should be sufficient to allow a knowledgeable individual with appropriate skills to reach the same conclusions based on a review of the relevant documentation.

• **Assignment Performance**—The IT assurance assignment must be scheduled with regard to the timing, availability, and other commitments and requirements of management and the auditee. Scheduling should also consider the needs of report users so that their timing requirements are met. In scheduling audit personnel, care must be taken to ensure that the correct personnel are available and that issues of continuity, skills and experience are addressed:
&ndash; Professional staff must be assigned to tasks that are within their knowledge and skills.

The work must be conducted with due care and appropriate consideration for:
&ndash; management and auditee issues and concerns, including timing and timeliness.
&ndash; IT audit performance must address the objectives and mandate of the audit.

• **Representations**—The IT audit and assurance professional will receive representations during the course of conducting the IT audit—some written and others oral. As part of the audit process, these representations should be documented and retained in the work-paper file. In addition, for attestation engagements, representations from the auditee should be obtained in writing to reduce possible misunderstandings. Matters that may appear in a representation letter include:
&ndash; A statement by the auditee acknowledging responsibility for the subject matter and, when applicable, the assertions
&ndash; A statement by the auditee acknowledging responsibility for the criteria, and where applicable, the assertions
&ndash; A statement by the auditee acknowledging responsibility for determining that the criteria are appropriate for the purposes
&ndash; A list of specific assertions about the subject matter based on the criteria selected
&ndash; A statement that all known matters contradicting the assertions have been disclosed to the IT audit and assurance professional
&ndash; A statement that all communications from regulators affecting the subject matter or the assertions have been disclosed to the IT audit and assurance professional
&ndash; A statement that the IT audit and assurance professional has been provided access to all relevant information and records, files, etc., pertaining to the subject matter
&ndash; A statement on any significant events that have occurred subsequent to the date of the audit report and prior to release of that report
&ndash; Other matters that the IT audit and assurance professional may deem relevant or appropriate

Frequently, a summary of all representations made during the assignment is prepared and signed prior to finalization of the audit or assurance work.

While the same degree of rigour is not essential in non-audit assurance engagements, the assurance professional should obtain representations from management on key issues.

Current ISACA IS Auditing Standards include the following performance standards:
• S1 Audit Charter
• S5 Planning
• S9 Irregularities and Illegal Acts
• S10 IT Governance
• S11 Use of Risk Assessment in Audit Planning
• S12 Audit Materiality
• S13 Using the Work of Other Experts
• S14 Audit Evidence
• S15 IT Controls
• S16 E-commerce

## Section 2600—Reporting Standards

The report produced by the IT audit and assurance professional will vary, depending on the type of assignment performed. Considerations include the level of assurance, whether the assurance professional was acting in an audit capacity, whether the assurance professional is providing a direct report on the subject matter or is reporting on assertions regarding the subject matter, and whether that report is based on work performed at the review level or the examination level.

Reporting standards address (1) types of reports, (2) the means of communication, and (3) information to be communicated.

At minimum, the IT audit and assurance professional's report and/or associated attachments should:
• Identify to whom the report is directed
• Identify the nature and objectives of the IT assurance assignment
• Identify the entity or portion thereof covered by the IT assurance report
• Identify the subject matter or assertions on which the IT audit and assurance professional is reporting
• Provide a description of the nature of the scope of the work, including a brief statement on matters that are not within the scope of the assignments as well as those that are, to remove any doubt about the scope
• State the time frame or period covered by the report
• State the period during which the IT assurance was performed
• Provide a reference to the applicable professional standards governing the IT assurance assignment and against which the IT assurance work was conducted
• Identify management assertions, if any
• Describe the responsibilities of management and the IT audit and assurance professional
• Identify the criteria against which the subject matter was evaluated
• State a conclusion on the level of assurance being provided (Depending on the type of assignment, this could range from an audit report to a review report where no assurance is provided.)
• State any reservations that the IT audit and assurance professional may have (These may include scope, timing, and inability to obtain sufficient information or conduct appropriate tests, and are particularly important in audit assignments.)
• State any restrictions on the distribution or use of the report
• State the date of the report
• State where the report was issued
• State who issued the report (name or organization of the IT auditor)
• Include the IT audit and assurance professional's signature on the written report

In addition, depending on the nature of the IT audit or assurance assignment, other information should be provided such as specific government directives, corporate policies or other information germane to the reader's understanding of the IT assurance assignment.

# The IS Audit Process

## Section 3000—IT Assurance Guidelines

Section 3000 addresses guidelines in the following areas:

| Section | Guideline Area |
| --- | --- |
| 3200 | Enterprise Topics |
| 3400 | IT Management Processes |
| 3600 | IT Audit and Assurance Processes |
| 3800 | IT Audit and Assurance Management |

Each section within the guidelines focuses on one of the following:
• IT issues and processes that the IT audit and assurance professional should understand and consider when determining the planning, scoping, execution and reporting of IT audit or assurance activities
• IT audit and assurance processes, procedures, methodologies and approaches that the IT audit and assurance professional should consider when conducting IT assurance activities

The guidelines are supported by references to additional ISACA resources.

## Section 3200—Enterprise Topics

Section 3200 addresses enterprise-wide issues that may impact the IT audit and assurance professional in the planning and performance of the IT audit and assurance mission. The guidelines provide the IT audit and assurance professional with an understanding of enterprise-wide issues such as executive actions, external events and decisions that impact the IT department and, hence, the IT audit and assurance planning, designing, executing and reporting processes. This understanding may be provided by executive and senior user management and can be obtained from within the IT department. In addition, relevant information may also be obtained from work performed by non-IT audit and assurance professionals, either as part of an integrated audit assignment or from the other audit findings and reports.

By gaining an understanding of the environment in which the IT function operates—whether a separate IT department or a technology function located within business units—the IT audit and assurance professional should also gain an appreciation for the business and political pressures the IT function must address. The IT audit and assurance professional also gains an appreciation for the perspectives from which the various stakeholders approach the IT services and assess the performance of IT service providers. Thus, the IT audit and assurance professional can put into context the various IT functions and initiatives.

In addition to the operational environment, the IT audit and assurance professional should also consider the control environment and the system of internal control.

Section 3200 addresses guidelines in the following areas:

| Section | Guideline Area |
| --- | --- |
| 3210 | Implication of Enterprise-wide Policies, Practices and Standards on the IT Function |
| 3230 | Implication of Enterprise-wide Assurance Initiatives on the IT Function |
| 3250 | Implication of Enterprise-wide Assurance Initiatives on IT Assurance Plans and Activities |
| 3270 | Additional Enterprise-wide Issues and Their Impact on the IT Function |

## Section 3400—IT Management Processes

Section 3400 addresses IT management. Guidelines in this section provide the IT audit and assurance professional with an understanding of various IT management and IT operations topics as a background to the planning and scoping of IT assurance activities. Guidance in this section may also provide the IT audit and assurance professional with direction or information that will be of assistance in conducting an audit and information on IT topics that the IT audit and assurance professional is likely to, or should expect to, encounter during the conduct of IT audit or assurance work.

IT management guidelines also provide the IT audit and assurance professional with insight into the practices and procedures of IT departments. As such, the section focuses on the planning, organization and strategizing of activities of IT departments; acquisition of information and information technology; implementation; support and delivery of IT services;

and the monitoring and improvement of IT practices and procedures to enhance security, control and shareholder value. The section provides the IT audit and assurance professional with information on common practices, issues and concerns as well as risks and pitfalls in each area, and approaches and methodologies management can use to enhance value. It also provides the IT audit and assurance professional with guidance on the types of controls that management is likely to or should implement.

Section 3400 addresses guidelines in the following areas:

| Section | Guideline Area |
|---------|----------------|
| 3410 | IT Governance (Mission, Goals, Strategy, Corporate Alignment, Reporting) |
| 3412 | Determining the Impact of Enterprise Initiatives on IT Assurance Activities |
| 3415 | Using the Work of Other Experts in Conducting IT Assurance Activities |
| 3420 | IT Project Management |
| 3425 | IT Information Strategy |
| 3427 | IT Information Management |
| 3430 | IT Plans and Strategy (Budgets, Funding, Metrics) |
| 3450 | IT Processes (Operations, Human Resources, Development, etc.) |
| 3470 | IT Risk Management |
| 3490 | IT Support of Regulatory Compliance |

## Section 3600—IT Audit and Assurance Processes

Section 3600 focuses on audit approaches, methodologies and techniques. It provides the IT audit and assurance professional with information on common practices, issues, concerns and pitfalls when employing various audit and assurance procedures, and guidance on how to plan and conduct the assurance activity to ensure success. It also provides the IT audit and assurance professional with specific guidance on testing controls.

The IT audit and assurance professional should recognize and appreciate the role of IT in the enterprise, and the relationships that exist between IT departments and enterprise operations and management.

When performing audit and assurance work, it is suggested that ISACA members indicate that 'The work was performed in accordance with ISACA audit and assurance standards'.

Section 3600 addresses guidelines in the following areas:

| Section | Guideline Area |
|---------|----------------|
| 3605 | Relying on the Work of Specialists and Others |
| 3607 | Integrating IT Audit and Assurance Work With Other Audit Activities |
| 3610 | Using COBIT in the IT Assurance Process |
| 3630 | Auditing IT General Controls (ITGCs) |
| 3650 | Auditing Application Controls |
| 3653 | Auditing Traditional Application Controls |
| 3655 | Auditing Enterprise Resource Planning (ERP) Systems |
| 3657 | Auditing Alternative Software Development Strategies |
| 3660 | Auditing Specific Requirements |
| 3661 | Auditing Government-specified Criteria |
| 3662 | Auditing Industry-specified Criteria |
| 3670 | Auditing With Computer-assisted Audit Techniques (CAATs) |
| 3680 | IT Auditing and Regulatory Reporting |
| 3690 | Selecting Items of Assurance Interest |

## Section 3800—IT Audit and Assurance Management

Section 3800 addresses IT audit and assurance management. Guidance in this section provides the IT audit and assurance professional with an understanding of information required to manage an IT audit assignment. The section commences with information about the creation and management of the IT audit or assurance function and follows with discussion of various

IT audit and assurance management topics. These topics include audit and assurance planning and scoping, then refining the initial scoping, putting information into a detailed IT audit plan and scope document that incorporates the IT audit or assurance objectives. Next, this section addresses managing the execution of the IT audit and assurance professional's work. The section provides guidance in documenting assurance work, and documenting and clearing findings and recommendations. The section also addresses effective assurance reporting considerations.

Section 3800 addresses guidelines in the following areas:

| Section | Guideline Area |
|---|---|
| 3810 | IT Audit or Assurance Function |
| 3820 | Planning and Scoping IT Audit and Assurance Objectives |
| 3830 | Planning and Scoping IT Audit and Assurance Work |
| 3835 | Planning and Scoping Risk Assessments |
| 3840 | Managing the IT Audit and Assurance Process Execution |
| 3850 | Integrating the Audit and Assurance Process |
| 3860 | Gathering Evidence |
| 3870 | Documenting IT Audit and Assurance Work |
| 3875 | Documenting and Confirming IT Audit and Assurance Findings |
| 3880 | Evaluating Results and Developing Recommendations |
| 3890 | Effective IT Audit and Assurance Reporting |
| 3892 | Reporting IT Audit and Assurance Recommendations |
| 3894 | Reporting on IT Advisory and Consultancy Reviews |

# 1.4 RISK ANALYSIS

Risk analysis is part of audit planning, and helps identify risks and vulnerabilities so the IS auditor can determine the controls needed to mitigate those risks.

In evaluating IT-related business processes applied by an organization, understanding the relationship between risk and control is important for IS audit and control professionals. IS auditors must be able to identify and differentiate risk types and the controls used to mitigate these risks. They must have knowledge of common business risks, related technology risks and relevant controls. They must also be able to evaluate the risk assessment and management techniques used by business managers, and to make assessments of risk to help focus and plan audit work. In addition to an understanding of business risk and control, IS auditors must understand that risk exists within the audit process.

There are many definitions of risk, reflecting that risk means different things to different people. Perhaps one of the most succinct definitions of risk used within the information security business world is provided by the *Guidelines for the Management of IT Security* published by the International Organization for Standardization (ISO):

> *"The potential that a given threat will exploit vulnerabilities of an asset (G.3) or group of assets and thereby cause harm to the organization."*
> *(ISO/IEC PDTR 13335-1)*

This definition is used commonly by the IT industry since it puts risk into an organizational context by using the concepts of assets and loss of value—terms that are easily understood by business managers.

Business risks are the likelihood of those threats that may negatively impact the assets, processes or objectives of a specific business or organization. The nature of these threats may be financial, regulatory or operational, and may arise as a result of the interaction of the business with its environment or as a result of the strategies, systems and particular technology, processes, procedures and information used by the business. The IS auditor is often focused toward high-risk issues associated with the confidentiality, availability or integrity of sensitive and critical information, and the underlying information systems and processes that generate, store and manipulate such information. In reviewing these types of risks, IS auditors will often assess the effectiveness of the risk management process that an organization uses.

The risk assessment process is characterized as an iterative life cycle that begins with identifying business objectives, information assets, and the underlying systems or information resources that generate/store, use or manipulate the assets (hardware, software, databases, networks, facilities, people, etc.) critical to achieving these objectives. The greatest degree of risk management effort may then be directed toward those considered most sensitive or critical to the organization. Once sensitive and/or critical information assets are identified, a risk assessment is performed to identify threats and determine the probability of occurrence, and the resulting impact and additional safeguards that would mitigate this impact to a level acceptable to management.

Next, during the risk mitigation phase, controls are identified for mitigating identified risks. These controls are risk-mitigating countermeasures that should prevent or reduce the likelihood of a risk event occurring, detect the occurrence of a risk event, minimize the impact, or transfer the risk to another organization.

The assessment of countermeasures should be performed through a cost-benefit analysis where controls to mitigate risks are selected to reduce risks to a level acceptable to management. This analysis process may be based on any of the following:
• The cost of the control compared to the benefit of minimizing the risk
• Management's appetite for risk (i.e., the level of residual risk that management is prepared to accept)
• Preferred risk-reduction methods (e.g., terminate the risk, minimize probability of occurrence, minimize impact, transfer the risk via insurance)

The final phase relates to monitoring performance levels of the risks being managed when identifying any significant changes in the environment that would trigger a risk reassessment, warranting changes to its control environment. It encompasses three processes—risk assessment, risk mitigation and risk reevaluation—in determining whether risks are being mitigated to a level acceptable to management. It should be noted that, to be effective, risk assessment should be an ongoing process in an organization that endeavors to continually identify and evaluate risks as they arise and evolve.

From the IS auditor's perspective, risk analysis serves more than one purpose:
• It assists the IS auditor in identifying risks and threats to an IT environment and IS system—risks and threats that would need to be addressed by management—and in identifying system-specific internal controls. Depending on the level of risk, this assists the IS auditor in selecting certain areas to examine.
• It helps the IS auditor in his/her evaluation of controls in audit planning.
• It assists the IS auditor in determining audit objectives.
• It supports risk-based audit decision making.

# 1.5 INTERNAL CONTROLS

Internal controls are normally composed of policies, procedures, practices and organizational structures which are implemented to reduce risks to the organization.

Internal controls are developed to provide reasonable assurance to management that the organization's business objectives will be achieved and risk events will be prevented, or detected and corrected. Internal control activities and supporting processes are either manual or driven by automated computer information resources. Internal controls operate at all levels within an organization to mitigate its exposures to risks that potentially could prevent it from achieving its business objectives. The board of directors and senior management are responsible for establishing the appropriate culture to facilitate an effective and efficient internal control system, and for continuously monitoring the effectiveness of the internal control system, although each individual within an organization must take part in this process.

There are two key aspects that controls should address: what should be achieved, and what should be avoided. Not only do internal controls address business/operational objectives, but they should also address undesired events through prevention, detection and correction.

Elements of controls that should be considered when evaluating control strength are classified as preventive, detective or corrective in nature.

**Exhibit 1.2** displays control categories, functions and usages.

| Exhibit 1.2—Control Classifications | | |
| --- | --- | --- |
| **Class** | **Function** | **Examples** |
| Preventive | • Detect problems before they arise.<br>• Monitor both operation and inputs.<br>• Attempt to predict potential problems before they occur and make adjustments.<br>• Prevent an error, omission or malicious act from occurring. | • Employ only qualified personnel.<br>• Segregate duties (deterrent factor).<br>• Control access to physical facilities.<br>• Use well-designed documents (prevent errors).<br>• Establish suitable procedures for authorization of transactions.<br>• Complete programmed edit checks.<br>• Use access control software that allows only authorized personnel to access sensitive files.<br>• Use encryption software to prevent unauthorized disclosure of data. |
| Detective | • Use controls that detect and report the occurrence of an error, omission or malicious act. | • Hash totals<br>• Check points in production jobs<br>• Echo controls in telecommunications<br>• Error messages over tape labels<br>• Duplicate checking of calculations<br>• Periodic performance reporting with variances<br>• Past-due account reports<br>• Internal audit functions<br>• Review of activity logs to detect unauthorized access attempts |
| Corrective | • Minimize the impact of a threat.<br>• Remedy problems discovered by detective controls.<br>• Identify the cause of a problem.<br>• Correct errors arising from a problem.<br>• Modify the processing system(s) to minimize future occurrences of the problem. | • Contingency planning<br>• Backup procedures<br>• Rerun procedures |

**Note:** A CISA candidate should know the differences between preventive, detective and corrective controls. An example of a question in the exam would be: Which of the following controls would **BEST** detect...

## 1.5.1 INTERNAL CONTROL OBJECTIVES

These types of controls include those controls related to the technology environment. Types of controls include:
• Internal accounting controls—Primarily directed at accounting operations such as the safeguarding of assets and the reliability of financial records.
• Operational controls—Directed at day-to-day operations, functions and activities to ensure that the operation is meeting the business objectives.
• Administrative controls—Concerned with operational efficiency in a functional area and adherence to management policies including operational controls. Described as supporting the operational controls specifically concerned with operating efficiency and adherence to organizational policy.

Internal control objectives are statements of the desired result or purpose to be achieved by implementing control activities (procedures).

For example, control objectives include:
• Safeguarding of IT assets
• Compliance with corporate policies and legal requirements
• Authorization and authentication
• Confidentiality
• Accuracy and completeness of data
• Reliability of process
• Availability of IT services
• Efficiency and economy of operations
• Change management process for IT and related systems

## 1.5.2 IS CONTROL OBJECTIVES

Internal control objectives apply to all areas, whether manual, automated, or a combination thereof (i.e., review of system logs). Therefore, conceptually, control objectives in an IS environment remain unchanged from those of a manual environment. However, the way these controls are implemented may be different. Thus, internal control objectives need to be addressed in a manner relevant to the specific IS-related processes.

IS control objectives may include:
• Safeguarding assets. Information on automated systems is secure from improper access and current.
• Ensuring integrity of general operating system (OS) environments, including network management and operations.
• Ensuring integrity of sensitive and critical application system environments, including accounting/financial and management information (information objectives) and customer data, through:
  – Authorization of the input. Each transaction is authorized and entered only once.
  – Validation of the input. Each input is validated and will not cause negative impact to the processing of transactions.
  – Accuracy and completeness of processing of transactions. All transactions are recorded and entered into the computer for the proper period.
  – Reliability of overall information processing activities
  – Accuracy, completeness and security of the output
  – Database integrity, availability and confidentiality
• Ensuring appropriate identification and authentication of users of IS resources (end users as well as infrastructure support).
• Ensuring the efficiency and effectiveness of operations (operational objectives).
• Complying with the users' requirements, organizational policies and procedures, and applicable laws and regulations (compliance objectives).
• Ensuring availability of IT services by developing efficient business continuity (BCP) and disaster recovery plans (DRP).
• Enhancing protection of data and systems by developing an incident response plan.
• Ensuring integrity and reliability of systems by implementing effective change management procedures.

Working through ISACA, ITGI publishes an IT governance and control framework incorporating good IT management practices—*Control Objectives for Information and related Technology* (CoBiT®). CoBiT is the leading framework for governance, control and assurance for information and related technology.

## 1.5.3 *CONTROL OBJECTIVES FOR INFORMATION AND RELATED TECHNOLOGY* (CoBiT)

CoBiT supports IT governance by providing a framework to ensure that IT is aligned with the business, IT enables the business and maximizes benefits, IT resources are used responsibly and IT risks are managed appropriately. CoBiT provides good practices across a domain and process framework, and presents activities in a manageable and logical structure. CoBiT's good practices represent the consensus of experts. CoBiT has a framework with a set of 34 IT processes grouped into four domains: Plan and Organize, Acquire and Implement, Deliver and Support, and Monitor and Evaluate. By adapting and using these practices from the relevant 34 IT processes, organizations can ensure that they have implemented a

system of IT governance and related controls as required for their IT environment. Supporting these IT processes are more than 200 detailed control objectives necessary for effective implementation. The component of control objectives in COBIT provides a reference framework on what controls should be in place while *COBIT® Control Practices: Guidance to Achieve Control Objectives for Successful IT Governance* facilitates the implementation of these controls.

COBIT's good practices are more strongly focused on control, less on execution. These practices will help optimize IT-enabled investments, ensure service delivery and provide a measure to judge against when things do go wrong.

For IT to be successful in delivering business requirements, management should put an internal control system or framework in place. The COBIT control framework contributes to these needs by:
• Making a link to the business requirements
• Organizing IT activities into a generally accepted process model
• Identifying the major IT resources to be leveraged
• Defining the management control objectives to be considered
The business orientation of COBIT consists of linking business goals to IT goals, providing metrics and maturity models to measure their achievement, and identifying the associated responsibilities of business and IT process owners.

In summary, to provide the information that the enterprise needs to achieve its objectives, IT resources need to be managed by a set of naturally grouped processes.

COBIT uses, as primary references, current major framework standards and regulations relating to IT. COBIT is directed to the management and staff of information services, control departments, audit functions and, most importantly, the business process owners using IT processes to ensure confidentiality, integrity and availability of sensitive and critical information. ITGI has also published the *IT Governance Implementation Guide* to facilitate enterprises in implementing IT governance using the COBIT framework. COBIT® *Quickstart*™ provides essentials of COBIT for small and medium enterprises. COBIT Online® provides all the components of COBIT on the Internet for users to adapt and customize COBIT components per their specific requirements. The online COBIT Foundation Course® and exam is an e-learning solution applicable to IT auditors, IT managers, IT quality professionals, IT leadership, IT developers, process practitioners and managers in IT service providing firms. They can be used to understand COBIT at a foundation level and help in the application of COBIT in practice.

COBIT provides a comprehensive framework for management and delivery of high-quality IT-based services. COBIT sets best practices for the process of value creation. Val IT adds best practices to unambiguously measure, monitor and optimize the realization of business value from investment in IT. Val IT complements COBIT from a business and financial perspective, and will help those with an interest in value delivery from IT. Val IT framework presents key management practices for three processes: value governance, portfolio management and investment management.

> **Note:** A CISA candidate will not be asked to specifically identify the COBIT process, the COBIT domains or the set of IT processes defined in each. However, candidates should know what frameworks are, what they do and why they are used by enterprises. Knowledge of the existence, structure and key principles of major standards and frameworks related to IT governance, assurance and security will also be advantageous. COBIT can be used as a supplemental study material in understanding control objectives and principles as detailed in this review material. Please refer to Appendix A for references between the CISA certification areas and the COBIT framework.

## 1.5.4 GENERAL CONTROLS

Controls include policies, procedures and practices (tasks and activities) established by management to provide reasonable assurance that specific objectives will be achieved.

General controls apply to all areas of the organization including IT infrastructure and support services. General controls include:
• Internal accounting controls that are primarily directed at accounting operations—controls that concern the safeguarding of assets and reliability of financial records

- Operational controls that concern day-to-day operations, functions and activities, and ensure that the operation is meeting the business objectives
- Administrative controls that are concern operational efficiency in a functional area and adherence to management policies (administrative controls support the operational controls specifically concerned with operating efficiency and adherence to organizational policies)
- Organizational security policies and procedures to ensure proper usage of information and technology assets
- Overall policies for the design and use of adequate documents and records (manual/automated) to help ensure proper recording of transactions—transactional audit trail
- Procedures and practices to ensure adequate safeguards over access to and use of assets and facilities
- Physical and logical security policies for all data centers and IT resources (e.g., servers and telecom infrastructure)

## 1.5.5 IS CONTROLS

Each general control procedure can be translated into an IS-specific control procedure. A well-designed information system should have controls built in for all its sensitive or critical functions. For example, the general procedure to ensure that adequate safeguards over access to assets and facilities can be translated into an IS-related set of control procedures, covering access safeguards over computer programs, data and computer equipment. The IS auditor should understand the basic control objectives that exist for all functions. IS control procedures include:
- Strategy and direction
- General organization and management
- Access to IT resources, including data and programs
- Systems development methodologies and change control
- Operations procedures
- Systems programming and technical support functions
- Quality assurance (QA) procedures
- Physical access controls
- Business continuity (BCP)/disaster recovery planning (DRP)
- Networks and communications
- Database administration
- Protection and detective mechanisms against internal and external attacks

The IS auditor should understand concepts regarding IS controls and how to apply them in planning an audit.

**Note:** The IS controls listed in this section should be considered by the CISA candidate within the related job practice area, i.e., Protection of Information Assets.

# 1.6 PERFORMING AN IS AUDIT

Auditing can be defined as a systematic process by which a qualified, competent, independent team or person objectively obtains and evaluates evidence regarding assertions about a process for the purpose of forming an opinion about and reporting on the degree to which the assertion is implemented.

IS audit can be defined as any audit that encompasses review and evaluation (wholly or partly) of automated information processing systems, related nonautomated processes and the interfaces among them.

To perform an audit, several steps are required. Adequate planning is a necessary first step in performing effective IS audits. To effectively use IS audit resources, audit organizations must assess the overall risks for the general and application areas and related services being audited, and then develop an audit program that consists of objectives and audit procedures to satisfy the audit objectives. The audit process requires the IS auditor to gather evidence, evaluate the strengths and weaknesses of controls based on the evidence gathered through audit tests, and prepare an audit report that presents those issues (areas of control weaknesses with recommendations for remediation) in an objective manner to management.

# The IS Audit Process

Audit management must ensure the availability of adequate audit resources and a schedule for performing the audits and, in the case of internal IS audit, for follow-up reviews on the status of corrective actions taken by management. The process of auditing includes defining the audit scope, formulating audit objectives, identifying audit criteria, performing audit procedures, reviewing and evaluating evidence, forming audit conclusions and opinions, and reporting to management after discussion with key process owners.

Auditee constraints may include:
• Recent employee turnover or unavailability
• Infringement on workload
• Overall lack of knowledge or documentation

To understand these constraints on the conduct of an audit, the IS auditor should have a good understanding of overall project management techniques. Often, these constraints can be minimized or avoided by adequate planning.

Project management techniques for managing and administering audit projects, whether automated or manual, include the following basic steps:
• Develop a detailed plan—The plan should chart out the necessary audit steps across a time line. Realistic estimates should be made of the time requirements for each task with proper consideration given to the availability of the auditee.
• Report project activity against the plan—There should be a reporting system in place to ensure that IS auditors can report their actual progress against planned audit steps.
• Adjust the plan and take corrective action—Actual accomplishments should be measured against the established plan on a continuous basis. Changes should be made in IS auditor assignments or in planned schedules as required.

## 1.6.1 CLASSIFICATION OF AUDITS

The IS auditor should understand the various types of audits that can be performed, internally or externally, and the audit procedures associated with each:
• **Financial audits**—The purpose of a financial audit is to assess the correctness of an organization's financial statements. A financial audit will often involve detailed, substantive testing, although increasingly, auditors are placing more emphasis on a risk- and control-based audit approach. This kind of audit relates to financial information integrity and reliability.
• **Operational audits**—An operational audit is designed to evaluate the internal control structure in a given process or area. IS audits of application controls or logical security systems are some examples of operational audits.
• **Integrated audits**—An integrated audit combines financial and operational audit steps. An integrated audit is also performed to assess the overall objectives within an organization, related to financial information and assets' safeguarding, efficiency and compliance. An integrated audit can be performed by external or internal auditors and would include compliance tests of internal controls and substantive audit steps.
• **Administrative audits**—These are oriented to assess issues related to the efficiency of operational productivity within an organization.
• **IS audits**—This process collects and evaluates evidence to determine whether the information systems and related resources adequately safeguard assets, maintain data and system integrity and availability, provide relevant and reliable information, achieve organizational goals effectively, consume resources efficiently, and have, in effect, internal controls that provide reasonable assurance that business, operational and control objectives will be met and that undesired events will be prevented, or detected and corrected, in a timely manner.
• **Specialized audits**—Within the category of IS audits, there are a number of specialized reviews that examine areas such as services performed by third parties. Because businesses are becoming increasingly reliant on third-party service providers, it is important that internal controls be evaluated in these environments. The Statement on Auditing Standards (SAS) 70, titled "Reports on the Processing of Transactions by Service Organizations," is a widely known auditing standard developed by the American Institute of Certified Public Accountants (AICPA). SAS 70 defines the professional standards used by a service auditor to assess the internal controls of a service organization. This type of audit has become increasingly relevant due to the current trend of outsourcing of financial and business processes to third-party service providers which, in some cases, may operate in different jurisdictions or even different countries. It should be noted that a Type 2 SAS 70 review is a more thorough variation of a regular SAS 70 review, which is often required in connection with regulatory reviews. Many other countries have their own equivalent of this standard. A SAS 70-type audit is important

because it represents that a service organization has been through an in-depth audit of their control activities, which generally include controls over information technology and related processes. SAS 70-type reviews provide guidance to enable an independent auditor (service auditor) to issue an opinion on a service organization's description of controls through a service auditor's report, which then can be relied on by the IS auditor of the entity that utilizes the services of the service organization.

• **Forensic audits**—Forensic auditing has been defined as auditing specialized in discovering, disclosing and following up on frauds and crimes. The primary purpose of such a review is the development of evidence for review by law enforcement and judicial authorities. In recent years, the forensic professional has been called on to participate in investigations related to corporate fraud and cybercrime. In cases where computer resources may have been misused, further investigation is necessary to gather evidence for possible criminal activity that can then be reported to appropriate authorities. A computer forensic investigation includes the analysis of electronic devices such as computers, phones, personal digital assistants (PDAs), disks, switches, routers, hubs and other electronic equipment. An IS auditor possessing the necessary skills can assist the information security manager in performing forensic investigations and conduct the audit of the systems to ensure compliance with the evidence collection procedures for forensic investigation. Electronic evidence is vulnerable to changes; therefore, it is necessary to handle electronic evidence with utmost care and controls should ensure that no manipulation can occur. Chain of custody for electronic evidence should be established to meet legal requirements.

Improperly handled computer evidence is subject to being ruled inadmissible by judicial authorities. The most important consideration for a forensic auditor is to make a bit-stream image of the target drive and examine that image without altering date stamps or other information attributable to the examined files. Further, forensic audit tools and techniques such as data mapping for security and privacy risk assessment, and the search for intellectual property for data protection, are also being used for prevention, compliance and assurance.

## 1.6.2 AUDIT PROGRAMS

Audit programs for financial, operational, integrated, administrative and IS audits are based on the scope and objective of the particular assignment. IS auditors often evaluate IT functions and systems from different perspectives such as security (confidentiality, integrity and availability), quality (effectiveness, efficiency), fiduciary (compliance, reliability), service and capacity. The audit work program is the audit strategy and plan—it identifies scope, audit objectives and audit procedures to obtain sufficient, relevant and reliable evidence to draw and support audit conclusions and opinions.

General audit procedures are the basic steps in the performance of an audit and usually include:
• Obtaining and recording an understanding of the audit area/subject
• A risk assessment and general audit plan and schedule
• Detailed audit planning
• Preliminary review of the audit area/subject
• Evaluating the audit area/subject
• Verifying and evaluating the appropriateness of controls designed to meet control objectives
• Compliance testing (tests of the implementation of controls, and their consistent application)
• Substantive testing (confirming the accuracy of information)
• Reporting (communicating results)
• Follow-up in cases where there is an internal audit function

The IS auditor must understand the procedures for testing and evaluating IS controls. These procedures could include:
• The use of generalized audit software to survey the contents of data files (including system logs)
• The use of specialized software to assess the contents of operating system database and application parameter files (or detect deficiencies in system parameter settings)
• Flow-charting techniques for documenting automated applications and business processes
• The use of audit logs/reports available in operation/application systems
• Documentation review
• Observation

The IS auditor should have a sufficient understanding of these procedures to allow for the planning of appropriate audit tests.

## 1.6.3 AUDIT METHODOLOGY

An audit methodology is a set of documented audit procedures designed to achieve planned audit objectives. Its components are a statement of scope, a statement of audit objectives and a statement of audit programs.

The audit methodology should be set up and approved by audit management to achieve consistency in the audit approach. This methodology should be formalized and communicated to all audit staff.

**Exhibit 1.3** lists the phases of a typical audit. An early and critical product of the audit process should be an audit program that is the guide for performing and documenting all the following audit steps, and the extent and types of evidential matter reviewed.

| Exhibit 1.3—Audit Phases | |
|---|---|
| **Audit Phase** | **Description** |
| Audit subject | • Identify the area to be audited. |
| Audit objective | • Identify the purpose of the audit. For example, an objective might be to determine whether program source code changes occur in a well-defined and controlled environment. |
| Audit scope | • Identify the specific systems, function or unit of the organization to be included in the review. For example, in the previous program changes example, the scope statement might limit the review to a single application system or to a limited period of time. |
| Preaudit planning | • Identify technical skills and resources needed.<br>• Identify the sources of information for test or review such as functional flow charts, policies, standards, procedures and prior audit workpapers.<br>• Identify locations or facilities to be audited. |
| Audit procedures and steps for data gathering | • Identify and select the audit approach to verify and test the controls.<br>• Identify a list of individuals to interview.<br>• Identify and obtain departmental policies, standards and guidelines for review.<br>• Develop audit tools and methodology to test and verify control. |
| Procedures for evaluating the test or review results | Organization-specific |
| Procedures for communication with management | Organization-specific |
| Audit report preparation | • Identify follow-up review procedures.<br>• Identify procedures to evaluate/test operational efficiency and effectiveness.<br>• Identify procedures to test controls.<br>• Review and evaluate the soundness of documents, policies and procedures. |

Although an audit program does not necessarily follow a specific set of steps, the IS auditor typically would follow, as a minimum course of action, sequential program steps to gain an understanding of the entity under audit, evaluate the control structure and test the controls.

Each audit department should design and approve an audit methodology as well as the minimum steps to be observed in any audit assignment.

All audit plans, programs, activities, tests, findings and incidents shall be properly documented in work papers.

The format and media of work papers can vary depending on specific needs of the department. IS auditors should particularly consider how to maintain the integrity and protection of audit test evidence in order to preserve their value as substantiation in support of audit results.

Work papers can be considered the bridge or interface between the audit objectives and the final report. Work papers should provide a seamless transition—with traceability and support for the work performed —from objectives to report and from report to objectives. In this context, the audit report can be viewed as a particular work paper.

| PRACTICE QUESTION |
| --- |
| **1-1**    Which of the following **BEST** describes the early stages of an IS audit? <br><br>      A.   Observing key organizational facilities <br>      B.   Assessing the IS environment <br>      C.   Understanding the business process and environment applicable to the review <br>      D.   Reviewing prior IS audit reports |
| *See answers and explanations to the practice questions at the end of the chapter. (page 68)* |

## 1.6.4 FRAUD DETECTION

The use of information technology for business has immensely benefited enterprises in terms of significantly increased quality of delivery of information. However, the widespread use of information technology and the Internet leads to risks that enable the perpetration of errors and frauds.

Management is primarily responsible for establishing, implementing and maintaining a framework and design of IT controls to meet the internal control objectives. A well-designed internal control system provides good opportunities for deterrence and/or timely detection of fraud. Internal controls may fail where such controls are circumvented by exploiting vulnerabilities or through management perpetrated weakness in controls or collusion among people.

Legislation and regulations relating to corporate governance cast significant responsibilities on management, auditors and the audit committee regarding detection and disclosure of any frauds, whether material or not.

IS auditors should observe and exercise due professional care (ISACA IS Auditing Standard S3) in all aspects of their work. IS auditors entrusted with assurance functions should ensure reasonable care while performing their work and be alert to the possible opportunities that allow fraud to materialize.

The presence of internal controls does not altogether eliminate fraud. IS auditors should be aware of the possibility and means of perpetrating fraud, especially by exploiting the vulnerabilities and overriding controls in the IT-enabled environment. IS auditors should have knowledge of fraud and fraud indicators, and be alert to the possibility of fraud and errors while performing an audit.

Besides instituting and maintaining a system of internal controls, management looks to IS auditors for assurance on the state of internal controls and their ability to deter and detect fraud, and recommendations from the IS auditors for improvement in internal controls.

During the course of regular assurance work, the IS auditor may come across instances or indicators of fraud. The IS auditor may, after careful evaluation, communicate the need for a detailed investigation to appropriate authorities. In the case of the auditor identifying a major fraud, or if the risk associated with the detection is high, audit management should also consider communicating in a timely manner to the audit committee.

Regarding fraud prevention, the IS auditor should be aware of potential legal requirements concerning the implementation of specific fraud detection procedures and reporting fraud to appropriate authorities.

## 1.6.5 RISK-BASED AUDITING

More organizations are moving to a risk-based audit approach that is usually adapted to develop and improve the continuous audit process. This approach is used to assess risk and to assist an IS auditor in making the decision to perform either compliance testing or substantive testing. It is important to stress that the risk-based audit approach efficiently assists the auditor in determining the nature and extent of testing.

Within this concept, inherent risk, control risk or detection risk should not be of major concern, despite some weaknesses. In a risk-based audit approach, IS auditors are not just relying on risk; they also are relying on internal and operational controls as well as knowledge of the company or the business. This type of risk assessment decision can help relate the cost-benefit analysis of the control to the known risk, allowing practical choices.

Business risks include concerns about the probable effects of an uncertain event on achieving established business objectives. The nature of these risks may be financial, regulatory or operational, and may also include risks derived from specific technology. For example, an airline company is subject to extensive safety regulations and economic changes, both of which impact the continuing operations of the company. In this context, the availability of IT service and its reliability are critical.

By understanding the nature of the business, IS auditors can identify and categorize the types of risks that will better determine the risk model or approach in conducting the audit. The risk model assessment can be as simple as creating weights for the types of risks associated with the business and identifying the risks in an equation. On the other hand, risk assessment can be a scheme where risks have been given elaborate weights based on the nature of the business or the significance of the risk. A simplistic overview of a risk-based audit approach can be seen in **exhibit 1.4**.

| Exhibit 1.4—Risk-based Audit Approach |
|---|

**Gather Information and Plan**

- Knowledge of business and industry
- Prior year's audit results
- Recent financial information
- Regulatory statutes
- Inherent risk assessments

**Obtain Understanding of Internal Control**

- Control environment
- Control procedures
- Detection risk assessment
- Control risk assessment
- Equate total risk

**Perform Compliance Tests**

- Identify key controls to be tested.
- Perform tests on reliability, risk prevention and adherence to organization policies and procedures.

**Perform Substantive Tests**

- Analytical procedures
- Detailed tests of account balances
- Other substantive audit procedures

**Conclude the Audit**

- Create recommendations.
- Write audit report.

| PRACTICE QUESTIONS |
|---|
| **1-2** In performing a risk-based audit, which risk assessment is completed initially by the IS auditor?<br><br>A. Detection risk assessment<br>B. Control risk assessment<br>C. Inherent risk assessment<br>D. Fraud risk assessment<br><br>**1-3** While developing a risk-based audit program, on which of the following would the IS auditor **MOST** likely focus?<br><br>A. Business processes<br>B. Critical IT applications<br>C. Operational controls<br>D. Business strategies |
| *See answers and explanations to the practice questions at the end of the chapter. (page 68)* |

## 1.6.6 AUDIT RISK AND MATERIALITY

Audit risk can be defined as the risk that information may contain a material error that may go undetected during the course of the audit. The IS auditor should also take into account, if applicable, other factors relevant to the organization: customer data, privacy, availability of provided services as well as corporate and public image as in the case of public organizations or foundations.

Audit risk can be categorized as:
• **Inherent risk**—The risk that an error exists that could be material or significant when combined with other errors encountered during the audit, assuming that there are no related compensating controls. Inherent risk can also be categorized as the susceptibility to a material misstatement in the absence of related controls. For example, complex calculations are more likely to be misstated than simple ones and cash is more likely to be stolen than an inventory of coal. Inherent risks exist independent of an audit and can occur because of the nature of the business.
• **Control risk**—The risk that a material error exists that will not be prevented or detected in a timely manner by the internal controls system. For example, the control risk associated with manual reviews of computer logs can be high because activities requiring investigation are often easily missed due to the volume of logged information. The control risk associated with computerized data validation procedures is ordinarily low if the processes are consistently applied.
• **Detection risk**—The risk that an IS auditor uses an inadequate test procedure and concludes that material errors do not exist when, in fact, they do. Detection of an error would not be determined during the risk assessment phase of an audit. However, identifying detection risk would better evaluate and assess the auditor's ability to test, identify and recommend the correction of material errors as the result of a test.
• **Overall audit risk**—The combination of the individual categories of audit risks assessed for each specific control objective. An objective in formulating the audit approach is to limit the audit risk in the area under scrutiny so the overall audit risk is at a sufficiently low level at the completion of the examination. Another objective is to assess and control those risks to achieve the desired level of assurance as efficiently as possible.

Audit risk is also used sometimes to describe the level of risk that the IS auditor is prepared to accept during an audit engagement. The auditor may set a target level of risk and adjust the amount of detailed audit work to minimize the overall audit risk.

> **Note:** Audit risk should not be confused with statistical sampling risk, which is the risk that incorrect assumptions are made about the characteristics of a population from which a sample is selected.

The word "material," when associated with any of these components of risks, refers to an error that should be considered significant to any party concerned with the item in question. Materiality considerations combined with an understanding of audit risk are essential concepts for planning areas to be audited as well as the specific tests to be performed in a given audit. The assessment of what is material is a matter of professional judgment and includes consideration of the effect on the organization as a whole, and errors, omissions, irregularities and illegal acts that may arise as a result of control weaknesses in the area being audited.

Specifically, this means that an internal control weakness or set of combined internal control weaknesses leaves the organization highly susceptible to the occurrence of a threat (e.g., financial loss, business interruption, loss of customer trust, economic sanction, etc.). The IS auditor should be concerned with assessing the materiality of the items in question through a risk-based audit approach to evaluating internal controls.

The IS auditor should have a good understanding of these audit risks when planning an audit. An audit sample may not detect every potential error in a population. However, by using proper statistical sampling procedures or a strong quality control process, the probability of detection risk can be reduced to an acceptable level.

Similarly, when evaluating internal controls, the IS auditor should realize that a given system may not detect a minor error. However, that specific error, combined with others, could become material to the overall system.

The concept of materiality requires sound judgment from the IS auditor. The IS auditor may detect a small error that could be considered significant at an operational level, but may not be viewed as significant to upper management. Materiality considerations combined with an understanding of audit risk are essential concepts for planning the areas to be audited and the specific test to be performed in a given audit.

Materiality can be more difficult for the IS auditor. For example, a logical security parameter setting that allows a programmer to access, without authorization, the source code for all programs might be a material error due to the potential pervasive impact on data integrity and accuracy. Similarly, access rights to only a few more insignificant programs might not be considered material to the IS auditor. Materiality is considered in terms of the total potential impact to the organization.

| PRACTICE QUESTIONS |
|---|
| 1-4    Which of the following types of audit risk assumes an absence of compensating controls in the area being reviewed?<br><br>A. Control risk<br>B. Detection risk<br>C. Inherent risk<br>D. Sampling risk |
| 1-5    An IS auditor performing a review of an application's controls finds a weakness in system software that could materially impact the application. The IS auditor should:<br><br>A. disregard these control weaknesses since a system software review is beyond the scope of this review.<br>B. conduct a detailed system software review and report the control weaknesses.<br>C. include in the report a statement that the audit was limited to a review of the application's controls.<br>D. review the system software controls as relevant and recommend a detailed system software review. |
| *See answers and explanations to the practice questions at the end of the chapter. (page 68)* |

## 1.6.7 RISK ASSESSMENT AND TREATMENT

### Assessing Security Risks

To develop a more complete understanding of audit risk, the IS auditor should also understand how the organization being audited approaches risk assessment and treatment.

Risk assessments should identify, quantify and prioritize risks against criteria for risk acceptance and objectives relevant to the organization. The results should guide and determine the appropriate management action, priorities for managing information security risks, and priorities for implementing controls selected to protect against these risks.

Risk assessment should include the systematic approach of estimating the magnitude of risks (risk analysis) and the process of comparing the estimated risks against risk criteria to determine the significance of the risks (risk evaluation).

Risk assessments should also be performed periodically to address changes in the environment, security requirements and in the risk situation (e.g., in the assets, threats, vulnerabilities, impacts), and when significant changes occur. These risk assessments should be undertaken in a methodical manner capable of producing comparable and reproducible results.

The information security risk assessment should have a clearly defined scope in order to be effective and should include relationships with risk assessments in other areas if appropriate.

The scope of a risk assessment can be either the whole organization, parts of the organization, an individual information system, specific system components, or services where this is practicable, realistic and helpful.

### Treating Security Risks

Before considering the treatment of a risk, the organization should decide the criteria for determining whether risks can be accepted. Risks may be accepted if, for example, it is assessed that the risk is low or that the cost of treatment is not cost-effective for the organization. Such decisions should be recorded.

Each of the risks identified in the risk assessment needs to be treated. Possible options for risk treatment include:
• Applying appropriate controls to reduce the risks.
• Knowingly and objectively accepting risks, providing they clearly satisfy the organization's policy and criteria for risk acceptance.
• Avoiding risks by not allowing actions that would cause the risks to occur.
• Transferring the associated risks to other parties, e.g. insurers or suppliers.

For those risks where the risk treatment decision has been to apply appropriate controls, controls should be selected to ensure that risks are reduced to an acceptable level, taking into account:
• Requirements and constraints of national and international legislation and regulations
• Organizational objectives
• Operational requirements and constraints
• Cost effectiveness (the need to balance the investment in implementation and operation of controls against the harm likely to result from security failures)

Controls can be selected from this standard or from other control sets, or new controls can be designed to meet the specific needs of the organization. It is necessary to recognize that some controls may not be applicable to every information system or environment, and might not be practical for all organizations.

Information security controls should be considered at the systems and projects requirements specification and design stage. Failure to do so can result in additional costs and less effective solutions, and, in a worst case scenario, the inability to achieve adequate security.

No set of controls can achieve complete security. Additional management action should be implemented to monitor, evaluate, and improve the efficiency and effectiveness of security controls to support the organization's aims.

# 1.6.8 RISK ASSESSMENT TECHNIQUES

When determining which functional areas should be audited, the IS auditor could face a large variety of audit subjects. Each of these subjects may represent different types of audit risks. The IS auditor should evaluate these various risk candidates to determine the high-risk areas that should be audited.

There are many risk assessment methodologies, computerized and noncomputerized, from which the IS auditor may choose. These range from simple classifications of high, medium and low, based on the IS auditor's judgment, to complex scientific calculations that provide a numeric risk rating.

One such risk assessment approach is a scoring system that is useful in prioritizing audits based on an evaluation of risk factors. The system considers variables such as technical complexity, level of control procedures in place and level of financial loss. These variables may or may not be weighted. The risk values are then compared to each other and audits are scheduled accordingly. Another form of risk assessment is judgmental, where an independent decision is made based on business knowledge, executive management directives, historical perspectives, business goals and environmental factors. A combination of techniques may be used as well. Risk assessment methods may change and develop over time to best serve the needs of the organization. The IS auditor should consider the level of complexity and detail appropriate for the organization being audited.

Using risk assessment to determine areas to be audited:
• Enables management to effectively allocate limited audit resources.
• Ensures that relevant information has been obtained from all levels of management, including boards of directors, IS auditors and functional area management. Generally, this information assists management in effectively discharging its responsibilities and ensures that the audit activities are directed to high-risk areas, which will add value for management.
• Establishes a basis for effectively managing the audit department.
• Provides a summary of how the individual audit subject is related to the overall organization as well as to the business plans.

# 1.6.9 AUDIT OBJECTIVES

Audit objectives refer to the specific goals that must be accomplished by the audit. In contrast, a control objective refers to how an internal control should function. An audit may, and generally does, incorporate several audit objectives.

Audit objectives often focus on substantiating that internal controls exist to minimize business risks, and that they function as expected. These audit objectives include assuring compliance with legal and regulatory requirements as well as the confidentiality, integrity, reliability and availability of information and IT resources. Audit management may give the IS auditor a general control objective to review and evaluate when performing an audit.

A key element in planning an IS audit is to translate basic and wide-ranging audit objectives into specific IS audit objectives. For example, in a financial/operational audit, an internal control objective could be to ensure that transactions are properly posted to the general ledger accounts. However, in the IS audit, the objective could be extended to ensure that editing features are in place to detect errors in the coding of transactions that may impact the account-posting activities.

The IS auditor must have an understanding of how general audit objectives can be translated into specific IS control objectives. Determining an audit's objectives is a critical step in planning an IS audit.

One of the basic purposes of any IS audit is to identify control objectives and the related controls that address the objective.

For example, the IS auditor's initial review of an information system should identify key controls. The IS auditor should then decide whether to test these controls for compliance. The IS auditor should identify both key general and application controls after developing an understanding, and documenting the business processes and the applications/functions that support these processes and general support systems. Based on that understanding, the IS auditor should identify the key control points.

Alternatively, an IS auditor may assist in assessing the integrity of financial reporting data-referred to as substantive testing-through computer-assisted audit techniques.

# 1.6.10 COMPLIANCE VS. SUBSTANTIVE TESTING

Compliance testing is evidence gathering for the purpose of testing an organization's compliance with control procedures. This differs from substantive testing in which evidence is gathered to evaluate the integrity of individual transactions, data or other information.

A compliance test determines if controls are being applied in a manner that complies with management policies and procedures. For example, if the IS auditor is concerned about whether production program library controls are working properly, the IS auditor might select a sample of programs to determine if the source and object versions are the same. The broad objective of any compliance test is to provide IS auditors with reasonable assurance that the particular control on which the IS auditor plans to rely is operating as the IS auditor perceived in the preliminary evaluation.

It is important that the IS auditor understands the specific objective of a compliance test and of the control being tested. Compliance tests can be used to test the existence and effectiveness of a defined process, which may include a trail of documentary and/or automated evidence-for example, to provide assurance that only authorized modifications are made to production programs.

A substantive test substantiates the integrity of actual processing. It provides evidence of the validity and integrity of the balances in the financial statements, and the transactions that support these balances. IS auditors could use substantive tests to test for monetary errors directly affecting financial statement balances, or other relevant data of the organization. Additionally, an IS auditor might develop a substantive test to determine if the tape library inventory records are stated correctly. To perform this test, the IS auditor might take a thorough inventory or might use a statistical sample, which will allow the IS auditor to develop a conclusion regarding the accuracy of the entire inventory.

There is a direct correlation between the level of internal controls and the amount of substantive testing required. If the results of testing controls (compliance tests) reveal the presence of adequate internal controls, then the IS auditor is justified in minimizing the substantive procedures. Conversely, if the control testing reveals weaknesses in controls that may raise doubts about the completeness, accuracy or validity of the accounts, substantive testing can alleviate those doubts.

Examples of compliance testing of controls where sampling could be considered include user access rights, program change control procedures, documentation procedures, program documentation, follow-up of exceptions, review of logs, software license audits, etc.

Examples of substantive tests where sampling could be considered include performance of a complex calculation (e.g., interest) on a sample of accounts or a sample of transactions to vouch for supporting documentation, etc.

The IS auditor could also decide during the preliminary assessment of the controls to include some substantive testing if the results of this preliminary evaluation indicate that implemented controls are not reliable or do not exist.

**Exhibit 1.5** shows the relationship between compliance and substantive tests, and describes the two categories of substantive tests.

> **Note:** The IS auditor should be knowledgeable on when to perform compliance tests or substantive tests.

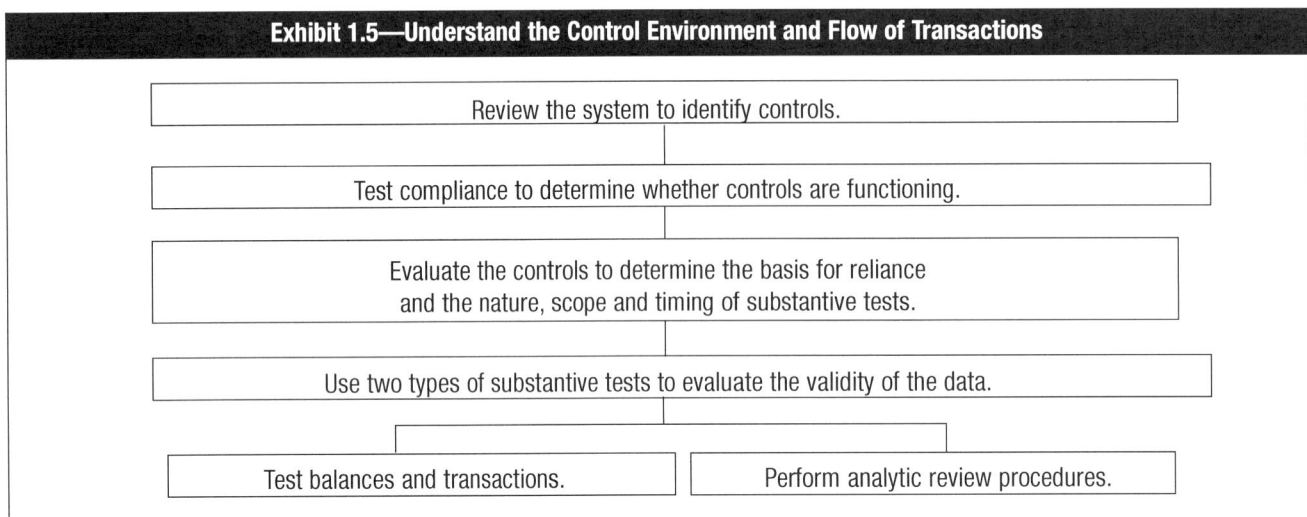

**Exhibit 1.5—Understand the Control Environment and Flow of Transactions**

Review the system to identify controls.

Test compliance to determine whether controls are functioning.

Evaluate the controls to determine the basis for reliance
and the nature, scope and timing of substantive tests.

Use two types of substantive tests to evaluate the validity of the data.

| Test balances and transactions. | Perform analytic review procedures. |

## 1.6.11 EVIDENCE

Evidence is any information used by the IS auditor to determine whether the entity or data being audited follows the established criteria or objectives, and supports audit conclusions. It is a requirement that the auditor's conclusions be based on sufficient, relevant and competent evidence. When planning the IS audit, the IS auditor should take into account the type of audit evidence to be gathered, its use as audit evidence to meet audit objectives and its varying levels of reliability.

Audit evidence may include the IS auditor's observations (presented to management), notes taken from interviews, material extracted from correspondence and internal documentation or contracts with external partners, or the results of audit test procedures. While all evidence will assist the IS auditor in developing audit conclusions, some evidence is more reliable than others. The rules of evidence and sufficiency as well as the competency of evidence must be taken into account as required by audit standards.

> **Note:** Knowledge of the evidence life cycle is difficult to test in the CISA exam because of the various laws and regulations governing the collection, protection and chain of custody of evidence. This topic, though relevant for the IS auditor, is not currently tested in the CISA exam.

Determinants for evaluating the reliability of audit evidence include:
- **Independence of the provider of the evidence**—Evidence obtained from outside sources is more reliable than from within the organization. This is why confirmation letters are used for verification of accounts receivable balances. Additionally, signed contracts or agreements with external parties could be considered reliable if the original documents are made available for review.
- **Qualifications of the individual providing the information/evidence**—Whether the providers of the information/evidence are inside or outside of the organization, the IS auditor should always consider the qualifications and functional responsibilities of the persons providing the information. This can also be true of the IS auditor. If an IS auditor does not have a good understanding of the technical area under review, the information gathered from testing that area may not be reliable, especially if the IS auditor does not fully understand the test.
- **Objectivity of the evidence**—Objective evidence is more reliable than evidence that requires considerable judgment or interpretation. An IS auditor's review of media inventory is direct, objective evidence. An IS auditor's analysis of the efficiency of an application, based on discussions with certain personnel, may not be objective audit evidence.
- **Timing of the evidence**—The IS auditor should consider the time during which information exists or is available in determining the nature, timing and extent of compliance testing and, if applicable, substantive testing. For example, audit evidence processed by dynamic systems, such as spreadsheets, may not be retrievable after a specified period of time if changes to the files are not controlled or the files are not backed up.

The IS auditor gathers a variety of evidence during the audit. Some evidence may be relevant to the objectives of the audit, while other evidence may be considered peripheral. The IS auditor should focus on the overall objectives of the review and not the nature of the evidence gathered.

The quality and quantity of evidence must be assessed by the IS auditor. These two characteristics are referred to by the International Federation of Accountants (IFAC) as competent (quality) and sufficient (quantity). Evidence is competent when it is both valid and relevant. Audit judgment is used to determine when sufficiency is achieved in the same manner that is used to determine the competency of evidence.

An understanding of the rules of evidence is important for IS auditors since they may encounter a variety of evidence types.

Gathering of evidence is a key step in the audit process. The IS auditor should be aware of the various forms of audit evidence and how evidence can be gathered and reviewed. The IS auditor should understand ISACA IS Auditing Standards S6 and S14, and should obtain evidence of a nature and sufficiency to support audit findings.

> **Note:** A CISA candidate, given an audit scenario, should be able to determine which type of evidence gathering technique would be best.

The following are techniques for gathering evidence:
- **Reviewing IS organization structures**—An organizational structure that provides an adequate separation or segregation of duties is a key general control in an IS environment. The IS auditor should understand general organizational controls and be able to evaluate these controls in the organization under audit. Where there is a strong emphasis on cooperative distributed processing or on end-user computing, IS functions may be organized somewhat differently than the classic IS organization which consists of separate systems and operations functions. The IS auditor should be able to review these organizational structures and assess the level of control they provide.
- **Reviewing IS policies and procedures**—An IS auditor should review whether appropriate policies and procedures are in place, determine whether personnel understand the implemented policies and procedures, and ensure that policies and procedures are being followed. The IS auditor should verify that management assumes full responsibility for formulating, developing, documenting, promulgating and controlling policies covering general aims and directives. Periodic reviews of policies and procedures for appropriateness should be carried out.
- **Reviewing IS standards**—The IS auditor should first understand the existing standards in place within the organization.
- **Reviewing IS documentation**—A first step in reviewing the documentation for an information system is to understand the existing documentation in place within the organization. This documentation could be held in hard copy form or stored electronically (e.g., document images stored on the internal corporate network). If the latter is the case, controls to preserve the document integrity should be evaluated by the IS auditor. The IS auditor should look for a minimum level of IS documentation. Documentation may include:
  – Systems development initiating documents (e.g., feasibility study)
  – Documentation provided by external application suppliers
  – Service level agreements (SLAs) with external IT providers
  – Functional requirements and design specifications
  – Tests plans and reports
  – Program and operations documents
    Program change logs and histories
  – User manuals
  – Operations manuals
  – Security-related documents (e.g., security plans, risk assessments)
  – BCPs
  – QA reports
  – Reports on security metrics
- **Interviewing appropriate personnel**—Interviewing techniques are an important skill for the IS auditor. Interviews should be organized in advance with objectives clearly communicated, follow a fixed outline and be documented by interview notes. An interview form or checklist prepared by an IS auditor is a good approach. The IS auditor should always remember that the purpose of such an interview is to gather audit evidence. Procedures to gather audit evidence include:

inquiry, observation, inspection, confirmation, performance and monitoring. Personnel interviews are discovery in nature and should never be accusatory. The IS auditor should verify the accuracy of the notes with the interviewee whether or not these notes would be necessary to support conclusions.

• **Observing processes and employee performance**—The observation of processes is a key audit technique for many types of review. The IS auditor should be unobtrusive while making observations and should document everything in sufficient detail to be able to present it, if required, as audit evidence at a later date. In some situations, the release of the audit report may not be timely enough to use this observation as evidence. This may necessitate the issuance of an interim report to management of the area being audited. The IS auditor may also wish to consider whether documentary evidence would be useful as evidence (e.g., photograph of a server room with doors fully opened).

All of these techniques for gathering evidence are part of an audit, but an audit is not considered only review work. An audit includes examination, which incorporates by necessity the testing of controls and audit evidence, and therefore includes the results of audit tests.

IS auditors should recognize that with systems development techniques such as computer-aided software engineering (CASE) or prototyping, traditional systems documentation will not be required or will be in an automated form rather than on paper. However, the IS auditor should look for documentation standards and practices within the IS organization.

The IS auditor should be able to review documentation for a given system and determine whether it follows the organization's documentation standards. In addition, the IS auditor should understand the current approaches to developing systems such as object orientation, CASE tools or prototyping, and how the documentation is constructed. The IS auditor should recognize other components of IS documentation such as database specifications, file layouts or self-documented program listings.

## 1.6.12 INTERVIEWING AND OBSERVING PERSONNEL IN ACTION

Observing personnel in the performance of their duties assists an IS auditor in identifying:
• **Actual functions**—Observation could be an adequate test to ensure that the individual who is assigned and authorized to perform a particular function is the person who is actually doing the job. It allows the IS auditor an opportunity to witness how policies and procedures are understood and practiced. Depending on the specific situation, the results of this type of test should be compared with the respective logical access rights.
• **Actual processes/procedures**—Performing a walk-through of the process/procedure allows the IS auditor to gain evidence of compliance and observe deviations, if any. This type of observation could prove to be useful for physical controls.
• **Security awareness**—Security awareness should be observed to verify an individual's understanding and practice of good preventive and detective security measures to safeguard the company's assets and data. This type of information could be complemented with an examination of previous and planned security training.
• **Reporting relationships**—Reporting relationships should be observed to ensure that assigned responsibilities and adequate segregation of duties are being practiced. Often, the results of this type of test should be compared with the respective logical access rights.

Interviewing information processing personnel and management should provide adequate assurance that the staff has the required technical skills to perform the job. This is an important factor that contributes to an effective and efficient operation.

## 1.6.13 SAMPLING

Sampling is used when time and cost considerations preclude a total verification of all transactions or events in a predefined population. The population consists of the entire group of items that need to be examined. The subset of population members used to perform testing is called a sample. Sampling is used to infer characteristics about a population, based on the characteristics of a sample.

**Note:** Increasing regulation of organizations has led to a major focus on the IS auditor's ability to verify the adequacy of internal controls through the use of sampling techniques. This has become necessary since many controls are transactional in nature, which can make it difficult to test the entire population. Although a candidate is not expected to become a sampling expert, it is important for the candidate to have a foundational understanding of the general principles of sampling and how to design a sample that can be relied on.

The two general approaches to audit sampling are statistical and nonstatistical:
1. **Statistical sampling**—An objective method of determining the sample size and selection criteria. Statistical sampling uses the mathematical laws of probability to: a) calculate the sampling size, b) select the sample items, and c) evaluate the sample results and make the inference. With statistical sampling, the IS auditor quantitatively decides how closely the sample should represent the population (assessing sample precision) and the number of times in 100 that the sample should represent the population (the reliability or confidence level). This assessment will be represented as a percentage. The results of a valid statistical sample are mathematically quantifiable.
2. **Nonstatistical sampling (often referred to as judgmental sampling)**—Uses auditor judgment to determine the method of sampling, the number of items that will be examined from a population (sample size) and which items to select (sample selection). These decisions are based on subjective judgment as to which items/transactions are the most material and most risky.

When using either statistical or nonstatistical sampling methods, the IS auditor should design and select an audit sample, perform audit procedures, and evaluate sample results to obtain sufficient, reliable, relevant and useful audit evidence. These methods of sampling require the IS auditor to use judgment when defining the population characteristics, and, thus are subject to the risk that the IS auditor will draw the wrong conclusion from the sample (sampling risk). However, statistical sampling permits the IS auditor to quantify the probability of error (confidence coefficient). To be a statistical sample, each item in the population should have an equal opportunity or probability of being selected. Within these two general approaches to audit sampling, there are two primary methods of sampling used by IS auditors—attribute sampling and variable sampling. Attribute sampling, generally applied in compliance testing situations, deals with the presence or absence of the attribute and provides conclusions that are expressed in rates of incidence. Variable sampling, generally applied in substantive testing situations, deals with population characteristics that vary, such as monetary values and weights (or any other measurement), and provides conclusions related to deviations from the norm.

Attribute sampling refers to three different but related types of proportional sampling:
1. **Attribute sampling** (also referred to as fixed sample-size attribute sampling or frequency-estimating sampling)—A sampling model that is used to estimate the rate (percent) of occurrence of a specific quality (attribute) in a population. Attribute sampling answers the question of "how many?" An example of an attribute that might be tested is approval signatures on computer access request forms.
2. **Stop-or-go sampling**—A sampling model that helps prevent excessive sampling of an attribute by allowing an audit test to be stopped at the earliest possible moment. Stop-or-go sampling is used when the IS auditor believes that relatively few errors will be found in a population.
3. **Discovery sampling**—A sampling model that can be used when the expected occurrence rate is extremely low. Discovery sampling is most often used when the objective of the audit is to seek out (discover) fraud, circumvention of regulations or other irregularities.

Variable sampling—also known as dollar estimation or mean estimation sampling—is a technique used to estimate the monetary value or some other unit of measure (such as weight) of a population from a sample portion. An example of variable sampling is a review of an organization's balance sheet for material transactions and an application review of the program that produced the balance sheet.

Variable sampling refers to a number of different types of quantitative sampling models:
1. **Stratified mean per unit**—A statistical model in which the population is divided into groups and samples are drawn from the various groups. Stratified mean sampling is used to produce a smaller overall sample size relative to unstratified mean per unit.
2. **Unstratified mean per unit**—A statistical model in which a sample mean is calculated and projected as an estimated total.
3. **Difference estimation**—A statistical model used to estimate the total difference between audited values and book (unaudited) values based on differences obtained from sample observations.

To perform attribute or variable sampling, the following statistical sampling terms need to be understood:
- **Confidence coefficient** (also referred to as confidence level or reliability factor)—A percentage expression (90 percent, 95 percent, 99 percent, etc.) of the probability that the characteristics of the sample are a true representation of the population. Generally, a 95 percent confidence coefficient is considered a high degree of comfort. If the IS auditor knows internal controls are strong, the confidence coefficient may be lowered. The greater the confidence coefficient, the larger the sample size.
- **Level of risk**—Equal to one minus the confidence coefficient. For example, if the confidence coefficient is 95 percent, the level of risk is five percent (100 percent minus 95 percent).
- **Precision**—Set by the IS auditor, it represents the acceptable range difference between the sample and the actual population. For attribute sampling, this figure is stated as a percentage. For variable sampling, this figure is stated as a monetary amount or a number. The higher the precision amount, the smaller the sample size and the greater the risk of fairly large total error amounts going undetected. The smaller the precision amount, the greater the sample size. A very low precision level may lead to an unnecessarily large sample size.
- **Expected error rate**—An estimate stated as a percent of the errors that may exist. The greater the expected error rate, the greater the sample size. This figure is applied to attribute sampling formulas but not to variable sampling formulas.
- **Sample mean**—The sum of all sample values, divided by the size of the sample. The sample mean measures the average value of the sample.
- **Sample standard deviation**—Computes the variance of the sample values from the mean of the sample. Sample standard deviation measures the spread or dispersion of the sample values.
- **Tolerable error rate**—Describes the maximum misstatement or number of errors that can exist without an account being materially misstated. Tolerable rate is used for the planned upper limit of the precision range for compliance testing. The term is expressed as a percentage. Precision range and precision have the same meaning when used in substantive testing.
- **Population standard deviation**—A mathematical concept that measures the relationship to the normal distribution. The greater the standard deviation, the larger the sample size. This figure is applied to variable sampling formulas but not to attribute sampling formulas.

Key steps in the construction and selection of a sample for an audit test include:
- Determining the objectives of the test
- Defining the population to be sampled
- Determining the sampling method, such as attribute vs. variable sampling
- Calculating the sample size
- Selecting the sample
- Evaluating the sample from an audit perspective

It is important to know that tools exist to analyze all of the data, not just those available through computer-assisted audit techniques.

> **Note:** The IS auditor should be familiar with the different types of sampling techniques and when it is appropriate to use each of them.

## 1.6.14 USING THE SERVICES OF OTHER AUDITORS AND EXPERTS

Due to the scarcity of IS auditors and the need for IT security specialists and other subject matter experts to conduct audits of highly specialized areas, the audit department or auditors entrusted with providing assurance may require the services of other auditors or experts. Outsourcing of IS assurance and security services is increasingly becoming a common practice. External experts could include experts in specific technologies such as networking, automated teller machine (ATM), wireless, systems integration and digital forensics, or subject matter experts such as specialists in a particular industry or area of specialization such as banking, securities trading, insurance, legal experts, etc.

When a part or all of IS audit services are proposed to be outsourced to another audit or external service provider, the following should be considered with regard to using the services of other auditors and experts:
- Restrictions on outsourcing of audit/security services provided by laws and regulations
- Audit charter or contractual stipulations

- Impact on overall and specific IS audit objectives
- Impact on IS audit risk and professional liability
- Independence and objectivity of other auditors and experts
- Professional competence, qualifications and experience
- Scope of work proposed to be outsourced and approach
- Supervisory and audit management controls
- Method and modalities of communication of results of audit work
- Compliance with legal and regulatory stipulations
- Compliance with applicable professional standards

Based on the nature of assignment, the following may also require special consideration:
- Testimonials/references and background checks
- Access to systems, premises and records
- Confidentiality restrictions to protect customer-related information
- Use of computer-assisted audit techniques (CAATs) and other tools to be used by the external audit service provider
- Standards and methodologies for performance of work and documentation
- Nondisclosure agreements

The IS auditor or entity outsourcing the services should monitor the relationship to ensure the objectivity and independence throughout the duration of the arrangement.

It is important to understand that often, even though a part of or the whole of the audit work may be delegated to an external service provider, the related professional liability is not necessarily delegated. Therefore, it is the responsibility of the IS auditor or entity employing the services of external service providers to:
- Clearly communicate the audit objectives, scope and methodology through a formal engagement letter.
- Put in place a monitoring process for regular review of the work of the external service provider with regard to planning, supervision, review and documentation. For example, review of the work papers of other IS auditors or experts to confirm the work was appropriately planned, supervised, documented and reviewed, and to consider the appropriateness and sufficiency of the audit evidence provided; or review of the report of other IS auditors or experts to confirm the scope specified in the audit charter, terms of reference or letter of engagement has been met, that any significant assumptions used by other IS auditors or experts have been identified, and the findings and conclusions reported have been agreed on by management.
- Assess the usefulness and appropriateness of reports of such external providers, and assess the impact of significant findings on the overall audit objectives.

> **Note:** The IS auditor should be familiar with ISACA's Auditing Standard on "Performance of Audit Work" (S6) and the IS Auditing Guideline "Using the Work of Other Auditors" (G1) focusing on the Rights of Access to the Work of Other Auditors or Experts.

## 1.6.15 COMPUTER-ASSISTED AUDIT TECHNIQUES

During the course of an audit, the IS auditor is to obtain sufficient, relevant and useful evidence to effectively achieve the audit objectives. The audit findings and conclusions should be supported by appropriate analysis and interpretation of the evidence. Today's information processing environments pose a significant challenge to the IS auditor to collect sufficient, relevant and useful evidence since the evidence exists on magnetic media.

CAATs are important tools for the IS auditor in gathering information from these environments. When systems have different hardware and software environments, data structure, record formats or processing functions, it is almost impossible for the auditors to collect evidence without a software tool to collect and analyze the records.

CAATs also enable IS auditors to gather information independently. CAATs provide a means to gain access and analyze data for a predetermined audit objective, and to report the audit findings with emphasis on the reliability of the records produced and maintained in the system. The reliability of the source of the information used provides reassurance on findings generated.

CAATs include many types of tools and techniques such as generalized audit software (GAS), utility software, debugging and scanning software, test data, application software tracing and mapping, and expert systems.

GAS refers to standard software that has the capability to directly read and access data from various database platforms, flat-file systems and ASCII formats. GAS provides IS auditors an independent means to gain access to data for analysis and the ability to use high-level, problem-solving software to invoke functions to be performed on data files. Features include mathematical computations, stratification, statistical analysis, sequence checking, duplicate checking and recomputations. The following functions are commonly supported by GAS:
- **File access**—Enables the reading of different record formats and file structures
- **File reorganization**—Enables indexing, sorting, merging and linking with another file
- **Data selection**—Enables global filtration conditions and selection criteria
- **Statistical functions**—Enables sampling, stratification and frequency analysis
- **Arithmetical functions**—Enables arithmetic operators and functions

| PRACTICE QUESTION |
| --- |
| 1-6    The **PRIMARY** use of generalized audit software (GAS) is to:<br><br>A.   test controls embedded in programs.<br>B.   test unauthorized access to data.<br>C.   extract data of relevance to the audit.<br>D.   reduce the need for transaction vouching. |
| *See answers and explanations to the practice questions at the end of the chapter. (page 68)* |

The effective and efficient use of software requires an understanding of its capabilities and limitations.

Utility software is a subset of software—such as report generators of the database management system—that provides evidence to auditors about system control effectiveness. Test data involve the auditors using a sample set of data to assess whether logic errors exist in a program and whether the program meets its objectives. The review of an application system will provide information about internal controls built in the system. The audit-expert system will give direction and valuable information to all levels of auditors while carrying out the audit because the query-based system is built on the knowledge base of the senior auditors or managers.

These tools and techniques can be used in performing various audit procedures:
- Tests of the details of transactions and balances
- Analytical review procedures
- Compliance tests of IS general controls
- Compliance tests of IS application controls
- Network and operating system (OS) vulnerability assessments
- Penetration testing
- Application security testing and source code security scans

The IS auditor should have a thorough understanding of CAATs, and know where and when to apply them. Please refer to G3: ISACA Guideline on Computer Assisted Audit Techniques.

An IS auditor should weigh the costs and benefits of CAATs before going through the effort, time and expense of purchasing or developing them. Issues to consider include:
- Ease of use, both for existing and future audit staff
- Training requirements
- Complexity of coding and maintenance
- Flexibility of uses
- Installation requirements
- Processing efficiencies (especially with a PC CAAT)
- Effort required to bring the source data into the CAATs for analysis

- Ensuring the integrity of imported data by safeguarding its authenticity
- Recording the time stamp of data downloaded at critical processing points to sustain the credibility of the review
- Obtaining permission to install the software on the auditee servers
- Reliability of the software
- Confidentiality of the data being processed

When developing CAATs, the following are examples of documentation to be retained:
- Online reports detailing high-risk issues for review
- Commented program listings
- Flowcharts
- Sample reports
- Record and file layouts
- Field definitions
- Operating instructions
- Description of applicable source documents

CAATs documentation should be referenced to the audit program, and clearly identify the audit procedures and objectives being served. When requesting access to production data for use with CAATs, the IS auditor should request read-only access. Any data manipulation by the IS auditor should be done to copies of production files in a controlled environment to ensure that production data are not exposed to unauthorized updating. Most of the CAATs provide for downloading production data from production systems to a standalone platform and then conducting analysis from the standalone platform, thereby insulating the production systems from any adverse impact.

## CAATs as a Continuous Online Audit Approach

An increasingly important advantage of CAATs is the ability to improve audit efficiency through continuous online auditing techniques. To this end, IS auditors must develop audit techniques that are appropriate for use with advanced computerized systems. In addition, they must be involved in the creation of advanced systems at the early stages of development and implementation, and must make greater use of automated tools that are suitable for their organization's automated environment. This takes the form of the continuous audit approach. (For more detailed information on continuous online auditing, see chapter 3, Systems and Infrastructure Life Cycle Management.)

# 1.6.16 EVALUATION OF AUDIT STRENGTHS AND WEAKNESSES

The IS auditor will review evidence gathered during the audit to determine if the operations reviewed are well controlled and effective. This is also an area that requires the IS auditor's judgment and experience. The IS auditor should assess the strengths and weaknesses of the controls evaluated and then determine if they are effective in meeting the control objectives established as part of the audit planning process.

A control matrix is often utilized in assessing the proper level of controls. Known types of errors that can occur in the area under review are placed on the top axis and known controls to detect or correct errors are placed on the side axis. Then, using a ranking method, the matrix is filled with the appropriate measurements. When completed, the matrix will illustrate areas where controls are weak or lacking.

In some instances, one strong control may compensate for a weak control in another area. For example, if the IS auditor finds weaknesses in a system's transaction error report, the IS auditor may find that a detailed manual balancing process over all transactions compensates for the weaknesses in the error report. The IS auditor should be aware of compensating controls in areas where controls have been identified as weak.

While a compensating control situation occurs when one stronger control supports a weaker one, overlapping controls are two strong controls. For example, if a data center employs a card key system to control physical access and a guard inside the door requires employees to show their card key or badge, an overlapping control exists. Either control might be adequate to restrict access, but the two complement each other.

Normally, a control objective will not be achieved by considering one control adequate. Rather, the IS auditor will perform a variety of testing procedures and evaluate how these relate to one another. Generally a group of controls, when aggregated together, may act as compensating controls, and thereby minimize the risk. An IS auditor should always review for compensating controls prior to reporting a control weakness.

The IS auditor may not find each control procedure to be in place but should evaluate the comprehensiveness of controls by considering the strengths and weaknesses of control procedures.

## *Judging the Materiality of Findings*

The concept of materiality is a key issue when deciding which findings to bring forward in an audit report. Key to determining the materiality of audit findings is the assessment of what would be significant to different levels of management. Assessment requires judging the potential effect of the finding if corrective action is not taken. A weakness in computer security physical access controls at a remote distributed computer site may be significant to management at the site, but will not necessarily be material to upper management at headquarters. However, there may be other matters at the remote site that would be material to upper management.

The IS auditor must use judgment when deciding which findings to present to various levels of management. For example, the IS auditor may find that the transmittal form for delivering tapes to the offsite storage location is not properly initialed or authorization evidenced by management as required by procedures. If the IS auditor finds that management otherwise pays attention to this process and that there have been no problems in this area, the IS auditor may decide that the failure to initial transmittal documents is not material enough to bring to the attention of upper management. The IS auditor might decide to discuss this only with local operations management. However, there may be other control problems that will cause the IS auditor to conclude that this is a material error because it may lead to a larger control problem in other areas. The IS auditor should always judge which findings are material to various levels of management and report them accordingly.

# 1.6.17 COMMUNICATING AUDIT RESULTS

The exit interview, conducted at the end of the audit, provides the IS auditor with the opportunity to discuss findings and recommendations with management. The objectives and scope of the audit can be discussed, and the IS audit process can be explained. During the exit interview, the IS auditor should:
• Ensure that the facts presented in the report are correct
• Ensure that the recommendations are realistic and cost-effective, and if not, seek alternatives through negotiation with auditee management
• Recommend implementation dates for agreed on recommendations

The IS auditor will frequently be asked to present the results of audit work to various levels of management. The IS auditor should have a thorough understanding of the presentation techniques necessary to communicate these results.

Presentation techniques could include the following:
• **Executive summary**—An easy-to-read, concise report that presents findings to management in an understandable manner. Most executive managers are not well versed in computer jargon; therefore, executive summaries should minimize the use of complex terminology. Findings and recommendations should be communicated from a business perspective. Detailed attachments can be more technical in nature since operations management will require the detail to correct the reported situations.
• **Visual presentation**—May include slides or computer graphics.

IS auditors should be aware that ultimately they are responsible to senior management and the audit committee of the board of directors. IS auditors should feel free to communicate issues or concerns to such management. An attempt to deny access by levels lower than senior management would limit the independence of the audit function.

Before communicating the results of an audit to senior management, the IS auditor should discuss the findings with the management staff of the audited entity. The goal of such a discussion would be to gain agreement on the findings and develop a course of corrective action. In cases where there is disagreement, the IS auditor should elaborate on the significance of the findings, risks and effects of not correcting the control weakness. Sometimes the auditee's management may request assistance from the IS auditor in implementing the recommended control enhancements. The IS auditor should communicate the difference between the IS auditor's role and that of a consultant, and give careful consideration to how assisting the auditee may adversely affect the IS auditor's independence.

Once agreement has been reached with the auditee, IS audit management should brief senior management of the audited organization. A summary of audit activities will be presented periodically to the audit committee. Audit committees typically are composed of individuals who do not work directly for the organization, and thus provide the auditors with an independent route to report sensitive findings.

## Audit Report Structure and Contents

Audit reports are the end product of the IS audit work. They are used by the IS auditor to report findings and recommendations to management. The exact format of an audit report will vary by organization; however, the skilled IS auditor should understand the basic components of an audit report and how it communicates audit findings to management.

> **Note:** The IS auditor should become familiar with the ISACA S7 Reporting and S8 Follow-up Activities standards.

There is no specific format for an IS audit report; the organization's audit policies and procedures will dictate the general format. Audit reports will usually have the following structure and content:
- An introduction to the report, including a statement of audit objectives, limitations to the audit and scope, the period of audit coverage, and a general statement on the nature and extent of audit procedures conducted and processes examined during the audit, followed by a statement on the IS audit methodology and guidelines
- A good practice is to include audit findings in separate sections. These findings can be grouped in sections by materiality and/or intended recipient.
- The IS auditor's overall conclusion and opinion on the adequacy of controls and procedures examined during the audit, and the actual potential risks identified as a consequence of detected deficiencies
- The IS auditor's reservations or qualifications with respect to the audit—This may state that the controls or procedures examined were found to be adequate or inadequate. The balance of the audit report should support that conclusion and the overall evidence gathered during the audit should provide an even greater level of support for the audit conclusions.
- Detailed audit findings and recommendations—The IS auditor would decide whether to include specific findings in an audit report. This should be based on the materiality of the findings and the intended recipient of the audit report. An audit report directed to the audit committee of the board of directors, for example, may not include findings that are important only to local management but have little control significance to the overall organization. The decision of what to include in various levels of audit reports depends on the guidance provided by upper management.
- A variety of findings, some of which may be quite material while others are minor in nature. The auditor may choose to present minor findings to management in an alternate format such as by memorandum.

The IS auditor, however, should make the final decision about what to include or exclude from the audit report. Generally, the IS auditor should be concerned with providing a balanced report, describing not only negative issues in terms of findings but positive constructive comments regarding improving processes and controls or effective controls already in place. Overall, the IS auditor should exercise independence in the reporting process.

Auditee management evaluates the findings, stating corrective actions to be taken and timing for implementing these anticipated corrective actions.

Management may not be able to implement all audit recommendations immediately. For example, the IS auditor may recommend changes to an information system that is also undergoing other changes or enhancements. The IS auditor should not necessarily expect that the other changes will be suspended until the IS auditor's recommendations are implemented. Rather, all may be implemented at once.

The IS auditor should discuss the recommendations and any planned implementation dates while in the process of releasing the audit report. The IS auditor must realize that various constraints—such as staff limitations, budgets or other projects—may limit immediate implementation. Management should develop a firm program for corrective actions. It is important to obtain a commitment from the auditee/management on the date by which the action plan will be implemented (the solution can be something which takes a long time for implementation) and the manner in which it will be done since the corrective action may bring certain risks that may be avoided if identified while discussing and finalizing the audit report. If appropriate, the IS auditor may want to report to upper management on the progress of implementing recommendations.

ISACA IS auditing standard S7 and the ISACA IS Auditing Guideline on Reporting (G20) state that the report should include all significant audit findings. When a finding requires explanation, the IS auditor should describe the finding, its cause and risk. When appropriate, the IS auditor should provide the explanation in a separate document and make reference to it in the report. For example, this approach may be appropriate for highly confidential matters. The IS auditor should also identify the organizational, professional and governmental criteria applied such as COBIT. The report should be issued in a timely manner to encourage prompt corrective action. When appropriate, the IS auditor should promptly communicate significant findings to the appropriate persons prior to the issuance of the report. Prior communication of significant findings should not alter the intent or content of the report.

> **Note:** The CISA candidate should review the detail from the IS Audit Guideline on "Reporting" (G20).

## 1.6.18 MANAGEMENT IMPLEMENTATION OF RECOMMENDATIONS

IS auditors should realize that auditing is an ongoing process. The IS auditor is not effective if audits are performed and reports issued, but no follow-up is conducted to determine whether management has taken appropriate corrective actions. IS auditors should have a follow-up program to determine if agreed on corrective actions have been implemented. Although IS auditors who work for external audit firms may not necessarily follow this process, they may achieve these tasks if agreed to by the audited entity.

The timing of the follow-up will depend on the criticality of the findings and would be subject to the IS auditor's judgment. The results of the follow-up should be communicated to appropriate levels of management.

The level of the IS auditor's follow-up review will depend on several factors. In some instances, the IS auditor may merely need to inquire as to the current status. In other instances, the IS auditor who works in an internal audit function may have to perform certain audit steps to determine whether the corrective actions agreed on by management have been implemented.

## 1.6.19 AUDIT DOCUMENTATION

Audit documentation should include, at a minimum, a record of the:
• Planning and preparation of the audit scope and objectives
• Description and/or walkthroughs on the scoped audit area
• Audit program
• Audit steps performed and audit evidence gathered
• Use of services of other auditors and experts
• Audit findings, conclusions and recommendations
• Audit documentation relation with document identification and dates

It is also recommended that documentation include:
• A copy of the report issued as a result of the audit work
• Evidence of audit supervisory review

Documents should include audit information that is required by laws and regulations, contractual stipulations and professional standards. Audit documentation is the necessary evidence supporting the conclusions reached, and hence should be clear, complete, easily retrievable and sufficiently comprehensible. Audit documentation is generally the property of the auditing entity and should be accessible only to authorized personnel under specific or general permission. Where access to audit documentation is requested by external parties, the auditor should obtain appropriate prior approval of senior management/client.

The IS auditor/IS audit department should also develop policies regarding custody, retention requirements and release of audit documentation.

> **Note:** The CISA candidate should be familiar with the detailed content of the IS Audit Guideline "Audit Documentation" (G8).

The documentation format and media are optional, but due diligence and best practices require that work papers are dated, initialed, page-numbered, relevant, complete, clear, self-contained and properly labeled, filed and kept in custody. Work papers may be automated. IS auditors should particularly consider how to maintain integrity and protection of audit test evidence to preserve their proof value in support of audit results.

Audit documentation or work papers can be considered the bridge or interface between the audit objectives and the final report. They should provide a seamless transition—with traceability and chargeability—from objectives to report and from report to objectives. The audit report, in this context, can be viewed as a set of particular work papers.

Audit documentation should support the finding and conclusions/opinion. Time of evidence sometimes will be crucial to supporting audit findings and conclusions. The IS auditor should take enough care to ensure that the evidence gathered and documented will be able to support audit findings and conclusions. An IS auditor should be able to prepare adequate working papers, narratives, questionnaires and understandable system flowcharts.

IS auditors are a scarce and expensive resource. Any technology capable of increasing the audit productivity is welcome. Automating work papers affects productivity directly and indirectly (granting access to other auditors, reusing documents or parts of them in recurring audits, etc.).

The quest for integrating work papers in the auditor's e-environment has resulted in all major audit and project management packages, CAATs and expert systems offering a complete array of automated documentation and import-export features.

ISACA IS Auditing Standards and Guidelines set forth many specifications about work papers, including how to use those of other auditors (previous or contractors), the need to document the audit plan, program and evidence, or the use of CAATs or sampling (G1 Using the Work of Other Auditors; G2 Audit Evidence Requirement; G3 Use of Computer Assisted Audit Techniques (CAATs); G8 Audit Documentation).

# 1.7 CONTROL SELF-ASSESSMENT

Control self-assessment (CSA) can be defined as a management technique that assures stakeholders, customers and other parties that the internal control system of the business is reliable. It also ensures that employees are aware of the risks to the business and they conduct periodic, proactive reviews of controls. It is a methodology used to review key business objectives, risks involved in achieving the business objectives and internal controls designed to manage these business risks in a formal, documented collaborative process.

In practice, CSA is a series of tools on a continuum of sophistication ranging from simple questionnaires to facilitated workshops, designed to gather information about the organization by asking those with a day-to-day working knowledge of an area as well as their managers. The basic tools used during a CSA project are the same whether the project is technical, financial or operational. These tools include management meetings, client workshops, worksheets, rating sheets and the CSA project approach. Like the continuum of tools used to gather information, there are diverse approaches to the levels below management that are queried; some organizations even include outsiders (such as clients or trading partners) when making CSA assessments.

The CSA program can be implemented by various methods. For small business units within organizations, it can be implemented by facilitated workshops where functional management and control professionals such as auditors can come together and deliberate how best to evolve a control structure for the business unit.

In the organizations with offices located at geographically dispersed locations, it may not be practical to organize facilitated workshops. In this case, a hybrid approach is needed. A questionnaire based on the control structure can be used. Operational managers can periodically complete the questionnaire, which can be analyzed and evaluated for effectiveness of the controls. However, a hybrid approach will be effective only if the analysis and readjustment of the questionnaire is performed using a life cycle approach, as shown in **exhibit 1.6**.

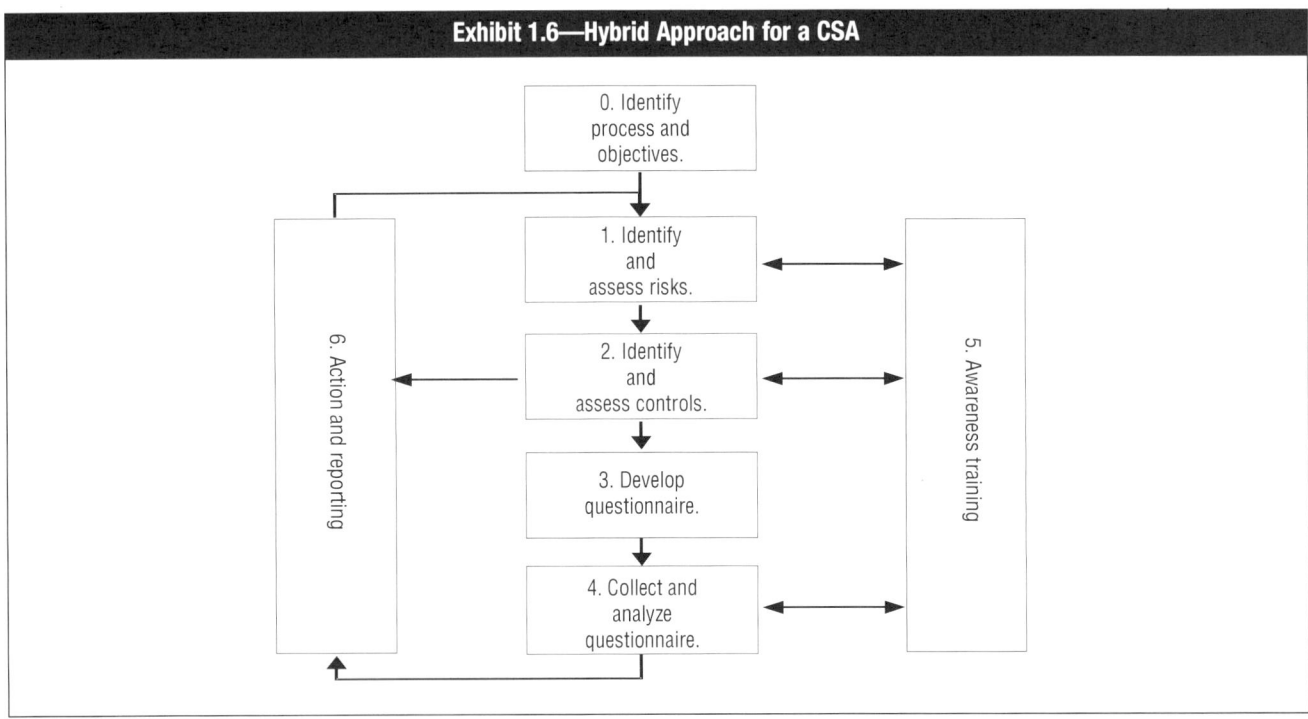

Exhibit 1.6—Hybrid Approach for a CSA

| PRACTICE QUESTION |
|---|

1-7  Which of the following is **MOST** effective for implementing a control self-assessment (CSA) within business units?

  A.  Informal peer reviews
  B.  Facilitated workshops
  C.  Process flow narratives
  D.  Data flow diagrams

*See answers and explanations to the practice questions at the end of the chapter (page 68).*

## 1.7.1 OBJECTIVES OF CSA

There are several objectives associated with adopting a CSA program. The primary objective is to leverage the internal audit function by shifting some of the control monitoring responsibilities to the functional areas. It is not intended to replace audit's responsibilities, but to enhance them. Clients such as line managers are responsible for controls in their environment; the managers also should be responsible for monitoring the controls. CSA programs also must educate management about control design and monitoring, particularly concentration on areas of high risk. These programs are not just policies requiring clients to comply with control standards. Instead, they offer a variety of support ranging from written suggestions outlining acceptable

control environments to in-depth workshops. When workshops are included in the program, an additional objective—the empowerment of workers to assess or even design the control environment—may be included in the program.

When employing a CSA program, measures of success for each phase (planning, implementation and monitoring) should be developed to determine the value derived from CSA and its future use. One critical success factor (CSF) is to conduct a meeting with the business unit representatives (including appropriate and relevant staff and management) to identify the business unit's primary objective-to determine the reliability of the internal control system. In addition, actions that increase the likelihood of achieving the primary objective should be identified.

A generic set of goals and metrics for each process, which can be used in designing and monitoring the CSA program, has been provided in CoBiT.

CoBiT is a governance and control framework that provides guidance in the development of the control assessment method. One could develop a CSA method by identifying the tasks and processes that are relevant to the business environment, and then defining the controls for relevant activities. A CSA questionnaire can be developed using the statements in the relevant control objectives of the identified IT processes. Various components of the CoBiT framework such as input-output matrix, RACI Chart, Goals and Metrics and Maturity Model can be converted in the form of a CSA questionnaire to assess each of the areas as required.

## 1.7.2 BENEFITS OF CSA

Some of the benefits of a CSA include the following:
• Early detection of risks
• More effective and improved internal controls
• Creation of cohesive teams through employee involvement
• Developing a sense of ownership of the controls in the employees and process owners, and reducing their resistance to control improvement initiatives
• Increased employee awareness of organizational objectives, and knowledge of risk and internal controls
• Increased communication between operational and top management
• Highly motivated employees
• Improved audit rating process
• Reduction in control cost
• Assurance provided to stakeholders and customers
• Necessary assurance given to top management about the adequacy of internal controls as required by the various regulatory agencies and laws such as the US Sarbanes-Oxley Act

## 1.7.3 DISADVANTAGES OF CSA

CSA does potentially contain several disadvantages which include:
• It could be mistaken as an audit function replacement
• It may be regarded as an additional workload (e.g., one more report to be submitted to management)
• Failure to act on improvement suggestions could damage employee morale
• Lack of motivation may limit effectiveness in the detection of weak controls

## 1.7.4 AUDITOR ROLE IN CSA

The auditor's role in CSAs should be considered enhanced when audit departments establish a CSA program. When these programs are established, auditors become internal control professionals and assessment facilitators. Their value in this role is evident when management takes responsibility and ownership for internal control systems under their authority through process improvements in their control structures, including an active monitoring component.

For an auditor to be effective in this facilitative and innovative role, the auditor must understand the business process being assessed. This can be attained via traditional audit tools such as a preliminary survey or walk-through. Also, the auditors must remember that they are the facilitators and the management client is the participant in the CSA process. For example, during a CSA workshop, instead of the auditor performing detailed audit procedures, the auditor will lead and guide the clients in assessing their environment by providing insight about the objectives of controls based on risk assessment. The managers, with a focus on improving the productivity of the process, might suggest replacement of preventive controls. In this case, the auditor is better positioned to explain the risks associated with such changes.

## 1.7.5 TECHNOLOGY DRIVERS FOR CSA

The development of techniques for empowerment, information gathering and decision making is a necessary part of a CSA program implementation. Some of the technology drivers include the combination of hardware and software to support CSA selection, and the use of an electronic meeting system and computer-supported decision aids to facilitate group decision making. Group decision making is an essential component of a workshop-based CSA where employee empowerment is a goal. In case of a questionnaire approach, the same principle applies for the analysis and readjustment of the questionnaire.

## 1.7.6 TRADITIONAL VS. CSA APPROACH

The traditional approach can be summarized as any approach in which the primary responsibility for analyzing and reporting on internal control and risk is assigned to auditors, and to a lesser extent, controller departments and outside consultants. This approach has created and reinforced the notion that auditors and consultants, not management and work teams, are responsible for assessing and reporting on internal control. The CSA approach, on the other hand, emphasizes management and accountability over developing and monitoring internal controls of an organization's sensitive and critical business processes.

A summary of attributes or focus that distinguishes each from the other is described in **exhibit 1.7**.

| Exhibit 1.7—Traditional and CSA Attributes | |
|---|---|
| **Traditional Historical** | **CSA** |
| Assigns duties/supervises staff | Empowered/accountable employees |
| Policy/rule-driven | Continuous improvement/learning curve |
| Limited employee participation | Extensive employee participation and training |
| Narrow stakeholder focus | Broad stakeholder focus |
| Auditors and other specialists | Staff at all levels, in all functions, are the primary control analysts. |
| Reporters | Reporters |

# 1.8 THE EVOLVING IS AUDIT PROCESS

The IS audit process must continually change to keep pace with innovations in technology. Topics to address these evolving changes include areas such as automated work papers, integrated auditing and continuous auditing.

## 1.8.1 AUTOMATED WORK PAPERS

Increasingly, audit teams are creating their audit work papers (risk analysis, audit programs, results, test evidences, conclusions, reports and other complementary information such as business information) in automated format, using specialized applications designed for this purpose.

Although auditors often use office automation packages such as text/word processors or spreadsheets, standard audit work paper packages are being implemented in more medium to large audit departments, and are proving to be useful and appropriate to help facilitate audit work.

In such cases, rules regarding integrity, confidentiality and availability of audit records should be applied that are equivalent to those required for hard copy or printed documents. Minimum controls that should be addressed include:
- Access to work papers (profiles and access rights)—i.e., no one should be authorized to change or delete audit records when audit work has been completed and a report issued, after audit management approval
- Audit trails-including when a document was changed, who performed the modification, automated update of a document version, when it was changed
- Automated features to provide and record approvals (e.g., by audit director, managers, etc.) of audit phases (audit program, conclusions, reports)
- Security and integrity controls regarding the operating system, databases and communication channels (e.g., server under audit control, corporate network, exporting documents, exclusive server)
- Backup and restore procedures
- Encryption techniques to provide confidentiality

## 1.8.2 INTEGRATED AUDITING

Dependence of business processes on information technology has necessitated that traditional financial and operational auditors develop an understanding of IT control structures, and IS auditors develop an understanding of the business control structures. Integrated auditing can be defined as the process whereby appropriate audit disciplines are combined to assess key internal controls over an operation, process or entity.

The integrated approach focuses on risk. A risk analysis assessment aims to understand and identify risks arising from the entity and its environment, including relevant internal controls. At this stage, the role of IT audit is typically to understand and identify risks under topical areas such as information management, IT infrastructure, IT governance and IT operations. Other audit specialists will seek to understand the organizational environment, business risks and business controls. A key element of the integrated approach is discussion of the risks arising among the whole audit team, with consideration of impact and likelihood.

Detailed audit work then focuses on the relevant controls in place to manage these risks. IT systems frequently provide a first line of preventive and detective controls, and the integrated audit approach depends on a sound assessment of their efficiency and effectiveness.

The integrated audit process typically involves:
- Identification of risks faced by the organization for the area being audited
- Identification of relevant key controls
- Review and understanding of the design of key controls
- Testing that key controls are supported by the IT system
- Testing that management controls operate effectively
- A combined report or opinion on control risks, design and weaknesses

The integrated audit demands a focus on business risk and a drive for creative control solutions. It is a team effort of auditors with different skill sets. Using this approach permits a single audit of an entity with one comprehensive report. An additional benefit is that this approach assists in staff development and retention by providing greater variety and the ability to see how all of the elements (functional and IT) mesh together to form the complete picture. See **exhibit 1.8** for an integrated auditing approach.

**Exhibit 1.8—An Integrated Audit**

Operational Audit

Financial Audit

IS Audit

The integrated audit concept has also radically changed the manner in which audits are looked on by the different stakeholders. Employees or process owners better understand the objectives of an audit as they are able to see the linkage between controls and audit procedures. Top management better understands the linkage between increased control effectiveness and corresponding improvements in the allocation and utilization of IT resources. Shareholders are able to better understand the linkage between the push for a greater degree of corporate governance and its impact on the generation of financial statements that can be relied on. All these developments have led to greater impetus for the growing popularity of integrated audits.

**Note:** This topic on Integrated Auditing, though important, is not specifically tested in the CISA exam.

## 1.8.3 CONTINUOUS AUDITING

The focus on increased effectiveness and efficiency of assurance, internal auditing and control has spurred the development of new studies and examination of new ideas concerning continuous auditing as opposed to more traditional periodic auditing reviews. Several research studies and documents addressing the subject carry different definitions of continuous auditing. All studies, however, recognize that a distinctive character of continuous auditing is the short time lapse between the facts to be audited, the collection of evidence and audit reporting.

Traditional financial reports and the traditional audit style sometime prove to be insufficient because they lack the essential element in today's business environment—updated information. Therefore, continuous auditing appears to be gaining more and more followers.

Some of the drivers of continuous auditing are a better monitoring of financial issues within a company, ensuring that real-time transactions also benefit from real-time monitoring, prevention of financial fraud and audit scandals such as Enron and WorldCom, and the use of software to determine that financial controls are proper. Continuous auditing involves a large amount of work because the company practicing continuous auditing will not provide one report at the end of a quarter, but will provide financial reports on a more frequent basis.

Continuous auditing is not a recent development. Traditional application systems may contain embedded audit modules. These would allow an auditor to trap predefined types of events, or to directly inspect abnormal or suspect conditions and transactions.

# The IS Audit Process

Most current commercial applications could be customized with such features. However, cost and other considerations and the technical skills that would be required to establish and operate these tools tend to limit the usage of embedded audit modules to specific fields and applications.

**Note:** This section on Continuous Auditing reflects the knowledge statement of continuous audit techniques (KS1.10); however, this is not specifically tested in the CISA exam.

To properly understand the implications and requirements of continuous auditing, a clear distinction has to be made between continuous auditing and continuous monitoring:
- **Continuous monitoring**—Provided by IS management tools and typically based on automated procedures to meet fiduciary responsibilities. For instance, real-time antivirus or intrusion detection systems (IDSs) may operate in a continuous monitoring fashion.
- **Continuous auditing**—"A methodology that enables independent auditors to provide written assurance on a subject matter using a series of auditors' reports issued simultaneously with, or a short period of time after, the occurrence of events underlying the subject matter" (from DeWayne L. Searcy and Jon B. Woodroof; "Continuous Auditing: Leveraging Technology," CICA/AICPA research report, May 2003). Continuous IS (and non-IS) auditing is typically completed using automated audit procedures.

Continuous auditing should be independent of continuous control or monitoring activities. When both continuous monitoring and auditing take place, continuous assurance can be established.

Efforts on the subject of continuous auditing often incorporate new IT developments, increased processing capabilities of current hardware and software, standards, and artificial intelligence (AI) tools. Continuous auditing attempts to facilitate the collection and analysis of data at the moment of the action. Data must be gathered from different applications working within different environments, transactions must be screened, the transaction environment has to be analyzed to detect trends and exceptions, and atypical patterns (i.e., a transaction with significantly higher or lower value than typical for a given business partner) must be exposed. If all of this must happen in real time, perhaps even before final sign-off of a transaction, it is mandatory to adopt and combine various top-level IT techniques. The IT environment is a natural enabler for the application of continuous auditing because of the intrinsic automated nature of its underlying processes.

Continuous auditing aims to provide a more secure platform to avoid fraud and a real-time process aimed at ensuring a high-level of financial control.

Prerequisites/preconditions for continuous auditing to succeed include:
- A high degree of automation
- An automated and highly reliable process in producing information about subject matter soon after or during the occurrence of events underlying the subject matter
- Alarm triggers to report timely control failures
- Implementation of highly automated audit tools that require the IS auditor to be involved in setting up the parameters
- Quickly informing IS auditors of the results of automated procedures, particularly when the process has identified anomalies or errors
- The quick and timely issuance of automated audit reports
- Technically proficient IS auditors
- Availability of reliable sources of evidence
- Adherence to materiality guidelines
- A change of mindset required for IS auditors to embrace continuous reporting
- Evaluation of cost factors

Simpler continuous auditing and monitoring tools are already built into many enterprise resource planning (ERP) packages and most operating system and network security packages. These environments, if appropriately configured and populated with rules, parameters and formulas, can output exception lists on request while operating against actual data. Therefore, they represent an instance of continuous auditing. The difficult but significant added value to using these features is that they postulate a definition of what would be a "dangerous" or exception condition. For instance, whether a set of granted IS access permissions is to be deemed risk-free will depend on having well-defined rules of segregation of duties. On the other hand, it may be much harder to decide if a given sequence of steps, taken to modify and maintain a database record, are pointing to a potential risk.

IT techniques that are used to operate in a continuous auditing environment must work at all data levels—single input, transaction and databases—and include:
• Transaction logging
• Query tools
• Statistics and data analysis (CAAT)
• Database management system (DBMS)
• Data warehouses, data marts, data mining
• Intelligent agents
• Embedded audit modules (EAM)
• Neural network technology
• Standards such as Extensible Business Reporting Language (XBRL)

Intelligent software agents may be used to automate the evaluation processes and allow for flexibility and dynamic analysis capabilities. The configuration and application of intelligent agents (sometimes referred to as bots), allows for continuous monitoring of systems settings and the delivery of alert messages when certain thresholds are exceeded or when certain conditions are met.

Full continuous auditing processes have to be carefully built into applications and work in layers. The auditing tools must operate in parallel to normal processing—capturing real-time data, extracting standardized profiles or descriptors, and passing the result to the auditing layers.

Continuous auditing has an intrinsic edge over point-in-time or periodic auditing because it captures internal control problems as they occur, preventing negative effects. Implementation can also reduce possible or intrinsic audit inefficiencies such as delays, planning time, inefficiencies of the audit process, overhead due to work segmentation, multiple quality or supervisory reviews, or discussions concerning the validity of findings.

Full top management support, dedication and extensive experience and technical knowledge are all necessary to accomplish this, while minimizing the impact on the underlying audited business processes. The auditing layers and settings may also need continual adjustment and updating. Besides difficulty and cost, continuous auditing has an inherent disadvantage in that internal control experts and auditors might be resistant to trust an automated tool in lieu of their personal judgment and evaluation. Also, mechanisms have to be put in place to eliminate false negatives and false positives in the reports generated by such audits so that the report generated continues to inspire stakeholders' confidence in the accuracy of the report.

The implementation of continuous auditing involves many factors; however, the task is not impossible. There is an increasing desire to provide auditing over information in a real-time environment (or as close to real time as possible).

| PRACTICE QUESTION |
|---|

**1-8** The **FIRST** step in planning an audit is to:

    A.  define audit deliverables.
    B.  finalize the audit scope and audit objectives.
    C.  gain an understanding of the business' objectives.
    D.  develop the audit approach or audit strategy.

**1-9** The approach an IS auditor should use to plan IS audit coverage should be based on:

    A.  risk.
    B.  materiality.
    C.  professional skepticism.
    D.  sufficiency of audit evidence.

**1-10** A company performs a daily backup of critical data and software files, and stores the backup tapes at an offsite location. The backup tapes are used to restore the files in case of a disruption. This is a:

    A.  preventive control.
    B.  management control.
    C.  corrective control.
    D.  detective control.

*See answers and explanations to the practice questions at the end of the chapter. (page 68)*

# 1.9 CHAPTER 1 QUICK REFERENCE REVIEW

| Chapter 1 Quick Reference Review |
| --- |
| Chapter 1 outlines the framework for performing IS auditing, specifically including those mandatory requirements regarding IS auditor mission and activity, as well as best practices to achieve a favorable IS auditing outcome.<br><br>CISA candidates should have a sound understanding of the following items, not only within the context of the present chapter, but also to correctly address questions in related subject areas. It is important to keep in mind that it is not enough to know these concepts from a definitional perspective. The CISA candidate must also be able to identify which elements among those presented may represent the greatest risk and which controls are most effective at mitigating this risk. |

- IS auditor roles and associated responsibilities, including expected audit outcomes: differences between IS auditing tasks within an assurance assignment and those within a consulting assignment.
- The need for audit independence and level of authority within the internal audit environment as opposed to an external context.
- Minimum planning audit requirements for an IS audit assignment, regardless of the specific or particular audit objective and scope.
- Understanding the required level of compliance with ISACA standards for IS auditing, as well as for ISACA guidelines.
- When planning audit work, the importance of clear identification of the audit approach related to controls defined as "general" versus auditing controls that are defined as "application controls."
- Scope, field work, application and execution of the concepts included in "audit risk" versus "business risk."
- The key role of requirements-compliant audit evidence when supporting the credibility of audit results and reporting.
- The reliance on electronic audit work papers and evidence.
- Purpose and planning opportunities of compliance testing versus substantive testing.
- Audit responsibility and level of knowledge when considering legal requirements affecting IT with an audit scope.
- The IS risk-oriented audit approach versus the complementary need for IS auditors to be acquainted with diverse IS standards and frameworks.
- Understanding the difference between the objectives of implemented controls and control procedures.

# 1.10 CHAPTER 1 CASE STUDIES

The following case studies are included as a learning tool to reinforce the concepts introduced in this chapter. Exam candidates should note that the CISA exam does not currently use this format for testing.

## 1.10.1 CASE STUDY A

### *Case Study A Scenario*

The IS auditor has been asked to perform preliminary work that will assess the readiness of the organization for a review to measure compliance with new regulatory requirements. These requirements are designed to ensure that management is taking an active role in setting up and maintaining a well-controlled environment, and accordingly will assess management's review and testing of the general IT control environment. Areas to be assessed include logical and physical security, change management, production control and network management, IT governance, and end-user computing. The IS auditor has been given six months to perform this preliminary work so sufficient time should be available. It should be noted that in previous years, repeated problems have been identified in the areas of logical security and change management so these areas will most likely require some degree of remediation. Logical security deficiencies noted included the sharing of administrator accounts and failure to enforce adequate controls over passwords. Change management deficiencies included improper segregation of incompatible duties and failure to document all changes. Additionally, the process for deploying operating system updates to servers was found to be only partially effective. In anticipation of the work to be performed by the IS auditor, the chief information officer (CIO) requested direct reports to develop narratives and process flows describing the major activities for which IT is responsible. These were completed, approved by the various process owners and the CIO, and then forwarded to the IS auditor for examination.

| CASE STUDY A QUESTIONS |
| --- |
| A1.    What should the IS auditor do **FIRST**?<br><br>    A.    Perform an IT risk assessment.<br>    B.    Perform a survey audit of logical access controls.<br>    C.    Revise the audit plan to focus on risk-based auditing.<br>    D.    Begin testing controls that the IS auditor feels are most critical. |
| A2.    When testing program change management, how should the sample be selected?<br><br>    A.    Change management documents should be selected at random and examined for appropriateness.<br>    B.    Changes to production code should be sampled and traced to appropriate authorizing documentation.<br>    C.    Change management documents should be selected based on system criticality and examined for appropriateness.<br>    D.    Changes to production code should be sampled and traced back to system-produced logs indicating the date and time of the change. |
| *See answers and explanations to the practice questions at the end of the chapter. (page 70)* |

*See answers and explanations to the practice questions at the end of the chapter. (page 70)*

## 1.10.2 CASE STUDY B

### *Case Study B Scenario*

An IS auditor is planning to review the security of a financial application for a large company with several locations worldwide. The application system is made up of a web interface, a business logic layer and a database layer. The application is accessed locally through a LAN and remotely through the Internet via a virtual private network (VPN) connection.

| CASE STUDY B QUESTIONS |
|---|
| B1. The **MOST** appropriate type of CAATs tool the auditor should use to test security configuration settings for the entire application system is:<br><br>    A.    generalized audit software (GAS).<br>    B.    test data.<br>    C.    utility software.<br>    D.    expert system. |
| B2. Given that the application is accessed through the Internet, how should the auditor determine whether to perform a detailed review of the firewall rules and VPN configuration settings?<br><br>    A.    Documented risk analysis<br>    B.    Availability of technical expertise<br>    C.    Approach used in previous audit<br>    D.    IS auditing guidelines and best practices |
| B3. During the review, if the auditor detects that the transaction authorization control objective cannot be met due to a lack of clearly defined roles and privileges in the application, the auditor should **FIRST**:<br><br>    A.    review the authorization on a sample of transactions.<br>    B.    immediately report this finding to upper management.<br>    C.    request that auditee management review the appropriateness of access rights for all users.<br>    D.    use a GAS to check the integrity of the database. |
| *See answers and explanations to the practice questions at the end of the chapter. (page 70)* |

## 1.10.3 CASE STUDY C

### *Case Study C Scenario*

An IS auditor has been appointed to carry out IS audits in an entity for a period of 2 years. After accepting the appointment the IS auditor noted that:

- The entity has an audit charter that detailed, among other things, the scope and responsibilities of the IS audit function and specifies the audit committee as the overseeing body for audit activity.
- The entity is planning a major increase in IT investment, mainly on account of implementation of a new ERP application, integrating business processes across units dispersed geographically. The ERP implementation is expected to become operational within the next 90 days. The servers supporting the business applications are hosted offsite by a third-party service provider.
- The entity has a new incumbent as Chief Information Security Officer (CISO); who reports to the Chief Financial Officer (CFO).
- The entity is subject to regulatory compliance requirements that require its management to certify the effectiveness of the internal control system as it relates to financial reporting. The entity has been recording consistent growth over the last two years at double the industry average. However, the entity has seen increased employee turnover as well.

| CASE STUDY C QUESTIONS |
|---|
| C1.   The **FIRST** priority of the IS auditor in Year 1 should be to study the:<br><br>   A.   previous IS audit reports and plan the audit schedule.<br>   B.   audit charter and plan the audit schedule.<br>   C.   impact of the new incumbent as CISO.<br>   D.   impact of the implementation of a new ERP on the IT environment and plan the audit schedule.<br><br>C2.   How should the IS auditor evaluate backup and batch processing within computer operations?<br><br>   A.   Plan and carry out an independent review of computer operations.<br>   B.   Rely on the service auditor's report of the service provider.<br>   C.   Study the contract between the entity and the service provider.<br>   D.   Compare the service delivery report to the service level agreement. |
| *See answers and explanations to the practice questions at the end of the chapter. (page 71)* |

# 1.11 ANSWERS TO PRACTICE QUESTIONS

1-1    **C**    Understanding the business process and environment applicable to the review is most representative of what occurs early on in the course of an audit. The other choices relate to activities actually occurring within this process.

1-2    **C**    Inherent risks exist independently of an audit and can occur because of the nature of the business. To successfully conduct an audit, it is important to be aware of the related business processes. To perform the audit the IS auditor needs to understand the business process, and by understanding the business process, the IS auditor better understands the inherent risks.

1-3    **A**    A risk-based audit approach focuses on the understanding of the nature of the business and being able to identify and categorize risk. Business risks impact the long-term viability of a specific business. Thus, an IS auditor using a risk-based audit approach must be able to understand business processes.

1-4    **C**    The risk of an error existing that could be material or significant when combined with other errors encountered during the audit, there being no related compensating controls, is the inherent risk. Control risk is the risk that a material error exists that will not be prevented or detected in a timely manner by the system of internal controls. Detection risk is the risk of an IS auditor using an inadequate test procedure that concludes that material errors do not exist, when they do. Sampling risk is the risk that incorrect assumptions are made about the characteristics of a population from which a sample is taken.

1-5    **D**    The IS auditor is not expected to ignore control weaknesses just because they are outside the scope of a current review. Further, the conduct of a detailed systems software review may hamper the audit's schedule and the IS auditor may not be technically competent to do such a review at this time. If there are control weaknesses that have been discovered by the IS auditor, they should be disclosed. By issuing a disclaimer, this responsibility would be waived. Hence, the appropriate option would be to review the systems software as relevant to the review and recommend a detailed systems software review for which additional resources may be recommended.

1-6    **C**    Generalized audit software facilitates direct access to and interrogation of the data by the IS auditor. The most important advantage of using GAS is that it helps in identifying data of interest to the IS auditor. GAS does not involve testing of application software directly. Hence, GAS helps in testing controls embedded in programs indirectly by testing data. GAS cannot identify unauthorized access to data if this information is not stored in the audit log file. However, this information may not always be available. Hence, this is not one of the primary reasons for using GAS. Vouching involves verification of documents. GAS could help in selecting transactions for vouching. Using GAS does not reduce transaction vouching.

1-7    **B**    Facilitated workshops work well within business units. Process flow narratives and data flow diagrams would not be as effective since they would not necessarily identify and assess all control issues. Informal peer reviews similarly would be less effective for the same reason.

1-8    **C**    The first step in audit planning is to gain an understanding of the business's mission, objectives and purpose, which in turn identifies the relevant policies, standards, guidelines, procedures, and organization structure. All other choices are dependent upon having a thorough understanding of the business's objectives and purpose.

1-9    **A**    Standard S5, Planning, establishes standards and provides guidance on planning an audit. It requires a risk-based approach.

1-10    **C**    A corrective control helps to correct or minimize the impact of a problem. Backup tapes can be used for restoring the files in case of damage of files, thereby reducing the impact of a disruption. Preventive controls are those that prevent problems before they arise. Backup tapes cannot be used to prevent damage to files and hence cannot be classified as a preventive control. Management controls modify processing systems to minimize a repeat occurrence of the problem. Backup tapes do not modify processing systems and hence do not fit the definition of a management control. Detective controls help to detect and report problems as they occur. Backup tapes do not aid in detecting errors.

# 1.12   ANSWERS TO CASE STUDY QUESTIONS

## ANSWERS TO CASE STUDY A QUESTIONS

A1.    **A**    An IT risk assessment should be performed first to ascertain which areas present the greatest risks and what controls mitigate those risks. Although narratives and process flows have been created, the organization has not yet assessed which controls are critical. All other choices would be undertaken after performing the IT risk assessment.

A2.    **B**    When testing a control, it is advisable to trace from the item being controlled to the relevant control documentation. When a sample is chosen from a set of control documents, there is no way to ensure that every change was accompanied by appropriate control documentation. Accordingly, changes to production code provide the most appropriate basis for selecting a sample. These sampled changes should then be traced to appropriate authorizing documentation. In contrast, selecting from the population of change management documents will not reveal any changes that bypassed the normal approval and documentation process. Similarly, comparing production code changes to system-produced logs will not provide evidence of proper approval of changes prior to their being migrated to production.

## ANSWERS TO CASE STUDY B QUESTIONS

B1.    **C**    When testing the security of the entire application system—including operating systems, database and application security—the auditor will most likely use a utility software that assists in reviewing the configuration settings. In contrast, the auditor might use GAS to perform a substantive testing of data and configuration files of the application.  Test data are normally used to check the integrity of the data and expert systems are used to inquire on specific topics.

B2.    **A**    In order to decide if the audit scope should include specific infrastructure components (in this case, the firewall rules and VPN configuration settings), the auditor should perform and document a risk analysis in order to determine which sections present the greatest risk and include these sections in the audit scope. The risk analysis may consider factors such as previous revisions to the system, related security incidents within the company or other companies of the same sectors, resources available to do the review and others. Availability of technical expertise and the approach used in previous audits may be taken into consideration; however, these should be of secondary importance. IS auditing guidelines and best practices provide a guide to the auditor on how to comply with IS audit standards, but by themselves they would not be sufficient to make this decision.

B3.    **A**    The auditor should first review the authorization on a sample of transactions in order to determine and be able to report the impact and materiality of this issue. Whether the auditor would immediately report the issue or wait until the end of the audit to report this finding will depend on the impact and materiality of the issue, which would require reviewing a sample of transactions. The use of GAS to check the integrity of the database would not help the auditor assess the impact of this issue.

## ANSWERS TO CASE STUDY C QUESTIONS

C1.  **D**  In terms of priority, as the implementation of the new ERP will have far reaching consequences on the way IS controls are configured in the system, the IS auditor should study the impact of implementation of the ERP and plan the audit schedule accordingly. Preferably, the IS auditor should discuss the audit plan with the external auditor and the internal audit division of the entity to make the audit more effective and useful for the entity.

C2.  **D**  The service delivery report which captures the actual performance of the service provider against the contractually agreed on levels provides the best and most objective basis for evaluation of the computer operations.  The Service Auditor's Report is likely to be more useful from a controls evaluation perspective for the external auditor of the entity.

# 1.13 SUGGESTED RESOURCES FOR FURTHER STUDY

Cascarino, Richard; *Auditor's Guide to Information Systems Auditing*, John Wiley & Sons Inc., USA, 2007

Davis, Chris; Mike Schiller; Kevin Wheeler; *IT Auditing Using Controls to Protect Information Assets*, McGraw Hill, USA, 2007

ISACA, *IS Standards, Guidelines and Procedures for Auditing and Control Professionals*, USA, 2007, *www.isaca.org/standards*

IT Governance Institute, *Control Objectives for Information and related Technology* (CoBIT) 4.1, USA, 2007

IT Governance Institute, *IT Control Objectives for Sarbanes-Oxley, 2nd Edition*, USA, 2006, *www.isaca.org/sox*

Senft, Sandra; Daniel P. Manson; Carol Gonzales; Frederick Gallegos; *Information Technology Control and Audit, 3rd Edition*, Auerbach, USA, 2008

*Note:* Publications in bold are stocked in the ISACA Bookstore. *Information Systems Control Journal* articles are available at *www.isaca.org/archives*. The articles are available online to ISACA members only during their first year of release, and then are opened to the public. All referenced *Journal* articles are available on the CISA Practice Question Database v9.

*Chapter 2:*

# IT Governance

# IT Governance

*Chapter 2:*

# IT Governance

# 2.1 INTRODUCTION

Knowledge of IT governance is fundamental to the work of the IS auditor. It forms the foundation for the development of sound control practices and mechanisms for management oversight and review.

## 2.1.1 OBJECTIVE

The objective of this area is to ensure that the CISA candidate understands and can provide assurance that the organization has the structure, policies, accountability mechanisms and monitoring practices in place to achieve the requirements of corporate governance of IT.

This area represents 15 percent of the CISA examination (approximately 30 questions).

## 2.1.2 TASKS

There are nine (9) tasks within the IT governance area:
T2.1  Evaluate the effectiveness of IT governance structure to ensure adequate board control over the decisions, directions and performance of IT, so it supports the organization's strategies and objectives.
T2.2  Evaluate IT organizational structure and human resources (personnel) management to ensure that they support the organization's strategies and objectives.
T2.3  Evaluate the IT strategy and process for their development, approval, implementation and maintenance to ensure that they support the organization's strategies and objectives.
T2.4  Evaluate the organization's IT policies, standards, procedures and processes for their development, approval, implementation and maintenance to ensure that they support the IT strategy and comply with regulatory and legal requirements.
T2.5  Evaluate management practices to ensure compliance with the organization's IT strategy, policies, standards and procedures.
T2.6  Evaluate IT resource investment, use and allocation practices to ensure alignment with the organization's strategies and objectives.
T2.7  Evaluate IT contracting strategies and policies and contract management practices to ensure that they support the organization's strategies and objectives.
T2.8  Evaluate risk management practices to ensure that the organization's IT-related risks are properly managed.
T2.9  Evaluate monitoring and assurance practices to ensure that the board and executive management receive sufficient and timely information about IT performance.

## 2.1.3 KNOWLEDGE STATEMENTS

There are fifteen (15) knowledge statements within the IT governance area:
KS2.1  Knowledge of the purpose of IT strategies, policies, standards and procedures for an organization and the essential elements of each
KS2.2  Knowledge of IT governance frameworks
KS2.3  Knowledge of the processes for the development, implementation and maintenance of IT strategies, policies, standards and procedures (e.g., protection of information assets, business continuity and disaster recovery, systems and infrastructure life cycle management, and IT service delivery and support)
KS2.4  Knowledge of quality management strategies and policies
KS2.5  Knowledge of organizational structure, roles and responsibilities related to the use and management of IT
KS2.6  Knowledge of generally accepted international IT standards and guidelines
KS2.7  Knowledge of enterprise IT architecture and its implications for setting long-term strategic directions
KS2.8  Knowledge of risk management methodologies and tools
KS2.9  Knowledge of the use of control frameworks (e.g., COBIT, COSO, ISO 17799)

KS2.10   Knowledge of the use of maturity and process improvement models (e.g., Capability Maturity Model [CMM], CobiT)
KS2.11   Knowledge of contracting strategies, processes and contract management practices
KS2.12   Knowledge of practices for monitoring and reporting of IT performance (e.g., balanced scorecards, key performance indicators [KPIs])
KS2.13   Knowledge of relevant legislative and regulatory issues (e.g., privacy, intellectual property, corporate governance requirements)
KS2.14   Knowledge of IT human resources (personnel) management
KS2.15   Knowledge of IT resource investment and allocation practices (e.g., portfolio management, return on investment [ROI])

## 2.1.4 RELATIONSHIP OF TASK TO KNOWLEDGE STATEMENTS

The task statements are what the CISA candidate is expected to know how to do. The knowledge statements delineate what the CISA candidate is expected to know in order to perform the tasks. The task and knowledge statements are approximately mapped in **exhibit 2.1** insofar as it is possible to do so. Note that although there is often overlap, each task statement will generally map to several knowledge statements.

| Exhibit 2.1—Tasks and Knowledge Statements Mapping ||
| Task Statements | Knowledge Statements |
|---|---|
| T2.1 Evaluate the effectiveness of IT governance structure to ensure adequate board control over the decisions, directions and performance of IT, so it supports the organization's strategies and objectives. | KS2.1 Knowledge of the purpose of IT strategies, policies, standards and procedures for an organization and the essential elements of each<br>KS2.2 Knowledge of IT governance frameworks<br>KS2.9 Knowledge of the use of control frameworks (e.g., CobiT, COSO, ISO 17799)<br>KS2.12 Knowledge of practices for monitoring and reporting of IT performance (e.g., balanced scorecards, key performance indicators [KPIs])<br>KS2.15 Knowledge of IT resource investment and allocation practices (e.g., portfolio management return on investment [ROI]) |
| T2.2 Evaluate IT organizational structure and human resources (personnel) management to ensure that they support the organization's strategies and objectives. | KS2.1 Knowledge of the purpose of IT strategies, policies, standards and procedures for an organization and the essential elements of each<br>KS2.5 Knowledge of organizational structure, roles and responsibilities related to the use and management of IT<br>KS2.14 Knowledge of IT human resources (personnel) management |
| T2.3 Evaluate the IT strategy and process for their development, approval, implementation and maintenance to ensure that they support the organization's strategies and objectives. | KS2.1 Knowledge of the purpose of IT strategies, policies, standards and procedures for an organization and the essential elements of each<br>KS2.3 Knowledge of the processes for the development, implementation and maintenance of IT strategies, policies, standards and procedures (e.g., protection of information assets, business continuity and disaster recovery, systems and infrastructure life cycle management, and IT service delivery and support)<br>KS2.4 Knowledge of quality management strategies and policies |

| Exhibit 2.1—Tasks and Knowledge Statements Mapping (*cont.*) ||
| Task Statements | Knowledge Statements |
| --- | --- |
| T2.4 Evaluate the organization's IT policies, standards, procedures and processes for their development, approval, implementation and maintenance to ensure that they support the IT strategy and comply with regulatory and legal requirements. | KS2.3 Knowledge of the processes for the development, implementation and maintenance of IT strategies, policies, standards and procedures (e.g., protection of information assets, business continuity and disaster recovery, systems and infrastructure life cycle management, and IT service delivery and support) <br><br> KS2.6 Knowledge of generally accepted international IT standards and guidelines <br><br> KS2.13 Knowledge of relevant legislative and regulatory issues (e.g., privacy, intellectual property, corporate governance requirements) |
| T2.5 Evaluate management practices to ensure compliance with the organization's IT strategy, policies, standards and procedures. | KS2.3 Knowledge of the processes for the development, implementation and maintenance of IT strategies, policies, standards and procedures (e.g., protection of information assets, business continuity and disaster recovery, systems and infrastructure life cycle management, and IT service delivery and support) <br><br> KS2.4 Knowledge of quality management strategies and policies <br><br> KS2.6 Knowledge of generally accepted international IT standards and guidelines <br><br> KS2.7 Knowledge of enterprise IT architecture and its implications for setting long-term strategic directions <br><br> KS2.9 Knowledge of the use of control frameworks (e.g., CoBiT, COSO, ISO 17799) <br><br> KS2.10 Knowledge of the use of maturity and process improvement models (e.g., CMM, CoBiT) <br><br> KS2.11 Knowledge of contracting strategies, processes and contract management practices <br><br> KS2.14 Knowledge of IT human resources (personnel) management <br><br> KS2.15 Knowledge of IT resource investment and allocation practices (e.g., portfolio management return on investment [ROI]) |
| T2.6 Evaluate IT resource investment, use and allocation practices to ensure alignment with the organization's strategies and objectives. | KS2.14 Knowledge of IT human resources (personnel) management |
| T2.7 Evaluate IT contracting strategies and policies and contract management practices to ensure that they support the organization's strategies and objectives. | KS2.11 Knowledge of contracting strategies, processes and contract management practices |

| Exhibit 2.1—Tasks and Knowledge Statements Mapping (*cont.*) | |
| --- | --- |
| **Task Statements** | **Knowledge Statements** |
| T2.8 Evaluate risk management practices to ensure that the organization's IT-related risks are properly managed. | KS2.3 Knowledge of the processes for the development, implementation and maintenance of IT strategies, policies, standards and procedures (e.g., protection of information assets, business continuity and disaster recovery, systems and infrastructure life cycle management, and IT service delivery and support)<br><br>KS2.8 Knowledge of risk management methodologies and tools |
| T2.9 Evaluate monitoring and assurance practices to ensure that the board and executive management receive sufficient and timely information about IT performance. | KS2.4 Knowledge of quality management strategies and policies<br><br>KS2.12 Knowledge of practices for monitoring and reporting of IT performance (e.g., balanced scorecards, key performance indicators [KPIs]) |

# 2.2 CORPORATE GOVERNANCE

Ethical issues, decision-making, and practices overall within an organization must be fostered through corporate governance practices. Corporate governance is defined as ethical corporate behavior by directors or others charged with governance in the creation and presentation of value for all stakeholders. The Organisation for Economic Co-operation and Development (OECD) States: "Corporate governance involves a set of relationships between a company's management, its board, its shareholders and other stakeholders. Corporate governance also provides the structure through which the objectives of the company are set, and the means of attaining those objectives and monitoring performance are determined. Good corporate governance should provide proper incentives for the board and management to pursue objectives that are in the interests of the company and its shareholders and should facilitate effective monitoring. *(OECD 2004, OECD Principles of Corporate Governance, p. 11)*"

With respect to public governance, the OECD states: "Good, effective public governance helps to strengthen democracy and human rights, promote economic prosperity and social cohesion, reduce poverty, enhance environmental protection and the sustainable use of natural resources, and deepen confidence in government and public administration." (OECD website on *Public Governance and Management*).

As part of this framework, rules should be established to manage and report on business risks. They should require organizations to have an internal control system to monitor risks when exploiting new and innovative ways to improve business. Simultaneously, this framework is a platform for the protection of stakeholders since it defines the responsibilities of the board of directors. In that way, shareholders, investors and other stakeholders will have defined duties and an adequate structure to decide about their investments, within a transparent framework. Thus, corporate governance seeks to strike a balance between the conflicting objectives of exploiting available opportunities to increase stakeholder value on the one hand, while keeping the organization's operations within the limits of regulatory requirements and social obligations on the other. Given the importance of social responsibility within corporate governance, the International Organization for Standardization (ISO) has launched the development of an International Standard (ISO 26000) providing voluntary guidelines for social responsibility (SR).

This framework is being increasingly mandated by government bodies of different countries in an effort to reduce the frequency and impact of inaccurate financial reporting, and provide greater transparency and accountability. Many of these government regulations include a requirement that senior management sign off on the adequacy of internal controls and include an assessment of organizational internal controls in the organization's financial reports.

# 2.3 MONITORING AND ASSURANCE PRACTICES FOR BOARD AND EXECUTIVE MANAGEMENT

IT governance is an inclusive term that encompasses information systems, technology and communication; business, legal and other issues; and all concerned stakeholders, directors, senior management, process owners, IT suppliers, users and auditors. Governance helps ensure the alignment of IT and enterprise objectives.

Enterprises are governed by generally accepted good or best practices, the assurance of which is provided by certain controls. From these practices flows the organization's direction, which indicates certain activities using the organization's resources. The results of these activities are measured and reported on, providing input to the cyclical revision and maintenance of controls.

IT is also governed by good or best practices that ensure that the organization's information and related technology support its business objectives, (i.e., strategic alignment), deliver value, that resources are used responsibly, that risks are managed appropriately, and that performance is measured.

Effective enterprise governance focuses individual and group expertise, and experience on specific areas where they can be most effective. Information technology, long considered only an enabler of an organization's strategy, is now regarded as an integral part of that strategy. Chief executive officers (CEOs), chief operating officers (COOs), chief financial officers (CFOs) and chief information officers (CIOs), and chief technology officers (CTOs) agree that strategic alignment between IT and enterprise objectives is a critical success factor. IT governance helps achieve this critical success factor by economically, efficiently and effectively deploying secure, reliable information and applied technology. Information technology is so critical to the success of enterprises that it cannot be relegated to either IT management or IT specialists, but must receive the attention of both in coordination with senior management.

Fundamentally, IT governance is concerned with two issues: that IT delivers value to the business and that IT risks are managed. The first is driven by strategic alignment of IT with the business. The second is driven by embedding accountability into the enterprise. Implementing the IT governance framework addresses these two issues by implementing practices that provide feedback on value delivery and risk management. The broad processes are:
• IT resource management which focuses on maintaining an updated inventory of all IT resources and addresses the risk management process.
• Performance measurement which focuses on ensuring that all IT resources perform as expected to deliver value to the business and also extends to identifying risks early on. This process is based on performance indicators that are optimized for value delivery and from which any deviation might lead to a materialization of risk.
• Compliance management which focuses on implementing processes that address legal and regulatory compliance requirements.

IT governance is the responsibility of the board of directors and executive management. It is an integral part of enterprise governance and consists of the leadership and organizational structures and processes that ensure that the organization's IT sustains and extends the organization's strategy and objectives (adapted from *Board Briefing on IT Governance, 2nd Edition*, ITGI, 2004).

A key element of IT governance is the alignment of business and IT, leading to the achievement of business value. This high-value goal can be achieved by aligning IT governance framework with best practices. Such a framework and practices should be composed of a variety of structures, processes and relational mechanisms. The key IT governance practices are IT strategy committee, risk management and standard IT balanced scorecard.

| PRACTICE QUESTION |
|---|
| 2-1    IT governance ensures that an organization aligns its IT strategy with: <br><br> A.    enterprise objectives. <br> B.    IT objectives. <br> C.    audit objectives. <br> D.    control objectives. |
| *See answers and explanations to the practice questions at the end of the chapter. (page 133)* |

Chapter 2:

# IT Governance

## 2.3.1 BEST PRACTICES FOR IT GOVERNANCE

Corporate governance is a set of responsibilities and practices used by an organization's management to provide strategic direction; thereby, ensuring that goals are achievable, risks are properly addressed and organizational resources are properly utilized (see **exhibit 2.2**).

---

**Exhibit 2.2—IT Governance Focus Areas**

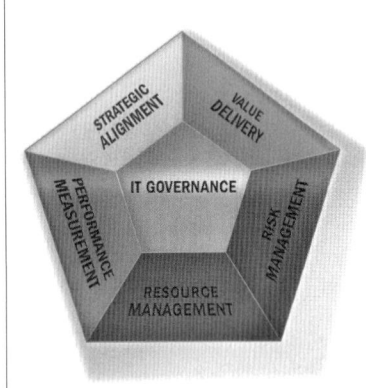

- **Strategic alignment** focuses on ensuring the linkage of business and IT plans; defining, maintaining and validating the IT value proposition; and aligning IT operations with enterprise operations.
- **Value delivery** is about executing the value proposition throughout the delivery cycle, ensuring that IT delivers the promised benefits against the strategy, concentrating on optimising costs and proving the intrinsic value of IT.
- **Resource management** is about the optimal investment in, and the proper management of, critical IT resources: applications, information, infrastructure and people. Key issues relate to the optimisation of knowledge and infrastructure.
- **Risk management** requires risk awareness by senior corporate officers, a clear understanding of the enterprise's appetite for risk, understanding of compliance requirements, transparency about the significant risks to the enterprise and embedding of risk management responsibilities into the organisation.
- **Performance measurement** tracks and monitors strategy implementation, project completion, resource usage, process performance and service delivery, using, for example, balanced scorecards that translate strategy into action to achieve goals measurable beyond conventional accounting.

---

IT governance integrates and institutionalizes good practices to ensure that the enterprise's IT supports the business objectives. IT governance enables the enterprise to take full advantage of its information, thereby maximizing benefits, capitalizing on opportunities and gaining competitive advantage. IT governance is a structure of relationships and processes used to direct and control the enterprise toward achievement of its goals by adding value while balancing risk vs. return over IT and its processes. Use of technology in all aspects of economic and social endeavors has created a critical dependency on information technology to initiate, record, move and manage all aspects of economic transactions, information and knowledge, creating a critical place for IT governance within enterprise governance.

The purpose of IT governance is to direct IT endeavors to ensure that IT's performance meets the objectives of aligning IT with the enterprise's objectives and the realization of promised benefits. Additionally, IT should enable the enterprise by exploiting opportunities and maximizing benefits. IT resources should be used responsibly, and IT-related risks should be managed appropriately.

IT governance has become significant due to a number of factors:
- Business managers and boards demanding a better return from IT investments (i.e., that IT deliver what the business needs to enhance stakeholder value)
- Concern over the generally increasing level of IT expenditure
- The need to meet regulatory requirements for IT controls in areas such as privacy and financial reporting (e.g., the US Sarbanes-Oxley Act, Basel II) and in specific sectors such as finance, pharmaceutical and healthcare
- The selection of service providers and the management of service outsourcing and acquisition
- Increasingly complex IT-related risks such as network security
- IT governance initiatives that include adoption of control frameworks and good practices to help monitor and improve critical IT activities to increase business value and reduce business risk
- The need to optimize costs by following, where possible, standardized rather than specially developed approaches
- The growing maturity and consequent acceptance of well-regarded frameworks
- The need for enterprises to assess how they are performing against generally accepted standards and their peers (benchmarking)

## *Audit Role in IT Governance*

IT governance, as one of the domains of enterprise governance, comprises the body of issues addressed in considering how IT is applied within the enterprise. IT is now intrinsic and pervasive within enterprises rather than being a separate function marginalized from the rest of the enterprise. How IT is applied within the enterprise will have an immense effect on whether the enterprise will attain its mission, vision or strategic goals. For this reason, an enterprise needs to evaluate its IT governance since it is an important part of the overall enterprise governance.

Audit plays a significant role in the successful implementation of IT governance within an organization. Audit is well positioned to provide leading practice recommendations to senior management to help improve the quality and effectiveness of the IT governance initiatives implemented.

As an entity that monitors compliance, audit helps ensure compliance with IT governance initiatives implemented within an organization. The continual monitoring, analysis and evaluation of metrics associated with IT governance initiatives require an independent and balanced view to ensure a qualitative assessment that subsequently facilitates the qualitative improvement of IT processes and associated IT governance initiatives.

Reporting on IT governance involves auditing at the highest level in the organization and may cross divisional, functional or departmental boundaries. The IS auditor should confirm that the terms of reference state the:
• Scope of the work, including a clear definition of the functional areas and issues to be covered
• Reporting line to be used, where IT governance issues are identified to the highest level of the organization
• IS auditor's right of access to information both within the organization and from third party service providers.

The organizational status and skill sets of the IS auditor should be considered for appropriateness with regard to the nature of the planned audit. Where this is found insufficient, the hiring of an independent third party to manage or perform the audit should be considered by an appropriate level of management.

In accordance with the defined role of the IS auditor, the following aspects related to IT governance need to be assessed:
• Alignment of the IS function with the organization's mission, vision, values, objectives and strategies
• Achievement of performance objectives established by the business (e.g., effectiveness and efficiency) by the IS function
• Legal, environmental, information quality, fiduciary, security, and privacy requirements
• The control environment of the organization
• The inherent risks within the IS environment

## 2.3.2 IT STRATEGY COMMITTEE

The creation of an IT strategy committee is an industry best practice. However, the IT strategy committee needs to broaden its scope to include not only advice on strategy when assisting the board in its IT governance responsibilities, but also to focus on IT value, risks and performance. This is a mechanism for incorporating IT governance into enterprise governance. As a committee of the board, it assists the board in overseeing the enterprise's IT-related matters by ensuring that the board has the internal and external information it requires for effective IT governance decision making.

Traditionally, organizations have had executive-level steering committees to deal with IT issues that are relevant organizationwide. There should be a clear understanding of both the IT strategy and steering levels. ITGI issued a document where a clear analysis is made (see **exhibit 2.3**).

**Note:** The Analysis of Steering Committee Responsibilities is information the CISA should know.

## 2.3.3 STANDARD IT BALANCED SCORECARD

The standard IT balanced scorecard is a process management evaluation technique that can be applied to the IT governance process in assessing IT functions and processes. The method goes beyond the traditional financial evaluation, supplementing it with measures concerning customer (user) satisfaction, internal (operational) processes and the ability to innovate. These

| Exhibit 2.3—Analysis of Steering Committee Responsibilities | | |
|---|---|---|
| **Level** | **IT Strategy Committee** | **IT Steering Committee** |
| **Responsibility** | • Provides insight and advice to the board on topics such as:<br>  – The relevance of developments in IT from a business perspective<br>  – The alignment of IT with the business direction<br>  – The achievement of strategic IT objectives<br>  – The availability of suitable IT resources, skills and infrastructure to meet the strategic objectives<br>  – Optimization of IT costs, including the role and value delivery of external IT sourcing<br>  – Risk, return and competitive aspects of IT investments<br>  – Progress on major IT projects<br>  – The contribution of IT to the business (i.e., delivering the promised business value)<br>  – Exposure to IT risks, including compliance risks<br>  – Containment of IT risks<br>  – Direction to management relative to IT strategy<br>  – Drivers and catalysts for the board's IT | • Decides the overall level of IT spending and how costs will be allocated<br>• Aligns and approves the enterprise's IT architecture<br>• Approves project plans and budgets, setting priorities and milestones<br>• Acquires and assigns appropriate resources<br>• Ensures projects continuously meet business requirements, including reevaluation of the business case<br>• Monitors project plans for delivery of expected value and desired outcomes, on time and within budget<br>• Monitors resource and priority conflict between enterprise divisions and the IT function as well as between projects<br>• Makes recommendations and requests for changes to strategic plans (priorities, funding, technology approaches, resources, etc.)<br>• Communicates strategic goals to project teams<br>• Is a major contributor to management's IT governance responsibilities governance practices |
| **Authority** | • Advises the board and management on IT strategy<br>• Is delegated by the board to provide input to the strategy and prepare its approval<br>• Focuses on current and future strategic IT issues | • Assists the executive in the delivery of the IT strategy<br>• Oversees day-to-day management of IT service delivery and IT projects<br>• Focuses on implementation |
| **Membership** | • Board members and specialist nonboard members | • Sponsoring executive<br>• Business executive (key users)<br>• CIO<br>• Key advisors as required (IT, audit, legal, finance) |

additional measures drive the organization toward optimum use of IT, which is aligned with the organization's strategic goals, while keeping all evaluation-related perspectives in balance. To apply to IT, a three-layered structure is used in addressing the four perspectives:

• Mission, for example:
  – Become the preferred supplier of information systems.
  – Deliver economic, effective and efficient IT applications and services.
  – Obtain a reasonable business contribution from IT investments.
  – Develop opportunities to answer future challenges.
• Strategies, for example:
  – Develop superior applications and operations.
  – Develop user partnerships and greater customer services.
  – Provide enhanced service levels and pricing structures.
  – Control IT expenses.
  – Provide business value to IT projects.
  – Provide new business capabilities.
  – Train and educate IT staff and promote excellence.
  – Provide support for research and development.
• Measures, for example:
  – Provide a balanced set of metrics (i.e., KPIs) to guide business-oriented IT decisions.

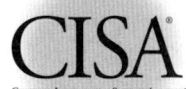
Use of an IT balanced scorecard is one of the most effective means to aid the IT strategy committee and management in achieving IT governance through proper IT and business alignment. The objectives are to establish a vehicle for management reporting to the board, foster consensus among key stakeholders about IT's strategic aims, demonstrate the effectiveness and added value of IT, and communicate IT's performance, risks and capabilities.

**Note:** The Standard IT Balanced Scorecard is information a CISA should know.

## 2.3.4 INFORMATION SECURITY GOVERNANCE

Within IT governance, information security governance should become a focused activity with specific value drivers: confidentiality, integrity, and availability of information, continuity of services and protection of information assets. Security has become a significant governance issue as a result of global networking, rapid technological innovation and change, increased dependence on IT, increased sophistication of threat agents and exploits, and extending the enterprise beyond its traditional boundaries. Therefore, information security should become an important and integral part of IT governance. Negligence in this regard will diminish an organization's capacity to take advantage of IT opportunities for business process improvement.

One of the major trends that global business is experiencing today is the outsourcing of in-house processes. Information security coverage extends beyond the geographic boundary of the organization's premises in onshoring and offshoring models being adopted by organizations. This trend has changed the way in which information security is managed.

### *Information Security Governance Overview*

Information can be defined as "data endowed with meaning and purpose." Today, information plays an increasingly important role in all aspects of our lives. Information has become an indispensable component of conducting business for virtually all organizations. In a growing number of companies, information is the business. This includes major players of the emerging knowledge society such as Google™, eBay®, Microsoft and countless others, large and small. Some might not think of software as information, but it is simply information for computers on how to operate or process something. In addition, a significant amount of data is created and distributed by end users without involving the IT organization.

Traditional organizations have undergone radical transformations in the "information age" as well. The graphic arts and printing industry, for example, deals almost entirely with information in digital form. Artwork and masters are no longer physical drawings or pieces of film but blocks of information stored on hard disks. Finally, many other organizations continue to strive for a "paperless" environment as well.

It would be difficult to find a business that has not been touched by information technology and is not dependent on the information it processes. Information systems have become pervasive in global society and business, and the dependence on these systems and the information they handle is arguably absolute.

The trend of escalating value of and dependence on information has increased exponentially over the last decade. There is every indication that this trend will continue into the foreseeable future.

During the same period, information crime and malicious acts have become the choice of a growing cadre of sophisticated criminals. Terrorists and others with enmity toward society have brazenly embraced the very IT they claim to despise to herald their view and advertise their horrid acts.

The majority of critical infrastructures in the developed world are controlled by the private sector. A preponderance of the task of protecting the information resources critical to survival falls squarely on corporate shoulders—coupled with often ineffective bureaucracies, countless conflicting jurisdictions and aging institutions unable to adapt to burgeoning global "information" crime..

To accomplish the task of adequate protection for information resources, the issue must be raised to a board-level activity as are other critical governance functions. The complexity, relevance and criticality of information security and its governance mandate that it be addressed and supported at the highest organizational levels.

Increasingly, those who understand the scope and depth of risks to information take the position that, as a critical resource, information must be treated with the same care, caution and prudence that any other asset essential to the survival of the organization and perhaps society itself would receive.

Until recently, the focus of protection has been on the IT systems that collect, process and store the vast majority of information rather than the information itself. But this approach is too narrow to accomplish the level of integration, process assurance and overall security that is now required. Information security takes the larger view that the content, information and the knowledge based on it must be adequately protected regardless of how it is handled, processed, transported or stored, and disposed of, particularly with tools that allow data and content to be shared easily over the Internet through blogs, newsfeeds, peer-to-peer or social networks, or web sites. Thus, the reach of protection efforts should encompass not only the process that generates the information but also the continued preservation of information generated as a result of the controlled processes.

IT security addresses the security of the technology and is typically driven from the CIO level. Information security related to privacy of information and information security itself addresses the universe of risks, benefits and processes involved with information, and must be driven by executive management (e.g., CEO, CFO, CTO, CIO) and supported by the board of directors.

The relentless advance of IT and the unparalleled ability to access, manipulate and use information has brought enormous benefits and opportunities to the global economy. It has also brought unparalleled new risks, ethical dilemmas, and a confounding patchwork of existing and pending laws and regulations. as well as social changes and related issues such as telecommuting and increased mobility.

Executive management is increasingly confronted by the need to stay competitive in the global economy and heed the promise of greater gains from the deployment of more information resources. But even as organizations reap those gains, the twin specters of increasing dependence on information and the systems that support it and advancing risks from a host of threats are forcing management to face difficult decisions about how to effectively address information security. In addition, scores of new and existing laws and regulations are increasingly demanding compliance and higher levels of accountability.

Information security governance is the responsibility of the board of directors and executive management, and must be an integral and transparent part of enterprise governance. Information security governance consists of the leadership, organizational structures and processes that safeguard information. As in the case of controls, nothing has changed with respect to the basic premise of information as an asset. What has changed is the platform and repositories used for collecting, processing and storing information. This explains why the board and executive management continue to be responsible and accountable for the organization's most valuable asset, which is information.

## IMPORTANCE OF INFORMATION SECURITY GOVERNANCE

From an organization's perspective, information security governance is increasingly critical as dependence grows on information and IT systems. For most organizations, information (and the knowledge based on it) has become increasingly one of their most important assets without which conducting business would not be possible. The systems and processes that handle this information have become truly pervasive throughout business and governmental organizations globally. This growing dependence of organizations on information and the systems that handle it, coupled with the risks, benefits and opportunities these resources present, have made information security governance an increasingly critical facet of overall governance. In addition to addressing legal and regulatory requirements, effective information security governance is simply good business and due diligence. Prudent management has come to understand that it provides a series of significant benefits including:
• Addressing the increasing exposure the organization and its management have to civil or legal liability as a result of inaccurate information provided to the public or to regulators, and the consequences of not exercising due care in the protection of private information (as demonstrated by leakage of credit card and other sensitive customer information)
• Providing assurance of policy and standards compliance
• Increasing predictability and reducing uncertainty of business operations by lowering risks to definable and acceptable levels
• Providing the structure and framework to optimize allocations of limited security resources
• Providing a level of assurance that critical decisions are not based on faulty information
• Providing a firm foundation for efficient and effective risk management, process improvement, and rapid incident response
• Providing accountability for safeguarding information during critical business activities such as mergers and acquisitions, business process recovery, and regulatory response

And finally, because new IT provides the potential for dramatically enhanced business performance, effective information security can add significant value to the organization by:
• Providing greater reliance on interactions with trading partners
• Improving trust in customer relationships
• Protecting the organization's reputation
• Enabling new and better ways to process electronic transactions

Information security covers all information processes, physical and electronic, regardless of whether they involve people and technology or relationships with trading partners, customers and third parties. Information security is concerned with all aspects of information and its protection at all points of its life cycle within the organization.

### OUTCOMES OF SECURITY GOVERNANCE

The basic outcomes of effective security governance should include:
1. Strategic alignment—Align information security with business strategy to support organizational objectives. To achieve alignment, the following should be accomplished:
   – Security requirements driven by enterprise requirements thoroughly developed to provide guidance on what must be done and a measure of when it has been achieved
   – Security solutions fit for enterprise processes that take into account the culture, governance style, technology and structure of the organization
   – Investment in information security aligned with the enterprise strategy and the well-defined threat, vulnerability and risk profile
2. Risk management—Manage and execute appropriate measures to mitigate risks and reduce potential impacts on information resources to an acceptable level. To achieve risk management, consider the following:
   – Collective understanding of the organization's threat, vulnerability and risk profile
   – Understanding of risk exposure and potential consequences of compromise including the regulatory, legal, operational and brand impacts
   – Awareness of risk management priorities based on potential consequences
   – Risk mitigation sufficient to achieve acceptable consequences from residual risk
   – Risk acceptance/deference based on an understanding of the potential consequences of residual risk
3. Value delivery—Optimize security investments in support of business objectives. To achieve value delivery, consider the following:
   – A standard set of security policies and practices (i.e., baseline security requirements following adequate and sufficient practices proportionate to risk)
   – Properly prioritized and distributed effort to areas with greatest impact and business benefit
   – Institutionalized and commoditized standards-based solutions
   – Complete solutions covering the organization and business processes as well as technology based on an understanding of the end-to-end business of the organization
   – A continuous improvement (i.e., Kaizen) culture based on the understanding that security is a process, not an event
4. Performance measurement—Measure, monitor and report on information security processes to ensure that SMART (Specific, Measurable, Achievable, Relevant and Time-bound) objectives are achieved. The following should be accomplished to achieve performance measurement:
   – A defined, agreed on and meaningful set of metrics properly aligned with strategic objectives
   – A measurement process that will help identify shortcomings and provide feedback on progress made in resolving issues
   – Independent assurance provided by external assessments and audits
5. Resource management—Utilize information security knowledge and infrastructure efficiently and effectively. To achieve resource management consider the following:
   – Ensure that knowledge is captured and available
   – Document security processes and practices
   – Develop security architecture(s) to define and utilize infrastructure resources efficiently
6. Process integration—This focuses on the integration of an organization's management assurance processes for security. Security activities are at times fragmented and segmented in silos with different reporting structures. This makes it difficult, if not impossible, to seamlessly integrate them. Process integration serves to improve overall security and operational efficiencies.

### Emerging Concepts of Business Process Assurance
Integration is a concept that consists of integrating all relevant assurance factors to ensure that processes operate as intended from end to end. To achieve integration, the following should be considered:
• Determine all organizational assurance functions.
• Develop formal relationships with other assurance functions.
• Coordinate all assurance functions for more complete security.
• Ensure that roles and responsibilities between assurance functions overlap.

## Effective Information Security Governance
Corporate governance is the set of responsibilities and practices exercised by the board and executive management to provide strategic direction, ensure that objectives are achieved, ascertain risks are managed appropriately and verify the enterprise's resources are used responsibly.

The strategic direction of the business will be defined by business goals and objectives. Information security must support business activities to be of value to the organization.

Information security governance is a subset of corporate governance that provides strategic direction for security activities and ensures that objectives are achieved. It ensures that information security risks are appropriately managed and enterprise information resources are used responsibly.

To achieve effective information security governance, management must establish and maintain a framework to guide the development and management of a comprehensive information security program that supports business objectives.

The governance framework will generally consist of:
• A comprehensive security strategy intrinsically linked with business objectives
• Governing security policies that address each aspect of strategy, controls and regulation
• A complete set of standards for each policy to ensure that procedures and guidelines comply with policy
• An effective security organizational structure void of conflicts of interest
• Institutionalized monitoring processes to ensure compliance and provide feedback on effectiveness

This framework in turn provides the basis for the development of a cost-effective information security program that supports the organization's business goals. Implementing a security program is covered in chapter 3, Systems Infrastructure Life Cycle Management. The objective of the program is a set of activities that provide assurance that information assets are given a level of protection commensurate with their value or the risk their compromise poses to the organization.

According to Julia Allen in *Governing for Enterprise Security*, "Governing for enterprise security means viewing adequate security as a non-negotiable requirement of being in business. If an organization's management—including boards of directors, senior executives and all managers—does not establish and reinforce the business need for effective enterprise security, the organization's desired state of security will not be articulated, achieved, or sustained. To achieve a sustainable capability, organizations must make enterprise security the responsibility of leaders at a governance level, not of other organizational roles that lack the authority, accountability and resources to act and enforce compliance."

### ROLES AND RESPONSIBILITIES OF SENIOR MANAGEMENT AND BOARDS OF DIRECTORS
Information security governance requires strategic direction and impetus. It requires commitment, resources and assigning responsibility for information security management as well as a means for the board to determine its intent has been met.

### Boards of Directors/Senior Management
Effective information security governance can be accomplished only by involvement of the board of directors and/or senior management in approving policy, appropriate monitoring and metrics as well as reporting and trend analysis.

Members of the board need to be aware of the organization's information assets and their criticality to ongoing business operations. This can be accomplished by periodically providing the board with the high-level results of comprehensive risk assessments and business impact analysis (BIA). It may also be accomplished by business dependency assessments of information resources. A result of these activities should include board members approving the assessment of key assets to be protected, and that protection levels and priorities are appropriate to a standard of due care.

The tone at the top must be conducive to effective security governance. It is unreasonable to expect lower-level personnel to abide by security measures if they are not exercised by senior management. Executive management endorsement of intrinsic security requirements provides the basis for ensuring that security expectations are met at all levels of the enterprise. Penalties for noncompliance must be defined, communicated and enforced from the board level down.

### Executive Management

Implementing effective security governance and defining the strategic security objectives of an organization is a complex, arduous task. As with any other major initiative, it must have leadership and ongoing support from executive management to succeed. Developing an effective information security strategy requires integration with and cooperation of business process owners. A successful outcome is the alignment of information security activities in support of business objectives. The extent to which this is achieved will determine the cost-effectiveness of the information security program in achieving the desired objective of providing a predictable, defined level of assurance for business processes and an acceptable level of impact from adverse events.

### Steering Committee

To some extent security affects all aspects of an organization. To be effective, security must be pervasive throughout the enterprise. To ensure that all stakeholders impacted by security considerations are involved, many organizations use a steering committee comprised of senior representatives of affected groups. This facilitates achieving consensus on priorities and trade-offs. It also serves as an effective communications channel and provides an ongoing basis for ensuring the alignment of the security program with business objectives. It can also be instrumental in achieving modification of behavior toward a culture more conducive to good security.

### Chief Information Security Officer

All organizations have a chief information security officer (CISO) whether anyone holds that title or not. It may be the CIO, CTO, CFO or, in some cases, the CEO even when there is an information security office or director in place. The scope and breadth of information security today is such that the authority required and the responsibility taken will inevitably make it a senior officer or top management responsibility. This could include a position such as a chief risk officer (CRO) or a chief compliance officer (CCO). Legal responsibility will, by default, extend up the command structure and ultimately reside with senior management and the board of directors. Failure to recognize this and implement appropriate governance structures can result in senior management being unaware of this responsibility and the attendant liability. It also usually results in a lack of effective alignment of business objectives and security activities. Increasingly, prudent management is elevating the position of information security officer to a senior management position (i.e., chief information security officer [CISO]), as organizations begin to understand their dependence on information and the growing threats to it.

### MATRIX OF OUTCOMES AND RESPONSIBILITIES

The relationship between the outcomes of effective security governance and management responsibilities are shown in **exhibit 2.4**. This matrix is not meant to be comprehensive but merely indicate some primary tasks and level for which management is responsible.

## 2.3.5 ENTERPRISE ARCHITECTURE

An area of IT governance receiving increasing attention is enterprise architecture (EA). Essentially, EA involves documenting an organization's IT assets in a structured manner to facilitate understanding, management and planning for IT investments. An EA often involves both a current state and optimized future state representation (e.g., a road map).

The current focus on EA is a response to the increasing complexity of IT, the complexity of modern organizations, and an enhanced focus on aligning IT with business strategy and ensuring that IT investments deliver real returns.

| Management Level | Strategic Alignment | Risk Management | Value Delivery | Performance Measurement | Resource Management | Process Assurance |
|---|---|---|---|---|---|---|
| **Exhibit 2.4—Relationships of Security Governance Outcomes to Management Responsibilities** | | | | | | |
| Board of directors | Require demonstrable alignment. | • Establish risk tolerance.<br>• Oversee a policy of risk management.<br>• Ensure regulatory compliance. | Require reporting of security activity costs. | Require reporting of security effectiveness. | Oversee a policy of knowledge management and resource utilisation. | Oversee a policy of assurance process integration. |
| Executive management | Institute processes to integrate security with business objectives. | • Ensure that roles and responsibilities include risk management in all activities.<br>• Monitor regulatory compliance. | Require business case studies of security initiatives. | Require monitoring and metrics for security activities. | Ensure processes for knowledge capture and efficiency metrics. | Provide oversight of all assurance functions and plans for integration. |
| Steering committee | • Review and assist security strategy and integration efforts.<br>• Ensure that business owners support integration. | Identify emerging risks, promote business unit security practices and identify compliance issues. | Review and advise on the adequacy of security initiatives to serve business functions. | Review and advise whether security initiatives meet business objectives. | Review processes for knowledge capture and dissemination. | • Identify critical business processes and assurance providers.<br>• Direct assurance integration efforts. |
| CISO/information security management | Develop the security strategy, oversee the security programme and initiatives, and liaise with business process owners for ongoing alignment. | • Ensure that risk and business impact assessments are conducted.<br>• Develop risk mitigation strategies.<br>• Enforce policy and regulatory compliance. | Monitor utilisation and effectiveness of security resources. | Develop and implement monitoring and metrics approaches, and direct and monitor security activities. | Develop methods for knowledge capture and dissemination, and develop metrics for effectiveness and efficiency. | • Liaise with other assurance providers.<br>• Ensure that gaps and overlaps are identified and addressed. |
| Audit executives | Evaluate and report on degree of alignment. | Evaluate and report on corporate risk management practices and results. | Evaluate and report on efficiency. | Evaluate and report on degree of effectiveness of measures in place and metrics in use. | Evaluate and report on efficiency or resource management. | Evaluate and report on effectiveness of assurance processes performed by different areas of management. |

Source: ITGI, *Information Security Governance: Guidance for Information Security Managers*, 2008. All rights reserved. Used by permission.

**Note:** **Exhibit 2.4** on Relationships of Security Governance Outcomes to Management Responsibilities is not specifically tested in the CISA exam but is information a CISA should be aware of.

The Framework for Enterprise Architecture, groundbreaking work in the field of EA, was first published by John Zachman in the late 1980s. The Zachman framework continues to be a starting point for many contemporary EA projects. Zachman reasoned that constructing IT systems had considerable similarities to building construction. In both cases there is a range of participants who become involved at differing stages of the project. In building construction, one moves from the abstract to the physical using models and representations (such as blueprints, floor plans and wiring diagrams). Similarly with IT, different artifacts (such as diagrams, flowcharts, data/class models and code) are used to convey different aspects of an organization's systems at progressively greater levels of detail.

The basic Zachman framework is described in **exhibit 2.5**.

The ultimate objective is to complete all cells of the matrix. At the outset of an EA project, most organizations will have difficulty providing details for every cell, particularly at the highest level.

| Exhibit 2.5—Zachman Framework for Enterprise Architecture | | | | | | |
|---|---|---|---|---|---|---|
| | Data | Functional (Application) | Network (Technology) | People (Organization) | Process (Workflow) | Strategy |
| Scope | | | | | | |
| Enterprise model | | | | | | |
| Systems model | | | | | | |
| Technology model | | | | | | |
| Detailed representation | | | | | | |

In attempting to complete an EA, organizations can address the challenge either from a technology perspective or a business process perspective.

Technology-driven EA attempts to clarify the complex technology choices faced by modern organizations. The idea is to provide guidance on issues such as whether and when to use advanced technical environments (e.g., J2EE or .NET) for application development, how to better connect intra- and interorganizational systems, how to "web enable" legacy and enterprise resource planning (ERP) applications (hopefully, without extensive rewrite), and whether to insource or outsource IT functions.

Business process-driven EA attempts to understand an organization in terms of its core value-adding and supporting processes. The idea is that by understanding processes, their constituent parts and the technology that supports them, business improvement can be obtained as aspects are progressively redesigned and replaced. The genesis for this type of thinking can be traced back to the work of Harvard professor Michael Porter-work first published in the 1980s and particularly his business value chain model. The effort to model business processes is being given extra impetus by a number of industrywide business models that are starting to emerge-e.g., the telecommunications industry's enhanced Telecom Operations Map (eTOM) and the Supply Chain Operations Reference (SCOR) model. The contents from a business process model can be mapped to upper tiers of the Zachman framework. Once the mapping is completed, an organization can consider the optimum mix of technologies needed to support its business processes.

The US federal government, with agencies collectively spending some US $60 billion on IT annually, is taking the subject of EA seriously. A US federal organization is required, by law, to develop an EA, and set up an EA governance structure that ensures that the EA is referenced and maintained in the planning and budgeting activities of all systems. The Federal Enterprise Architecture (FEA) has been developed to guide this process. The FEA is described (on the FEA web site, *www.feapmo.gov*) as "a business and performance based framework to support cross-agency collaboration, transformation and government-wide improvement." The FEA has a hierarchy of five reference models:

- **Performance reference model**—A framework to measure the performance of major IT investments and their contribution to program performance
- **Business reference model**—A function-driven framework that describes the functions and subfunctions performed by the government, independent of the agencies that actually perform them
- **Service component reference model**—A functional framework that classifies the service components that support business and performance objectives
- **Technical reference model**—A framework that describes how technology supports the delivery, exchange and construction of service components
- **Data reference model**—While still being developed, this will describe the data and information that support program and business line operations

The documentation on corporate architecture and the FEA are primarily used for maintaining and describing technological coherence, continually describing and evaluating the technology that is being managed by the IS department.

Relevant aspects of IT governance regarding the management of an IS department are the processes for selection and/or the methodologies that are used to change strategic technologies. This relevant topic affects management decisions and is subject to great business risks.

# 2.4 INFORMATION SYSTEMS STRATEGY

## 2.4.1 STRATEGIC PLANNING

Strategic planning from an IS standpoint relates to the long-term direction an organization wants to take in leveraging information technology for improving its business processes.

Under the responsibility of top management, factors to consider include identifying cost-effective IT solutions in addressing problems and opportunities that confront the organization, and developing action plans for identifying and acquiring needed resources. In developing strategic plans, generally three to five years in duration, organizations should ensure that the plans are fully aligned and consistent with the overall organizational goals and objectives. IS department management, along with the IS steering committee and the strategy committee (which provides valuable strategic input related to stakeholder value), plays a key role in the development and implementation of the plans.

Effective IT strategic planning involves a consideration of the organization's demand for IT and its IT supply capacity. Determining IT demand will involve a systematic consideration of the organization's strategic intentions, how these translate into specific objectives and business initiatives, and what IT capabilities will be needed to support these objectives and initiatives. In assessing IT capabilities, the existing system's portfolio should be reviewed in terms of functional fit, cost and risk. IT supply planning involves assessing the organization's technical IT infrastructure and key support processes (e.g., software development and maintenance practices, security administration and help desk services) to determine whether expansion or improvement is necessary. It is important that the strategic planning process encompasses not just the delivery of new systems and technology, but considers the returns being achieved from investment in existing "lights on" IT and the decommisioning of legacy systems. In many organizations, spending on existing IT systems, infrastructure and support services accounts for 85 percent or more of total annual IT spending. The strategic IT plan should balance the cost of maintenance of existing systems against the cost of new initiatives or systems to support the business strategies.

The IS auditor should pay full attention to the importance of IT strategic planning, taking management control practices into consideration. In addition, the IT governance objective requires that IT strategic plans be synchronized with the overall business strategy. An IS auditor must focus on the importance of a strategic planning process or planning framework. Particular attention should be paid to the need to note requirements to translate operational or tactical plans from the business and IT strategies, contents of strategic plans, requirements for updating and communicating plans, and monitoring and evaluation requirements. The IS auditor should consider how the CIO or senior IT management are involved in the creation of the overall business strategy. A lack of involvement of IT in the creation of the business strategy indicates that there is a risk that the IT strategy and plans will not be aligned with the business strategy.

| PRACTICE QUESTIONS |
|---|
| 2-2    Which of the following would be included in an IS strategic plan? |

    A.    Specifications for planned hardware purchases
    B.    Analysis of future business objectives
    C.    Target dates for development projects
    D.    Annual budgetary targets for the IS department

*See answers and explanations to the practice questions at the end of the chapter. (page 133)*

| PRACTICE QUESTIONS |
| --- |

2-3    Which of the following **BEST** describes an IT department's strategic planning process?

    A.    The IT department will have either short-range or long-range plans depending on the organization's broader plans and objectives.
    B.    The IT department's strategic plan must be time- and project-oriented, but not so detailed as to address and help determine priorities to meet business needs.
    C.    Long-range planning for the IT department should recognize organizational goals, technological advances and regulatory requirements.
    D.    Short-range planning for the IT department does not need to be integrated into the short-range plans of the organization since technological advances will drive the IT department plans much quicker than organizational plans.

*See answers and explanations to the practice questions at the end of the chapter. (page 133)*

# 2.4.2 STEERING COMMITTEE

The organization's senior management should appoint a planning or steering committee to oversee the IS function and its activities. A high-level steering committee for information technology is an important factor in ensuring that the IS department is in harmony with the corporate mission and objectives. Although not a common practice, it is highly desirable that a member of the board of directors who understands the risks and issues is responsible for information technology, and is chair of this committee. The committee should include representatives from senior management, user management and the IS department.

The committee's duties and responsibilities should be defined in a formal charter. Members of the committee should know IS department policies, procedures and practices. Each member should have the authority to make decisions within the group for his/her respective areas.

Such a committee typically serves as a general review board for major IS projects and should not become involved in routine operations. Primary functions performed by this committee include:
• Review the long- and short-range plans of the IS department to ensure that they are in accordance with the corporate objectives.
• Review and approve major acquisitions of hardware and software within the limits approved by the board of directors.
• Approve and monitor major projects and the status of IS plans and budgets, establish priorities, approve standards and procedures, and monitor overall IS performance.
• Review and approve sourcing strategies for select or all IS activities, including insourcing or outsourcing, and the globalization or offshoring of functions.
• Review adequacy of resources and allocation of resources in terms of time, personnel and equipment.
• Make decisions regarding centralization vs. decentralization and assignment of responsibility.
• Support development and implementation of an enterprisewide information security management program.
• Report to the board of directors on IS activities.

**Note:** Consider that the responsibilities will vary from organization to organization and that these responsibilities listed are the most common of the steering committee. The CISA candidate should know the purpose of the IS steering committee and it's major responsibilities. Also please note that many organizations may refer to this committee with a different name. The IS auditor needs to identify the group that performs the above functions

The IS steering committee should receive the appropriate management information from IS departments, user departments and the audit to effectively coordinate and monitor the organization's IS resources. The committee should monitor performance and institute appropriate action to achieve desired results. The committee should meet regularly and report to senior management. Formal minutes of the IS steering committee meetings should be maintained to document the committee's activities and decisions.

# 2.5 POLICIES AND PROCEDURES

Policies and procedures reflect management guidance and direction in developing controls over information systems and related resources.

## 2.5.1 POLICIES

Policies are high-level documents that represent the corporate philosophy of an organization and the strategic thinking of senior management and business process owners. To be effective, policies must be clear and concise. Management must create a positive control environment by assuming responsibility for formulating, developing, documenting, promulgating and controlling policies covering general goals and directives. Management should take the steps necessary to ensure that employees affected by a specific policy receive a full explanation of the policy and understand its intent. In addition, policies may also apply to third parties and outsourcers, who will need to be bound to follow the policies through contracts or statements of work (SOW).

In addition to corporate policies that set the tone for the organization as a whole, individual divisions and departments should define lower-level policies. The lower-level policies should be consistent with the corporate-level policies. These would apply to the employees and operations of these units, and would focus at the operational level.

A top-down approach to the development of lower-level policies is desired in instances where they are derived from corporate policies since consistency is facilitated across the organization. However, some organizations begin by defining operational-level policies as immediate priorities. These companies view this as being the more cost-effective approach since these policies are often derived and implemented as the result of risk assessments. This is a bottom-up approach in which corporate policies are a subsequent development and a synthesis of existing operational policies. This approach may seem more practical but it leaves room for inconsistency and conflicting situations in policies.

Management should review all policies periodically. Policies need to be updated to reflect new technology, changes in environment (e.g. regulatory compliance requirements) and significant changes in business processes in exploiting information technology for efficiency and effectiveness in productivity or competitive gains. Policies formulated must support achievement of business objectives and implementation of IS controls. However, management must be responsive to the needs of the customers and change policies that may hinder customer satisfaction or the organization's ability to achieve business objectives. Just as every control is formulated to meet a control objective, the broad policies at a higher level and the detailed policies at a lower level need to be in alignment with the business objectives.

IS auditors should understand that policies are a part of the audit process and test the policies for compliance. IS controls should flow from the enterprise's policies, and IS auditors should use policies as a benchmark for evaluating compliance. However, if policies exist that hinder the achievement of business objectives, these policies must be identified and reported for improvement. The IS auditor should also consider the extent to which the policies apply to third parties or outsourcers the extent to which third parties or outsourcers comply with the policies, and whether the policies of the third parties or utsourcers are in conflict with the enterprise's policies.

## *Information Security Policy*

A security policy communicates a coherent security standard to users, management and technical staff. A security policy for information and related technology is a first step toward building the security infrastructure for technology-driven organizations. Policies will often set the stage in terms of what tools and procedures are needed for the organization. Security policies must balance the level of control with the level of productivity. Also, the cost of a control should never exceed the expected benefit to be derived. In designing and implementing these policies, the organizational culture will play an important role. The security policy must be approved by senior management, and should be documented and communicated, a appropriate, to all employees, service providers and business partners (i.e., suppliers). The security policy should be used by IS auditors as a reference framework for performing various IS audit assignments. The adequacy and appropriateness of the security policy could also be an area of review for the IS auditor.

The information security policy guides the entire organization by defining what needs to be protected, responsibilities for protection, and the strategy that will be followed for protection.

The information security policy provides management the direction and support for information security in accordance with business requirements, and relevant laws and regulations. Management should set a clear policy direction in line with business objectives, and demonstrate support for and commitment to information security through the issuance and maintenance of an information security policy for the organization.

### INFORMATION SECURITY POLICY DOCUMENT

The information security policy document should state management's commitment and set out the organization's approach to managing information security. The policy document should contain:
• A definition of information security, its overall objectives and scope, and the importance of security as an enabling mechanism for information sharing (see introduction)
• A statement of management intent, supporting the goals and principles of information security in line with the business strategy and objectives
• A framework for setting control objectives and controls, including the structure of risk assessment and risk management
• A brief explanation of the security policies, principles, standards and compliance requirements of particular importance to the organization including:
  – Compliance with legislative, regulatory and contractual requirements
  – Security education, training and awareness requirements
  – Business continuity management
  – Consequences of information security policy violations
• A definition of general and specific responsibilities for information security management, including reporting information security incidents
• References to documentation which may support the policy; e.g., more detailed security policies, standards, and procedures for specific information systems or security rules with which users should comply

This information security policy should be communicated throughout the organization to users in a form that is accessible and understandable to the intended reader. The information security policy might be a part of a general policy document and may be suitable for distribution to third parties and outsourcers of the organization as long as care is taken not to disclose sensitive organizational information. All employees having access to information assets should be required to sign off on their understanding and willingness to comply with the information security policy at the time they are hired and on a regular basis thereafter (e.g., annually) to account for policy changes over time.

Depending upon the need and appropriateness, organizations may document information security policies as a set of policies. Generally, the following policy groups are addressed:
• **High-level Information Security Policy:** This policy should include statements on confidentiality, integrity and availability.
• **Data Classification Policy:** This policy should describe the classifications, levels of control at each classification and responsibilities of all potential users including ownership.

- **Acceptable Usage Policy:** There must be a comprehensive policy that includes information for all information resources (HW/SW, networks, Internet, etc.) and describes the organizational permissions for the usage of IT and information-related resources.
- **End-User Computing Policy:** This policy describes the parameters and usage of desktop tools by users.
- **Access Control Policies:** This policy describes the method for defining and granting access to users to various IT resources.

## REVIEW OF THE INFORMATION SECURITY POLICY

The information security policy should be reviewed at planned intervals or when significant changes occur to ensure its continuing suitability, adequacy and effectiveness. The information security policy should have an owner who has approved management responsibility for the development, review and evaluation of the security policy. The review should include assessing opportunities for improvement to the organization's information security policy and approach to managing information security in response to changes to the organizational environment, business circumstances, legal conditions or technical environment.

The maintenance of the information security policy should take into account the results of these reviews. There should be defined management review procedures, including a schedule or period for the review.

The input to the management review should include:
- Feedback from interested parties
- Results of independent reviews
- Status of preventive, detective and corrective actions
- Results of previous management reviews
- Process performance and information security policy compliance
- Changes that could affect the organization's approach to managing information security, including changes to the organizational environment; business circumstances; resource availability; contractual, regulatory and legal conditions; or technical environment
- Usage of the consideration of outsourcers or offshore of IT or business functions
- Trends related to threats and vulnerabilities
- Reported information security incidents
- Recommendations provided by relevant authorities

The output from management review should include any decisions and actions related to:
- Improvement of the organization's approach to managing information security and its processes
- Improvement of control objectives and controls
- Improvement in the allocation of resources and/or responsibilities

A record of management reviews should be maintained and management approval for the revised policy should be obtained.

---

**Note:** This review is performed by management to address the changes in environmental factors. While reviewing the policies the IS auditor needs to assess the following:
- Basis on which the policy has been defined—generally, it is based on a risk management process
- Appropriateness of these policies
- Contents of policies
- Exceptions to the policies—clearly noting in which area the policies do not apply and why—e.g., password policies may not be compatible with legacy applications)
- Policy approval process
- Policy implementation process
- Effectiveness of implementation of policies
- Awareness and training
- Periodic review and update process

---

## 2.5.2 PROCEDURES

Procedures are detailed steps defined and documented for implementing policies. They must be derived from the parent policy and must implement the spirit (intent) of the policy statement. Procedures must be written in a clear and concise manner so they may be easily and properly understood by those governed by them. Procedures document business processes (administrative and operational) and the embedded controls. Procedures are formulated by process owners as an effective translation of policies.

Generally, procedures are more dynamic than their respective parent policies. Procedures must reflect the regular changes in business focus and environment. Therefore, frequent reviews and updates of procedures are essential if they are to be relevant. Auditors review procedures to identify/evaluate and, thereafter, test controls over business processes. The controls embedded in procedures are evaluated to ensure that they fulfill necessary control objectives while making the process as efficient and practical as possible. Where operational practices do not match documented procedures or where documented procedures do not exist, it is difficult (for management and auditors) to identify controls and ensure that they are in continuous operation.

One of the most critical aspects related to procedures is that they should be well known by the people they govern. A procedure that is not thoroughly known by the personnel who are to use it is, essentially, ineffective. Therefore, attention should be paid to deployment methods and automation of mechanisms to store, distribute and manage IT procedures.

An independent review is necessary to ensure that policies and procedures have been properly documented, understood and implemented. The reviewer should maintain independence at all times and not be influenced by anyone in the group being reviewed. Evidence of work performed should be adequate and provide the reviewer with a level of confidence that the work was performed in compliance with established policies and procedures. Reviews can be performed as part of an IT function implementing the Software Engineering Institute's Capability Maturity Model (CMM), Information Technology Infrastructure Library (ITIL) or ISO standards.

# 2.6 RISK MANAGEMENT

Risk management is the process of identifying vulnerabilities and threats to the information resources used by an organization in achieving business objectives and deciding what countermeasures (safeguards or controls), if any, to take in reducing risk to an acceptable level (i.e., residual risk), based on the value of the information resource to the organization.

Effective risk management begins with a clear understanding of the organization's appetite for risk. This drives all risk management efforts and, in an IT context, impacts future investments in technology, the extent to which IT assets are protected and the level of assurance required. Risk management encompasses identifying, analyzing, evaluating, treating, monitoring and communicating the impact of risk on IT processes. Having defined risk appetite and identified risk exposure, strategies for managing risk can be set and responsibilities clarified. Depending on the type of risk and its significance to the business, management and the board may choose to:
• Avoid—e.g., where feasible, choose not to implement certain activities or processes that would incur risk (i.e., eliminate the risk by eliminating the cause).
• Mitigate—e.g., lessen the probability or impact of the risk by defining, implementing, and monitoring appropriate controls.
• Transfer (deflect, or allocate)—e.g., share risk with partners or transfer via insurance coverage, contractual agreement, or other means.
• Accept—i.e., formally acknowledge the existence of the risk and monitor it.

In other words, risk can be avoided, reduced, transferred, or accepted. An organization can also choose to reject risk by ignoring it, which can be dangerous.

## 2.6.1 DEVELOPING A RISK MANAGEMENT PROGRAM

To develop a risk management program:
- **Establish the purpose of the risk management program**—The first step is to determine the organization's purpose for creating a risk management program. The program's purpose may be to reduce the cost of insurance or reduce the number of program-related injuries. By determining its intention before initiating risk management planning, the organization can define key performance indicators (KPIs) and evaluate the results to determine its effectiveness. Typically, the executive director, with the board of directors, sets the tone for the risk management program.
- **Assign responsibility for the risk management plan**—The second step is to designate an individual or team responsible for developing and implementing the organization's risk management program. While the team is primarily responsible for the risk management plan, a successful program requires the integration of risk management within all levels of the organization. Operations staff and board members should assist the risk management committee in identifying risks and developing suitable loss control and intervention strategies.

## 2.6.2 RISK MANAGEMENT PROCESS

The first step in the process is the identification and classification of information resources or assets that need protection because they are vulnerable to threats. The purpose of the classification may be either to prioritize further investigation and identify appropriate protection (simple classification based on asset value), or to enable a standard model of protection to be applied (classification in terms of criticality and sensitivity). Examples of typical assets associated with information and IT include:
- Information and data
- Hardware
- Software
- Services
- Documents
- Personnel

Other more traditional business assets for consideration are buildings, stock of goods (inventory), and cash and intangible assets such as goodwill or image/reputation.

The next step in the process is to assess threats and vulnerabilities associated with the information resource and the likelihood of their occurrence. In this context, threats are any circumstances or events with the potential to cause harm to an information resource such as destruction, disclosure, modification of data and/or denial of service. Common classes of threats are:
- Errors
- Malicious damage/attack
- Fraud
- Theft
- Equipment/software failure

Threats occur because of vulnerabilities associated with use of information resources. Vulnerabilities are characteristics of information resources that can be exploited by a threat to cause harm. Examples of vulnerabilities are:
- Lack of user knowledge
- Lack of security functionality
- Poor choice of passwords
- Untested technology
- Transmission of unprotected communications

The result of any of these threats occurring is called an impact and can result in a loss of one sort or another. In commercial organizations, threats usually result in a direct financial loss in the short term or an ultimate (indirect) financial loss in the long term. Examples of such losses include:
- Direct loss of money (cash or credit)

- Breach of legislation
- Loss of reputation/goodwill
- Endangering of staff or customers
- Breach of confidence
- Loss of business opportunity
- Reduction in operational efficiency/performance
- Interruption of business activity

Once the elements of risk have been established, they are combined to form an overall view of risk. A common method of combining the elements is to calculate the vulnerability of impact X (probability of occurrence related to a particular information resource) for each threat to give a measure of overall risk. The risk is proportional to the value of the loss/damage and the estimated frequency of the threat.

Once risks have been identified, existing controls can be evaluated or new controls designed to reduce the vulnerabilities to an acceptable level of risk. These controls are referred to as countermeasures or safeguards. They could be actions, devices, procedures or techniques (i.e., people, processes or products). The strength of a control can be measured in terms of its inherent or design strength and the likelihood of its effectiveness. Elements of controls that should be considered when evaluating control strength include whether the controls are preventive, detective or corrective, manual or programmed, and formal (i.e., documented in procedure manuals and evidence of their operation is maintained) or *ad hoc*.

The remaining level of risk, once controls have been applied, is called residual risk and can be used by management to further reduce risk by identifying those areas in which more control is required. An acceptable level of risk target can be established by management (appetite for risk.) Risks in excess of this level should be reduced by the implementation of more stringent controls. Risks below this level should be evaluated to determine whether an excessive level of control is being applied and if cost savings can be made by removing these excessive controls. Final acceptance of residual risks takes into account:
- Organizational policy
- Risk identification and measurement
- Uncertainty incorporated in the risk assessment approach
- Cost and effectiveness of implementation

It is important to realize that IT risk management needs to operate at multiple levels, including:
- **The operational level**—At the operational level, one is concerned with risks that could compromise the effectiveness and efficiency of IT systems and supporting infrastructure, the ability to bypass system controls, the possibility of loss or unavailability of key resources (e.g., systems, data, communications, personnel, premises), and failure to comply with laws and regulations.
- **The project level**—Risk management needs to focus on the ability to understand and manage project complexity, and, if this is not done effectively, the consequent risk that project objectives will not be met.
- **The strategic level**—The risk focus shifts to considerations such as how well the IT capability is aligned with the business strategy, how it compares with that of competitors and the threats (as well as the opportunities) posed by technological change.

The identification, evaluation and management of IT risks at various levels will be the responsibility of different individuals and groups within the organization. However, these individuals and groups should not operate separately since risks at one level or in one area may also impact another. A major system malfunction could impair an organization's ability to deliver customer service or deal with suppliers, and it could have strategic implications that require top management attention. Similarly, problems with a major project could have strategic implications. Also, as projects deliver new IT systems and infrastructure, the new operational risk environment needs to be considered.

In summary, the risk management process should achieve a cost-effective balance between the application of security controls as countermeasures and the significant threats. Some of the threats are related to security issues that can be extremely sensitive for some industries.

# 2.6.3 RISK ANALYSIS METHODS

This section discusses qualitative, semiquantitative and quantitative risk management methods, and the advantages and limitations of the latter.

## *Qualitative Analysis Methods*

Qualitative risk analysis methods use word or descriptive rankings to describe the impacts or likelihood. They are the simplest and most frequently used methods. They are normally based on checklists and subjective risk ratings such as high, medium or low.

Such approaches lack the rigor that is customary for accounting and management.

## *Semiquantitative Analysis Methods*

In semiquantitative analysis, the descriptive rankings are associated with a numeric scale. Such methods are frequently used when it is not possible to utilize a quantitative method or to reduce subjectivity in qualitative methods.

## *Quantitative Analysis Methods*

Quantitative analysis methods use numeric values to describe the likelihood and impacts of risks, using data from several types of sources such as historic records, past experiences, industry practices and records, statistical theories, testing, and experiments.

Many quantitative risk analysis methods are currently used by military, nuclear, chemical, financial and other areas. The following selections describe concepts related to quantitative methods.

Many organizations refer to quantitative risk analysis, which expresses risks in numeric (e.g., monetary) terms. A quantitative risk analysis is generally performed during a business impact analysis (BIA). The main problem within this process is the valuation of information assets. Different individuals may assign different values to the same asset, depending on the relevance of information to the individuals. In the case of technology assets, it is not the cost of the asset that is considered but also the cost of replacement and the value of information processed by that asset.

### PROBABILITY AND EXPECTANCY
Most of these methods are based on "classical" statistical theories of probability and expectancy.

Most natural phenomena are affected by many variables. Even in cases where scientific laws are found for their behavior, actual behavior fluctuates around predicted values. Artificial variables, e.g., social, economic, technological (such as voting attitudes), gross domestic product (GDP) or boiler temperature, are subject to a similar so-called stochastic behavior. The occurrence of a storm cannot be accurately predicted, but (if enough scientific historical information is available to establish a trend or pattern) the probability of an occurrence can be assigned with a certain accuracy and confidence level.

An event's probability is denoted by $p$ ($0 \leq p \leq 1$). The value of an asset that will be affected by the referred event is denoted by v. The expected loss or gain is equal to v x $p$ (the value times the probability that the event takes place).

### ANNUAL LOSS EXPECTANCY METHOD
Annual loss expectancy (ALE) simplifies the assignment of value (v) and probability (p) in a manner that is easier to quantify.

With this approach, a custom-made worksheet can be easily created and adapted to the specific currency, asset materiality and time horizons desired (**exhibit 2.6**).

**Exhibit 2.6—Annual Loss Expectancy Approach**

| Frequency | Asset Value (US $) | | | | | | | |
|---|---|---|---|---|---|---|---|---|
| | **1** | **100** | **1,000** | **10,000** | **100,000** | **1 million** | **10 million** | **1 billion** |
| 1 minute | 526 | 52,560 | 525,600 | | | | | |
| 1 hour | 9 | 876 | 8,760 | 87,600 | 876,000 | | | |
| 1 day | | 37 | 365 | 3,650 | 36,500 | 36,500 | | |
| 1 week | | 5 | 52 | 521 | 5,214 | 52,143 | 521,429 | |
| 1 month | | 1 | 12 | 120 | 1,200 | 12,000 | 120,000 | |
| 3 months | | | 4 | 40 | 400 | 4,000 | 40,000 | |
| 1 year | | | 1 | 10 | 100 | 1,000 | 10,000 | 1,000,000 |
| 5 years | | | | 2 | 20 | 200 | 2,000 | 200,000 |
| 10 years | | | | 1 | 10 | 100 | 1,000 | 100,000 |
| 20 years | | | | 1 | 5 | 50 | 500 | 50,000 |
| 50 years | | | | | 2 | 20 | 200 | 20,000 |
| 100 years | | | | | 1 | 10 | 100 | 10,000 |
| 300 years | | | | | | 3 | 33 | 3,333 |

(Figures are rounded to US $1,000)

Management and IS auditors should keep in mind certain considerations including:
- Risk management (RM) should be applied to IT functions throughout the company.
- RM is a senior management responsibility.
- Quantitative RM is preferred over qualitative approaches.
- Quantitative RM always faces the challenge of estimating risks (their probability) and relies on subjectivity and qualitative approaches.
- Quantitative RM provides more objective (and traceable) assumptions.
- The real complexity or the apparent sophistication of the methods or packages used should not be a substitute for business commonsense or professional diligence.
- Special care should be taken to ensure that adequate consideration is given to dealing with very high impact events, even if probability expressed as a frequency of occurrence over time is low. Recent tragedies such as terrorist attacks and natural disasters have demonstrated the potential for unpredicted catastrophic events to occur. Determining future probabilities based on past experience is often difficult, particularly where events are unprecedented, so it is prudent to assume that the worst case could occur. While to some extent distasteful, organizations need to be realistic in making plans to deal with loss of key staff, loss of premises, loss of systems, and widespread loss, corruption or theft of data. It should be noted that if an event is assessed as having a probability of one in 50 years, this does not mean it will take 50 years from now to happen—it could happen tomorrow.

# 2.7 IS MANAGEMENT PRACTICES

IS management practices reflect the implementation of policies and procedures developed for various IS-related management activities. In most organizations, the IS department is a service (support) department. The traditional role of a service department is to help production (line) departments conduct their operations more effectively and efficiently. Today, however, IS has become an integral part of every facet of the operations of an organization. Its importance continues to grow year after year, and there is little likelihood of a reversal of this trend. IS auditors must understand and appreciate the extent to which a well-managed IS department is crucial to achieving the organization's objectives.

Management activities to review the policy/procedure formulations and their effectiveness within the IS department would include practices such as personnel management, sourcing and IT change management.

## 2.7.1 HUMAN RESOURCE MANAGEMENT

Human resource management relates to organizational policies and procedures for recruiting, selecting, training and promoting staff, measuring staff performance, disciplining and staff, succession planning, and staff retention. The effectiveness of these activities, as they relate to the IS function, impacts the quality of staff and the performance of IS duties.

> **Note:** The IS auditor should be aware of human resource management issues but this information is not tested in the CISA exam due to its subjectivity and organizational specific subject matter.

### Hiring

An organization's hiring practices are important to ensure that the most effective and efficient staff are chosen and that the company is in compliance with legal recruitment requirements. Some of the common controls would include:
- Background checks (e.g., criminal, financial, professional, references).
- Confidentiality agreements
- Employee bonding to protect against losses due to theft, mistakes and neglect (Note: Employee bonding is not always an accepted practice all over the world; in some countries, it is not legal.)
- Conflict of interest agreements
- Codes of professional conduct/ethics
- Noncompete agreements

Control risks include:
- Staff may not be suitable for the position they are recruited to fill.
- Reference checks may not be carried out.
- Temporary staff and third-party contractors may introduce uncontrolled risks.
- Lack of awareness of confidentiality requirements may lead to the compromise of the overall security environment.

### Employee Handbook

Employee handbooks, distributed to all employees at time of hire, should explain items such as:
- Security policies and procedures
- Acceptable and unacceptable conduct
- Organizational values and ethics code
- Company expectations
- Employee benefits
- Vacation (holiday) policies
- Overtime rules
- Outside employment
- Performance evaluations
- Emergency procedures
- Disciplinary actions for:
  – Excessive absence
  – Breach of confidentiality and/or security
  – Noncompliance with policies

In general, there should be a published code of conduct for the organization that specifies the responsibilities of all employees.

### Promotion Policies

Promotion policies should be fair and equitable and understood by employees. Policies should be based on objective criteria and consider an individual's performance, education, experience and level of responsibility.

The IS auditor should ensure that the IS organization has well-defined policies and procedures for promotion, and is adhering to them.

## Training

Training should be provided on a regular basis to all employees based on the areas where employee expertise is lacking. Training is particularly important for IS professionals, given the rapid rate of change of technology and products. Training not only assures more effective and efficient use of IS resources, but also strengthens employee morale. Training must be provided when new hardware and/or software is being implemented. Training should also include relevant management training, project management and technical training.

Cross-training is having more than one individual properly trained to perform a specific job or procedure. This practice has the advantage of decreasing dependence on one employee and can be part of succession planning. It also provides a backup for personnel in the event of absence for any reason and, thereby, provides for continuity of operations. However, in using this approach, it would be prudent to have first assessed the risks of any person knowing all parts of a system and what exposure this may cause.

## Scheduling and Time Reporting

Proper scheduling provides for more efficient operation and use of computing resources. Time reporting allows management to monitor the scheduling process. Management can then determine whether staffing is adequate and if the operation is running efficiently. It is important that the information being entered or recorded into such a system is accurate.

Time reporting can be an excellent source of information for IT governance purposes. One of the scarcest resources in IT is time, and its proper reporting will definitely help to better manage this finite resource. This input can be useful for cost allocation, key goal indicator (KGI) and key performance indicator (KPI) measurement, and activities analysis (e.g., how many hours the organization dedicates to application changes vs. new developments).

## Employee Performance Evaluations

Employee assessment/performance evaluations must be a standard and regular feature for all IS staff. The human resources (HR) department should ensure that IS managers and IS employees set mutually agreed on goals and expected results. Assessment can be set against these goals only if the process is objective and neutral.

Salary increments, performance bonuses and promotions should be based on performance. The same process can also allow the organization to gauge employee aspirations and satisfaction, and identify problems.

## Required Vacations

A required vacation (holiday) ensures that once a year, at a minimum, someone other than the regular employee will perform a job function. This reduces the opportunity to commit improper or illegal acts. During this time, it may be possible to discover fraudulent activity as long as there has been no collusion between employees to cover possible discrepancies.

Job rotation provides an additional control (to reduce the risk of fraudulent or malicious acts) since the same individual does not perform the same tasks all the time. This provides an opportunity for an individual other than the regularly assigned person to perform the job and notice possible irregularities.

> **Note:** The IS auditor should be familiar with ways to mitigate internal fraud. Mandatory leave is such a control measure.

## Termination Policies

Written termination policies should be established to provide clearly defined steps for employee separation. It is important that policies be structured to provide adequate protection for the organization's computer assets and data. Termination practices should address voluntary and involuntary (e.g., immediate) terminations. For certain situations, such as involuntary terminations under adverse conditions, an organization should have clearly defined and documented procedures for escorting the terminated employee from the premises. In all cases, however, the following control procedures should be applied:
• Return of all access keys, ID cards and badges—to prevent easy physical access
• Deletion/revocation of assigned logon IDs and passwords—to prohibit system access

• Notification to appropriate staff and security personnel regarding the employee's status change to "terminated"
• Arrangement of the final pay routines—to remove the employee from active payroll files
• Performance of a termination interview—to gather insight on the employee's perception of management
• Return of all company property

## 2.7.2 SOURCING PRACTICES

Sourcing practices relate to the way in which the organization will obtain the IS functions required to support the business. Organizations can perform all the IS functions in-house (known as "insourcing") in a centralized fashion, or outsource all functions across the globe. The sourcing strategy should consider each IS function and determine which approach allows the IS function to meet the enterprise's goals.

Delivery of IS functions can include:
• **Insourced**—Fully performed by the organization's staff
• **Outsourced**—Fully performed by the vendor's staff
• **Hybrid**—Performed by a mix of the organization's and vendor's staffs; can include joint ventures/supplemental staff

IS functions can be performed across the globe, taking advantage of time zones and arbitraging labor rates, and can include:
• **Onsite**—Staff work onsite in the IS department
• **Offsite**—Also known as nearshore, staff work at a remote location in the same geographic area
• **Offshore**—Staff work at a remote location in a different geographic region

The organization should evaluate their IS functions and determine the most appropriate method of delivering the IS functions, giving consideration to the following:
• Is this a core function for the organization?
• Does this function have specific knowledge, processes and staff critical to meeting its goals and objectives, and that cannot be replicated externally or in another location?
• Can this function be performed by another party or in another location for the same or lower price, with the same or higher quality, and without increasing risk?
• Does the organization have experience managing third parties or using remote/offshore locations to execute IS or business functions?

On completion of the sourcing strategy, the IS steering committee should review and approve the strategy. At this point, if the organization has chosen to use outsourcing, a rigorous process should be followed, including the following steps:
• Define the IS function to be outsourced.
• Describe the service levels required and minimum metrics to be met.
• Know the desired level of knowledge, skills and quality of the expected service provider desired.
• Know the current in-house cost information to compare with third-party bids.
• Conduct due dilligence reviews of potential service providers.

Using this information, the organization can perform a detailed analysis of the service provider bids and determine whether outsourcing will allow the organization to meet their goals in a cost-effective manner, with limited risk.

The same process should be considered when an organization chooses to "globalize" or take their IS functions offshore.

## Outsourcing Practices and Strategies

Outsourcing practices relate to contractual agreements under which an organization hands over control of part or all of the functions of the IS department to an external party. Most IS departments utilize information resources from a wide array of vendors and, therefore, need a defined outsourcing process for effectively managing contractual agreements with these vendors.

The contractor provides the resources and expertise required to perform the agreed service. Outsourcing is becoming increasingly important in many organizations. The IS auditor must be aware of the various forms outsourcing can take and the associated risks.

The specific objectives for IT outsourcing vary from organization to organization. Typically, though, the goal is to achieve lasting, meaningful improvement in business processes and services through corporate restructuring, to take advantage of a vendor's core competencies. As with the decision to downsize or rightsize, the decision to outsource services and products requires management to revisit the control framework on which it can rely.

Reasons for embarking on outsourcing include:
• A desire to focus on core activities
• Pressure on profit margins
• Increasing competition that demands cost savings
• Flexibility with respect to both organization and structure

The services provided by a third party can include:
• Data entry
• Design and development of new systems in the event that the in-house staff does not have the requisite skills or is otherwise occupied in higher-priority tasks, or in the event of a one-time task in which case there is no need to recruit additional in-house skilled staff
• Maintenance of existing applications to free in-house staff to develop new applications
• Conversion of legacy applications to new platforms-for example, a specialist company may web-enable the front end of an old application.
• Operating the help desk or the call center
• Operations processing

Possible advantages of outsourcing include:
• Commercial outsourcing companies can achieve economies of scale through the deployment of reusable component software.
• Outsourcing vendors are likely to be able to devote more time and focus more effectively and efficiently on a given project than in-house staff.
• Outsourcing vendors are likely to have more experience with a wider array of problems, issues and techniques than in-house staff.
• The act of developing specifications and contractual agreements using outsourcing services is likely to result in better specifications than if developed only by in-house staff.
• As vendors are highly sensitive to time-consuming diversions and changes, feature creep or scope creep is substantially less likely with outsourcing vendors.

Possible disadvantages and business risks of outsourcing include:
• Costs exceeding customer expectations
• Loss of internal IS experience
• Loss of control over IS
• Vendor failure (going concern)
• Limited product access
• Difficulty in reversing or changing outsourced arrangements
• Deficient compliance with legal and regulatory requirements
• Contract terms not being met
• Lack of loyalty of contractor personnel toward the customer
• Disgruntled customers/employees as a result of the outsource arrangement
• Service costs not being competitive over the period of the entire contract
• Obsolescence of vendor IT systems
• Failure of either company to receive the anticipated benefits of the outsourcing arrangement
• Damage to either or both companies' reputations due to project failures

• Lengthy, expensive litigation
• Risk of loss or leakage of information or processes

Some of the ways that these risks can be reduced are by:
• Establishing measurable, partnership-enacted shared goals and rewards
• Using multiple suppliers or withholding a piece of business as an incentive
• Performing periodic competitive reviews and benchmarking/benchtrending
• Implementing short-term contracts
• Forming a cross-functional contract management team
• Including contractual provisions to consider as many contingencies as can reasonably be foreseen-for example:
  – Incorporating service quality expectations, including usage of the Capability Maturity Model (CMM), Information Technology Infrastructure Library (ITIL) or International Organization for Standardization (ISO) methodologies
  – Ensuring adequate contractual consideration of access control/security administration, whether vendor- or owner-controlled
  – Ensuring that violation reporting and follow-up are required by the contract, and any requirements for owner notification and cooperation with any investigations
  – Ensuring that change/version control and testing requirements are contractually required for the implementation and production phases
  – Ensuring that the parties responsible and the requirements for network controls are adequately defined, and any necessary delineation of these responsibilities established
  – Stating specific, defined performance parameters that must be met; for example, minimum processing times for transactions or minimum hold times for contractors
  – Incorporating capacity management criteria
  – Providing contractual provisions for making changes to the contract
  – Providing a clearly defined dispute escalation and resolution process
  – Ensuring that the contract indemnifies the company from damages caused by the organization responsible for the outsourced services
  – Requiring confidentiality agreements protecting both parties
  – Incorporating clear, unambiguous "right to audit" provisions, providing the right to audit vendor operations (e.g., access to facilities, access to records, right to make copies, access to personnel, provision of computerized files) as they relate to the contracted services
  – Ensuring that the contract adequately addresses business continuity and disaster recovery provisions, and appropriate testing
  – Establishing that the confidentiality, integrity, and availability (sometimes referred to as the CIA triad) of organization-owned data must be maintained, and clearly establishing the ownership of the data
  – Requiring that the vendor comply with all relevant legal and regulatory requirements, including those enacted after contract initiation
  – Establishing ownership of intellectual property developed by the vendor on behalf of the customer
  – Establishing clear warranty and maintenance periods
  – Providing software escrow provisions
  – Establishing clear roles and responsibilities between the parties
  – Requiring that the vendor follow the organization's policies, including their information security policy, unless the vendor's policies have been agreed to in advance by the organization
  – Requiring the vendor to identify all subcontract relationships and requiring the organization's approval to change subcontractors

Outsourcing requires management to actively manage the relationship and the outsourced services. Since the outsourcing agreement is governed by the contract terms, the contract with the outsourced service provider should include a description of the means, methods, processes and structure accompanying the offer of IS services and products, and the control of quality. The formal or legal character of these agreements depends on the relationship between the parties and the demands placed by principals on those performing the engagement.

Once the outsourcer has been selected, the IS auditor should regularly review the contract and service levels to ensure that they are appropriate. In addition, the IS auditor could review the outsourcer's documented procedures and results of their

quality programs-which could include, for example, CMM, ITIL and ISO methodologies. These quality programs require regular audits to certify that the process and procedures meet the quality standard.

Outsourcing is not only a cost decision; it is a strategic decision that has significant control implications for management. Quality of service, guarantees of continuity of service, control procedures, competitive advantage and technical knowledge are issues that need to be part of the decision to outsource IT services. Choosing the right supplier is extremely important, particularly when outsourcing is a long-term strategy. The compatibility of suppliers in terms of culture and personnel is an important issue that should not be overlooked by management.

The decision to outsource a particular service currently within the organization demands proper attention to contract negotiations. A well-balanced contract and service level agreement (SLA) is of great importance for quality purposes and future cooperation between the concerned parties.

SLAs are a contractual means of helping the IS department manage information resources under the control of a vendor. SLAs stipulate and commit a vendor to a required level of service and support options. This includes providing for a guaranteed level of system performance regarding downtime or uptime, and a specified level of customer support. Software or hardware requirements are also stipulated. SLAs also provide for penalty provisions and enforcement options for services not provided, and may include incentives such as bonuses or gain-sharing for exceeding service levels.

Above all, the SLA should serve as an instrument of control. Where the outsourcing vendor is from another country, the organization should be aware of cross-border legislation.

## Globalization Practices and Strategies

Many organizations have chosen to globalize their IS functions in addition to outsourcing functions. The globalization of IS functions is performed for many of the same reasons cited for outsourcing; however, the organization may choose not to outsource the function. Globalizing IS functions requires management to actively oversee the remote or offshore locations.

Where the organization performs functions in-house, it may choose to move the IS functions offsite or offshore. The IS auditor can assist in this process by ensuring that IS management considers the following risks and audit concerns when defining the globalization strategy and completing the subsequent transition to remote offshore locations:
- **Legal, regulatory and tax issues**—Operating in a different country or region may introduce new risks about which the organization may have limited knowledge.
- **Continuity of operations**— Business continuity and disaster recovery may not be adequately provided for and tested.
- **Personnel**—Needed modifications to personnel policies may not be considered.
- **Telecommunication issues**—Network controls and access from remote or offshore locations may be subject to more frequent outages or a larger number of security exposures.
- **Cross-border and cross-cultural issues**—Managing people and processes across multiple time zones, languages and cultures may present unplanned challenges and problems.

## Outsourcing and Third-party Audit Reports

One method for the IS auditor to have assurance of the controls implemented by a service provider requires the provider periodically provide a third-party audit report. These reports cover the range of issues related to confidentiality, integrity and availability of data. The third-party auditor should be found acceptable by the IS auditor and the auditee, and must be agreed on in advance. For certain industries, third-party audits may fall under regulatory oversight and control. For example, there is legislation such as SAS 70-an audit guide by the American Institute of Certified Public Accountants (AICPA) in the US (and similar legislation mandated in the UK and Canada) for providing third-party audit reports.

SAS 70 is designed to provide guidance to service auditors engaged to issue reports on a service organization's controls that may be part of a user organization's information systems in the context of a financial statement. It also provides guidance to external auditors engaged to audit the financial statements of entities that use service organizations.

The second method could be to allow periodic review by the user's auditor of the organization. This may not be acceptable to the vendor since it involves providing time and resources for each audit.

## Governance in Outsourcing

Outsourcing is the mechanism that allows organizations to transfer the delivery of services to third parties. Fundamental to outsourcing is accepting that, while service delivery is transferred, accountability remains firmly with the management of the client organization-which must ensure that the risks are properly managed and there is continued delivery of value from the service provider. Transparency and ownership of the decision-making process must reside within the purview of the client.

The decision to outsource is a strategic, not merely a procurement, decision. The organization that outsources is effectively reconfiguring its value chain by identifying those activities that are core to its business, retaining them and making noncore activities candidates for outsourcing. Understanding this in the light of governance is key, not only because well-governed organizations have been shown to increase shareholder value, but more importantly, because organizations are competing in an increasingly aggressive, global and dynamic market.

Establishing and retaining competitive and market advantage requires the organization to be able to respond effectively to competition and changing market conditions. Outsourcing can support this, but only if the organization understands which parts of its business truly create competitive advantage. Disaggregating these and giving them to a third party must, in themselves, become core competencies because outsourcing is a strategic mechanism that allows the organization to constantly focus its efforts and expertise. As a strategic resource, outsourcing must be governed accordingly. This is not just about purchasing, but about effective management and ensuring that both parties benefit.

Governance of outsourcing is the set of responsibilities, roles, objectives, interfaces and controls required to anticipate change and manage the introduction, maintenance, performance, costs and control of third-party provided services. It is an active process that the client and service provider must adopt to provide a common, consistent and effective approach that identifies the necessary information, relationships, controls and exchanges among many stakeholders across both parties.

The decision to outsource and subsequently successfully manage that relationship demands effective governance. Most people who conduct outsourcing contracts include basic control and service execution provisions; however, one of the main objectives of the outsourcing governance process, as defined in the outsourcing contract, is to ensure continuity of service at the appropriate levels, profitability and added value to sustain the commercial viability of both parties. Experience has shown that many companies make assumptions about what is included in the outsource proposition. Whereas it is neither possible nor cost-effective to contractually define every detail and action, the governance process provides the mechanism to balance risk, service demand, service provision and cost.

The governance of outsourcing extends both parties' (i.e., client and supplier) responsibilities into:
• Ensuring contractual viability through continuous review, improvement and benefit gain to both parties
• Inclusion of an explicit governance schedule to the contract
• Management of the relationship to ensure that contractual obligations are met through SLAs and operating level agreements (OLAs)
• Identification and management of all stakeholders, their relationships and expectations
• Establishment of clear roles and responsibilities for decision making, issue escalation, dispute management, demand management and service delivery
• Allocation of resources, expenditure and service consumption in response to prioritized needs
• Continuous evaluation of performance, cost, user satisfaction and effectiveness
• Ongoing communication across all stakeholders

## Current Outsourcing Governance Approaches

The increasing size of the technology solution space is driven by the pace of technological evolution. Acquiring, training and retaining qualified staff is becoming more expensive in an increasingly global, dynamic and mobile economy. Investing in costly technology implementation and training is seen as less of an organizational core activity than is the ability to work effectively across the value chain by integrating the outsourcing of services where appropriate.

Although the term 'business alignment' is often used, what it encompasses is not always clear. In the widest sense, it involves making the services provided by the corporate IT function more closely reflect the requirements and desires of the business users. When organizations recognize what is core to their business, which services provide them differential advantage and then outsource the activities that support these services, business alignment can be achieved. If the degree to which this alignment is approached is to be understood, the implication is that SLAs and OLAs must be established, monitored and measured in terms of performance and user satisfaction. Business alignment should be driven by the service end user.

Governance should be preplanned and built into the contract as part of the service cost optimization. The defined governance processes should evolve as the needs and conditions of the outsourcing relationship adapt to changes to service demand and delivery, and to technology innovation.

## Capacity and Growth Planning

Given the strategic importance of IT in companies and the constant change in technology, capacity and growth planning are essential. This activity must be reflective of the long- and short-range business plans and must be considered within the budgeting process. Changes in capacity should not only reflect changes in the underlying infrastructure, but also in the number of staff available to support the organization. A lack of appropriately qualified staff may delay projects that are critical to the organization or result in not meeting agreed on service levels. This is what leads some organizations to choose outsourcing as a solution for growth.

## Third-party Service Delivery Management

Every organization using the services of third parties should have a service delivery management system in place to implement and maintain the appropriate level of information security and service delivery in line with third-party service delivery agreements.

The organization should check the implementation of agreements, monitor compliance with the agreements and manage changes to ensure that the services delivered meet all requirements agreed to with the third party.

### SERVICE DELIVERY

Security controls, service definitions and delivery levels included in the third-party service delivery agreement should be implemented, operated and maintained by the third party.

Service delivery by a third party should include the agreed on security arrangements, service definitions and aspects of service management. In case of outsourcing arrangements, the organization should plan the necessary transitions (of information, data center facilities, and anything else that needs to be moved) and ensure that security is maintained throughout the transition period.

The organization should ensure that the third party maintains sufficient service capability together with workable plans designed to ensure that agreed on service continuity levels are maintained following major service failures or disaster.

### MONITORING AND REVIEW OF THIRD-PARTY SERVICES

The services, reports and records provided by the third party should be regularly monitored and reviewed, and audits should be carried out regularly. Monitoring and review of third-party services should ensure that the information security terms and conditions of the agreements are being adhered to, and information security incidents and problems are managed properly. This should involve a service management relationship and process between the organization and the third party to:
• Monitor service performance levels to check adherence to the agreements
• Review service reports produced by the third party and arrange regular progress meetings as required by the agreements
• Provide information about information security incidents and review of this information by the third party and the organization as required by the agreements and any supporting guidelines and procedures
• Review third-party audit trails and records of security events, operational problems, failures, tracing of faults and disruptions related to the service delivered
• Resolve and manage any identified problems

The responsibility for managing the relationship with a third party should be assigned to a designated individual or service management team. In addition, the organization should ensure that the third party assigns responsibilities for checking for compliance and enforcing the requirements of the agreements. Sufficient technical skills and resources should be made available to monitor whether requirements of the agreement, in particular the information security requirements, are being met. Appropriate action should be taken when deficiencies in the service delivery are observed.

The organization should maintain sufficient overall control and visibility into all security aspects for sensitive or critical information or information processing facilities accessed, processed or managed by a third party. The organization should ensure that they retain visibility in security activities such as change management, identification of vulnerabilities and information security incident reporting/response through a clearly defined reporting process, format and structure. When outsourcing, the organization needs to be aware that the ultimate responsibility for information processed by an outsourcing party remains with the organization.

### MANAGING CHANGES TO THIRD-PARTY SERVICES
Changes to the provision of services, including maintaining and improving existing information security policies, procedures and controls, should be managed taking into account the criticality of business systems and processes involved and reassessing risks.

The process of managing changes to a third-party service needs to take into account:
- Changes made by the organization to implement:
  - Enhancements to the current services offered
  - Development of any new applications and systems
  - Modifications or updates of the organization's policies and procedures
  - New controls to resolve information security incidents and to improve security
  - Updates to policies, including the IT security policy
- Changes in third-party services to implement:
  - Changes and enhancements to networks
  - Use of new technologies
  - Adoption of new products or newer versions/releases
  - New development tools and environments
  - Changes to physical location of service facilities
  - Change of vendors or subcontractors

## Service Improvement and User Satisfaction
SLAs set the baseline by which outsourcers perform the IS function. In addition, organizations can set service improvement expectations into the contracts with associated penalties and rewards. Examples of service improvements include:
- Reductions in the number of help desk calls
- Reductions in the number of system errors
- Improvements to system availability

Service improvements should be agreed on by users and IT with the goals of improving user satisfaction and attaining business objectives. User satisfaction should be monitored by interviewing and surveying users.

## Industry Standards/Benchmarking
Industry standards/benchmarking provide a means of determining the level of performance provided by similar information processing facility environments. These standards or benchmarking statistics can be obtained from vendor user groups, industry publications and professional associations. Examples include ISO 9000 and the Software Engineering Institute's CMM. Outsourcing organizations must adhere to a well-defined set of standards that can be relied on by their clients.

## 2.7.3 ORGANIZATIONAL CHANGE MANAGEMENT

Organizational change management is managing involves use of a defined and documented process to identify and apply technology improvements at the infrastructure and application level that are beneficial to the organization and involve all levels of the organization impacted by the changes. This level of involvement and communication will ensure that the IS department fully understands the users' expectations and changes are not resisted or ignored by users once implemented.

The IS department is the focal point for such changes by leading or facilitating change in the organization. This includes staying abreast of technology changes that could lead to significant business process improvements and obtaining senior management commitment for the changes or project that will be required at the user level.

Once senior management support is obtained to move forward with the changes or project, the IS department can begin working with each functional area and their management to obtain support for the changes. In addition, the IS department will need to develop a communication process which is directed at the end users to update them on the changes, the impact and benefit of the changes, and provide a method for obtaining user feedback and involvement.

User feedback should be obtained throughout the project, including validation of the business requirements and training on and testing of the new or changed functionality.

## 2.7.4 FINANCIAL MANAGEMENT PRACTICES

Financial management is a critical element of all business functions. In a cost-intensive computer environment, it is imperative that sound financial management practices are in place.

The user-pays scheme, a form of chargeback, can improve application and monitoring of IS expenses and available resources. In this scheme the costs of IS services—including staff time, computer time and other relevant costs—are charged back to the end users based on a standard (uniform) formula or calculation.

Chargeback is a joint responsibility of IS management and user management. Chargeback provides IS personnel and users with a tool to measure the effectiveness and efficiency of the service provided by the information processing facility.

### IS Budgets

IS management, like all other departments, must develop a budget.

A budget allows for forecasting, monitoring and analyzing financial information. The budget allows for an adequate allocation of funds, especially in an IS environment where expenses can be cost-intensive. The IS budget should be linked to short- and long-range IT plans.

## 2.7.5 QUALITY MANAGEMENT

Quality management is the means by which IS department-based processes are controlled, measured and improved. Processes in this context are defined as a set of tasks that, when properly performed, produce the desired results. Areas of control for quality management may include the following:
• Software development, maintenance and implementation
• Acquisition of hardware and software
• Day-to-day operations
• Service management
• Security
• HR management
• General administration

The development and maintenance of defined and documented processes by the IS department is evidence of effective governance of information resources. Insistence on the observance of processes and related process management techniques is key to the effectiveness and efficiency of the IS organization. Various standards have emerged to assist IS organizations in achieving these results. Quality standards are increasingly being used to assist IS organizations in achieving an operational environment that is predictable, measurable, repeatable and certified for their information technology resources.

> **Note:** The IS auditor should be aware of Quality Management. However, the CISA exam does not test specifics on any ISO standards.

A prominent standard receiving wide recognition and acceptance is the ISO 9001:2000 Quality Management Systems, which replaces the ISO 9000, 9001, 9002 and 9003 standards enacted to govern the management of quality. This standard is becoming the most important quality management system standard and is representative of those standards that are most likely to be met in practice. This applies to all types of organizations and can be service- or product-oriented in complying with the needs of organizational objectives, customers and clients.

The quality management system is based on a set of documents, manuals and records.

A key practice by an organization that wants to develop a quality management system that meets ISO 9001:2000 requirements is to perform a gap analysis against the requirements in the standard. The analysis allows for improvements in the company's processes to fill the gaps and comply with the standards. After successfully meeting requirements of an internal process audit of the quality management system, an ISO certificate is issued and recorded in a registry. The certification is valid until the next audit. However, registration is not required. Companies may choose to voluntarily comply with the standards.

IS auditors should be concerned with whether functions and processes are documented and practiced, as defined, to produce desired results. Since there is a cost associated with developing and implementing process management techniques, an IS auditor would be most concerned with defined and documented IT-related processes for critical business functions. Toward this end, an auditor would recommend development of a process improvement program that prioritizes required actions, establishes a plan to accomplish required actions and commits resources to execute the plan.

Additionally, IS auditors are concerned with whether up-to-date process documentation exists for managing the following functions within the IS organization:
• Computer operations
• Service management
• Systems software acquisition, implementation and maintenance
• Hardware acquisition and maintenance
• Application software acquisition/development and maintenance
• Management reporting
• Physical and logical security
• Short- and long-term planning
• Time reporting
• HR management

In practice, ISO 9001:2000 can directly impact an IS audit due to the strength of clauses in the standard. Three examples are:
• **Quality manual**—ISO 9001:2000 requires (clause 4.2.2) that a quality manual containing documented procedures or reference to them be established and maintained for the quality management system and the processes involved.
• **HR**—ISO 9001:2000 (clause 6.2) requires that personnel performing work affecting quality shall be competent on the basis of appropriate education, training, skills and experience. The establishment and maintenance of suitably trained and experienced personnel is a mandatory and auditable aspect of the quality management system, and directly impinges, for example, on the training of IS personnel.
• **Purchasing**—ISO 9001:2000 (clause 7.4) provides strong control on purchasing, including supplier evaluation, using defined and documented processes that can include the entire standard. When applied, for example, to the buying of outsourced IS services, meeting the conditions for ISO 9001:2000 purchasing controls would satisfy most of the needs of an IS audit.

The standard also requires a set of mandatory quality records be maintained. Records are used to demonstrate the existence and efficacy of the system, and are reviewed during internal and external audits.

The ISO 9001:2000 quality management system process audits (internal and external) include the provision of IS services as they affect the quality performance of the entire organization. The application of ISO quality standards should not excuse management from performing periodic risk assessment regarding IT services.

(See chapter 3, Systems and Infrastructure Life Cycle Management, for a discussion of software quality development processes—such as ISO 9126 and the software CMM sofftware developed by the Software Engineering Institute at Carnegie Mellon University—for evaluating quality of software products.)

Another ISO Standard, ISO/IEC 20000-1:2005 (Information Technology—Service Management), is more specific regarding service management than the ISO 9001:2000 standard due to its nature as a generic quality management standard.

ISO 20000 is tightly aligned with ITIL V2 Service Support and Service Delivery disciplines and promotes the adoption of an integrated process approach to the effective delivery of IT Services. It consists of a set of auditable statements and requirements against which an organization can be assessed and audited.

**Exhibit 2.7** outlines the content of ISO/IEC 20000-1:2005 (the numbers beside the headings depict the sections of the standard). It extends ITIL V2 Service Delivery and Service Support in some areas, introduces a new grouping of processes and challenges overall management practices that are common within other relevant ISO standards (like ISO 9001:2000).

| Exhibit 2.7—ISO/IEC 20000-1:2005 |
|---|

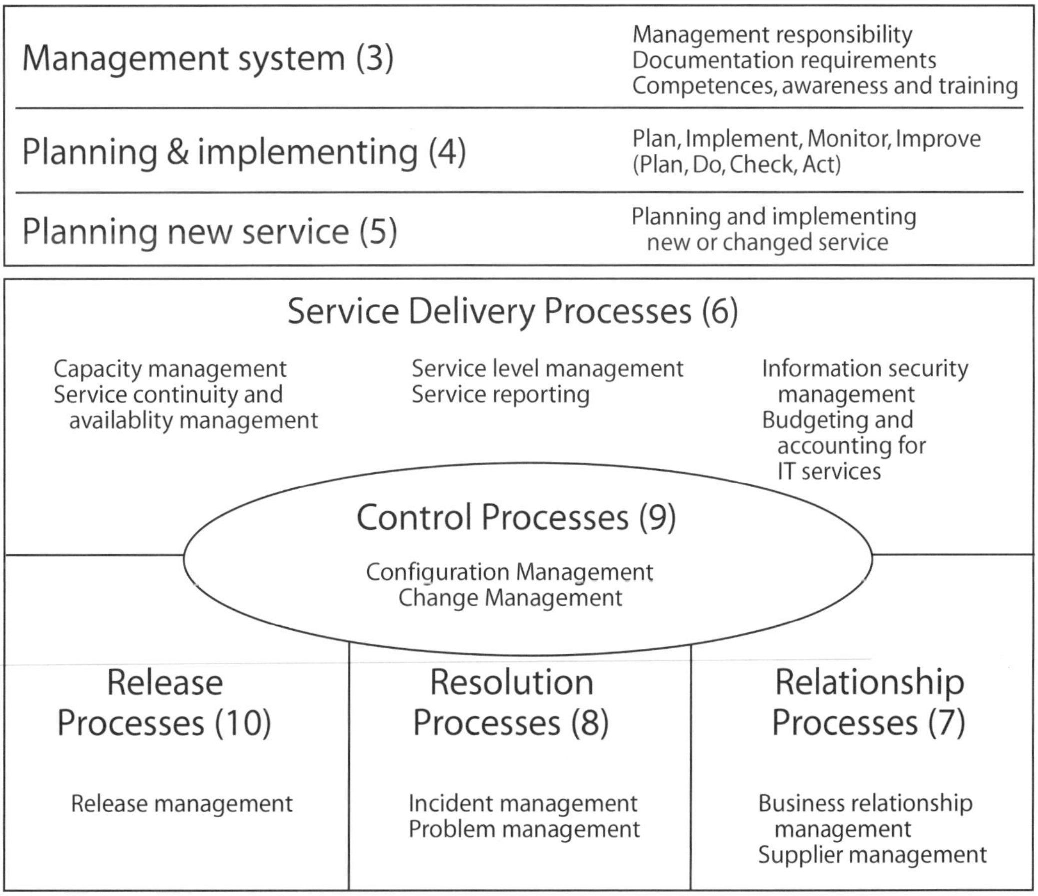

| | |
|---|---|
| **Management system (3)** | Management responsibility<br>Documentation requirements<br>Competences, awareness and training |
| **Planning & implementing (4)** | Plan, Implement, Monitor, Improve<br>(Plan, Do, Check, Act) |
| **Planning new service (5)** | Planning and implementing<br>new or changed service |

**Service Delivery Processes (6)**

| | | |
|---|---|---|
| Capacity management<br>Service continuity and<br>availablity management | Service level management<br>Service reporting | Information security<br>management<br>Budgeting and<br>accounting for<br>IT services |

**Control Processes (9)**

Configuration Management
Change Management

| **Release Processes (10)** | **Resolution Processes (8)** | **Relationship Processes (7)** |
|---|---|---|
| Release management | Incident management<br>Problem management | Business relationship<br>management<br>Supplier management |

## 2.7.6 INFORMATION SECURITY MANAGEMENT

Information security management provides the lead role to ensure that the organization's information and the information processing resources under its control are properly protected. This would include leading and facilitating the implementation of an organizationwide IT security program which includes the development of business impact analysis (BIA), business continuity (BCPs) and disaster recovery plans (DRPs) related to IS department functions in support of the organization's critical business processes. A major component in establishing such programs is the application of risk management principles to assess the risks to IT assets, mitigate these risks to an appropriate level as determined by management and monitor the remaining residual risks.

See chapter 5, Protection of Information Assets, for more details on information security management.

| PRACTICE QUESTIONS |
| --- |
| **2-4** The **MOST** important responsibility of a data security officer in an organization is: <br><br> A. recommending and monitoring data security policies. <br> B. promoting security awareness within the organization. <br> C. establishing procedures for IT security policies. <br> D. administering physical and logical access controls. <br><br> **2-5** Which of the following is **MOST** likely to be performed by the security administrator? <br><br> A. Approving the security policy <br> B. Testing application software <br> C. Ensuring data integrity <br> D. Maintaining access rules |
| *See answers and explanations to the practice questions at the end of the chapter. (page 133)* |

## 2.7.7 PERFORMANCE OPTIMIZATION

Measuring IT performance is a dynamic process. This is significant because today's business environment is complex and changing. Traditional measurement systems may give misleading signals to program managers, especially in trying to assess information technology's contribution to the organization's mission. Frequently, many variables affect the performance of information systems. Performance measurement of IT may be a statutory requirement.

Performance optimization is a process driven by performance indicators. These indicators are defined based on the complexity of an organization's business operations and processes, its strategic IT solution, and the primary corporate strategic objectives of IT implementation. Optimization refers to the process of improving the productivity of information systems to the highest level possible without unnecessary, additional investment in the IT infrastructure.

The broad phases of performance measurement are:
• Establishing and updating performance measures
• Establishing accountability for performance measures
• Gathering and analyzing performance data
• Reporting and using performance information

Caveats of performance measurement include:
• **Model**—A model is first built or established to evaluate the performance and alignment with the business objectives.
• **Measurement error**—Conventional measures do not properly account for the true inputs and outputs.
• **Lags**—Time lags between expense and benefit are not properly accounted for in current measures.
• **Redistribution**—IT is used to redistribute the source of costs in firms; there is no difference in total output, only in the means of getting it.

• **Mismanagement**—The lack of explicit measures of the value of information makes resources vulnerable to misallocation and overconsumption by managers. As a result proper performance measurement techniques will play an increasing role for program managers and investment review boards.

There are generally five ways to use performance measures:
1. Measure products/services
2. Manage products/services
3. Ensure accountability
4. Make budget decisions
5. Optimize performance

An effective performance management system should have leadership, a conceptual framework, effective internal and external communication, accountability for the results, and intelligence for decision makers. Rewards, compensation and recognition should be linked to performance measures. It is also important to share results and progress with employees, customers and stakeholders.

COBIT management guidelines are primarily designed to meet the needs of IT management for performance measurement. Goals, metrics and maturity models are provided for each of the 34 IT processes. These are generic and action-oriented for the purpose of addressing the following types of management concerns:
• **Performance measurement**—What are the indicators of good performance?
• **IT control profiling**—What is important? What are the critical success factors for control?
• **Awareness**—What are the risks of not achieving our objectives?
• **Benchmarking**—What do others do? What is measured and compared?

From a control perspective, management guidelines address the key issue of determining the right level of control for IT such that it supports the objectives of the enterprise.

Performance measures should be short and focused, complementing the high-level control guidance provided by the COBIT framework which states that IT enables the business by delivering information the business needs. COBIT management guidelines concentrate on defining action-oriented and generic guidelines for management required to maintain control over the enterprise's information and related processes and technology. These guidelines may include the use of maturity models for strategic choice and benchmark comparison, and goals and metrics.

| PRACTICE QUESTION |
| --- |
| 2-6      An IS auditor should ensure that IT governance performance measures: |
|      A.     evaluate the activities of IT oversight committees. |
|      B.     provide strategic IT drivers. |
|      C.     adhere to regulatory reporting standards and definitions. |
|      D.     evaluate the IT department. |
| *See answers and explanations to the practice questions at the end of the chapter. (page 133)* |

# 2.8 IS ORGANIZATIONAL STRUCTURE AND RESPONSIBILITIES

An IS department can be structured in different ways. One such format is shown in **exhibit 2.8**. The organization chart depicted includes functions related to security, applications development and maintenance, technical support for network and systems administration, and operations. The organizational structure shows the IS department typically headed by an IT manager/director or, in large organizations, by a CIO.

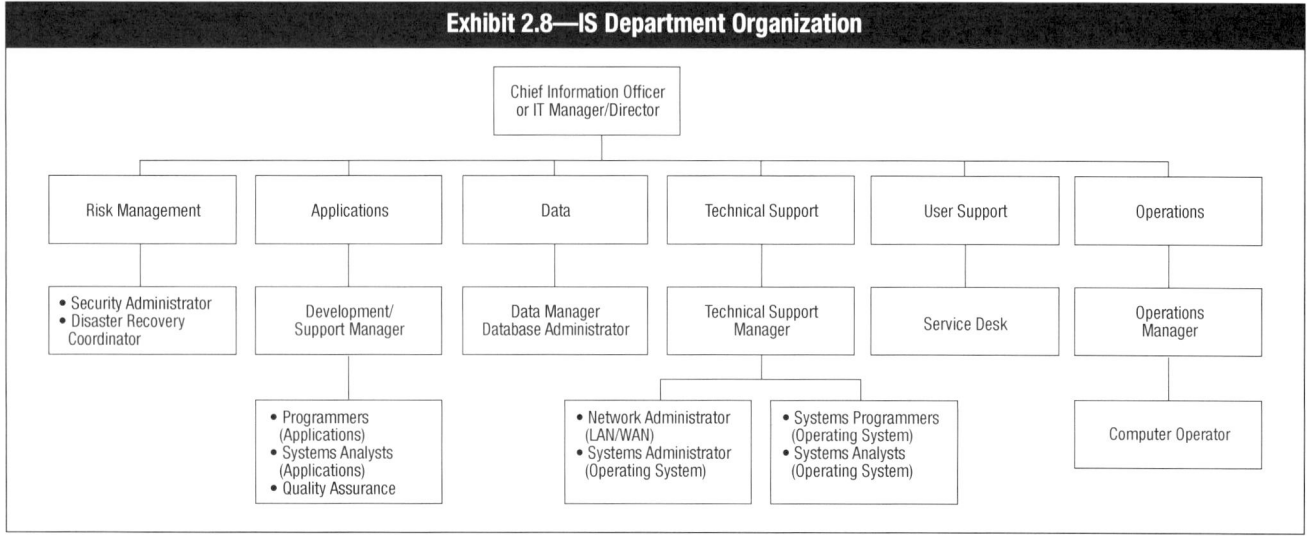

**Exhibit 2.8—IS Department Organization**

**Note:** The CISA exam does not test specific job responsibilities since they might vary within organizations. However, universally known responsibilities such as the business owners, information security functions and executive management might be tested, especially when testing access controls and data ownership. The IS auditor should be familiar with separation of duties.

## 2.8.1 IS ROLES AND RESPONSIBILITIES

Organizational structure charts are important items for all employees to have since they provide a clear definition of the department's hierarchy and authorities. Additionally, job descriptions provide IS department employees a clear direction regarding their roles and responsibilities. The IS auditor should spend time in an auditee's area to observe and determine whether the job description and structures are adequate. Generally, the following IS functions should be reviewed:

• **Systems development manager**—Responsible for programmers and analysts who implement new systems and maintain existing systems
• **Service desk (help desk)**—In today's IS environment more and more companies find it important to have a service desk function. It is a unit within an organization that responds to technical questions and problems faced by users. Most software companies have service desks. Questions and answers can be delivered by telephone, fax, e-mail or instant messaging. Service desk personnel may use third-party help desk software that enables them to quickly find answers to common questions. A procedure to record the problems reported, solved and escalated should be in place for analysis of the problems/questions. It helps in monitoring the user groups and improving the software/information processing facility (IPF) services.

Service desk/support administration includes the following activities:
– Acquiring hardware/software (HW/SW) on behalf of end users
– Assisting end users with HW/SW difficulties
– Training end users to use HW/SW and databases

– Answering queries of end users
– Monitoring technical developments and informing end users of pertinent developments
– Determining the source of problems with production systems and initiating corrective actions
– Informing end users of problems with HW/SW or databases that could affect their control of the installation of HW/SW upgrades
– Initiating changes to improve efficiency
• **End user**—Responsible for operations related to business application services; used to distinguish the person for whom the product (generally application level) was designed from the person who programs, services or installs applications. It is worth noting that there is a small distinction between the terms "end user" and "user." End user is slightly more specific and refers to someone who will access a business application, as stated above. The term user is broader and could refer to administrative accounts and accounts to access platforms.
• **End-user support manager**—Responsible as a liaison between the IS department and the end users
• **Data management**—Responsible for the data architecture in larger IT environments and tasked with managing data as a corporate asset
• **Quality assurance (QA) manager**—Responsible for negotiating and facilitating quality activities in all areas of information technology
• **Information security management**—This is a function that generally needs to be separate from the IS department and headed by a CISO. The CISO may report to the CIO or have a dotted-line (indirect reporting) relationship to the CIO. Even when the security officer reports to the CIO there is a possibility of conflict since the goals of the CIO are to efficiently provide continuous IT services whereas the CISO may be less interested in cost reduction if this impacts the quality of protection.

## Vendor and Outsourcer Management

With the increase in outsourcing, including the use of multiple vendors, dedicated staff may be required to manage the vendors and outsourcers including performing the following functions:
• Act as the prime contact for the vendor and outsourcer within the IS function.
• Provide direction to the outsourcer on issues and escalate internally within the organization and IS function.
• Monitor and report on the service levels to management.
• Review changes to the contract due to new requirements and obtain IS approvals.

## Infrastructure Operations and Maintenance

An **operations manager** is responsible for computer operations personnel, including all the staff required to run the computer IPF efficiently and effectively (e.g., computer operators, librarians, schedulers and data control personnel). The IPF includes the computer, peripherals, magnetic media and the data stored on the media. It constitutes a major asset investment and impacts the organization's ability to function effectively. The computer room should be secured and only authorized personnel should have access. No one except operations personnel should have access to the IPF. Within computer operations, management controls can be subdivided into three categories related to physical security, data security and processing controls.

The **control group** is responsible for the collection, conversion and control of input, and the balancing and distribution of output to the user community. The supervisor of the control group usually reports to the IPF operations manager. The input/output control group should be in a separate area where only authorized personnel are permitted since they handle sensitive data.

## Media Management

**Media management** is required to record, issue, receive and safeguard all program and data files that are maintained on removable media. Depending on the size of the organization, this function may be assigned to a full-time individual or a member of operations who also performs other duties.

This is a crucial function. Therefore, many organizations provide additional support for this function through the use of software which assists in maintaining inventory and movement of media. The use of this software also helps to maintain version control and configuration management of the programs.

## Data Entry

Data entry is critical to the information processing activity. Data entry can include batch entry or online entry.

In most organizations personnel in user departments do their own data entry online. In many online environments data are captured from the original source (e.g., electronic data interchange [EDI] input documents, data captured from bar codes for time management, departmental store inventory.). The user department and the system application must have controls in place to ensure that data are validated, accurate, complete and authorized.

With the advancement of technology and need to acquire data at its origination, automated systems for data acquisition are being deployed by organizations. These systems include barcode readers, or systems that are referred as SCADA (Supervisory Control and Data Acquisition). The term SCADA usually refers to centralized systems which monitor and control entire sites, or complexes of systems spread out over large areas (on the scale of kilometers or miles). Most site control is performed automatically by remote terminal units (RTUs) or by programmable logic controllers (PLCs). Host control functions are usually restricted to basic site overriding or supervisory level intervention. For example, measuring and controlling the extraction of oil from oil rigs or controlling the temperature and flow of water, etc.

Data acquisition begins at the RTU or PLC level and includes meter readings and equipment status reports that are communicated to SCADA as required. Data are then compiled and formatted in such a way that a control room operator using human machine interfacing (HMI) networks can make supervisory decisions to adjust or override normal RTU (PLC) controls. Data may also be fed to a history log, often built on a commodity database management system, to allow trending and other analytical auditing.

## Systems Administration

The **systems administrator** is responsible for maintaining major multiuser computer systems, including local area networks (LANs), wireless local area networks (WLANs), wide area networks (WANs), personal area networks (PANs), storage area networks (SANs), intranets and extranets, and mid-range and mainframe systems. Typical duties include:
- Adding and configuring new workstations and peripherals.
- Setting up user accounts
- Installing systemwide software
- Performing procedures to prevent/detect/correct the spread of viruses
- Allocating mass storage space

Small organizations may have just one systems administrator, whereas larger enterprises usually have a team of systems administrators. Some mainframe-centric organizations may refer to a systems administrator as a systems programmer.

## Security Administration

Security administration begins with management's commitment. Management must understand and evaluate security risks, and develop and enforce a written policy that clearly states the standards and procedures to be followed. The duties of the **security administrator** should be defined in the policy. To provide adequate segregation of duties, this individual should be a full-time employee who may report directly to the infrastructure director. However, in a small organization, it may not be practical to hire a full-time individual for this position. The individual performing the function should ensure that the various users are complying with the corporate security policy and controls are adequate to prevent unauthorized access to the company assets (including data, programs and equipment). The security administrator's functions usually include:
- Maintaining access rules to data and other IT resources
- Maintaining security and confidentiality over the issuance and maintenance of authorized user IDs and passwords
- Monitoring security violations and taking corrective action to ensure that adequate security is provided
- Periodically reviewing and evaluating the security policy and suggesting necessary changes to management
- Preparing and monitoring the security awareness program for all employees
- Testing the security architecture to evaluate the security strengths and detect possible threats
- Working with compliance, risk management and audit functions to ensure that security is appropriately designed and updated based on audit feedback or testing

## Quality Assurance

**Quality assurance personnel** usually perform two distinct tasks:
- **Quality assurance** (QA)—Helps the IS department to ensure that personnel are following prescribed quality processes. For example, QA will help to ensure that programs and documentation adhere to the standards and naming conventions.
- **Quality control** (QC)—Responsible for conducting tests or reviews to verify and ensure that software is free from defects and meets user expectations. This could be done at various stages of the development of an application system, but it must be done before the programs are moved into production.

The QA group is in charge of developing, promulgating and maintaining standards for the IS function. They also provide training in QA standards and procedures. The QA group can also assist by periodically checking the accuracy and authenticity of the input, processing and output of various applications.

To enable this group to play an effective role, the group should be independent. In some organizations this group may be a part of the control group. In smaller organizations it may not be possible to have a separate QA group, in which case individuals may possess more than one role. However, under no circumstances should an individual perform a quality review of his/her own work. Additionally, the review should not be performed by an individual whose role would conflict e.g., a systems programmer performing a quality review of application system changes.

## Database Administration

The **database administrator** (DBA), as custodian of an organization's data, defines and maintains the data structures in the corporate database system. The DBA must understand the organization, and user data and data relationship (structure) requirements. This position is responsible for the security of the shared data stored on database systems. The DBA is responsible for the actual design, definition and proper maintenance of the corporate databases. The DBA usually reports directly to the director of the IPF. The DBA's role includes:
- Specifying the physical (computer-oriented) data definition
- Changing the physical data definition to improve performance
- Selecting and implementing database optimization tools
- Testing and evaluating programmer and optimization tools
- Answering programmer queries and educating programmers in the database structures
- Implementing database definition controls, access controls, update controls and concurrency controls
- Monitoring database usage, collecting performance statistics and tuning the database
- Defining and initiating backup and recovery procedures

The DBA has the tools to establish controls over the database and the ability to override these controls. The DBA also has the capability of gaining access to all data, including production data. It is usually not practical to prohibit or completely prevent access to production data by the DBA. Therefore, the IS department must exercise close control over database administration through:
- Segregation of duties
- Management approval of DBA activities
- Supervisor review of access logs and activities
- Detective controls over the use of database tools

## Systems Analyst

**Systems analysts** are specialists who design systems based on the needs of the user and are usually involved during the initial phase of the system development life cycle (SDLC). These individuals interpret the needs of the user and develop requirements and functional specifications as well as high-level design documents. These documents enable programmers to create a specific application.

## Security Architect

**Security architects** evaluate security technologies; design security aspects of the network topology, access control, identity management and other security systems; and establish security policies and security requirements. One may argue that systems analysts perform the same role; however, the set of skills required are quite different. The deliverables (e.g., program specifications vs. policies, requirements, architecture diagrams) are different as well. Security architects should also work with compliance, risk management and audit functions to incorporate their requirements and recommendations for security into the security policies and architecture.

## Applications Development and Maintenance

**Applications staff** are responsible for developing and maintaining applications. Development can include developing new code or changing the existing setup or configuration of the application. Staff develop the programs or change the application setup that will ultimately run in a production environment. Therefore, management must ensure that staff cannot modify production programs or application data. Staff should work in a test-only environment and turn their work to another group to move programs and application changes into the production environment.

## Infrastructure Development and Maintenance

**Infrastructure staff** are responsible for maintaining the systems software, including the operating system. This function may require staff to have broad access to the entire system. IS management must closely monitor activities by requiring that electronic logs capture this activity and are not susceptible to alteration. Infrastructure staff should only have access to the system libraries of the specific software that they maintain. Usage of domain administration and superuser accounts should be tightly controlled and monitored.

## Network Management

Today many organizations have widely dispersed IPFs. They may have a central IPF, but they also make extensive use of:
• **LANs**—Local area networks at branches and remote locations
• **WANs**—Wide area networks (WANs), where LANs may be interconnected for ease of access by authorized personnel from other locations
• **Wireless networks**—Established through personal digital assistants (PDAs) and other mobile devices

**Network administrators** are responsible for key components of this infrastructure (routers, switches, firewalls, network segmentation, performance management, remote access, etc.). Because of geographical dispersion, each LAN may need an administrator. Depending on the policy of the company, these administrators can report to the director of the IPF or, in a decentralized operation, may report to the end-user manager. This position is responsible for technical and administrative control over the LAN. This includes ensuring that transmission links are functioning correctly, backups of the system are occurring, and software/hardware purchases are authorized and installed properly. In smaller installations this person may be responsible for security administration over the LAN. The LAN administrator should have no application programming responsibilities but may have systems programming and end-user responsibilities.

## 2.8.2 SEGREGATION OF DUTIES WITHIN IS

Actual job titles and organizational structures vary greatly from one organization to another depending on the size and nature of the business. However, it is important for an IS auditor to obtain information to assess the relationship among various job functions, responsibilities and authorities in assessing adequate segregation of duties. Segregation of duties avoids the possibility that a single person could be responsible for diverse and critical functions in such a way that errors or misappropriations could occur and not be detected in a timely manner and in the normal course of business processes. Segregation of duties is an important means by which fraudulent and/or malicious acts can be discouraged and prevented.

Duties that should be segregated include:
• Custody of the assets
• Authorization
• Recording transactions

If adequate segregation of duties does not exist, the following could occur:
• Misappropriation of assets
• Misstated financial statements
• Inaccurate financial documentation (i.e., errors or irregularities)
• Improper use of funds or modification of data could go undetected
• Unauthorized or erroneous changes or modification of data and programs may not be detected

When duties are segregated, access to the computer, production data library, production programs, programming documentation, and operating system and associated utilities can be limited, and potential damage from the actions of any one person is, therefore, reduced. The IS and end-user departments should be organized to achieve an adequate segregation of duties. See **exhibit 2.9** for a guideline of the job responsibilities that should not be combined.

| Exhibit 2.9—Segregation of Duties Control Matrix | | | | | | | | | | | | |
|---|---|---|---|---|---|---|---|---|---|---|---|---|
| | Control Group | Systems Analyst | Application Programmer | Help Desk and Support Manager | End User | Data Entry | Computer Operator | Database Administrator | Network Administrator | Systems Administrator | Security Administrator | Systems Programmer | Quality Assurance |
| Control Group | | X | X | X | | X | X | X | X | X | | X | |
| Systems Analyst | X | | | X | X | | X | | | | X | | X |
| Application Programmer | X | | | X | X | X | X | X | X | X | X | X | X |
| Help Desk and Support Manager | X | X | X | | X | X | | X | X | X | | X | |
| End User | | X | X | X | | | X | X | X | | | X | X |
| Data Entry | X | | X | X | | | X | X | X | X | X | X | |
| Computer Operator | X | X | X | | X | X | | X | X | X | X | X | |
| Database Administrator | X | | X | X | X | X | X | | X | X | | X | |
| Network Administrator | X | | X | X | X | X | X | X | | | | | |
| System Administrator | X | | X | X | | X | X | X | | | | X | |
| Security Administrator | | X | X | | | X | X | | | | | X | |
| Systems Programmer | X | | X | X | X | X | X | X | | X | X | | X |
| Quality Assurance | | X | X | | X | | | | | | | X | |

X—Combination of these functions may create a potential control weakness.

**Note:** The segregation of duties control matrix (**exhibit 2.9**) is not an industry standard, but a guideline indicating which positions should be separated and which require compensating controls when combined. The matrix illustrates potential segregation of duties issues and should not be viewed or used as an absolute; rather, it should be used to help identify potential conflicts so proper questions may be asked to identify compensating controls.

In actual practice, functions and designations may vary in different enterprises. Further, depending on the nature of the business processes and technology deployed, risks may vary. However, it is important for an IS auditor to understand the functions of each of the designations specified in the manual. IS auditors need to understand the risk of combining functions as indicated in the segregation of duties control matrix. In addition, depending on the complexity of the applications and systems deployed, an automated tool may be required to evaluate the actual access a user has against the segregation of duties matrix. Most tools come with a predefined segregation of duties matrix which must be tailored to an organization's IT and business processes, including any additional functions or risks that are not included in the delivered segregation of duties matrix.

Compensating controls are internal controls that are intended to reduce the risk of an existing or potential control weakness when duties cannot be appropriately segregated. The organization structure and roles should be taken into account in determining the appropriate controls for the relevant environment. For example, an organization may not have all the positions described in the matrix or one person may be responsible for many of the roles described. The size of the IT department may also be an important factor that should be considered; i.e., certain combinations of roles in an IT department of a certain size should never be used. However, if for some reason combined roles are required, then compensating controls should be described.

The purpose of segregation of duties is to reduce or eliminate business risks through the identification of compensating controls. The magnitude and probability of risk can be assessed from low to high, depending on the organization.

| PRACTICE QUESTIONS |
| --- |

2-7    Which of the following tasks may be performed by the same person in a well-controlled information processing computer center?

    A.    Security administration and change management
    B.    Computer operations and system development
    C.    System development and change management
    D.    System development and systems maintenance

2-8    Which of the following is the **MOST** critical control over database administration?

    A.    Approval of DBA activities
    B.    Segregation of duties
    C.    Review of access logs and activities
    D.    Review of the use of database tools

*See answers and explanations to the practice questions at the end of the chapter. (page 133)*

# 2.8.3 SEGREGATION OF DUTIES CONTROLS

Several control mechanisms can be used to enforce segregation of duties. The controls are described in the following sections.

## Transaction Authorization

Transaction authorization is the responsibility of the user department. Authorization is delegated to the degree that it relates to the particular level of responsibility of the authorized individual in the department. Periodic checks must be performed by management and audit to detect the unauthorized entry of transactions.

## Custody of Assets

Custody of corporate assets must be determined and assigned appropriately. The data owner usually is assigned to a particular user department, and his/her duties should be specific and in writing. The owner of the data has responsibility for determining authorization levels required to provide adequate security while the administration group is often responsible for implementing and enforcing the security system.

## Access to Data

Controls over access to data are provided by a combination of physical, system and application security in the user area and the IPF. The physical environment must be secured to prevent unauthorized personnel from accessing the various tangible devices connected to the central processing unit and, thereby, permitting access to data. System and application security are additional layers that may prevent unauthorized individuals from gaining access to corporate data. Access to data from external connections is a growing concern since the advent of the Internet. Therefore, IS management has added responsibilities to protect information assets from unauthorized access.

Access control decisions are based on organizational policy and two generally accepted standards of practice—segregation (separation) of duties and least privilege. Controls for effective use must not disrupt the usual work flow more than necessary or place too much burden on administrators, auditors or authorized users. Further access must be conditional and access controls must adequately protect all of the organization's resources.

Policies establish levels of sensitivity—such as top secret, secret, confidential and unclassified—for data and other resources. These levels should be used for guidance on the proper procedures for handling information resources. The levels may be also used as a basis for access control decisions. Individuals are granted access to only those resources at or below a specific level of sensitivity. Labels are used to indicate the sensitivity level of electronically stored documents. Policy-based controls may be characterized as either mandatory or discretionary.

For further detail, refer to chapter 5, to the Mandatory and Discretionary Access Controls section under the topic of Importance of Information Assets.

### AUTHORIZATION FORMS

Managers of user departments must provide IS with formal authorization forms (either hard copy or electronic) that define the access rights of their employees. In other words, managers must define who should have access to what. Authorization forms must be evidenced properly with management-level approval. Generally, all users should be authorized with specific system access via formal request of management. In large companies or in those with remote sites, signature authorization logs should be maintained and formal requests should be compared to the signature log. Access privileges should be reviewed periodically to ensure that they are current and appropriate to the user's job functions.

### USER AUTHORIZATION TABLES

The IS department should use the data from the authorization forms to build and maintain user authorization tables. These will define who is authorized to update, modify, delete and/or view data. These privileges are provided at the system, transaction or field level. In effect, these are user access control lists. These authorization tables must be secured against unauthorized access by additional password protection or data encryption. A control log should record all user activity and appropriate management should review this log. All exception items should be investigated.

## Compensating Controls for Lack of Segregation of Duties

In a small business where the IS department may only consist of four to five people, compensating control measures must exist to mitigate the risk resulting from a lack of segregation of duties. Before relying on system generated reports or functions as compensating controls, the IS auditor should carefully evaluate the reports, applications and related processes for appropriate controls, including testing and access controls to make changes to the reports or functions. Compensating controls would include:
- **Audit trails**—Audit trails are an essential component of all well-designed systems. Audit trails help the IS and user departments as well as the IS auditor by providing a map to retrace the flow of a transaction. Audit trails enable the user and IS auditor to recreate the actual transaction flow from the point of origination to its existence on an updated file. In the absence of adequate segregation of duties, good audit trails may be an acceptable compensating control. The IS auditor should be able to determine who initiated the transaction, time of day and date of entry, type of entry, what fields of information it contained, and what files it updated.

• **Reconciliation**—Reconciliation is ultimately the responsibility of the user. In some organizations limited reconciliation of applications may be performed by the data control group with the use of control totals and balance sheets. This type of independent verification increases the level of confidence that the application processed successfully and the data are in proper balance.

• **Exception reporting**—Exception reporting should be handled at the supervisory level and should require evidence, such as initials on a report, noting that the exception has been handled properly. Management should also ensure that exceptions are resolved in a timely manner.

• **Transaction logs**—A transaction log may be manual or automated. An example of a manual log is a record of transactions (grouped or batched) before they are submitted for processing. An automated transaction log or journal provides a record of all transactions processed and is maintained by the computer system.

• **Supervisory reviews**—Supervisory reviews may be performed through observation and inquiry or remotely.

• **Independent reviews**—Independent reviews are carried out to compensate for mistakes or intentional failures in following prescribed procedures. These reviews are particularly important when duties in a small organization cannot be appropriately segregated. Such reviews will help detect errors or irregularities.

---

### PRACTICE QUESTIONS

**2-9**  When a complete segregation of duties cannot be achieved in an online system environment, which of the following functions should be separated from the others?

  A.  Origination
  B.  Authorization
  C.  Recording
  D.  Correction

**2-10**  In a small organization where segregation of duties is not practical, an employee performs the function of computer operator and application programmer. Which of the following controls should the IS auditor recommend?

  A.  Automated logging of changes to development libraries
  B.  Additional staff to provide segregation of duties
  C.  Procedures that verify that only approved program changes are implemented
  D.  Access controls to prevent the operator from making program modifications

*See answers and explanations to the practice questions at the end of the chapter. (pages 133-134)*

---

# 2.9 AUDITING IT GOVERNANCE STRUCTURE AND IMPLEMENTATION

While many conditions concern the IS auditor when auditing the IS function, some of the more significant indicators of potential problems include:
• Unfavorable end-user attitudes
• Excessive costs
• Budget overruns
• Late projects
• High staff turnover
• Inexperienced staff
• Frequent HW/SW errors
• An excessive backlog of user requests
• Slow computer response time
• Numerous aborted or suspended development projects
• Unsupported or unauthorized HW/SW purchases

---

- Frequent HW/SW upgrades
- Extensive exception reports
- Exception reports that were not followed up
- Poor motivation
- Lack of succession plans
- A reliance on one or two key personnel
- Lack of adequate training

## 2.9.1 REVIEWING DOCUMENTATION

The following documents should be reviewed:
- **IT strategies, plans and budgets**—They provide evidence of planning and management's control of the IS environment and alignment with the business strategy.
- **Security policy documentation**—This documentation provides the standard for compliance. The documentation should state the position of the organization with regard to any and all security risks. The documentation should identify who is responsible for the safeguarding of company assets, including programs and data, and it should state the preventive measures to be taken to provide adequate protection and actions to be taken against violators. For this reason, this part of the policy document should be treated as confidential.
- **Organization/functional charts**—These charts provide the IS auditor with an understanding of the reporting line within a particular department or organization. The charts illustrate a division of responsibility and give an indication of the degree of segregation of duties within the organization.
- **Job descriptions**—These descriptions define the functions and responsibilities of positions throughout the organization. Job descriptions provide an organization with the ability to group similar jobs in different grade levels to ensure fair compensation for the workforce. Furthermore, job descriptions give an indication of the degree of segregation of duties within the organization and may help identify possible conflicting duties. Job descriptions should identify the position to which these personnel report. The IS auditor should then verify that the levels of reporting relationships are based on sound business concepts and do not compromise the segregation of duties.
- **Steering committee reports**—These reports provide documented information regarding new system projects. The reports are reviewed by upper management and disseminated among the various business units.
- **System development and program change procedures**—These procedures provide a framework within which to undertake system development or program change.
- **Operations procedures**—These procedures describe the responsibilities of the operations staff. Performance measurement procedures are generally embedded in operational procedures and are periodically reported to senior management/steering committees. IS auditors should ensure that these procedures are embedded in operational procedures.
- **HR manuals**—These manuals provide the rules and regulations (determined by an organization) that specify how employees should conduct themselves.
- **QA procedures**—These procedures provide framework and standards that can be followed by the IS department.

The various documents reviewed should be further assessed to determine if:
- They were created as management authorized and intended
- They are current and up to date

## 2.9.2 REVIEWING CONTRACTUAL COMMITMENTS

There are various phases to computer hardware, software and IS service contracts, including:
- Development of contract requirements and service levels
- Contract bidding process
- Contract selection process
- Contract acceptance
- Contract maintenance
- Contract compliance

Each of these phases should be supported by legal documents, subject to the authorization of management. The IS auditor should verify management participation in the contracting process and ensure a proper level of timely contract compliance review. The IS auditor may wish to perform a separate compliance review on a sample of such contracts.

In reviewing a sample of contracts, the IS auditor should evaluate the adequacy of the following terms and conditions:
• Service levels
• Right to audit or third party audit reporting
• Software escrow
• Penalties for noncompliance
• Adherence to security policies and procedures
• Protection of customer information
• Contract change process
• Contract termination and any associated penalties

---

**Note:** An IS auditor should be familiar with the request for proposal (RFP) process and know what needs to be reviewed in an RFP. It is also important to be noted that a CISA should know, from a governance perspective, the evaluation criteria and methodology of an RFP, and the requirements to meet organizational standards.

---

# 2.10 CHAPTER 2 QUICK REFERENCE REVIEW

**Chapter 2 Quick Reference Review**

Chapter 2 addresses the need for IT governance. An IS auditor must be able to understand and provide assurance that the organization has the structure, policies, accountability mechanisms and monitoring practices in place to achieve the requirements of corporate governance of IT. For an IS auditor, knowledge of IT governance forms the foundation for evaluating control practices and mechanisms for management oversight and review.

CISA candidates should have a sound understanding of the following items, not only within the context of the present chapter, but also to correctly address questions in related subject areas. It is important to keep in mind that it is not enough to know these concepts from a definitional perspective. The CISA candidate must also be able to identify which elements among those presented may represent the greatest risk and which controls are most effective at mitigating this risk.

- An objective of corporate governance is to resolve the conflicting objectives of exploiting available opportunities to increase stakeholder value while keeping the organization's operations within the limits of regulatory requirements and social obligations. Applied to IT, governance helps ensure the alignment of IT and enterprise objectives. IT governance is concerned with two issues: that IT delivers value to the business and that IT risks are managed. The first is driven by strategic alignment of IT with the business. The second is driven by establishing accountability into the enterprise. IT governance is the responsibility of the board of directors and executive management, and the key IT governance practices for executive management include an IT strategy committee, a risk management process and a standard IT balanced scorecard.
- An IT strategy committee monitors IT value, risks and performance and provides information to the board to support decision making on IT strategies. The IS auditor must evaluate the effectiveness of IT governance structure to ensure adequate board control over the decisions, direction and performance of IT, so it supports the organization's strategies and objectives.
- A key aspect of IT governance is the governance of information security. Information is an organization's most valuable asset and must be adequately protected regardless of how it is handled, processed, transported, stored, or disposed. Information security includes all information processes, physical and electronic, regardless of whether they involve people, technology, or relationships with trading partners, customers and third parties. It ensures that information security risks are appropriately managed and enterprise information resources are used responsibly.
- The governance of information security should be executed and supported with information security strategies, policies and organization structure. Information security governance must be the responsibility of the board of directors/senior management to approve policy and penalties for noncompliance and the mandate of information security may be delegated to a chief information security officer.
- IT governance is concerned with ensuring that IT delivers value to the business and, to ensure this governance, an information systems strategy should be in place that contains the long-term direction of how information technology will be used to support and improve the business. Effective IT strategic planning involves consideration of the organization's demand for IT and its IT supply capacity. The strategy is governed by a steering committee and the strategy is guided and controlled by policies and procedures, including the information security policy. Strategies, polices and procedures should be evaluated by the IS auditor to determine the importance placed on the planning process; the involvement of senior IT management in the overall business strategy; and policy compliance, relevance and applicability to third parties.
- IT governance encompasses minimizing IT risks to the organization. Risk management is the process of identifying vulnerabilities and threats to the information resources used by an organization in achieving business objectives and deciding what countermeasures (safeguards or controls), if any, to take in reducing risk to an acceptable level (i.e., residual risk), based on the value of the information resource to the organization. This process begins with understanding the organization's appetite for risk and then determining the risk exposure on its IT assets. From this identification, risk management strategies and responsibilities are defined. Depending on the type of risk and its significance to the business, management risk can be avoided, reduced, transferred, accepted or rejected. The result of a risk occurring is called an impact and can result in losses, such as financial, legal, reputation, efficiency.

- Risks are measured using a qualitative analysis (defining risks in terms of high/medium/low); semiqualitative analysis (defining risks according to a numeric scale) or quantitative analysis (applying several values to risk, including financial, and calculating the risk's probability and impact). Once risks have been identified, existing or new controls are designed and measured for their strength and likelihood of effectiveness. Controls may be preventive, detective, or corrective; manual or programmed; and formal (i.e., documented) or *ad hoc*. Residual risk can be used by management to determine which areas require more control and whether the benefits of such controls outweigh the costs. This entire process of IT risk management needs to be managed at multiple levels in the organization, including the operational, project and strategic levels, and should form part of the IS management practice.
- Key management processes that will shape the effectiveness of an IS department and outline controls on strategy and use of resources are human resource management, change management, financial practices, quality management, information security management and performance optimization practices. Management's control and governance of the IS environment can be evaluated based on the review of its organizational structure. Charts should provide a clear definition of the department's hierarchy and authorities, and the specific roles and responsibilities. The structure should define the role of each area in the IS department and indicate appropriate segregation of duties within the IS department.
- The purpose of segregation of duties is to reduce or eliminate business risks through the identification of compensating controls. Specifically, the duties that should be segregated are: custody of the assets, authorization and recording transactions. If combined roles are required, then compensating controls should be described. In summary, for an IS auditor, all IS management practices should be evaluated to determine management's governance over IT, including documentation regarding IT strategies, budgets, policies and procedures; control over information security; as well as the structure of the IS department because each of these elements illustrates how effective an organization is at ensuring that IT delivers value to the business and IT risks are managed.

# 2.11 CHAPTER 2 CASE STUDY

The following case studies are included as a learning tool to reinforce the concepts introduced in this chapter. Exam candidates should note that the CISA exam does not currently use this format for testing.

## 2.11.1 CASE STUDY A

### *Case Study A Scenario*

An IS auditor has been asked to review the draft of an outsourcing contract and SLA, and recommend any changes or point out any concerns prior to these documents being submitted to senior management for final approval. The agreement includes outsourcing support of Windows and UNIX server administration, and network management to a third party. Servers will be relocated to the outsourcer's facility that is located in another country, and connectivity will be established using the Internet. Operating system software will be upgraded on a semiannual basis but it will not be escrowed. All requests for addition or deletion of user accounts will be processed within three business days. Intrusion detection software will be continuously monitored by the outsourcer and the customer notified by e-mail if any anomalies are detected. New employees hired within the last three years were subject to background checks. Prior to that time there was no policy in place. A right to audit clause is in place but 24-hour notice is required prior to an onsite visit. If the outsourcer is found to be in violation of any of the terms or conditions of the contract, the outsourder will have 10 business days to correct the deficiency. The outsourcer does not have an IS auditor but is audited by a regional public accounting firm.

| CASE STUDY A QUESTIONS |
| --- |

A1.    Which of the following should be of **MOST** concern to the IS auditor?

      A.    User account changes are processed within three business days.
      B.    Twenty-four hour notice is required prior to an onsite visit.
      C.    The outsourcer does not have an IS audit function.
      D.    Software escrow is not included in the contract.

A2.    Which of the following would be the **MOST** significant issue to address if the servers contain personally identifiable customer information that is regularly accessed and updated by end users?

      A.    The country in which the outsourcer is based prohibits the use of strong encryption for transmitted data.
      B.    The outsourcer limits its liability if it took reasonable steps to protect the customer data.
      C.    The outsourcer did not perform background checks for employees hired over three years ago.
      D.    System software is only upgraded once every six months.

*See answers and explanations to the practice questions at the end of the chapter. (page 135)*

## 2.11.2 CASE STUDY B

## *Case Study B Scenario*

An organization has implemented an integrated application for supporting business processes. It has also entered into an agreement with a vendor for application maintenance and providing support to the users and system administrators. This support will be provided by a remote vendor support center using a privileged user ID with O/S level super user authority having read and write access to all files. The vendor will use this special user ID to log ointo the system for troubleshooting and implementing application updates (patches). Due to the volume of transactions, activity logs are only maintained for 90 days.

| CASE STUDY B QUESTIONS |
| --- |
| B1.  Which of the following is a **MAJOR** concern for the IS auditor?<br><br>  A.  User activity logs are only maintained for 90 days.<br>  B.  The special user ID will access the system remotely.<br>  C.  The special user ID can alter activity log files.<br>  D.  The vendor will be testing and implementing patches on servers.<br><br>B2.  Which of the following actions would be **MOST** effective in reducing the risk that the privileged user account may be misused?<br><br>  A.  The special user ID should be disabled except when maintenance is required.<br>  B.  All usage of the special user account should be logged.<br>  C.  The agreement should be modified so that all support is performed onsite.<br>  D.  All patches should be tested and approved prior to implementation. |
| *See answers and explanations to the practice questions at the end of the chapter. (page 135)* |

## 2.11.3 CASE STUDY C

### *Case Study C Scenario*

An IS auditor was asked to review alignment between IT and business goals for a small financial institution. The IS auditor requested various information including business goals and objectives and IT goals and objectives. The IS auditor found that business goals and objectives were limited to a short bulleted list, while IT goals and objectives were limited to slides used in meetings with the CIO (the CIO reports to the CFO). It was also found in the documentation provided that over the past two years, the risk management committee (composed of senior management) only met on three occasions, and no minutes of what was discussed were kept for these meetings. When the IT budget for the upcoming year was compared to the strategic plans for IT, it was noted that several of the initiatives mentioned in the plans for the upcoming year were not included in the budget for that year.

| CASE STUDY C QUESTIONS |
| --- |
| C1.    Which of the following should be of **GREATEST** concern to the IS auditor?<br>     A.    Strategy documents are informal and incomplete.<br>     B.    The risk management committee seldom meets and does not keep minutes .<br>     C.    Budgets do not appear adequate to support future IT investments.<br>     D.    The CIO reports to the CFO.<br><br>C2.    Which of the following would be the **MOST** significant issue to address?<br><br>     A.    The prevailing culture within IT.<br>     B.    The lack of information technology policies and procedures.<br>     C.    The risk management practices as compared to peer organizations.<br>     D.    The reporting structure for IT. |
| *See answers and explanations to the practice questions at the end of the chapter. (page 135)* |

## 2.11.4 CASE STUDY D

## *Case Study D Scenario*

An IS Auditor is auditing the IT governance practices for an organization. During the course of the work, it is noted that the organization does not have a full time chief Information officer (CIO). The organization chart of the entity provides for an information systems manager reporting to the chief financial officer (CFO), who in turn reports to the board of directors. The board plays a major role in monitoring IT initiatives in the entity and the CFO communicates on a frequent basis the progress of IT initiatives. From reviewing the segregation of duties matrix, it is apparent that application programmers are only required to obtain approval from the data base administrator (DBA) to directly access production data. It is also noted that the application programmers have to provide the developed program code to the program librarian, who then migrates it to production. Information systems audits are carried out by the internal audit department, which reports to the CFO at the end of every month, as part of business performance review process;, the financial results of the entity are reviewed in detail and signed off by the business managers for correctness of data contained therein.

| CASE STUDY D QUESTIONS |
| --- |

D1.   Given the circumstances described, what would be of **GREATEST** concern from an IT governance perspective?

    A.   The organization does not have a full-time CIO.
    B.   The organization does not have an IT steering committee.
    C.   The board of the organization plays a major role in monitoring IT initiatives.
    D.   The information systems manager reports to the CFO.


D2.   Given the case, what would be of **GREATEST** concern from a segregation of duties perspective?

    A.   Application programmers are required to obtain approval only from the DBA for direct write access to data.
    B.   Application programmers are required to turn over the developed program code to the program librarian for migration to production.;
    C.   The internal audit department reports to the CFO.
    D.   Business performance reviews are required to be signed off only by the business managers.

D3.   Which of the following would **BEST** address data integrity from a mitigating control standpoint?

    A.   Application programmers are required to obtain approval from DBA for direct access to data.
    B.   Application programmers are required to hand over the developed program codes to the program librarian for transfer to production.
    C.   The internal audit department reports to the CFO.
    D.   Business performance results are required to be reviewed and signed off by the business managers.

*See answers and explanations to the practice questions at the end of the chapter. (page 136)*

# 2.12 ANSWERS TO PRACTICE QUESTIONS

2-1    **A**    IT governance ensures that the organization aligns its IT strategy with the enterprise/business objectives. Choices B, C and D are too limited.

2-2    **B**    IS strategic plans must address the needs of the business and meet future business objectives. Hardware purchases may be outlined, but not specified, and neither budget targets nor development projects are relevant choices. Choices A, C and D are not strategic items.

2-3    **C**    Long-range planning for the IT department should recognize organizational goals, technological advances and regulatory requirements. Typically, the IT department will have long-range and short-range plans that are consistent and integrated with the organization's plans. These plans must be time- and project-oriented and address the organization's broader plans toward attaining its goals.

2-4    **A**    A data security officer's prime responsibility is recommending and monitoring data security policies. Promoting security awareness within the organization is one of the responsibilities of a data security officer, but it is not as important as recommending and monitoring data security policies. The IT department, not the data security officer, is responsible for establishing procedures for IT security policies recommended by the data security officer and for the administration of physical and logical access controls.

2-5    **D**    Maintaining access rules over data and IT resources is one of the primary functions of the security administrator. Approving the security policy is the responsibility of senior management. Maintaining and implementing the security policy is the responsibility of the security administrator. Testing application software is the function of the programmer or user. Ensuring data integrity is the responsibility of the user and processing controls built into the application.

2-6    **A**    Evaluating the activities of boards and committees providing oversight is an important aspect of governance and should be measured. Choices B, C and D are irrelevant to the evaluation of IT governance performance measures.

2-7    **D**    It is common for system development and maintenance to be undertaken by the same person. In both, the programmer requires access to the source code in the development environment, but should not be allowed access in the production environment. Choice A is incorrect because the roles of security administration and change management are incompatible functions. The level of security administration access rights could allow changes to go undetected. Computer operations and system development (choice B) are incompatible since it would be possible for an operator to run a program that he/she had amended. Choice C is incorrect because the combination of system development and change control would allow program modifications to bypass change control approvals.

2-8    **B**    Segregation of duties will prevent combination of conflicting functions. This is a preventive control and it is the most critical control over database administration. Approval of DBA activities does not prevent the combination of conflicting functions. Review of access logs and activities is a detective control. If DBA activities are improperly approved, review of access logs and activities may not reduce the risk. Reviewing the use of database tools does not reduce the risk since this is only a detective control and does not prevent combination of conflicting functions.

2-9    **B**    Authorization should be separated from all aspects of record keeping (origination, recording and correction). Such a separation enhances the ability to detect the recording of unauthorized transactions.

2-10   **C**   In smaller organizations it generally is not appropriate to recruit additional staff to achieve a strict segregation of duties. The IS auditor must look at alternatives. Of the choices, C is the only practical one that has an impact. The IS auditor should recommend processes that detect changes to production source and object code, such as code comparisons, so the changes can be reviewed by a third party on a regular basis. This would be a compensating control process. Choice A, involving logging of changes to development libraries, would not detect changes to production libraries. Choice D is in effect requiring a third party to do the changes, which may not be practical in a small organization.

# 2.13 ANSWERS TO CASE STUDY QUESTIONS

## ANSWERS TO CASE STUDY A QUESTIONS

A1.   **A**   Three business days to remove the account of a terminated employee would create an unacceptable risk to the organization. In the intervening time significant damage could be done. In contrast, some degree of advance notice prior to an onsite visit is generally accepted within the industry. Also, not every outsourcer will have its own internal audit function or IS auditor. Software escrow is primarily of importance when dealing with custom application software where there is a need to store a copy of the source code with a third party. Operating system software for generally available commercial operating systems would not require software escrow.

A2.   **A**   Since connectivity to the servers is over the Internet, the prohibition against strong encryption will place any transmitted data at risk. The limitation of liability is a standard industry practice. Although the failure to perform background checks for employees hired more than three years ago is of importance, it is not as significant an issue. Upgrading system software once every six months does not present any significant exposure.

## ANSWERS TO CASE STUDY B QUESTIONS

B1.   **C**   Because the super user ID has read and write access to all files, there is no way to ensure that the activity logs are not modified to hide unauthorized activity by the vendor. Remote access is not a major concern as long as the connection is made over an encrypted line, and testing and implementing patches on servers is part of vendor- provided support. Although 90-day retention of logs may not be sufficient in some business situations, it is not as major a concern as is the fact that the vendor has the ability to alter the activity logs.

B2.   **A**   The **MOST** effective and practical control in this situation is to lock the special user account when it is not needed. The account should be opened only when vendor needs access for support and closed immediately after use. All activities should be logged and reviewed for appropriateness. The other choices are not as effective or practical in reducing the risk.

## ANSWERS TO CASE STUDY C QUESTIONS

C1.   **B**   The fact that the risk management committee seldom meets and when it does meet, no minutes are taken, is the greatest concern. Because senior management is not meeting regularly to discuss key risk issues, and minutes are not captured which would provide for follow up, analysis and commitment, this indicates a serious lack of governance. The other options are not as serious in their potential impact on the organization.

C2.   **B**   The absence of policies and procedures makes it difficult if not impossible to implement effective IT governance. Other issues are secondary by comparison.

## ANSWERS TO CASE STUDY D QUESTIONS

D1.  **D**  The information systems manager should ideally report to the board of directors or the chief executive officer (CEO) to provide a sufficient degree of independence. The reporting structure that requires the iInformation sSystems manager to report to the CFO is not a desirable situation and could lead to the compromise of certain controls.

D2.  **A**  The application programmers should obtain approval from the business owners before accessing data. DBAs are only custodians of the data and should only provide access that is authorized by the data owner.

D3.  **D**  Sign-off on data contained in the financial results by the business managers at the end of the month would detect any significant discrepancies that would result from tampering of data through inappropriate direct access of data without the approval or knowledge of the business managers.

# 2.14 SUGGESTED RESOURCES FOR FURTHER STUDY

De Haes, Steven; David Gilmore; Wim Van Grembergen; Gary Hardy; Alan Simmons; Paul A. Williams; *IT Governance Domain Practices and Competencies Series*, IT Governance Institute, USA, 2005

IT Governance Institute, *Board Briefing on IT Governance, 2nd Edition*, USA, 2003

IT Governance Institute, CoBiT 4.1, USA, 2007

Ramos, Michael J.; *How to Comply with Sarbanes-Oxley Section 404, 3rd Edition*, John Wiley & Sons Inc., USA, 2008

SABSA® (Sherwood Applied Business Security Architecture), Enterprise Security Architecture, UK, 2008, *www.sabsa-institute.org*

Simmonds, Alan; David Gilmore; *Governance of Outsourcing*, IT Governance Institute, USA, 2005

Tarantino, Anthony; *Manager's Guide to Compliance*, John Wiley & Sons Inc., USA, 2006

*Note:* Publications in bold are stocked in the ISACA Bookstore. *Information Systems Control Journal* articles are available at *www.isaca.org/archives*. The articles are available online to ISACA members only during their first year of release, and then are opened to the public. All referenced *Journal* articles are available on the CISA Practice Question Database v9.

Page intentionally left blank

*Chapter 3:*

# Systems and Infrastructure Life Cycle Management

# Systems and Infrastructure Life Cycle Management

# Systems and Infrastructure Life Cycle Management

# Systems and Infrastructure Life Cycle Management

# Systems and Infrastructure Life Cycle Management

# 3.1 INTRODUCTION

The Systems and Infrastructure Lifecycle Management chapter provides the IS auditor with an overview of key processes and methodologies used by organizations when creating and changing application systems and infrastructure components. Due to the wide variety of topics addressed in this area, the IS auditor is encouraged to research these topics in greater detail using the list of references cited at the end of this chapter.

## 3.1.1 OBJECTIVE

The objective of this area is to ensure that the CISA candidate understands and can provide assurance that the management practices for the development/acquisition, testing, implementation, maintenance, and disposal of systems and infrastructure will meet the organization's objectives.

**This area represents 16 percent of the CISA examination (approximately 32 questions).**

## 3.1.2 TASKS

There are 10 tasks within the systems and infrastructure life cycle management area:

T3.1   Evaluate the business case for the proposed system development/acquisition to ensure that it meets the organization's business goals.

T3.2   Evaluate the project management framework and project governance practices to ensure that business objectives are achieved in a cost-effective manner while managing risks to the organization.

T3.3   Perform reviews to ensure that a project is progressing in accordance with project plans, is adequately supported by documentation and status reporting is accurate.

T3.4   Evaluate proposed control mechanisms for systems and/or infrastructure during specification, development/acquisition and testing to ensure that they will provide safeguards and comply with the organization's policies and other requirements.

T3.5   Evaluate the processes by which systems and/or infrastructure are developed/acquired and tested to ensure that the deliverables meet the organization's objectives.

T3.6   Evaluate the readiness of the system and/or infrastructure for implementation and migration into production.

T3.7   Perform postimplementation review of systems and/or infrastructure to ensure that they meet the organization's objectives and are subject to effective internal control.

T3.8   Perform periodic reviews of systems and/or infrastructure to ensure that they continue to meet the organization's objectives and are subject to effective internal control.

T3.9   Evaluate the process by which systems and/or infrastructure are maintained, to ensure the continued support of the organization's objectives, and are subject to effective internal control.

T3.10   Evaluate the process by which systems and/or infrastructure are disposed to ensure that they comply with the organization's policies and procedures.

## 3.1.3 KNOWLEDGE STATEMENTS

There are 18 knowledge statements within the systems and infrastructure life cycle management area:

KS3.1   Knowledge of benefits management practices (e.g., feasibility studies, business cases)

KS3.2   Knowledge of project governance mechanisms (e.g., steering committee, project oversight board)

KS3.3   Knowledge of project management practices, tools and control frameworks

KS3.4   Knowledge of risk management practices applied to projects

KS3.5   Knowledge of project success criteria and risks

KS3.6   Knowledge of configuration, change and release management in relation to development and maintenance of systems and/or infrastructure

# Systems and Infrastructure Life Cycle Management

KS3.7   Knowledge of control objectives and techniques that ensure the completeness, accuracy, validity and authorization of transactions and data within IT systems applications

KS3.8   Knowledge of enterprise architecture related to data, applications and technology (e.g., distributed applications, web-based applications, web services, n-tier applications)

KS3.9   Knowledge of requirements analysis and management practices (e.g., requirements verification, traceability, gap analysis)

KS3.10  Knowledge of acquisition and contract management processes (e.g., evaluation of vendors, preparation of contracts, vendor management, escrow)

KS3.11  Knowledge of system development methodologies and tools, and an understanding of their strengths and weaknesses (e.g., agile development practices, prototyping, rapid application development [RAD], object-oriented design techniques)

KS3.12  Knowledge of quality assurance methods

KS3.13  Knowledge of the management of testing processes (e.g., test strategies, test plans, test environments, entry and exit criteria)

KS3.14  Knowledge of data conversion tools, techniques and procedures

KS3.15  Knowledge of system and/or infrastructure disposal procedures

KS3.16  Knowledge of software and hardware certification, and accreditation practices

KS3.17  Knowledge of postimplementation review objectives and methods (e.g., project closure, benefits realization, performance measurement)

KS3.18  Knowledge of system migration and infrastructure deployment practices

---

**Note:** Hardware and software are integrated components of the technical infrastructure that make it possible for the business applications to operate. Considering them as separate audit items is an intrinsic scope limitation and requires a cautious definition of the audit objectives since these must generally be framed around their efficiency and effectiveness in supporting the business. The need to separately evaluate a component, or a class of components, must be determined as a part of the overall system assessment. A specific audit would be necessary, for example, if a preceding higher-level review concluded that a component, or a class of components, weakens the overall system performance. Furthermore, IS auditors must be aware of other synergies and restrictions that affect a specific component's functionality in the overall system since these have a direct impact on the significance of the audit results.

---

## 3.1.4 RELATIONSHIP OF TASK TO KNOWLEDGE STATEMENTS

The task statements are what the CISA candidate is expected to know how to do. The knowledge statements delineate what the CISA candidate is expected to know in order to perform the tasks. The task and knowledge statements are approximately mapped in **exhibit 3.1** insofar as it is possible to do so. The candidate will not be tested on which knowledge statements map to individual task statements. Note that although there is often overlap, each task statement will generally map to several knowledge statements.

| Exhibit 3.1—Tasks and Knowledge Statements Mapping | |
|---|---|
| **Task Statements** | **Knowledge Statements** |
| T3.1  Evaluate the business case for the proposed system development/acquisition to ensure that it meets the organization's business goals. | KS3.1   Knowledge of benefits management practices (e.g., feasibility studies, business cases)<br>KS3.5   Knowledge of project success criteria and risks<br>KS3.8   Knowledge of enterprise architecture related to data, applications and technology (e.g., distributed applications, web-based applications, web services, n-tier applications)<br>KS3.10  Knowledge of acquisition and contract management processes (e.g., evaluation of vendors, preparation of contracts, vendor management, escrow) |

# Systems and Infrastructure
# Life Cycle Management

| Exhibit 3.1—Tasks and Knowledge Statements Mapping (*cont.*) | |
|---|---|
| **Task Statements** | **Knowledge Statements** |
| T3.1 (*cont.*) | KS3.11 Knowledge of system development methodologies and tools, and an understanding of their strengths and weaknesses (e.g., agile development practices, prototyping, rapid application development [RAD], object-oriented design techniques)<br><br>KS3.12 Knowledge of quality assurance methods<br><br>KS3.13 Knowledge of the management of testing processes (e.g., test strategies, test plans, test environments, entry and exit criteria)<br><br>KS3.14 Knowledge of data conversion tools, techniques and procedures<br><br>KS3.16 Knowledge of software and hardware certification, and accreditation practices<br><br>KS3.18 Knowledge of system migration and infrastructure deployment practices |
| T3.2 Evaluate the project management framework and project governance practices to ensure that business objectives are achieved in a cost-effective manner while managing risks to the organization. | KS3.2 Knowledge of project governance mechanisms (e.g., steering committee, project oversight board)<br><br>KS3.3 Knowledge of project management practices, tools and control frameworks<br><br>KS3.4 Knowledge of risk management practices applied to projects<br><br>KS3.5 Knowledge of project success criteria and risks<br><br>KS3.7 Knowledge of control objectives and techniques that ensure the completeness, accuracy, validity and authorization of transactions and data within IT systems applications<br><br>KS3.9 Knowledge of requirements analysis and management practices (e.g., requirements verification, traceability, gap analysis)<br><br>KS3.10 Knowledge of acquisition and contract management processes (e.g., evaluation of vendors, preparation of contracts, vendor management, escrow)<br><br>KS3.16 Knowledge of software and hardware certification, and accreditation practices<br><br>KS3.18 Knowledge of system migration and infrastructure deployment practices |
| T3.3 Perform reviews to ensure that a project is progressing in accordance with project plans, is adequately supported by documentation and status reporting is accurate. | KS3.1 Knowledge of benefits management practices (e.g., feasibility studies, business cases)<br><br>KS3.2 Knowledge of project governance mechanisms (e.g., steering committee, project oversight board)<br><br>KS3.3 Knowledge of project management practices, tools and control frameworks<br><br>KS3.4 Knowledge of risk management practices applied to projects<br><br>KS3.5 Knowledge of project success criteria and risks<br><br>KS3.9 Knowledge of requirements analysis and management practices (e.g., requirements verification, traceability, gap analysis) |

| Exhibit 3.1—Tasks and Knowledge Statements Mapping (*cont.*) | |
|---|---|
| **Task Statements** | **Knowledge Statements** |
| T3.3 (*cont.*) | KS3.10 Knowledge of acquisition and contract management processes (e.g., evaluation of vendors, preparation of contracts, vendor management, escrow) |
| | KS3.11 Knowledge of system development methodologies and tools, and an understanding of their strengths and weaknesses (e.g., agile development practices, prototyping, rapid application development [RAD], object-oriented design techniques) |
| | KS3.12 Knowledge of quality assurance methods |
| | KS3.13 Knowledge of the management of testing processes (e.g., test strategies, test plans, test environments, entry and exit criteria) |
| | KS3.14 Knowledge of data conversion tools, techniques and procedures |
| | KS3.16 Knowledge of software and hardware certification, and accreditation practices |
| | KS3.18 Knowledge of system migration and infrastructure deployment practices |
| T3.4 Evaluate proposed control mechanisms for systems and/or infrastructure during specification, development/ acquisition and testing to ensure that they will provide safeguards and comply with the organization's policies and other requirements. | KS3.1 Knowledge of benefits management practices (e.g., feasibility studies, business cases) |
| | KS3.2 Knowledge of project governance mechanisms (e.g., steering committee, project oversight board) |
| | KS3.3 Knowledge of project management practices, tools and control frameworks |
| | KS3.4 Knowledge of risk management practices applied to projects |
| | KS3.5 Knowledge of project success criteria and risks |
| | KS3.6 Knowledge of configuration, change and release management in relation to development and maintenance of systems and/or infrastructure |
| | KS3.7 Knowledge of control objectives and techniques that ensure the completeness, accuracy, validity and authorization of transactions and data within IT systems applications |
| | KS3.8 Knowledge of enterprise architecture related to data, applications and technology (e.g., distributed applications, web-based applications, web services, n-tier applications) |
| | KS3.9 Knowledge of requirements analysis and management practices (e.g., requirements verification, traceability, gap analysis) |
| | KS3.10 Knowledge of acquisition and contract management processes (e.g., evaluation of vendors, preparation of contracts, vendor management, escrow) |
| | KS3.11 Knowledge of system development methodologies and tools, and an understanding of their strengths and weaknesses (e.g., agile development practices, prototyping, rapid application development [RAD], object-oriented design techniques) |

# Systems and Infrastructure Life Cycle Management

| Exhibit 3.1—Tasks and Knowledge Statements Mapping (*cont.*) | |
|---|---|
| **Task Statements** | **Knowledge Statements** |
| T3.4 (*cont.*) | KS3.12 Knowledge of quality assurance methods<br>KS3.13 Knowledge of the management of testing processes (e.g., test strategies, test plans, test environments, entry and exit criteria)<br>KS3.14 Knowledge of data conversion tools, techniques and procedures<br>KS3.16 Knowledge of software and hardware certification, and accreditation practices |
| T3.5 Evaluate the processes by which systems and/or infrastructure are developed/acquired and tested to ensure that the deliverables meet the organization's objectives. | KS3.1 Knowledge of benefits management practices (e.g., feasibility studies, business cases)<br>KS3.2 Knowledge of project governance mechanisms (e.g., steering committee, project oversight board)<br>KS3.3 Knowledge of project management practices, tools and control frameworks<br>KS3.4 Knowledge of risk management practices applied to projects<br>KS3.6 Knowledge of configuration, change and release management in relation to development and maintenance of systems and/or infrastructure<br>KS3.9 Knowledge of requirements analysis and management practices (e.g., requirements verification, traceability, gap analysis)<br>KS3.10 Knowledge of acquisition and contract management processes (e.g., evaluation of vendors, preparation of contracts, vendor management, escrow)<br>KS3.11 Knowledge of system development methodologies and tools, and an understanding of their strengths and weaknesses (e.g., agile development practices, prototyping, rapid application development [RAD], object-oriented design techniques)<br>KS3.12 Knowledge of quality assurance methods<br>KS3.13 Knowledge of the management of testing processes (e.g., test strategies, test plans, test environments, entry and exit criteria)<br>KS3.14 Knowledge of data conversion tools, techniques and procedures<br>KS3.16 Knowledge of software and hardware certification, and accreditation practices |
| T3.6 Evaluate the readiness of the system and/or infrastructure for implementation and migration into production. | KS3.2 Knowledge of project governance mechanisms (e.g., steering committee, project oversight board)<br>KS3.3 Knowledge of project management practices, tools and control frameworks<br>KS3.4 Knowledge of risk management practices applied to projects<br>KS3.6 Knowledge of configuration, change and release management in relation to development and maintenance of systems and/or infrastructure |

| Exhibit 3.1—Tasks and Knowledge Statements Mapping (*cont.*) | |
| --- | --- |
| **Task Statements** | **Knowledge Statements** |
| T3.6 (*cont.*) | KS3.7 Knowledge of control objectives and techniques that ensure the completeness, accuracy, validity and authorization of transactions and data within IT systems applications<br><br>KS3.8 Knowledge of enterprise architecture related to data, applications and technology (e.g., distributed applications, web-based applications, web services, n-tier applications)<br><br>KS3.9 Knowledge of requirements analysis and management practices (e.g., requirements verification, traceability, gap analysis)<br><br>KS3.10 Knowledge of acquisition and contract management processes (e.g., evaluation of vendors, preparation of contracts, vendor management, escrow)<br><br>KS3.11 Knowledge of system development methodologies and tools, and an understanding of their strengths and weaknesses (e.g., agile development practices, prototyping, rapid application development [RAD], object-oriented design techniques)<br><br>KS3.12 Knowledge of quality assurance methods<br><br>KS3.13 Knowledge of the management of testing processes (e.g., test strategies, test plans, test environments, entry and exit criteria)<br><br>KS3.14 Knowledge of data conversion tools, techniques and procedures<br><br>KS3.16 Knowledge of software and hardware certification, and accreditation practices<br><br>KS3.18 Knowledge of system migration and infrastructure deployment practices |
| T3.7 Perform postimplementation review of systems and/or infrastructure to ensure that they meet the organization's objectives and are subject to effective internal control. | KS3.1 Knowledge of benefits management practices (e.g., feasibility studies, business cases)<br><br>KS3.2 Knowledge of project governance mechanisms (e.g., steering committee, project oversight board)<br><br>KS3.3 Knowledge of project management practices, tools and control frameworks<br><br>KS3.4 Knowledge of risk management practices applied to projects<br><br>KS3.5 Knowledge of project success criteria and risks<br><br>KS3.7 Knowledge of control objectives and techniques that ensure the completeness, accuracy, validity and authorization of transactions and data within IT systems applications<br><br>KS3.8 Knowledge of enterprise architecture related to data, applications and technology (e.g., distributed applications, web-based applications, web services, n-tier applications)<br><br>KS3.9 Knowledge of requirements analysis and management practices (e.g., requirements verification, traceability, gap analysis) |

# Systems and Infrastructure
# Life Cycle Management

| Exhibit 3.1—Tasks and Knowledge Statements Mapping (*cont.*) | |
|---|---|
| **Task Statements** | **Knowledge Statements** |
| T3.7 (*cont.*) | KS3.10 Knowledge of acquisition and contract management processes (e.g., evaluation of vendors, preparation of contracts, vendor management, escrow) <br> KS3.11 Knowledge of system development methodologies and tools, and an understanding of their strengths and weaknesses (e.g., agile development practices, prototyping, rapid application development [RAD], object-oriented design techniques) <br> KS3.12 Knowledge of quality assurance methods <br> KS3.13 Knowledge of the management of testing processes (e.g., test strategies, test plans, test environments, entry and exit criteria) <br> KS3.14 Knowledge of data conversion tools, techniques and procedures <br> KS3.16 Knowledge of software and hardware certification, and accreditation practices <br> KS3.17 Knowledge of postimplementation review objectives and methods (e.g., project closure, benefits realization, performance measurement) <br> KS3.18 Knowledge of system migration and infrastructure deployment practices |
| T3.8 Perform periodic reviews of systems and/or infrastructure to ensure that they continue to meet the organization's objectives and are subject to effective internal control. | KS3.4 Knowledge of risk management practices applied to projects <br> KS3.6 Knowledge of configuration, change and release management in relation to development and maintenance of systems and/or infrastructure <br> KS3.7 Knowledge of control objectives and techniques that ensure the completeness, accuracy, validity and authorization of transactions and data within IT systems applications <br> KS3.8 Knowledge of enterprise architecture related to data, applications and technology (e.g., distributed applications, web-based applications, web services, n-tier applications) <br> KS3.11 Knowledge of system development methodologies and tools, and an understanding of their strengths and weaknesses (e.g., agile development practices, prototyping, rapid application development [RAD], object-oriented design techniques) <br> KS3.12 Knowledge of quality assurance methods <br> KS3.13 Knowledge of the management of testing processes (e.g., test strategies, test plans, test environments, entry and exit criteria) <br> KS3.14 Knowledge of data conversion tools, techniques and procedures <br> KS3.16 Knowledge of software and hardware certification, and accreditation practices <br> KS3.17 Knowledge of postimplementation review objectives and methods (e.g., project closure, benefits realization, performance measurement) |

# Systems and Infrastructure Life Cycle Management

| Exhibit 3.1—Tasks and Knowledge Statements Mapping (*cont.*) | |
|---|---|
| **Task Statements** | **Knowledge Statements** |
| T3.9 Evaluate the process by which systems and/or infrastructure are maintained to ensure the continued support of the organization's objectives, and are subject to effective internal control. | KS3.4 Knowledge of risk management practices applied to projects<br><br>KS3.6 Knowledge of configuration, change and release management in relation to development and maintenance of systems and/or infrastructure<br><br>KS3.7 Knowledge of control objectives and techniques that ensure the completeness, accuracy, validity and authorization of transactions and data within IT systems applications<br><br>KS3.8 Knowledge of enterprise architecture related to data, applications and technology (e.g., distributed applications, web-based applications, web services, n-tier applications)<br><br>KS3.11 Knowledge of system development methodologies and tools, and an understanding of their strengths and weaknesses (e.g., agile development practices, prototyping, rapid application development [RAD], object-oriented design techniques)<br><br>KS3.12 Knowledge of quality assurance methods<br><br>KS3.13 Knowledge of the management of testing processes (e.g., test strategies, test plans, test environments, entry and exit criteria)<br><br>KS3.14 Knowledge of data conversion tools, techniques and procedures<br><br>KS3.16 Knowledge of software and hardware certification, and accreditation practices<br><br>KS3.17 Knowledge of postimplementation review objectives and methods (e.g., project closure, benefits realization, performance measurement) |
| T3.10 Evaluate the process by which systems and/or infrastructure are disposed to ensure that they comply with the organization's policies and procedures. | KS3.4 Knowledge of risk management practices applied to projects<br><br>KS3.6 Knowledge of configuration, change and release management in relation to development and maintenance of systems and/or infrastructure<br><br>KS3.8 Knowledge of enterprise architecture related to data, applications and technology (e.g., distributed applications, web-based applications, web services, n-tier applications)<br><br>KS3.15 Knowledge of system and/or infrastructure disposal procedures<br><br>KS3.16 Knowledge of software and hardware certification, and accreditation practices<br><br>KS3.18 Knowledge of system migration and infrastructure deployment practices |

# Systems and Infrastructure Life Cycle Management

# 3.2 BUSINESS REALIZATION

Business realization of projects is a compromise among major factors such as cost, quality, speed, reliability and dependability. Strategy-makers perform a comprehensive study and evaluate which factors are "qualifying" or "winning" and then compare those factors with strengths, weaknesses and competencies of their services.

## 3.2.1 PORTFOLIO/PROGRAM MANAGEMENT

A program can be seen as a group of projects and time-bound tasks that are closely linked together through common objectives, a common budget, and intertwined schedules and strategies. Like projects, programs have a limited time frame (start and end date) and organizational boundaries. A differentiator is that programs are more complex, usually have a longer duration, a higher budget, higher risks associated with them and are of higher strategic importance in nature.

A typical program might be the implementation of a large-scale enterprise resource planning (ERP) system, such as SAP, including technology infrastructure, operations, organizational realignment, business process reengineering (BPR) and optimization, training, and development. Mergers and acquisitions (M&As) might serve as an example for non-IS-related programs.

In either case, the objective of program management is the successful execution of programs including, but not limited to, management of:
• Program scope, program financials (costs, resources, cash flow, etc.), program schedules, and program objectives and deliverables
• Program context and environment
• Program communication and culture
• Program organization

To make autonomous projects possible on one side and make use of synergies between projects on the other side, a specific program organization is required. Typical program roles are the program owner, the program manager, the program team and the program office. The program owner role is to be differentiated from the project owner role. Typical communication structures in a program are program owner's meetings and program team meetings.

Methodology and processes used in program management are very similar to those in project management and run in parallel to each other. However, they must not be mixed up, and have to be handled and carried out separately.

To formally start a program, some form of written assignment from the program sponsor (owner) to the program manager and program team is required. Since programs most often emerge from projects, such an assignment is of paramount importance to set the program context and boundaries as well as formal management authority.

To deal with program management, project portfolio management and project management, an organization requires specific and well-designed structures such as expert pools, a project management office and project portfolio groups. These organizations make use of specific integrative tools such as project management guidelines, standard project plans and project management marketing instruments.

The project management office, as an owner of the project management and program management process, must be a permanent structure and adequately staffed to provide professional support in these areas to maintain current and develop new procedures and standards.

Its objective is to improve project and program management quality and secure project success, but it can only focus on activities and tasks that come out of this area (project management process), and not on project or program content. So an auditor has to differentiate between auditing project content and/or procedural aspects of a program or project.

# Systems and Infrastructure Life Cycle Management

A project portfolio is defined as all the projects being carried out in an organization at a given point in time (snapshot). In contrast to program management in which all relevant projects are closely coupled, this is not a requirement in a project portfolio. Projects of a program belong to the company's project portfolio.

But even loosely coupled projects need management. The objectives of project portfolio management are:
• Optimization of the results of the project portfolio (not of the individual projects)
• Prioritizing and scheduling projects
• Resource coordination (internal and external)
• Knowledge transfer throughout the projects

A project portfolio database is mandatory for project portfolio management. It must include project data such as owner, schedules, objectives, project type, status, cost, etc. Project portfolio management requires specific project portfolio reports. Typical project portfolio reports are a project portfolio bar chart, a profit vs. risk matrix, a project portfolio progress graph, etc.

## 3.2.2 BUSINESS CASE DEVELOPMENT AND APPROVAL

An important consideration in any IT project, whether it be the development of a new system or the investment in new infrastructure, is the business case. It has been increasingly recognized that the achievement of business benefits should drive projects.

A business case provides the information required for an organization to decide whether a project should proceed. Depending on organizations and often the size of the investment, the development of a business is either the first step in a project or a precursor to the commencement of a project.

The initial business case would normally derive from a feasibility study undertaken as part of project initiation/planning. This is an early study of a problem to assess if a solution is feasible. The feasibility study will normally scope the problem, identify and explore a number of solutions, and make a recommendation on what action to take. Part of the work in developing options is to calculate and outline a business case for each solution as an aspect of comparison.

The business case should be of sufficient detail to describe the justification for setting up and continuing a project. The business case should provide the reasons for the project and answer the question, "Why?"

The business case should also be a key element of the decision process throughout the life cycle of any project. If at any stage the business case is thought to no longer be valid, through increased costs or through reduction in the anticipated benefits, the project sponsor or steering committee should consider whether the project should proceed. In a well-planned project, there will be decision points, sometimes called "stage gates" or "kill points," at which the business case is formally reviewed to ensure that it is still valid. If the business case changes during the course of an IT project, the project should be reapproved through the departmental planning and approval process.

## 3.2.3 BENEFITS REALIZATION TECHNIQUES

Increasingly, a majority of business benefits are obtained through changes enabled by technology. With the evolution of the application of IT from straight automation of work, through information management to business transformation, the realization of benefits has become a more complex task. Benefits do not just happen when the new technology is delivered; they occur throughout the business cycle.

A planned approach to benefits realization is required, looking beyond project cycles to longer term cycles that consider the benefits. Benefits rarely happen according to plan. Organizations have to keep checking and adjusting strategies.

This concept is called benefits management or benefits realization. It requires:
• Describing benefits management or benefits realization
• Assigning a measure and target

- Establishing a tracking/measuring regimen
- Documenting the assumption
- Establishing key responsibilities for realization
- Validating the benefits predicted in the business
- Planning the benefit that is to be realized

Benefits realization is a continuous process that must be managed just like any business process. Gateway assessment, involving assessment of the benefits realization processes and the business case, should be a key element of benefits realization processes. Enterprisewide benefits realization studies should be aggregated and lessons learned should fine-tune the enterprise benefit realization process.

Benefits realization often includes a postimplementation review 6-18 months after the implementation of systems. Time must be allowed for teething problems to be eradicated and for the project benefits to accrue as users become familiar with the new processes and procedures.

Benefits realization must be part of governance and management of projects. There must be business sponsorship of projects. Project governance structures should involve the project and the functional line organization, all governance decisions about the project should be driven through the business case, and there must be periodic review of benefits.

# 3.3 PROJECT MANAGEMENT STRUCTURE

Today, many approaches to project management exist. Some are focused on software development, others have a more general approach; some concentrate on a holistic and systemic view, others provide a very detailed workflow including templates for document creation. Some of the most prominent *de facto* standards and organizations are the Project Management Body of Knowledge (PMBOK®) (i.e., IEEE standard 1490) from the Project Management Institute (PMI), Projects in a Controlled Environment (PRINCE2), and International Project Management Association (IPMA). Since there are significant differences in scope, content and wording in each of these standards, an auditor has to become familiar with the standard used prior to getting involved such a project.

Although each *modus operandi* has its own pros and cons, several elements are common across all project management methodologies.

## 3.3.1 GENERAL ASPECTS

IS projects may be initiated from within any part of the organization, including the IS department.

A project is always a time-bound effort. A project can be complex and involve an element of risk, with a specific objective, deliverable, and start and end dates; and is divisible into explicit phases (see the description of a system development life cycle [SDLC] in this chapter). A project can be perceived not only as a group of complex tasks, but also as a social system and/or a temporary organization (in contrast to the relatively stable structures of the permanent organization).

Project management should be a business process of the project-oriented organization. The project management process begins with the project charter and ends with the completion of the project. Project management contains subprocesses such as project initiation, project planning, project coordination, project execution, project controlling, project discontinuity management (including the management of stage boundaries) and project closing. These subprocesses of project management are related to one another, and their objective is to consider the project's goal, scope, schedule, resources, costs, organization, culture and context (i.e., pre- and postproject phases, project environments, other projects, etc.).

The complexity of project management requires a careful and explicit design of the project management process, which is normally done for a business process but sometimes neglected for the project management process. Thus, all design issues applicable for business process engineering should also be applied for the project management process.

# Systems and Infrastructure Life Cycle Management

## 3.3.2 PROJECT CONTEXT AND ENVIRONMENT

A project context can be divided into a time and a social context. In the analysis of the content's dimension of the context, the following must be taken into account:
• Importance of the project in the organization
• Connection between the organization's strategy and the project
• Relationship between the project and other projects
• Connection between the project to the underlying business case

Since there are normally several projects running at the same time, the relationships between those projects have to be investigated to identify common objectives for the business organization, identify and manage risks, and identify resource connections. A common approach to address these issues is to establish a project portfolio management and/or a program management structure.

A project, by definition, has a specific start date and end date. The project's time context, however, should consider the phase before project startup and the phase after project closing. During the preproject phase there are key activities and objectives identified that must be considered in project planning. A thorough transfer of information about the negotiations and decisions as well as necessary knowledge from the preproject phase is critical. The expectations in regard to the postproject phase also influence the tasks to be fulfilled and the strategy for designing the project environment relationships.

Because a project represents a social system, it is also necessary to consider its relationships to its own social environments. The design of the project environment relationships is a project management activity. The objective is to determine all relevant environments for the project, which will have a significant influence on overall project planning and project success.

## 3.3.3 PROJECT ORGANIZATIONAL FORMS

Three major forms of organizational alignment for project management can be observed:
1. Influence project organization
2. Pure project organization
3. Matrix project organization

In influence project organization, the project manager has only a staff function without formal management authority. The project manager is only allowed to advise peers and team members as to which activities should be completed.

In a pure project organization, the project manager has formal authority over those taking part in the project. Often, this is bolstered by providing a special working area for the project team that is separated from their normal office space.

In a matrix project organization, management authority is shared between the project manager and the department heads. For an auditor, it is important to understand these organizational forms and their implications on controls in project management activities.

Requests for major projects should be submitted to and prioritized by the IS steering committee. A project manager should be identified and appointed by the IS steering committee. The project manager, who does not need to be an IS staff member, should be given complete operational control over the project and be allocated the appropriate resources, including IS professionals and other staff from user departments, for the successful completion of the project. IS auditors may be included in the project team as control experts. They may also provide an independent, objective review to ensure that the level of involvement (commitment) of the responsible parties is appropriate. In such cases, IS auditors are not performing an audit, but are participating on the project in an advisory role. An example of such a structure is shown in **exhibit 3.2**.

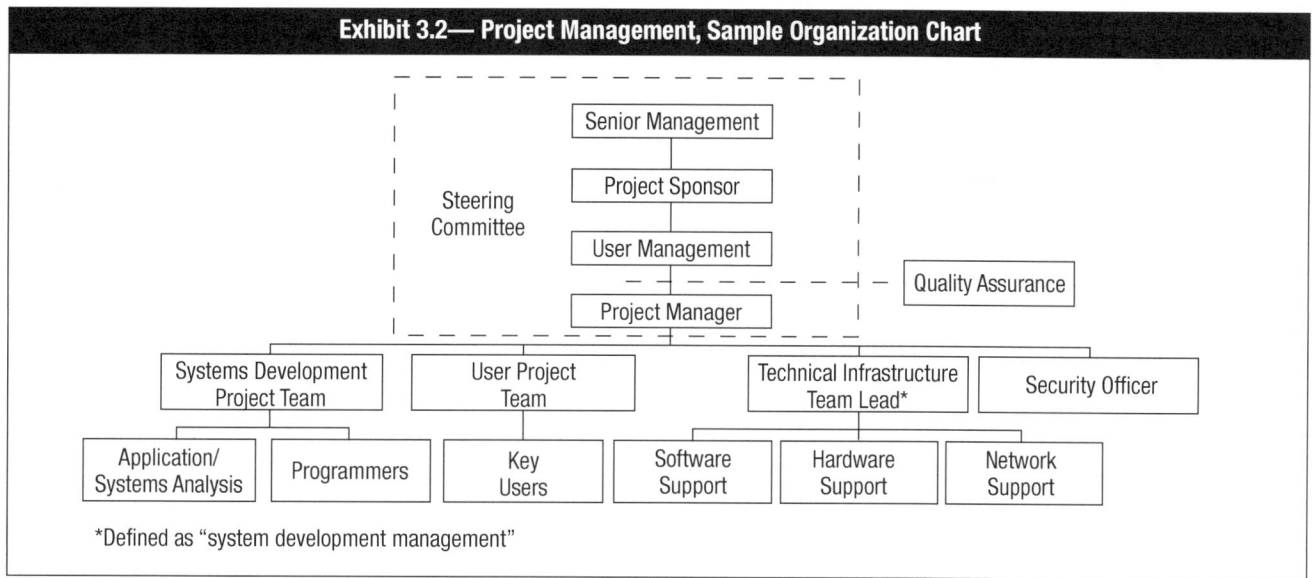

**Exhibit 3.2— Project Management, Sample Organization Chart**

*Defined as "system development management"

## 3.3.4 PROJECT COMMUNICATION AND CULTURE

Depending on the size and complexity of the project and the affected parties, communication when initiating the project management project process may be achieved by:
• One-on-one meetings
• Kick-off meetings
• Project start workshops
• A combination of the three

One-on-one meetings and a project start workshop help to facilitate two-way communication between the project team members and the project manager. A kick-off meeting may be used by the project manager to inform the team of what has to be done for the project.

An optimal method to ensure communication is open and clear among the project team is to use a project start workshop to obtain cooperation from all team members and buy-in from stakeholders. This helps to develop a big picture of the project and the project culture.

As an independent social system, each project has its own culture that defines its norms and rules of engagement. The project culture cannot be described, but manifests itself in the applied project management techniques, including project planning, forms of communication, etc. The dynamics of the project culture provide for a common understanding of what is seen as desirable and serves as an orientation for the team.

Methods for developing a project culture include the establishment of a project mission statement, a project name and logo, project team meeting rules and communication protocol, and project-specific social events.

## 3.3.5 PROJECT OBJECTIVES

A project needs clearly defined results that are specific, measurable, achievable, relevant and time-bound (SMART). A holistic project view ensures the consideration and consolidation of all closely coupled objectives. These objectives are broken down into main objectives, additional objectives and nonobjectives.

Main objectives shall always be directly coupled with business success (see the Key Business Drivers section later in this chapter). Additional objectives are objectives that are not directly related to the main results of the project but may contribute to project success (e.g., business unit reorganization in a software development project).

# Systems and Infrastructure Life Cycle Management

Nonobjectives add clarity to scope and make project boundaries become clearer, shaping the contours of the deliverables and supporting all parties to gain a clear understanding of what has to be done and to avoid any ambiguities.

A commonly accepted approach to define project objectives is to start off with an object breakdown structure (OBS). It represents the individual components of the solution and their relationships to each other in a hierarchical manner, either graphically or in a table. An OBS can help, especially when dealing with nontangible project results such as organizational development, to ensure that a material deliverable is not overlooked.

After the OBS has been compiled or a solution is defined, a work breakdown structure (WBS) is designed to structure all the tasks that are necessary to build up the elements of the OBS during the project. The WBS represents the project in terms of manageable and controllable units of work, serves as a central communications tool in the project, and forms the baseline for cost and resource planning.

In contrast to the OBS, the WBS does not include basic elements of the solution to build, but shows individual work packages (WPs) instead. The structuring of the WBS is process-oriented and in phases. The level of detail of the WBS serves as the basis for the negotiations of objectives between the project sponsor, project manager and the project team members.

Detailed specifications regarding the WBS can be found in WPs. Each WP must have a distinct owner and a list of main objectives, and may have a list of additional objectives and nonobjectives. The WP specifications should include dependencies on other WPs and a definition of how to evaluate performance and goal achievement. An example of a WBS is shown in **exhibit 3.3**.

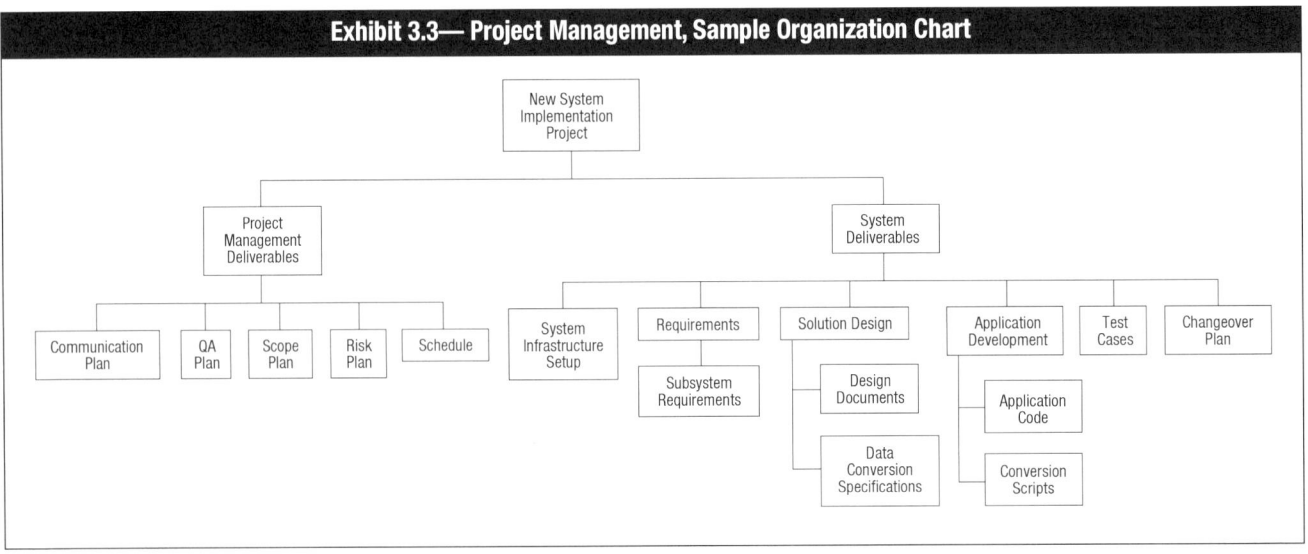

**Exhibit 3.3— Project Management, Sample Organization Chart**

To support communications, to-do lists are often used. A to-do list is a list of actions to be carried out in relation to work packages with assigned responsibilities and deadlines. The to-do list aids the individual project team members in operational planning as well as in making agreements.

These to-do lists are most typically compiled into a project schedule at the initiating phase of a project, and are used in the controlling phase of the project to monitor and track the progress and completion of the WPs. Project schedules are living documents and should indicate the tasks for a WP, the start and finish dates, percentage completed, task dependencies, and resource names of individuals planned to work on those tasks. A project schedule will also indicate the stage boundaries explained in section 3.2.2.

## 3.3.6 ROLES AND RESPONSIBILITIES OF GROUPS AND INDIVIDUALS

To achieve a successful completion and implementation of any new system, it is advisable that the audit function have an active part, where appropriate, in the life cycle development of the business application. This will facilitate efforts to ensure that proper controls are designed and implemented in the new system (e.g., continuous concurrent controls for paperless

e-commerce systems). Additionally, there are other key stakeholders who should be involved in the system's design, development and implementation. The various roles and responsibilities of groups/individuals that may be involved in the development process are summarized as follows:

- **Senior management**—Demonstrates commitment to the project and approves the necessary resources to complete the project. This commitment from senior management helps ensure involvement by those needed to complete the project.
- **User management**—Assumes ownership of the project and resulting system, allocates qualified representatives to the team, and actively participates in business process redesign, system requirements definition, test case development, acceptance testing and user training. User management should review and approve system deliverables as they are defined and implemented. User management is concerned particularly with the following questions:
  - Are the required functions available in the software?
  - How reliable is the software?
  - How efficient is the software?
  - Is the software easy to use?
  - How easy is it to transfer or adapt old data from preexisting software to this environment?
  - How easy is it to transfer the software to another environment?
  - Is it possible to add new functions?
  - Does it meet regulatory requirements?
- **Project steering committee**—Provides overall direction and ensures appropriate representation of the major stakeholders in the project's outcome. The project steering committee is ultimately responsible for all deliverables, project costs and schedules. This committee should be comprised of a senior representative from each business area that will be significantly impacted by the proposed new system or system modification. Each member must have the authority to make decisions related to system designs that will affect their respective departments. Generally, a project sponsor who would assume the overall ownership and accountability of the project will chair the steering committee. The project manager should also be a member of this committee and, in some cases, may serve as its chair. The project steering committee performs the following functions:
  - Reviews project progress regularly (for example, semimonthly or monthly) and holds emergency meetings when required.
  - Serves as coordinator and advisor. Members of the committee should be available to answer questions and make user-related decisions about system and program design.
  - Takes corrective action. The committee should evaluate progress and take action or make recommendations regarding personnel changes on the project team, managing budgets or schedules, changes in project objectives, and the need for redesign. The committee should be available to address risks and issues that are escalated and cannot be resolved at the project level. The project manager should have the ability to escalate such matters and rely on the project steering committee to resolve risks and project issues for the benefit of the organization. In some cases, the committee may recommend that the project be halted.
- **Project sponsor**—Provides funding for the project and works closely with the project manager to define the critical success factors (CSFs) and metrics for measuring the success of the project. It is crucial that success is translated to measurable and quantifiable terms. Data and application ownership are assigned to a project sponsor. A project sponsor is typically the senior manager in charge of the primary business unit that the application will support.
- **Systems development management**—Provides technical support for hardware and software environments by developing, installing and operating the requested system. This area also provides assurance that the system is compatible with the organization's computing environment and strategic IT direction, and assumes operating support and maintenance activities after installation.
- **Project manager**—Provides day-to-day management and leadership of the project, ensures that it remains in line with the overall direction, ensures appropriate representation of the affected departments, ensures the project adheres to local standards, ensures that deliverables meet the quality expectations of key stakeholders, resolves interdepartmental conflicts, and monitors and controls costs and the project timetable. The project manager may also facilitate the definition of the scope of the project, manage the budget of the project, and control the activities via a project schedule. This person can be an end user, a member of the systems development team or a professional project manager. Where projects are staffed by personnel dedicated to the project, the project manager will have a line responsibility for such personnel.
- **Systems development project team**—Completes assigned tasks, communicates effectively with users by actively involving them in the development process, works according to local standards and advises the project manager of necessary project plan deviations.

- **User project team**—Completes assigned tasks, communicates effectively with the systems developers by actively involving themselves in the development process as subject matter experts (SMEs), works according to local standards and advises the project manager of expected and actual project plan deviations.
- **Security officer**—Ensures that system controls and supporting processes provide an effective level of protection, based on the data classification set in accordance with corporate security policies and procedures; consults throughout the life cycle on appropriate security measures that should be incorporated into the system; reviews security test plans and reports prior to implementation; evaluates security-related documents developed in reporting the system's security effectiveness for accreditation; and periodically monitors the security system's effectiveness during its operational life.
- **Quality assurance (QA)**—Personnel who review results and deliverables within each phase and at the end of each phase, and confirm compliance with requirements. Their objective is to ensure that the quality of the project by measuring the adherence of the project staff to the organization's software development life cycle (SDLC), advise on deviations, and propose recommendations for process improvements or greater control points when deviations occur. The points where reviews occur depend on the SDLC methodology used, the structure and magnitude of the system, and the impact of potential deviations. Additionally, focus may include a review of appropriate, process-based activities related to either project management or the use of specific software engineering processes within a particular life cycle phase. Such a focus is crucial to completing a project on schedule and within budget, and in achieving a given software process maturity level (see the Software Process Improvement Practices section of this chapter).

Specific objectives of the QA function include:
- Ensuring the active and coordinated participation by all relevant parties in the revision, evaluation and dissemination of standards, management guidelines and procedures
- Maintaining the agreed on systems development methodology
- Reviewing and evaluating large system projects at significant development milestones, and making appropriate recommendations for improvement
- Establishing, enhancing and maintaining a stable, controlled environment for the implementation of changes within the production software environment
- Defining, establishing and maintaining a standard, consistent and well-defined testing methodology for computer systems
- Reporting to management on systems that are not performing as defined or designed

It is essential for the IS auditor to understand the systems development, acquisition and maintenance methodology in use and to identify potential vulnerabilities and points requiring control. If controls are lacking (either as a result of the organizational structure or of the software methods used) or the process is disorderly, it is the IS auditor's role to advise the project team and senior management of the deficiencies. It may also be necessary to advise those engaged in development and acquisition activities of appropriate controls or processes to implement and follow.

> **Note:** The CISA candidate should be familiar with general roles and responsibilities of groups or individuals involved in the systems development process.

# 3.4 PROJECT MANAGEMENT PRACTICES

Project management is the application of knowledge, skills, tools and techniques applied to a broad range of activities to achieve a stated objective such as meeting the defined user requirements and deadlines for an IS project. Project management knowledge and practices are best described in terms of their component processes of initiating, planning, executing, controlling and closing a project. Overall characteristics of successful project planning are that it is a risk-based management process and iterative in nature. Project management techniques also provide systematic quantitative and qualitative approaches to software size estimating, scheduling, allocating resources and measuring productivity.

There are numerous project management techniques and tools available to assist the project manager in controlling the time and resources utilized in the development of a system. The techniques and tools may vary from a simple manual effort to a more elaborate computerized process. The size and complexity of the project may require different approaches. These tools typically provide assistance in areas described in the following sections.

# Systems and Infrastructure Life Cycle Management

There are three elements or dimensions of a project that should always be taken into account:
1. **Time/duration**—How long will it take to complete the project?
2. **Cost/resources**—How much will it cost?
3. **Deliverables**—What has to be done?

**Exhibit 3.4** shows these relationships. From a heuristic view, the area of the triangle surrounding the project remains constant but the length of the sides may vary. The behavior of the triangle greatly depends on its degree of freedom. If, for example, constraints exist in two dimensions (e.g., project must end by a certain date and must not exceed a specified budget), the remaining side, reflecting the deliverables, is automatically determined. If there is only one constraint (e.g., specified deliverables must be accomplished), then one dimension can be freely chosen (e.g., time frame of the project) and the last dimension has to compensate (cost and resources).

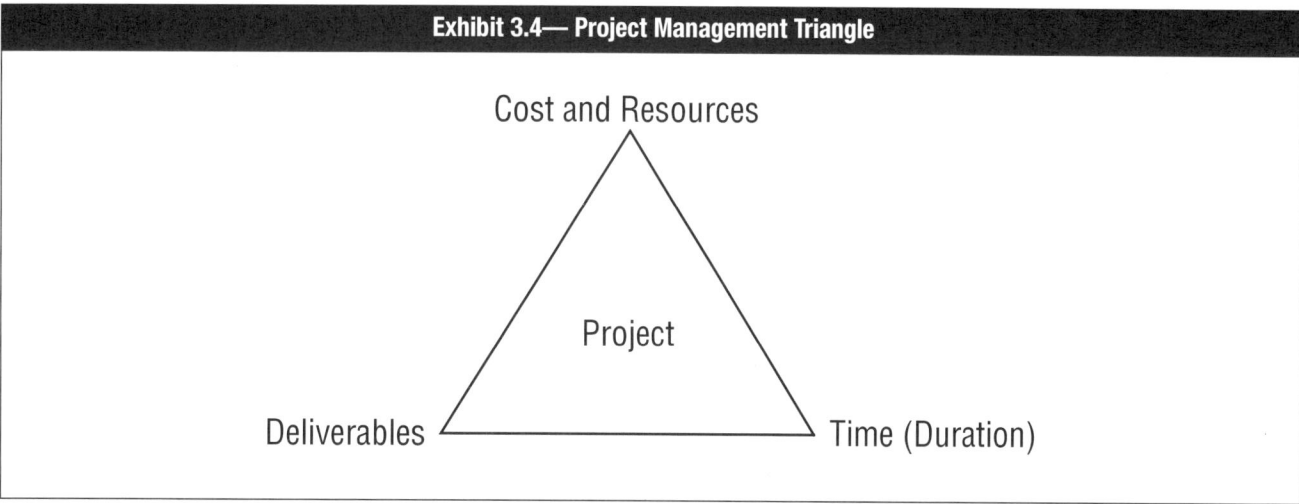

It is important to recognize that if a project is completely set up (all dimensions are agreed on), the area is fixed. If, for example, feature or scope creep emerges (more deliverables), either the cycle time of the project will have to increase or costs/resource utilization will increase.

An important element of the deliverables that is also considered during the management of time and resources is the quality of the deliverables. The parameters for quality of the deliverables may be specified clearly by the project steering committee or project sponsor, or the project manager may have to elicit this from user management. In either case, the project manager must have a clear and documented view of the quality expectations for the deliverables of the project steering committee, sponsor and users.

## 3.4.1 INITIATION OF A PROJECT

A project will be initiated by a project manager or sponsor gathering the information required to gain approval for the project to be created. This will often be compiled into terms of reference or a project charter that states the objective of the project, the stakeholders in the system to be produced, and the project manager and sponsor. Approval of a project initiation document (PID) or a project request document (PRD) is the authority for a project to begin.

## 3.4.2 PROJECT PLANNING

Software development/acquisition or maintenance projects have to be planned and controlled.

The project manager needs to determine:
• The various tasks that need to be performed to produce the expected business application system
• The sequence or the order in which these tasks need to be performed

# Systems and Infrastructure Life Cycle Management

- The duration or the time window for each task
- The priority of each task
- The IT resources, which are available and required to perform these tasks
- Budget or costing for each of these tasks
- Source and means of funding

Realistically, on other than small projects, detailed task-level planning can only occur toward the end of a phase or iteration for the upcoming phase or iteration, or through to the next decision gate. Until work products and decisions are produced during the current phase or iteration, insufficient information is available to understand in detail what will occur in the next phrase/iteration.

There are some techniques that are useful in creating a project plan and monitoring its progress throughout the execution of the project to provide continual support for making it successful. Some of these techniques are addressed in the following sections.

## Software Size Estimation

Software size estimation relates to methods of determining the relative physical size of the application software to be developed. It can be used as a guide to the allocation of resources, estimates of time and cost required for its development, and as a comparison of the total effort required by the resources available.

Traditionally, software sizing has been done using single-point estimations such as source lines of code (SLOC). For complex systems, this technique has not worked well since it does not support more than one parameter in different types of programs, which in turn affects the cost, schedule and quality metrics.

Metrics are identified to measure the development effort. The first step is to identify resources (e.g., people with requisite skills, development tools, facilities) for software development. This will help in estimating and budgeting software development resources.

## Lines of Source Code

The traditional and simplest method in measuring size by counting the number of lines of source code, measured in SLOC, or measuring in thousands of lines of code, which is referred to as kilo lines of code (KLOC). An alternative but similar metric is thousand delivered source instructions (KDSI). This method has proved to be superior when using structured programming languages such as BASIC or COBOL. The variance in estimation by two different application developers is found to be minimal. However, recent technologies have introduced development tools that have added one more dimension to the estimation factor. Thus, the problem with SLOC is that the estimation becomes less reliable using just a one-factor metric vs. the code length. Several multifactor methods have been developed.

The advantages of SLOC and KLOC are that they could be unambiguously defined for any language used for mainframe and allow quantitative measure. The disadvantage of using the SLOC and KLOC approach for software sizing is that it has little meaning to the customer or end user and is not applicable for newer software development tools, which are nontextual.

Current technologies now take the form of more abstract representations such as diagrams, objects, spreadsheet cells, database queries and graphical user interface (GUI) widgets. These technologies are more closely related to "functionality" deliverables rather than "work" or lines that need to be created.

## Function Point Analysis

The function point analysis (FPA) technique was developed in the late 1970s and has become widely used for estimating complexity in developing large business applications.

# Systems and Infrastructure Life Cycle Management

The results of FPA are a measure of the size of an information system based on the number and complexity of the inputs, outputs, files, interfaces and queries with which a user sees and interacts. This is an indirect measure of software size and the process by which it is developed vs. direct size-oriented measures such as SLOC counts.

Function points (FPs) are computed by first completing a table (see **exhibit 3.5**) to determine whether a particular entry is simple, average or complex. Five FP count values are defined, including the number of user inputs, user outputs, user inquiries, files and external interfaces.

| Exhibit 3.5—Computing Function Point Metrics | | | | | |
|---|---|---|---|---|---|
| Measurement Parameter | Count | Weighting Factor | | | |
| | | Simple | Average | Complex | Results |
| Number of user inputs | | x 3 | 4 | 6 | =_____ |
| Number of user outputs | | x 4 | 5 | 7 | =_____ |
| Number of user inquiries | | x 3 | 4 | 6 | =_____ |
| Number of files | | x 7 | 10 | 15 | =_____ |
| Number of external interfaces | | x 5 | 7 | 10 | =_____ |
| Count total: | | | | | |
| Note: Organizations that use FP methods develop criteria for determining whether a particular entry is simple, average or complex. | | | | | |

Upon completion of the table entries, the count total in deriving the function point is computed through an algorithm that takes into account complexity adjustment values (i.e., rating factors) based on responses to questions related to issues such as reliability, criticality, complexity, reusability, changeability and portability. Function points derived from this equation are then used in a manner analogous to SLOC counts as a measure for cost, schedule, productivity and quality metrics (e.g., productivity = FP/person-month, quality = defects/FP, and cost = $/FP).

FPA is a more refined and useful five-factor method. It is an indirect measurement (FP) of the software size. It is based on the number and complexity of inputs, outputs, files, interfaces and queries.

**Note:** The CISA candidate should be familiar with the use of Function Point Analysis; however, the exam does not test the specifics on how to perform calculations.

## FPA Feature Points

In most standard applications, lists of functions are identified and the corresponding effort is estimated. In web-enabled applications, the development effort depends on the number of screens (forms), number of images, type of images (static or animated), features to be enabled, interfaces and cross-referencing that is required. Thus, from the point of view of web applications, the effort would include all that is mentioned under function point estimation, plus the features that need to be enabled for different types of user groups. The measurement would involve identification or listing of features, access rules, links, storage, etc.

FPA behaves reasonably well in estimating business applications, but not as well for other types of software (i.e., OS, process control, communications, engineering, etc.). Other estimation methods include the constructive cost model (COCOMO 2), De Marco and Walston-Felix.

## Cost Budgets

A system development project should be analyzed with a view toward estimating the amount of effort that will be required to carry out each task. The estimates for each task should contain some or all of the following elements:
• Personnel hours by type (e.g., system analyst, programmer, clerical)
• Machine hours (predominantly computer time as well as duplication facilities, office equipment and communication equipment)

# Systems and Infrastructure Life Cycle Management

• Other external costs such as third-party software, licensing of tools for the project, consultant or contractor fees, training costs, certification costs (if required), and occupation costs (if extra space is required for the project)

Having established a best estimate of expected work efforts by task (i.e., actual hours, minimum/maximum) for personnel, costs budgeting now becomes a two-step process to:
1. Obtain a phase-by-phase estimate of human and machine effort by summing the expected effort for the tasks within each phase
2. Extend the summation of effort expressed in hours by the appropriate hourly rate to obtain a phase-by-phase estimate of systems development expenditure

Other costs may require tenders or quotes.

## Software Cost Estimation
Cost estimation is a consequence of software size estimation. This is a necessary step in properly scoping a project.

There are automated techniques for cost estimation of projects at each phase of system development. To use these products, a system is usually divided into main components, and a set of cost drivers is established. Components include:
• Source code language
• Execution time constraints
• Main storage constraints
• Data storage constraints
• Computer access
• The target machine used for development
• The security environment
• Staff experience

Once all the drivers are defined, the program will develop cost estimates of the system and total project.

## Scheduling and Establishing the Time Frame
While budgeting involves totaling the human and machine effort involved in each task, scheduling involves establishing the sequential relationship among tasks. This is achieved by arranging tasks according to:
• Earliest start date, by considering the logical sequential relationship among tasks and attempting to perform tasks in parallel, wherever possible
• Latest expected finish date, by considering the estimate of hours per the budget and the expected availability of personnel or other resources, and allowing for known, elapsed-time considerations (e.g., holidays, recruitment time, full-time/part-time employees)

The schedule can be graphically represented using various techniques such as Gantt charts or Program Evaluation Review Technique (PERT) diagrams. At key points/milestones within the project, the budget and schedule should be revisited to verify compliance and identify variances. Any variances to the budget and schedule should be analyzed to determine the cause and corrective action to take in minimizing or eliminating the total project variance. Variances and the variance analysis should be reported to management on a timely basis.

## Critical Path Methodology
All project management techniques compute what is called a critical path. Since a project consists of an ordered set of independent activities, it can be represented as a network where activities are shown as branches connected at nodes immediately preceding and immediately following activities.

A path through the network is any set of successive activities, which goes from the beginning to the end of the project. Associated with each activity in the network is a single number that best estimates the amount of time that the activity will consume. Differences in the way this number is obtained distinguish the major variants of the technique.

# Systems and Infrastructure Life Cycle Management

A critical path, then, is one whose sum of activity time is longer than that for any other path through the network. This path is important because, if everything goes according to schedule, its length gives the shortest possible completion time for the overall project.

In addition, a slack time can be associated with each activity in the project. This is the difference between the latest possible completion time of each activity that will not delay the completion of the overall project and the earliest possible completion time based on all predecessor activities. Activities on a critical path have zero slack time and, conversely, activities with zero slack time are on a critical path.

The critical path and associated slack times for a project are found by simply working forward through the network (forward pass), computing the earliest possible completion time for each activity, until the earliest possible completion time for the total project is found. Then by working backward through the network, the latest completion time for each activity is found, the slack time computed and the critical path identified. This procedure is usually programmed into a computer for easy calculation and what-if scenarios.

Computation of the critical path proceeds by first assuming a nominal time for each activity. Activities on the critical path become candidates for crashing or for a reduction in their time by payment of a premium for early completion. By successively relaxing activities on the critical path, a curve (S-Curve) showing total project costs vs. time to completion can be obtained.

Most critical path methodology (CPM) packages facilitate the analysis of resource utilization per time unit (e.g., day, week, etc.) and resource leveling, which is a way to level off alternative peaks and valleys. Resource peaks and valleys (e.g., management, hiring, firing, and/or overtime and idle resource costs) are expensive. A constant, base resource utilization is preferable.

There are few, if any, scientific (algorithmic) resource-leveling methods available, but there is a battery (which CPM packages offer) of efficient heuristic methods that yield satisfactory results.

## Gantt Charts

Gantt charts can be constructed to aid in scheduling the activities (tasks) needed to complete a project. The charts show when an activity should begin and when it should end. The charts also show which activities can be in progress concurrently and which activities must be completed sequentially. The charts can also reflect the resources assigned to each task and by what percent allocation. Gantt charts aid in identifying what will happen if an activity is either completed early or late by illustrating a baseline. Progress of a project can be shown on the charts and can reveal if the project is behind, ahead or on schedule by creating a baseline project plan. Gantt charts can also be used to track the achievement of milestones or significant accomplishments for the project such as the end of a project phase or completion of a key deliverable.

## Program Evaluation Review Technique

PERT is a network management technique often used in system development projects based on well-defined events and activities for specific project management tasks. A diagram illustrating use of the PERT network management technique is shown in **exhibit 3.6**, where events are points in time or milestones for starting and completing activities. Activities are the time-based estimates shown from the arrows to an event signifying a task's completion.

To determine a task's completion, three estimates are shown for completing each activity. The first is the most optimistic time (if everything went well) and the third is the pessimistic or worst-case scenario. The second is the most likely scenario. This estimate is based on experience attained from projects similar in size and scope. To calculate the PERT time estimate for each given task, the following calculation is applied:

$$[\text{Optimistic} + \text{Pessimistic} + 4(\text{most likely})]/6$$

# Systems and Infrastructure Life Cycle Management

Using PERT, a critical path is also derived. The critical path is the longest path through the network (only one critical path in a network). The critical path is the route along which the project is shortened (accelerated) or lengthened (delayed). In **exhibit 3.6**, the critical path is A, C, E, F, H and I.

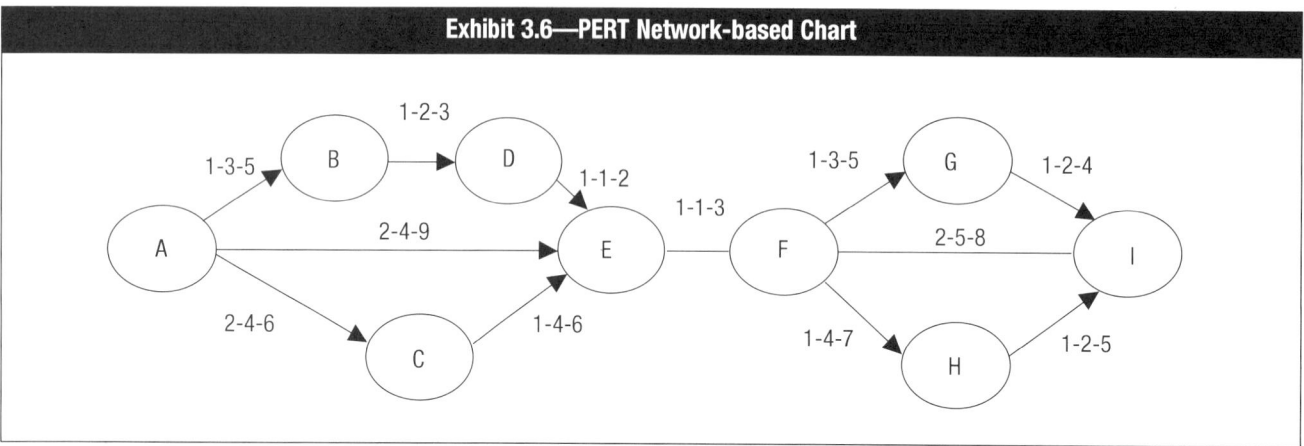

**Exhibit 3.6—PERT Network-based Chart**

When designing a PERT network for system development projects, the first step is to identify all the activities and related events/milestones of the project and their relative sequence. For example, an event or result may be the completion of the operational feasibility study or the point at which the user accepts the detailed design. The analyst must be careful not to overlook any activity. Additionally, some activities such as analysis and design must be preceded by others before program coding can be begin. The list of activities determines the detail of the PERT network. The analyst may prepare many diagrams that provide increasingly more detailed time estimates.

## Timebox Management

Timebox management is another project management technique for defining and deploying software deliverables within a relatively short and fixed period of time, and with predetermined specific resources. There is a need to balance software quality and meet the delivery requirements within the timebox or time window. The project manager has some degree of flexibility and uses discretion in scoping requirements. The timebox management approach is well suited for any kind of end-user computing objectives, using techniques such as prototyping or rapid application development-type approaches in which key features to be delivered for specified application are considered. The key features may include making provision for interfaces for future integrations as may be decided at the time of design or preparation of functional specification for the application under this project. The major advantage of this approach is that it prevents project cost overruns and delays from scheduled delivery. The project does not necessarily eliminate the need for a quality process. The design and development phase is shortened due to the use of newer developmental tools and techniques. Preparation of test cases and testing requirements are easily written down as a result of end-user participation. System test and user acceptance testing are normally performed together.

## 3.4.3 GENERAL PROJECT MANAGEMENT

General project management involves automated techniques to handle proposals and cost estimations, and to monitor, predict and report on performance with recommended action items. Many of these techniques are provided as decision support systems (DSS) for planning and controlling project resources. These automated techniques may perform functions ranging from staffing requirements to budgeting systems. These automated systems often incorporate PERT and CPM techniques.

## Documentation

Automated documentation tools handle the production, validation and maintenance of system and program documentation. These tools often allow the user to input program and system parameters without having to learn high-powered word processing functions. The package then generates program narratives and flowcharts from the input provided by the user.

# Systems and Infrastructure Life Cycle Management

## *Office Automation*

Office automation reduces programmer involvement in functions that are required to comply with policies and staff meetings. Types of office automation tools that have proven to be useful and productive are electronic mail systems, voice-messaging systems, automated time management tools (such as electronic calendars and date reminders), automated libraries, and filing and retrieval systems.

## 3.4.4 PROJECT CONTROLLING

The controlling activities of a project include management of scope, resource usage and risk. It is important that new requirements for the project are documented and, if approved, allocated appropriate resources. Control of change during a project ensures that projects are completed within stakeholders' expectations of time, use of funds and quality objectives.

## 3.4.5 CLOSING A PROJECT

A project should have a finite life so, at some point, it is closed and the new or modified system is handed over to the users and/or system support staff. At this point there may still be some issues that need to be resolved—ownership of which needs to be assigned. The project sponsor should be satisfied that the system produced is acceptable and ready for delivery. Custody of contracts may need to be assigned, and documentation archived or passed on to those who will need it. It is also common at this stage to survey the project team, development team, users and other stakeholders to identify any lessons learned that can be applied to future projects including content-related criteria such as performance fulfillment, fulfillment of additional objectives, adherence to the schedule and costs, and process-related criteria such as quality of the project teamwork and relationships to relevant environments.

Review may be done in a formal process such as a postproject review in which lessons learned and an assessment of project management processes used are documented and referenced, in the future, by other project managers or users working on projects of similar size and scope. A postimplementation review, in contrast, is typically completed once the project has been in use (or in "production") for some time—long enough to realize its business benefits and costs, and measure the project's overall success and impact on the business units.

# 3.5 BUSINESS APPLICATION DEVELOPMENT

Companies often commit significant IT resources (e.g., people, applications, facilities and technology) to develop, acquire and maintain application systems that are critical to the effective functioning of key business processes. These systems in turn often control critical information assets and should be considered assets that need to be effectively managed and controlled. IT processes for managing and controlling these IT resources and other such activities are part of a life cycle process with defined phases applicable to business application development, deployment, maintenance and retirement. In this process, each step or phase in the life cycle is an incremental step that lays the foundation for the next phase, for effective management control in building and operating business application systems.

The implementation process for business applications, commonly referred to as an SDLC, begins when an individual application is initiated as a result of one or more of the following situations:
• A new opportunity that relates to a new or existing business process
• A problem that relates to an existing business process
• A new opportunity that will enable the organization to take advantage of technology
• A problem with the current technology

All of these situations are tightly coupled with key business drivers. Key business drivers, in this context, can be defined as the attributes of a business function that drive the behavior and implementation of that business function to achieve the strategic business goals of the company.

# Systems and Infrastructure Life Cycle Management

Thus, all critical business objectives (as a breakdown of the corporate strategy) have to be translated into key business drivers for all parties involved in business operations during an SDLC project. If, and only if, this is done in a SMART way, the decomposition of general requirements into a specific scorecard will provide objective evidence to measure the business value of an application and allow for prioritizing of requirements.

Benefits of using this approach are that all affected parties will have a common and clear understanding of the objectives and how they contribute to business support. Additionally, conflicting key business drivers (e.g., cost vs. functionality) and mutually dependent key business drivers can be detected and resolved in early stages of an SDLC project.

Business applications should be initiated using well-defined procedures or activities as part of a defined process to communicate business needs to management. These procedures often require detailed documentation identifying the need or problem, specifying the desired solution and relating the potential benefits to the organization. All internal and external factors affected by the problem and their effect on the corporation should be identified.

A risk in any software development project is that the final outcome may not meet all requirements. Problems due to translation errors arise when initially defining the requirements for interim products. The waterfall model and variants of the model normally involve a life cycle verification approach that ensures that potential mistakes are corrected early and not solely during final acceptance testing. The verification and validation model, sometimes called the V-model, also emphasizes the relationship between development phases and testing levels (see **exhibit 3.7**). The most granular testing—the unit test—occurs immediately after programs have been written. Following this model, testing occurs to validate the detailed design. System testing relates to the architectural specification of the system while final user-acceptance testing references the requirements.

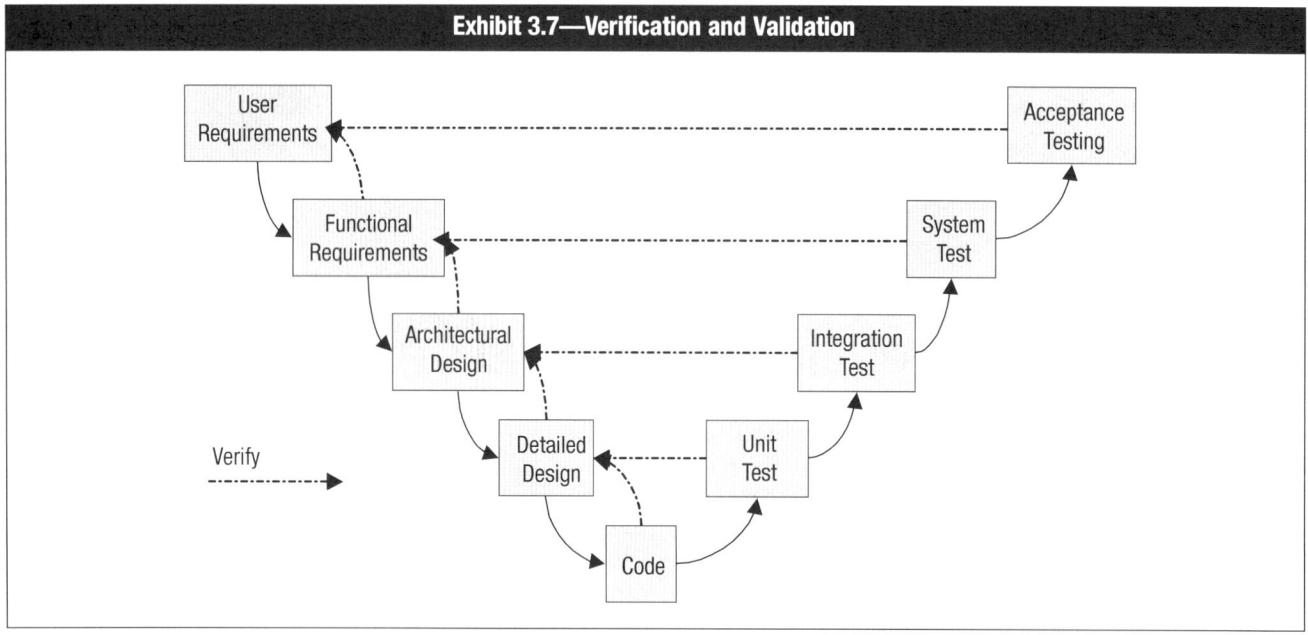

**Exhibit 3.7—Verification and Validation**

From an IS auditor's perspective, such an an approach with defined life cycle phases and specific points for review and evaluation provides the following advantages:
• The IS auditor's influence is significantly increased when there are formal procedures and guidelines identifying each phase in the business application life cycle and the extent of auditor involvement.
• The IS auditor can review all relevant areas and phases of the systems development project, and report independently to management on the adherence to planned objectives and company procedures.
• The IS auditor can identify selected parts of the system and become involved in the technical aspects on the basis of their skills and abilities.
• The IS auditor can provide an evaluation of the methods and techniques applied through the development phases of the business application life cycle.

# Systems and Infrastructure Life Cycle Management

Any business application system developed will fall under one of two major categories: organization-centric (management information system [MIS], ERP, customer relationship management [CRM], supply chain management [SCM], etc.) and end-user-centric computing. The organization-centric application continues to use the SDLC approach while an end-user-centric application takes an alternate development approach. The objective of an organization-centric application is to collect, collage, store, archive and share information to business users and various applicable support functions on a need-to-know basis. Thus, sales data are made available to accounts, administration, governmental levy payment departments, etc. Regulatory levy fulfillment (i.e., tax compliance) is also addressed by the presence of organization-centric applications. The objective of an end-user-centric application is to provide different views of data for their performance optimization. This objective includes DSS, geographic information systems (GIS), techniques, etc. Most of these applications are developed using alternate development approaches.

## 3.5.1 TRADITIONAL SDLC APPROACH

Over the years business application development has occurred largely through the use of the traditional SDLC phases shown in **exhibit 3.8**. Also referred to as the waterfall technique, this life cycle approach is the oldest and most widely used for developing business applications. The approach is based on a systematic, sequential approach to software development (largely of business applications) that begins with a feasibility study and progresses through requirements definition, design, development, implementation and postimplementation. The series of steps or phases have defined goals and activities to perform with assigned responsibilities, expected outcomes and target completion dates.

| Exhibit 3.8—Traditional Systems Development Life Cycle Approach | |
|---|---|
| **SDLC Phase** | **General Description** |
| Phase 1—Feasibility | Determine the strategic benefits of implementing the system either in productivity gains or in future cost avoidance, identify and quantify the cost savings of a new system, and estimate a payback schedule for costs incurred in implementing the system. Further, intangible factors such as readiness of the business users and maturity of the business processes will also be considered and assessed. This business case provides the justification for proceeding to the next phase. |
| Phase 2—Requirements | Define the problem or need that requires resolution and define the functional and quality requirements of the solution system. This can be either a customized approach or vendor-supplied software package, which would entail following a defined and documented acquisition process. In either case, the user needs to be actively involved. |
| Phase 3A—Design | Based on the requirements defined, establish a baseline of system and subsystem specifications that describe the parts of the system, how they interface, and how the system will be implemented using the chosen hardware, software and network facilities. Generally, the design also includes program and database specifications, and will address any security considerations. Additionally, a formal change control process is established to prevent uncontrolled entry of new requirements into the development process. |
| Phase 3B—Selection | Based on the requirements defined, prepare a request for proposal from suppliers of packaged systems. In addition to the functionality requirements, there will be operational, support and technical requirements, and these, together with considerations of the suppliers' financial viability and provision for escrow, will be used to select the packaged system that best meets the organization's total requirements. |
| Phase 4A—Development | Use the design specifications to begin programming and formalizing supporting operational processes of the system. Various levels of testing also occur in this phase to verify and validate what has been developed. This would generally include all unit and system testing, as well as several iterations of user acceptance testing. |

| Exhibit 3.8—Traditional Systems Development Life Cycle Approach (*cont.*) | |
|---|---|
| **SDLC Phase** | **General Description** |
| Phase 4B—Configuration | Configure the system, if it is a packaged system, to tailor it to the organization's requirements. This is best done through the configuration of system control parameters, rather than changing program code. Modern software packages are extremely flexible, making it possible for one package to suit many organizations simply by switching functionality on or off and setting parameters in tables. There may be a need to build interface programs that will connect the acquired system with existing programs and databases. |
| Phase 5—Implementation | Establish the actual operation of the new information system, with the final iteration of user acceptance testing and user sign-off conducted in this phase. The system also may go through a certification and accreditation process to assess the effectiveness of the business application in mitigating risks to an appropriate level and providing management accountability over the effectiveness of the system in meeting its intended objectives and in establishing an appropriate level of internal control. |
| Phase 6—Postimplementation | Following the successful implementation of a new or extensively modified system, implement a formal process that assesses the adequacy of the system and projected cost-benefit or ROI measurements *vis-à-vis* the feasibility stage findings and deviations. In so doing, IS project and end-user management can provide lessons learned and/or plans for addressing system deficiencies as well as recommendations for future projects regarding system development and project management processes followed. |

This approach works best when a project's requirements are likely to be stable and well defined. It facilitates the determination of a system architecture relatively early in the development effort. Another type of software development approach is the iterative approach where business requirements are developed and tested in iterations until the entire application is designed, built and tested. This approach is useful in web applications where prototypes of screens are necessary to aid in the completion of requirements and design.

As purchased packages have become more common, the design and development phases of the traditional life cycle are being replaced with the selection and configuration phases.

> **Note:** The CISA candidate should be familiar with the SDLC phases. An example of a CISA question would be: What should the IS auditor look for when reviewing the feasibility study?

Some examples of measurements of CFSs for an e-commerce project could include productivity, quality, economic value and customer service shown in **exhibit 3.9**.

| Exhibit 3.9—Measurements of Critical Success Factors | |
|---|---|
| Productivity | Dollars spent per user<br>Number of transactions per month<br>Number of transactions per user |
| Quality | Number of discrepancies<br>Number of disputes<br>Number of occurrences of fraud/misuse detection |
| Economic value | Total processing time reduction<br>Monetary value of administration costs |
| Customer service | Turnaround time for customer question handling<br>Frequency of useful communication to users |

# Systems and Infrastructure Life Cycle Management

The primary advantage of this approach is that it provides a template into which methods for the requirements (i.e., definition, design, programming, etc.) can be placed. However, some of the problems encountered with this approach include:

- Unanticipated events. Since real projects rarely follow the sequential flow prescribed, iteration always occurs and creates problems in implementing the approach.
- The difficulty of obtaining an explicit set of requirements from the customer/user as the approach requires
- Managing requirements and convincing the user about the undue or unwarranted requirements in the system functionality, which may lead to friction in the project
- The necessity of customer/user patience, which is required since under this approach a working version of the system's programs will not be available until late in the project's life cycle
- A changing business environment that alters or changes the customer's/user's requirements before they are delivered

Moreover, the actual phases for each project may vary depending on whether a developed or acquired solution is chosen. For example, system maintenance efforts may not require the same level of detail or number of phases as new applications. The phases and deliverables should be decided during the early planning stages of the project.

## 3.5.2 INTEGRATED RESOURCE MANAGEMENT SYSTEMS

A growing number of organizations—public and private—are shifting from separate groups of interrelated applications to a fully integrated corporate solution. Such solutions are often marketed as ERP solutions. Many vendors, mainly from Europe and the US, have been focusing on this market and offer packages with commercial names such as SAP, PeopleSoft, Oracle Financials, SSG (Baan) or J.D. Edwards.

This type of software implementation is much larger than a regular software acquisition project. The acquisition and implementation of an ERP system impacts the way the corporation does business, its entire control environment, technological direction and internal resources. Generally speaking, a corporation is required to convert management philosophies, policies and practices to those of the integrated software solution providers, notwithstanding the numerous customization options. In this respect such a solution will either impair or enhance information technology's ability to support the organization's mission and goals. When considering a change of this magnitude, it is imperative that a thorough impact and risk assessment be conducted. When implementing an ERP solution or any off-the-shelf software, the business unit has the option of implementing a "vanilla" solution (referring to the generic ice cream flavor). This term simply means implementing and configuring the new system as-is, out-of-the-box and not developing any additional functionality or customizations to suit the specific gaps in business processes. Instead, the business opts to change business processes to suit the industry standard as dictated by the software solution. While this decision results in less software design, development and testing work than does customization, it does imply greater change in the business units to work differently. Due to the large costs of customization in software development, maintenance and continuous upgrading and patching, it is not usually recommended by software vendors.

Because of the magnitude of the risks involved, it is imperative that senior management assess and approve all changes in the system's architecture, technological direction, migration strategies and IS budgets.

## 3.5.3 DESCRIPTION OF TRADITIONAL SDLC PHASES

As shown in **exhibit 3.8**, the traditional SDLC approach is made up of six distinct phases, each with a defined set of activities and outcomes. The following section describes in detail each phase's purpose and relationship to prior phases, the general activities performed, and expected outcomes.

### *Feasibility Study*

Once the determination has been made to move forward with a project, an analysis begins to clearly define the need and to identify alternatives for addressing the need. This analysis is known as the feasibility study.

A feasibility study is concerned with analyzing the benefits and solutions for the identified problem area. Study includes development of a business case, which determines the strategic benefits of implementing the system either in productivity

gains or in future cost avoidance; identifies and quantifies the cost savings of the new system; and estimates a payback schedule for the cost incurred in implementing the system or shows the projected return on investment (ROI). Intangible benefits such as improved morale may also be identified; however, benefits should be quantified wherever possible.

Within the feasibility study, the following typically are addressed:
• Define a time frame for the implementation of the required solution.
• Determine an optimum alternative risk-based solution for meeting business needs and general information resource requirements (e.g., whether to develop or acquire a system). Information resources, as defined in COBIT, include people, applications, technology, facilities and data organized and managed through a given set of IT processes (e.g., SDLC and rapid application development [RAD]) for the acquisition and implementation domain.
• Determine if an existing system can correct the situation with slight or no modification (e.g., workaround).
• Determine if a vendor product offers a solution to the problem.
• Determine the approximate cost to develop the system to correct the situation.
• Determine if the solution fits the business strategy.

Factors impacting whether to develop or acquire a system include:
• The date the system needs to be functional
• The cost to develop the system as opposed to buying it
• The resources, staff (availability and skill sets) and hardware required to develop the system or implement a vendor solution
• In a vendor system, the license characteristics (e.g., yearly renewal, perpetual) and maintenance costs
• The other systems needing to supply information to or use information from the vendor system that will need the ability to interface with the system
• Compatibility with strategic business plans
• Compatibility with the organization's IT infrastructure
• Likely future requirements for changes to functionality offered by the system

The result of the completed feasibility study should be some type of a comparative report that shows the results of criteria analyzed (e.g., costs, benefits, risks, resources required and organizational impact) and includes alternatives/solutions.

Closely related to a feasibility study is the development of an impact assessment. An impact assessment is a study of the potential future effects of a development project on current projects and resources. The resulting document should list the pros and cons of pursuing a specific course of action.

## *Requirements Definition*

Requirements definition is concerned with identifying and specifying the business requirements of the system chosen for development during the feasibility study. Requirements include descriptions of what a system should do, how users will interact with a system, conditions under which the system will operate, and the information criteria the system should meet. The COBIT framework defines information criteria that should be incorporated in system requirements to address issues associated with effectiveness, efficiency, confidentiality, integrity, availability, compliance and reliability. The requirements definition phase deals with the issues that are sometimes called nonfunctional requirements.

To accomplish the above in the requirements definition phase you should:
• Identify and consult stakeholders to determine their expectations.
• Analyze requirements to detect and correct conflicts and determine priorities.
• Identify system bounds and how the system should interact with its environment.
• Convert user requirements into system requirements (e.g., an interactive user interface prototype that demonstrates the screen look and feel).
• Record requirements in a structured format. Historically, requirements have been recorded in a written requirements specification, possibly supplemented by some schematic models. Commercial requirements management tools now are available that allow requirements and related information to be stored in a multiuser database.
• Verify that requirements are complete, consistent, unambiguous, verifiable, modifiable, testable and traceable. Because of the high cost of rectifying requirements' problems in downstream development phases, effective requirements reviews have a large payoff.

• Resolve conflicts between stakeholders.
• Resolve conflicts between the requirements set and the resources that are available.

The users in this process specify their information resource needs, nonautomated as well as automated, and how they wish to have them addressed by the system (e.g., access controls, regulatory requirements, management information needs and interface requirements).

From this interactive process, a general preliminary design of the system may be developed and presented to user management for their review, modification, approval and endorsement. A project schedule is created for developing, testing and implementing the system. Also, commitments are obtained from the system's developers and affected user departments to contribute the necessary resources to complete the project.

It is important to note that all concerned management and user groups must be actively involved in the requirements definition phase to prevent problems such as expending resources on a system that will not satisfy the business requirements. User involvement is necessary to obtain commitment and full benefit from the system. Without management sponsorship, clearly defined requirements and user involvement, the benefits may never be realized.

An important tool in the creation of a general preliminary design is the use of entity relationship diagrams (ERDs).

## Entity Relationship Diagrams

An ERD depicts a system's data and how these data interrelate. An ERD can be used as a requirements analysis tool to obtain an understanding of the data a system needs to capture and manage. In this case, the ERD represents a logical data model. An ERD can also be used later in the development cycle as a design tool that helps document the actual database schema that will be implemented. Used in this way, the ERD represents a physical data model.

As the name suggests, the essential components of an ERD are entities and relationships.

Entities are groupings of like data elements or instances that may represent actual physical objects or logical constructs. An entity is described by attributes, which are properties or characteristics common to all or some of the instances of the entity. Particular attributes, either singularly or in combination, form the keys of an entity. An entity's primary key uniquely identifies each instance of the entity. Entities are represented on ERDs as rectangular boxes with an identifying name.

Relationships depict how two entities are associated (and in some cases how instances of the same entity are associated). The classic way of depicting a relationship is via a diamond with connecting lines to each related entity. The name in the diamond describes the nature of the relationship. The relationship may also specify the foreign key attributes that achieve the association between/among the entities. A foreign key is one or more attributes held in one entity that map to the primary key of a related entity.

IS auditors are involved at this stage to determine whether adequate security requirements have been defined to address, at a minimum, the confidentiality, integrity and availability requirements of the system. This includes whether adequate audit trails are defined as part of the system since these affect the auditor's ability to identify issues for proper follow-up.

> **Note:** If the result of the decision to develop/acquire is to buy a vendor-supplied software package, the user must be actively involved in the package evaluation and selection process.

## Software Acquisition

Software acquisition is not a phase in the standard SDLC. However, if a decision was reached to acquire rather than develop software, software acquisition is the process that should occur after the requirements definition phase. The decision is generally based on various factors such as the cost differential between development and acquisition, availability of generic software, and the time gap between development and acquisition.

# Systems and Infrastructure Life Cycle Management

The feasibility study should contain documentation that supports the decision to acquire the software. Depending on the software required there could be three cases:

1. Software is required for a generic business process for which vendors are available and software can be implemented without customization.
2. The vendor's software needs to be customized to suit business processes.
3. Software needs to be developed by the vendor.

A project team with participation by technical support staff and key users should be created to write a request for proposal (RFP) or invitation to tender (ITT). An RFP needs to be prepared separately for each case referred to previously.

The invitation to respond to an RFP should be widely distributed to appropriate vendors and, if possible, posted via a public procurement medium (Internet or newspaper). This process allows the business to determine which of the responding vendors' products offers the best solution at the most cost-effective price.

The RFP should include the areas shown in **exhibit 3.10**.

| Exhibit 3.10—RFP Contents | |
|---|---|
| **Item** | **Description** |
| Product vs. system requirements | The chosen vendor's product should come as close as possible to meeting the defined requirements of the system. If no vendor's product meets all of the defined requirements, the project team, especially the users, will have to decide whether to accept the deficiencies. An alternative to living with a product's deficiencies is for the vendor or the purchaser to make customized changes to the product. |
| Customer references | Project management should check vendor-supplied references to validate the vendor's claims of product performance and completion of work by the vendor. |
| Vendor viability/financial stability | The vendor supplying or supporting the product should be reputable and, therefore, should be able to provide evidence of financial stability. A vendor may not be able to prove financial stability; if the product is new, the vendor presents a substantially higher risk to the organization. |
| Availability of complete and reliable documentation | The vendor should be willing and able to provide a complete set of system documentation for review prior to acquisition. The level of detail and precision found in the documentation may be an indicator of the detail and precision utilized within the design and programming of the system itself. |
| Vendor support | The vendor should have available a complete line of support products for the software package. This may include a 24-hour, seven-day-a-week help line, onsite training during implementation, product upgrades, automatic new version notification and onsite maintenance, if requested. |
| Source code availability | The source code should be received either from the vendor initially or there should be provisions for acquiring the source code in the event that the vendor goes out of business. Usually, these clauses are part of a software escrow agreement in which a third party holds the software in escrow should such an event occur. The acquiring company should ensure that product updates and program fixes are included in the escrow agreement. |
| Number of years of experience in offering the product | More years indicate stability and familiarity with the business that the product supports. |
| A list of recent or planned enhancements to the product, with dates | A short list suggests the product is not being kept current. |
| Number of client sites using the product with a list of current users | A larger number suggests wide acceptance of the product in the marketplace. |
| Acceptance testing of the product | Such testing is crucial in determining whether the product really satisfies the system requirements. This is allowed before a purchasing commitment must be made. |

# Systems and Infrastructure Life Cycle Management

It is worth noting that a distinction can sometimes be drawn between ITT and RFP with respect to their applicability. When the product and related services are known in advance, a user organization will prefer an ITT so they can obtain the best combination of price and services. This is more applicable where procurement of hardware, network, database, etc. is involved. When the requirement is more toward a solution and related support and maintenance, an organization generally prefers an RFP, where the capability, experience and approach can be measured against the requirement. This is more applicable in system integration projects such as ERP and supply chain management (SCM) that involve delivery or escrowing of source code.

Often, prior to the development of an RFP, organizations will develop a request for information (RFI) to solicit software development vendors for advice in addressing problems with existing systems. Information obtained in this manner may be used to develop an RFP.

> **Note:** The CISA candidate should be familiar with the RFP process.

The project team needs to carefully examine and compare the vendors' responses to the RFP. This comparison should be done using an objective method such as a scoring and ranking methodology. After the RFP responses have been examined, the project team may be able to identify a single vendor whose product satisfies most or all of the stated requirements in the RFP. Other times, the team may narrow the list to two or three acceptable candidates (i.e., short list of vendors). In evaluating the best-fit solution and vendor against the given set of business requirements and conditions, a suitable methodology of evaluation should be adopted. The methodology should ensure objective, equitable and fair comparison of the products/vendors (e.g., a gap analysis to find out the differences between requirements and software, the parameters required to modify, etc.).

It is important to keep in mind the minimum and recommended requirements to use the software including:
• Required hardware such as memory, disk space, and server or client characteristics
• Operating system versions and patch levels supported
• Additional tools such as import and export tools
• Databases supported

Often it is likely that more than one product/vendor fits the requirements with advantages and disadvantages with respect to each other. To resolve such a situation, agenda-based presentations should be requested from the short-listed vendors. The agenda-based presentations are scripted business scenarios that are designed to show how the vendor will perform certain critical business functions. Vendors are typically invited to demonstrate their product and follow the sample business scenarios given to them to prepare. It is highly recommended that adequate participation from various user groups is included when evaluating the product/vendor's fit and the system's ease of use. The project team thus has an opportunity to check the intangible issues such as the vendor's knowledge of the product and the vendor's ability to understand the business issue at hand. With each short-listed vendor demonstrating their product following a scripted document, this also enables the project team to evaluate and finalize the product/vendor with knowledge and objectivity built into the process. The finalist vendor candidate is then requested to organize site visits to confirm the findings from the agenda-based presentations and check the system in a live environment. Once the finalist is confirmed, a conference room pilot needs to be conducted. The conference room pilot will enable the project team to understand the system with a hands-on session and identify the areas that need certain customizations or workarounds.

Additionally, for the short list of vendors remaining, it can be beneficial for the project team to talk to current users of each of the potential products. If it can be arranged and can be cost-justified, an onsite visit can be even more beneficial. Whenever possible, the companies chosen should be those that use the products in a manner similar to the way the company will use the products. The IS auditor should encourage the project team to contact current users. The information obtained from these discussions or visits validates statements made in the vendor's proposal and can determine which vendor is selected.

The discussions with the current users should concentrate on each vendor's:
• **Reliability**—Are the vendor's deliverables (enhancements or fixes) dependable?
• **Commitment to service**—Is the vendor responsive to problems with its product? Does the vendor deliver on time?
• **Commitment to providing training, technical support and documentation for its product**—What is the level of customer satisfaction?

Upon completing the activities cited, vendor presentations and final evaluations, the project team can make a product selection. The reasons for making a particular choice should be documented.

The last step in the acquisition process is to negotiate and sign a contract for the chosen product. Appropriate legal counsel should review the contract prior to its signing.
The contract should contain the following items:
• Specific description of deliverables and their costs
• Commitment dates for deliverables
• Commitments for delivery of documentation, fixes, upgrades, new release notifications and training
• Commitments for data migration
• Allowance for a software escrow agreement, if the deliverables do not include source code
• Description of the support to be provided during installation/customization
• Criteria for user acceptance
• Provision for a reasonable acceptance testing period, before the commitment to purchase is made
• Allowance for changes to be made by the purchasing company
• Maintenance agreement
• Allowance for copying software for use in business continuity efforts and for test purposes
• Payment schedule linked to actual delivery dates

Managing the contract should also involve a major level of effort to ensure that deployment efforts are controlled, measured and improved on, where appropriate. This may include regular status reporting requirements. Additionally, the milestones and metrics to be reported against should be agreed on with the vendor.

IS auditors are involved in the software acquisition process to determine whether an adequate level of security controls has been considered prior to any agreement being reached. If security controls are not part of the software, it may become difficult to ensure data integrity for the information that will be processed through the system. Risks involved with the software package include inadequate audit trails, password controls and overall security of the application. Because of these risks, the IS auditor should ensure that these controls are built into the software application.

## Design
Based on the general preliminary design and user requirements defined in the requirements definition phase, a detailed design can be developed. Generally, a programming and analyst team is assigned the tasks of defining the software architecture depicting a general blueprint of the system and then detailing or decomposing the system into its constituent parts such as modules and components. This approach is an enabler for effective allocation of resources to design and for defining how the system will satisfy all its information requirements. Depending on the complexity of the system, several iterations in defining system-level specifications may be needed to get down to the level of detail necessary to start development activities such as coding.

### USER INVOLVEMENT IN THE DESIGN
Once business processes have been documented and it is understood how those processes might be executed in the new system, involvement of users during the design phase is limited. Given the technical discussion that usually occurs during a design review, end-user participation in the review of detailed design work products is normally not appropriate. However, developers should be able to explain how the software architecture will satisfy system requirements and outline the rationale for key design decisions. Choices of particular hardware and software configurations may have cost implications of which stakeholders need to be aware and control implications that are of interest to the IS auditor.

Key design phase activities include:
• Developing system flowcharts and entity relationship models to illustrate how information will flow through the system
• Determining the use of structured design techniques (which are processes to define applications through a series of data or process flow diagrams) that show various relationships from the top level down to the details
• Describing inputs and outputs such as screen designs and reports. If a prototyping tool is going to be used, they most often are used in the screen design and presentation process (via online programming facilities) as part of an integrated development environment.

# Systems and Infrastructure Life Cycle Management

- Determining processing steps and computation rules when addressing functional requirement needs
- Determining data file or database system file design
- Preparing program specifications for various types of requirements or information criteria defined
- Developing test plans for the various levels of testing:
  - Unit (program)
  - Subsystem (module)
  - Integration (system)
  - Interface with other systems
  - Loading and initializing files
  - Stress
  - Security
  - Backup and recovery
- Developing data conversion plans to convert data and manual procedures from the old system to the new system. Detailed conversion plans will alleviate implementation problems that arise due to incompatible data, insufficient resources or staff who are unfamiliar with the operations of the new system.

## SOFTWARE BASELINING
The software design phase represents the optimum point for software baselining to occur. The term software baseline means the cutoff point in the design and is also referred to as design freeze. User requirements are reviewed, item by item, and considered in terms of time and cost. The changes are undertaken after taking into account various risks, and change does not occur without undergoing formal strict procedures for approval based on a cost-benefit impact analysis. Failure to adequately manage the requirements for a system through baselining can result in a number of risks. Foremost among these risks is scope creep—the process through which requirements change during development. Empirical studies have shown that a typical project can experience at least a 25 percent change in requirements throughout development, resulting in an increase in the effort and costs required for development.

Software baselining also relates to the point when formal establishment of the software configuration management process occurs. At this point, software work products are established as configuration baselines with version numbers. This would include, for example, functional requirements, specifications and test plans. All of these work products are configuration items, and are identified and brought under formal change management control. This process will be used throughout the application system's life cycle, where SDLC procedures for analysis, design, development, testing and deployment are enforced on new requirements or changes to existing requirements.

## END OF DESIGN PHASE
After the detailed design has been completed, including user approvals and software baselining, the design is distributed to the system developers for coding.

## IS AUDITOR INVOLVEMENT
The IS auditor involvement is primarily focused on whether an adequate system of controls is incorporated into system specifications and test plans, and whether continuous online auditing functions are built into the system (particularly for e-commerce applications and other types of paperless environments). Additionally, the IS auditor is interested in evaluating the effectiveness of the design process itself (such as in the use of structured design techniques, prototyping and test plans, and software baselining) to establish a formal software change process that effectively freezes the inclusion of any changes to system requirements without a formal review and approval process.

The key documents coming out of this phase include system, subsystem, program and database specifications, test plans, and a defined and documented formal software change control process.

# Systems and Infrastructure Life Cycle Management

## Development

The development phase uses the detailed design developed in the previous step to begin coding, moving the system one step closer to a final physical software product. Responsibilities in this phase rest primarily with programmers and systems analysts who are building the system. Key activities performed in a test/development environment include:
• Coding and developing program and system-level documents
• Debugging and testing the programs developed
• Developing programs to convert data from the old system for use on the new system
• Creating user procedures to handle transition to the new system
• Training selected users on the new system since their participation will be needed
• Ensuring modifications are documented, and applied accurately and completely to vendor-acquired software to ensure that future updated versions of the vendor's code can be applied

### PROGRAMMING METHODS AND TECHNIQUES

To enhance the quality of programming activities and future maintenance capabilities, program coding standards should be applied. Program coding standards are essential to reading and understanding code, simply and clearly, without having to refer back to design specifications. Elements of program coding standards may include methods and techniques for internal (source code level) documentation, methods for data declaration, and an approach to statement construction and techniques for input/output (I/O). The programming standards applied are an essential control since they serve as a method of communicating among members of the program team, and between the team and users during system development. Program coding standards minimize system development setbacks resulting from personnel turnover, provide the material needed to use the system effectively, and are required for program maintenance and modifications.

Additionally, traditional structured programming techniques should be applied in developing quality and easily maintained software products. They are a natural progression from the top-down structuring design techniques previously described. Like the design specifications, structured application programs are easier to develop, understand and maintain since they are divided into subsystems, components, modules, programs, subroutines and units. Generally, the greater extent to which each software item described performs a single, dedicated function (cohesion) and retains independence from other comparable items (coupling), the easier it is to maintain and enhance a system since it is easier to determine where and how to apply a change, and reduce the chances of unintended consequences.

### ONLINE PROGRAMMING FACILITIES (INTEGRATED DEVELOPMENT ENVIRONMENT)

To facilitate effective use of structured programming methods and techniques, an online programming facility should be applied as part of an integrated development environment (IDE). This allows programmers to code and compile programs interactively with a remote computer or server from a terminal or a client's PC workstation. Through this facility, programmers can enter, modify and delete programming codes as well as compile, store and list programs (source and object) on the computer. The online facilities can also be used by non-IS staff to update and retrieve data directly from computer files.

In general, an online programming facility allows faster program development and the application of standards and structured programming techniques. Online systems improve the programmer's problem-solving abilities, but online systems create opportunities for errors resulting from unauthorized access. Access control software should be used to help reduce the risk. Today, online programming facilities are used on PC workstations. The program library is on a server, such as a mainframe library management system, but the modification/development and testing are performed on the workstation. This approach can lower the development costs, maintain rapid response time, and expand the programming resources and aids available (e.g., editing tools, programming languages, debugging aids). From the perspective of control, this approach introduces the potential weaknesses of:
• The proliferation of multiple versions of programs
• Reduced program and processing integrity through the increased potential for unauthorized access and updating
• The possibility that valid changes could be overwritten by other changes

# Systems and Infrastructure Life Cycle Management

## PROGRAMMING LANGUAGES

Application programs must first be coded in statements, instructions or a programming language that is easy for a programmer to write and that can be read by the computer. These statements will then be translated by the language translator/compiler into a binary machine code or machine language (i.e., source code into object code) that the computer can execute.

Programming languages commonly used for developing application programs are:
• High-level, general-purpose programming languages such as COBOL and the C programming language
• Object-oriented languages for business purposes such as C++, 00-Basic, Eifell and JAVA
• IDEs such as Visual Studio or Jbuilder
• Hypertext Markup Language (HTML)
• Scripting languages such as SH(SHELL), PERL, TCL, Python, JAVAScript and VB Script. In web development, scripting languages are used commonly to write common gateway interface (CGI) scripts that are used to extend the functionality of web server application software (e.g., to interface with search engines, create dynamic web pages and respond to user input).
• Low-level assembler languages designed for a specific processor type
• Fourth-generation, high-level programming languages (4GLs), which consist of a DBMS, embedded database manager, and a nonprocedural report and screen generation facility. 4GLs provide fast iteration through successive designs. Examples of 4GLs include FOCUS, Natural and dBase.
• Decision support or expert systems languages (Express, LISP and PROLOG)

## PROGRAM DEBUGGING

Many programming bugs are detected during the system development process, after a programmer runs a program in the test environment. The purpose of debugging programs during system development is to ensure that all program abends (unplanned ending of a program due to programming errors) and program coding flaws are detected and corrected before the final program goes into production. A debugging tool is a program that will assist a programmer in fine-tuning, fixing or debugging the program under development. Compilers have some potential to provide feedback to a programmer, but they are not considered debugging tools. These tools fall into three main categories:
• Logic path monitors, which report on the sequence of events achieved by the program, thus providing the programmer with clues on logic errors
• Memory dumps, which provide a picture of the internal memory's content at one point in time. This is often produced at the point where the program is aborted, providing the programmer with clues on inconsistencies in data or parameter values. A variant, called a trace, will do the same at different stages in the program execution to show the evolution of such things as counters and registers.
• Output analyzers, which help check results of program execution for accuracy. This is achieved by comparing expected results with the actual results.

## TESTING

Testing is an essential part of the development process that verifies and validates that a program, subsystem or application performs the functions for which it has been designed. Testing also determines whether the units being tested operate without any malfunction or adverse effect on other components of the system.

The variety of development methodologies and organizational requirements provide for a large range of testing schemes or levels. Each set of tests is performed with a different set of data and under the responsibility of different people or functions. The IS auditor can play a preventive or detective role in the testing process.

# Systems and Infrastructure Life Cycle Management

**3-1** To assist in testing a core banking system being acquired, an organization has provided the vendor with sensitive data from its existing production system. An IS auditor's **PRIMARY** concern is that the data should be:

A. sanitized.
B. complete.
C. representative.
D. current.

*See answers and explanations to the practice questions at the end of the chapter. (page 274)*

## ELEMENTS OF A SOFTWARE TESTING PROCESS

To guide the testing process and help ensure that all facets of the system function as expected, basic elements for application software testing activities have been defined and include:

- **Test plan**—Developed early in the life cycle and refined until the actual testing phase, test plans identify the specific portions of the system to be tested. Test plans may include a categorization of types of deficiencies that can be found during the test. Categories of such deficiencies may be system defects, incomplete requirements, designs, specifications, or errors in the test case itself. Test plans also specify severity levels of problems found as well as guidelines on identifying the business priority. The tester determines the severity of the problem found during testing. Based on the severity level, the problem may be fixed or remain. Often, cosmetic problems with the interface are classified as a lower severity and may not be fixed if time constraints become an issue for the project manager. This would be an example of a low severity, low priority defect. However, a defect in a report output such as spelling, look and feel, or specific content may also be a low severity defect for the tester, but a high priority defect for the business and thus would warrant resolution within the time constraints of the project. The project sponsor, end-user management and the project manager decide early in the test phase on the severity definitions. Test plans also identify test approaches, such as the two reciprocal approaches, to software testing:
  - Bottom up—Begin testing of atomic units, such as programs or modules, and work upward until a complete system testing has taken place. The advantages are:
    - No need for stubs or drivers
    - Can be started before all programs are complete
    - Errors in critical modules are found early
  - Top down—Follow the opposite path, either in depth-first or breadth-first search order. The advantages are:
    - Tests of major functions and processing are conducted early
    - Interface errors can be detected sooner
    - Confidence is raised in the system since programmers and users actually see a working system

Generally, most application testing of large systems follows a bottom-up testing approach that involves several levels of testing (e.g., unit or program, subsystem/integration, system, etc.):

- **Conduct and report test results**—Describe resources needed for testing, including personnel involved and information resources/facilities used during the test as well as actual vs. expected test results. Results reported, along with the test plan, should be retained as part of the system's permanent documentation.
- **Address outstanding issues**—Identify errors and irregularities from the actual tests conducted. When such problems occur, the specific tests in question have to be redesigned in the test plan until acceptable conditions occur when the tests are redone.

## TESTING CLASSIFICATIONS

The following tests relate, to varying degrees, to the above approaches that can be performed based on the size and complexity of the modified system:

- **Unit testing**—The testing of an individual program or module. Unit testing uses a set of test cases that focus on the control structure of the procedural design. These tests ensure that the internal operation of the program performs according to specification.
- **Interface or integration testing**—A hardware or software test that evaluates the connection of two or more components that pass information from one area to another. The objective is to take unit-tested modules and build an integrated structure dictated by design. The term integration testing is also used to refer to tests that verify and validate the functioning of the application under test with other systems, where a set of data is transferred from one system to another.

- **System testing**—A series of tests designed to ensure that modified programs, objects, database schema, etc., which collectively constitute a new or modified system, function properly. These test procedures are often performed in a nonproduction test/development environment by software developers designated as a test team. The following specific analyses may be carried out during system testing:
  - Recovery testing—Checking the system's ability to recover after a software or hardware failure
  - Security testing—Making sure the modified/new system includes provisions for appropriate access controls and does not introduce any security holes that might compromise other systems
  - Stress/volume testing—Testing an application with large quantities of data to evaluate its performance during peak hours
  - Volume testing—Studying the impact on the application by testing with an incremental volume of records to determine the maximum volume of records (data) that the application can process
  - Stress testing—Studying the impact on the application by testing with an incremental number of concurrent users/services on the application to determine the maximum number of concurrent users/services the application can process
  - Performance testing—Comparing the system's performance to other equivalent systems using well-defined benchmarks
- **Final acceptance testing**—After the system staff is satisfied with their initial and/or system tests, the modified system is ready for the acceptance testing, which occurs during the implementation phase. During this testing phase, the defined methods of testing to apply should be incorporated into the organization's QA methodology. QA activities should proactively encourage that adequate levels of testing be performed on all software development projects. Final acceptance testing has two major parts: quality assurance testing (QAT) focusing on technical aspects of the application, and user acceptance testing (UAT) focusing on functional aspect of the application. QAT and UAT have different objectives and therefore should not be combined.

In the case of acquired software, after attending to the changes during testing by the vendor, the accepted version should be controlled and used for implementation. In the absence of controls, the risk of introducing malicious patches/Trojan horse programs is very high.

QAT focuses on the documented specifications and the technology employed. It verifies that the application works as documented by testing the logical design and the technology itself. It also ensures that the application meets the documented technical specifications and deliverables. QAT is performed primarily by the IS department. The participation of the end user is minimal and on request. QAT does not focus on functionality testing.

> **Note:** The CISA candidate should be familiar with the need for coding standards and with details on QA activities, Software QA plan and the Application QA function.

UAT supports the process of ensuring that the system is production-ready and satisfies all documented requirements. The methods include:
- Definition of test strategies and procedures
- Design of test cases and scenarios
- Execution of the tests
- Utilization of the results to verify system readiness

Acceptance criteria are defined criteria that a deliverable must meet to satisfy the predefined needs of the user. A UAT plan must be documented for the final test of the completed system. The tests are written from a user's perspective and should test the system in a manner as close to production as possible. For example, tests may be based around typical predefined, business process scenarios. If new business processes have been developed to accommodate the new or modified system they should also be tested at this point. A key aspect of testing should also include testers seeking to verify that supporting processes integrate into the application in an acceptable fashion. Successful completion would generally enable a project team to hand over a complete integrated package of application and supporting procedures.

Ideally, UAT should be performed in a secure testing or staging environment. A secure testing environment where both source and executable code are protected helps to ensure that unauthorized or last-minute changes are not made to the system without going through the standard system maintenance process. The nature and extent of the tests will be dependent on the magnitude and complexity of the system change.

# Systems and Infrastructure Life Cycle Management

Even though packaged systems are tested by the vendor prior to distribution, these systems and any subsequent changes should be tested thoroughly by the end user and the system maintenance staff. These supplemental tests will help ensure that programs function as designed by the vendor and the changes do not interact adversely with existing systems.

Many organizations rely on integrated test facilities. Test data usually are processed in production-like systems. This confirms the behavior of the new application or modules in real-life conditions. These conditions include peak volume and other resource-related constraints. In this environment, IS will perform their tests with a set of fictitious data whereas client representatives use extracts of production data to cover the most possible scenarios as well as some made-up data for scenarios that would not be tested by the production data. In some organizations where a subset of production data is used in a test environment, such production data may be obfuscated to scramble the data such that the confidential nature of data is obscured from the tester. This is often the case where the acceptance testing is done by team members who, under usual circumstances, would not have access to such production data.

On completion of acceptance testing, the final step is usually a certification and accreditation process. This process includes evaluating program documentation and testing effectiveness. The process will result in a final decision for deploying the business application system. For information security issues, the evaluation process includes reviewing security plans, the risk assessments performed and test plans, and the evaluation process results in an assessment of the effectiveness of the security controls and processes to be deployed. Generally involving security staff and the business owner of the application, this process provides some degree of accountability to the business owner regarding the state of the system that he/she will accept for deployment.

> **Note:** The CISA candidate should be familiar with details on independent certification, evaluating confidentiality for security certification and benchmarking.

When the tests are completed, the IS auditor should issue an opinion to management as to whether the system meets the business requirements, has implemented appropriate controls, and is ready to be migrated to production. This report should specify the deficiencies in the system that need to be corrected and should identify and explain the risk(s) that the organization is taking by implementing the new system.

| PRACTICE QUESTION |
|---|
| 3-2 An IS auditor is performing a project review to identify whether a new application has met business objectives. Which of the following test reports offers the **MOST** assurance that business objectives are met? |
|    A.   User acceptance |
|    B.   Performance |
|    C.   Sociability |
|    D.   Penetration |
| *See answers and explanations to the practice questions at the end of the chapter. (page 274)* |

## OTHER TYPES OF TESTING

Other types of testing include:

- **Alpha and beta testing**—A very early version of the application system (or software product) that may not contain all of the features that are planned for the final version. Typically, software goes through two stages of testing before it is considered finished. The first stage, called alpha testing, is often performed only by users within the organization developing the software (i.e., systems testing). The second stage, called beta testing, a form of user acceptance testing, generally involves a limited number of external users. Beta testing is the last stage of testing, and normally involves sending the product to beta test sites outside the development environment for real-world exposure. Beta testing is often preceded by a round of testing called alpha testing.
- **Pilot testing**—A preliminary test that focuses on specific and predetermined aspects of a system. It is not meant to replace other testing methods but rather to provide a limited evaluation of the system.
- **White box testing**—Assesses the effectiveness of software program logic. Specifically, test data are used in determining procedural accuracy or conditions of a program's specific logic paths (i.e., applicable to unit and integration testing). However, testing all possible logic paths in large information systems is not feasible and cost-prohibitive, and therefore is used on a select basis only.

- **Black box testing**—An integrity-based form of testing associated with testing components of an information system's "functional" operating effectiveness without regard to any specific internal program structure. Applicable to integration (interface) and user acceptance testing processes.
- **Function/validation testing**—It is similar to system testing but is often used to test the functionality of the system against the detailed requirements to ensure that the software that has been built is traceable to customer requirements (i.e., Are we building the right product?).
- **Regression testing**—The process of rerunning a portion of a test scenario or test plan to ensure that changes or corrections have not introduced new errors. The data used in regression testing should be the same as the data used in the original.
- **Parallel testing**—This is the process of feeding test data into two systems—the modified system and an alternative system (possibly the original system)—and comparing the results.
- **Sociability testing**—The purpose of these tests is to confirm that the new or modified system can operate in its target environment without adversely impacting existing systems. This should cover not only the platform that will perform primary application processing and interfaces with other systems but, in a client server or web development, changes to the desktop environment. Multiple applications may run on the user's desktop, potentially simultaneously, so it is important to test the impact of installing new dynamic link libraries (DLLs), making operating system registry or configuration file modifications, and possibly extra memory utilization.

### AUTOMATED APPLICATION TESTING
Automated testing techniques are used in this process. For example, test data generators can be used to systematically generate random data that can be used to test programs. The generators work by using the field characteristics, layout and values of the data. In addition to test data generators, there are interactive debugging aids and code logic analyzers available to assist in the testing activities.

## *Implementation*
During the implementation phase, the actual operation of the new information system is established and tested. Final UAT is conducted in this environment. The system may also go through a certification and accreditation process to assess the effectiveness of the business application at mitigating risks to an appropriate level, and provide management accountability over the effectiveness of the system in meeting its intended objectives and establishing an appropriate level of internal control.

After a successful full-system testing, the system is ready to migrate to the production environment. The programs have been tested and refined; program procedures and production schedules are in place; all necessary data have been successfully converted and loaded into the new system; and the users have developed procedures and been fully trained in the use of the new system. A date (or dates) for system migration is determined and production turnover takes place.

Planning for the implementation should commence well in advance of the actual implementation date, and a formal implementation plan should be constructed in the design phase and revised accordingly as development progresses. Each step in setting up the production environment should be stipulated, including who will be responsible, how the step will be verified and the backout procedure if problems are experienced. If the new system will interface with other systems or is distributed across multiple platforms, some final commissioning tests of the production environment may be desirable to prove end-to-end connectivity. If such tests are run, care will be needed to ensure test transactions do not remain in production databases or files.

In the case of acquired software, the implementation project should be coordinated by user management with the help of IS management, if required. The total process should not be delegated to the vendor—to avoid possible unauthorized changes or introduction of malicious code by the vendor's employees/representatives.

Once operations are established the next step is to perform site acceptance testing, which is a full-system test conducted on the actual operations environment. UAT supports the process of ensuring that the system is production-ready and satisfies all documented requirements. Details on UAT methods and procedures are discussed earlier in the Testing Phase section of this chapter.

## IMPLEMENTATION PLANNING

Once developed and ready for operation, the new system delivered by the project will need an efficient support structure. It is not enough to just set up a support structure defining some roles and naming some people to fulfill these roles. These people will need to acquire new skills. Workload has to be distributed, in order for the right people to support the right issues, thus new processes have to be developed while respecting the specificities of IT department requirements. Additionally, an infrastructure dedicated for support staff has to be made available.

For these and other reasons, setting up a support structure normally is a project in itself and requires planning, a methodology and best practices adaptation from past experiences.

The objective of such a project is to develop and establish the to-be support structure for the new technical infrastructure.

The main goals are to:
• Provide appropriate support structures for first-, second- and third-line support teams
• Provide a single point of contact (SPOC)
• Provide roles and skills definitions with applicable training plans

Often the project sponsor's company operates and supports a legacy solution and will implement a new system environment based on new system architecture. The existing support procedures and the organizational units will have to maintain the future system to provide the appropriate level of support for the new platform as well as for the old one.

One of the major challenges, therefore, is to manage the phases from build to integrate, to migrate, and to transfer for the existing system and the new system. The reason for this is that the migration normally is not a big-bang scenario, but a step-by-step transition of the affected services. Further, the implemented processes for a legacy environment might be different from what may be implemented with the new platform and must be considered in the field of people change management.

To achieve significant improvement of the situation, it is necessary to address some important questions such as:
• How can the existing support staff be involved in the setup of the new project without neglecting the currently running system?
• What is the gap of "know-how" that must be addressed in the training plan?
• How big is the difference from the current legacy environment operation to the operation of the new platform?

Such a project shall include the following requirements and expectations:
• There should be a smooth transition from the existing platform to the new platform, without any negative effect on users of the system.
• There should be maximum employment of the existing support staff to operate the new system environment and keep the effort of new hires at a minimum level.

### Phase 1—Develop To-be Support Structures
Phase 1 includes the development of:
*Gap Analysis*
One possible approach to determine the gap—the differences between the current support organization and the future one—is to schedule workshops with the appropriate staff members who are dealing with system operational tasks or support processes at present. This should also include representatives of the current help desk unit.

The gap analysis should be based on the results of those workshops and on the data that can be gathered from self-assessment where representatives of all present support units will take part. The fields of the gap analysis should include processes, skills, tools, headcount, services of the help desk and interfaces to other organizational units.

*Role Definitions*
The role definitions for the required roles have to be provided in detail.

# Systems and Infrastructure Life Cycle Management

**Phase 2—Establish Support Functions**

Phase 2 includes the development of:

*Service Level Agreement*

The SLA should at least consider the following items:
- Operating time
- Support time
- Mean time between failures (MTBF)
- Mean time to repair (MTTR)
- Response time (technical)

At least these basic key attributes must be measurable with a reasonable effort.

*Implementation Plan/Knowledge Transfer Plan*

Due to best practices, the transfer should follow the shadowing and relay-baton method. This approach is the best suitable concept to transfer knowledge, but also to transfer responsibility in a transparent way. The metaphor of the relay-baton expresses exactly what must be achieved.

*Training Plans*

After the roles and responsibilities are defined, they will be documented in the form of a chart to achieve a clear and easy-to-read overview.

The training plans for the staff should show all of the required training in terms of:
- Content
- Scheduling information
- Duration
- Delivery mechanism (classroom and/or web-based)
- Train-the-trainer concept

The plan should consider the role definitions and skill profiles for the new to-be structure, and the results of the gap analysis. The plan takes into account that the staff who need to be trained must still run the current system, and therefore a detailed coordination with the daily business tasks is maintained.

The following list gives an impression about work tasks defined to fulfill the overall project goal:
- Collate existing support structure documentation.
- Review the existing IT organization model.
- Define the new support organization structure.
- Define the new support processes.
- Map the new process to the organization model.
- Execute the new organization model.
- Establish support functions.
- Develop communications material for support staff.
- Conduct briefing and training sessions.
- Review mobilization progress.
- Transfer to new organization structure.
- Review of items above.

## END-USER TRAINING

The goal of a training plan is to ensure that the end user can become self-sufficient in the operation of the system.

One of the most important keys in end-user training is to ensure that training is considered early in the development process.

The training project plan must start early in the development process. A strategy can be developed that would take into consideration the timing, extent and delivery mechanisms for training the end users. The timing of the delivery of training is very important. If training is delivered too early, the users will forget much of the training by the time the system actually goes into production. If it is delivered too late, there will not be enough time to obtain feedback from the pilot group and implement the necessary changes into the main training program. The training should be piloted using a cross-section of users to determine how best to customize the training to the different user groups. Following the pilot, the training approach can be adjusted as necessary, based on the feedback received from the pilot group. Separate classes should be developed for individuals who will assist in the training process. These train-the-trainer classes also provide useful feedback for improving the content of the training program.

Training classes should be customized to address skill level and needs of users based on their role within the organization.

To create the training strategy, the organization must name a training administrator who should identify users who need to be trained with respect to their specific job functions. Consideration should be given to the following format and delivery mechanisms:
• Case studies
• Role-based training
• Lecture and breakout sessions
• Modules at different experience levels
• Practical sessions on how to use the system
• Remedial computer training (if needed)
• Online sessions on the web or on a CD-ROM

It is important to have a library of cases or tests, including user errors and the system response to those errors.

The training administrator needs to record student information in a database or spreadsheet, including student feedback for improving the training course.

## DATA CONVERSION

A data conversion (also known as data porting) is required if the source and target systems utilize different field formats or sizes, file/database structures, or coding schemes. For example, a number may be stored as text, floating point, or as binary-coded-decimal. Another example is it the colors red, green, and blue are represented using the codes "R", "G" and "B" on the existing system and "RD", GN" and "BL" in the new system. Note that this conversion would also address a change in the field size. Conversions are often necessary when the source and target systems are on different hardware and/or operating system platforms, and where different file or database structures (e.g., relational database, flat files, VSAM, etc.) are used. Another example of different coding schemes requiring conversion would be the use of different character representation schemes on the two systems (e.g. ASCII versus EBCDIC).

Existing data may be converted in the existing operating environment and then transmitted to the new system, transmitted to a converted third-party system (e.g., a vendor's system) and, then transferred to the new system, or transferred and then converted on the new system. Conversions may be performed with existing system utilities (on either system), vendor tools, or one-time-use applications specifically written for the conversion project.

The objective of data conversion is to convert existing data into the new required format, coding and structure while preserving the meaning and integrity of the data. The data conversion process must provide some means, such as audit trails and logs, which allow for the verification of the accuracy and completeness of the converted data. This verification of accuracy and completeness may be performed through a combination of manual processes, system utilities, vendor tools, and one-time-use special applications.

A large-scale data conversion can potentially become a project within a project since considerable analysis, design and planning will be required. Among the steps necessary for a successful data conversion are:
• Determining what data should be converted programmatically and what, if any, should be converted manually
• Performing any necessary data cleansing ahead of conversion

# Systems and Infrastructure Life Cycle Management

- Identifying the methods to be used to verify the conversion, such as automated file comparisons, comparing record counts and control totals, accounting balances, and individual data items on a sample basis
- Establishing the parameters for a successful conversion. For example, is 100 percent consistency between the old and new systems necessary, or will some differences within defined ranges be acceptable?
- Scheduling the sequence of conversion tasks
- Designing audit trail reports to document the conversion, including data mappings and transformations
- Designing exception reports that will record any items that cannot be converted automatically
- Establishing responsibility for verifying and signing off on individual conversion steps and accepting the overall conversion. Typically, this would be the responsibility of the system owner.
- Developing and testing conversion programs, including functionality and performance
- Performing one or more conversion dress rehearsals to familiarize persons with the sequence of events and their roles, and test the conversion process end-to-end with real data
- Outsourcing the conversion process should be controlled with a proper agreement covering nondisclosure
- Running the actual conversion with all necessary personnel onsite, or, at least, able to be contacted

A successful data conversion delivers the new system on time, on budget, and with the required quality. Additionally, a project should carefully planned and utilize appropriate methodologies and tools to minimize the risk of:
- the disruption of routine operations
- violation of the security and confidentiality of data
- conflicts and contention between legacy and migrated operations
- data inconsistencies and loss of data integrity during the migration process

In large or complex environments, it is best if the implementation of the future architecture follows an evolutionary model with multiple releases. Each release could contain one or more existing systems and its components could coexist with existing systems during the migration.

The data model and the new application model should be stored in an enterprise repository. Using a repository allows a simulation of the migration scenario and promotes information traceability during the project. An enterprise repository enables an overview of the reengineering and data migration process (e.g., which modules and entities are in which stage such as in service or already migrated). These models will be managed due to the processes described in the following sections.

**Refining the Migration Scenario**
In order to determine the scope of the implementation project, module analysis should be undertaken to identify the affected functional modules and data entities. The plan of the implementation project should be refined based on this information and an analysis of business requirements.

The next step is to develop a migration screenplay. This is a detailed plan (listing of tasks to be executed) for the production deployment of the new system. Within this plan decision points are defined to make "go" or "no-go" decisions. The following processes require decision points:
- Support migration process—A support process to administer the enterprise repository must be implemented. Since this repository should be used on completion of the project to manage the software components of the new architecture, this process should be capable of supporting future development processes. The enterprise repository administration and report generation supports the migration by supporting the reverse-engineering of changes in the legacy architecture and facilitating the creation of impact analysis reports.
- Migration infrastructure—The project develops specifications for the infrastructure of the migration project. This approach ensures consistency and increases confidence in the functionality of the fallback scenario. The migration project team completes a high-level analysis of the legacy and new data models to establish links between them that will be refined later. The migration infrastructure is the basis for specifying the following components:
  - Data redirector (temporary adapters)—Best practices suggest the staged deployment of applications to minimize the end-user impact of their implementation and limit the risk by having a fallback scenario with minimum impact. For this reason, an infrastructure component is needed to handle distributed data on different platforms within distributed applications. The design of a data redirector on the new architecture corresponds to service-oriented architectures and should cover features such as access to the not-yet-migrated legacy data during run time (e.g., extended binary-coded

decimal interchange code [EBCDIC]—American standard code for information interchange [ASCII] conversion), data consistency due to the usage of standards such as X/Open XA interface, and a homogeneous new architecture.
- Data conversion components—The need to create an enterprise data model to eliminate data redundancies and inconsistencies often is identified. For this reason, infrastructure components to transform the legacy data model to the new data model must be provided. These components can be described as follows:
- Unload components, to execute the unload of the data off the legacy database
- Transfer components, to execute the data transfer from the legacy system to the new system
- Load components, to execute the load of the data into the new database

The acquisition of software packages such as ERP and document management should be carried out as soon as the software evaluation is done. The data conversion plan should be based on the available databases and migration tools provided by the selected vendor(s).

### Fallback (Rollback) Scenario
Not all new system deployments go as planned. To mitigate the risk of downtime for mission-critical systems, best practices dictate that the tools and applications required to reverse the migration are available prior to attempting the production cutover. Some or all of these tools and applications may need to be developed as part of the project.

Components have to be delivered that can back out all changes and restore data to the original applications in the case of nonfunctioning new applications. Two types of components should be considered as part of a fallback contingency plan:
1. The first consists of unload components to execute the unloading of the data from the new data structures, transfer components for the data conversion, and load components to execute the loading of the data into the legacy data structures.
2. The second consists of a log component to log the data modifications within the new data model during runtime within the service layer, transfer components for the data conversion, and load components to execute the load of the data into the legacy data structures.

The decision on which method to use for data conversion has to be made as part of the implementation project and should be based on the following criteria:
• Transaction volume
• Change degree of the data model

In summary, data conversion projects begin with planning and preparation. The first step is to understand the new system's data structure. This is done by reading the software user guides, analyzing the entity relationship diagrams, understanding the relationships between data elements and reviewing definitions of key terms (such as entity and record) in the new system.

The next step in data conversion is to review the decisions on how business processes should be conducted in the new system. Changes are identified and the output of this exercise is a translation of what the new data terminology is compared to the current definitions of data. In this step, the project team identifies how current data are defined in the new system. Following this step, a data cleanup is completed to eliminate inconsistencies in the current database, if possible, and duplications of data sets, if applicable. The rules of translation are defined and documented with the objective of ensuring the business processes executed in the new system yield results that maintain data integrity and relationships.

Data conversion rules are programmed by the software development team. Data conversion scripts are created to convert the data from the old database to the new database. The data conversion scripts are tested on a discreet selection of data. This is referred to as program or unit testing. Following the sign-off of data conversion scripts by programmers, the scripts are run on a test copy of the production database. The values of data are tested by executing tests including business process tests. Users and developers complete cycles of testing until conversion scripts are fine-tuned. After testing has been completed, the next step is to promote the converted database to production. Data conversion can also be done manually, and is often done so, if the conversion of data requires interpretation or if the conditions or data conversion rules are too complex for programming.

The key points to be taken into consideration in a data conversion project are to ensure:
• Completeness of data conversion—i.e., the total number of records from the source database is transferred to the new database (assuming the number of fields is the same)

# Systems and Infrastructure
# Life Cycle Management

- Integrity of data—i.e., the data are not altered manually, mechanically or electronically by person, program, substitution or overwriting in the new system. Integrity thus, includes error creep due to transposition, transcription, transrecord, transfield, transfile, translibrary, etc.
- Confidentiality of data under conversion—i.e., data are backed up before conversion for future reference or any emergency that might arise out of data conversion program management. Unauthorized copy or too many copies can lead to misuse, abuse or theft of data from the system.
- Consistency of data—i.e., the field/record called for from the new application should be consistent with that of the original application. This should enable consistency in repeatability of the testing exercise.
- Continuity—i.e., the new application should be able to continue with newer records as addition (append) and help in ensuring seamless business continuity.
- The last copy of the data before conversion from the old platform and the first copy of the data after conversion to the new platform should be maintained separately in the archive for any future reference

## CHANGEOVER (GO-LIVE OR CUTOVER) TECHNIQUES

Changeover refers to an approach to shift users from using the application from the existing (old) system to the replacing (new) system. This is possible only after testing the new system with respect to its program and relevant data. This is sometimes called the go-live technique since it enables the start of the new system. This approach is also called the cutover technique since it helps in cutting out from the older system and moving over to the newer system.

This technique can be achieved in three different ways. See **exhibits 3.11**, **3.12** and **3.13**.

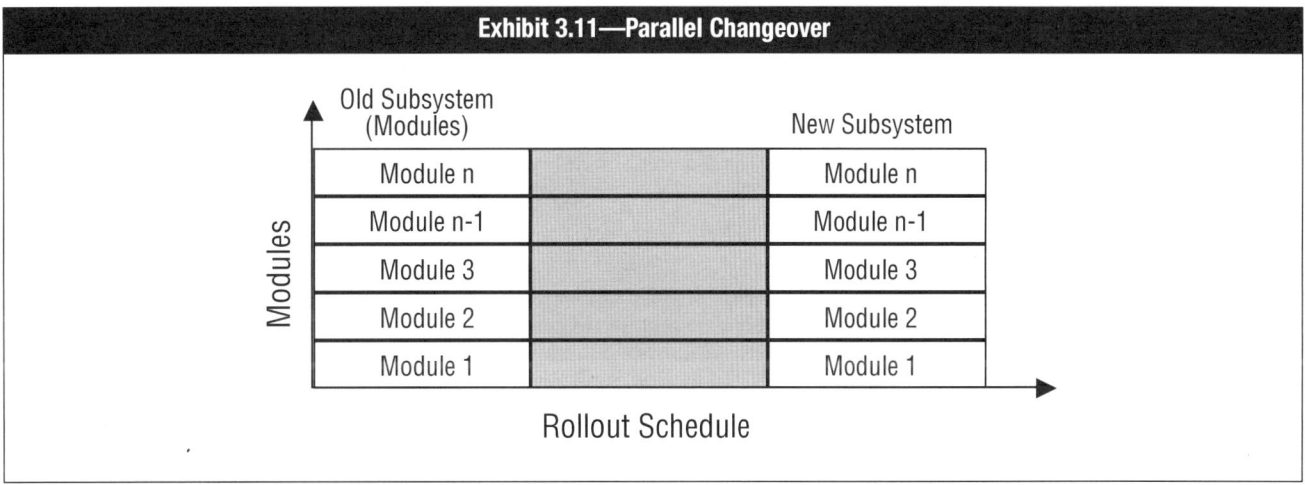

**Exhibit 3.11—Parallel Changeover**

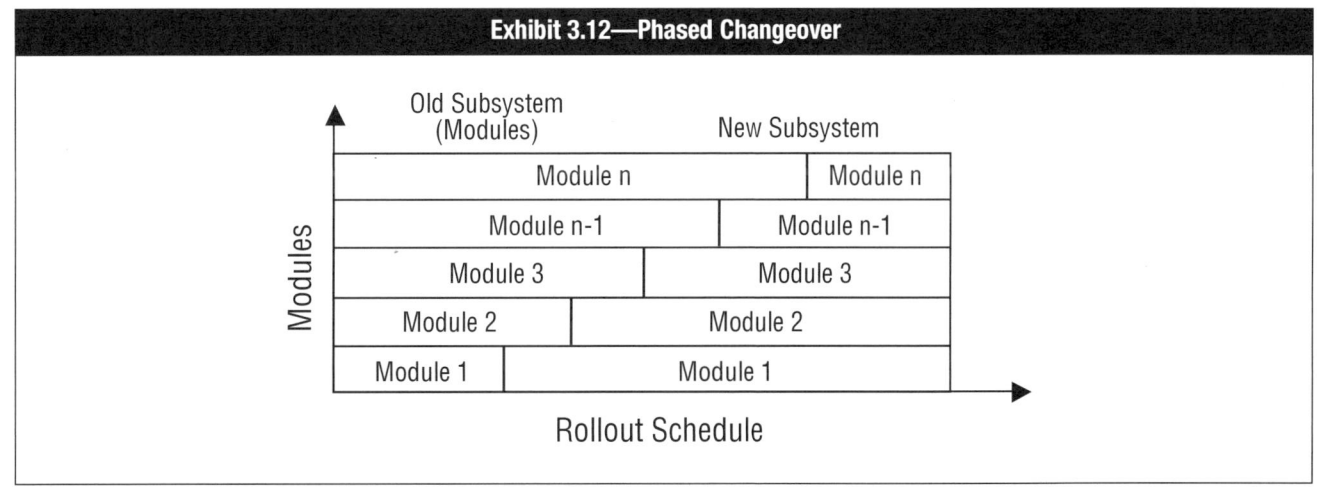

**Exhibit 3.12—Phased Changeover**

# Systems and Infrastructure Life Cycle Management

| Exhibit 3.13—Abrupt Changeover |
|---|

Old Subsystem (Modules) — New Subsystem

| Module n | Module n |
|---|---|
| Module n-1 | Module n-1 |
| Module 3 | Module 3 |
| Module 2 | Module 2 |
| Module 1 | Module 1 |

Modules (vertical axis) — Rollout Schedule (horizontal axis)

## Parallel Changeover

This technique includes running the old system, then running both the old and new systems in parallel, and finally fully changing over to the new system after gaining confidence in the working of the new system. With this approach the users will have to use both systems during the period of overlap. This will minimize the risk of using the newer system, and at the same time help in identifying problems, issues or any concerns that the user comes across in the newer system in the beginning. After a period of overlap the user gains confidence and assurance in relying on the newer system. At this point, the use of the older system is discontinued and the new system becomes totally operational.

## Phased Changeover

In this approach the older system is broken into deliverable modules. Initially, the first module of the older system is phased out using the first module of the newer system. Then, the second module of the older system is phased out, using the second module of the newer system and so forth until reaching the last module. Thus, the changeover from the older system to the newer system takes place in a preplanned, phased manner.

Some of the risk areas that may exist in the phased changeover include:
• Resource challenges (both on the IT side—to be able to maintain two unique environments such as hardware, operating systems, databases and code; and on the operations side—to be able to maintain user guides, procedures and policies, definitions of system terms, etc.)
• Extension of the project life cycle to cover two systems
• Change management for requirements and customizations to maintain ongoing support of the older system

## Abrupt Changeover

In this approach the newer system is changed over from the older system on a cutoff date and time, and the older system is discontinued once changeover to the new system takes place.

Changeover to the newer system involves four major steps or activities:
1. Conversion of files and programs; test running on test bed
2. Installation of new hardware, operating system, application system and the migrated data
3. Training employees or users in groups
4. Scheduling operations and test running for go-live or changeover

Some of the risk areas that are possible to exist in the changeover include:
• Asset safeguarding
• Data integrity
• System effectiveness
• System efficiency
• Change management challenges (depending on the configuration items considered)
• Records being the same (duplicate or erroneous records may exist if data cleansing is not done)

# Systems and Infrastructure Life Cycle Management

## CERTIFICATION/ACCREDITATION

Certification is a process by which an assessor organization, performs comprehensive assessment of management, operational, and technical controls in an information system. The assessor examines the level of compliance in meeting certain requirements such as standards, policies, processes, procedures, work instructions and guidelines—requirements made in support of accreditation. The goal is to determine the extent to which controls are implemented correctly, operating as intended, and producing the desired outcome with respect to meeting the system's security requirements. The results of a certification are used to reassess the risks and update the system security plan, thus providing the factual basis for an authorizing official to render an accreditation decision.

Accreditation is the official management decision (given by a senior official) to authorize operation of an information system and to explicitly accept the risk to the organization's operations, assets, or individuals based on the implementation of an agreed-u pon set of requirements and security controls. Security accreditation provides a form of quality control and challenges managers and technical staff at all levels to implement the most effective security controls possible in an information system, given mission requirements, and technical, operational, and cost/schedule constraints.

By accrediting an information system, a senior official accepts responsibility for the security of the system and is fully accountable for any adverse impact to the organization if a breach of security occurs. Thus, responsibility and accountability are core principles that characterize accreditation.

> **Note:** The CISA candidate should be familiar with the auditor's role in the certification process.

## *Postimplementation Review*

Following the successful implementation of a new or extensively modified system, it is beneficial to verify the system has been properly designed and developed, and proper controls have been built into the system. A postimplementation review should meet the following objectives:
• Assess the adequacy of the system:
  – Does the system meet user requirements and business objectives?
  – Have access controls been adequately defined and implemented?
• Evaluate the projected cost benefits or ROI measurements.
• Develop recommendations that address the system's inadequacies and deficiencies.
• Develop a plan for implementing the recommendations.
• Assess the development project process:
  – Were the chosen methodologies, standards and techniques followed?
  – Were appropriate project management techniques used?

It is important to note that, for a postimplementation review to be effective, the information to be reviewed should be identified during the project startup phase, and collected during each stage of the project. For instance, the project manager might establish certain checkpoints to measure effectiveness of software development processes and accuracy of software estimates during the project execution. Business measurements should also be established up front and collected before the project begins and after the project is implemented (for examples of such measurements see **exhibit 3.9**).

It is also important to allow a sufficient number of business cycles to be executed in the new system to realize the new system's actual return on investment.

A postproject review should be performed jointly by the project development team and appropriate end users. Typically, the focus of this type of internal review is to assess and critique the project process, whereas a postimplementation review has the objective of assessing and measuring the value the project has on the business (business realization).

Alternatively, an independent group not associated with the project implementation (internal or external audit) can perform a postimplementation review. The IS auditors performing this review should be independent of the system development process. Therefore, IS auditors involved in consulting with the project team on the development of the system should not perform this review. Unlike internal project team reviews, postimplementation reviews performed by IS auditors have a tendency to concentrate on the control aspects of the system development and implementation processes.

It is important that all audit involvement in the development project be thoroughly documented in the audit workpapers to support the IS auditor's findings and recommendations. This audit report and documentation should be reused during maintenance and changes to validate, verify and test the impact of any changes made to the system. The system should periodically undergo a review to ensure the system is continuing to meet business objectives in a cost-effective manner and that control integrity still exists.

> **Note:** The CISA candidate should be familiar with issues related to dual authorization within the postimplementation review and with those related to reviewing results of live processing.

## 3.5.4 RISKS ASSOCIATED WITH SOFTWARE DEVELOPMENT

There are many potential risks that can occur when designing and developing software systems.

One type of risk is business risk relating to the likelihood that the new system may not meet the users' business needs, requirements and expectations. For example, the business requirements that were to be addressed by the new system are still unfulfilled, and the process has been a waste of resources. In such a case, even if the system is implemented, it will most likely be underutilized and not maintained, making it obsolete in a short period of time.

Another risk is project risk, where the project activities to design and develop the system exceed the limits of the financial resources set aside for the project and, as a result, it may be completed late, if ever. There are many potential risks that can occur when designing and developing software systems. Software project risks exist at multiple levels:
• Within the project—e.g., risks associated with not identifying the right requirements to deal with the business problem or opportunity that the system is meant to address, and not managing the project to deliver within time and cost constraints
• With suppliers—e.g., failure to clearly communicate requirements and expectations, resulting in suppliers delivering late, at over expected cost and/or with deficient quality
• Within the organization—e.g., stakeholders not providing needed inputs or committing resources to the project, and changing organizational priorities and politics
• With the external environment—e.g., impacts on the projects caused by the actions and changing preferences of customers, competitors, government/regulators and economic conditions

The foremost cause of these problems is a lack of discipline in managing the software development process such as the use of a methodology inappropriate to the system being developed. In such instances, organizations are not providing the infrastructure and support necessary to help projects avoid these problems. Success on projects, if occurring, is not repeatable, and SDLC activities are not defined and followed adequately. However, with effective management, SDLC management activities can be controlled, measured and improved.

The IS auditor should be aware that merely following an SDLC management approach does not ensure the successful completion of a development project. The IS auditor should also review the management discipline over a project related to the following:
• The project meeting cooperative goals and objectives
• Project planning, including effective estimates of resources and time
• Control over scope creep where a software baseline does not exist—meaning that requirements can be added into the software design and the development process is uncontrolled
• Management tracking over software design and development activities
• Senior management support provided to the software project's design and development efforts
• Periodic review and risk analysis in each project phase

(See also the Process Improvement Practices section later in this chapter.)

## 3.5.5 USE OF STRUCTURED ANALYSIS, DESIGN AND DEVELOPMENT TECHNIQUES

Closely associated with the traditional, classic SDLC approach is the use of structured analysis, design and development techniques. These techniques provide a framework for representing the data and process components of an application using various graphical notations at different levels of abstraction, until it reaches the abstraction level that enables programmers to code the system. Early on, for example, the following activities occur in defining the requirements for a new system:
• Develop system context diagrams (e.g., high-level business process flow schema).
• Perform hierarchical data flow/control flow decomposition.
• Develop control transformations.
• Develop minispecifications.
• Develop data dictionaries.
• Define all external events—inputs from external environment.
• Define single transformation data flow diagrams (DFDs) from each external event.

The next level provides greater detail for building the system, including developing system flowcharts, inputs/outputs, processing steps and computations, and program and data file or database specifications. It should be noted that representation of functions is developed in a modularized top-down fashion. This enables programmers to systematically develop and test modules in a linear fashion.

Auditors should be particularly concerned with whether the processes under a structured approach are well defined, documented and followed when using the traditional SDLC approach to business application development.

# 3.6 ALTERNATIVE APPLICATION DEVELOPMENT APPROACHES

In the face of increasing system complexity and the need to implement new systems more quickly to achieve benefits before the business changes, software development practitioners have adopted new ways of organizing software projects that vary, or in some cases radically depart from, the traditional waterfall model described above. In addition there has been continued evolution in the thinking about how best to analyze, design and construct software systems and in the information technologies available to perform these activities.

# 3.7 ALTERNATE FORMS OF SOFTWARE PROJECT ORGANIZATION

The following section describes different approaches to organizing a software project, depending on imperatives such as time to delivery, scale of the system, clarity of requirements and maturity of the technology being used.

An IS auditor must understand that the basic steps outlined in the traditional approach exist to some extent on nearly all software development projects. However, the sequencing of steps, the number of times they will be repeated, their duration and the methods employed vary considerably from project to project. It is also true that the thinking on how best to organize a software project has evolved over time as the complexity of systems has increased.

# Systems and Infrastructure Life Cycle Management

Given these factors, other approaches an IS auditor may encounter include:
- **Incremental or progressive development**—The system is built in stages or releases rather than being delivered in its entirety in one implementation. The separate releases are often still large undertakings that operate as discrete subprojects. The usual practice is to deliver the basic system architecture in the first release. Subsequent releases expand the system in terms of functionality, range of users or usage locations.
- **Iterative development**—This involves building the system in iterations or increments, with feedback occurring after each increment to facilitate any necessary adjustment of project plans and software development products. Iterative development is now commonly regarded as best practice. Iterative development is considered the most appropriate way of dealing with the complexities and associated risks found in contemporary software development projects. Sources of complexity include:
  - Range and scope of systems now developed—e.g., e-business, customer relationship management (CRM), supply-chain integration, online transaction processing and online analytical processing
  - Rate of business change, which increases requirements instability
  - Diverse architectural decisions that need to be made—e.g., whether the system should have a graphical user interface (GUI) or browser interface; which firewall to use; which web server to use; whether to use an application server or other integration/control middleware; which DBMS to use; which communication and data definition protocols to use; whether to organize the system in terms of subsystems and modules, classes and objects, or employ a component architecture; and where should business logic reside?
  - A need to integrate new systems with legacy systems in various cases

Within the overall category of an iterative life cycle there are a number of variants. These include:
- **Evolutionary development**—Prototyping is used to build a working model that is used to elicit/verify requirements and explore design issues. Eventually the prototype is hardened so it can be implemented into production, or perhaps the system is recoded based on learning from the prototype.
- **Spiral development**—A series of prototypes is used to develop a solution to the point of detailed design, build and test. The solution spirals out from the initial limited prototype to become progressively more expansive and detailed. Formal risk analysis precedes each prototype and the various prototypes (depending on the iteration reached) form the basis for producing software development products, including the requirement's specification, system design and test plans.
- **Agile development**—The project is broken down into relatively short, time-boxed iterations. From the earliest iterations the emphasis is to produce actual working functionality, although software release may not always coincide with completion of an increment. In early increments, a tracer-bullet approach may be used in which a piece of functionality is developed in a manner consistent with the intended development approach and architecture. The functionality extends from the user interface, through intermediate layers, to stored data and back. Once confidence is obtained in the intended architecture, the focus in later iterations shifts to delivering the functionality with the greatest return on investment.

## 3.7.1 AGILE DEVELOPMENT

The term "agile development" refers to a family of similar development processes that espouse a nontraditional way of developing complex systems. One of the first agile processes emerged in the early 1990s—Scrum.

Scrum aims to move planning and directing tasks from the project manager to the team, leaving the project manager to work on removing the obstacles to the team, achieving their objectives. Scrum is a project management approach that fits well with other agile techniques. Other agile processes have since emerged such as Extreme Programming (XP), Crystal, Adaptive Software Development, Feature Driven Development, and Dynamic Systems Development Method. These processes are termed "agile" because they are designed to flexibly handle changes to the system being developed or the project that is performing the development.

Agile development processes have a number of common characteristics:
- The use of small, time-boxed subprojects or iterations, as shown in **exhibit 3.14**. In this instance, each iteration forms the basis for planning the next iteration.
- Replanning the project at the end of each iteration (referred to as a "sprint" in Scrum), including reprioritizing requirements, identifying any new requirements and determining within which release delivered functionality should be implemented

# Systems and Infrastructure Life Cycle Management

• Relatively greater reliance, compared to traditional methods, on tacit knowledge—the knowledge in people's heads—as opposed to external knowledge that is captured in project documentation

• A heavy influence on mechanisms to effectively disseminate tacit knowledge and promote teamwork. Therefore, teams are kept small in size, comprise both business and technical representatives, and are located physically together. Team meetings to verbally discuss progress and issues occur daily, but with strict time limits.

• At least some of the agile methods stipulate pair-wise programming (two persons code the same part of the system) as a means of sharing knowledge and as a quality check.

• A change in the role of the project manager, from one primarily concerned with planning the project, allocating tasks and monitoring progress to that of a facilitator and advocate. Responsibility for planning and control devolves to the team members.

Agile development does not ignore the concerns of traditional software development but approaches them from a different perspective:

• Agile development only plans for the next iteration of development in detail, rather than planning subsequent development phases far out in time.

• Agile development's adaptive approach to requirements does not emphasize managing a requirements baseline.

• Agile development's focus is to quickly prove an architecture by building actual functionality vs. formally defining, early on, software and data architecture in increasingly more detailed models and descriptions.

• Agile development assumes limits to defect testing but attempts to validate functions through a frequent-build test cycle, and correct problems in the next subproject before too much time and cost are incurred.

• Agile development does not emphasize defined and repeatable processes but instead performs and adapts its development based on frequent inspections.

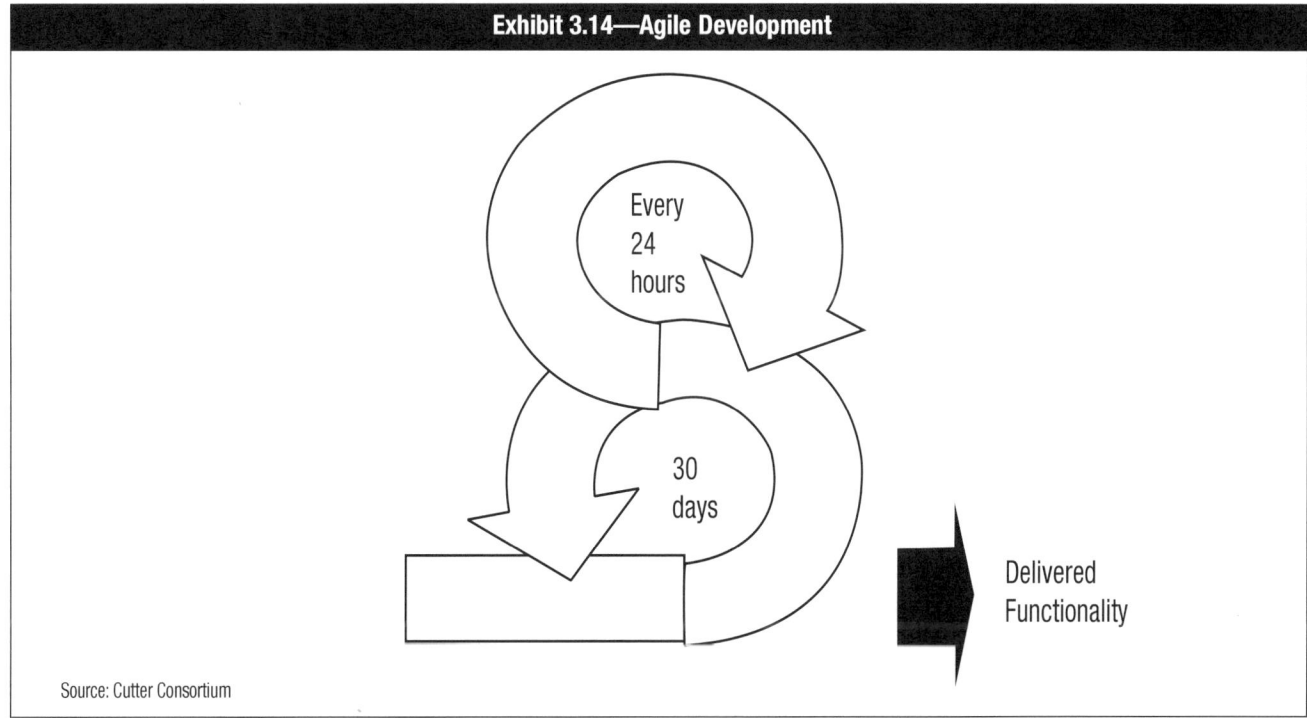

**Exhibit 3.14—Agile Development**

Every 24 hours

30 days

Delivered Functionality

Source: Cutter Consortium

## 3.7.2 PROTOTYPING

Prototyping, also known as heuristic or evolutionary development, is the process of creating a system through controlled trial and error procedures to reduce the level of risks in developing the system. That is, it enables the developer and customer to understand and react to risks at each evolutionary level (using prototyping as a risk reduction mechanism). It combines the best features of classic SDLC by maintaining the systematic stepwise approach but incorporates it into an iterative framework that more realistically reflects the real world.

Moreover, prototyping reduces the time to deploy systems primarily by using faster development tools, such as fourth-generation techniques, that allow a user to see a high-level view of the workings of the proposed system within a short period of time. A typical software development environment that supports fourth-generation techniques may include some or all of the following tools: nonprocedural languages for database query, report generation, data manipulation, screen interaction and definition, high-level graphics capability, and configuration management.

The initial emphasis during the development of the prototype is usually placed on the reports and screens, which are the system aspects most used by the end users. This allows the end user to see a working model of the proposed system within a short time.

There are two basic methods or approaches to prototyping:
1. Build the model to create the design (i.e., the mechanism for defining requirements). Then, based on that model, develop the system design with all the performance, quality and maintenance features needed.
2. Gradually build the actual system that will operate in production using a 4GL that has been determined to be appropriate for the system being built.

The problem with the first approach is that there can be considerable pressure to implement an early prototype. Often, users observing a working model cannot understand why the early prototype has to be refined further. The fact that the prototype has to be expanded to handle transaction volumes, client-server network connectivity, and backup and recovery procedures, and provide for auditability and control is not often understood.

The second approach typically works with small applications using 4GL tools. However, for larger efforts, it is necessary to develop a design strategy for the system even if a 4GL is used. The use of 4GL techniques alone will cause the same difficulties (e.g., poor quality, poor maintainability and low user acceptance) encountered when developing business applications using conventional approaches.

Another overall disadvantage of prototyping is that it often leads to functions or extras being added to the system that are not included in the initial requirements document. All major enhancements beyond the initial requirements document should be reviewed to ensure they meet the strategic needs of the organization and are cost-effective. Otherwise, the final system may be functionally rich but inefficient.

A potential risk with prototyped systems is that the finished system will have poor controls. By focusing mainly on what the user wants and what the user sees, system developers may miss some of the controls that come out of the traditional system development approach such as backup recovery, security and audit trails.

Change control often becomes much more complicated with prototyped systems. Changes in designs and requirements happen so quickly that they are seldom documented or approved, and the system can escalate to a point of not being maintainable.

Although the IS auditor should be aware of the risks associated with prototyping, the IS auditor should also be aware that this method of system development can provide the organization with significant time and cost savings.

## 3.7.3 RAPID APPLICATION DEVELOPMENT

RAD is a methodology that enables organizations to develop strategically important systems quickly while reducing development costs and maintaining quality. This is achieved by using a series of proven application development techniques within a well-defined methodology. These techniques include the use of:
• Small, well-trained development teams
• Evolutionary prototypes
• Integrated power tools that support modeling, prototyping and component reusability
• A central repository
• Interactive requirements and design workshops
• Rigid limits on development time frames

RAD supports the analysis, design, development and implementation of individual application systems. However, RAD does not support the planning or analysis required to define the information needs of the enterprise as a whole or of a major business area of the enterprise. RAD provides a means for developing systems faster while reducing cost and increasing quality. This is done by automating large portions of the SDLC, imposing rigid limits on development time frames and reusing existing components. The RAD methodology has four major stages:

1. The concept definition stage defines the business functions and data subject areas that the system will support and determines the system scope.
2. The functional design stage uses workshops to model the system's data and processes and build a working prototype of critical system components.
3. The development stage completes the construction of the physical database and application system, builds the conversion system, and develops user aids and deployment work plans.
4. The deployment stage includes final-user testing and training, data conversion and the implementation of the application system.

RAD uses prototyping as its core development tool no matter which underlying technology is used. In contrast, object-oriented software development (OOSD) and data-oriented system development (DOSD) use continuously developing models but have a focus on content solution space (e.g., how to best address the problem to make the code reusable and maintainable) and can be applied using a traditional waterfall approach. It should also be noted that BPR attempts to convert an existing business process rather than make dynamic changes.

# 3.8 ALTERNATIVE DEVELOPMENT METHODS

This section describes different techniques of understanding, designing and constructing a software system. The choice of a particular method will be driven by considerations such as organizational policy, developer knowledge and preference, and the technology being used.

> **Note:** The selection of one of the methods described in this section is generally independent of the selection of a project organization model. An object-oriented approach to design and coding could be utilized on a project organized into distinct phases as in the waterfall model of software development just as it could be with a so-called agile project where each short iteration delivers working software.

## 3.8.1 DATA-ORIENTED SYSTEM DEVELOPMENT

DOSD is a method for representing software requirements by focusing on data and their structure. There are institutions such as stock exchanges and service providers such as airlines, telephone companies, etc., that generate time-dependent data. These data are provided to some of their subscribers offline—e.g., stock exchange to stock brokers to subbrokers; airline to their travel agencies, etc. These data are in preknown or prespecified formats, which may come in CD-ROM or be downloadable through File Transfer Protocol (FTP) in comma separated value (CSV) or ASCII or other specified formats. The user organization develops its own application (platform variant, development tool variant) and can use these data directly in their applications to issue tickets or initiate the purchase or sale of stocks to their clients. The major advantage of this data-oriented system development approach is that it eliminates data transformation errors such as porting, conversion, transcription and transposition. It is generally combined with another development technique that considers processing issues to develop a suitable business solution.

## 3.8.2 OBJECT-ORIENTED SYSTEM DEVELOPMENT

OOSD is the process of solution specification and modeling where data and procedures can be grouped into an entity known as an object. An object's data are referred to as its attributes and its functionality is referred to as its methods. This contrasts with the traditional structured SDLC approach which considers data separately from the procedures that act on

them (e.g., program and database specifications). Proponents of OOSD claim the combination of data and functionality is aligned with how humans conceptualize everyday objects.

OOSD is a programming technique, not a software development methodology. One can do OOSD while following any of the widely diverse set of software methodologies: waterfall, iterative, software engineering, agile and even pure hacking (and prototyping). A particular programming language, or use of a particular programming technique, does not imply or require use of a particular software development methodology.

Objects usually are created from a general template called a class. The template contains the characteristics of the class without containing the specific data that need to be inserted into the template to form the object. Classes are the basis for most design work in objects. Classes are either superclasses (i.e., root or parent classes) with a set of basic attributes or methods, or subclasses which inherit the characteristics of the parent class and may add (or remove) functionality as required. In addition to inheritance, classes may interact through sharing data, referred to as aggregate or component grouping, or sharing objects. Aggregate classes interact through messages, which are requests for services from one class (called a client) to another class (called a server). The ability of two or more objects to interpret a message differently at execution, depending on the superclass of the calling object, is termed polymorphism.

The first objected-oriented language, Simula67, was released in 1967. Smalltalk emerged during the 1970s as the first commercial object-oriented language. Then followed a series of languages that were either object-oriented from inception (e.g., Eiffel) or had been modified to include object-oriented capabilities (e.g., C++, Object Pascal, and Ada95). The emergence of Java during the late 1990s provided a significant boost to the acceptance of object technology.

To realize the full benefits of using object-oriented programming, it is necessary to employ object-oriented analysis and design approaches. Dealing with objects should permit analysts, developers and programmers to consider larger logical chunks of a system and clarify the programming process.

The major advantages of OOSD are as follows:
• The ability to manage an unrestricted variety of data types
• Provision of a means to model complex relationships
• The capacity to meet the demands of a changing environment

A significant development in OOSD has been the decision by some of the major players in object-oriented development to join forces and merge their individual approaches into a unified approach using the Unified Modeling Language (UML). UML is a general-purpose notational language for specifying and visualizing complex software for large object-oriented projects. This signals a maturation of the object-oriented development approach. While object-orientation is not yet pervasive, it can accurately be said to have entered the computing mainstream. Applications that use object-oriented technology are:
• Web applications
• E-business applications
• CASE for software development
• Office automation for e-mail and work orders
• Artificial intelligence
• Computer-aided manufacturing (CAM) for production and process control

## 3.8.3 COMPONENT-BASED DEVELOPMENT

Component-based development can be regarded as an outgrowth of object-oriented development. Component-based development means assembling applications from cooperating packages of executable software that make their services available through defined interfaces (i.e., enabling pieces of programs called objects, to communicate with one another regardless of which programming language they were written in or what operating system they are running). The basic types of components are:
• **In-process client components**—These components must run from within a container of some kind such as a web browser; they cannot run on their own.

- **Stand-alone client components**—Applications that expose services to other software can be used as components. Well-known examples are Microsoft's Excel and Word.
- **Stand-alone server components**—Processes running on servers that provide services in standardized ways can be components. These are initiated by remote procedure calls or some other kind of network call. Technologies supporting this include Microsoft's Distributed Component Object Model (DCOM), Object Management Group's Common Object Request Broker Architecture (CORBA) and Sun's Java through Remote Method Invocation (RMI).
- **In-process server components**—These run on servers within containers. Examples include Microsoft's Transaction Server (MTS) and Sun's Enterprise Java Beans (EJB).

> **Note:** The CISA candidate will not be tested on vendor-specific products or services in the CISA exam.

A number of different component models have emerged. Microsoft has its Component Object Model (COM). MTS when combined with COM allows developers to create components that can be distributed in the Windows environment. COM is the basis for ActiveX technologies, with ActiveX Controls being among the most widely used components. Alternative component models include the CORBA Component Model and Sun's EJB.

> **Note:** COM/DCOM, CORBA (itself a standard, not a specific product) and RMI are sometimes referred to as distributed object technologies. As the name suggests, they allow objects on distributed platforms to interact. These technologies, among others, are also termed middleware. Middleware is a broad term, but a basic definition is software that provides run-time services whereby programs/objects/components can interact with one another.

Tool developers are supporting one or another of these standards with powerful, visual tools now available for designing and testing component-based applications. Industry "heavy weights," such as Microsoft and Rational are supporting component-based development. Additionally, a growing number of commercially available application servers now support MTS or EJB. There is a growing market for third-party components. A primary benefit of component-based development is the ability to buy proven, tested software from commercial developers. The range of components available has increased. The first components were simple in concept—e.g., buttons and list boxes. Components now provide much more diverse functionality. Databases are now available on the web to search for commercial components.

Components play a significant role in web-based applications. Applets are required to extend static HTML, Active X controls or Java. Both technologies are compatible with component development. Component-based development:
- Reduces development time. If an application system can be assembled from prewritten components and only code for unique parts of the system needs to be developed, then this should prove faster than writing the entire system from scratch.
- Improves quality. Using prewritten components means a significant percentage of the system code has been tested already.
- Allows developers to focus more strongly on business functionality. An outcome of component-based development and its enabling technologies is to further increase abstraction already achieved with high-level languages, databases and user interfaces. Developers are shielded from low-level programming details.
- Promotes modularity. By encouraging or forcing impassable interfaces between discrete units of functionality, it encourages modularity.
- Simplifies reuse. It avoids the need to be conversant with procedural or class libraries, allowing cross-language combination and allowing reusable code to be distributed in an executable format—i.e., no source is required. (To date, however, large-scale reuse of business logic has not occurred.)
- Reduces development cost. Less effort needs to be expended on design and build. Instead, the cost of software components can be spread across multiple users.
- Supports multiple development environments. Components written in one language can interact with components written in other languages or running on other machines.
- Allows a happy compromise between build and buy options. Instead of buying a complete solution, which perhaps does not entirely fit requirements, it could be possible to purchase only needed components and incorporate these into a customized system.

In considering these advantages, there needs to be attention given to software integration early and continuously during the development process. No matter how efficient component-based development is, if system requirements are poorly defined or the system fails to adequately address business needs, the project will not be successful.

# Systems and Infrastructure Life Cycle Management

## 3.8.4 WEB-BASED APPLICATION DEVELOPMENT

Web-based application development is an important emerging software development approach designed to achieve easier and more effective integration of code modules within and between enterprises. Historically, software written in one language on a particular platform has used a dedicated application programming interface (API). The use of specialized APIs has caused difficulties in integrating software modules across platforms. Technologies such as CORBA and COM that use remote procedure calls (RPCs) have been developed to allow real-time integration of code across platforms. However, using these RPC approaches for different APIs still remains complex. Web-based application development and associated Extensible Markup Language (XML) technologies are a recent development designed to further facilitate and standardize code module and program integration.

The other problem that web-based application development seeks to address is to avoid the need to perform redundant computing tasks with the inherent need for redundant code. One obvious example of this is a change of address notification from a customer. Instead of having to update details separately in multiple databases—e.g., contact management, accounts receivable and credit control—it is preferable that a common update process executed to update the address includes details in the multiple places required. Web services are intended to make this relatively easy to achieve.

Web application development is different than traditional third- or fourth-generation program developments in many ways— from the languages and programming techniques used, to the methodologies (or lack thereof) used to control the development work, to the way the users test and approve the development work. But the risks of application development remain the same. For example, buffer overflows had been a risk since computer programming was invented (for example, truncation issues with first generation computer programs), but they really hit the headlines only when they could be exploited by almost anyone, almost anywhere in the world, courtesy of the Internet.

As with traditional program development, a risk-based approach should be taken in the assessment of web application vulnerabilities: identify the business goals and supporting IT goals related to the development, then identify what can go wrong. One's previous experience can be used to identify risks related to inadequate specifications, poor coding techniques, inadequate documentation, inadequate QC and QA (including testing inadequacies), lack of proper change control and controls over promotion into production, and so on, and put these in the context of the web application languages, development processes and deliverables (perhaps with the support of best practice material/literature on web applications development). The focus should be on application development risks, the associated business risks and technical vulnerabilities, and how these could materialize and be controlled/addressed. Some controls will look the same for all application development activity, but many will need to reflect the way the development activity is taking place in the area under review.

With web-based application development, an XML language known as Simple Object Access Protocol (SOAP) is used to define APIs. SOAP will work with any operating system and programming language that understands XML. SOAP is simpler than using an RPC-based approach, with the advantage that modules are coupled loosely. Therefore a change to one component does not normally require changes to other components.

The second key component of web development is the Web Services Description Language (WSDL), which is also based on XML. WSDL is used to identify the SOAP specification that is to be used for the code module API and the formats of the SOAP messages used for input and output to the code module. The WSDL is also used to identify the particular web service accessible via a corporate intranet or across the Internet by being published to a relevant intranet or Internet web server.

The final component of web services is another XML-based language—Universal Description, Discovery and Integration (UDDI). UDDI is used to make an entry in a UDDI directory, which acts as an electronic directory accessible via a corporate intranet or across the Internet, and allows interested parties to learn of the existence of available web services.

Standards for SOAP, WSDL and UDDI have been accepted by the World Wide Web Consortium. A number of current or soon-to-be-released software products and development environments, including Microsoft's .Net family of products, support web services. However, some important standards such as those addressing security and transaction management are yet to be defined. Other issues such as charging for use of commercially developed web services also need to be addressed.

## 3.8.5 REENGINEERING

Reengineering is a process of updating an existing system by extracting and reusing design and program components. This process is used to support major changes in the way an organization operates. A number of tools are now available to support this process.

## 3.8.6 REVERSE ENGINEERING

Reverse engineering is the process of taking apart an application, a software application or a product to see how it functions and to use that information to develop a similar system. This process can be carried out in several ways:
• Decompiling object or executable code into source code and using it to analyze the program
• Utilizing the reverse-engineered application as a black box test and unveiling its functionality using test data

The major advantages of reverse engineering are:
• Faster development and reduced SDLC duration
• The creation of an improved system using the reverse-engineered application drawbacks

The IS auditor should be aware of the following risks:
• Software license agreements often contain clauses prohibiting the licensee from reverse engineering the software so that any trade secrets or programming techniques are not compromised.
• Decompilers are relatively new tools with functions that depend on specific computers, operating systems and programming languages. Any change in one of these components will require developing or purchasing a new decompiler.

# 3.9 INFRASTRUCTURE DEVELOPMENT/ ACQUISITION PRACTICES

The physical architecture analysis, the definition of a new one and the necessary road map to move from one to the other is a critical task for an IT department. Its impact is not only economic, but also technological, since it decides many other choices downstream such as operational procedures, training needs, installation issues and total cost of ownership (TCO).

Conflicting requirements such as evolving toward a services-based architecture, legacy hardware considerations, secure data access independent of data location, zero data loss, 24x7 availability and many others ensure no single platform satisfies all these requirements equally. Thus physical architecture analysis cannot be based solely on price or isolated features. A formal, reasoned choice must be made.

In this first section, steps are listed to arrive at the choice of a good physical architecture and the way to define a possible road map for supporting the migration of the technical architecture to the new one to reach the following goals:
• To successfully analyze the existing architecture
• To design a new architecture which takes into account the existing architecture as well as a company's particular constraints/requirements such as:
  – Reduced costs
  – Increased functionality
  – Minimum impact on daily work
  – Security and confidentiality issues
  – Progressive migration to the new architecture
• To write the functional requirements of this new architecture
• To develop a proof of concept based on these functional requirements:
  – To characterize price, functionality and performance
  – To identify additional requirements that will be used later

Chapter 3:

# Systems and Infrastructure Life Cycle Management

The resulting requirements will be documents and drawings describing the reference infrastructure that will be used by all projects downstream. With these requirements in hand, these projects can begin to start implementation.

The requirements are validated using a proof of concept. The proof of concept is a test-bed implementation of the physical architecture. It saves money because any problems are detected and corrected early, when they are cheaper to correct, and it gives confidence to the teams that the requirements correctly instruct potential vendors on the requirements they have to meet.

The main objective of the second section is to plan the physical implementation of the required technical infrastructure to set up the future environment (normally production, test and development environment). This task will cover the procurement activities such as contracting partners, setting up the SLAs, and developing installation plans and installation test plans. A well-designed selection process must be ensured, taking analytical results and intuition into account and guaranteeing alignment and commitment for implementation success. Due to the heterogeneous nature of the infrastructure found, it is necessary to develop a clear implementation plan (including deliverables, delivery times, test plans, etc.). It is also necessary to plan the coexistence of the old and new system, to avert possible mistakes during the installation and go-live phase (see **exhibit 3.15**).

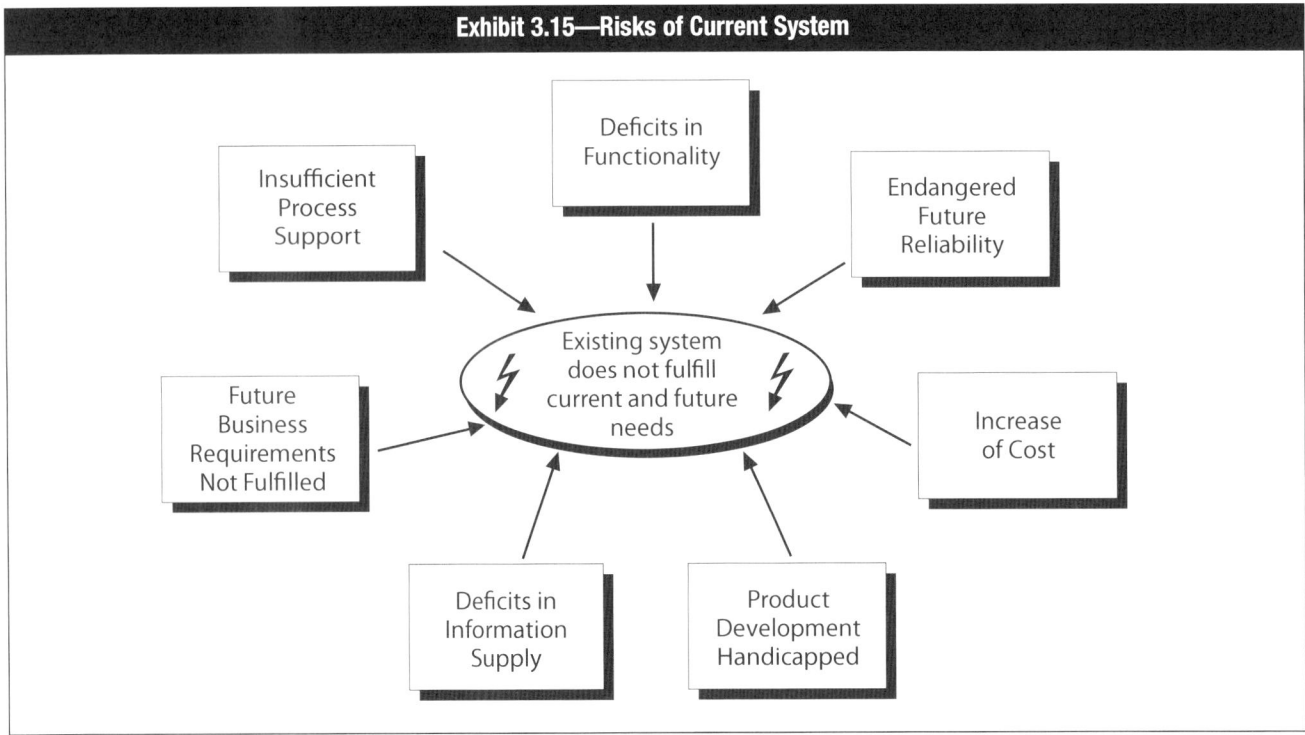

**Exhibit 3.15—Risks of Current System**

Thus, information and communication technologies (ICT) departments often face these requirements. The suggested solution must:
• Ensure alignment of the IT systems with corporate standards
• Provide appropriate levels of security
• Integrate with current IT systems
• Consider IT industry trends
• Provide future operational flexibility to support business processes
• Allow for projected growth in infrastructure without major upgrades
• Include technical architecture considerations for information security, secure storage, etc.
• Ensure cost-effective, day-to-day operational support
• Allow the usage of standardized hardware and software
• Maximize ROI, cost transparency and operational efficiency

# Systems and Infrastructure Life Cycle Management

## 3.9.1 PROJECT PHASES OF PHYSICAL ARCHITECTURE ANALYSIS

**Exhibit 3.16** shows the project phases to physical architecture analysis and, in the background, the time at which the vendor selection process may start.

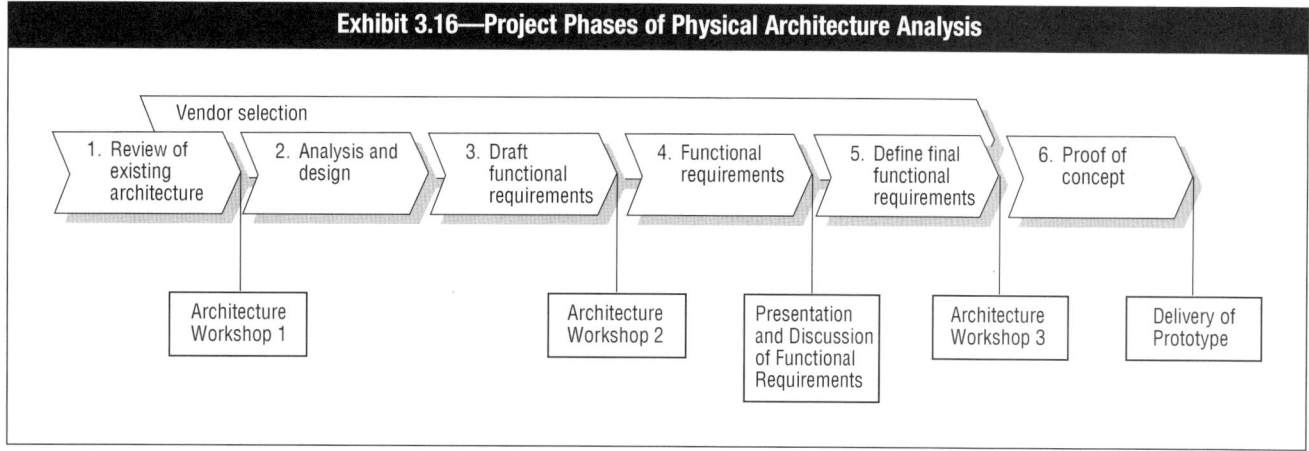

Exhibit 3.16—Project Phases of Physical Architecture Analysis

### Review of Existing Architecture

To start the process, the latest documents about the existing architecture must be reviewed. Participants of the first workshop will be specialists of the ICT department in all areas directly impacted by physical architecture. Examples are server, storage, security and overall IT infrastructure.

Special care must be taken in characterizing all the operational constraints that impact physical architecture such as:
• Ground issues
• Size limits
• Weight limits
• Current supply
• Physical security issues

The output of the first workshop is a list of components of the current infrastructure and constraints defining the target physical architecture.

### Analysis and Design

After reviewing the existing architecture, the design of the actual physical architecture has to be analyzed, and again compared against best practices and business requirements.

### Draft Functional Requirements

With the first physical architecture design in hand, the first (draft) of functional requirements is composed. This material is the input for the next step and the vendor selection process.

### Vendor and Product Selection

While the draft functional requirements are written, the vendor selection process is started in parallel. This process is described in detail later in this chapter.

### Writing Functional Requirements

After finishing the draft functional requirements and feeding the second part of this project, the functional requirements document is written, which will be introduced at the second architecture workshop with staff from all affected parties. The results will be discussed and a list of the requirements that need to be refined or added will be composed.

# Systems and Infrastructure Life Cycle Management

This is the last checkpoint before the sizing and the proof of concept (POC) starts, although the planning of the POC starts after the second workshop. With the finished functional requirements, the proof of concept phase begins.

## Proof of Concept

Establishing a POC is highly recommended to prove that the selected hardware and software are able to meet all expectations, including security requirements. The deliverable of the POC should be a running prototype, including the associated document and test protocols describing the tests and their results.

To start, the POC should be based on the results of the procurement phase described below in this section. For this purpose, a subset of the target hardware is used. The software to run the POC can be either test versions or software already supplied by the vendor; therefore, additional costs are expected to be minimal.

To keep costs low, most elements of the framework are implemented in a simplified form. They will be extended to their final form in later phases.

The prototype should demonstrate the following features:
• The basic setup of the core security infrastructure
• Correct functionality of auditing components
• Basic but functional implementation of security measures as defined
• Secured transactions
• Characterization in terms of installation constraints and limits (server size, server current consumption, server weight, server room physical security)
• Performance
• Resiliency
• Funding and costing model

Related implementation projects which prepare the ground for deployment should also be part of the POC since they will be used in the same way as they are used in the production physical architecture. At the end of this phase a last workshop is held where the production sizing and layout is adapted to include POC conclusions.

## 3.9.2 PLANNING IMPLEMENTATION OF INFRASTRUCTURE

To ensure the quality of the results, it is necessary to use a phased approach to fit the whole puzzle together. It is also fundamental to set up the communication processes to other projects like the one described earlier. Through these different phases the components are fit together, and a clear understanding of the available and contactable vendors is established by using the selection process during the procurement phase and beyond. Furthermore, it is necessary to select the scope of key business and technical requirements to prepare the next steps, which include the development of the delivery, installation and test plans. Moreover, to ensure a future proven solution, it is integral to choose the right partners with the right skills in place. In the course of the project, it is necessary to use a phased approach, shown in **exhibit 3.17**.

As shown in **exhibit 3.17**, the requirements analysis is not part of this process but constantly feeds results into the process. If a Gantt chart is produced with these phases, most likely some phases likely overlap; therefore, the different phases must be considered an iterative process.

During the four different phases it is necessary to fit all the components together to get prepared for projects downstream (e.g., data migration).

# Systems and Infrastructure Life Cycle Management

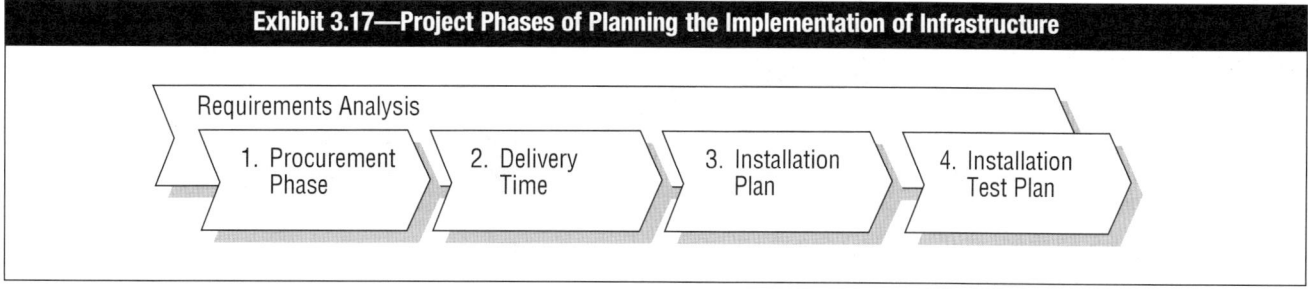

**Exhibit 3.17—Project Phases of Planning the Implementation of Infrastructure**

## Procurement Phase

During the procurement phase the communication processes is established with the analysis project to get an overview of the chosen solution and determine the quantity structure of the deliverables. The requirements statements are also produced.

Additionally, the procurement process begins the service-level management process. During these activities, the preferred partners are invited to the negotiations process and the deliverables, contracts and SLAs are signed (See **exhibit 3.18**).

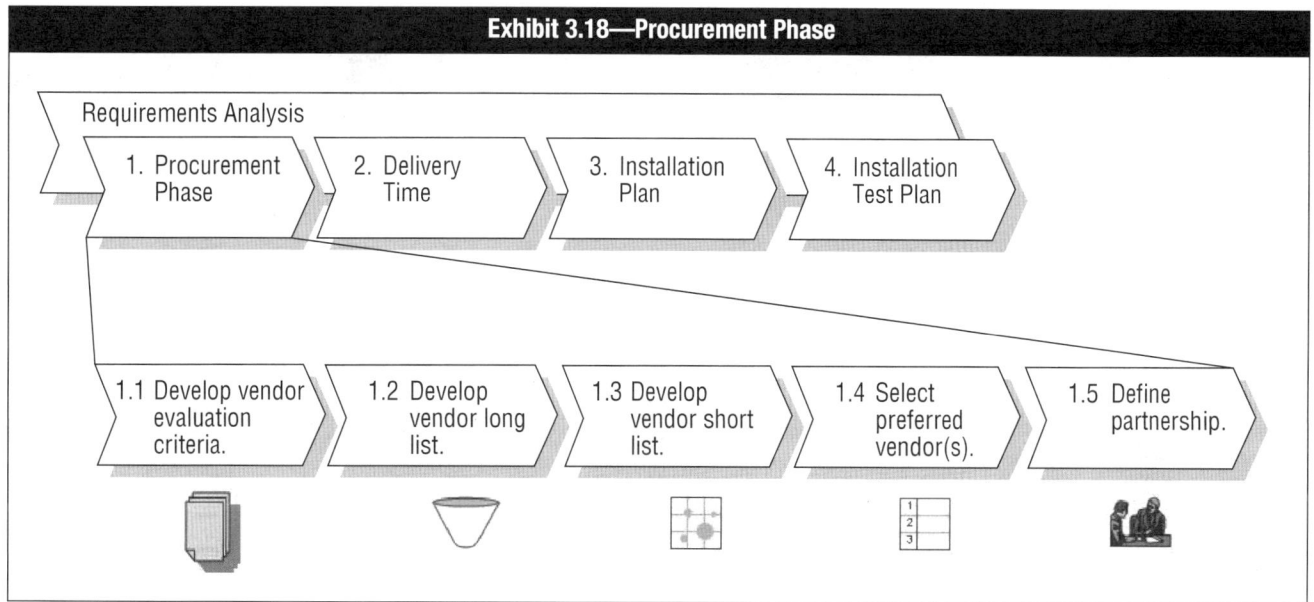

**Exhibit 3.18—Procurement Phase**

## Delivery Time

During the delivery time phase, the delivery plan is developed (see **exhibit 3.19**). This phase overlaps in some parts with the procurement phase.

The delivery plan should include topics such as priorities, goals and nongoals, key facts, principles, communication strategies, key indicators, progress on key tasks, and responsibilities.

## Installation Plan

During the installation planning phase, the installation plan is developed in cooperation with all affected parties (see **exhibit 3.20**).

An additional step is to review the plan with the involved parties and of course with those responsible for the integration projects. This is an iterative process.

# Systems and Infrastructure Life Cycle Management

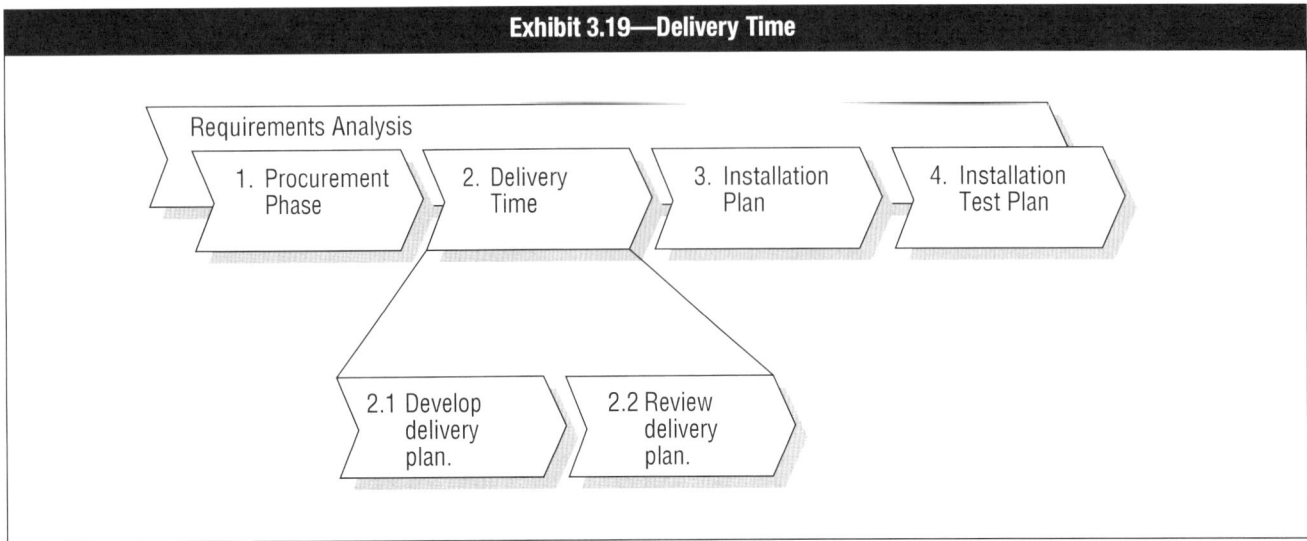

Exhibit 3.19—Delivery Time

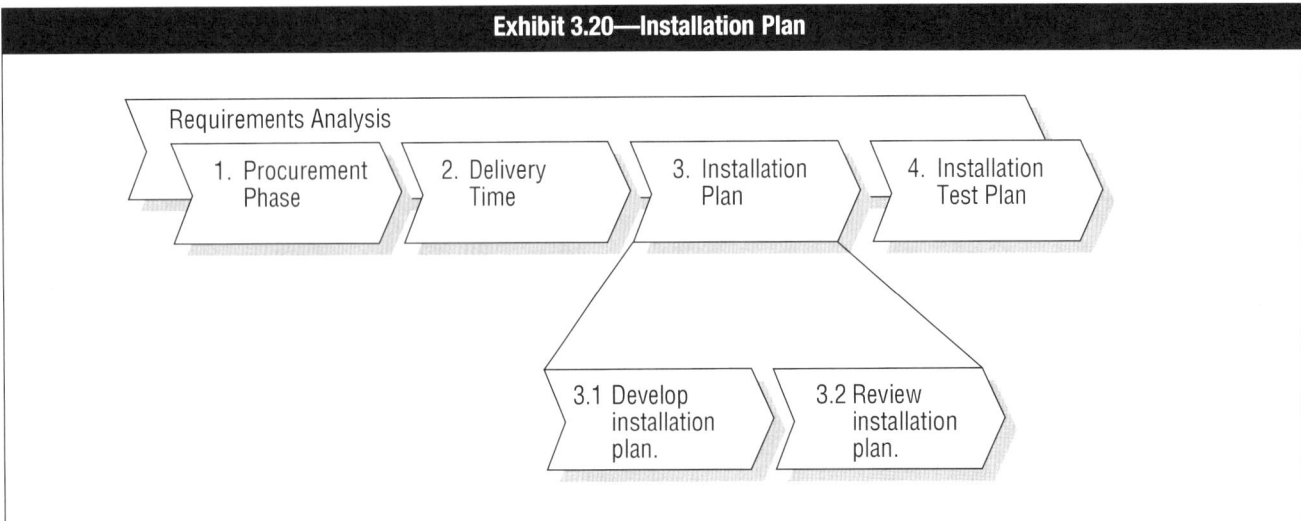

Exhibit 3.20—Installation Plan

## Installation Test Plan

Based on the known dependencies of the installation plan, the test plan is developed (see **exhibit 3.21**).

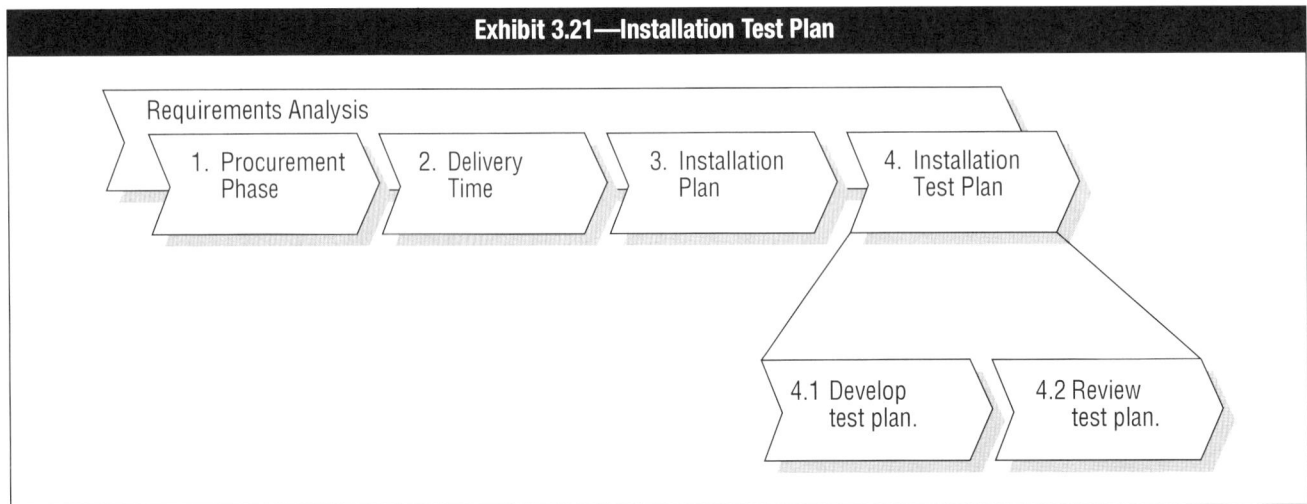

Exhibit 3.21—Installation Test Plan

# Systems and Infrastructure Life Cycle Management

The test plan includes test cases, basic requirements' specifications, definition of the processes and, as far as possible, measurement information for the applications and the infrastructure.

## 3.9.3 CRITICAL SUCCESS FACTORS

Critical success factors of planning the implementation include:
• To avoid delays, the appropriate skilled staff must attend workshops and participate for the entire project duration.
• The documentation needed for carrying out the work needs to be ready at project initiation.
• Decision makers must be involved at all steps to ensure all necessary decisions can be made quickly.
• To get prepared, part one of the project (Analysis of Physical Architecture) must be completed and the needed infrastructure decisions must be made.

## 3.9.4 HARDWARE ACQUISITION

Selection of a computer hardware and software environment frequently requires the preparation of specifications for distribution to hardware/software (HW/SW) vendors and criteria for evaluating vendor proposals. The specifications are sometimes presented to vendors in the form of an invitation to tender (ITT), also known as a request for proposal (RFP).

The specifications must define, as completely as possible, the usage, tasks and requirements for the equipment needed, and must include a description of the environment where that equipment will be used.

When acquiring a system, the IT or specifications, should include the following:
• Organizational descriptions indicating whether the computer facilities are centralized or decentralized, and distributed or outsourced
• Information processing requirements such as:
  – Major existing application systems and future application systems
  – Workload and performance requirements
  – Processing approaches (e.g., online/batch, client-server, real-time databases, continuous operation)
• Hardware requirements such as:
  – Central processing unit (CPU) speed
  – Disk space requirements
  – Memory requirements
  – Number of CPUs required
  – Peripheral devices (e.g., sequential devices such as tape drives; direct access devices such as magnetic disk drives, printers, compact disc drives, digital video disc drives, Universal Serial Bus [USB] peripherals, and secure digital multimedia cards [SD/MMC])
  – Data preparation/input devices that accept and convert data for machine processing
  – Direct entry devices (e.g., terminals, point-of-sale terminals or automated teller machines)
  – Networking capability (e.g., Ethernet connections, modems and integrated services digital network [ISDN] connections)
  – Number of terminals or nodes the system needs to support
• System software applications such as:
  – Operating system software (current version and any required upgrades)
  – Compilers
  – Program library software
  – Database management software and programs
  – Communications software
  – Access control software
• Support requirements such as:
  – System maintenance (for preventive, detective [fault reporting] or corrective purposes)
  – Training (user and technical staff)
  – Backups (daily and disaster backups)

- Adaptability requirements such as:
  - HW/SW upgrade capabilities
  - Compatibility with existing HW/SW platforms
  - Changeover to other equipment capabilities
- Constraints such as:
  - Staffing levels
  - Existing hardware capacity
  - Delivery dates
- Conversion requirements such as:
  - Test time for the HW/SW
  - System conversion facilities
  - Cost/pricing schedule

## Acquisition Steps

When purchasing (acquiring) HW/SW from a vendor, consideration should be given to the following:
- Testimonials or visits with other users
- Provisions for competitive bidding
- Analysis of bids against requirements
- Comparison of bids against each other using predefined evaluation criteria
- Analysis of the vendor's financial condition
- Analysis of the vendor's capability to provide maintenance and support (including training)
- Review of delivery schedules against requirements
- Analysis of HW/SW upgrade capability
- Analysis of security and control facilities
- Evaluation of performance against requirements
- Review and negotiation of price
- Review of contract terms (including right to audit clauses)
- Preparation of a formal written report summarizing the analysis for each of the alternatives and justifying the selection based on benefits and cost

The criteria and data used for evaluating vendor proposals should be properly planned. The following are some of the criteria that should be considered in the evaluation process:
- **Turnaround time**—The time that the help desk or vendor takes to fix a problem from the moment it is logged in
- **Response time**—The time a system takes to respond to a specific query by the user
- **System reaction time**—The time taken for logging into a system or getting connected to a network
- **Throughput**—The quantity of useful work made by the system per unit of time. Throughput can be measured in instructions per second or some other unit of performance. When referring to a data transfer operation, throughput measures the useful data transfer rate and is expressed in kilobits per second (kbps), megabits per second (Mbps), and gigabits per second (Gbps).
- **Workload**—The capacity to handle the required volume of work, or the volume of work that the vendor's system can handle in a given time frame
- **Compatibility**—The capability of an existing application to run successfully on the newer system supplied by the vendor
- **Capacity**—The capability of the newer system to handle a number of simultaneous requests from the network for the application and the volume of data that it can handle from each of the users
- **Utilization**—The system availability time vs. the system downtime

When performing an audit of this area, the IS auditor should:
- Determine if the acquisition process began with a business need and whether the hardware requirements for this need were considered in the specifications.
- Determine if several vendors were considered and whether the comparison between them was done according to the aforementioned criteria.

## 3.9.5 SYSTEM SOFTWARE ACQUISITION

Every time a technological development has allowed for increased computing speeds or new capabilities, these have been absorbed immediately by the demands placed on computing resources by more ambitious applications. Consequently, improvements have led to decentralized, interconnected open systems through functions bundled in operating system software to meet these needs. For example, network management and connectivity are features now found in most operating systems.

It is IS management's responsibility to be aware of these new capabilities because they generally are enablers which improve business processes, and provide expanded application services to businesses and customers in a more effective way. Therefore, it is important that organizations stay current by applying the latest version or release of system software to remain competitive. If the version or release is not current, the organization risks being dependent on software that may become obsolete and is no longer supported by the software vendor. Additionally, because of the open interconnected systems, an organization may become susceptible to security exposures if they do not apply the latest software versions and periodic updates/patches.

As a result, business operations could be seriously impeded or disrupted if changes are not made. Software vendors usually advise their customers of the latest system upgrades or fixes aimed at reducing security-related vulnerabilities. In addition, noncurrent software may be incompatible with new technological requirements that may accompany the introduction of new applications.

Short- and long-term plans should document IS management's plan for migrating to newer, more efficient and more effective operating systems and related systems software.

When selecting new system software, a number of business and technical issues must be considered including:
• Business, functional and technical needs and specifications
• Cost and benefit(s)
• Obsolescence
• Compatibility with existing systems
• Security
• Demands on existing staff
• Training and hiring requirements
• Future growth needs
• Impact on system and network performance

## 3.9.6 SYSTEM SOFTWARE IMPLEMENTATION

System software implementation involves identifying features, configuration options and controls for a standard configuration to apply across the organization. Additionally, implementation involves testing the software in a nonproduction environment and obtaining some form of certification and accreditation to place the approved operating system software into production.

## 3.9.7 SYSTEM SOFTWARE CHANGE CONTROL PROCEDURES

All test results should be documented, reviewed and approved by technically qualified subject matter experts prior to production use.

Change control procedures are designed to ensure that changes are authorized and do not disrupt processing. This requires that IS management and personnel are aware of and involved in the system software change process. Change control procedures should ensure that changes impacting the production systems (particularly in relation to the impact of failure during installation) have been assessed appropriately, and that appropriate recovery/backout (rollback) procedures exist so that the impact of any failure during installation can be minimized. For example, change control could be accomplished by having a

configuration management system in place for maintaining prior OS versions or prior states when applying security patches related to high-risk security issues (see chapter on protection of information assets). In many situations, IS management will have to apply patches to implement vendor solutions related to security issues affecting network- and host-based systems. Change control procedures should also ensure that all appropriate members of the management team who could be affected by the change have been properly informed and have made an assessment of the impact of the change in each area.

# 3.10 INFORMATION SYSTEMS MAINTENANCE PRACTICES

System maintenance practices refer primarily to the process of managing change to application systems while maintaining the integrity of both the production source and executable code.

Once a system is moved into production, it seldom remains static. Change is an expected event in all systems regardless of whether they are vendor-supplied or internally developed. Reasons for change in normal operations include IT/business changes, changes in classification related to either sensitivity or criticality, audits, and adverse incidents such as intrusions and viruses.

To control the ongoing maintenance of the system, a standard process for performing and recording changes is necessary. This process, which mirrors the organization's SDLC process, should include steps to ensure that the system changes are appropriate to the needs of the organization, appropriately authorized, documented, thoroughly tested and approved by management. The process typically is established in the design phase of the application when application system requirements are baselined.

## 3.10.1 CHANGE MANAGEMENT PROCESS OVERVIEW

The change management process begins with authorizing changes to occur. For this purpose, a methodology should exist for prioritizing and approving system change requests. Change requests are initiated from end users as well as operational staff and system development/maintenance staff. In any case, authorization needs to be obtained from appropriate levels of end-user and systems management (e.g., a change control group, configuration control boards). For acquired systems, for example, a vendor may distribute periodic updates, patches or new release levels of the software. User and systems management should review such changes. Determination should be made as to whether the changes are appropriate for the organization or will negatively affect the existing system.

Users should convey system change requests to the system management using some type of formal correspondence such as a standard change request form, memo or e-mail message. At a minimum, the user request should include the requestor's name, date of request, date the change is needed, priority of the request, a thorough description of the change request, a description of any anticipated effects on other systems or programs, and fallback procedures in case the changes cause the failure of the system. The user could also provide a reason for the change, a cost justification analysis and the expected benefits of the change. In addition, the request should provide evidence that it has been reviewed and authorized by user management. A signature on the request form or memo typically provides this evidence.

Change requests should be in a format that ensures all changes are considered for action and allows the system management staff to easily track the status of the request. This is usually done by assigning a unique control number to each request and entering the change request information into a computerized system. This can also be performed manually. With detailed information regarding each request, management can identify those requests that have been completed and are still in progress or have not been addressed yet. Management can also use this information to help ensure that user requests are addressed in a timely manner. See **exhibit 3.22**.

# Systems and Infrastructure
# Life Cycle Management

**Exhibit 3.22—Sample Change Request Form**

# Request for Change (RFC) Document

## 1. Contents

This document shall ensure that at least every major change will be applied to the customer's mission or business critical systems in a controlled environment. It shall support all affected parties in gaining a more reliable and resilient infrastructure. Thus the usage of this document is mandatory for all involved personnel and—during normal operations—has to be distributed in a timely manner to provide the setup of a Forward Schedule of Change. In the rare condition of emergency changes it must be belatedly completed for documentation purpose. **In either case there must always be a formal approval for the RFC covered in this document.**

## 2. Usage Guidance

The following figure shall illustrate the usage guideline for this document.

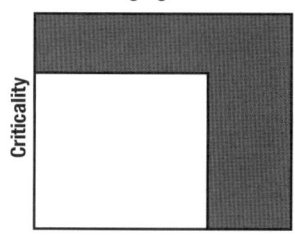

Penetration

The area shaded in red symbolizes the mandatory usage of the formal RFC procedure depicted in this document.

If the proposed change has an impact either on critical/vital systems (criticality is high) or on a massive amount of systems (wide penetration) then it is—by definition—a major change and the RFC procedure has to be executed.

In any other case (area left blank on the figure) it is up to the requestor to choose a formal approach or to bypass the RFC procedure.

To follow best practices a list shall by compiled which itemizes all systems classified as either critical or vital so the requestor will know what to head for (criticality). Additionally a percentage rate of affected systems shall be published as a threshold so that it is clearly defined when to execute the RFC procedure (penetration).

## 3. General RFC data

| RFC title: | | | | ID: | |
|---|---|---|---|---|---|
| Standard/emergency change: | | □ / □ | | | |
| Issued by: | | Intended recipient: | | | |
| Issued on: | | Scheduled for: | | | |
| Checked by: | | Checked on: | | Status: | |
| Postimplementation review done by: | | Postimplementation review done on: | | Status: | |

▨ Has to be filled out by the requestor
▧ Has to be filled out by the approver

## 4. Scope of Change

| Abstract |
|---|
| |
| **Detailed description (please be as specific as you can)** |
| |
| **Expected benefit** |
| |

**Exhibit 3.22—Sample Change Request Form (cont.)**

**Implementation checklist/release management**

| Task | Description | Responsible Party |
|------|-------------|-------------------|
|      |             |                   |

**Affected configuration items (CIs) and impact on CIs**

| | | |
|---|---|---|
| | | |

| | Yes/No | Description |
|---|---|---|
| Performance degradation | ☐ ☐ | |
| Redundancy loss | ☐ ☐ | |
| Service disruption | ☐ ☐ | |
| Reboot | ☐ ☐ | |
| Downtime | ☐ ☐ | |
| Fallback/backout implemented | ☐ ☐ | |

**Associated costs (hardware, software, manpower, etc.)**

| |
|---|
| |

### 5. Approval/Rejection

**Comments**

| |
|---|
| |

### 6. Postimplementation Review

**Summary**

| |
|---|
| |

| | Yes/No | Description |
|---|---|---|
| Does the implementation meet your expectations? | ☐ ☐ | |
| Are there any deviations from the outlined procedure above? | ☐ ☐ | |
| Is the documentatin/configuration management database up to date? | ☐ ☐ | |
| Are the stakeholders informed of the change? | ☐ ☐ | |

All requests for changes and related information should be maintained by the system maintenance staff as part of the system's permanent documentation.

Maintenance records of all program changes should exist either manually or automatically. Several library management software products provide this type of audit trail. The maintenance information usually consists of the programmer ID, time and date of change, project or request number associated with the change, and before and after images of the lines of code that were changed.

This process becomes even more important when the programmer who creates the program is also the operator. In this case it is assumed the IS department is either small or there are few applications being processed. Change management procedures must be closely followed since segregation of duties cannot be established in this environment. It requires user management to pay more attention to changes and upgrades made by the programmer, and proper authorization must be given to the programmer before putting any change into production. In addition to the manual process of management approving changes before the programmer can submit them into production, management could have automated change control software installed to prevent unauthorized program changes. By doing this the programmer is no longer responsible for migrating changes into production. The change control software becomes the operator that migrates programmer changes into production based on approval by management.

Programmers should not have write, modify or delete access to production data. Depending on the type of information in production, programmers may not even have read-only access (or access to customer credit card numbers, US Social Security numbers/national ID numbers, or other sensitive information that may require added security).

## Deploying Changes
After the end user is satisfied with the system test results and the adequacy of the system documentation, approval should be obtained from user management. User approval could be documented on the original change request or in some other fashion (memo or e-mail); however, evidence that verifies user approval should be retained by the system maintenance staff.

## Documentation
To ensure the effective utilization and future maintenance of a system, it is important that all relevant system documentation be updated. Due to tight time constraints and limited resources, thorough updates to documentation are often neglected. Documentation requiring revision may consist of program and/or system flowcharts, program narratives, data dictionaries, entity relationship models, data flow diagrams (DFDs), operator run books and end-user procedural manuals.

Procedures should be in place to ensure that documentation stored offsite for disaster recovery purposes is updated. This documentation is often overlooked and may be out of date.

## Testing Changed Programs
Changed programs should be tested and certified with the same discipline as newly developed systems to ensure that the changes perform the intended functions. In addition, if the risk analysis determines it is necessary, additional testing would be required to ensure:
• Existing functionality is not damaged by the change
• System performance is not degraded because of the change
• No security exposures have been created because of the change

## Auditing Program Changes
In evaluating whether procedures for program changes are adequate, the IS auditor should ensure that controls are in place to protect production application programs from unauthorized changes. The control objectives are as follows:
• Access to program libraries should be restricted.
• Supervisory reviews should be conducted.
• Change requests should be approved and documented.
• Potential impact of changes should be assessed.

# Systems and Infrastructure Life Cycle Management

- The change request should be documented on a standard form, paying particular attention to the following:
  - The change specifications should be adequately described, a cost analysis developed and a target date established.
  - The change form should be signed by the user to designate approval.
  - The change form should be reviewed and approved by programming management.
  - The work should be assigned to an analyst, programmer and programming group leader for supervision.
- A sample of program changes made during the audit period should be selected and traced to the maintenance form to determine whether the changes are authorized, check that the form has appropriate approvals, and compare the date on the form with the date of production update for agreement.
- If an independent group updates the program changes in production, the IS auditor should determine before the update whether procedures exist to ensure possession of the change request form. (This is accomplished by watching the groups perform their jobs.)

## Emergency Changes

There may be times when emergency changes are required to resolve system problems and enable critical "production job" processing to continue. Procedures should primarily exist in the application's operations manual to ensure emergency fixes can be performed without compromising the integrity of the system. This typically involves the use of special logon IDs (i.e., emergency IDs) that grant a programmer/analyst temporary access to the production environment during these emergency situations. The use of emergency IDs should be logged and carefully monitered since their use grants someone powerful privileges. Emergency fixes should be completed using after-the-fact, follow-up procedures which ensure that all normal change management controls are retroactively applied. Changes done in this fashion are held in a special emergency library from where they should be moved through the change management process into normal production libraries in an expeditious manner. IS auditors need to pay particular attention that emergency changes are handled in an appropriate manner.

## Deploying Changes Back Into Production

Once user management has approved the change, the modified programs can be moved into the production environment. A group that is independent of computer programming should perform the migration of programs from test to production. Groups such as computer operations, QA or a designated change control group should perform this function.

To ensure that only authorized individuals have the ability to migrate programs into production, proper access restrictions must be in place. Such restrictions can be implemented through the use of operating system security or an external security package.

Distributed systems, such as point-of-sale systems, offer an additional challenge in ensuring that changed programs are rolled out to all nodes. The rollout may be performed over a significant period of time to enable:
- Controls to be exercised over conversion of data
- Training of staff who will be using the changed software
- Support to be provided to users of the changed system
- Reduction of the risk associated with changing all nodes at the same time, and having to back them all out if something goes wrong

In view of the time it may take to roll out changes to the whole distributed system, controls must ensure that all nodes are eventually updated. If changes are made on a regular basis, it may be necessary to check regularly that all nodes are running the same versions of all software.

## Change Exposures (Unauthorized Changes)

An unauthorized change to application system programs can occur for several reasons:
- The programmer has access to production libraries containing programs and data including object code.
- The user responsible for the application was not aware of the change (no user signed the maintenance change request approving the start of the work).
- A change request form and procedures were not formally established.
- The appropriate management official did not sign the change form approving the start of the work.

# Systems and Infrastructure Life Cycle Management

• The user did not sign the change form signifying acceptance before the change was updated.
• The changed source code was not properly reviewed by the appropriate programming personnel.
• The appropriate management official did not sign the change form approving the program for update to production.
• The programmer put in extra code for personal benefit (i.e., committed fraud).
• Changes received from the acquired software vendor were not tested or the vendor was allowed to load the changes directly into production/site. This happens in cases of distributed processing sites, such as point-of-sale, banking applications, ATM networks, etc.

## 3.10.2 CONFIGURATION MANAGEMENT

Because of the difficulties associated with exercising control over programming maintenance activities, some organizations implement configuration management systems. In a configuration management system, maintenance requests must be formally documented and approved by a change control group (e.g., configuration control boards). In addition, careful control is exercised over each stage of the maintenance process via checkpoints, reviews and sign-off procedures. From an audit perspective, effective use of this software provides important evidence of management's commitment to careful control over the maintenance process.

Configuration management involves procedures throughout the software life cycle (from requirements analysis to maintenance) to identify, define and baseline software items in the system and thus provide a basis for problem management, change management and release management.

Configuration management is important because software is subject to ongoing change both during and after development. The process involves identification of the items that are likely to change (called configuration items). These include things such as programs, documentation and data. Once an item is developed and approved, it is handed over to a configuration management team for safekeeping and assigned a reference number. Once baselined in this way, an item should only be changed through a formal change control process.

Checking in is the process of moving an item to the controlled environment. When a change is required (and supported by a change control form), the item will be checked out by the configuration manager. Once the change is made, it can be checked using a different version number. The process of checking out also prevents or manages simultaneous code edits.

For configuration management to work, management must support the concept of software configuration management. The configuration management process is implemented by developing and following a configuration management plan and operating procedures. This plan should not be limited to just the software developed, but should also include all system documentation, test plans and procedures. As part of the software configuration management task, the maintainer performs the following task steps:
1. Develop the configuration management plan.
2. Baseline the code and associated documents.
3. Analyze and report on the results of configuration control.
4. Develop the reports that provide configuration status information.
5. Develop release procedures.
6. Perform configuration control activities such as identification and recording of the request.
7. Update the configuration status accounting database.

In many cases, commercial software products will be used to automate the manual processes. Such tools should allow control to be maintained for applications software from the outset of system analysis and design to running live.

Configuration management tools will support change management and release management by providing automated support for the following:
1. Identification of items affected by a proposed change to assist with impact assessment
2. Recording configuration items affected by authorized changes
3. Implementation of changes in accordance with authorization records
4. Registering of configuration item changes when authorized changes and releases are implemented

5. Recording of baselines that are related to releases (with known consequences) to which an organization would revert if an implemented change fails

A new version of the system (or builds) should only be built from the baselined items. In many cases, software tools are available to automate the process of preparing a release to avoid human errors and resource costs.

# 3.11 SYSTEM DEVELOPMENT TOOLS AND PRODUCTIVITY AIDS

System development tools and productivity aids include code generators, CASE applications and fourth-generation languages. These tools and aids are addressed in the following sections.

## 3.11.1 CODE GENERATORS

Code generators are tools, often incorporated with CASE products, that generate program code based on parameters defined by a systems analyst or on data/entity flow diagrams developed by the design module of a CASE product. These products allow most developers to implement software programs with efficiency. As noted, the IS auditor should be aware of nontraditional origins of source code.

## 3.11.2 COMPUTER-AIDED SOFTWARE ENGINEERING

Application development efforts require collecting, organizing and presenting a substantial amount of data at the application, systems and program levels. A substantial amount of the application development effort involves translating this information into program logic and code for subsequent testing, modification and implementation. This often is a time consuming process but it is necessary to develop, use and maintain computer applications.

CASE is the use of automated tools to aid in the software development process. Their use may include the application of software tools for software requirements analysis, software design, code production, testing, document generation and other software development activities.

CASE products are generally divided into three categories:
1. **Upper CASE**—These products are used to describe and document business and application requirements. This information includes data object definitions and relationships, and process definitions and relationships.
2. **Middle CASE**—These products are used for developing the detailed designs. These include screen and report layouts, editing criteria, data object organization and process flows. When elements or relationships change in the design, it is necessary to make only minor alterations to the automated design and all other relationships are automatically updated.
3. **Lower CASE**—These products are involved with the generation of program code and database definitions. These products use detailed design information, programming rules and database syntax rules to generate program logic, data file formats or entire applications.

CASE tools are available for mainframe, minicomputer and microcomputer environments. These tools can provide higher quality systems more quickly. CASE products enforce a uniform approach to system development, facilitate storage and retrieval of documents, and reduce the manual effort in developing and presenting system design information. This power of automation changes the nature of the development process by eliminating or combining some steps and altering the means of verifying specifications and applications.

The IS auditor needs to recognize the changes in the development process brought on by CASE. Some CASE systems allow a project team to produce a complete system from the DFDs and data elements without any traditional source code. In these situations, the DFDs and data elements become the source code.

# Systems and Infrastructure Life Cycle Management

The IS auditor should gain assurances that approvals are obtained for the appropriate specifications, users continue to be involved in the development process, and investments in CASE tools yield benefits in quality and speed. Other key issues the IS auditor needs to consider with CASE include the following:

- CASE tools help in the application design process, but do not ensure that the design, programs and system are correct or that they fully meet the needs of the organization.
- CASE tools should complement and fit into the application development methodology, but there needs to be a methodology in place for CASE to be effective. The methodology should be understood and used effectively by the organization's software developers.
- The integrity of data moved between CASE products or between manual and CASE processes needs to be monitored and controlled.
- Changes to the application should be reflected in stored CASE product data.
- Just like a traditional application, application controls need to be designed.
- The CASE repository (the database that stores and organizes the documentation, models and other outputs from the different phases) needs to be secured on a need-to-know basis. Strict version control should be maintained on this database.

The IS auditor may also become a user of CASE tools as several features facilitate the audit process. DFDs, which may be the product of upper and middle CASE tools, may be used as an alternative to other flowcharting techniques. IS auditors whose IS departments are moving into CASE are using CASE-generated documentation as part of the audit. Some are even experimenting with the use of CASE tools to create audit documentation. In addition, CASE tools can be used to develop interrogation software and embedded audit modules (EAMs). Repository reports should be used to gain an understanding of the system and to review controls over the development process.

## 3.11.3 FOURTH-GENERATION LANGUAGES

While a standard definition of a 4GL does not exist, the common characteristics of 4GLs are the following:
- **Nonprocedural language**—Most 4GLs do not obey the procedural paradigm of continuous statement execution and subroutine call and control structures. Instead, they are event-driven and make extensive use of object-oriented programming concepts such as objects, properties and methods.

For example, a COBOL programmer who wants to produce a report sorted in a given sequence must first open and read the data file, then sort the file and finally produce the report. A typical 4GL treats the report as an object with properties, such as input file name and sort order, and methods such as sort file and print report.

Care should be taken when using 4GLs. Unlike traditional languages, the 4GLs can lack the lower level detail commands necessary to perform certain types of data intensive or online operations. These operations are usually required when developing major applications. For this reason, the use of 4GLs as development languages should be weighed carefully against traditional languages such as COBOL.
- **Environmental independence (portability)**—Many 4GLs are portable across computer architectures, operating systems and telecommunications monitors. Some 4GLs have been implemented on mainframe processors and microcomputers.
- **Software facilities**—These facilities include the ability to design or paint retrieval screen formats, develop computer-aided training routines or help screens, and produce graphical outputs.
- **Programmer workbench concepts**—The programmer has access through the terminal to easy filing facilities, temporary storage, text editing and operating system commands. This type of a workbench approach is closely associated with the CASE application development approach. It is often referred to as an integrated development environment.
- **Simple language subsets**—4GLs generally have simple language subsets that can be used by less-skilled users in an information center.

4GLs are often classified in the following ways:
- **Query and report generators**—These specialized languages can extract and produce reports (audit software). Recently, more powerful languages have been produced that can access database records, produce complex online outputs and be developed in an almost natural language.

- **Embedded database 4GLs**—These depend on self-contained database management systems. This characteristic often makes them more user-friendly but also may lead to applications that are not integrated well with other production applications. Examples include FOCUS, RAMIS II and NOMAD 2.
- **Relational database 4GLs**—These high-level language products are usually an optional feature on a vendor's DBMS product line. These allow the applications developer to make better use of the DBMS product, but they often are not end-user-oriented. Examples include SQL+, MANTIS and NATURAL.
- **Application generators**—These development tools generate lower-level programming languages (3GLs) such as COBOL and C. The application can be further tailored and customized. Data processing development personnel, not end users, use application generators.

# 3.12 PROCESS IMPROVEMENT PRACTICES

Business processes require improvements, which are accomplished with practices and techniques addressed in the following sections.

## 3.12.1 BUSINESS PROCESS REENGINEERING AND PROCESS CHANGE PROJECTS

"Business is in reality the underlying business process and successful business implies efficient underlying business processes. Everything else that constitutes a business is losing its uniqueness and becoming a commodity. It is no wonder that the efficient organization of processes is becoming a boardroom agenda the world over." (Nagaraj, N.S.; *Business Process Management—An Emerging Trend*, Infosys/SETLabs, India, 2001)

A business process can be seen as a sociotechnical system, that is "a set of interrelated work activities characterized by specific inputs and value-added tasks that produce specific customer-focused outputs. Business processes consist of horizontal work flows that cut across several departments or functions." (Seth, Vikram; William King; *Organizational Transformation through Business Process Reengineering*, Prentice Hall, USA, 1998)

The generic model shown in **exhibit 3.23** is a very basic description of a process (some form of information enters the process, is processed, and the outcome is measured against the goal or objective of the process). The level of detail needed (e.g., breakdown of subprocesses to activities) depends highly on the complexity of the process, the knowledge of the affected staff and the company's requirements regarding audit functionality (performance and compliance) of the process and shall fit into an existing quality management system (e.g., ISO 9001:2000).

Any output produced by a process must be bound to a business objective and adhere to defined corporate standards. Monitoring of effectiveness (goal achievement), efficiency (minimum effort) and compliance must be done on a regular basis and shall be included in management reports for review under the plan-do-check-act (PDCA) cycle.

BPR is the process of responding to competitive and economic pressures, and customer demands to survive in the current business environment. This is usually done by automating system processes so that there are fewer manual interventions and manual controls. BPR achieved with the help of implementing an ERP system is often referred to as package-enabled reengineering (PER). Advantages of BPR are usually experienced where the reengineering process appropriately suits the business needs. BPR has increased in popularity as a method for achieving the goal of cost savings through streamlining operations and gaining the advantages of centralization within the same process.

The steps in a successful BPR are:
- Define the areas to be reviewed.
- Develop a project plan.
- Gain an understanding of the process under review.
- Redesign and streamline the process.
- Implement and monitor the new process.
- Establish a continuous improvement process.

# Systems and Infrastructure Life Cycle Management

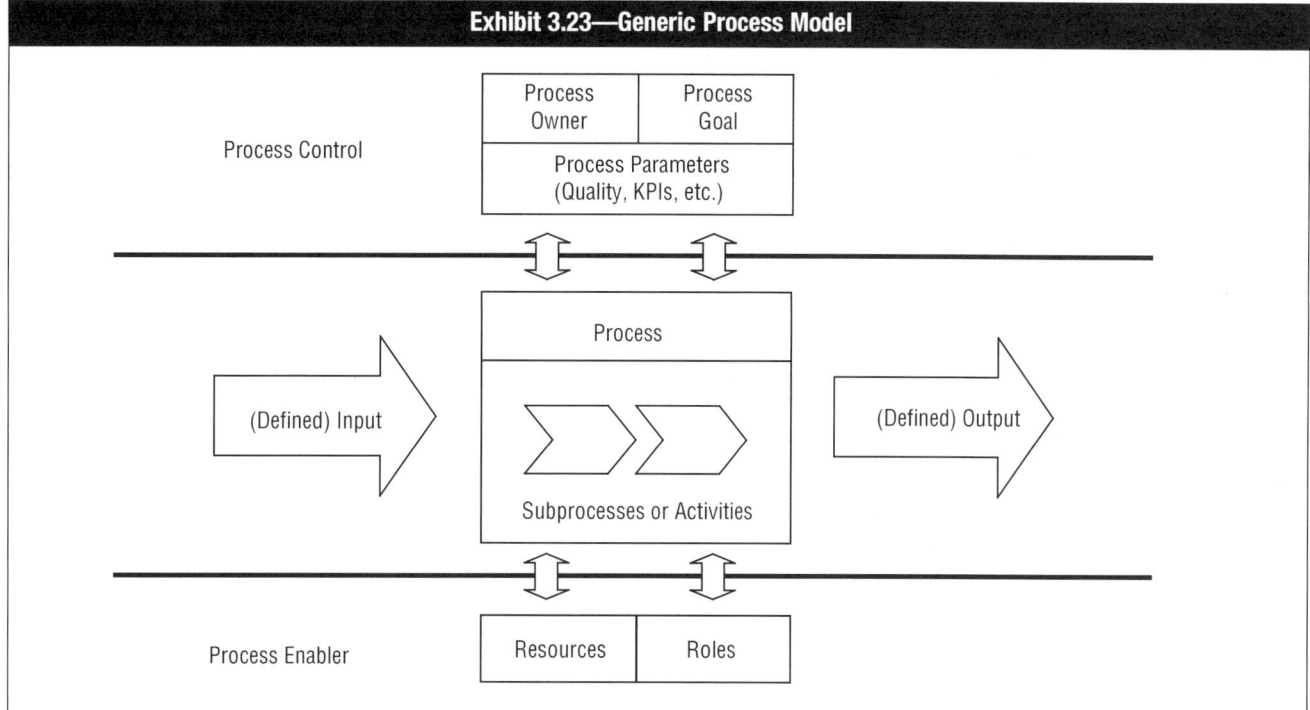

**Exhibit 3.23—Generic Process Model**

Process Control

Process Owner | Process Goal

Process Parameters (Quality, KPIs, etc.)

(Defined) Input

Process

Subprocesses or Activities

(Defined) Output

Process Enabler

Resources | Roles

As a reengineering process takes hold, new results begin to emerge:
• New business priorities based on value and customer requirements
• A concentration on process as a means of improving product, service and profitability
• New approaches to organizing and motivating people inside and outside the enterprise
• New approaches to the use of technologies in developing, producing and delivering goods and services
• New approaches to the use of information as well as powerful and more accessible information technologies
• Refined roles for suppliers including outsourcing, joint development, quick response, just-in-time inventory and support
• Redefined roles for clients and customers, providing them with more direct and active participation in the enterprise's business process

A successful BPR/process change project requires the project team to perform the following for the existing processes:
• Process decomposition to the lowest level required for effectively assessing a business process (typically referred to as an elementary process), which is a unit of work performed with a definitive input and output
• Identification of customers, process-based managers or process owners responsible for processes from beginning to end
• Documentation of the elementary process-related profile information including:
  – Duration
  – Trigger (which triggers the process to act)
  – Frequency
  – Effort
  – Responsibility (process owner)
  – Input and output
  – External interfaces
  – System interaction
  – Risk and control information
  – Performance measurement information
  – Identified problematic areas and their root causes

The existing baseline processes must be documented—preferably in the form of flowcharts and related profile documents—so the baseline processes can be compared to the processes after reengineering.

The newly designed business processes inevitably involve changes in the way(s) of doing business, and could impact the finances, philosophy and personnel of the organization, its business partners and customers.

Throughout the change process, the BPR team must be sensitive to organization culture, structure, direction and the components of change. Management must also be able to predict and/or anticipate issues and problems, and offer appropriate resolutions that will accelerate the change process.

BPR teams can be used to facilitate and assist the staff in transitioning into the reengineered business processes. BPR professionals are valuable in monitoring progress toward the achievement of the strategic plan of the organization.

A major concern in BPR is that key controls may be reengineered out of a business process. The IS auditor's task is to identify the existing key controls and evaluate the impact of removing these controls. If the controls are key preventive controls, the IS auditor must ensure that management is aware of the removal of the control and management is willing to accept the potential material risk of not having that preventive control.

## BPR Methods and Techniques

Applying BPR methods and techniques to a process creates an immediate environment for change and provides consistency of results.

### BENCHMARKING PROCESS

Benchmarking is about improving business processes. It is defined as a continuous, systematic process for evaluating the products, services and work processes of organizations recognized as representing best practices for the purpose of organizational improvement.

The steps listed below are followed generally by the benchmarking team in a benchmarking exercise:
1. **Plan**—In the planning stage, critical processes are identified for the benchmarking exercise. The benchmarking team should identify the critical processes and understand how they are measured, what kinds of data are needed and how the data need to be collected.
2. **Research**—The team should collect baseline data about its own processes before collecting these data about others. The next step is to identify the benchmarking partners through sources such as business newspapers and magazines, quality award winners, trade journals, etc.
3. **Observe**—The next step is to collect data and visit the benchmarking partner. There should be an agreement with the partner organization, a data collection plan and a method to facilitate proper observation.
4. **Analyze**—This step involves summarizing and interpreting the data collected, and analyzing the gaps between an organization's process and its partner's process. Converting key findings into new operational goals will be the goal of this stage.
5. **Adapt**—Adapting the results of benchmarking can be the most difficult step. In this step, the team needs to translate the findings into a few core principles and work down from principles to strategies to action plans.
6. **Improve**—Continuous improvement is the key focus in a benchmarking exercise. Benchmarking links each process in an organization with an improvement strategy and organizational goals.

## BPR Audit and Evaluation Techniques

When reviewing an organization's business process change (reengineering) efforts, IS auditors must determine whether:
- The organization's change efforts are consistent with the overall culture and strategic plan of the organization
- The reengineering team is making an effort to minimize any negative impact the change might have on the organization's staff
- The BPR team has documented lessons to be learned after the completion of the BPR/process change project

The IS auditor would also provide a statement of assurance or conclusion with respect to the objectives of the audit.

| PRACTICE QUESTIONS |
| --- |

**3-3** When conducting a review of business process reengineering, an IS auditor found that a key preventive control had been removed. In this case the IS auditor should:

    A.   inform management of the finding and determine whether management is willing to accept the potential material risk of not having that preventive control.

    B.   determine if a detective control has replaced the preventive control during the process and, if it has not, report the removal of the preventive control.

    C.   recommend that this and all control procedures that existed before the process was reengineered be included in the new process.

    D.   develop a continuous audit approach to monitor the effects of the removal of the preventive control.

**3-4** During which of the following steps in the business process reengineering should the benchmarking team visit the benchmarking partner?

    A.   Observation

    B.   Planning

    C.   Analysis

    D.   Adaptation

*See answers and explanations to the practice questions at the end of the chapter. (page 274)*

# 3.12.2 ISO 9126

ISO 9126 provides the definition of the characteristics and associated quality evaluation process to be used when specifying the requirements for, and evaluating the quality of, software products throughout their life cycle. Attributes evaluated include:
- **Functionality**—The set of attributes that bears on the existence of a set of functions and their specified properties. The functions are those that satisfy stated or implied needs.
- **Reliability**—The set of attributes that bears on the capability of software to maintain its level of performance under stated conditions for a stated period of time.
- **Usability**—The set of attributes that bears on the effort needed for use and on the individual assessment of such use by a stated or implied set of users.
- **Efficiency**—The set of attributes that bears on the relationship between the level of performance of the software and the amount of resources used under stated conditions.
- **Maintainability**—The set of attributes that bears on the effort needed to make specified modifications.
- **Portability**—The set of attributes that bears on the ability of software to be transferred from one environment to another.

# 3.12.3 SOFTWARE CAPABILITY MATURITY MODEL

The Capability Maturity Model (CMM) for Software, developed by Carnegie Mellon's Software Engineering Institute and various industry and government affiliates in the early 1990s, is a process maturity model or framework that helps organizations improve their software life cycle processes. The model is particularly adept at enabling organizations to prevent excessive project schedule delays and cost overruns by providing the appropriate infrastructure and support necessary to help projects avoid these problems.

Based on quality process management principles of five maturity levels, the CMM was designed to guide software organizations in selecting process improvement strategies by determining current process maturity and identifying the few issues most critical to software quality and process improvement. This enables organizations to focus on a limited set of activities to steadily improve software process capability. The five maturity levels attainable by software organizations include:
1. **Initial**—Characterized as *ad hoc*, where success depends solely on individual effort

# Systems and Infrastructure Life Cycle Management

2. **Repeatable**—Disciplined management processes are established to plan and track cost, schedule and functionality, and provide oversight over a software project, and create a learning environment where successfully defined and applied processes can be repeated successfully on other projects of similar size and scope.
3. **Defined**—Lessons learned from the prior phase provide the impetus to develop a standard software process across the organization. This includes both management and software engineering activities, documented, standardized and integrated into an institutionalized standard software process applicable to all software development projects.
4. **Managed**—Once processes are well-defined and applied, the organization has reached a point where it can develop and apply quantitative managed control over its software development processes. This provides a greater degree of precision and control over software projects for improving software productivity and reaching zero-defect goals.
5. **Optimizing**—When an organization has attained the ability to quantitatively and successfully control its software projects, it is in a position to use continuous process improvement strategies in applying innovative solutions and state-of-the-art technologies to its software processes.

Improvement activities performed by an organization occur at maturity levels two through five. The starting point is one level above the organization's current maturity level.

## 3.12.4 CAPABILITY MATURITY MODEL INTEGRATION

Following the release and successful adoption of CMM for Software, other models were developed for disciplines such as systems engineering, integrated product development, etc. The Capability Maturity Model Integration (CMMI) was conceived as a means of combining the various models into a set of integrated models. CMMI also describes five levels of maturity, although the descriptions of what constitutes each level differ from those used in the original CMM. Supporters of CMMI see it as less directly aligned with the traditional waterfall approach toward development and better aligned with contemporary software development practices including:
• Iterative development
• Early definition of architecture
• Model-based design notation
• Component-based development
• Demonstration-based assessment of intermediate development products
• Use of scalable, configurable processes

The maturity models, CMM and CMMI, are useful to evaluate a computer center management, evaluate the development function management process, and implement and measure the IT change management process.

## 3.12.5 ISO 15504

ISO/IEC TR 15504, also known as Software Process Improvement and Capability Determination (SPICE), is based on CMM and Bootstrap and was first published in 1998 (ISO/IEC TR 15504:1998). Based on the feedback of more than 4,000 assessments, it has been republished during 2003-2005 as a full international standard.

SPICE, like CMMI, serves as a baseline for process improvement and process benchmarking, and leverages the adoption of best practices. Reference models for SPICE include:
• **Software life cycle processes**—Through ISO/IEC 12207 AMD1/2
• **System life cycle processes**—Through ISO/IEC 15288
• **Human-centered life cycle processes**—Through ISO 18529
• **Component-based development processes**—Through the OOSPICE project
• **IT service management processes**—Through a SPICE user group initiative
• **Quality management system processes**—Through SPICE for 9000 (S9K)
• **Automotive embedded software**—Through Automotive SPICE (an initiative of The Procurement Forum and The SPICE User Group with the major European car manufacturers)
• **Medical device software**—Through the Medi SPICE initiative

# Systems and Infrastructure
# Life Cycle Management

The process capability dimensions are slightly different from those used in CMMI (see **exhibit 3.24**).

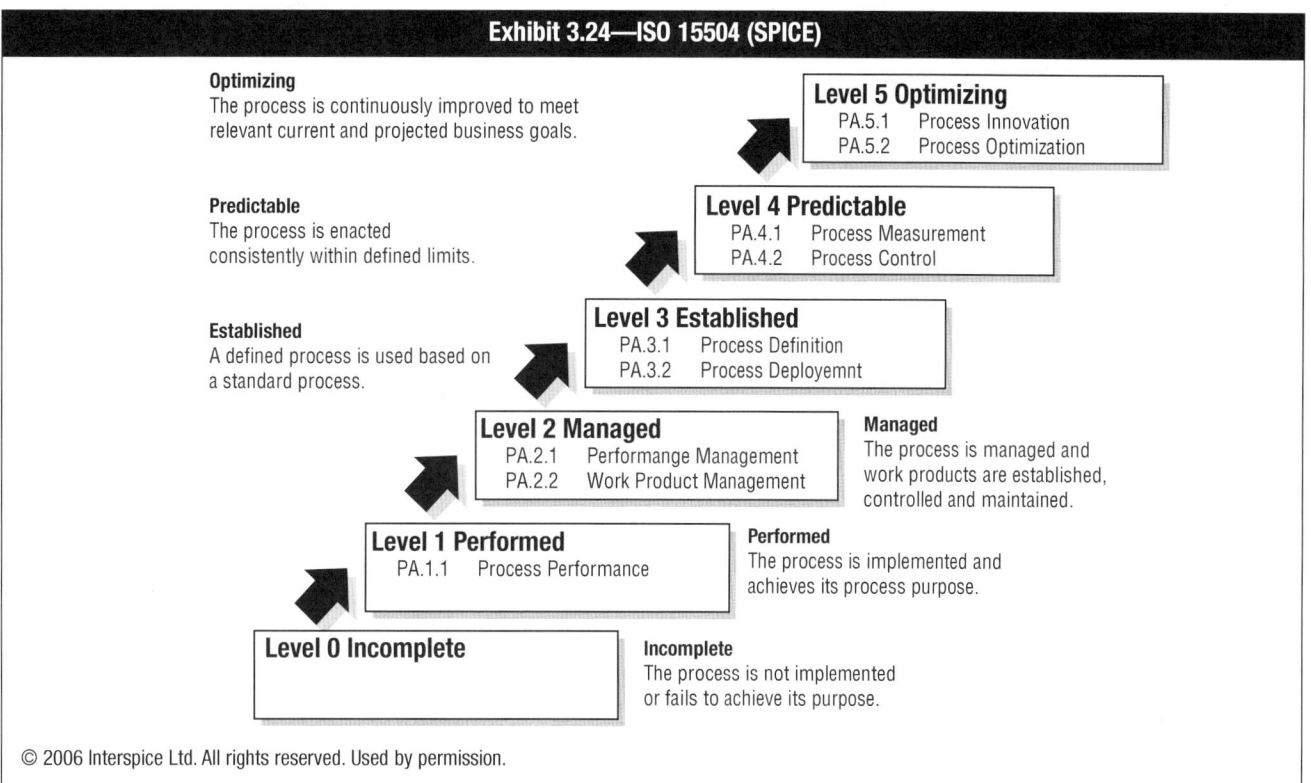

**Exhibit 3.24—ISO 15504 (SPICE)**

**Optimizing**
The process is continuously improved to meet relevant current and projected business goals.

**Level 5 Optimizing**
PA.5.1    Process Innovation
PA.5.2    Process Optimization

**Predictable**
The process is enacted consistently within defined limits.

**Level 4 Predictable**
PA.4.1    Process Measurement
PA.4.2    Process Control

**Established**
A defined process is used based on a standard process.

**Level 3 Established**
PA.3.1    Process Definition
PA.3.2    Process Deployemnt

**Level 2 Managed**
PA.2.1    Performange Management
PA.2.2    Work Product Management

**Managed**
The process is managed and work products are established, controlled and maintained.

**Level 1 Performed**
PA.1.1    Process Performance

**Performed**
The process is implemented and achieves its process purpose.

**Level 0 Incomplete**

**Incomplete**
The process is not implemented or fails to achieve its purpose.

A very good heuristic approach to clarify how these levels can be seen is depicted in **exhibit 3.25**.

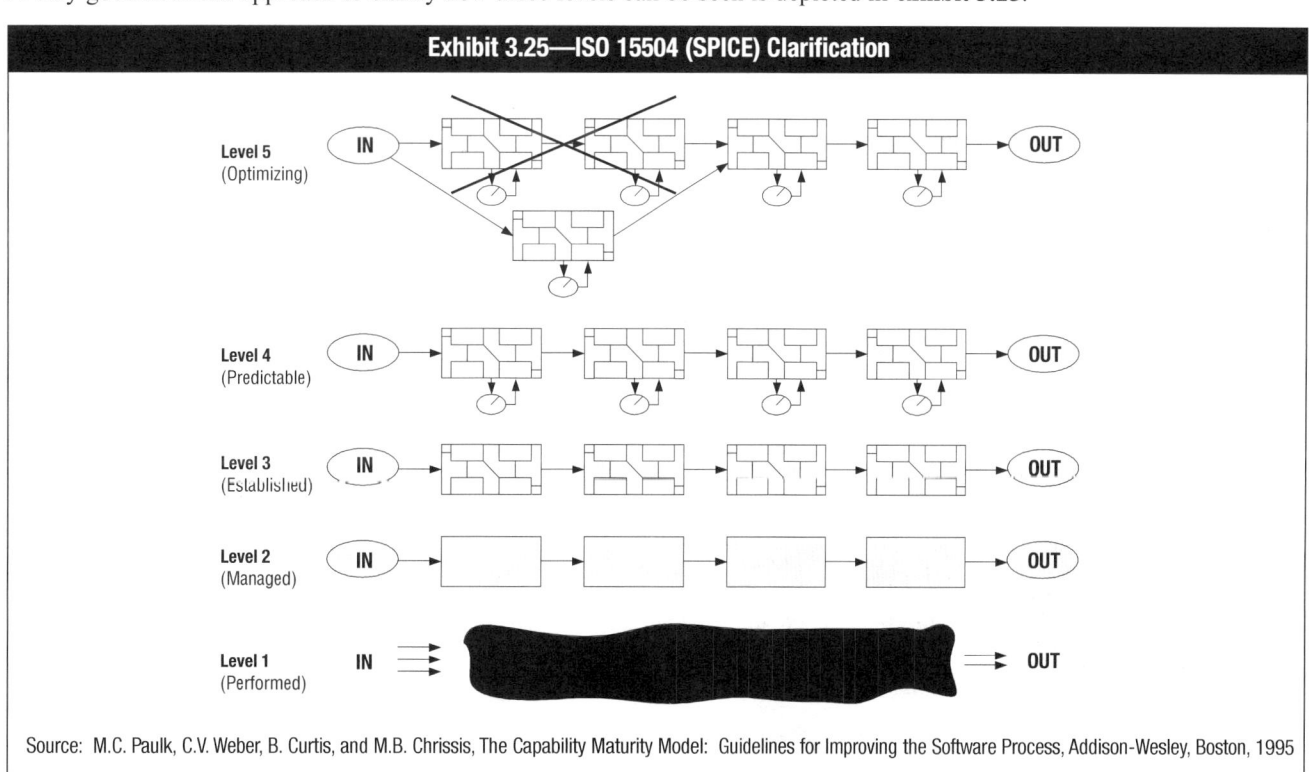

**Exhibit 3.25—ISO 15504 (SPICE) Clarification**

Level 5 (Optimizing)

Level 4 (Predictable)

Level 3 (Established)

Level 2 (Managed)

Level 1 (Performed)

IN    OUT

Source: M.C. Paulk, C.V. Weber, B. Curtis, and M.B. Chrissis, The Capability Maturity Model: Guidelines for Improving the Software Process, Addison-Wesley, Boston, 1995

# 3.13 APPLICATION CONTROLS

Application controls refer to the transactions and data relating to each computer-based application system; therefore, they are specific to each application. The objectives of application controls, which may be manual or programmed, are to ensure the completeness and accuracy of the records and the validity of the entries made therein resulting from manual and programmed processing. Application controls are controls over input, processing and output functions. They include methods for ensuring that:
• Only complete, accurate and valid data are entered and updated in a computer system
• Processing accomplishes the correct task
• Processing results meet expectations
• Data are maintained

Application controls may consist of edit tests, totals, reconciliations, and identification and reporting of incorrect, missing or exception data. Automated controls should be coupled with manual procedures to ensure proper investigation of exceptions.

These controls help ensure data accuracy, completeness, validity, verifiability and consistency, thus achieving data integrity and data reliability. Implementation of these controls helps ensure system integrity, that applicable system functions operate as intended, and that information contained by the system is relevant, reliable, secure and available when needed.

The IS auditor's tasks include the following:
• Identifying the significant application components and the flow of transactions through the system, and gaining a detailed understanding of the application by reviewing the available documentation and interviewing appropriate personnel
• Identifying the application control strengths, and evaluating the impact of the control weaknesses on the development of a testing strategy by analyzing the accumulated information
• Testing the controls to ensure their functionality and effectiveness by applying appropriate audit procedures
• Evaluating the control environment by analyzing the test results and other audit evidence to determine that control objectives were achieved
• Considering the operational aspects of the application to ensure its efficiency and effectiveness by comparing the system with efficient programming standards, analyzing procedures used and comparing them to management's objectives for the system

## 3.13.1 INPUT/ORIGINATION CONTROLS

Input control procedures must ensure that every transaction to be processed is received, processed and recorded accurately and completely. These controls should ensure that only valid and authorized information is input and that these transactions are only processed once. In an integrated systems environment, output generated by one system is the input for another system.

Therefore, the edit checks, validations and access controls of the system generating the output must be reviewed as input/origination controls for the system receiving that information as its own data input.

### *Input Authorization*

Input authorization verifies that all transactions have been authorized and approved by management.

Authorization of input helps ensure that only authorized data are entered into the computer system for processing by applications. Authorization can be performed online at the time when the data are entered into the system. A computer-generated report listing the items requiring manual authorization may also be generated. It is important that controls exist throughout processing to ensure that the authorized data remain unchanged. This can be accomplished through various accuracy and completeness checks incorporated into an application's design.

Types of authorization include:
• **Signatures on batch forms or source documents**—Provide evidence of proper authorization.
• **Online access controls**—Ensure that only authorized individuals may access data or perform sensitive functions.

# Systems and Infrastructure Life Cycle Management

- **Unique passwords**—Necessary to ensure that access authorization cannot be compromised through use of another individual's authorized data access. Individual passwords also provide accountability for data changes. (See chapter 5, Protection of Information Assets, for controls regarding password use and structure.)
- **Terminal or client workstation identification**—Used to limit input to specific terminals or workstations as well as to individuals. Terminals or client workstations in a network can be configured with a unique form of identification such as serial number or computer name that is authenticated by the system.
- **Source documents**—The forms used to record data. A source document may be a piece of paper, a turnaround document or an image displayed for online data input. A well-designed source document achieves several purposes. It increases the speed and accuracy with which data can be recorded, controls work flow, facilitates preparation of the data in machine-readable form for pattern recognition devices, increases the speed and accuracy with which data can be read, and facilitates subsequent reference checking.

Ideally, source documents should be preprinted forms to provide consistency, accuracy and legibility. Source documents should include standard headings, titles, notes and instructions. Source document layouts should:
- Emphasize ease of use and readability
- Group similar fields together to facilitate input
- Provide predetermined input codes to reduce errors
- Contain appropriate cross-reference numbers or a comparable identifier to facilitate research and tracing
- Use boxes to identify field size errors
- Include an appropriate area for management to document authorization

All source documents should be appropriately controlled. If source documents are not prenumbered, procedures should be established to ensure that all source documents have been input and taken into account.

## Batch Controls and Balancing

Batch controls group input transactions to provide control totals. The batch control can be based on total monetary amount, total items, total documents or hash totals.

Batch header forms are a data preparation control. All input forms should be clearly identified with the application name and transaction codes. Where possible, preprinted and prenumbered forms, with transaction identification codes and other constant data items, are recommended. This would help ensure that all pertinent data have been recorded on the input forms and can reduce data recording/entry errors.

Types of batch controls include:
- **Total monetary amount**—Verification that the total monetary value of items processed equals the total monetary value of the batch documents. For example, the total monetary value of the sales invoices in the batch agrees with the total monetary value of the sales invoices processed.
- **Total items**—Verification that the total number of items included on each document in the batch agrees with the total number of items processed. For example, the total number of units ordered in the batch of invoices agrees with the total number of units processed.
- **Total documents**—Verification that the total number of documents in the batch equals the total number of documents processed. For example, the total number of invoices in a batch agrees with the total number of invoices processed.
- **Hash totals**—Verification that the total (albeit meaningless in itself) for a predetermined numeric field that exists for all documents in a batch agrees with the total calculated by the system. For example, a total of customer account numbers is meaningless, but can be manually calculated and compared to the system calculated value.

Batch balancing can be performed through manual or automated reconciliation. Batch totaling must be combined with adequate follow-up procedures. Adequate controls should exist to ensure that each transaction creates an input document, all documents are included in a batch, all batches are submitted for processing, all batches are accepted by the computer, batch reconciliation is performed, procedures for the investigation and timely correction of differences are followed, and controls exist over the resubmission of rejected items.

Types of batch balancing include:
- **Batch registers**—These registers enable manual recording of batch totals and subsequent comparison with system reported totals.
- **Control accounts**—Control account use is performed through an initial edit file to determine batch totals. The data are then processed to the master file, and a reconciliation is performed between the totals processed during the initial edit file and the master file.
- **Computer agreement**—Computer agreement with batch totals is performed through the input of batch header details that record the batch totals; the system compares these to calculated totals, either accepting or rejecting the batch.

## Error Reporting and Handling

Input processing requires that controls be identified to verify data are accepted into the system correctly and input errors are recognized and corrected.

Data conversion error corrections are needed during the data conversion process. Errors can occur due to duplication of transactions and inaccurate data entry. These errors can, in turn, greatly impact the completeness and accuracy of the data. Corrections to data should be processed through normal data conversion processes and should be verified, authorized and reentered into the system as a part of normal processing.

Input error handling can be processed by:
- **Rejecting only transactions with errors**—Only transactions containing errors would be rejected; the rest of the batch would be processed.
- **Rejecting the whole batch of transactions**—Any batch containing errors would be rejected for correction prior to processing.
- **Holding the batch in suspense**—Any batches containing errors would not be rejected; however, the batch would be held in suspense, pending correction.
- **Accepting the batch and flagging error transactions**—Any batch containing errors would be processed; however, those transactions containing errors would be flagged for identification, enabling subsequent error correction.

Input control techniques include:
- **Transaction log**—Contains a detailed list of all updates. The log can be either manually maintained or provided through automatic computer logging. A transaction log can be reconciled to the number of source documents received to verify that all transactions have been input.
- **Reconciliation of data**—Controls whether all data received are properly recorded and processed
- **Documentation**—Written evidence of user, data entry and data control procedures
- **Error correction procedures**—These include:
  - Logging of errors
  - Timely corrections
  - Upstream resubmission
  - Approval of corrections
  - Suspense file
  - Error file
  - Validity of corrections
- **Anticipation**—The user or control group anticipates the receipt of data
- **Transmittal log**—Documents transmission or receipt of data
- **Cancellation of source documents**—Procedures to cancel source documents such as by punching with holes or marking them to avoid duplicate entry

## Batch Integrity in Online or Database Systems

Online systems also require control over input. Batches may be established by the time of day, the specific terminal or the individual inputting the data. A supervisor should then review the online batch and release it to the system for processing. This method is preferred over review of the output by the same person preparing the input.

# Systems and Infrastructure Life Cycle Management

## 3.13.2 PROCESSING PROCEDURES AND CONTROLS

Processing procedures and controls ensure the reliability of application program processing. IS auditors need to understand the procedures and controls that can be exercised over processing to evaluate what exposures are covered by these controls and what exposures remain.

### *Data Validation and Editing Procedures*

Procedures should be established to ensure that input data are validated and edited as close to the time and point of origination as possible. Preprogrammed input formats ensure that data are input to the correct field in the correct format. If input procedures allow supervisor overrides of data validation and editing, automatic logging should occur. A management individual who did not initiate the override should review this log.

Data validation identifies data errors, incomplete or missing data and inconsistencies among related data items. Front-end data editing and validation can be performed if intelligent terminals are used.

Edit controls are preventive controls that are used in a program before data are processed. If the edit control is not in place or does not work effectively, the preventive control measures do not work effectively. This may cause processing of inaccurate data. **Exhibit 3.26** describes various types of data validation edits.

| Exhibit 3.26—Data Validation Edits and Controls | |
|---|---|
| **Edits** | **Description** |
| Sequence check | The control number follows sequentially and any out-of-sequence or duplicated control numbers are rejected or noted on an exception report for follow-up purposes. For example, invoices are numbered sequentially. The day's invoices begin with 12001 and end with 15045. If any invoice larger than 15045 is encountered during processing, that invoice would be rejected as an invalid invoice number. |
| Limit check | Data should not exceed a predetermined amount. For example, payroll checks should not exceed US $4,000. If a check exceeds US $4,000, the data would be rejected for further verification/authorization. |
| Range check | Data should be within a predetermined range of values. For example, product type codes range from 100 to 250. Any code outside this range should be rejected as an invalid product type. |
| Validity check | Programmed checking of the data validity in accordance with predetermined criteria. For example, a payroll record contains a field for marital status and the acceptable status codes are M or S. If any other code is entered, the record should be rejected. |
| Reasonableness check | Input data are matched to predetermined reasonable limits or occurrence rates. For example, a widget manufacturer usually receives orders for no more than 20 widgets. If an order for more than 20 widgets is received, the computer program should be designed to print the record with a warning indicating that the order appears unreasonable. |
| Table lookups | Input data comply with predetermined criteria maintained in a computerized table of possible values. For example, the input clerk enters a city code of 1 to 10. This number corresponds with a computerized table that matches the code to a city name. |
| Existence check | Data are entered correctly and agree with valid predetermined criteria. For example, a valid transaction code must be entered in the transaction code field. |
| Key verification | The keying process is repeated by a separate individual using a machine that compares the original keystrokes to the repeated keyed input. For example, the worker number is keyed twice and compared to verify the keying process. |
| Check digit | A numeric value that has been calculated mathematically is added to data to ensure that the original data have not been altered or an incorrect, but valid, value substituted. This control is effective in detecting transposition and transcription errors. For example, a check digit is added to an account number so it can be checked for accuracy when it is used. |
| Completeness check | A field should always contain data rather than zeros or blanks. A check of each byte of that field should be performed to determine that some form of data, not blanks or zeros, is present. For example, a worker number on a new employee record is left blank. This is identified as a key field and the record would be rejected, with a request that the field be completed before the record is accepted for processing. |
| Duplicate check | New transactions are matched to those previously input to ensure that they have not already been entered. For example, a vendor invoice number agrees with previously recorded invoices to ensure that the current order is not a duplicate and, therefore, the vendor will not be paid twice. |
| Logical relationship check | If a particular condition is true, then one or more additional conditions or data input relationships may be required to be true and consider the input valid. For example, the hire date of an employee may be required to be more than 16 years past his/her date of birth. |

# Systems and Infrastructure Life Cycle Management

## Processing Controls

Processing controls ensure the completeness and accuracy of accumulated data. They ensure that data in a file/database remain complete and accurate until changed as a result of authorized processing or modification routines. The following are processing control techniques that can be used to address the issues of completeness and accuracy of accumulated data:

- **Manual recalculations**—A sample of transactions may be recalculated manually to ensure that processing is accomplishing the anticipated task.
- **Editing**—An edit check is a program instruction or subroutine that tests the accuracy, completeness and validity of data. It may be used to control input or later processing of data.
- **Run-to-run totals**—Run-to-run totals provide the ability to verify data values through the stages of application processing. Run-to-run total verification ensures that data read into the computer were accepted and then applied to the updating process.
- **Programmed controls**—Software can be used to detect and initiate corrective action for errors in data and processing. For example, if the incorrect file or file version is provided for processing, the application program could display messages instructing that the proper file and version be used.
- **Reasonableness verification of calculated amounts**—Application programs can verify the reasonableness of calculated amounts. The reasonableness can be tested to ensure appropriateness to predetermined criteria. Any transaction that is determined to be unreasonable may be rejected pending further review.
- **Limit checks on calculated amounts**—An edit check can provide assurance, through the use of predetermined limits, that calculated amounts have been keyed correctly. Any transaction exceeding the limit may be rejected for further investigation.
- **Reconciliation of file totals**—Reconciliation of file totals should be performed on a routine basis. Reconciliations may be performed through the use of a manually maintained account, a file control record or an independent control file.
- **Exception reports**—An exception report is generated by a program that identifies transactions or data that appear to be incorrect. These items may be outside a predetermined range or may not conform to specified criteria.

## Data File Control Procedures

File controls should ensure that only authorized processing occurs to stored data. Types of controls over data files are shown in **exhibit 3.27**.

| Exhibit 3.27—Data File Controls | |
|---|---|
| **Method** | **Description** |
| Before and after image reporting | Computer data in a file prior to and after a transaction is processed can be recorded and reported. The before and after images make it possible to trace the impact transactions have on computer records. |
| Maintenance error reporting and handling | Control procedures should be in place to ensure that all error reports are properly reconciled and corrections are submitted on a timely basis. To ensure segregation of duties, error corrections should be reviewed properly and authorized by personnel who did not initiate the transaction. |
| Source documentation retention | Source documentation should be retained for an adequate time period to enable retrieval, reconstruction or verification of data. Policies regarding the retention of source documentation should be enforced. Originating departments should maintain copies of source documentation and ensure that only authorized personnel have access. When appropriate, source documentation should be destroyed in a secure, controlled environment. |
| Internal and external labeling | Internal and external labeling of removable storage media is imperative to ensure that the proper data are loaded for processing. External labels provide the basic level of assurance that the correct data medium is loaded for processing. Internal labels, including file header records, provide assurance that the proper data files are used and allow for automated checking. |
| Version usage | It is critical that the proper version of a file be used as well as the correct file, for processing to be correct. For example, transactions should be applied to the most current database, while restart procedures should use earlier versions. |
| Data file security | Data file security controls prevent unauthorized access by unauthorized users that may have access to the application to alter data files. These controls do not provide assurances relating to the validity of data, but ensure that unauthorized users who may have access to the application cannot alter stored data improperly. |
| One-for-one checking | Individual documents agree with a detailed listing of documents processed by the computer. It is necessary to ensure that all documents have been received for processing. |

# Systems and Infrastructure
# Life Cycle Management

| Exhibit 3.27—Data File Controls (*cont.*) | |
|---|---|
| **Method** | **Description** |
| Prerecorded input | Certain information fields are preprinted on blank input forms to reduce initial input errors. |
| Transaction logs | All transaction input activity is recorded by the computer. A detailed listing, including date of input, time of input, user ID and terminal location, can then be generated to provide an audit trail. It also permits operations personnel to determine which transactions have been posted. This will help to decrease the research time needed to investigate exceptions and decrease recovery time if a system failure occurs. |
| File updating and maintenance authorization | Proper authorization for file updating and maintenance is necessary to ensure that stored data are safeguarded adequately, correct and up to date. Application programs may contain access restrictions in addition to the overall system access restrictions. The additional security may provide levels of authorization as well as an audit trail of file maintenance. |
| Parity checking | Data transfers in a computer system are expected to be made in a relatively error-free environment. However, when programs or vital data are transmitted, additional controls are needed. Transmission errors are controlled primarily by error-detecting or correcting codes. The former is used more often because error-correcting codes are costly to implement and are unable to correct all errors. Generally, error detection methods such as a check bit and redundant transmission are adequate. Redundancy checking is a common error-detection routine. A transmitted block of data containing one or more records or messages is checked for the number of characters or patterns of bits contained in it. If the numbers or patterns do not conform to predetermined parameters, the receiving device ignores the transmitted data and instructs the user to retransmit. Check bits are often added to the transmitted data by the telecommunications control unit and may be applied either horizontally or vertically. These checks are similar to the parity checks normally applied to data characters within on-premises equipment. A parity check on a single character generally is referred to as a vertical or column check, and a parity check on all the equivalent bits is known as a horizontal, longitudinal or row check. Use of both checks greatly improves the possibilities of detecting a transmission error, which may be missed when either of those checks is used alone. |

Data files, or indeed database tables, generally fall into four categories:
- **System control parameters**—The entries in these files change the workings of the system and may alter controls exercised by the system; for example, the tolerance allowed before an exceptional transaction is reported or blocked. Any change to these files should be controlled in a similar way to program changes.
- **Standing data**—These include data, such as supplier/customer names and addresses, that do not frequently change and are referred to during processing. These data should be authorized before entry or maintenance. Input controls may include a report of changed data that is checked and approved. Audit trails may log all changes.
- **Master data/balance data**—Running balances and totals that are updated by transactions should not be capable of adjustment except under strict approval and review controls. Audit trails are important here since there may be financial reporting implications for the change.
- **Transaction files**—These are controlled using validation checks, control totals, exception reports, etc.

| **PRACTICE QUESTION** |
|---|
| 3-5    A hash total of employee numbers is part of the input to a payroll master file update program. The program compares the hash total with the corresponding control total. What is the purpose of this procedure?<br><br>    A.    Verify that employee numbers are valid<br>    B.    Verify that only authorized employees are paid<br>    C.    Detect errors in payroll calculations<br>    D.    Detect the erroneous update of records |
| *See answers and explanations to the practice questions at the end of the chapter. (page 274)* |

# 3.13.3 OUTPUT CONTROLS

Output controls provide assurance that the data delivered to users will be presented, formatted and delivered in a consistent and secure manner.

Output controls include:
- **Logging and storage of negotiable, sensitive and critical forms in a secure place**—Negotiable, sensitive or critical forms should be properly logged and secured to provide adequate safeguards against theft or damage. The form log should be routinely reconciled to have inventory on hand, and any discrepancies should be properly researched.
- **Computer generation of negotiable instruments, forms and signatures**—The computer generation of negotiable instruments, forms and signatures should be properly controlled. A detailed listing of generated forms should be compared to the physical forms received. One should properly account for all exceptions, rejections and mutilations.
- **Report distribution**—Output reports should be distributed according to authorized distribution parameters, which may be automated or manual. Operations personnel should verify that output reports are complete and delivered according to schedule. All reports should be logged prior to distribution. In most environments, processing output is spooled to a buffer or print spool on completion of job processing, where it waits for an available printer. Controls over access to the print spools are important to prevent reports from being deleted accidentally from print spools or directed to a different printer. In addition, changes to the output print priority can delay printing of critical jobs. Access to distributed reports can compromise confidentiality. Therefore, physical distribution of reports should be controlled adequately. Reports containing sensitive data should be printed under secure, controlled conditions. Secure output drop-off points should be established. Output disposal should also be secured adequately to ensure that no unauthorized access can occur. Reports that are distributed electronically through the computer system also need to be considered. Logical access to these reports should also be controlled carefully and subject to authorization. When distributed manually, assurance should be provided that sensitive reports are properly distributed. Such assurance should include the recipient signing a log as evidence of receipt of output (i.e., manual nonrepudiation).
- **Balancing and reconciling**—Data processing application program output should be balanced routinely to the control totals. Audit trails should be provided to facilitate the tracking of transaction processing and the reconciliation of data.
- **Output error handling**—Procedures for reporting and controlling errors contained in the application program output should be established. The error report should be timely and delivered to the originating department for review and error correction.
- **Output report retention**—A record retention schedule should be adhered to firmly. Any governing legal regulations should be included in the retention policy.
- **Verification of receipt of reports**—To provide assurance that sensitive reports are properly distributed, the recipient should sign a log as evidence of receipt of output.

The IS auditor should be aware of existing concerns regarding record-retention policies for the organization and address legal requirements. Output can be restricted to particular IT resources or devices (e.g., a particular printer)

## 3.13.4 BUSINESS PROCESS CONTROL ASSURANCE

In an integrated application environment, controls are embedded and designed into the application that supports the processes. Business process control assurance involves evaluating controls at the process and activity level. These controls may be a combination of management, programmed and manual controls. In addition to evaluating general controls that affect the processes, business process owner-specific controls—such as establishing proper security and segregation of duties, periodic review and approval of access, and application controls within the business process—are evaluated.

Specific matters to consider in the business process control assurance are:
- Process maps
- Process controls
- Assessing business risks within the process
- Benchmarking with best practices
- Roles and responsibilities
- Activities and tasks
- Data restrictions

# Systems and Infrastructure
# Life Cycle Management

# 3.14 AUDITING APPLICATION CONTROLS

The IS auditor's tasks include the following:
- Identifying the significant application components and the flow of transactions through the system, and gaining a detailed understanding of the application by reviewing the available documentation and interviewing appropriate personnel
- Identifying the application control strengths and evaluating the impact of the control weaknesses to develop a testing strategy by analyzing the accumulated information
- Reviewing application system documentation to provide an understanding of the functionality of the application. In many cases—mainly in large sysems or packaged software—it is not feasible to review the whole application documentation. Thus a selected review should be performed. If an application is vendor supplied, technical and user manuals should be reviewed. Any changes to applications should be documented properly. The following documentation should be reviewed to gain an understanding of an application's development:
  - **System development methodology documents**—These documents include cost-benefit analysis and user requirements.
  - **Functional design specifications**—This document provides a detailed explanation of the application. An understanding of key control points should be noted during review of the design specifications.
  - **Program changes**—Documentation of any program change should be available for review. Any change should provide evidence of authorization and should be cross-referenced to source code.
  - **User manuals**—A review of the user manuals provides the foundation for understanding how the user is utilizing the application. Often control weaknesses can be noted from the review of this document.
  - **Technical reference documentation**—This documentation includes any vendor-supplied technical manuals for purchased applications in addition to any in-house documentation. Access rules and logic usually are included in these documents.

## 3.14.1 FLOW OF TRANSACTIONS THROUGH THE SYSTEM

A transaction flowchart provides information regarding key processing controls. Points where transactions are entered, processed and posted should be reviewed for control weaknesses.

## 3.14.2 RISK ASSESSMENT MODEL TO ANALYZE APPLICATION CONTROLS

Risk assessment, as discussed in chapter 1, provides information relating to the inherent risk of an application.

A risk assessment model can be based on many factors, which may include a combination of the following:
- The quality of internal controls
- Economic conditions
- Recent accounting system changes
- Time elapsed since last audit
- Complexity of operations
- Changes in operations/environment
- Recent changes in key positions
- Time in existence
- Competitive environment
- Assets at risk
- Prior audit results
- Staff turnover
- Transaction volume
- Regulatory agency impact
- Monetary volume
- Sensitivity of transactions
- Impact of application failure

# Systems and Infrastructure Life Cycle Management

## 3.14.3 OBSERVING AND TESTING USER PERFORMING PROCEDURES

Some of the user procedures that should be observed and tested include.
- **Separation of duties**—Ensures that no individual has the capability of performing more than one of the following processes: origination, authorization, verification or distribution. Observation and review of job descriptions and review of authorization levels and procedures may provide information regarding the existence and enforcement of separation of duties.
- **Authorization of input**—Evidence of input authorization can be achieved via written authorization on input documents or with the use of unique passwords. One may test this by looking through a sampling of input documents for proper authorization or reviewing computer-access rules. Supervisor overrides of data validation and editing should be reviewed to ensure that automatic logging occurs. This override activity report should be tested for evidence of managerial review. Excessive overrides may indicate the need for modification of validation and editing routines to improve efficiency.
- **Balancing**—Performed to verify that run-to-run control totals and other application totals are reconciled on a timely basis. This may be tested by independent balancing or reviewing past reconciliations.
- **Error control and correction**—In the form of reports that provide evidence of appropriate review, research, timely correction and resubmission. Input errors and rejections should be reviewed prior to resubmission. Managerial review and authorization of corrections should be evidenced. Testing of this effort can be achieved by retabulating or reviewing past error corrections.
- **Distribution of reports**—Critical output reports should be produced and maintained in a secure area and distributed in an authorized manner. The distribution process can be tested by observation and review of distribution output logs. Access to online output reports should be restricted. Online access may be tested through a review of the access rules or by monitoring user output.
- **Review and testing of access authorizations and capabilities**—Access control tables provide information regarding access levels by individuals. Access should be based on job descriptions and should provide for a separation of duties. Testing can be performed through the review of access rules to ensure that access has been granted as management intended.

Activity reports provide details, by user, of activity volume and hours. Activity reports should be reviewed to ensure that activity occurs only during authorized hours of operation.

Violation reports indicate any unsuccessful and unauthorized access attempts. Violation reports should indicate the terminal location, date and time of attempted access. These reports should evidence managerial review. Repeated unauthorized access violations may indicate attempts to circumvent access controls. Testing may include review of follow-up procedures.

## 3.14.4 DATA INTEGRITY TESTING

Data integrity testing is a set of substantive tests that examines accuracy, completeness, consistency and authorization of data presently held in a system. It employs testing similar to that used for input control. Data integrity tests will indicate failures in input or processing controls. Controls for ensuring the integrity of accumulated data in a file can be exercised by regularly checking data in the file. When this checking is done against authorized source documentation, it is common to check only a portion of the file at a time. Since the whole file is regularly checked in cycles, the control technique is often referred to as cyclical checking.

Two common types of data integrity tests are relational and referential integrity tests. Relational integrity tests are performed at the data element and record-based levels. Relational integrity is enforced through data validation routines built into the application or by defining the input condition constraints and data characteristics at the table definition in the database stage. Sometimes it is a combination of both.

Referential integrity tests define existence relationships between entities in a database that needs to be maintained by the DBMS. It is required for maintaining interrelation integrity in the relational data model. Whenever two or more relations are related through referential constraints (primary and foreign key), it is necessary that references be kept consistent in the event of insertions, deletions and updates to these relations. Database software generally provides various built-in automated procedures for checking and ensuring referential integrity. Referential integrity checks involve ensuring that all references to a primary key from another file (i.e., a foreign key) actually exist in their original file. In nonpointer databases (e.g., relational), referential integrity checks involve making sure that all foreign keys exist in their original table.

# Systems and Infrastructure Life Cycle Management

## 3.14.5 DATA INTEGRITY IN ONLINE TRANSACTION PROCESSING SYSTEMS

In multiuser transaction systems, it is necessary to manage parallel user access to stored data typically controlled by a DBMS, and deliver fault tolerance. Of particular importance are four online data integrity requirements known collectively as the ACID principle:

- **Atomicity**—From a user perspective, a transaction is either completed in its entirety (i.e., all relevant database tables are updated) or not at all. If an error or interruption occurs, all changes made up to that point are backed out.
- **Consistency**—All integrity conditions in the database are maintained with each transaction, taking the database from one consistent state into another consistent state.
- **Isolation**—Each transaction is isolated from other transactions and hence each transaction only accesses data that are part of a consistent database state.
- **Durability**—If a transaction has been reported back to a user as complete, the resulting changes to the database survive subsequent hardware or software failures.

This type of testing is vital in today's vast array of online Internet-accessible, multiuser DBMSs.

## 3.14.6 TEST APPLICATION SYSTEMS

Testing the effectiveness of application controls involves analyzing computer application programs, testing computer application program controls, or selecting and monitoring data process transactions. Testing controls by applying appropriate audit procedures is important to ensure their functionality and effectiveness. Methods and techniques for each category are described in **exhibit 3.28**.

| Exhibit 3.28—Testing Application Systems | | | |
|---|---|---|---|
| **Analyzing Computer Application Programs** | | | |
| **Technique** | **Description** | **Advantages** | **Disadvantages** |
| Snapshot | • Records flow of designated transactions through logic paths within programs | • Verifies program logic | • Requires extensive knowledge of the IS environment |
| Mapping | • Identifies specific program logic that has not been tested, and analyzes programs during execution to indicate whether program statements have been executed | • Increases efficiency by identifying unused code<br>• Identifies potential exposures | • Cost of software |
| Tracing and tagging | • Tracing shows the trail of instructions executed during an application. Tagging involves placing an indicator on selected transactions at input and using tracing to track them. | • Provides an exact picture of sequence of events, and is effective with live and simulated transactions | • Requires extensive amounts of computer time, an intimate knowledge of the application program and additional programming to execute trace routines |

# Systems and Infrastructure Life Cycle Management

| Exhibit 3.28—Testing Application Systems (cont.) | | | |
|---|---|---|---|
| **Analyzing Computer Application Programs** | | | |
| **Technique** | **Description** | **Advantages** | **Disadvantages** |
| Test data/deck | • Simulates transactions through real programs | • May use actual master files or dummies<br>• Source code review is unnecessary.<br>• Can be used on a surprise basis<br>• Provides objective review and verification of program controls and edits<br>• Initial use can be limited to specific program functions minimizing scope and complexity.<br>• Requires minimal knowledge of the IS environment | • Difficult to ensure that the proper program is checked<br>• Risk of not including all transaction scenarios<br>• Requires good knowledge of application systems<br>• Does not test master file and master file records |
| Base-case system evaluation | • Uses test data sets developed as part of a comprehensive testing of programs<br>• Verifies correct system operations before acceptance, as well as periodic revalidation | • Comprehensive testing verification and compliance testing | • Extensive effort to maintain data sets<br>• Close cooperation is required among all parties. |
| Parallel operation | • Processes actual production data through existing and newly developed programs at the same time and compares results, and is used to verify changed production prior to replacing existing procedures | • Verifies new system before discontinuing the old one | • Added processing costs |
| Integrated testing facility | • Creates a fictitious file in the database with test transactions processed simultaneously with live data | • Periodic testing does not require separate test process. | • Need for careful planning<br>• Need to isolate test data from production data |
| Parallel simulation | • Processes production data using computer programs that simulate application program logic | • Eliminates need to prepare test data | • Programs must be developed. |
| Transaction selection programs | • Use audit software to screen and select transactions input to the regular production cycle | • Independent of production system<br>• Controlled by the auditor<br>• Requires no modification to production systems | • Cost of development and maintenance |

# Systems and Infrastructure Life Cycle Management

| Exhibit 3.28—Testing Application Systems (cont.) | | | |
|---|---|---|---|
| **Analyzing Computer Application Programs** | | | |
| **Technique** | **Description** | **Advantages** | **Disadvantages** |
| Embedded audit data collection | • Software embedded in host computer applications screens. It selects input transactions and generated transactions during production. Usually, it is developed as part of system development. Types include:<br>– Systems control audit review file (SCARF)—Auditor determines reasonableness of tests incorporated into normal processing. It provides information for further review.<br>– Sample audit review file (SARF)—Randomly selects transactions to provide representative file for analysis | • Provides sampling and productions statistics | • High cost of development and maintenance<br>• Auditor independence issues |
| Extended records | • Gathers all data that have been affected by a particular program | • Records are put into one convenient file. | • Adds to data storage costs and overhead, and to system development costs |

To facilitate the evaluation of application system tests, an IS auditor may also want to use generalized audit software (GAS). This is particularly useful when specific application control weaknesses are discovered that affect, for example, updates to master file records and certain error conditions on specific transaction records. Additionally, GAS can be used to perform certain application control tests, such as parallel simulation, in comparing expected outcomes to live data.

## 3.14.7 CONTINUOUS ONLINE AUDITING

Continuous online auditing is becoming increasingly important in today's e-business world since it provides a method for the IS auditor to collect evidence on system reliability while normal processing takes place. The approach allows IS auditors to monitor the operation of such a system on a continuous basis and gather selective audit evidence through the computer. If the selective information collected by the computer technique is not deemed serious or material enough to warrant immediate action, the information is stored in separate audit files for verification by the IS auditor at a later time. The continuous audit approach cuts down on needless paperwork and leads to the conduct of an essentially paperless audit. In such a setting, an IS auditor can report directly through the microcomputer on significant errors or other irregularities that may require immediate management action. This approach reduces audit cost and time.

Continuous audit techniques are important IS audit tools, particularly when they are used in time-sharing environments that process a large number of transactions but leave a scarce paper trail. By permitting IS auditors to evaluate operating controls on a continuous basis without disrupting the organization's usual operations, continuous audit techniques improve the security of a system. When a system is misused by someone withdrawing money from an inoperative account, a continuous audit technique will report this withdrawal in a timely fashion to the IS auditor. Thus the time lag between the misuse of the system and the detection of that misuse is reduced. The realization that failures, improper manipulation and lack of controls will be detected on a timely basis by the use of continuous audit procedures gives IS auditors and management greater confidence in a system's reliability.

# Systems and Infrastructure Life Cycle Management

## 3.14.8 ONLINE AUDITING TECHNIQUES

There are five types of automated evaluation techniques applicable to continuous online auditing:

1. **Systems Control Audit Review File and Embedded Audit Modules (SCARF/EAM)**—The use of this technique involves embedding specially written audit software in the organization's host application system so the application systems are monitored on a selective basis.
2. **Snapshots**—This technique involves taking what might be termed pictures of the processing path that a transaction follows, from the input to the output stage. With the use of this technique, transactions are tagged by applying identifiers to input data and recording selected information about what occurs for the auditor's subsequent review.
3. **Audit hooks**—This technique involves embedding hooks in application systems to function as red flags and to induce IS auditors to act before an error or irregularity gets out of hand.
4. **Integrated test facility (ITF)**—In this technique, dummy facilities are set up and included in an auditee's production files. The IS auditor can make the system either process live transactions or test transactions during regular processing runs, and have these transactions update the records of the dummy entity. The operator enters the test transactions simultaneously with the live transactions that are entered for processing. The auditor then compares the output with the data that have been independently calculated to verify the correctness of the computer-processed data.
5. **Continuous and intermittent simulation (CIS)**—During a process run of a transaction, the computer system simulates the instruction execution of the application. As each transaction is entered, the simulator decides whether the transaction meets certain predetermined criteria and, if so, audits the transaction. If not, the simulator waits until it encounters the next transaction that meets the criteria.

In **exhibit 3.29**, the relative advantages and disadvantages of the various concurrent audit tools are presented.
The use of each of the continuous audit techniques has advantages and disadvantages. Their selection and implementation depends, to a large extent, on the complexity of an organization's computer systems and applications, and the IS auditor's ability to understand and evaluate the system with and without the use of continuous audit techniques. In addition, IS auditors must recognize that continuous audit techniques are not a cure for all control problems and that the use of these techniques provides only limited assurance that the information processing systems examined are operating as they were intended to function.

| Exhibit 3.29—Concurrent Audit Tools—Advantages and Disadvantages | | | | | |
|---|---|---|---|---|---|
| | **SCARF/EAM** | **ITF** | **Snapshots** | **CIS** | **Audit Hooks** |
| Complexity | Very high | High | Medium | Medium | Low |
| Useful when: | Regular processing cannot be interrupted. | It is not beneficial to use test data. | An audit trail is required. | Transactions meeting certain criteria need to be examined. | Only select transactions or processes need to be examined. |

# 3.15 AUDITING SYSTEMS DEVELOPMENT, ACQUISITION AND MAINTENANCE

The IS auditor's tasks in system development, acquisition and maintenance generally include the following:

• Meet with key systems development and user project team members to determine the main components, objectives and user requirements of the system to identify the areas that require controls.
• Discuss the selection of appropriate controls with systems development and user project team members to determine and rank the major risks to and exposures of the system.
• Discuss references to authoritative sources with systems development and user project team members to identify controls to mitigate the risks to and exposures of the system.

# Systems and Infrastructure Life Cycle Management

- Evaluate available controls and participate in discussions with systems development and user project team members to advise the project team regarding the design of the system and implementation of controls.
- Periodically meet with systems development and user project team members, and review the documentation and deliverables to monitor the systems development process to ensure that controls are implemented, user and business requirements are met, and the systems development/acquisition methodology is being followed. Also review and evaluate the application system audit trails to ensure that documented controls are in place to address all security, edit and processing controls. Audit trails are tracking mechanisms that can help IS auditors ensure program change accountability. Tracking information in a change management system includes:
  - History of all work order activity (date of work order, programmer assigned, changes made and date closed)
  - History of logons and logoffs by programmers
  - History of program deletions
- Participate in postimplementation reviews.
- Review appropriate documentation, discuss with key personnel, and use observation to evaluate system maintenance standards and procedures to ensure their adequacy.
- Discuss and examine supporting records to test system maintenance procedures to ensure that they are being applied as described in the standards.
- Analyze test results and other audit evidence to evaluate the system maintenance process to determine whether control objectives were achieved.
- Identify and test existing controls to determine the adequacy of production library security to ensure the integrity of the production resources.

## 3.15.1 PROJECT MANAGEMENT

Throughout the project management process the IS auditor should analyze the associated risks and exposures inherent in each phase of the SDLC and ensure that the appropriate control mechanisms are in place to minimize these risks in a cost-effective manner. Caution should be exercised to avoid recommending controls that cost more to administer than the associated risks the controls are designed to minimize.

When reviewing the SDLC process, the IS auditor should obtain documentation from the various phases and attend project team meetings, offering advice to the project team throughout the system development process. The IS auditor should also assess the project team's ability to produce key deliverables by the promised dates.

Typically, the IS auditor should review the adequacy of the following project management activities:
- Levels of oversight by project committee/board
- Risk management methods within the project
- Issue management
- Cost management
- Processes for planning and dependency management
- Reporting processes to senior management
- Change control processes
- Stakeholder management involvement
- Sign-off process—At a minimum, signed approvals from systems development and user management responsible for the cost of the project and/or use of the system

Additionally, adequate and complete documentation of all phases of the SDLC process should be evident. Typical types of documentation may include, but should not be limited to, the following:
- Objectives defining what is to be accomplished during that phase
- Key deliverables by phases with project personnel assigned direct responsibilities for these deliverables
- A project schedule with highlighted dates for the completion of key deliverables
- An economic forecast for that phase, defining resources and the cost of the resources required to complete the phase

## 3.15.2 FEASIBILITY STUDY

The IS auditor should perform the following functions:
- Review the documentation produced in this phase for reasonableness.
- Determine whether all cost justifications/benefits are verifiable and present them, showing the anticipated benefits to be realized.
- Identify and determine the criticality of the need.
- Determine if a solution can be achieved with systems already in place. If not, review the evaluation of alternative solutions for reasonableness.
- Determine the reasonableness of the chosen solution.

## 3.15.3 REQUIREMENTS DEFINITION

The IS auditor should perform the following functions:
- Obtain the detailed requirements definition document and verify its accuracy through interviews with the relevant user departments.
- Identify the key team members on the project team and verify all affected user groups have appropriate representation.
- Verify that project initiation and cost have received proper management approval.
- Review the conceptual design specifications (transforms [e.g., Laplace transform], data descriptions) to ensure that they address the needs of the user.
- Review the conceptual design to ensure that control specifications have been defined.
- Determine whether a reasonable number of vendors received a proposal covering the project scope and user requirements.
- Review the UAT specification.
- Determine whether the application is a candidate for the use of an embedded audit routine. If so, request that the routine be incorporated in the conceptual design of the system.

| PRACTICE QUESTION |
|---|
| 3-6     When auditing the requirements phase of a software acquisition, the IS auditor should:<br><br>A.    assess the feasibility of the project timetable.<br>B.    assess the vendor's proposed quality processes.<br>C.    ensure that the best software package is acquired.<br>D.    review the completeness of the specifications. |
| *See answers and explanations to the practice questions at the end of the chapter. (page 274)* |

## 3.15.4 SOFTWARE ACQUISITION PROCESS

The IS auditor should perform the following functions:
- Analyze the documentation from the feasibility study to determine whether the decision to acquire a solution was appropriate.
- Review the RFP to ensure that it covers the items listed in this section.
- Determine whether the selected vendor is supported by RFP documentation.
- Attend agenda-based presentations and conference room pilots to ensure that the system matches the vendor's response to the RFP.
- Review the vendor contract prior to its signing to ensure that it includes the items listed.
- Ensure the contract is reviewed by legal counsel before it is signed.

| PRACTICE QUESTION |
| --- |

3-7  An organization decides to purchase a software package instead of developing it. In such a case, the design and development phases of a traditional software development life cycle (SDLC) would be replaced with:

    A.   selection and configuration phases.
    B.   feasibility and requirements phases.
    C.   implementation and testing phases.
    D.   nothing; replacement is not required.

*See answers and explanations to the practice questions at the end of the chapter. (page 274)*

## 15.5 DETAILED DESIGN AND DEVELOPMENT

The IS auditor should perform the following functions:
- Review the system flowcharts for adherence to the general design. Verify that appropriate approvals were obtained for any changes and all changes were discussed and approved by appropriate user management.
- Review the input, processing and output controls designed into the system for appropriateness.
- Interview the key users of the system to determine their understanding of how the system will operate, and assess their level of input into the design of screen formats and output reports.
- Assess the adequacy of audit trails to provide traceability and accountability of system transactions.
- Verify the integrity of key calculations and processes.
- Verify that the system can identify and process erroneous data correctly.
- Review the quality assurance results of the programs developed during this phase.
- Verify that all recommended corrections to programming errors were made and the recommended audit trails or EAMs were coded into the appropriate programs.

| PRACTICE QUESTION |
| --- |

3-8  User specifications for a project using the traditional SDLC methodology have not been met. An IS auditor looking for a cause should look in which of the following areas?

    A.   Quality assurance
    B.   Requirements
    C.   Development
    D.   User training

*See answers and explanations to the practice questions at the end of the chapter. (page 275)*

## 3.15.6 TESTING

Testing is crucial in determining the user requirements have been validated, the system is performing as anticipated and internal controls work as intended. Therefore it is essential that the IS auditor be involved in reviewing this phase and perform the following:
- Review the test plan for completeness; indicate evidence of user participation such as user development of test scenarios and/or user sign-off of results; and consider rerunning critical tests.
- Reconcile control totals and converted data.
- Review error reports for their precision in recognizing erroneous data and resolution of errors.
- Verify cyclical processing for correctness (month-end, year-end processing, etc.).
- Interview end users of the system for their understanding of new methods, procedures and operating instructions.
- Review system and end-user documentation to determine its completeness and verify its accuracy during the test phase.
- Review parallel testing results for accuracy.
- Verify that system security is functioning as designed by developing and executing access tests.
- Review unit and system test plans to determine whether tests for internal controls are planned and performed.

# Systems and Infrastructure Life Cycle Management

• Review the user acceptance testing and ensure that the accepted software has been delivered to the implementation team. The vendor should not be able to replace this version.
• Review procedures used for recording and following through on error reports.

## 3.15.7 IMPLEMENTATION PHASE

This phase is initiated only after a successful testing phase. The system should be installed according to the organization's change control procedures. The IS auditor should verify that appropriate sign-offs have been obtained prior to implementation and perform the following:
• Review the programmed procedures used for scheduling and running the system along with system parameters used in executing the production schedule.
• Review all system documentation to ensure its completeness and that all recent updates from the testing phase have been incorporated.
• Verify all data conversion to ensure that they are correct and complete before implementing the system in production.

## 3.15.8 POSTIMPLEMENTATION REVIEW

After the new system has stabilized in the production environment, a postimplementation review should be performed. Prior to this review, it is important that sufficient time be allowed for the system to stabilize in production. In this way, any significant problems will have had a chance to surface.

The IS auditor should perform the following functions:
• Determine if the system's objectives and requirements were achieved. During the postimplementation review, careful attention should be paid to the end users' utilization and overall satisfaction with the system. This will indicate whether the system's objectives and requirements were achieved.
• Determine if the cost benefits identified in the feasibility study are being measured, analyzed and accurately reported to management.
• Review program change requests performed to assess the type of changes required of the system. The type of changes requested may indicate problems in the design, programming or interpretation of user requirements.
• Review controls built into the system to ensure that they are operating according to design. If an EAM was included in the system, use this module to test key operations.
• Review operators' error logs to determine if there are any resource or operating problems inherent within the system. The logs may indicate inappropriate planning or testing of the system prior to implementation.
• Review input and output control balances and reports to verify that the system is processing data accurately.

## 3.15.9 SYSTEM CHANGE PROCEDURES AND THE PROGRAM MIGRATION PROCESS

Following implementation and stabilization, a system enters into the ongoing development or maintenance stage. This phase continues until the system is retired. The phase involves those activities required to either correct errors in the system or enhance the capabilities of the system. In this regard, the IS auditor should consider the following:
• The use and existence of a methodology for authorizing, prioritizing and tracking system change requests from the user
• Whether emergency change procedures are addressed in the operations manuals
• Whether change control is a formal procedure for the user and the development groups
• Whether the change control log ensures all changes shown were resolved
• The user's satisfaction with the turnaround—timeliness and cost—of change requests
• The adequacy of the security access restrictions over production source and executable modules
• The adequacy of the organization's procedures for dealing with emergency program changes
• The adequacy of the security access restrictions over the use of the emergency logon IDs

# Systems and Infrastructure Life Cycle Management

For a selection of changes on the change control log:
- Determine whether changes to requirements resulted in appropriate change-development documents such as program and operations documents.
- Determine whether changes were made as documented.
- Determine whether current documentation reflects the changed environment.
- Evaluate the adequacy of the procedures in place for testing system changes.
- Review evidence (test plans and test results) to ensure that procedures are carried out as prescribed by organizational standards.
- Review the procedures established for ensuring executable and source code integrity.
- Review production executable modules and verify there is one and only one corresponding version of the program source code.

Additionally, the IS auditor should review the overall change management process for possible improvements in acknowledgement, response time, response effectiveness and user satisfaction with the process.

# 3.16 BUSINESS APPLICATION SYSTEMS

To develop effective audit programs, the IS auditor must obtain a clear understanding of the application system under review. Some types of application systems and the related processes are described in the following sections.

Numerous financial and operational functions are computerized for the purpose of improving efficiency and increasing the reliability of information. These applications range from traditional (including general ledger, accounts payable and payroll) to industry-specific (such as bank loans, trade clearing and material requirements planning). Given their unique characteristics, computerized application systems add complexity to audit efforts. These characteristics may include limited audit trails, instantaneous updating and information overload. Application systems may reside in the various environments that follow.

## 3.16.1 ELECTRONIC COMMERCE

Electronic commerce (e-commerce) is one of the most popular e-business implementations. It is the buying and selling of goods online, usually via the Internet. Typically, a web site will advertise goods and services, and the buyer will fill in a form on the web site to select the items to be purchased and provide delivery and payment details or banking services such as transfers and payment orders. The web site may gather details about customers and offer other items that may be of interest. The cost of a brick-and-mortar store is avoided and the savings are often a benefit to the customers, sometimes leading to spectacular growth. The term e-business includes buying and selling online as well as other aspects of online business such as customer support or relationships between businesses.

E-commerce, as a general model, uses technology to enhance the processes of commercial transactions among a company, its customers and business partners. The used technology can include the Internet, multimedia, web browsers, proprietary networks, ATMs and home banking, and the traditional approach to EDI. However, the primary area of growth in e-commerce is through the use of the Internet as an enabling technology.

### E-commerce Models
E-commerce models include the following:
- **Business-to-consumer (B-to-C) relationships**—The greatest potential power of e-commerce comes from its ability to redefine the relationship with customers in creating a new convenient, low-cost channel to transact business. Companies can tailor their marketing strategies to an individual customer's needs and wants. As more of its business shifts online, a company will have an enhanced ability to track how its customers interact with it.
- **Business-to-business (B-to-B) relationships**—The relationship among the selling services of two or more businesses opens up the possibility of reengineering business processes across the boundaries that have traditionally separated external entities from each other. Because of the ease of access and the ubiquity of the Internet, for example, companies can build

business processes (order processing, payments and after-sale service) that combine previously separated activities. The result is a faster, higher-quality and lower-cost set of transactions. The market has even created a subdivision of B-to-B called business-to-small business (B-to-SB) relationships.

- **Business-to-employee (B-to-E) relationships**—Web technologies also assist in the dissemination of information to and among an organization's employees.
- **Business-to-government (B-to-G) relationships**—Covers all transactions between companies and government organizations. Currently this category is in its infancy, but it could expand quite rapidly as governments use their own operations to promote awareness and growth of e-commerce. In addition to public procurement, administrations may also offer the option of electronic interchange for such transactions as VAT returns and the payment of corporate taxes.
- **Consumer-to-government (C-to-G) relationships**—This has not yet emerged. However, in the wake of a growth of both the business-to-consumer and business-to-government categories, governments may extend electronic interaction to such areas as welfare payments and self-assessed tax returns.
- **Exchange-to-exchange (X-to-X) relationships**—This is the linkage of multiple B-to-B relationships/marketplaces. X-to-X is the next logical step beyond B-to-B or B-to-SB. Organizations are able to purchase supplies/products at competitive prices.

## E-commerce Architectures

There are a large number of choices to be made in determining an appropriate e-commerce architecture. Initially, e-commerce architectures were either two-tiered (i.e., client browser and web server) or three-tiered (i.e., client browser, web server and database server). With increasing emphasis on integrating the web channel with a business' internal legacy systems and the systems of its business partners, company systems now typically will run on different platforms, running different software and with different databases. It is also the case that, in addition to supporting browser connections, companies eventually may move in the direction of supporting connections from active content clients, mobile phones or other wireless devices, and host-to-host connections.

The challenge of integrating diverse technologies within and beyond the business has increasingly led companies to move to component-based systems that utilize a middleware infrastructure based around an application server. This supports current trends in the evolution of software development—build systems from proven quality and cataloged components, just as hardware is built. While this is yet to be fully realized, component models—notably Microsoft's COM and Sun's Enterprise Java Beans—are widely used.

E-components that one could expect to see in a B-to-C system would include marketing, sales and customer service components (e.g., personalization, membership, product catalog, customer ordering, invoicing, shipping, inventory replacement, online training and problem notification).

Application servers will support a particular component model and provide services (such as data management, security and transaction management) either directly or through connection to another service or middleware product such as MQSeries. A number of major software vendors (including BEA, IBM, Oracle, Sybase and Linux) offer application servers. Microsoft offers their version of an application server (Microsoft Transaction Server) as a standard part of the NT server operating system.

Application servers in conjunction with other middleware products provide for multitiered systems (i.e., a business transaction can span multiple platforms and software layers). For example, a system's presentation layer typically will consist of a browser or other client application. A web server will be used to manage web content and connections; business logic and other services will be provided by the application server; and one or more database(s) will be used for data storage.

Databases play a key role in most e-commerce systems, maintaining data for web site pages, accumulating customer information, and possibly storing click-stream data for analyzing web site usage. To provide full functionality and achieve back-end efficiencies, an e-commerce system may involve connections to in-house legacy systems—accounting, inventory management or possibly an ERP system—or business partner systems. Thus, further business logic and data persistence tiers are added.

# Systems and Infrastructure Life Cycle Management

Note that, for security reasons, persistent customer data should not be stored on web servers that are exposed directly to the Internet. Extensible Markup Language (XML) is also likely to form an important part of an organization's overall e-commerce architecture. While originally conceived as a technique to facilitate electronic publishing, XML was quickly seized on as a medium that could store and enclose any kind of structured information so it could be passed between different computing systems. XML has emerged as a key means of exchanging a wide variety of data on the web and elsewhere. In addition to basic XML, a variety of associated standards has been and is continuing to be developed. Some of these include:

- **Extensible Stylesheet Language (XSL)**—Defines how an XML document is to be presented; e.g., on a web page
- **XML Query (XQuery)**—Deals with querying XML format data
- **XML Encryption**—Deals with encrypting, decrypting and digitally signing XML documents

A particularly important offshoot of XML is Web Services. Web Services represents a way of using XML format information to remotely invoke processing. Since a Web Services message can contain both an XML document and a corresponding schema defining the document, in theory it is self-describing and assists in achieving the goal of "loose coupling." If the format of a Web Services message changes, the receiving Web Services will still work, provided the accompanying schema is updated. This advantage, combined with the support for Web Services that has emerged from major software industry players such as IBM, Microsoft and SUN, means Web Services is emerging as the key middleware to connect distributed web systems.

Cooperation between the different software vendors means that Web Services can interoperate, irrespective of hardware, operating system and programming language. Development environments such as Microsoft's .NET and Sun's ONE (Open Net Environment) support Web Services.

It will be necessary to reach some agreement on metadata definitions for Web Services to serve as a means of enabling cooperative processing across organizational boundaries. (Metadata are data about data, and the term is referred to in Web Services' standards as ontology). Web Services may be successfully called and the resulting XML data may be successfully parsed by the calling program, but to use these data effectively it is necessary to understand the business meaning of the data. This is similar to previous attempts at interorganizational computing (such as EDI), where it was necessary to agree in advance on electronic document formats and meanings. Efforts are now underway to define industry-standard XML documents and standard business process definitions that can be represented in XML.

## E-commerce Risks

E-commerce, as any other form of commerce, depends on the existence of a level of trust between two parties. For example, the Internet presents a challenge between the buyer and seller, similar to those a catalog or direct-mail retailer faces. The challenges are proving to the buyer that the seller is who they say they are, proving to the buyer that their personal information such as credit card numbers (and other personally identifiable information) remains confidential and that the seller cannot later refute the occurrence of a valid transaction. Therefore some of the most important elements at risk are:

- **Confidentiality**—Potential consumers are concerned about providing unknown vendors with personal (sometimes sensitive) information for a number of reasons including the possible theft of credit card information from the vendor following a purchase. Connecting to the Internet via a browser requires running software on the computer that has been developed by someone unknown to the organization. Moreover, the medium of the Internet is a broadcast network, which means that whatever is placed on it is routed over wide-ranging and essentially uncontrolled paths.
- **Integrity**—Data, both in transit and in storage, could be susceptible to unauthorized alteration or deletion (i.e., hacking or the e-business system itself could have design or configuration problems).
- **Availability**—The Internet holds out the promise of doing business on a 24-hour, seven-day-a-week basis. Hence high availability is important with any system's failure becoming immediately apparent to customers or business partners.
- **Authentication and nonrepudiation**—The parties to an electronic transaction should be in a known and trusted business relationship, which requires that they prove their respective identities before executing the transaction in preventing man-in-the-middle attacks (i.e., preventing the seller from being an impostor). Then, after the fact, there must be some manner of ensuring that the transacting parties cannot deny that the transaction was entered into and the terms on which it was completed.

# Systems and Infrastructure Life Cycle Management

- **Power shift to customers**—The Internet gives consumers unparalleled access to market information and generally makes it easier to shift between suppliers. Firms participating in e-business need to make their offerings attractive and seamless in terms of service delivery. This will involve not only system design, but also reengineering of business processes. Back-end support processes need to be as efficient as possible because, in many cases, doing business over the Internet forces down prices (e.g., online share brokering). To avoid losing their competitive advantage of doing business online, firms need to enhance their services, differentiate from the competition and build additional value. Hence the drive to personalize web sites by targeting content based on analyzed customer behavior and allowing direct contact with staff through instant messaging technology and other means.

It is important to take into consideration the importance of security issues that extend beyond confidentiality objectives.

## E-commerce Requirements

Some e-commerce requirements include:
- Build a business case (IT as an enabler).
- Develop a clear business purpose.
- Use technology to first improve costs.
- Build business case around the four Cs: customers, costs, competitors and capabilities.

Other requirements for e-commerce involve the following:
- **Top-level commitment**—Because of the breadth of changes required (i.e., business processes, company culture, technology and customer boundaries), e-commerce cannot succeed without a clear vision and strong commitment from the top of the organization.
- **Business process reconfiguration**—Technology is not the key innovation needed to make e-commerce work, but it is the ingenuity needed to envision how that technology can enable the company to fundamentally reconfigure some of its basic business processes. This requires thinking that is outside-the-box and outside-the-walls; i.e., looking outside of the organization and understanding what customers are doing and how changes in the overall process can create new value for them.
- **Links to legacy systems**—Organizations must take seriously the requirement to accelerate response times, provide real interaction to customers and customize responses to individual customers. Specifically, in applying enterprise application integration (EAI), organizations must create online interfaces and make sure those interfaces communicate with existing databases and systems for customer service and order processing. A term often referred to in establishing this communication is middleware, which is defined as independent software and services that distributed business applications use to share computing resources across heterogeneous technologies. A range of middleware technologies—message brokers, gateways, process managers, data transformation software and file transfer—are likely to be deployed to create an integration infrastructure. Increasingly, integration will be viewed not as a responsibility of an individual application development team, but as something to be managed across the organization using a standard approach and technologies.

## E-commerce Audit and Control Issues (Best Practices)

When reviewing the adequacy of contracts in e-commerce applications, audit and control professionals should assess applicable use of the following items:
- Security mechanisms and procedures that, taken together, constitute a security architecture for e-commerce (e.g., Internet firewalls, public key infrastructure [PKI], encryption, certificates and password management)
- Firewall mechanisms that are in place to mediate between the public network (the Internet) and an organization's private network
- A process whereby participants in an e-commerce transaction can be identified uniquely and positively (e.g., process of using some combination of public and private key encryption and certifying key pairs)
- Digital signatures so the initiator of an e-commerce transaction can be uniquely associated with it. Attributes of digital signatures include:
  - The digital signature is unique to the person using it.
  - The signature can be verified.
  - The mechanism for generating and affixing the signature is under the sole control of the person using it.
  - The signature is linked to data in such a manner that if the data are changed, the digital signature is invalidated.

# Systems and Infrastructure Life Cycle Management

- Infrastructure to manage and control public key pairs and their corresponding certificates which include:
  - **Certificate authority (CA)**—Attests, as trusted provider of the public/private key pairs, to the authenticity of the owner (entity or individual) to whom a public/private key pair has been given. The process involves a CA who makes a decision to issue a certificate based on evidence or knowledge obtained in verifying the identity of the recipient. On verifying the identity of the recipient, the CA signs the certificate with its private key for distribution to the user. On receipt, the user will decrypt the certificate with the CA's public key (e.g., commercial CAs such as Verisign provide public keys on web browsers). The ideal CA is authoritative (someone that the user trusts) for the name or key space it represents.
  - **Registration authority (RA)**—An optional entity separate from a CA that would be used by a CA with a very large customer base. CAs use RAs to delegate some of the administrative functions associated with recording or verifying some or all of the information needed by a CA to issue certificates or certification revocation lists (CRLs) and to perform other certificate management functions. However, with this arrangement, the CA retains sole responsibility for signing either digital certificates or CRLs. If an RA is not present in the established PKI structure, the CA is assumed to have the same set of capabilities as defined for an RA.
  - **Certification revocation list**—Instrument for checking the continued validity of the certificates. If a certificate is compromised, if the holder is no longer authorized to use the certificate or if there is a fault in binding the certificate to the holder, the certificate must be revoked and taken out of circulation as rapidly as possible and all parties in the trust relationship must be informed. The CRL is usually a highly controlled online database through which subscribers and administrators may determine the status of a target partner's certificate.
  - **Certification practice statement (CPS)**—A detailed set of rules governing the certificate authority's operations. The CPS provides an understanding of the value and trustworthiness of certificates issued by a given CA, the terms of the controls that an organization observes, the method used to validate the authenticity of certificate applicants and the CA's expectations of how its certificates may be used.
- Procedures in place to control changes to an e-commerce presence
- E-commerce application logs which are monitored by responsible personnel. This includes operating system logs and console messages, network management messages, firewall logs and alerts, router management messages, intrusion detection alarms, application and server statistics, and system integrity checks.
- Methods and procedures to recognize security breaches when they occur (network and host-based intrusion detection systems [IDSs])
- Features in e-commerce applications to reconstruct the activity performed by the application
- Protection in place to ensure that data collected about individuals are not disclosed without their consent nor used for purposes other than that for which they are collected
- Means to ensure confidentiality of data communicated between customers and vendors (safeguarding resources such as through encrypted Secure Sockets Layer [SSL])
- Mechanisms to protect the presence of e-commerce and supporting private networks from computer viruses and to prevent them from propagating viruses to customers and vendors
- Features within the e-commerce architecture to keep all components from failing and allow them to repair themselves, if they should fail
- Plan and procedure to continue e-commerce activities in the event of an extended outage of required resources for normal processing
- Commonly understood set of practices and procedures to define management's intentions for the security of e-commerce
- Shared responsibility within an organization for e-commerce security
- Communications from vendors to customers about the level of security in an e-commerce architecture
- Regular program of audit and assessment of the security of e-commerce environments and applications to provide assurance that controls are present and effective

# Systems and Infrastructure Life Cycle Management

## 3.16.2 ELECTRONIC DATA INTERCHANGE

In use for more than 20 years, EDI is one of the first e-commerce applications used among business partners for transmitting business transactions between organizations with dissimilar computer systems. EDI involves the exchange and transmittal of business documents (such as invoices, purchase orders and shipping notices) in a standard, machine-processable format. The transmissions use standard formats such as specific record types and field definitions. The process works by translating data from a business application into a standard format, transmitting the data over communication lines to a trading partner, and then retranslating using the trading partner's application.

The benefits associated with the adoption of EDI include:
• Less paperwork
• Fewer errors during the exchange of information
• Improved information flow, database-to-database and company-to-company
• No unnecessary rekeying of data
• Fewer delays in communication
• Improved invoicing and payment processes

Since EDI replaces the traditional paper document exchange—such as purchase orders, invoices or material release schedules—the proper controls and edits need to be built within each company's application system to allow this communication to take place.

### General Requirements

An EDI system requires communications software, translation software and access to standards. Communications software moves data from one point to another, flags the start and end of an EDI transmission, and determines how acknowledgments are transmitted and reconciled. Translation software helps build a map and shows how the data fields from the application correspond to elements of an EDI standard. Later, it uses this map to convert data back and forth between the application and EDI formats.

To build a map, an EDI standard appropriate for the kind of EDI data to be transmitted is selected. For example, there are specific standards for invoices, purchase orders, advance shipping notices, etc.

The final step is to write a partner profile that tells the system where to send each transaction and how to handle errors and exceptions.

In summary, components of an EDI process include system software and application systems. EDI system software includes transmission, translation and storage of transactions initiated by or destined for application processing. EDI is also an application system in that the functions it performs are based on business needs and activities. The applications, transactions and trading partners supported will change over time, and the intermixing of transactions, purchase orders, shipping notices, invoices and payments in the EDI process makes it necessary to include application processing procedures and controls in the EDI process.

# Systems and Infrastructure Life Cycle Management

In reviewing EDI, IS auditors need to be aware of the two approaches related to EDI, the traditional proprietary version of EDI used by large companies and the development of EDI through the publicly available commercial infrastructure offered through the Internet. The difference between the approaches relates to cost, where use of a public commercial infrastructure such as the Internet provides significantly reduced costs vs. development of a customized proprietary approach. From a security standpoint, risks associated with not having a completely trustworthy relationship arise in addressing Internet security and risks.

## Traditional EDI

Moving data in a batch transmission process through the traditional EDI process generally involves three functions within each trading partner's computer system:

1. **Communications handler**—Process for transmitting and receiving electronic documents between trading partners via dial-up lines, public-switched network, multiple dedicated lines or a value-added network (VAN). VANs use computerized message switching and storage capabilities to provide electronic mailbox services similar to a post office. The VAN receives all the outbound transactions from an organization, sorts them by destination and passes them to recipients when they log on to check their mailbox and receive transmissions. VANs may also perform translation and verification services. VANs specializing in EDI applications also provide technical support, help desk and troubleshooting assistance for EDI and telecommunications problems. VANs help in configuration of software, offer upgrades to telecommunications connectivity, provide data and computer security, audit and trace transactions, recover lost data, and confirm service reliability and availability.

2. **EDI interface**—Interface function that manipulates and routes data between the application system and the communications handler. The interface consists of two components:
   - **EDI translator**—This device translates the data between the standard format (ANSI X12) and a trading partner's proprietary format.
   - **Application interface**—This interface moves electronic transactions to or from the application systems and performs data mapping. Data mapping is the process by which data are extracted from the EDI translation process and integrated with the data or processes of the receiving company. The EDI interface may generate and send functional acknowledgments, verify the identity of partners and check the validity of transactions by checking transmission information against a trading partner master file. Functional acknowledgments are standard EDI transactions that tell the trading partners that their electronic documents were received. Different types of functional acknowledgments provide various levels of detail and can, therefore, act as an audit trail for EDI transactions.

3. **Application system**—The programs that process the data sent to, or received from, the trading partner. Although new controls should be developed for the EDI interface, the controls for existing applications, if left unchanged, are usually unaffected.

Application-initiated transactions (such as purchase orders from the purchasing system) are passed to a common application interface for storage and interpretation. All outbound transactions are formatted according to an externally defined standard and batched by destination and transaction type by the translator. The batches of transactions, like functional groups, are routed to the communications processor for transmission. This entire process is reversed for inbound transactions, including invoices destined for the purchasing and accounts payable systems. Controls need to recognize and deal with error conditions and provide feedback on the process for the EDI system to be considered well-managed.

The EDI process can be accomplished in a number of ways. The best method depends on the volume and number of transactions being processed. Many organizations begin by using a microcomputer with a modem to receive and transmit transactions. The transactions to be sent are keyed into the microcomputer from computer or manually prepared documents. The transactions received are printed and then processed in the same way that a transaction received in the mail would be processed. Elimination of the manual intervention and use of paper documents is appropriate when the transaction volume is expected to be high enough to justify the added costs of developing and maintaining the customized application interfaces.

## Web-based EDI

Web-based EDI has come into prominence because of the following:
- Internet-through-Internet service providers offer a generic network access (i.e., not specific to EDI) for all computers connected to the Internet, whereas VAN services have typically used a proprietary network or a network gateway linked with a specific set of proprietary networks. The result is a substantially reduced cost to EDI applications.

# Systems and Infrastructure Life Cycle Management

- Its ability to attract new partners via web-based sites to exchange information, take orders and link the web site to back-end order processing and financial systems via EDI
- New security products available to address issues of confidentiality, authentication, data integrity and nonrepudiation of origin and return
- Improvements in the x.12 EDI formatting standard

Web-based EDI trading techniques aim to improve the interchange of information between trading partners, suppliers and customers by bringing down the boundaries that restrict how they interact and do business with each other. For example, the use of Internet service provider (ISP)-related services can provide functions similar to those of the more traditional VANs, but with a much broader array of available services (i.e., processing transactions of all types through the Internet). This is beneficial particularly for smaller organizations wanting to enter the e-commerce EDI market since ISPs have a ready network infrastructure of servers offering e-mail, web services and the network of routers, and modems attached to a permanent, high-speed Internet "backbone" connection.

## 3.16.3 EDI RISKS AND CONTROLS

The hybrid nature of EDI adds a new dimension to the design and auditing of the EDI process. The traditional procedures for managed and controlled implementation of system software—such as requirements definition, version and release identification, testing and limited implementation with a fallback strategy—apply to software used for EDI. In addition, there are issues and risks unique to EDI.

Foremost of these risks is transaction authorization. Since the interaction between parties is electronic, there is no inherent authentication occurring. Computerized data can look the same no matter what the source and do not include any distinguishing human element or signature.

Where responsibilities of trading partners are not clearly defined by a trading partner agreement, there could be uncertainty related to specific, legal liability. Therefore, it is important that, to protect both parties, any agreement be codified legally in what is known as a trading partner agreement. Another risk is the loss of business continuity. Corruption of EDI applications, whether done innocently or deliberately, could affect every EDI transaction undertaken by a company. This would have a negative impact on both customer and vendor relations. In an extreme situation, it could ultimately affect the ability of a company to stay in business.

Additional security risks include:
- Unauthorized access to electronic transactions
- Deletion or manipulation of transactions prior to or after establishment of application controls
- Loss or duplication of EDI transmissions
- Loss of confidentiality and improper distribution of EDI transactions while in the possession of third parties

## 3.16.4 CONTROLS IN EDI ENVIRONMENT

Security risks can be addressed by enforcing general controls and establishing an added layer of application control procedures over the EDI process that can take over where traditional application controls leave off. These controls need to secure the current EDI activity as well as historical activities that may be called on to substantiate business transactions should a dispute arise.

To protect EDI transmissions, the EDI process should include the following electronic measures:
- Standards should be set to indicate that the message format and content are valid to avoid transmission errors.
- Controls should be in place to ensure that standard transmissions are properly converted for the application software by the translation application.
- The receiving organization must have controls in place to test the reasonableness of messages received. This should be based on a trading partner's transaction history or documentation received that substantiates special situations.

# Systems and Infrastructure Life Cycle Management

- Controls should be established to guard against manipulation of data in active transactions, files and archives. Attempts to change records should be recorded by the system for management review and attention.
- Procedures should be established to determine messages are only from authorized parties and transmissions are properly authorized.
- Direct or dedicated transmission channels among the parties should exist to reduce the risk of tapping into the transmission lines.
- Data should be encrypted using algorithms agreed on by the parties involved.
- Electronic signatures should be in the transmissions to identify the source and destination.
- Message authentication codes should exist to ensure that what is sent is received.

The EDI process needs the ability to detect and deal with transactions that do not conform to the standard format or are from/to unauthorized parties. Options for handling detected errors include requesting retransmissions or manually changing the data.

The critical nature of many EDI transactions, such as orders and payments, requires that there be positive assurances that the transmissions were complete. The transactions need to be successfully passed from the originating computer application to the destination organization. Methods for providing these assurances include internal batch total checking, run-to-run and transmission record count balancing, and use of special acknowledgment transactions for functional acknowledgments.

Organizations desiring to exchange transactions using EDI are establishing a new business relationship. This business relationship needs to be defined so both parties can conduct business in a consistent and trusting manner. This relationship usually is defined in a legal document called a trading partner agreement. The document should define the transactions to be used, responsibilities of both parties in handling and processing the transactions, as well as the written business terms and conditions associated with the transactions.

The evolving nature of EDI means that the transaction standards are evolving too, particularly with use of the Internet as an available, low-cost primary business transaction delivery system. Specifically, not all trading partners desire or need to use the current standard. As a result, the EDI process needs to adapt to changes in standards and be able to support multiple versions of the standard as ISPs offer more EDI-related services (vs. VANs).

Other issues relate to many organizations with a large, installed base of applications that need to be retrofit to accommodate EDI. In addition, not all transactions will be processed through EDI. Some transactions will continue to be processed in the traditional way. The application processing control procedures must be modified to include the EDI transaction processing and the dual sources/destinations.

## *Receipt of Inbound Transactions*

Controls should ensure that all inbound EDI transactions are accurately and completely received (communication phase), translated (translation phase) and passed to an application (application interface phase) as well as processed only once.

The control considerations for receipt of inbound transactions are as follows:
- Use appropriate encryption techniques when using public Internet infrastructures for communication in assuring confidentiality, authenticity and integrity of transactions.
- Perform edit checks to identify erroneous, unusual or invalid transactions prior to updating an application.
- Perform additional computerized checking to assess transaction reasonableness, validity, etc. (Consider expert system front ends for complex comparisons.)
- Log each inbound transaction on receipt.
- Use control totals on receipt of transactions to verify the number and value of transactions to be passed to each application; reconcile totals between applications and with trading partners.
- Segment count totals built into the transaction set trailer by the sender.
- Control techniques in the processing of individual transactions such as check digits on control fields, loop or repeat counts.
- Ensure the exchange of control totals of transactions sent and received between trading partners at predefined intervals.
- Maintain a record of the number of messages received/sent and validate these with the trading partners from time to time.
- Arrange for security over temporary files and data transfer to ensure that inbound transactions are not altered or erased between time of transaction receipt and application updates.

## Outbound Transactions

Controls should ensure that only properly authorized outbound transactions are processed. This includes the objectives that outbound EDI messages are initiated on authorization, that they contain only preapproved transaction types and that they are sent only to valid trading partners.

The control considerations for outbound transactions are as follows:
- Controlling the set up and change of trading partner details
- Comparing transactions with trading partner transaction profiles
- Matching the trading partner number to the trading master file (prior to transmission)
- Limiting the authority of users within the organization to initiate specific EDI transactions
- Segregating initiation and transmission responsibilities for high-risk transactions
- Documenting management sign-off on programmed procedures and subsequent changes
- Logging all payment transactions to a separate file which is reviewed for authorization before transmission
- Segregating duties within the transaction cycle, particularly where transactions are automatically generated by the system
- Segregating access to different authorization processes in a transaction cycle
- Reporting large (value) or unusual transactions for review prior to or after transmission
- Logging outbound transactions in a secure temporary file until authorized and due for transmission
- Requiring paperless authorization which would establish special access to authorization fields (probably two levels, requiring the intervention of different users) within the computer system

## Auditing EDI

The IS auditor must evaluate EDI to ensure that all inbound EDI transactions are received and translated accurately, passed to an application, and processed only once.

To accomplish this goal, IS auditors must review the following:
- Internet encryption processes put in place to ensure authenticity, integrity, confidentiality and nonrepudiation of transactions
- Edit checks to identify erroneous, unusual or invalid transactions prior to updating the application
- Additional computerized checking to assess transaction reasonableness and validity
- Each inbound transaction to ensure that it is logged on receipt
- The use of control totals on receipt of transactions to verify the number and value of transactions to be passed to each application and reconcile totals between applications and with trading partners
- Segment count totals built into transaction set trailers by the sender
- Transaction set count totals built into the functional group headers by the sender
- Batch control totals built into the functional group headers by the sender
- The validity of the sender against trading partner details by:
  - Using control fields within an EDI message at either the transaction, function, group or interchange level (often within the EDI header, trailer or control record)
  - Using VAN sequential control numbers or reports (if applicable)
  - Sending an acknowledgment transaction to inform the sender of message receipt. The sender should then match this against a file/log of EDI messages sent.

EDI audits also involve:
- Audit monitors—Devices can be installed at EDI workstations to capture transactions as they are received. Such transactions can be stored in a protected file for use by the auditor. Consideration should be given to storage requirements for voluminous amounts of data.
- Expert systems—Within the context of utilizing the computer system for internal control checks, consideration should be given to having audit monitors evaluate the transactions received. Based upon judgmental rules, the system can determine the audit significance of such transactions and provide a report for the auditor's use.

As use of EDI becomes more widespread, additional methods for auditing transactions will be developed. It is important to stay current with these developments.

# Systems and Infrastructure Life Cycle Management

| PRACTICE QUESTION |
|---|
| 3-10    Which of the following procedures should be implemented to help ensure the completeness of inbound transactions via electronic data interchange (EDI)? <br><br>      A.    Segment counts built into the transaction set trailer <br>      B.    A log of the number of messages received, periodically verified with the transaction originator <br>      C.    An electronic audit trail for accountability and tracking <br>      D.    Matching acknowledgement transactions received to the log of EDI messages sent |
| *See answers and explanations to the practice questions at the end of the chapter. (page 275)* |

## 3.16.5 ELECTRONIC MAIL

Electronic mail (e-mail), may be the most heavily used feature of the Internet or LANs in an organization. At the most basic level, the e-mail process can be divided into two principal components:
• Mail servers—hosts that deliver, forward and store mail
• Clients—interface with users and allow users to read, compose, send and store e-mail messages

E-mail messages are sent in the same way as most Internet data. When a user sends an e-mail message it is first broken up by the transmission control protocol (TCP) protocol into Internet protocol (IP) packets. Those packets are then sent to an internal router (a router that is inside the user's network) that examines the address. Based on the address, the router decides whether the mail is to be delivered to someone on the same network or to someone outside of the network. If the mail goes to someone on the same network, the mail is delivered to them. If the mail is addressed to someone outside the network, it may pass through a firewall, which is a computer that shields the network from the broader Internet so intruders cannot break into the network. The firewall keeps track of messages and data going into and out of the network—to and from the Internet. The firewall also can prevent certain packets from getting through it.

Once out on the Internet, the message is sent to an Internet router. The router examines the address, determines where the message should be sent and sends the message on its way. A gateway at the receiving network receives the e-mail message. This gateway uses TCP to reconstruct the IP packets into a full message. The gateway then translates the message into the protocol the target network uses and sends it on its way. The message may be required to also pass through a firewall on the receiving network. The receiving network examines the e-mail address and sends the message to a specific mailbox.

A user can also attach binary files such as pictures, videos, sounds and executable files to the e-mail message. To do this, the user must encode the file in a way that will allow it to be sent across the network. The receiver will have to decode the file once it is received. There are a variety of different encoding schemes that can be used. Some e-mail software packages automatically do the encoding for the user and the decoding on the receiving end.

When a user sends e-mail to someone on the Internet or within a closed network, that message often has to travel through a series of networks before it reaches the recipient. These networks might use different e-mail formats. Gateways perform the job of translating e-mail formats from one network to another so the messages can make their way through all the networks. An e-mail message is made up of binary data, usually in the ASCII text format. ASCII is a standard that allows any computer, regardless of its operating system or hardware, to read the text. ASCII code describes the characters that users see on their computer screens.

## *Security Issues of E-mail*

Some security issues involved in e-mails are as follows:
• Flaws in the configuration of the mail server application may be used as the means of compromising the underlying server and the attached network.
• Denial-of-service (DoS) attacks may be directed to the mail server, denying or hindering valid users from using the mail server.
• Sensitive information transmitted unencrypted between mail server and e-mail client may be intercepted.

- Information within the e-mail may be altered at some point between the sender and recipient.
- Viruses and other types of malicious code may be distributed throughout an organization via e-mail.
- Users may send inappropriate, proprietary or other sensitive information via e-mail leading to a legal exposure.

## Standards for E-mail Security

To improve e-mail security, organizations should:
- Address the security aspects of the deployment of a mail server through maintenance and administration standards
- Ensure that the mail server application is deployed, configured and managed to meet the security policy and guidelines laid down by management
- Consider the implementation of encryption technologies to protect user authentication and mail data

In e-mail security, a digital signature authenticates a transmission from a user in an untrusted network environment. A digital signature is a sequence of bits appended to a digital document. Like a handwritten signature, its authenticity can be verified. But unlike a handwritten signature, it is unique to the document being signed. Digital signatures are another application of public key cryptography. Digital signatures are a good method of securing e-mail transmissions in that:
- The signature cannot be forged.
- The signature is authentic and encrypted.
- The signature cannot be reused (a signature on one document cannot be transferred to another document).
- The signed document cannot be altered; any alteration to the document (whether or not it has been encrypted) renders the signature invalid.

On the receiver's end of an e-mail transmission, a user needs the public key to decrypt the message as well as a digital signature verification program to verify the signature. Digital signatures are based on a procedure called message digesting, which computes a short, fixed-length number called a digest for any message of any length. Several different messages may have the same digest, but it is extremely difficult to produce any of them from the digest. A message digest is a 128-bit, cryptographically strong, one-way hash function of the message. It is similar to a checksum in that it compactly represents the message and is used to detect changes in the message. One also can think of the message digest as a fingerprint of the message. The message digest authenticates the user's message in such a way that if it were altered, the message would be considered corrupted and in some cases unreadable.

In using digital signatures for securing e-mail messages, there are two different types of encryption techniques used to ensure secure messages. Messages can be secured with a symmetric (secret key) key management system using Data Encryption Standard (DES) or with a public key (asymmetric) management system using RSA—a public key cryptosystem developed by R. Rivest, A. Shamir and L. Adleman (see chapter 5, Protection of Information Assets, section 5.3).

Organizations should employ their network infrastructure to protect their mail server(s) through appropriate use of firewalls, routers and IDSs.

(See chapter 4, Protection of Information Assets, for a more detailed discussion of this topic.)

## 3.16.6 POINT-OF-SALE SYSTEMS

Point-of-sale (POS) systems enable the capture of data at the time and place that sales transactions occur. The most common payment instruments to operate with POS are credit and debit cards, which are associated with banks accounts. POS terminals may have attached peripheral equipment—such as optical scanners to read bar codes and magnetic card readers for credit or debit cards or electronic readers for smart cards—to improve the efficiency and accuracy of the transaction recording process.

POS systems may be online to a central computer owned by a financial institution or a cards administrator, or may use local processors/microcomputers owned by a business to hold the transactions for a specified period after which they are sent to the main computer for batch processing.

# Systems and Infrastructure Life Cycle Management

It is most important for the IS auditor to determine whether any credit cardholder information is stored on the local POS system such as credit card numbers, personal identification numbers (PINs), etc. Any such information, if stored on the POS system, should be encrypted using strong encryption methods.

## 3.16.7 ELECTRONIC BANKING

Banking organizations have been remotely delivering electronic services to consumers and businesses for years. Electronic funds transfer (EFT) (including small payments and corporate cash management systems), publicly accessible automated machines for currency withdrawal and retail account management are global fixtures.

Continuing technological innovation and competition among existing banking organizations and new market entrants has allowed for a much wider array of electronic banking products and services for retail and wholesale banking customers. However, the increased worldwide acceptance of the Internet as a delivery channel for banking products and services provides new business opportunities as well as new risks.

Major risks associated with banking activities are strategic, reputational, operational (including security—sometimes called transactional—and legal risks), credit, price, foreign exchange, interest rate and liquidity. Electronic banking (e-banking) activities do not raise risks that were not already identified in traditional banking, but e-banking increases and modifies some of these traditional risks. The core business and the IT environment are tightly coupled, thereby influencing the overall risk profile of e-banking.

In particular, from the perspective of the IS auditor, the main issues are strategic, operational and reputational risk since these are directly related to threats to reliable data flow and operational risk, and are certainly heightened by the rapid introduction and underlying technological complexity of electronic banking.

Banks should have a risk management process to enable them to identify, measure, monitor and control their technology risk exposure. Risk management of new technologies has three essential elements:
- Risk management is the responsibility of the board of directors and senior management. They are responsible for developing the bank's business strategy and establishing an effective risk management methodology. They need to possess the knowledge and skills to manage the bank's use of electronic banking and all related risks. The board should take an explicit, informed and documented strategic decision as to whether and how the bank is to provide e-banking services. The initial decision should include the specific accountabilities, policies and controls to address risks, including those arising in a cross-border context. The board should review, approve and monitor e-banking technology-related projects that have a significant impact on the bank's risk profile and ensure that adequate controls are identified, planned and implemented.
- Implementing technology is the responsibility of IT senior management members. They should have the skills to effectively evaluate e-banking technologies and products, and ensure that they are appropriately installed and documented.
- Measuring and monitoring risk is the responsibility of members of operational management. They should have the skills to effectively identify, measure, monitor and control risks associated with e-banking. The board of directors should receive regular reports on the technologies employed, the risks assumed and how those risks are managed.

### *Risk Management Challenges in E-banking*

E-banking presents a number of risk management challenges:
- The speed of change relating to technological and service innovation in e-banking is unprecedented. Currently, banks are experiencing competitive pressure to roll out new business applications in very compressed time frames. This competition intensifies the management challenge to ensure that adequate strategic assessment, risk analysis and security reviews are conducted prior to implementing new e-banking applications.
- Transactional e-banking web sites and associated retail and wholesale business applications are typically integrated as much as possible with legacy computer systems to allow more straight-through processing of electronic transactions. Such straight-through automated processing reduces opportunities for human error and fraud inherent in manual processes, but it also increases dependence on sound system design and architecture as well as system interoperability and operational scalability.

- E-banking increases banks' dependence on information technology, thereby increasing the technical complexity of many operational and security issues and furthering a trend toward more partnerships, alliances and outsourcing arrangements with third parties such as ISPs, telecommunication companies and other technology firms.
- The Internet is ubiquitous and global by nature. It is an open network accessible from anywhere in the world by unknown parties. Messages are routed through unknown locations and via fast-evolving wireless devices. Therefore, the Internet significantly magnifies the importance of security controls, customer authentication techniques, data protection, audit trail procedures and customer privacy standards.

## Risk Management Controls for E-banking

Effective risk management controls for electronic banking include the following 14 controls divided among three categories:
- **Board and management oversight:**
  1. Effective management oversight of e-banking activities
  2. Establishment of a comprehensive security control process
  3. Comprehensive due diligence and management oversight process for outsourcing relationships and other third-party dependencies
- **Security controls:**
  4. Authentication of e-banking customers
  5. Nonrepudiation and accountability for e-banking transactions
  6. Appropriate measures to ensure segregation of duties
  7. Proper authorization controls within e-banking systems, databases and applications
  8. Data integrity of e-banking transactions, records and information
  9. Establishment of clear audit trails for e-banking transactions
  10. Confidentiality of key bank information
- **Legal and reputational risk management:**
  11. Appropriate disclosures for e-banking services
  12. Privacy of customer information
  13. Capacity, business continuity and contingency planning to ensure availability of e-banking systems and services
  14. Incident response planning

# 3.16.8 ELECTRONIC FINANCE

Electronic finance (e-finance) is changing the face of the financial services industry and enabling new providers to emerge within and across countries, including online banks, brokerages and companies that allow consumers to compare financial services such as mortgage loans and insurance policies. Nonfinancial entities are also entering the market, including telecommunication and utility companies that offer payment and other services. Vertically integrated financial service companies are growing rapidly and creating synergies by combining brand names, distribution networks and financial service production.

Advantages of this approach to consumers are:
- Lower costs
- Increased breadth and quality
- Widening access to financial services
- A-synchrony (time-decoupled)
- A-topy (location-decoupled)

By using credit scoring and other data mining techniques, providers can create and tailor products over the Internet without much human input and at a very low cost. Providers can better stratify their customer base through analysis of Internet-collected data and allow consumers to build preference profiles online. This not only permits personalization of information and services, it also allows more personalized pricing of financial services and more effective identification of credit risks. At the same time, the Internet allows new financial service providers to compete more effectively for customers because it does not distinguish between traditional brick-and-mortar providers of financial services and those without physical presence. All these forces are delivering large benefits to consumers at the retail and commercial levels. These mechanisms should be used within privacy law statements (regarding confidentiality and authorization) to gather diverse user information and set up profiles.

## 3.16.9 PAYMENT SYSTEMS

There are two types of parties involved in all payment systems—the issuers and the users. An issuer is an entity that operates the payment service. An issuer holds the items that the payments represent (e.g., cash held in regular bank accounts). The users of the payment service perform two main functions—making payments and receiving payments—and therefore can be described as a payer or a payee, respectively.

### *Electronic Money Model*

The objective of electronic money systems is to emulate physical cash. An issuer attempts to do this by creating digital certificates, which are then purchased (or withdrawn) by the users who redeem (deposit) them with the issuer at a later date. In the interim, certificates can be transferred among users to trade for goods or services.

For the certificates to take on some of the attributes of physical cash, certain techniques are used so that when a certificate is deposited, the issuer cannot determine the original withdrawer of the certificate. This provides the electronic certificates with unconditional untraceability.

Electronic money systems can be hard to implement in practice due to the overheads of solving the "double spending" problem (i.e., preventing a user from depositing the same money twice).

Some advantages of electronic money systems are:
• The payer does not need to be online (generally) at the time of the purchase (since the electronic money can be stored on the payer's computer).
• The payer can have unconditional untraceability (albeit at the expense of lost interest on deposits).

### *Electronic Checks Model*

Electronic check systems model real-world checks quite well and are thus relatively simple to understand and implement. A user writes an electronic check, which is a digitally signed instruction to pay. This is transferred (in the course of making a purchase) to another user, who then deposits the electronic check with the issuer. The issuer will verify the payer's signature on the payment and transfer the funds from the payer's account to the payee's account.

Some advantages of electronic check systems are:
• Easy to understand and implement
• The availability of electronic receipts, allowing users to resolve disputes without involving the issuer
• No need for payer to be online to create a payment

These systems are usually fully traceable, which is an advantage for certain law enforcement, tax collection and marketing purposes, but a disadvantage for those concerned about privacy.

### *Electronic Transfer Model*

Electronic transfer systems are the simplest of the three payment models. The payer simply creates a payment transfer instruction, signs it digitally and sends it to the issuer. The issuer then verifies the signature on the request and performs the transfer. This type of system requires the payer to be online, but not the payee.

Some advantages of electronic transfer systems are:
• Easy to understand and implement
• The payee does not need to be online—a considerable advantage in some circumstances (e.g., paying employee wages)

## 3.16.10 INTEGRATED MANUFACTURING SYSTEMS

Integrated manufacturing systems (IMS) have a long history and, accordingly, there has been quite a diverse group of models and approaches.

Some of the integrated manufacturing systems include bill of materials (BOM), BOM processing (BOMP), manufacturing resources planning (MRP), computer-assisted design (CAD), computer-integrated (or computer-intensive) manufacturing (CIM) and manufacturing accounting and production (MAP).

Original IMSs were based on BOM and BOMP, and usually supported by a hierarchical DBMS.

Evolution toward further integration with other business functions (e.g., recording of raw materials, work-in-process and finished goods transactions, inventory adjustments, purchases, supplier management, sales, accounts payable, accounts receivable, goods received, inspection, invoices, cost accounting, maintenance, etc.) led to MRP (initially standing for material requirements processing, now for manufacturing resources planning), which is a family of widely used standards and standard-based packages.

An early and successful integration was IBM Manufacturing Accounting and Production Information Control System (MAPICS). The most popular MRP standard is MRP II—the current version (2004) being MRP III—supporting just-in-time (JIT) and lean manufacturing. There is also MRP IV, which is meant to support SCM.

MRP is a typical module of most ERP packages such as SAP, Navision or J.D. Edwards, and is usually integrated in modern CRM and SCM systems.

CAD, computer-assisted engineering (CAE) and CAM—the latter including computerized numeric control (CNC)—have led to CIM. CIM is frequently used to run huge lights-out plants, with a significant portion of consumer goods is being manufactured in these environments.

The importance for the IS auditor lies in the high number of systems and applications using these technologies. The larger the scale of integration, the more auditor attention is required.

Highly integrated CIM projects require the same attention from the auditor as the ERPs previously mentioned in this chapter. They are major undertakings that should be based on comprehensive feasibility studies and subject to top management approval and close supervision.

Continuity planning is also a primary area that should be reviewed by the IS auditor.

## 3.16.11 ELECTRONIC FUNDS TRANSFER

As the Internet continues to transform commercial transactions, the method of payment is one bothersome concept that will take on an increasingly significant role in the relationship between seller and buyer. The underlying goal of the automated environment is to wring out costs inherent in the business processes. EFT is the exchange of money via telecommunications without currency actually changing hands. In other words, EFT is the electronic transfer of funds between a buyer, seller and his/her respective financial institution. EFT refers to any financial transaction that transfers a sum of money from one account to another electronically. EFT allows parties to move money from one account to another account, replacing traditional check writing and cash collection procedures. EFT services have been available for two decades. With the increased interest in Internet business, more and more consumers and businesses have begun to utilize EFT services. In the settlement between parties, EFT transactions usually function via an internal bank transfer from one party's account to another or via a clearinghouse network. Usually, transactions originate from a computer at one institution (location) and are transmitted to a computer at another institution (location) with the monetary amount recorded in the respective organization's accounts. Because of the potential high volume of money being exchanged, these systems may be in an extremely high-risk category. Therefore access security and authorization of processing are important controls. Regarding EFT transactions, central bank requirements should be reviewed for application in these processes.

# Systems and Infrastructure
# Life Cycle Management

## 3.16.12 CONTROLS IN AN EFT ENVIRONMENT

Because of the potential high volume of money being exchanged, these systems may be in an extremely high-risk category and security in an EFT environment becomes extremely critical.

Security includes the methods used by the customer to gain access to the system, the communications network and the host or application processing site. Individual consumer access to the EFT system is generally controlled by a plastic card and a PIN. Both items are required to initiate a transaction. The IS auditor should review the physical security of unissued plastic cards, the procedures used to generate PINs, the procedures used to issue cards and PINs, and the conditions under which the consumer uses the access devices.

Security in an EFT environment ensures that:
• All the equipment and communication linkages are tested to effectively and reliably transmit and receive data
• Each party uses security procedures that are reasonably sufficient for affecting the authorized transmission of data and for protecting business records and data from improper access
• There are guidelines set for the receipt of data and to ensure that the receipt date and time for data transmitted are the date and time the data have been received
• On receipt of data, the receiving party will immediately transmit an acknowledgment or notification to communicate to the sender that a successful transmission occurred
• Data encryption standards are set
• Standards for unintelligible transmissions are set
• Regulatory requirements for enforceability of electronic data transmitted and received are explicitly stated

Access to commercial EFT systems generally does not require a plastic card, but the IS auditor should ensure that reasonable identification methods are required. The communications network should be designed to provide maximum security. Data encryption is recommended for all transactions; however, the IS auditor should determine any conditions under which the PIN might be accessible in a clear mode.

An EFT switch involved in the network is also an audit concern. An EFT switch is the facility that provides the communication linkage for all equipment in the network. The IS auditor should review the contract with the switch and the third-party audit of the switch operations. If a third-party audit has not been performed, the auditor should consider visiting the switch location.

At the application processing level, the IS auditor should review the interface between the EFT system and the application systems that process the accounts from which funds are transferred. Availability of funds or adequacy of credit limits should be verified before funds are transferred. Unfortunately, this is not always the case. Because of the penalties for failure to make a timely transfer, the IS auditor should review backup arrangements or other methods used to ensure continuity of operations. Since EFT reduces the flow of paper and consequently reduces normal audit trails, the IS auditor should determine that alternative audit trails are available.

## 3.16.13 INTEGRATED CUSTOMER FILE

Integrated customer files provide details regarding all business relationships a customer maintains with an organization. The integration aids in customer analysis and marketing. An example of an integrated file is an integrated banking customer file. The file includes data regarding the customer loans, checking accounts, savings accounts and any certificates of deposit.

## 3.16.14 OFFICE AUTOMATION

Currently, many offices use a variety of electronic devices and techniques to aid in the conduct of business. Word processors, automated spreadsheets and e-mail are used daily in many offices. LANs link local offices and computers to facilitate the use of these technologies. Office automation devices and networks may contain sensitive data; however, access controls and security are frequently weak or nonexistent.

## 3.16.15 AUTOMATED TELLER MACHINE

An ATM is a specialized form of the POS terminal that is designed for the unattended use by a customer of a financial institution. These machines customarily allow a range of banking and debit operations—especially financial deposits and cash withdrawals. ATMs are usually located in uncontrolled areas to facilitate easy access to customers after hours. This facility can be within a bank, across local banks and in banks outside a region. They are becoming known as retail EFT networks, transferring information and money over communication lines. Therefore the system must provide high levels of logical and physical security for both the customer and the machinery. The ATM architecture has a physical network layer, a switch and a communication layer connecting the various ATM POS terminals.

Recommended internal control guidelines for ATMs, apart from what has been provided for any EFT, include the following:
• Written policies and procedures covering personnel, security controls, operations, disaster recovery credit and check authorization, floor limits, override, settlement, and balancing
• Reconciliation of all general ledger accounts related to retail EFTs and review of exception items and suspense accounts
• Procedures for PIN issuance and protection during storage
• Procedures for the security of PINs during delivery and the restriction of access to a customer's account after a small number of unsuccessful attempts
• Systems should be designed, tested and controlled to preclude retrieval of stored PINs in any nonencrypted form. Application programs and other software containing formulas, algorithms and data used to calculate PINs must be subject to the highest level of access for security purposes.
• Controls over plastic card procurement should be adequate with a written agreement between the card manufacturer and the bank that details control procedures and methods of resolution to be followed if problems occur.
• Controls and audit trails of the transactions that have been made in the ATM. This should include internal registration in the ATM, either in internal paper or digital media, depending on regulation or laws in each country and on the hosts that are involved in the transaction.

### *Audit of ATM*
To perform an audit of ATMs, the IS auditor should:
• Review measures to establish proper customer identification and maintenance of their confidentiality
• Review file maintenance and retention system to trace transactions
• Review exception reports to provide an audit trail
• Review daily reconciliation of ATM transactions including:
  – Review segregation of duties in the opening of ATM and recount of deposit
  – Review the procedures made for the retained cards
• Review encryption key change management procedures

## 3.16.16 COOPERATIVE PROCESSING SYSTEMS

Cooperative processing systems are divided into segments so different parts may run on different independent computer devices. The system divides the problem into units that are processed in a number of environments and communicates the results among them to produce a solution to the total problem. The system must be designed to minimize and maintain the integrity of communication between the component parts and to use the most appropriate processor for each of the problem units.

## 3.16.17 INTERACTIVE VOICE RESPONSE

In telephony, interactive voice response (IVR), is a phone technology that allows a computer to detect voice and touch tones using a normal phone call. The caller uses the telephone keypad to select from preset menu choices provided by the IVR. The IVR system then responds with pre-recorded or dynamically generated audio to further direct callers or route the caller to a customer service representative. IVR systems can be used to control almost any function where the interface can be

broken down into a series of simple menu choices. IVR systems generally scale well to handle large call volumes. Controls over such systems must be in place to prevent unauthorized individuals from entering system—level commands that may permit them to change or rerecord menu options.

## 3.16.18 PURCHASE ACCOUNTING SYSTEM

Financial transactions frequently go through more than one system when processed. In a department store, a sale is first processed in the sales accounting system, then processed by the accounts receivable system (if the purchase was by credit card) and, for either cash or credit sales, through the inventory system (when they are linked). That same sale might trigger the purchase accounting system to replace depleted inventory. Eventually the transactions become part of the general ledger system since all transactions are recorded somewhere in that system. For the integration of systems to be effective, processing of transactions must be complete, accurate and timely. If it is not, a ripple effect impairs the integrity of the data.

Purchase accounting systems process the data for purchases and payments. Since purchases automatically lead to payments, if purchases are properly contracted, partial control over payments exists. Additional controls over payments are still needed to ensure that each payment was made for goods and services received, that the same purchases were not paid for twice and that they were, indeed, paid. Most purchase accounting systems perform three basic accounting functions:
1. **Accounts payable processing**—Recording transactions in the accounts payable records
2. **Goods received processing**—Recording details of goods received, but not yet invoiced
3. **Order processing**—Recording goods ordered, but not yet received

The computer may be involved in each of these activities, and the extent to which they are computerized determines the complexity of the purchase accounting system.

## 3.16.19 IMAGE PROCESSING

Image processing is computer manipulation of images. Some of the many algorithms used in image processing include convolution (on which many others are based), fast Fourier transform (FFT), discrete cosine transform (DCT), thinning (or skeletonization), edge detection and contrast enhancement. These are usually implemented in software but may also use special-purpose hardware for speed.

An imaging system stores, retrieves and processes graphic data, such as pictures, charts and graphs, instead of or in addition to text data. The storage capacities must be enormous and most image systems include optical disk storage. In addition to optical disks, the systems include high-speed scanning, high-resolution displays, rapid and powerful compression, communications functions, and laser printing. The systems include techniques that can identify levels of shades and colors that cannot be differentiated by the human eye. These systems are expensive and companies do not invest in them lightly.

Most businesses that perform image processing obtain benefits from using the imaging system. Examples of potential benefits are:
• Item processing (e.g., signature storage and retrieval)
• Immediate retrieval via a secure optical storage medium
• Increased productivity
• Improved control over paper files
• Reduced deterioration due to handling
• Enhanced disaster recovery procedures

Imaging systems are the fastest growth area of the micrographics industry and are an outgrowth of microfilm and microfiche, which have, in the past, been heavily used in paper-intensive fields such as insurance and banking. Not surprisingly, these same fields were the first to incorporate imaging systems into their standard operations. The replacement of paper documents with electronic images can have a significant impact on the way that an institution does business. Many of the traditional audit and security controls for paper-based systems may be reduced or absent in electronic documentation workflow. New controls must be developed and designed into the automated process to ensure that information image files cannot be altered, erased or lost.

Risk areas that management should address when installing imaging systems and that IS auditors should be aware of when reviewing an institution's controls over imaging systems include:

- **Planning**—The lack of careful planning in selecting and converting paper systems to document imaging systems can result in excessive installation costs, the destruction of original documents and the failure to achieve expected benefits. Critical issues include converting existing paper storage files and integration of the imaging system into the organization workflow and electronic media storage to meet audit and document retention legal requirements.
- **Audit**—Imaging systems may change or eliminate the traditional controls as well as the checks and balances inherent in paper-based systems. Audit procedures may have to be redesigned and new controls designed into the automated process.
- **Redesign of workflow**—Institutions generally redesign or reengineer workflow processes to benefit from imaging technology.
- **Scanning devices**—Scanning devices are the entry point for image documents and a significant risk area in imaging systems. Scanning operations can disrupt workflow if the scanning equipment is not adequate to handle the volume of documents or the equipment breaks down. The absence of controls over the scanning process can result in poor quality images, improper indexing, and incomplete or forged documents being entered into the system. Factors that should be considered in an imaging system are quality control over the scanning and indexing process, the scanning rate of the equipment, the storage of images, equipment backup and the experience level of personnel scanning the document. Procedures should be in place to ensure that original documents are not destroyed before determining that a good image has been captured.
- **Software security**—Security controls over image system documents are critical to protect institutions and customer information from unauthorized access and modifications. The integrity and reliability of the imaging system database is related directly to the quality of controls over access to the system.
- **Training**—Inadequate training of personnel scanning the documents can result in poor-quality document images and indexes, and the early destruction of original documents. The installation and use of imaging systems can be a major change for department personnel. They must be trained adequately to ensure quality control over the scanning and storage of imaging documents as well as the use of the system to maximize the benefits of converting to imaging systems.

## 3.16.20 ARTIFICIAL INTELLIGENCE AND EXPERT SYSTEMS

Artificial intelligence is the study and application of the principles by which:
- Knowledge is acquired and used
- Goals are generated and achieved
- Information is communicated
- Collaboration is achieved
- Concepts are formed
- Languages are developed

Artificial intelligence fields include, among others:
- Expert systems
- Natural and artificial (such as programming) languages
- Neural networks
- Intelligent text management
- Theorem proving
- Abstract reasoning
- Pattern recognition
- Voice recognition
- Problem solving
- Machine translation of foreign languages

Two main programming languages that have been developed for artificial intelligence are LISP and PROLOG.

Expert systems are in an area of artificial intelligence that is becoming increasingly popular. They perform a specific function or are prevalent in certain industries. An expert system allows the user to specify certain basic assumptions or formulas and then uses these assumptions or formulas to analyze arbitrary events. Based on the information used as input to the system, a conclusion is produced.

*Chapter 3:*

# Systems and Infrastructure Life Cycle Management

The use of expert systems has many potential benefits within an organization including:
- Capturing the knowledge and experience of individuals before they leave the organization
- Sharing knowledge and experience in areas where there is limited expertise
- Facilitating consistent and efficient quality decisions
- Enhancing personnel productivity and performance
- Automating highly (statistically) repetitive tasks (help desk, score credits, etc.)
- Operating in environments where a human expert is not available (e.g., medical assistance on board of a ship, satellites, etc.)

Expert systems are comprised of the primary components shown in **exhibit 3.30**, called shells, when they are not populated with particular data, and the shells are designed to host new expert systems.

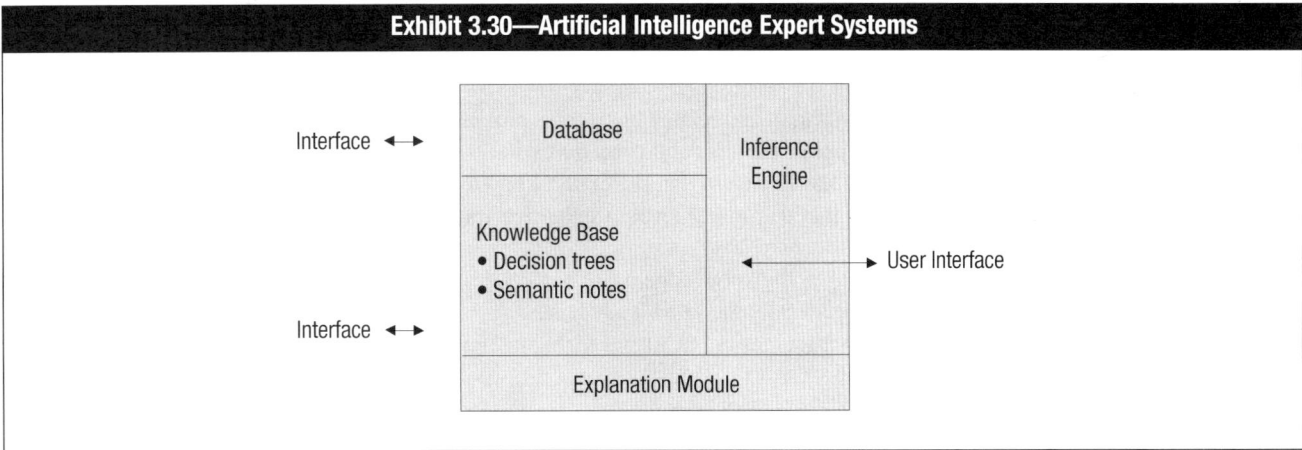

Key to the system is the knowledge base (KB), which contains specific information or fact patterns associated with particular subject matter and the rules for interpreting these facts. The KB interfaces with a database in obtaining data to analyze a particular problem in deriving an expert conclusion. The information in the KB can be expressed in several ways:
- **Decision trees**—Using questionnaires to lead the user through a series of choices, until a conclusion is reached. Flexibility is compromised because the user must answer the questions in an exact sequence.
- **Rules**—Expressing declarative knowledge through the use of if-then relationships. For example, if a patient's body temperature is over 39°C (102.2°F) and their pulse is under 60, then they might be suffering from a certain disease.
- **Semantic nets**—Consist of a graph in which the nodes represent physical or conceptual objects and the arcs describe the relationship between the nodes. Semantic nets resemble a data flow diagram and make use of an inheritance mechanism to prevent duplication of data.

Additionally, the inference engine shown is a program that uses the KB and determines the most appropriate outcome based on the information supplied by the user. In addition, an expert system includes the following components:
- **Knowledge interface**—Allows the expert to enter knowledge into the system without the traditional mediation of a software engineer
- **Data interface**—Enables the expert system to collect data from nonhuman sources, such as measurement instruments in a power plant

An explanation module that is user-oriented in addressing the problem is analyzed, and the expert conclusion reached is also provided.

Expert systems are gaining acceptance and popularity as audit tools. Specifically related to IS auditing, expert systems have been developed to facilitate the IS auditor in auditing such areas as operating systems, online software environments, access control products and microcomputer environments. These tools can take the form of a series of well-designed questionnaires or actual software that integrates and reports on system parameters and data sets. Other accounting- and auditing-related applications for expert systems include audit planning, internal control analysis, account attribute analysis, quality review, accounting decisions, tax planning and user training.

Consistent with standard systems development methodologies, stringent change control procedures should be followed since the basic assumptions and formulas may need to be changed as more expertise is gained. As with other systems, access should be on a need-to-know basis.

The IS auditor should be knowledgeable about the various AI and expert system applications used within the organization.

The IS auditor needs to be concerned with the controls relevant to these systems when used as an integral part of an organization's business process or mission critical functions, and the level of experience or intelligence used as a basis for developing the software. This is critical since errors produced by AI systems may have a more severe impact than those produced by traditional systems. This is true especially of intelligent systems that facilitate health care professionals in the diagnosis and treatment of injuries and illnesses. Error loops/routines should be designed into these systems.

Specifically, the IS auditor should:
• Understand the purpose and functionality of the system
• Assess the system's significance to the organization and related businesses processes as well as the associated potential risks
• Review the adherence of the system to corporate policies and procedures
• Review the decision logic built into the system to ensure that the expert knowledge or intelligence in the system is sound and accurate. The IS auditor should ensure that the proper level of expertise was used in developing the basic assumptions and formulas.
• Review procedures for updating information in the knowledge base
• Review security access over the system, specifically the knowledge base
• Review procedures to ensure that qualified resources are available for maintenance and upgrading

## 3.16.21 BUSINESS INTELLIGENCE

Business intelligence (BI) is a broad field of IT that encompasses the collection and dissemination of information to assist decision making and assess organizational performance.

Investments in BI technology can be applied to enhance understanding of a wide range of business questions. Some typical areas in which BI is applied for measurement and analysis purposes include:
• Process cost, efficiency and quality
• Customer satisfaction with product and service offerings
• Customer profitability, including determination of which attributes are useful predictors of customer profitability
• Staff and business unit achievement of key performance indicators
• Risk management—e.g., by identifying unusual transaction patterns and accumulation of incident and loss statistics

The interest in BI as a distinct field of IT activity is being spurred by a number of factors:
• **The increasing size and complexity of modern organizations**—The result of this is that even fundamental business questions cannot be properly answered without establishing serious BI capability.
• **Pursuit of competitive advantage**—Most organizations have, for many years, automated their basic, high-volume activities. Significant organizationwide IT investment such as ERP systems is now common place. Many companies have or are now investing in Internet technology as a means of distributing product/service and supply chain integration. However, utilization of IT to maintain and extend a firm's knowledge capital represents a new opportunity to use technology to gain an advantage over competitors.
• **Legal requirements**—Legislation such as the US Sarbanes-Oxley Act and the US Patriot Act exist to enforce the need for companies to have an understanding of the "whole of business." Financial institutions must now be able to report on all accounts/instruments that their customers have and all transactions against those accounts/instruments, including any suspicious transaction patterns.

To deliver effective BI, organizations need to design and implement (progressively, in most cases) a data architecture. A complete data architecture consists of two components:
• The enterprise data flow architecture (EDFA)
• A logical data architecture

# Systems and Infrastructure Life Cycle Management

An optimized enterprise data flow architecture is depicted in **exhibit 3.31**. Explanations of the various layers/components of this data flow architecture follow:

- **Presentation/desktop access layer**—This is where end users directly deal with information. This layer includes familiar desktop tools such as Microsoft Access and Excel, direct querying tools, reporting and analysis suites offered by vendors such as Cognos and Business Objects, and possibly purpose-built applications such as balanced scorecards and digital dashboards. Power users will have the ability to build their own queries and reports while other users will interact with the data in predefined ways. Increasingly, users are being provided with more than static reporting capabilities by embellishing reports with parameterization and drill-down capabilities and presenting data in visual formats as a supplement or replacement of textual/tabular data presentation.

- **Data source layer**—Enterprise information derives from a number of sources:
  - Operational data—Data captured and maintained by an organization's existing systems, and usually held in system-specific databases or possibly flat files
  - External data—Data provided to an organization by external sources. This could include data such as customer demographics and market share information.
  - Nonoperational data—Information needed by end users that is not currently maintained in a computer-accessible format

- **Core data warehouse**—This is where all the data (or at least the majority) of interest to an organization is captured and organized to assist reporting and analysis. DWs are normally instituted as large relational databases. While there is not unanimous agreement, many pundits suggest the warehouse should hold fully normalized data to give it the flexibility to deal with complex and changing business structures. A properly constituted DW should support three basics forms of inquiry:
  - Drilling up and drilling down—Using dimensions of interest to the business, it should be possible to aggregate data (e.g., sum store sales to get region sales and ultimately national sales) as well as drill down (e.g., break store sales down to counter sales). Attributes available at the more granular levels of the warehouse can also be used to refine the analysis (e.g., analyze sales by product).
  - Drill across—Use common attributes to access a cross-section of information in the warehouse such as sum sales across all product lines by customer and groups of customers according to length of association with the company (and/or other attribute of interest)
  - Historical analysis—The warehouse should support this by holding historical, time-variant data. An example of historical analysis would be to report monthly store sales and then repeat the analysis using only customers who were preexisting at the start of the year in order to separate the effect of new customers from the ability to generate repeat business with existing customers.

- **Data mart layer**—Data marts represent subsets of information from the core DW selected and organized to meet the needs of a particular business unit or business line. Data marts may be relational databases or some form of online analytical processing (OLAP) data structure (also known as a data cube). OLAP technologies and some variants (e.g., relational OLAP [ROLAP]), allow users to "slice and dice" data presented in terms of standardized measures (i.e., numerical facts) and dimensions (i.e., business hierarchies). Data marts have a simplified structure compared to the normalized DW. If using a relational database, a popular structure is the star schema which involves a fact table connected to denormalized dimension tables. A simplified structure assists the business user's understanding of the data and relationships. If data are held in relational tables, less joins (connections between tables) are necessary when querying, and the need to become familiar with a large number of code values is avoided if full text descriptions are used instead.

- **Data staging and quality layer**—This layer is responsible for data copying, transformation into DW format and quality control. It is particularly important that only reliable data get loaded to the core DW. This layer needs to be able to deal with problems periodically thrown up by operational systems such as changes to account number formats and reuse of old account and customer numbers (when the DW still holds information on the original entity).

- **Data access layer**—This layer operates to connect the data storage and quality layer with data stores in the data source layer and, in the process, avoiding the need to know exactly how these data stores are organized. Technology now permits SQL access to data even if it is not stored in a relational database.

- **Data preparation layer**—This layer is concerned with the assembly and preparation of data for loading into data marts. The usual practice is to precalculate the values that are loaded into OLAP data repositories to increase access speed. Specialist data mining also normally requires preparation of data. Data mining is concerned with exploring large volumes of data to determine patterns and trends of information. Data mining often identifies patterns that are counterintuitive due to the number and complexity of data relationships. Data quality needs to be very high to not corrupt the results.

**Exhibit 3.31—Data Flow Architecture Sample**

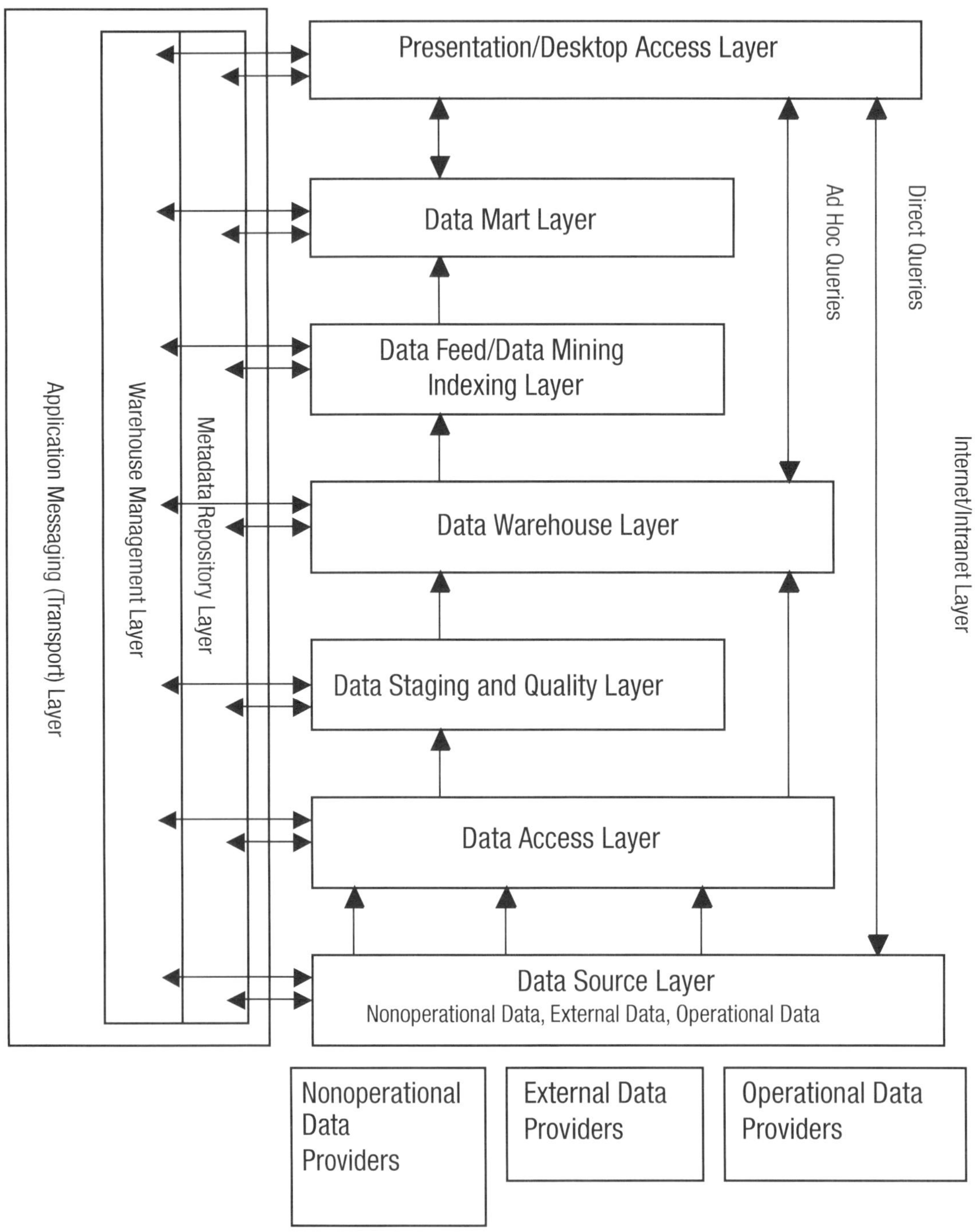

```
Presentation/Desktop Access Layer

Data Mart Layer

Data Feed/Data Mining
Indexing Layer

Data Warehouse Layer

Data Staging and Quality Layer

Data Access Layer

Data Source Layer
Nonoperational Data, External Data, Operational Data
```

Application Messaging (Transport) Layer

Warehouse Management Layer

Metadata Repository Layer

Ad Hoc Queries

Direct Queries

Internet/Intranet Layer

Nonoperational Data Providers

External Data Providers

Operational Data Providers

- **Metadata repository layer**—Metadata are data about data. The information held in the metadata layer needs to extend beyond data structure names and formats to provide detail on business purpose and context. The metadata layer should be comprehensive in scope, covering data as they flow between the various layers, including documenting transformation and validation rules. Ideally, information in the metadata layer can be directly sourced by software operating in the other layers, as required.
- **Warehouse management layer**—The function of this layer is the scheduling of the tasks necessary to build and maintain the DW and populate data marts. This layer is also involved in the administration of security.
- **Application messaging layer**—This layer is concerned with transporting information between the various layers. In addition to business data, this layer encompasses generation, storage and targeted communication of control messages.
- **Internet/intranet layer**—This layer is concerned with basic data communication. Included here are browser-based user interfaces and TCP/IP networking.

The construction of the logical data architecture for an enterprise is a major undertaking that would normally be undertaken in stages. One reason for separating logical data model determination by business domain is that different parts of large business organizations will often deal with different transaction sets, customers and products.

Ultimately, the data architecture needs to be structured to accommodate the needs of the organization in the most efficient manner. Factors to consider include the types of transactions in which the organization engages, the entities that participate in or form part of these transactions (e.g., customers, products, staff and communication channels), and the dimensions (hierarchies) that are important to the business (e.g., product and organization hierarchies).

With modern DWs, storage capacity is not really an issue. Therefore the goal should be to obtain the most granular or atomic data possible. The lowest level data are most likely to have attributes that can be used for analysis purposes that would be lost if summarized data are loaded.

Various analysis models used by data architects/analysts follow:
- **Context diagrams**—Outline the major processes of an organization and the external parties with which the business interacts.
- **Activity or swim-lane diagrams**—Deconstruct business processes.
- **Entity relationship diagrams**—Depict data entities and how they relate. These data analysis methods obviously play an important part in developing an enterprise data model. However, it is also crucial that knowledgeable business operatives are involved in the process. This way proper understanding can be obtained of the business purpose and context of the data. This also mitigates the risk of the replication of suboptimal data configurations from existing systems and databases into the DW.

## *Business Intelligence Governance*

To maximize the value an organization obtains from its BI initiatives, an effective BI governance process needs to be in place.

An important part of the governance process involves determining which BI initiatives to fund, what priority to assign to initiatives and how to measure their ROI. This is particularly important since the investment needed to build BI infrastructure, such as a DW, is considerable. Additionally, the scope and complexity of an organizationwide DW means that, realistically, it must be built in stages.

A recommended practice in the area of BI funding governance is to establish a business/IT advisory team that allows different functional perspectives to be represented, recommends investment priorities and establishes cross-organizational benefit measures. Final funding decisions should rest with a technology steering committee that comprises senior management.

A further important part of overall BI governance is data governance. Aspects to be considered here include establishing standard definitions for data, business rules and metrics, identifying approved data sources, and establishing standards for data reconciliation and balancing.

# Systems and Infrastructure Life Cycle Management

## 3.16.22 DECISION SUPPORT SYSTEM

A decision support system (DSS) is an interactive system that provides the user with easy access to decision models and data from a wide range of sources in order to support semistructured decision-making tasks typically for business purposes. It is an informational application that is designed to assist an organization in making decisions through data provided by business intelligence tools (in contrast to an operational application which collects the data in the course of normal business operations). Typical information that a decision support application might gather and present would be:
• Comparative sales figures between one week and the next
• Projected revenue figures based on new product sales assumptions
• The consequences of different decision alternatives given past experience in the described context

A DSS may present information graphically and may include an expert system or artificial intelligence. Further, it may be aimed at business executives or some other group of knowledge workers.

Characteristics of a DSS are:
• Aims at solving less-structured, underspecified problems that senior managers face
• Combines the use of models or analytic techniques with traditional data access and retrieval functions
• Emphasizes flexibility and adaptability to accommodate changes in the environment and the decision-making approach of the users

### *Efficiency vs. Effectiveness*
A principle of DSS design is to concentrate less on efficiency (i.e., performing tasks quickly and reducing costs) and more on effectiveness (i.e., performing the right task). Therefore DSSs are often developed using 4GL tools that are less efficient, but allow for flexible and easily modified systems.

### *Decision Focus*
A DSS is often developed with a specific decision or well-defined class of decisions to solve; therefore, some commercial software packages that claim to be DSS are nothing more than a DSS generator (tools with which to construct a DSS).

### *DSS Frameworks*
Frameworks are generalizations about a field that help put many specific cases and ideas into perspective. The G. Gorry-M.S. Morton framework is the most complete knowledge- and system-control-related IS model, and it is based on problem classification into structured and unstructured types as well as the time horizon of the decisions. This framework characterizes DSS activities along two dimensions:
1. The degree of structure in the decision process being supported
2. The management level at which decision making takes place

This framework also portrays all IS efforts as addressing distinct types of problems, depending on the above two factors.

The management-level dimension is broken into three parts:
1. Operational control
2. Management control
3. Strategic planning

The decision-structure dimension is also broken into three parts:
1. Structured
2. Semistructured
3. Unstructured

The degree to which a problem or decision is structured corresponds roughly to the extent to which it can be automated or programmed.

Another DSS framework is the Sprague-Carson framework that is initiated with an effort to create family trees—which is a generalization of the structure of a DSS. This framework suggests that every DSS has data, a model and a dialog generator subsystem. This framework emphasizes the importance of data management in DSS work. This framework also stresses the importance of interactive user interfaces in a DSS. The generation and management of these interfaces require appropriate software and hardware—the dialog management system. In general, a system must offer more than one interface and might need to provide a tailored interface for each user.

## Design and Development

Prototyping is the most popular approach to DSS design and development. Prototyping usually bypasses the usual requirement definition. System requirements evolve through the user's learning process. The benefits of prototyping include the following:
• Learning is explicitly incorporated into the design process because of the iterative nature of the system design.
• Feedback from design iterations is rapid to maintain an effective learning process for the user.
• The user's expertise in the problem area helps the user suggest system improvements.
• The initial prototype must be inexpensive to create.

## Implementation and Use

It is difficult to implement a DSS because of its discretionary nature. Using a DSS to solve a problem represents a change in behavior on the part of the user. Implementing a DSS is an exercise in changing an organization's behavior. The main challenge is to get the users to accept the use of software. The following are the steps involved in changing behavior:
• **Unfreezing**—This step alters the forces acting on individuals such that the individuals are distracted sufficiently to change. Unfreezing is accomplished either through increasing the pressure for change or by reducing some of the threats of or resistance to change.
• **Moving**—This step presents a direction of change and the actual process of learning new attitudes.
• **Refreezing**—This step integrates the changed attitudes into the individual's personality.

## Risk Factors

Developers should be prepared for eight implementation risk factors:
1. Nonexistent or unwilling users
2. Multiple users or implementers
3. Disappearing users, implementers or maintainers
4. Inability to specify purpose or usage patterns in advance
5. Inability to predict and cushion impact on all parties
6. Lack or loss of support
7. Lack of experience with similar systems
8. Technical problems and cost-effectiveness issues

## Implementation Strategies

To plan for the risks and prevent them from occurring:
• Divide the project into manageable pieces.
• Keep the solution simple.
• Develop a satisfactory support base.
• Meet user needs and institutionalize the system.

## Assessment and Evaluation

The true test of a DSS lies in whether it improves a manager's decision making, which is something not easily measured. A DSS also rarely results in cost displacements such as a reduction in staff or other expenses. In addition, because a DSS is evolutionary in nature, it lacks neatly defined completion dates.

Using an incremental approach to DSS development reduces the need for evaluation. By developing one step at a time and achieving tangible results at the end of each step, the user does not need to make extensive commitments of time and money at the beginning of the development process.

The DSS designer and user should use broad evaluation criteria. These criteria should include:
• Traditional cost-benefit analysis
• Procedural changes, more alternatives examined, less time consumed in making the decision
• Evidence of improvement in decision making
• Changes in the decision process

Some common trends in DSS usage include:
• The need for more accurate information by managers is an important motivator for DSS development.
• Few DSS evaluations use the traditional cost-benefit analysis.
• End users are usually the motivators in developing a DSS.
• Development staff members are drawn largely from functional area staff or the planning department, not from the IS department.
• Users perceive flexibility as the most important factor influencing the system's success.
• Few DSS projects are being developed today for third-generation software facilities, user-oriented fourth-generation languages and planning languages predominate.
• Planning, evaluation and training for DSS projects traditionally have been performed quite poorly.

## DSS Common Characteristics
Some of the common characteristics of DSSs include:
• Oriented toward decision making
• Usually based on 4GL
• Surfable
• Linkable
• Drill-down
• Semaphores (signals to automatically alert when a decision needs to be made)
• Time series analysis
• What if (refers to scenario modeling; i.e., determining what is the end result of a changing variable or variables)
• Sensitivity analysis
• Goal-seeking
• Excellent graphic presentations
• Dynamic graphic, data editing
• Simulation

## DSS Trends
DSS trends include:
• Gradual improvement and sharpening of skills in the development and implementation of a traditional DSS
• Advances in database and graphics capabilities for microcomputers
• Exploratory work in such fields as expert systems, a DSS to support group decision making and visual interactive modeling

# 3.16.23 CUSTOMER RELATIONSHIP MANAGEMENT

The emerging customer-driven business trend is to be focused on the wants and needs of the customers. With the customer's expectations constantly increasing, these objectives are becoming more difficult to achieve. This emphasizes the importance of focusing on information relating to transaction data, preferences, purchase patterns, status, contact history, demographic information and service trends of customers rather than on products.

# Systems and Infrastructure Life Cycle Management

All these factors lead to customer relationship management (CRM), which is an optimum combination of strategy, tactics, processes, skill sets and technology. CRM has become a strategic success factor for all types of business, and its proficiency has a significant impact on profitability.

The customer expectations are increasing tremendously which, in turn, raises the expectation of service levels. Therefore, the customer-centered applications focus on CRM processes emphasizing the customer, rather than marketing, sales or any other function. The new business model will have an integration of telephony, web and database technologies, and interenterprise integration capabilities. Also, this model spreads to the other business partners who can share information, communicate and collaborate with the organization with the seamless integration of web-enabled applications and without bothering their local network and other configurations.

It is possible to distinguish between operational and analytical CRM. Operational CRM is concerned with maximizing the utility of the customer's service experience while also capturing useful data about the customer interaction. Analytical CRM seeks to analyze information captured by the organization about its customers and their interactions with the organization into information that allows greater value to be obtained from the customer base. Among uses of analytical CRM are increasing customer product holdings or "share of customer wallet," moving customers into higher margin products, moving customers to lower-cost service channels, increasing marketing success rates and making pricing decisions.

# 3.16.24 SUPPLY CHAIN MANAGEMENT

Supply chain management (SCM) is about linking the business processes between the related entities such as the buyer and the seller. The link is provided to all the connected areas such as managing logistics and the exchange of information, services and goods among supplier, consumer, warehouse, wholesale/retail distributors and the manufacturer of goods.

SCM has become a focal point and is seen as a new area in strategic management because of the shift in the business scenario at the advent of global competition, proliferation of the Internet, the instantaneous transmission of information and web presence in all spheres of business activities. SCM is all about managing the flow of goods, services and information among suppliers, manufacturers, wholesalers, distributors, stores, consumers and end users.

SCM shifts the focus since all the entities in the supply chain can work collaboratively and in a real-time mode, thus reducing, to a great extent, the inventory. The JIT concept becomes possible, and the cycle time is reduced with an objective toward reducing the unwanted inventory. Seasonal (e.g., both availability and demand) and regional (e.g., preferences in size, shape, quantity, etc.) factors are addressed.

Stock levels of nonmoving items are significantly reduced, and there is an automated flow of supply and demand. Also, the intrinsic costs and errors associated with the manual means such as fax, input of data, delay and inaccurate orders, can be avoided.

EDI, which is extensively used in SCM, is not a new technology but the cost associated with it was not affordable to many until recent years. Now, availability of inexpensive, web-based SCM solutions has opened this area for everybody. However, there is a need for significant change in the business model.

# Systems and Infrastructure Life Cycle Management

# 3.17 CHAPTER 3 QUICK REFERENCE REVIEW

## Chapter 3 Quick Reference Review

Chapter 3 addresses the need for systems and infrastructure life cycle management. For an IS auditor to provide assurance that the an organization's objectives are being met by the management practices of its systems and infrastructure, it is important to understand how an organization evaluates, develops, implements, maintains and disposes of its IT systems and related components.

CISA candidates should have a sound understanding of the following items, not only within the context of the present chapter, but also to correctly address questions in related subject areas. It is important to keep in mind that it is not enough to know these concepts from a definitional perspective. The CISA candidate must also be able to identify which elements among those presented may represent the greatest risk and which controls are most effective at mitigating this risk.

- The management practice of business realization is the process by which an organization evaluates technology solutions to business problems. A feasibility study scopes the problem, outlines possible solutions, and makes a recommendation. A business case is normally derived from the feasibility study, and contains information for the decision-making process on whether it should be undertaken; if so, it becomes the basis for a project.
- A project portfolio is defined as all of the projects (related or unrelated) being carried out in an organization at a given point in time. A program can be seen as a group of projects that are linked together through common objectives, strategies, budgets and schedules. Portfolios, programs and projects are controlled by a project management office— which governs the processes of project management—but are not typically involved in the management of their content.
- A project may be initiated from within any part of the organization, including the IS department. A project is time bound, with specific start and end dates, a specific objective and a set of deliverables. Throughout the project life cycle, the business case should be reviewed at predefined decision points. The project management processes of project initiation, planning, execution, controlling and project closing are management practices that outline how an organization works to implement projects.
- Using object and work breakdown structures, the project activities are identified and sized using a variety of techniques, such as estimating lines of code (LOC) to feature and function point analysis. The project manager then determines the optimal schedule by using the critical path method. The schedule can be depicted in several forms, such as the Gantt chart or PERT diagrams.
- A project must take into account the duration, costs and deliverables (quality) of the project. These elements are managed throughout the controlling and execution phases of the project. At project closing, a postproject review is conducted to assess lessons learned of the project management process.
- The management practices of specific IT processes are used for managing and controlling IT resources and activities. These IT processes form a life cycle, referred to as the SDLC. A project may follow various forms of an SDLC, such as the waterfall method, the iterative method, or other alternative methods of development, testing and implementation.
- The major phases of an SDLC include feasibility study, requirements definition, design, development, testing, training, implementation, certification and postimplementation review. The phases for each project depend on whether a solution is developed or acquired.
- Development processes may differ based on the use of system development tools and productivity aids including code generators, CASE applications and fourth-generation languages.
- In addition to selecting phases and deliverables, risks to the project should also be identified. Business risks include risks that the new system may not meet the users' business requirements, whereas project risks are risks where the project activities exceed the limits of the project budget.
- As part of its IT processes, an organization uses management practices to identify and analyze the needs for IT infrastructure. The physical architecture analysis, the definition of a new architecture and the necessary road map to move from architecture to another is a critical task for an IT department.
- As an organization's systems and infrastructure change, management practices are used to manage that change (change management and configuration management). Change management processes include the governance on how a change is submitted, documented and approved.

- The IS auditor should understand the systems development, acquisition and maintenance methodology in use and to identify potential vulnerabilities and points requiring control. It is the IS auditor's role to advise the project team and senior management of the deficiencies within each of these processes.
- Other components of an organization's management practices for its systems and architecture include business process reengineering, e-commerce and business intelligence. In each of these IT practices, input control procedures are evaluated to determine that they permit only valid, authorized and unique transactions. In an integrated application environment, controls are embedded and designed into the application that supports the processes. Business process control assurance involves evaluating controls at the process and activity level. These controls may be a combination of management of programmed and manual controls.

# 3.18 CHAPTER 3 CASE STUDIES

The following case studies are included as a learning tool to reinforce the concepts introduced in this chapter. Exam candidates should note that the CISA exam does not currently use this format for testing.

## 3.18.1 CASE STUDY A

### *Case Study A Scenario*

A major retailer asked the IS auditor to review their readiness for complying with credit card company requirements for protecting cardholder information. The IS auditor subsequently learned the following information. The retailer uses wireless point-of-sale (POS) registers that connect to application servers located at each store. These registers use wired equivalent protection (WEP) encryption. The application server, usually located in the middle of the store's customer service area, forwards all sales data over a frame relay network to database servers located at the retailer's corporate headquarters, and using strong encryption over an Internet virtual private network (VPN) to the credit card processor for approval of the sale. Corporate databases are located on a protected screened subset of the corporate local area network. Additionally, weekly aggregate sales data by product line is copied from the corporate databases to magnetic media and mailed to a third party for analysis of buying patterns. It was noted that the retailer's database software has not been patched in over two years. This is because vendor support for the database package was dropped due to management's plans to eventually upgrade to a new ERP system.

| CASE STUDY A QUESTIONS |
|---|
| A1. Which of the following would present the **MOST** significant risk to the retailer?<br><br>    A.   Wireless point-of-sale registers use WEP encryption.<br>    B.   Databases patches are severely out-of-date.<br>    C.   Credit cardholder information is sent over the Internet.<br>    D.   Aggregate sales data are mailed to a third party. |
| A2. Based on the case study, which of the following controls would be the **MOST** important to implement?<br><br>    A.   Store application servers should be located in a secure area.<br>    B.   Point-of-sale registers should use two-factor authentication.<br>    C.   Wireless access points should use MAC address filtering.<br>    D.   Aggregate sales data sent offsite should be encrypted. |
| *See answers and explanations to the practice questions at the end of the chapter. (page 276)* |

# Systems and Infrastructure Life Cycle Management

## 3.18.2 CASE STUDY B

### *Case Study B Scenario*

A large industrial concern has begun a complex IT project, with ERP, to replace the main component systems of its accounting and project control departments. Sizeable customizations were anticipated and are being carried out with a phased approach of partial deliverables. These deliverables are released to users for pilot usage on real data and actual projects. In the meanwhile, detailed design and programming of the next phase begins. After a period of initial adjustment, the pilot users start experiencing serious difficulties. In spite of positive test results, already stabilized functionalities began to have intermittent problems; transactions hang during execution and, more and more frequently, project data get corrupted in the database. Additional problems show up—errors already corrected started occurring again and functional modifications already tested tend to present other errors. The project, already late, is now in a critical situation. The IS auditor, after collecting the evidence, requests an immediate meeting with the head of the project steering committee to communicate findings and suggest actions capable to improve the situation.

| CASE STUDY B QUESTIONS |
| --- |
| B1.    The IS auditor should indicate to the head of the project steering committee that:<br><br>    A.    the observed project problems are a classic example of loss of control of project activities and loose discipline in following procedures and methodologies. A new project leader should be appointed.<br>    B.    relays due to an underestimation of project efforts have led to failures in the versioning and modification control procedures.  New programming and system resources must be added to solve the root problem.<br>    C.    the problems are due to excessive system modifications after each delivery phase. The procedure for control of modifications must be tightened and made more selective.<br>    D.    the nature of initial problems is such as to lead to doubts regarding the adequacy and reliability of the platform. An immediate technical review of the hardware and software platform (parameters, configuration) is necessary.<br><br>B2.    In order to contribute more directly to solve the situation, the IS auditor should:<br><br>    A.    research the problems further to identify root causes and define appropriate countermeasures.<br>    B.    review the validity of the functional project specifications as the basis for an improved software baselining definition.<br>    C.    propose to be included in the project team as a consultant for the quality control of deliverables.<br>    D.    contact the project leader and discuss the project plans and recommend redefining the delivery schedule using the PERT methodology. |
| *See answers and explanations to the practice questions at the end of the chapter. (page 276)* |

# 3.19 ANSWERS TO PRACTICE QUESTIONS

3-1     **A**     Test data should be sanitized to prevent sensitive data from leaking to unauthorized persons.

3-2     **A**     User acceptance testing (UAT) is performed to ensure that the application meets predefined user needs. Therefore, it is the most relevant from the list provided to determine whether business objectives have been met. Performance testing is an element of systems testing that is performed to ensure that the application functions properly. This includes both business and nonbusiness requirements and is, therefore, less relevant than UAT to determine whether business requirements have been met. Sociability testing confirms that the new application can operate in the target environment without adversely impacting existing systems, not whether business objectives have been met. Penetration testing uses the tools and techniques available to a hacker to identify weaknesses in configurations and security settings. If the business objectives include security, penetration testing may identify whether this has been met. However, this will apply only for some applications, so penetration testing is less relevant than UAT.

3-3     **A**     Management should be informed immediately to determine whether they are willing to accept the potential material risk of not having that preventive control in place. The existence of a detective control instead of a preventive control usually increases the risks that a material problem may occur. Often during a BPR, many nonvalue-added controls will be eliminated. This is good, unless they increase the business and financial risks. The IS auditor may wish to monitor or recommend that management monitor the new process, but this should be done only after management has been informed and accepts the risk of not having the preventive control in place.

3-4     **A**     During the observation stage, the team collects data and visits the benchmarking partner. In the planning stage, the team identifies the critical processes for the benchmarking purpose. The analysis stage involves summarizing and interpreting the data collected and analyzing the gaps between an organization's process and its partner's process. During the adaptation step, the team needs to translate the findings into a few core principles and work down from principles to strategies to action plans.

3-5     **D**     A hash total is a technique to improve the accuracy of transaction processing by detecting omissions or errors. Hash totals do not verify that all payroll records are posted to the master file. Hash totals do not verify that only authorized employees are paid. Hash totals do not detect errors in payroll calculations.

3-6     **D**     The purpose of the requirements phase is to specify the functionality of the proposed system; therefore, the IS auditor would concentrate on the completeness of the specifications. The decision to purchase a package from a vendor would come after the requirements have been completed. Therefore, choices B and C are incorrect. Choice A is incorrect because a project timetable normally would not be found in a requirements document.

3-7     **A**     With purchase packages becoming more common, the design and development phases of the traditional life cycle have become replaceable with selection and configuration phases. A request for proposal from the supplier of packaged systems is called for and evaluated against predefined criteria for selection, before a decision is made to purchase the software. Thereafter, it is configured to meet the organization's requirement. The other phases of the SDLC such as feasibility study, requirements definition, implementation and postimplementation remain unaltered.

3-8    **C**    Obviously, project management has failed to either set up or verify controls that provide for software or software modules under development that adhere to those specifications made by the users. Quality assurance has its focus on formal aspects of software development such as adhering to coding standards or a specific development methodology. Functionality issues in terms of business usability are out of scope and are, in this case, part of the development phase. To fail at user specifications implies that requirements engineering has been done to describe the users' demands. Otherwise, there would not be a baseline of specifications to check against. Depending on the chosen approach (traditional waterfall, RAD, etc.), these discrepancies might show up during user training or acceptance testing—whatever occurs first.

3-9    **D**    The main change when using thin client architecture is making the servers critical to the operation; therefore, the probability that one of them fails is increased and, as a result, the availability risk is increased (option D). Because the other elements do not need to change, the other listed risks are not increased (choices A, B and C).

3-10    **A**    Control totals built into the trailer record of each segment is the only option that will ensure all individual transactions sent are received completely. The other options provide supporting evidence, but their findings are either incomplete or not timely.

# 3.20 ANSWERS TO CASE STUDY QUESTIONS

## ANSWERS TO CASE STUDY A QUESTIONS

A1.  **A**  Use of WEP encryption would present the most significant risk since WEP uses a fixed secret key that is easy to break. Transmission of credit cardholder information by wireless registers would be susceptible to interception, and would present a very serious risk. With regard to the unpatched database servers, since they are located on a screened subnet, this would mitigate the risk to the organization. Sending credit cardholder data over the Internet would be less of a risk since strong encryption is being utilized. Since the sales data being sent to the third party are aggregate data, no cardholder information should be included.

A2.  **A**  Locating application servers in an unsecured area creates a significant risk to the retailer since cardholder data may be retained on the server. Additionally, physical access to the server could potentially allow programs or devices to be installed that would capture sensitive cardholder information. Two-factor authentication for POS registers is not as critical. In the case of MAC address filtering, this can be spoofed. Since sales data are aggregate, no cardholder information should be at risk when the data are sent to a third party.

## ANSWERS TO CASE STUDY B QUESTIONS

B1.  **D**  The IS auditor has only found symptoms but not the root causes of the severe problems, so attributing them to limited skills of the project leader is inappropriate. The sequence of negative events seems to show that a fundamental problem is causing serious technical difficulties and, as a consequence of delays, negatively impacting the versioning and change control procedures. It is necessary, however, to ascertain immediately the nature of the suspect problems so if such a problem exists and is not found, it could severely affect the final result or kill the project. Also, the added technical programming resources are an inappropriate remedy because the development methodology is RAD, which is based on using a small group of skilled and experienced technical resources. The initial effects of the problem (intermittent block of transactions, corrupted data) lead to the suspicion that a problem might exist in the setup and configuration of the technical hardware/software environment or platform.

B2.  **A**  The only appropriate action is additional research, even if the apparently technical nature of the problem renders it unlikely that the auditor may find it alone. Functional project specifications should be executed by users and systems analysts, and not by the auditor. To propose to be project consultant for quality would not bring about an essential contribution since quality is a formal characteristic, whereas in the current case the problem is a substantial system instability. To contact the project leader and redesign the schedule of deliveries would not solve the problem. Furthermore, the definition of real causes may sensibly alter the project environment, invalidating in this way all preceding plans and projections.

# 3.21 SUGGESTED RESOURCES FOR FURTHER STUDY

**Bonham, Stephen S.;** *IT Project Portfolio Management,* **Artech House Inc., USA, 2005**

**Doss, George M.; Michael Wallace; Larry Webber;** *IS Project Management Handbook 2006 Edition,* **Aspen Publishers Inc., USA, 2006**

Information Processing Limited, "Software Testing and Software Development Lifecycles," *www.ipl.com/pdf/p0821.pdf*

**IT Governance Institute,** *Measuring and Demonstrating the Value of IT,* **USA, 2005**

**Maizlish, Bryan; Robert Handler;** *IT Portfolio Management Step-by-Step: Unlocking the Business Value of Technology,* **John Wiley & Sons, USA, 2005**

**Natan, Ron Ben;** *Implementing Database Security and Auditing,* **Elsevier Digital Press, USA, 2005**

**Wysocki, Robert K.;** *Effective Project Management,* **Wiley Publishing Inc., USA, 2007**

*Note:* Publications in bold are stocked in the ISACA Bookstore. *Information Systems Control Journal* articles are available at *www.isaca.org/archives.* The articles are available online to ISACA members only during their first year of release, and then are opened to the public. All referenced *Journal* articles are available on the CISA Practice Question Database v9.

Page intentionally left blank

*Chapter 4:*

# IT Service Delivery and Support

# IT Service Delivery and Support

# 4.1 INTRODUCTION

IT service management practices are important to provide assurance to users as well as management that the expected level of service will be delivered. Service level expectations are derived from the organization's business objectives. IT service delivery includes IS operations, IT services and management of IS and the groups responsible for supporting them.

## 4.1.1 OBJECTIVE

The objective of this area is to ensure that the CISA candidate understands and can provide assurance that the IT service management practices will ensure the delivery of the level of services required to meet the organization's objectives.

**This area represents 14 percent of the CISA examination (approximately 28 questions).**

## 4.1.2 TASKS

There are seven (7) tasks within the IT service delivery and support area:

T4.1    Evaluate service-level management practices to ensure that the level of service from internal and external service providers is defined and managed.

T4.2    Evaluate operations management to ensure that IT support functions effectively meet business needs.

T4.3    Evaluate data administration practices to ensure the integrity and optimization of databases.

T4.4    Evaluate the use of capacity and performance monitoring tools and techniques to ensure that IT services meet the organization's objectives.

T4.5    Evaluate change, configuration and release management practices to ensure that changes made to the organization's production environment are adequately controlled and documented.

T4.6    Evaluate problem and incident management practices to ensure that incidents, problems or errors are recorded, analyzed and resolved in a timely manner.

T4.7    Evaluate the functionality of the IT infrastructure (e.g., network components, hardware, system software) to ensure that it supports the organization's objectives.

## 4.1.3 KNOWLEDGE STATEMENTS

There are 11 knowledge statements within the IT service delivery and support area:

KS4.1    Knowledge of service-level management practices

KS4.2    Knowledge of operations management best practices (e.g., workload scheduling, network services management, preventive maintenance)

KS4.3    Knowledge of systems performance monitoring processes, tools and techniques (e.g., network analyzers, system utilization reports, load balancing)

KS4.4    Knowledge of the functionality of hardware and network components (e.g., routers, switches, firewalls, peripherals)

KS4.5    Knowledge of database administration practices

KS4.6    Knowledge of the functionality of system software, including operating systems, utilities and database management systems

KS4.7    Knowledge of capacity planning and monitoring techniques

KS4.8    Knowledge of processes for managing scheduled and emergency changes to the production systems and/or infrastructure including change, configuration, release and patch management practices

KS4.9    Knowledge of incident/problem management practices (e.g., help desk, escalation procedures and tracking)

KS4.10   Knowledge of software licensing and inventory practices

KS4.11   Knowledge of system resiliency tools and techniques (e.g., fault-tolerant hardware, elimination of single point of failure and clustering)

## 4.1.4 RELATIONSHIP OF TASK TO KNOWLEDGE STATEMENTS

The task statements are what the CISA candidate is expected to know how to do. The knowledge statements delineate what the CISA candidate is expected to know in order to perform the tasks. The task and knowledge statements are approximately mapped in **exhibit 4.1** insofar as it is possible to do so. Note that although there is often overlap, each task statement will generally map to several knowledge statements.

| Exhibit 4.1—Tasks and Knowledge Statements Mapping | |
|---|---|
| **Task Statements** | **Knowledge Statements** |
| T4.1 Evaluate service-level management practices to ensure that the level of service from internal and external service providers is defined and managed. | KS4.1 Knowledge of service-level management practices<br>KS4.2 Knowledge of operations management best practices (e.g., workload scheduling, network services management, preventive maintenance)<br>KS4.3 Knowledge of systems performance monitoring processes, tools and techniques (e.g., network analyzers, system utilization reports, load balancing)<br>KS4.6 Knowledge of the functionality of system software, including operating systems, utilities and database management systems<br>KS4.7 Knowledge of capacity planning and monitoring techniques<br>KS4.9 Knowledge of incident/problem management practices (e.g., help desk, escalation procedures and tracking)<br>KS4.11 Knowledge of system resiliency tools and techniques (e.g., fault-tolerant hardware, elimination of single point of failure and clustering) |
| T4.2 Evaluate operations management to ensure that IT support functions effectively meet business needs. | KS4.2 Knowledge of operations management best practices (e.g., workload scheduling, network services management, preventive maintenance)<br>KS4.3 Knowledge of systems performance monitoring processes, tools and techniques (e.g., network analyzers, system utilization reports, load balancing)<br>KS4.4 Knowledge of the functionality of hardware and network components (e.g., routers, switches, firewalls, peripherals)<br>KS4.5 Knowledge of database administration practices<br>KS4.6 Knowledge of the functionality of system software, including operating systems, utilities and database management systems<br>KS4.8 Knowledge of processes for managing scheduled and emergency changes to the production systems and/or infrastructure including change, configuration, release and patch management practices<br>KS4.9 Knowledge of incident/problem management practices (e.g., help desk, escalation procedures and tracking)<br>KS4.10 Knowledge of software licensing and inventory practices<br>KS4.11 Knowledge of system resiliency tools and techniques (e.g., fault-tolerant hardware, elimination of single point of failure and clustering) |

# IT Service Delivery and Support

| Exhibit 4.1—Tasks and Knowledge Statements Mapping (*cont.*) | |
| --- | --- |
| **Task Statements** | **Knowledge Statements** |
| T4.3 Evaluate data administration practices to ensure the integrity and optimization of databases. | KS4.5 Knowledge of database administration practices<br>KS4.6 Knowledge of the functionality of system software, including operating systems, utilities and database management systems<br>KS4.7 Knowledge of capacity planning and monitoring techniques<br>KS4.11 Knowledge of system resiliency tools and techniques (e.g., fault-tolerant hardware, elimination of single point of failure and clustering) |
| T4.4 Evaluate the use of capacity and performance monitoring tools and techniques to ensure that IT services meet the organization's objectives. | KS4.1 Knowledge of service-level management practices<br>KS4.2 Knowledge of operations management best practices (e.g., workload scheduling, network services management, preventive maintenance)<br>KS4.3 Knowledge of systems performance monitoring processes, tools and techniques (e.g., network analyzers, system utilization reports, load balancing)<br>KS4.7 Knowledge of capacity planning and monitoring techniques<br>KS4.11 Knowledge of system resiliency tools and techniques (e.g., fault-tolerant hardware, elimination of single point of failure and clustering) |
| T4.5 Evaluate change, configuration and release management practices to ensure that changes made to the organization's production environment are adequately controlled and documented. | KS4.2 Knowledge of operations management best practices (e.g., workload scheduling, network services management, preventive maintenance)<br>KS4.8 Knowledge of processes for managing scheduled and emergency changes to the production systems and/or infrastructure including change, configuration, release and patch management practices<br>KS4.9 Knowledge of incident/problem management practices (e.g., help desk, escalation procedures and tracking) |
| T4.6 Evaluate problem and incident management practices to ensure that incidents, problems or errors are recorded, analyzed and resolved in a timely manner. | KS4.1 Knowledge of service-level management practices<br>KS4.2 Knowledge of operations management best practices (e.g., workload scheduling, network services management, preventive maintenance)<br>KS4.3 Knowledge of systems performance monitoring processes, tools and techniques (e.g., network analyzers, system utilization reports, load balancing)<br>KS4.8 Knowledge of processes for managing scheduled and emergency changes to the production systems and/or infrastructure including change, configuration, release and patch management practices<br>KS4.9 Knowledge of incident/problem management practices (e.g., help desk, escalation procedures and tracking) |

| Exhibit 4.1—Tasks and Knowledge Statements Mapping (*cont.*) | |
|---|---|
| **Task Statements** | **Knowledge Statements** |
| T4.7 Evaluate the functionality of the IT infrastructure (e.g., network components, hardware, system software) to ensure that it supports the organization's objectives. | KS4.2 Knowledge of operations management best practices (e.g., workload scheduling, network services management, preventive maintenance)<br><br>KS4.3 Knowledge of systems performance monitoring processes, tools and techniques (e.g., network analyzers, system utilization reports, load balancing)<br><br>KS4.4 Knowledge of the functionality of hardware and network components (e.g., routers, switches, firewalls, peripherals)<br><br>KS4.5 Knowledge of database administration practices<br><br>KS4.6 Knowledge of the functionality of system software, including operating systems, utilities and database management systems<br><br>KS4.11 Knowledge of system resiliency tools and techniques (e.g., fault-tolerant hardware, elimination of single point of failure and clustering) |

# 4.2 INFORMATION SYSTEMS OPERATIONS

IS operations are in charge of the daily support of an organization's IS hardware and software environment. This function is particularly important when very large and centralized computing tasks are regularly executed for business purposes, producing output or updating situations. The organization of IS operations varies depending on the size of the computer environment and workload.

IS operations generally include the following functional areas:
• Management of IS operations
• Infrastructure support including computer operations
• Technical support/help desk
• Scheduling
• Controlling input/output of data
• Quality assurance
• Change control and release management
• Configuration management
• Program library management system and control software
• Problem management procedures
• Performance monitoring and management
• Capacity monitoring and planning
• Management of physical and environmental security
• Information security management

## 4.2.1 MANAGEMENT OF IS OPERATIONS

IS management has the overall responsibility for all operations within the IS department. This area will involve allocation of resources (planning and organization), adherence to standards and procedures (delivery and support), and monitoring of IS operation processes (monitoring and evaluation). Therefore, operations management functions include:
• Resource allocation—IS management is responsible for ensuring that the necessary resources are available to perform the planned activities within the IS function.

- Standards and procedures—IS management is responsible for establishing the necessary standards and procedures for all operations in accordance with the overall business strategies and policies.
- Process monitoring—IS management is responsible for monitoring and measuring the effectiveness and efficiency of IS operation processes, so that the processes will be improved over time.

## *Control Functions*

Management control functions include the following:
- Ensuring that detailed schedules exist for each operating shift
- Planning to ensure the most efficient and effective use of an operation's resources
- Authorizing and monitoring IT resource usage based on corporate policy
- Ensuring that security vulnerabilities (internal and external) are identified and resolved in a timely manner
- Detecting intrusion attempts
- Resolving information security events, incidents and problems in a timely manner
- Reviewing and authorizing changes to the operations schedules
- Reviewing and authorizing changes to the network, system and applications
- Monitoring operations to ensure compliance with standards
- Review log from all IT systems to detect critical system events and establish accountability of IT operations
- Ensuring that IS processing can recover in a timely manner from minor and major disruptions of operations
- Ensuring the confidentiality, integrity and availability of the data
- Monitoring system performance and resource usage to optimize computer resource utilization
- Anticipating equipment replacement/capacity to maximize current job throughput and strategically plan future acquisitions
- Monitoring the environment and security of the facility to maintain proper conditions for equipment performance
- Ensuring that changes to hardware and software do not cause undue disruption to normal processing
- Maintaining job accounting reports and other audit records
- Limiting logical and physical access to computer resources to those who require and are authorized to use it

| PRACTICE QUESTION |
|---|
| **4-1**    Which one of the following provides the **BEST** method for determining the level of performance provided by similar information processing facility environments? <br><br>     A.    User satisfaction <br>     B.    Goal accomplishment <br>     C.    Benchmarking <br>     D.    Capacity and growth planning |
| *See answers and explanations to the practice questions at the end of the chapter. (page 357)* |

# 4.2.2 IT SERVICE MANAGEMENT

IT service management (ITSM) is a concept that comprises processes and procedures for efficient and effective delivery and support of various IT functions. It focuses on fine-tuning IT services to meet the changing demands of the enterprise, and to measure and demonstrate improvements in the quality of IT services offered with a reduction in the cost of service in the long term.

ITSM focuses on the business deliverables and covers infrastructure management of IT applications that support and deliver these IT services.

IT support services are service desk, incident management, problem management, configuration management, change management and release management. IT delivery services include service-level management, IT financial management, capacity management, IT service continuity management and availability management. Though each management area is a separate process by itself, each process is highly interdependent on one or more other process.

IT services can be better managed with a service level agreement (SLA) and the services offered form a basis for such agreements. There is a possibility of a gap between customer expectations and the services offered, and this is narrowed by the SLA, which completely defines the nature, type, time and other relevant information for the services being offered.

For example, when a complaint is received, the service desk looks for an available solution from the information stored after classifying and storing it as an incident. Repeated incidents or major incidents may lead to problems that call for the problem management process. If changes are needed, the change management group of the process/program can provide a supporting role after consulting the configuration management group.

Any required change, whether it originated as a solution to a problem, an enhancement or for any other reason, goes through the change management process. The cost-benefit and feasibility studies are reviewed before the changes are accepted and approved. The risk of the changes should be studied and a fallback plan should be developed. The change may be for one configuration item or for multiple items, and the change management process invokes the configuration management process.

For example, the software could comprise different systems, each containing different programs and each program having different modules. The configuration can be maintained at the system level, the program level or the module level. The organization may have a policy saying that any changes made at the system level will be released as a new version. It may also decide to release a new version, if it involves changes at the program level for yet another application.

The releases, whether major or minor, will have a unique identity. Sometimes, the minor or small fixes may trigger some other problem. Fully tested, major releases may not have such problems. Because of testing time, space and other constraints, it is also possible to have a partial release, which is known as a delta release. The delta release contains only those items that have undergone changes since the last release.

The releases are controlled, and in the event of any problems in the new release, one should be able to back out completely and restore the system to its previous state. Suitable contingency plans may also be developed, if it is not completely restorable. These plans are developed before the new release is implemented.

Service management metrics should be captured and appropriately analyzed so that this information can be used to enhance the quality of service.

Service management comprises a configuration management database (CMDB), which stores each of the configuration items based on the solutions that are offered, the problems that are resolved or an SLA that is drawn.

ITSM implementation within an enterprise can be certified against the ISO/IEC 20000 standard. ITSM is an approach based on the Information Technology Infrastructure Library (ITIL) framework, which was developed by the UK government to demonstrate the industry best practice processes and procedures. Customer operation performance centre (COPC) is another best practice similar to that of the ITIL process. The key difference between ITIL and COPC is that ITIL provides input for establishing an infrastructure that is expected to meet all the delivery requirements, while COPC provides requirements that an infrastructure should deliver. Thus, IT service delivery expectations can be written using COPC and used to measure IT deliveries against these expectations.

## Service Level

The IS department is a service organization for end users. As such, the success of the IS department is dependent upon satisfying users and adhering to targets set within SLAs. Characteristics that should be considered in the delivery of these services include accuracy, completeness, timeliness and proper distribution of output related to application processing. Many tools are available to monitor the efficiency and effectiveness of services provided by IS personnel. These tools include:

• **Exception reports**—These automated reports identify all applications that did not successfully complete or otherwise malfunctioned. An excessive number of exceptions may indicate:
  – Poor application design, development or testing
  – Inadequate operation instructions

- Inadequate operations support
- Inadequate operator training or performance monitoring
- Inadequate sequencing of tasks
- Inadequate system configuration
- Inadequate capacity management

- **Job rerun reports**—Additional automated reports that should be considered with this process are the job rerun reports that identify all application reruns and restarts. These would provide additional, useful information regarding IS operations since most abnormal job terminations result in job restarts.
- **Operator problem reports**—These manual reports are used by operators to log computer operations problems and their resolutions. Operator responses should be reviewed by IS management to determine whether operator actions were appropriate or whether additional training should be provided to operators.
- **Output distribution reports**—These reports identify all application reports generated and their distribution locations. The reports can be either manual or automated and can be helpful for tracking lost, delayed or misrouted reports.
- **Console logs**—These automated reports identify most of the activities performed on the computer. Because of size and complexity, it is difficult to use console logs to monitor computer activity. Generally, IS management uses one of the immediately preceding, function-specific reports. Many installations produce such logs of all computer activity automatically, such as IBM's system management facility (SMF) log for the OS or the dump and temp logs in the UNIX system. Programs have been developed which analyze the system log to report on specifically defined items. Using this software, the auditor can carry out tests to ensure that:
- Only approved programs access sensitive data
- Software utilities that can alter data files and program libraries are used only for authorized purposes
- Approved programs are run only when scheduled and, conversely, that unauthorized runs do not take place
- The correct data file generation is accessed for production purposes
- Data files are adequately protected

In addition, there are software packages similar to the console logs that provide facilities to log and report all incidents of terminal logon/logoff, hardware malfunctions, system start-up and shutdown, and job start or stop.
- **Operator work schedules**—These schedules are generally maintained manually by IS management to assist in human resource planning. By ensuring proper staffing of operation support personnel, IS management is assured that service requirements of end users will be met. This is especially important during critical or heavy computer usage periods. These schedules should be flexible enough to allow for proper cross-training and emergency staffing requirements.

Many IS departments define the level of service that they will guarantee to users of the IS facilities. This level of service is often documented in SLAs. It is particularly important to define service levels where there is a contractual relationship between the IS department and the end user or customer. SLAs are often tied to chargeback systems, in which a certain percentage of the cost is apportioned from the end-user department to the IS department. When functions of the IS department are performed by a third party, it is important to have an outsourcing SLA.

Service levels are often defined to include hardware and software performance targets (such as user response time and hardware availability) but can also include a wide range of other performance measures. Such measures might include financial performance measures (such as year-to-year incremental cost reduction), human relationship measures (such as resource planning, staff turnover, development or training) or risk management measures (compliance with control objectives). The IS auditor should be aware of the different types of measures available and should ensure that they are comprehensive and include risk, security and control measures as well as efficiency and effectiveness measures.

Where the functions of an IS department are outsourced, the IS auditor should ensure that a provision is made for independent audit reports that cover all essential areas and provide full audit access.

| PRACTICE QUESTIONS |
|---|

**4-2** When reviewing a service level agreement (SLA) for an outsourced computer center an IS auditor should **FIRST** determine that:

   A.   the cost proposed for the services is reasonable.
   B.   security mechanisms are specified in the agreement.
   C.   the services in the agreement are based on an analysis of business needs.
   D.   audit access to the computer center is allowed under the agreement.

**4-3** A university's IT department and financial services office (FSO) have an existing service level agreement (SLA) that requires availability during each month to exceed 98 percent. The FSO has analyzed availability and noted that it has exceeded 98 percent for each of the last 12 months, but has averaged only 93 percent during month-end closing. Which of the following options **BEST** reflects the course of action the FSO should take?

   A.   Renegotiate the agreement.
   B.   Inform IT that the agreement is not meeting the required availability standard.
   C.   Acquire additional computing resources.
   D.   Streamline the month-end closing process.

*See answers and explanations to the practice questions at the end of the chapter. (page 357)*

# 4.2.3 INFRASTRUCTURE OPERATIONS

Computer operators are responsible for the accurate and efficient operation of scheduled computer jobs.

The operator's tasks include:
• Running jobs/macros/programs
• Restarting computer applications after an abnormal termination has been investigated and resolved by the responsible department
• Facilitating timely backup of computer files
• Monitoring the information processing facility (IPF) for unauthorized entry
• Monitoring and reviewing the extent of adherence to documented job schedules as established by IS and business management
• Participating in tests of disaster recovery plans (DRPs)
• Monitoring the performance, capacity, availability and failure of information resources
• Facilitating troubleshooting and incident handling

Procedures detailing instructions for operational tasks and procedures coupled with appropriate IS management oversight are necessary parts of the IS control environment.

This documentation should include:
• Operator procedures based on computer and peripheral equipment operating instructions and job flows
• Procedures for rectifying machine or program failures and the escalation of unresolved failures
• Instructions for output report distribution
• Procedures for obtaining files from the offline library and returning files to the library
• Procedures for reporting run delays
• Procedures for reporting and correcting equipment failures and job processing delays
• Procedures for recovering archived files

## *Lights-out Operations (Automated Unattended Operations)*

Lights-out operations is the automation of key computer room operations, whereby tasks can take place without human intervention.

The types of tasks being automated with the use of system operations software are:
• Job scheduling
• Console operation
• Report balancing and distribution
• Rerun/restart activities
• Tape mounting and management
• Storage device management
• Environmental monitoring
• Physical and data security

Advantages of lights-out operations are:
• Cost containment and/or reduction in IS operations
• Continuous operations (24/7)
• Reduced number of system errors and interruptions

## Input/Output Control Function

Input/output (I/O) control personnel are responsible for ensuring that information is processed accurately and completely, in accordance with IS management's intent and authorization.

In online environments where processing is generally paperless, the interrelationships among various applications are most appropriately dealt with during the processing control portion of an applications review. For example, in a paperless order-processing system, where customers initiate orders and are invoiced via electronic data interchange (EDI) or web-based applications, 20 or more separate applications may be involved. In such a system, the output from one system is the input for the next system in sequence. In general, while the individual applications are maintained online, files may be transferred among the applications using a batch file transfer mechanism controlled by the job scheduler. If the standard job schedule is modified, it is conceivable that input files expected by a successor application may not be present and may cause production processing to halt. It should be noted that these transfers of data between applications are frequently being built as online interfaces (similar to web services-oriented architectures); however, the batch transfer method is still relevant.

Based on the aforementioned example and on the responsibilities associated with this position, it should be clear that the typical tasks performed by I/O control personnel include ensuring that:
• Input is processed accurately and in a timely manner
• Output is produced in the proper format and is distributed to the appropriate people through the proper transmission methods. With the availability of communication facilities, the high-speed printing of paper documents (i.e., reports, bills, etc.) is often accomplished at a remote and sometimes separate location from the IS department and files may be stored for remote retrieval (e.g., File Transfer Protocol [FTP]) by other organizations.
• Output that will become input to the next system in sequence is accurate and complete, and produced in a timely manner
• Correct files were used during processing
• Proper actions were taken by operators
• No evidence exists indicating that data were altered in an unauthorized manner

## Job Accounting

Job accounting applications are designed to monitor and record the use of IS resources. Accounting becomes increasingly important in connection with SLAs where the client often divides the cost among applicable business units. Information recorded by these applications (such as the performance and utilization of the central processing unit (CPU), secondary storage media and terminal connect time) is used by IS management to perform activities that include:
• Matching resource utilization with associated users for billing purposes
• Optimizing hardware performance by changing or tuning system software defaults

## Scheduling

Scheduling is a major function within the IS department. The schedule includes the jobs that must be run, the sequence of job execution and the conditions that cause program execution. Low-priority jobs can also be scheduled, if time becomes available. Automated job scheduling software provides control over the scheduling process since job information is set up once—reducing the possibility of errors; job dependencies can be defined and software can provide security over access to production data.

High-priority jobs should be given optimal resource availability while maintenance functions such as backup and system reorganization should, if possible, be performed during nonpeak times. Schedules provide a means of keeping customer demand at a manageable level and permit unexpected or on-request jobs to be processed without unnecessary delay.

Job scheduling procedures are necessary to ensure that IS resources are utilized optimally based on processing requirements. Applications are increasingly required to be continually available; therefore, job scheduling (maintenance or long processing times) represents a greater challenge than before.

### JOB SCHEDULING SOFTWARE

Job scheduling software is system software used by installations that process a large number of batch routines. The scheduling software sets up daily work schedules and automatically determines which jobs are to be submitted to the system for processing.

The following are some of the advantages of using job scheduling software:
• Job information is set up only once, reducing the probability of an error.
• Job dependencies are defined so that if a job fails, subsequent jobs relying on its output will not be processed.
• Records are maintained of all job successes and failures.
• Reliance on operators is reduced.

# 4.2.4 MONITORING USE OF RESOURCES

Computer resources, like any other organizational asset, should be used in a manner that benefits the entire organization. This includes providing information to authorized personnel when and where it is needed, and at a cost that is identifiable and auditable. Computer resources include hardware, software, telecommunications, networks and data.

Controls over these resources are sometimes referred to as general controls. Effective control over computer resources is critical because of the reliance on computer processing in managing the business.

## Process of Incident Handling

Incident management is one of the critical processes in IT service management. IT needs to be attended to on a periodic basis to better serve the customer. Incident management focuses on providing increased continuity of service by reducing or removing the adverse effect of disturbances to IT services, and covers almost all nonstandard operations of IT services—thereby defining the scope to include virtually any nonstandard event which, in certain organizations, may include a request for an ink cartridge. In addition to incident initiation, other steps include classification of incidents, assignment to specialists, resolution and incident closure.

It is essential for any incident handling process to prioritize items after determining the impact and urgency. For example, there could be a situation where a service request from the chief information officer (CIO) for a printer problem arrives at the same time as a request from the technology team to attend to a server crash. IS management should have parameters in place for assigning the priority of these incidents, considering both the urgency and impact.

Unresolved incidents are escalated based on the criteria set by IS management. Incident management is reactive and its objective is to respond to and resolve issues as quickly as possible. Formal service level agreements are sometimes in place to define acceptable ranges for various incident management statistics.

## Problem Management

Problem management aims to resolve issues through the investigation and in-depth analysis of a major incident, or several incidents that are similar in nature, in order to identify the root cause. One of the standard methodologies for this analysis is the development of an Ishikawa (or cause-and-effect) diagram through brainstorming by the affected parties.

Once a problem is identified and analysis has identified a root cause, the condition becomes a "known error." A workaround can then be developed to address the error state and prevent future occurrences of the related incidents.

Problem management and incident management are related but have different methods and objectives. Problem management's objective is to reduce the number and/or severity of incidents, while incident management's objective is to return the effected business process back to its "normal state" as quickly as possible, minimizing the impact on the business. Effective problem management can show a significant improvement in the quality of service of an IS organization.

## Detection, Documentation, Control, Resolution and Reporting of Abnormal Conditions

Because of the highly complex nature of software, hardware and their interrelationships, a mechanism should exist to detect and document any abnormal conditions that could lead to the identification of an error. This documentation generally takes the form of an automated or manual log.

Errors that should be entered in the log include:
- Program errors
- System errors
- Operator errors
- Network errors
- Telecommunication errors
- Hardware errors

Examples of items that should appear in an error log entry include:
- Error date
- Error resolution description
- Error code
- Error description
- Source of error
- Escalation date and time
- Initials of the individual responsible for maintaining the log
- Initials of the individual responsible for closing the log entry
- Department/center responsible for error resolution
- Status code of problem resolution (i.e., problem open, problem closed pending some future specified date, or problem irresolvable in current environment)
- Narrative of the error resolution status

For control purposes, the ability to add to the error log should not be restricted. The ability to update the error log, however, should be restricted to authorized individuals, and the updates should be traceable. Proper segregation of duties requires that the ability to close an error log entry be assigned to a different individual than the one responsible for maintaining or initiating the error log entry.

Management should be aware of any communication devices or access mechanisms (i.e., modems, virtual private networks [VPNs], etc.) the organization has that permit vendors to remotely access production systems to troubleshoot and correct an issue. With devices that have dial-in capabilities, it is possible to bypass security features and make modifications without the knowledge of IS management. This also provides a method for unauthorized individuals to gain access and compromise the organization's computing resources.

IS management should ensure that the incident and problem management mechanisms are properly maintained and monitored and that outstanding errors are being adequately addressed and resolved in a timely manner.

IS management should develop operations documentation to ensure that procedures exist for the escalation of unresolved problems to a higher level of IS management. While there are many reasons why a problem may remain outstanding for a long period of time, it should not be acceptable for a problem to remain unresolved indefinitely. The primary risk resulting from lack of attention to unresolved problems is the interruption of business operations. An unresolved hardware or software problem could potentially corrupt production data. Problem escalation procedures should be well-documented. IS management should ensure that the problem escalation procedures are being adhered to properly. Problem escalation procedures generally include:
• Names/contact details of individuals who can deal with specific types of problems
• Types of problems that require urgent resolution
• Problems that can wait until normal working hours

Problem resolution should be communicated to appropriate systems, programming, operations and user personnel to ensure that problems are resolved in a timely manner. The IS auditor should examine problem reports and logs to ensure that they are resolved in a timely manner and are assigned to the individuals or groups most capable of resolving the problem.

The departments and positions responsible for problem resolution should be part of problem management documentation. This documentation must be maintained properly to be useful.

## 4.2.5 SUPPORT/HELP DESK

The responsibility of the technical support function is to provide specialist knowledge of production systems to identify and assist in system change/development and problem resolution. In addition, it is technical support's responsibility to apprise management of current technologies that may benefit overall operations.

Procedures covering the tasks to be performed by the technical support personnel must be established in accordance with an organization's overall strategies and policies.

The support functions include:
• Determining the source of computer problems and taking appropriate corrective actions
• Initiating problem reports as required and ensuring that problems are resolved in a timely manner
• Obtaining detailed knowledge of the operating system and other systems software
• Answering inquiries regarding specific systems
• Controlling the installation of vendor and system software to improve its efficiency and customize the system based on organizational requirements and computer configuration
• Providing technical support for computerized telecommunications processing
• Maintaining documentation of vendor software, including issuance of new releases and problem fixes, as well as documentation of utilities and systems developed in house
• Communicating with IS operations to signal abnormal patterns in calls or application behavior

The primary purpose of the help desk is to service the user. The help desk personnel must ensure that all hardware and software problems that arise are fully documented and escalated based on the priorities established by management. In many organizations, the help desk function means different things. However, the basic function of the help desk is to perform the following:
• Document problems that arise from users and initiate problem resolution
• Prioritize the issues and forward them to the appropriate managers
• Follow up on unresolved problems
• Close out resolved problems, noting proper authorization to close out the problem by the user

# IT Service Delivery and Support

## 4.2.6 CHANGE MANAGEMENT PROCESS

Change control procedures are a part of the more encompassing function referred to as change management and are established by IS management to control the movement of applications (composed of jobs or programs) from the test environment, where development and maintenance occurs, to the quality assurance (QA) environment, where thorough testing occurs, and finally to the production environment. For software changes, the portion of the program change control mechanism that describes the actions to be performed by IS operations personnel after a job or program has passed user acceptance testing (UAT) and is to be moved from the staging environment to the production environment may be referred to as migration procedures.

The procedures associated with this migration process ensure that:
• System, operations and program documentation are complete, up to date and in compliance with the established standards
• Job preparation, scheduling and operating instructions have been established
• System and program test results have been reviewed and approved by user and project management
• Data file conversion, if necessary, has occurred accurately and completely as evidenced by review and approval by user management
• System conversion has occurred accurately and completely as evidenced by review and approval by user management
• All aspects of jobs turned over have been tested, reviewed and approved by control/operations personnel
• The risks of adversely affecting the business operation are reviewed and a rollback plan is developed to back out the changes, if necessary

Apart from program change control, standardized methods and procedures for change management are needed to ensure and maintain agreed-on levels in quality service. These methods are aimed at minimizing the adverse impact of any probable incidents triggered by change that may arise.

This is achieved by formalizing and documenting the process of change request, authorization, testing, implementation and communication to the users. Change requests are often categorized into emergency changes, major changes and minor changes, and may have different change management procedures in place for each type of change.

| PRACTICE QUESTION |
|---|
| **4-4** Which of the following is the **MOST** effective method for an IS auditor to use in testing the program change management process? <br><br> A. Trace from system-generated information to the change management documentation <br> B. Examine change management documentation for evidence of accuracy <br> C. Trace from the change management documentation to a system-generated audit trail <br> D. Examine change management documentation for evidence of completeness |
| *See answers and explanations to the practice questions at the end of the chapter. (page 357)* |

See chapter 3, Systems and Infrastructure Life Cycle Management, for details on program change controls.

## 4.2.7 PROGRAM LIBRARY MANAGEMENT SYSTEMS

Program library management systems provide several functional capabilities to facilitate effective and efficient management of the data center software inventory. The inventory may include application and system software program code, job control statements that identify resources used and processes to be performed, and processing parameters which direct processing.

Some of the library management system's traits/capabilities are:
• **Integrity**—Each source program is assigned a modification number and version number, and each source statement is associated with a creation date. Security over program libraries, job control language sets and parameter files provided is through the use of passwords, encryption, data compression facilities and automatic backup creation.

- **Update**—Library management systems facilitate the addition, modification, deletion, resequencing and editing of library members. For example, they can create an automated backup copy prior to changes being made and maintain an audit trail of all library activities.
- **Reporting**—Lists of additions, deletions and modifications, the library catalog and library member attributes can be prepared for management and auditor review.
- **Interface**—Library software packages may interface with the operating system, job scheduling system, access control system and online program management.

## 4.2.8 LIBRARY CONTROL SOFTWARE

Library control software should be used to separate test libraries from production libraries in both mainframe and client-server environments. The major objective of library software is to ensure that all program changes have been authorized. The minimum control necessary is authorization of the changes to the production library, which requires an authorization procedure between the test and production libraries. Authorization controls should be applied to test and production libraries. The risk of not having controls over both libraries is that a test version may contain unauthorized or fraudulent code, which could be used in the next change to production. After being linked to a production object, the fraudulent code would be placed into production.

The primary purpose of library control software is to:
- Prohibit programmers from accessing production source and object libraries
- Prohibit batch program updating
- Require the control group or operator to release source code and place it in the programmer's library
- Require the programmer to turn over changed source code to the control group or operator who updates the object library or promotes the code to a test region
- Require the control group or operator to update the object library with the production version after testing is completed
- Allow read-only access to source code
- Enforce a program naming convention that allows test programs to be readily distinguished from production programs
- Implement job control language screening controls to prevent inadvertent production execution of a test program
- Allow changes to be backed out
- Enforce programming/coding standards

### Executable and Source Code Integrity

Each production executable module should have one corresponding source module. Whenever a modified program is moved into the production source code library, a compiled executable version of that program should be moved to the production library. Usually this is a manual procedure performed by the group or individual responsible for program migration. An optimal procedure would be one that automatically creates an executable code module whenever a source module is moved into production. This procedure is sometimes a feature of library management software products that are available for many computer platforms.

Ensuring the executable/source integrity is vital to the proper control of production systems because it provides assurance that an improper version of a program is not being executed. The time stamp of the source's module should not be later than that of the corresponding executable module. At no time should the user or application programmer have access to production source code or load libraries in the production environment. This is used to minimize the risk of inappropriate segregation of duties in the change control process. However, in a user-controlled application development, users develop their own applications independent of the IS department. Users work with stand-alone PCs or in environments that are similar to the production environment to meet their business needs. IS auditors must review with special care those applications developed by autonomous end users since they may not adhere to corporate standards or may not be as securely configured. This is especially important when the applications can be used to access or manipulate central databases. Without the guidance of professional developers, users can potentially develop flawed programs, generate misleading management reports, adversely affect system performance or even introduce errors into databases.

## Source Code Comparison

Source code comparison software is an effective and easy-to-use method for tracing changes to programs. The IS auditor must first obtain and store (on tape or disk) a control copy of the source code of the program to be tested. At some future date (in a week or month) the IS auditor runs the source comparison program to trace differences between the control and current source. A list of additions, deletions and changes is generated and used by the auditor to review change authorization approval documentation such as work requests, approvals and program testing.

Source comparison programs allow the IS auditor to examine source-program changes and controls without assistance or possible misleading information from IS personnel. This technique, however, does not detect a source program that has been changed and restored to its original form during the time between control copy acquisition and the comparison run. Methods for detecting changes to load modules should be used with the source code comparison software.

A manual review of the changes in the source code is considered valid but not totally effective. The IS auditor obtains a source listing of the module being audited and scans the program for audit trail statements. For example, programmers can enter evidence of a program change (e.g., a date, initials, work request number). In addition, deleted or changed code need not be physically removed from the source code; deleted or original source lines can be changed to comments. The programmer can also write a history or narrative of changes in the remarks section allowed in most programming language rules of coding.

## 4.2.9 RELEASE MANAGEMENT

Software release management is the process through which software is made available to users. The term "release" is used to describe a collection of authorized changes. The release will typically consist of a number of problem fixes and enhancements to the service. A release consists of the new or changed software required. Releases are often divided into:
• **Major releases**—Normally contain a significant change or addition to new functionality. A major upgrade or release usually supersedes all preceding minor upgrades. Grouping together a number of changes facilitates more comprehensive testing and planned user training. Large organizations typically have a predefined timetable for implementing major releases throughout the year (e.g., quarterly). Smaller organizations may have only one release during the year or numerous releases if the organization is quickly growing.
• **Minor software releases**—Upgrades, normally containing small enhancements and fixes. A minor upgrade or release usually supersedes all preceding emergency fixes. Minor releases are generally used to fix small reliability or functionality problems that cannot wait until the next major release. The entire release process should be followed for the preparation and implementation of minor releases, but it is likely to take less time because the development, testing and implementation activities do not require as much time as major releases do.
• **Emergency software releases**—Normally containing the corrections to a small number of known problems. Emergency releases are fixes that require implementation as quickly as possible to prevent significant user downtime to business-critical functions. Depending upon the required urgency of the release, limited testing and release management activities are executed prior to implementation. Such changes should be avoided whenever possible because they increase the risk of errors being introduced.

Many new system implementations will involve phased delivery of functionality and thus require multiple releases. In addition, planned releases will offer an ongoing process for system enhancement.

The main roles and responsibilities in release management should be defined to ensure that everyone understands their role and level of authority and those of others involved in the process. The organization should decide the most appropriate approach, depending on the size and nature of the systems, the number and frequency of releases required, and any special needs of the users, for example, if a phased rollout is required over an extended period of time. All releases should have a unique identifier that can be used by configuration management.

Planning a release involves:
• Gaining consensus on the release's contents
• Agreeing to the phasing over time and by geographical location, business unit and customers
• Producing a high-level release schedule
• Planning resource levels (including staff overtime)

• Agreeing on roles and responsibilities
• Producing back-out plans
• Developing a quality plan for the release
• Planning acceptance of support groups and the customer

## 4.2.10 QUALITY ASSURANCE

QA personnel verify that system changes are authorized, tested and implemented in a controlled manner prior to being introduced into the production environment. With the assistance of librarian software, personnel also oversee the proper maintenance of program versions and source code to object integrity.

See chapter 3, System Development Life Cycle, for more details on QA and on specific objectives of the QA function.

## 4.2.11 INFORMATION SECURITY MANAGEMENT

Information security management, which ensures continuous IT operation and security of business process and data, is a critical part of IS operations. Information security management includes various security processes to protect the information assets. Information security processes should be integrated in all IT operation processes.

Information security management includes:
• Performing risk assessments on information assets
• Performing business impact analyses (BIAs)
• Developing and enforcing information security policy, procedures and standards
• Conducting security assessments on a regular basis
• Implementing a formal vulnerability management process

See chapter 5, Protection of Information Assets, for more details on information security management.

## 4.2.12  MEDIA SANITIZATION PROGRAM

An effective media sanitization program establishes the controls, techniques and processes necessary to preserve the confidentiality of sensitive information stored on media to be reused, transported, or discarded. In general, sanitization involves the eradication of information recorded on storage media to the extent of providing reasonable assurance that residual content can not be salvaged or restored. From an organizational perspective, identifying media types sanctioned for information storage is essential to detect unauthorized storage media being used to record organizational information, and more importantly, to target sanitization techniques for approved storage media in proportion to the information sensitivity level and protection requirements. Given this conceptual approach, the probability of unauthorized access to information can be significantly reduced to acceptable risk levels as a direct consequence of senior management sponsorship, coupled with its commitment to providing sufficient resources to achieve organization objectives.

# 4.3 INFORMATION SYSTEMS HARDWARE

This section provides an introduction to hardware platforms that make up the enterprise systems of today's organizations. The section describes the basic concepts of and history behind the different types of computers developed, and the advances in information technology that have occurred. Also discussed are the key audit considerations such as capacity management, system monitoring, maintenance of hardware and typical steps in the acquisition of new hardware.

**Note:** Vendor-specific terminology is used within this manual for illustrative purposes only. Candidates will not be examined on the components of vendor-specific hardware offerings or on vendor-specific terminology unless this terminology has become generalized and is used globally.

## 4.3.1 COMPUTER HARDWARE COMPONENTS AND ARCHITECTURES

The hardware components of computer systems include differing interdependent components performing specific functions, which can be classified as either processing or input/output components.

### *Processing Components*

The main hardware component of a computer is the CPU. The CPU consists of an arithmetic logic unit (ALU), a control unit and an internal memory. The control unit consists of electrical circuits that control/direct all operations in the computer system. The ALU performs mathematical and logical operations. The internal memory is that part of the memory within the CPU, such as the registry, that is used for processing transactions.

Other key components of a computer include random access memory (RAM) and read-only memory (ROM). The computer usually requires permanent storage devices, such as the hard disk, in order to operate.

### *Input/Output Components*

The I/O components are used to pass instructions/information to the computer and to collect or record the output generated by the computer. Some components, such as the keyboard, are input-only devices, while others, such as the touch screen, are both input and output devices. Printers are an example of an output-only device.

### *Types of Computers*

Computers can be categorized following several criteria, mainly based on their processing power, size and architecture. These categories include:

- **Supercomputers**—Very large and expensive computers with the highest processing speed, designed to be used for specialized purposes or fields that require extensive processing power (e.g., complex mathematical or logical calculations). They are typically dedicated to a few specific specialized system or application programs.
- **Mainframes**—Large, general-purpose computers that are made to share their processing power and facilities with thousands of internal or external users. Mainframes accomplish this by executing a large variety of tasks almost simultaneously. The range of capabilities of these computers is extensive. A mainframe computer often has its own proprietary operating system that can support background (batch) and real-time (online) programs operating parallel applications. Mainframes have traditionally been the main data processing and data warehousing resource of large concerns and, as such, have long been protected by a number of the early security and control tools.

  Many of the best practices that grew out of the mainframe environment have been adopted, to some degree, in other types of computer environments. For example, mainframes have segregation built into their system hardware controls. This allows the software components (e.g., programming languages, link libraries, stored system programs, databases, user application programs, etc.) to be kept in separate areas and to execute in a fully separate environment. This protects the operating system from the risk of interference (integrity and availability) and the business application from the risk of unexpected disclosures (confidentiality). Additionally, mainframe systems, through separate logical partitions on the physical unit, provide a built-in mechanism for controlled access to the production (live) and development (test) environments and to their associated databases, storage and peripherals, thus preserving management programs (integrity).
- **Midrange computers and servers**—Multiprocessing systems capable of supporting up to hundreds of simultaneous users. In size and power, midrange computers and servers can be comparable to small mainframes. Midrange computers and servers have many of the control features of mainframes such as memory management, segregation of work folders, command language, etc. Their capabilities are also comparable to small mainframes in terms of speed for processing data and execution of client programs, but they cost much less than mainframes. Their operating systems and system software base components are often commercial products. The higher-end devices generally utilize UNIX and, in many cases, are used as database servers while smaller devices are more likely to utilize the Windows operating system and be used as application servers and file/print servers.
- **Microcomputers** (personal computers [PCs])—Small computer systems referred to as PCs or workstations that are designed for individual users, inexpensively priced and based on microprocessor technology which places an entire CPU on a chip. Their use includes office automation functions such as word processing, spreadsheets and e-mail; database management; interaction with web-based applications; and others such as personal graphics, voice, imaging, design, web access and entertainment. Although designed as single-user systems, these computers are commonly linked together to form a network.

In terms of power, at the high end, the distinction between personal computers and workstations has faded. Both, generally, come with a large, high-resolution graphics screen, up to 1 gigabyte (GB) of volatile memory (RAM), and support for network interface and communications cards. Both also have large and very fast magnetic disk storage devices, referred to as disk drives, that can permanently hold much more information than RAM (currently in the range of 30 to 250 GB); however, for network usage, a special type of workstation, called a diskless workstation, comes without a disk drive. The most common operating systems used by business organizations for PCs are different versions of UNIX and Microsoft Windows.
- **Thin client computers**—These are personal computers that are generally configured with minimal hardware features (e.g., diskless workstation) with the intent being that most processing occurs at the server level using software, such as Microsoft Terminal Services or Citrix Presentation Server, to access a suite of applications.
- **Notebook/laptop computers**—Lightweight (under 10 pounds/5 kilograms) personal computers that are easily transportable and are powered by a normal AC connection or by a rechargeable battery pack. Similar to the desktop variety of personal computers in capability, they have similar CPUs, memory capacity and disk storage capacity, but the battery pack makes them less vulnerable to power failures.

This family of computers has evolved to fulfill the requirements of portable mobile computing, where users can carry their e-mail, office suite and relevant business client applications in their laptop and work from any place within the workplace's premises or remotely as a telecommuter. The computers use common operating systems such as different versions of UNIX and Microsoft Windows. Being portable, these are vulnerable to theft. Devices may be stolen to obtain information contained therein and hijack connectivity, either within an internal local area network (LAN) or remotely.
- **Personal Digital Assistants (PDAs)**—Handheld devices that enable its users to use a small computing device the size of a calculator as a personal organizer and planner. Some of its uses include a scheduler, a phone and address book, creating and tracking to-do lists, an expense manager, and an assortment of other office automation-based functions. Such devices can also combine computing, telephone/fax and networking features together so they can be used anytime and anywhere. A common hardware peripheral feature differing from other computers is that most PDAs are pen-based devices. Besides being operable through a keyboard, they accept character input from a stylus, writing or pointing on the screen, and incorporate handwriting recognition features. PDAs are also capable of interfacing with PCs to back up or transfer important information. Likewise, information from a PC can be downloaded to a PDA.

## Common Characteristics of Different Types of Computers

All types of computers can support color monitors with high-quality resolution, high-speed wired or wireless network connections, and peripheral devices such as audio cards, CD-ROMs, DVDs, tape drives, printers, high-speed modems and customized devices for workstations.

With the increases in speed, processing power and storage capacity and the reduction in size of computers in general, the distinctions between categories and types of usage are now less important than in the past. For instance, some mainframes use UNIX operating systems; mainframes and minicomputers are often used as database servers; and minicomputers, either stand-alone or in cluster configuration, are used as the main shared computing resource by average-sized companies. All levels of machines may have more than one processor. Additionally, through the CPU, all types of computers can perform several tasks nearly simultaneously using the following principles:
- **Multitasking**—The capability of running two or more applications concurrently—often performed by allocating processing time to each application. Each individual task is segmented and each segment is run during a (short) time lapse. Subsequently, execution of the concurrent application is suspended. The processing rights are then given to a different task that runs its own segment, and so on, for all active tasks that need to compute. In this way, each task appears to be continuous without any disruption felt by the user of the system.
- **Multiprocessing**—The linking of more than one processor sharing the same memory to execute programs simultaneously
- **Multiusing**—Also referred to as timesharing. By using multitasking capabilities, multiple users can access and use a computer operating system simultaneously. All mainframes, midrange and microcomputer systems designed for client-server computing are multiuser-based environments. For example, in personal computer environments, workstations access servers through a network so that the computing load is shared among workstations and servers.
- **Multithreading**—When a program is executed there are several subactivities taking place within the program. These subactivities are performed through separate predefined or identified threads and produce output for the entire activity seamlessly. The advantages are the optimization of the processor and the minimization of idle time within the processor.

- **Grid computing**—Some applications are designed to utilize the processing power of many, in some cases even thousands, of computers to execute programs that require enormous amounts of processing power similar to that found in supercomputers. By utilizing unused processing cycles on many computers that are accessible over the Internet, this technique known as grid computing, allows resource intensive tasks to be performed by arrays of ordinary machines.

## Common Computer Roles

In today's distributed environment, there are many different devices used in delivering application services. The following are some of the most common devices encountered.

- **Print servers**—Businesses of all sizes require that printing capability be made available to users across multiple sites and domains. Generally, a network printer is configured based upon where the printer is physically located and who within the organization needs to use it. Print servers allow businesses to consolidate printing resources for cost-savings.
- **File servers**—Provide for organizationwide access to files and programs. Document repositories can be centralized to a few locations within the organization and controlled with an access-control matrix. Group collaboration and document management are easier when a document repository is used, rather than dispersed storage across multiple workstations.
- **Application (program) servers**—Application servers typically host the software programs that provide application access to client computers, including the processing of the application business logic and communication with the application's database. Consolidation of applications and licenses in servers enables centralized management and a more secure environment.
- **Web servers**—Provide information and services to external customers and internal employees through web pages. A web server is normally accessed by its universal resource locator (URL), e.g., *http://www.isaca.org*.
- **Proxy servers**—Provide an intermediate link between users and resources. As opposed to direct access, proxy servers will access services on a user's behalf. Depending on the services being proxied, a proxy server may render more secure and faster response than direct access.
- **Database servers**—Store raw data and act as a repository. The servers concentrate on storing information rather than presenting it to be usable. Application servers and web servers use the data stored in database servers and process the data into usable information.
- **Appliances (specialized devices)**—Provide a specific service and normally would not be capable of running other services. As a result, the devices are significantly smaller and faster, and very efficient. Capacity and performance demands require certain services to be run on appliances instead of generic servers. A few examples of appliances commonly found in today's environments are:
  - **Firewall**—A specific device that inspects all traffic going between segments and applies security policies to help ensure a secure network. An effective firewall implementation depends on the quality of the security policies written and their compliance with best practices.
  - **Intrusion detection system (IDS)**—Listens to all incoming and outgoing traffic to deduce and warn of potentially malicious connections.
  - **Intrusion prevention system (IPS)**—Actively attempts to prevent intrusion by monitoring traffic and identifying irregular usage patterns.
  - **Switches**—Data link-level devices that can divide and interconnect network segments and help to reduce collision domains in Ethernet-based networks
  - **Routers**—Devices used to link two or more physically separate network segments. The network segments linked by a router remain logically separate and can function as independent networks.
  - **VPNs**—VPNs provide remote access to enterprise IT resources or can link two or more physically separate networks through a security tunnel. A Secure Sockets Layer-virtual private network (SSL-VPN) provides clientless remote access only through an Internet browser.
  - **Load balancer**—A load balancer distributes traffic across several different devices to increase the performance and availability of IT services

## Universal Serial Bus

Adapters, cables, CD-RW, DVD-RW, scanners, webcams, flash readers, hard drives, hubs, switches and video phones can be connected to the computer using a USB connection.

USB ports overcome the limitations of the serial and parallel ports in terms of speed and the actual number of connections that can be made. USB ports can transfer data up to 480 Mbps, and up to 127 devices can be connected to a single computer. It is easy to connect the devices using a USB connection since they are hot-swappable, i.e., system restart is not required to initialize the device.

Most operating systems recognize when a USB device is connected and assign each device an address. The operating system will then attempt to load the necessary device drivers by searching for the appropriate drivers already on the system. If none is found, the user will then be prompted to manually install appropriate drivers for the device.

## Memory Cards/Flash Drives
Because of the versatility of computers and computing devices, alternative devices have emerged to store information. There has been a complete shift in the shape, size and capacity of the storage devices, with some memory cards as small as postage stamps. They are nonvolatile, removable, small, easy to erase and store, and are often referred to as flash memory cards.

They can be used in cameras, music players, PDAs, mobile phones, printers, navigation systems and many other portable devices.

One type of device combines flash memory with a USB plug to form a so-called "flash drive" or "pen drive," which is quite useful and reliable in storing information. Despite the fact that encryption is available for flash drives, they pose serious security issues:
• Anyone using such a device can copy gigabytes of data without necessarily leaving any trace.
• A flash drive that is lost or stolen could expose gigabytes of confidential information.
• Some of the bootable USB devices bypass security checks and completely expose the system.

## Radio Frequency Identification
Radio Frequency Identification (RFID) uses radio waves to identify tagged objects within a limited radius. A tag consists of a microchip and an antenna. The microchip stores information along with an ID to identify a product. The other part of the tag is the antenna, which transmits the information to the RFID reader.

The power needed to drive the tag can be derived in two modes. The first mode, used in passive tags, draws power from the incident radiation arriving from the reader. The second mode, used in active tags, derives its electrical supply from batteries attached to it and is capable of utilizing higher frequencies and achieving longer communication distances. An active tag is reusable and expensive, and can contain more data. A passive tag is generally the opposite.

A passive tag can generally contain about two kilobytes (KB) of data, which is enough to store basic information about an item. As noted above, an active tag can contain more information. Tags can be used to identify an item based on either direct product identification or carrier identification. In the case of the latter, an article's ID is manually fed into the system (e.g., using a bar code) and is used along with strategically placed radio frequency readers to track and locate the item. RFID has many applications, including logistics, inventory, article surveillance, item tracking and even monitoring livestock.

## Write Once and Read Many
The advancement of computing technology requires bigger and faster storage media with affordable price tags. In the past, a floppy disc and a small capacity hard drive would be sufficient to do almost everything. This no longer holds true. In recent years, many new technologies have been introduced, including those in the write once and read many (WORM) category. It is worth mentioning that this type of WORM is totally unrelated to the malicious code also referred to as a "worm."

There are several types of WORM media, including:
• **Compact discs (CDs)**—Commonly found in audio and computer stores, they have multipurpose storage functions. A CD can be used to store audio, known as music CDs, or data files. CDs can store up to 80 minutes of audio or 700 megabytes (MB) of data and are the most common form of WORM media. CDs are common and affordable, and the number used can be varied to fit the needs of the user.

- **Compact disc recordable/rewritable (CD-R or CD-RW)**—A type of CD that can be written with consumer appliances such as CD-R/CD-RW writers. Compact disc media that is rewritable (RW) is technically not WORM media.
- **Digital video disc (DVD)**—Refers to optical disc storage technology. DVDs are essentially a larger capacity (several GB), faster CD that can hold cinema-like video, better-than-CD audio, still photos and computer data. DVDs aim to encompass home entertainment, computers and business information with a single, digital format. They have replaced laserdisc, are well on the way to replacing videotape and video game cartridges, and could eventually replace audio CD and CD-ROM.
- **Digital video disc-high density (DVD-HD, Blu-Ray)**—The most recent generation of digital video disc technology, this format utilizes a blue-violet laser that, because of its shorter wavelength, can write more information on a single disk (15-25 GB).

## 4.3.2 HARDWARE MAINTENANCE PROGRAM

To ensure proper operation, hardware must be routinely cleaned and serviced. Maintenance requirements vary based on complexity and performance workloads (e.g., processing requirements, terminals access and number of applications running). In any event, maintenance should be scheduled to closely coincide with vendor-provided specifications. The hardware maintenance program is designed to document the performance of this maintenance.

Information typically maintained by this program includes:
- Reputable service company information for each IS hardware resource requiring routine maintenance
- Maintenance schedule information
- Maintenance cost information
- Maintenance performance history information such as planned versus unplanned, executed and exceptional

IS management should monitor, identify and document any deviations from vendor maintenance specifications as well as provide supporting arguments for this deviation.

When performing an audit of this area, the IS auditor should:
- Ensure that a formal maintenance plan has been developed and approved by management
- Identify maintenance costs that exceed budget or are excessive. These overages may be an indication of a lack of adherence to maintenance procedures or of upcoming changes to hardware. Proper inquiry and follow-up procedures should be performed.

## 4.3.3 HARDWARE MONITORING PROCEDURES

The following are typical procedures and reports for monitoring the effective and efficient use of hardware:
- **Availability reports**—These reports indicate the time periods during which the computer is in operation and available for utilization by users or other processes. A key concern addressed by this report is excessive IS unavailability, referred to as downtime. This unavailability may indicate inadequate hardware facilities, excessive operating system maintenance, the need for preventive maintenance, inadequate environmental facilities (e.g., power supply or air conditioning) or inadequate training for operators.
- **Hardware error reports**—These reports identify CPU, I/O, power and storage failures. These reports should be reviewed by IS operations management to ensure that equipment is functioning properly, to detect failures and to initiate corrective action. The IS auditor should be aware that a sure attribution of an error in hardware or software is not necessarily easy and immediate. Reports should be checked for intermittent or recurring problems, which might indicate difficulties in properly diagnosing the errors.
- **Utilization reports**—These automated reports document the use of the machine and peripherals. Software monitors are used to capture utilization measurements for processors, channels and secondary storage media (such as disk and tape drives). Depending on the operating system, resource utilization for multiuser computing environments found in mainframe/large-scale computers should average in the 85 to 95 percent range, with allowances for utilization occasionally reaching 100 percent and falling below 70 percent. Trends provided by utilization reports can be used by IS management to predict whether more or fewer processing resources are required.
- **Asset/fleet management reports**—Inventory of network-connected equipment such as PCs, servers, routers and other devices

If utilization is routinely above the 95 percent level, IS management may consider reviewing user and application patterns to free space, upgrading computer hardware, and/or investigating where savings can be made by eliminating nonessential processing or moving processing that is less critical to time periods that are in less demand (such as during the night). If the utilization is routinely below the 85 percent level, there is a need to determine whether hardware exceeds processing requirements. A decision to allow lower utilization could depend on the need to keep resources free for backup purposes.

## 4.3.4 CAPACITY MANAGEMENT

Capacity management is the planning and monitoring of computing resources to ensure that the available resources are used efficiently and effectively. This requires that the expansion or reduction of resources takes place in parallel with the overall business growth or reduction. The capacity plan should be developed based on input from both user and IS management to ensure that business goals are achieved in the most efficient and effective way. This plan should be reviewed and updated at least annually.

Capacity planning should include projections substantiated by past experience, considering the growth of existing business as well as future expansions. The following information is key to the successful completion of this task:

- CPU utilization
- Computer storage utilization
- Telecommunications and WAN bandwidth utilization
- I/O channel utilization
- Number of users
- New technologies
- New applications
- SLAs

The IS auditor must realize that the amount and distribution of these requirements has an intrinsic flexibility. Specialized resources of a given class may have an impact on the requirements for other classes. For example, the proper use of more "intelligent" terminals may consume less processor power and less communications bandwidth than other terminals. Consequently, the above information is strictly related to type and quality of used or planned system components.

One element in capacity management is deciding on whether to host the organization's applications distributed across a number of small servers or consolidated onto a few large servers. Consolidating applications on a few large servers (also known as application stacking) often allows the organization to make better overall use of the resources, but on the other hand, it increases the impact of a server outage and it affects more applications when the server has to be shut down for maintenance.

Larger organizations often have hundreds, if not thousands, of servers which are arrayed in groups referred to as server farms.

| PRACTICE QUESTION | |
|---|---|
| 4-5 | The key objective of capacity planning procedures is to ensure that: |
| | A. available resources are fully utilized. |
| | B. new resources will be added for new applications in a timely manner. |
| | C. available resources are used efficiently and effectively. |
| | D. utilization of resources does not drop below 85 percent. |
| *See answers and explanations to the practice questions at the end of the chapter. (page 357)* | |

# 4.4 IS ARCHITECTURE AND SOFTWARE

The architecture of most computers can be viewed as a number of layers of circuitry and logic, arranged in a hierarchical structure that interacts with the computer's operating system. At the base of the hierarchy is the computer hardware, which includes some hard-coded instructions (firmware). The next level up in the hierarchy comprises the nucleus functions.

Functions of the nucleus relate to basic processes associated with the operating system, which include:
• Interrupt handling
• Process creation/destruction
• Process state switching
• Dispatching
• Process synchronization
• Interprocess communication
• Support of I/O processes
• Support of the allocation and reallocation/release of memory

The nucleus is a highly privileged area where access by most users is restricted. Above the nucleus are various operating system processes that support users. These processes, referred to as system software, are a collection of computer programs used in the design, processing and control of all computer applications used to operate and maintain the computer system. Comprised of system utilities and programs, the system software ensures the integrity of the system, controls the flow of programs and events in the computer, and manages the interfaces with the computer. Software developed for the computer must be compatible with its operating system. Examples include:
• Access control software
• Data communications software
• Database management software
• Program library management systems
• Tape and disk management systems
• Network management software
• Job scheduling software
• Utility programs

Some or all of the above may be built into the operating system.

## 4.4.1 OPERATING SYSTEMS

Before discussion of the various forms of system software, the most significant system software related to a computer—its operating system (OS)—needs to be further addressed. The operating system contains programs that interface between the user, processor and applications software. It provides the primary means of managing the sharing and use of computer resources such as processors, real memory (e.g., RAM), auxiliary memory (e.g., disk storage) and I/O devices.

An operating system:
• Defines user interfaces
• Permits users to share hardware
• Permits users to share data
• Schedules resources among users
• Informs users of any errors that occur with the processor, I/O or programs
• Permits recovery from system errors
• Communicates between the operating system and application programs, allocating memory to processors and making the memory available upon the completion of a process
• Allows system file management
• Allows system accounting management

As mentioned, one of the roles of operating systems is to manage the computer resources and processing. These resources include:
• **I/O devices**—Keyboards, mouse devices, scanners, webcams, microphones, displays, terminal screens, printers and speakers
• **Memory**—The internal storage directly attached to and/or a part of the CPU
• **Internal and/or external storage**—Disk drives, diskettes, tape drives, CD-ROMs, CD-RWs, DVD-Rs and DVD-RWs
• **CPU time**—The available time for instructional processing in the CPU
• **Networks**—The communication channels connecting I/O devices and computer processors

Most modern operating systems have also expanded the basic OS functionalities to include capabilities for a more efficient operation of system and applications software. For example, all modern operating systems possess a virtual storage memory capability which allows programs to use and reference a range of addresses greater than the real memory. The system hardware and software operate the mapping between virtual and real addresses, the latter being only a subset of "segments" of the full range of virtual addresses. A full image of the virtual memory is normally kept on high-speed, addressable, permanent storage (hard disk); the segments containing the addresses being referenced are loaded onto the internal working storage, with newer segments overwriting the oldest unused segments. This technique of mapping parts of a large slower memory to a faster and smaller working memory is used between various levels of "cached memory" within modern systems.

In some systems and specialized devices, program code can be stored in various types of ROM and very high-speed control storage, called firmware, instead of being loaded and run in the main computer memory. Firmware storage provides increased processing speed for frequently used functions and the ability to be configured and changed more easily than custom-designed computer circuits.

Operating systems vary in the resources managed, comprehensiveness of management and techniques used to manage resources. The type of computer, its intended use, and normal, expected attached devices and networks influence the OS requirements, characteristics and complexity. For example, a single-user microcomputer operating in stand-alone mode needs an operating system capable of cataloging files and loading programs to be effective.

A mainframe computer handling large volumes of transactions for consolidation and distribution requires an operating system capable of managing extensive resources and many concurrent operations, in terms of application input and output, with a very high degree of reliability. For example, the z/OS operating system from IBM has been engineered specifically to complement this environment.

A server with multiple users interacting with data and programs from database servers and middleware connections to legacy mainframe applications, requires an operating system that can accommodate multiprocessing, multitasking and multithreading. It must be able to share disk space (files) and CPU time among multiple users and system processes as well as manage connections to devices on the network. For example, the UNIX operating system is designed to specifically address this type of environment.

A microcomputer in a networked environment functioning as a server with specialized functions (applications, DBMS, directory/file storage, etc.) also has the ability to interact with data and programs of multiple users to provide services to client workstations throughout the network.

## Software Control Features or Parameters

Various OS software products provide parameters and options for the tailoring of the system and activation of features such as activity logging. Parameters are important in determining how a system runs because they allow a standard piece of software to be customized to diverse environments.

Software control parameters deal with:
• Data management
• Resource management
• Job management
• Priority setting

Parameter selections should be appropriate to the organization's workload and control environment structure. The most effective means of determining how controls are functioning within an operating system is to review the software control features and/or parameters.

Improper implementation and/or monitoring of operating systems can result in undetected errors and corruption of the data being processed as well as lead to unauthorized access and inaccurate logging of system usage.

# IT Service Delivery and Support

## Software Integrity Issues

OS integrity is a very important requirement and ability of the operating system, and it involves utilizing specific hardware and software features to:
- Protect itself from deliberate and inadvertent modification
- Ensure that privileged programs cannot be interfered with by user programs
- Provide for effective process isolation to ensure that:
  - Multiple processes running concurrently will not interfere by accident or by design with each other and are protected from writing into each other's memory (e.g., changing instructions, sharing resources, etc.)
  - Enforcement of least privilege where processes have no more privilege than needed to perform functions and modules call on more privileged routines only if, and for as long as, needed

A hardware-supported feature specifically designed for this purpose is the ability to operate in two different control states:

- **The supervisory/administrator state**—Refers to the most privileged operating mode available on a computer's operating system and is typically restricted to the nucleus of the system. This state permits complete unrestricted access to all resources such as memory, hardware devices and privileged instructions. Generally, system software programs such as system utilities operate in this state and, therefore, must be strictly controlled. A user directly allowed to achieve, or otherwise achieving, supervisory state privileges could bypass any kind of security mechanisms in place and virtually own the system and its resources. The IS auditor will need to ascertain the degree to which system administrators are able to operate in a totally unrestricted mode. In some cases, user IDs may be granted access rights that are equivalent to system administration superuser accounts.
- **The general user state**—Refers to the most constrained processor mode, where the user can only access their own virtual address space and must rely on code running in a privileged OS area to perform other privileged functions such as opening a communication line; that is, the system allows the general user code to call upon functions running in a privileged area, and these act on behalf of the requestor. However, this delegation of functions is strictly controlled so that the user's request may not cause interference or damage to other processes, devices or the system itself.

Therefore, to maintain system and data integrity, it is necessary to correctly and consistently define, enforce and monitor the operating environment and the granted permissions. IS management is responsible for the implementation of appropriate authorization techniques to prevent nonprivileged users from gaining the ability to execute privileged instructions and thus take control of the entire machine.

For example, IBM mainframe z/OS systems are customized at system generation (SYSGEN) time. When these systems are started (initial program load), important options and parameters are read from information kept in a key system directory (referred to as the SYS1.PARMLIB partitioned data set). The directory specifies critical initialization parameters used to meet the data center's installation requirements (i.e., other system software activated for job scheduling, security, activity logging, etc). These options, if uncontrolled, provide a nonprivileged user a way to gain access to the operating system's supervisory state. Audit and control professionals should review system configuration directories/files in all operating systems for control options used to protect the supervisory state.

Likewise, PC-based client-server Windows, UNIX and LINUX operating systems have special system configuration files and directories. The existence of program flaws or errors in configuring, controlling and updating the systems to the latest security patches makes them vulnerable to being compromised by perpetrators. Important Windows system options and parameters are set in special system configuration files, referred to as a "registry." Therefore, the registry is an important aspect of IS auditing. Noting any changes that take place in the registry is crucial for maintaining the integrity, confidentiality and availability of the systems. In UNIX-based operating systems, the same issues are present. Critical system configuration files and directories related to the nucleus (kernel) operations, system start-up, network file sharing and other remote services should be appropriately secured and checked for correctness.

## Activity Logging and Reporting Options

Computer processing activity can be logged for analysis of system functions. The following are some of the areas that can be analyzed based on the activity log:

- Data file versions used for production processing
- Program accesses to sensitive data
- Programs scheduled and run
- Utilities or service aids usage
- Operating system operation to ensure that the integrity of the operating system has not been compromised due to improper changes to system parameters and libraries
- Databases to:
  - Evaluate the efficiency of the database structure
  - Assess database security
  - Validate the DBA documentation
  - Determine whether the organization's standards have been followed
- Access control to:
  - Evaluate the access controls over critical data files/bases and programs
  - Evaluate security facilities that are active in communications systems, DBMSs and applications

Many intruders will attempt to alter logs to hide their activities. Secure logging is also needed to preserve evidence authenticity should the logs be required for legal/court use. For secure logging it is, therefore, important that logs be stored on removable media that are then placed in a physically secure area. Although WORM media could be used, their capacity may not be sufficient.

# 4.4.2 ACCESS CONTROL SOFTWARE

Access control software is designed to prevent unauthorized access to data, unauthorized use of system functions and programs, and unauthorized updates/changes to data, and to detect or prevent unauthorized attempts to access computer resources. For more details on access control software, see chapter 5, Protection of Information Assets.

# 4.4.3 DATA COMMUNICATIONS SOFTWARE

Data communications software is used to transmit messages or data from one point to another, which may be local or remote. For example, a database request from an end user is actually transmitted from that user's terminal to an online application, then to a DBMS in the form of messages handled by data communications software. Likewise, responses back to the user are handled in the same manner (i.e., from the DBMS to the online application and back to the user's terminal). Some of the common transmission codes are EBCDIC, ASCII and Unicode.

A typical simple data communications system has three components:
1. The transmitter (source)
2. The transmission path (channel or line)
3. The receiver

A one-way communication is said to exist when communication flows in one direction only. In a two-way communication, both ends may simultaneously operate as source and receiver, with data flowing over the same channel in both directions. The data communications system is concerned only with the correct transmission between two points. It does not operate on the content of the information.

A data communication system is divided into multiple functional layers. At each layer, software interfaces with hardware to provide a specific set of functions. All data communication systems have at least a physical layer and a data link layer. (Refer to the Network Standards and Protocols section later in this chapter for a discussion regarding data communication layers.)

Communication-based applications operate in local area network (LAN) and and wide area network (WAN) environments to support:
• Electronic funds transfer (EFT) systems
• Office information systems
• Customer electronic services/electronic data interchange (EDI)
• Electronic message systems, including bulletin boards and e-mail

The data communication system interfaces with the operating system, application programs, database systems, telecommunication address method systems, network control system, job scheduling system and operator consoles.

## 4.4.4 DATA MANAGEMENT

Data management capabilities are enabled by the system software components that enact and support the definition, storage, sharing and processing of user data, and deal with file management activities.

### *File Organization*
User and system data are usually partitioned into manageable units—data files. Some data management file organizations that may be supported include:
• **Sequential**—One record is processed after another, from the beginning to the end of a file.
• **Direct random access**—Records are addressed individually based on a key not related to the data (e.g., record number).

The decision on which access method to use is often based on the platform utilized. For nonmainframe environments, the access method will usually be nondiscretionary, based on the database that is utilized.

Commercial and open systems make available structured software interfaces for data storage and management purposes. Database management systems can be linked to these interfaces and are available for high-level user application programming.

## 4.4.5 DATABASE MANAGEMENT SYSTEM

DBMS system software aids in organizing, controlling and using the data needed by application programs. A DBMS provides the facility to create and maintain a well-organized database. Primary functions include reduced data redundancy, decreased access time and basic security over sensitive data.

DBMS data are organized in multilevel schemes, with basic data elements such as the fields (e.g., a Social Security number could be a field) at the lowest level. The levels above each field have differing properties depending on the architecture of the database.

The DBMS can include a data dictionary that identifies the fields, their characteristics and their use. Active data dictionaries require entries for all data elements and assist application processing of data elements such as providing validation characteristics or print formats. Passive dictionaries are only a repository of information that can be viewed or printed. A DBMS can control user access at the following levels:
• User and the database
• Program and the database
• Transaction and the database
• Program and data field
• User and transaction
• User and data field

The following are some of the advantages of a DBMS:
• Data independence for application systems
• Ease of support and flexibility in meeting changing data requirements

• Transaction processing efficiency
• Reduction of data redundancy
• Ability to maximize data consistency
• Ability to minimize maintenance cost through data sharing
• Opportunity to enforce data/programming standards
• Opportunity to enforce data security
• Availability of stored data integrity checks
• Facilitation of terminal users' *ad hoc* access to data, especially through designed query language/application generators

## DBMS Architecture

Data elements required to define a database are called metadata. This includes data about data elements used to define logical and physical fields, files, data relationships, queries, etc. There are three types of metadata: conceptual schema, external schema and internal schema. If any of these elements is missing from the data definition maintained within the DBMS, the DBMS may not be adequate to meet the users' needs.

## Detailed DBMS Metadata Architecture

Within each level, there is a data definition language (DDL) component for creating the schema representation necessary for interpreting and responding to the user's request. At the external level, a DBMS will typically accommodate multiple DDLs for several application programming languages compatible with the DBMS. The conceptual level will provide appropriate mappings between the external and internal schemas. External schemas are location independent of the internal schema.

## Data Dictionary/Directory System

A data dictionary/directory system (DD/DS) helps define and store source and object forms of all data definitions for external schemas, conceptual schemas, the internal schema and all associated mappings. The data dictionary contains an index and description of all of the items stored in the database. The directory describes the location of the data and the access method.

DD/DS provides the following functional capabilities:
• A data definition language processor, which allows the database administrator to create or modify a data definition for mappings between external and conceptual schemas
• Validation of the definition provided to ensure the integrity of the metadata
• Prevention of unauthorized access to, or manipulation of, the metadata
• Interrogation and reporting facilities that allow the DBA to make inquiries on the data definition

DD/DS can be used by several DBMSs; therefore, using one DD/DS could reduce the impact of changing from one DBMS to another DBMS. Some of the benefits of using DD/DS include:
• Enhancing documentation
• Providing common validation criteria
• Facilitating programming by reducing the needs for data definition
• Standardizing programming methods

## Database Structure

There are three major types of database structure: hierarchical, network and relational. Of these, the first two were mainly used prior to 1990 and have been mostly replaced by relational databases. A description of these models follows:
• **Hierarchical database model**—In this model there is a hierarchy of parent and child data segments. To create links between them, this model uses parent-child relationships. These are 1:N (one-to-many) mappings between record types represented by logical trees, as shown in **exhibit 4.2**. A child segment is restricted to having only one parent segment, so data duplication is necessary to express relationships to multiple parents. Subordinate segments are retrieved through the parent segment. Reverse pointers are not allowed. When the data relationships are hierarchical, the database is easy to implement, modify and search.

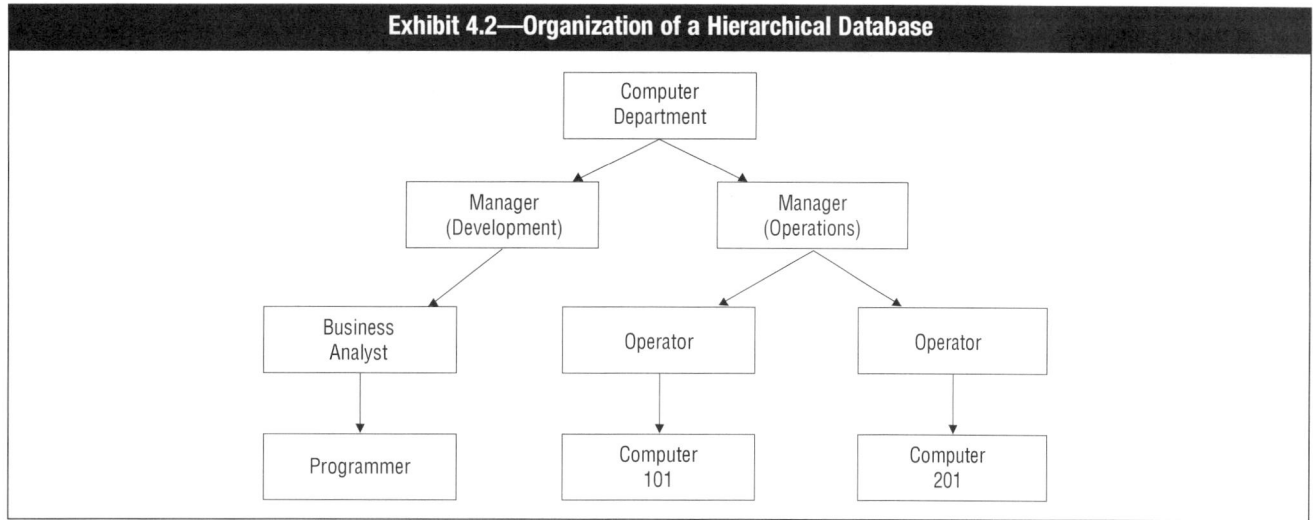

**Exhibit 4.2—Organization of a Hierarchical Database**

- **Network database model**—In the network model, the basic data modeling construct is called a set. A set is formed by an owner record type, a member record type and a name. A member record type can have that role in more than one set, so a multiowner relationship is allowed. An owner record type can also be a member or owner in another set. Usually, a set defines a 1:N relationship, although one-to-one (1:1) is permitted. A disadvantage of the network model is that such structures can be extremely complex and difficult to comprehend, modify or reconstruct in case of failure. This model is rarely used in current environments. See **exhibit 4.3**. The hierarchical and network models do not support high-level queries. The user programs have to navigate the data structures.

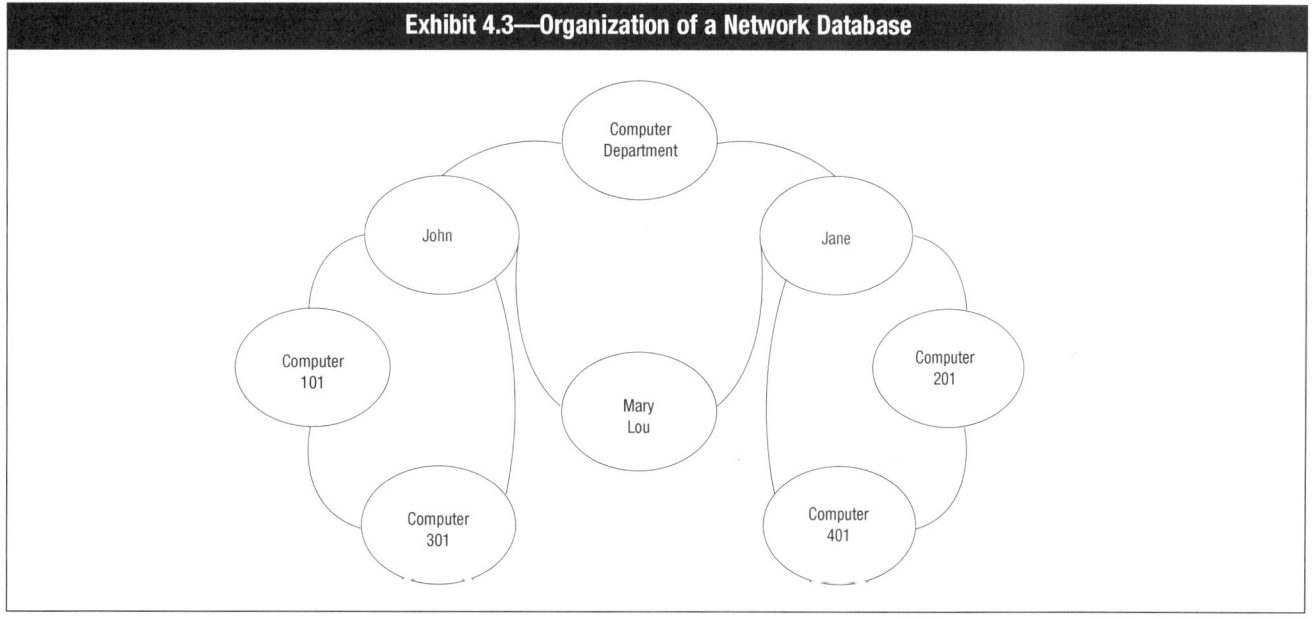

**Exhibit 4.3—Organization of a Network Database**

- **Relational database model**—An example of a relational database can be seen in **exhibit 4.4**. The relational model is based on the set theory and relational calculations. A relational database allows the definition of data structures, storage/retrieval operations and integrity constraints. In such a database, the data and relationships among these data are organized in tables. A table is a collection of rows, also known as tuples, and each tuple in a table contains the same columns. Columns, called domains or attributes, correspond to fields. Tuples are equal to records in a conventional file structure. An example of a relational database can be seen in **exhibit 4.4**.

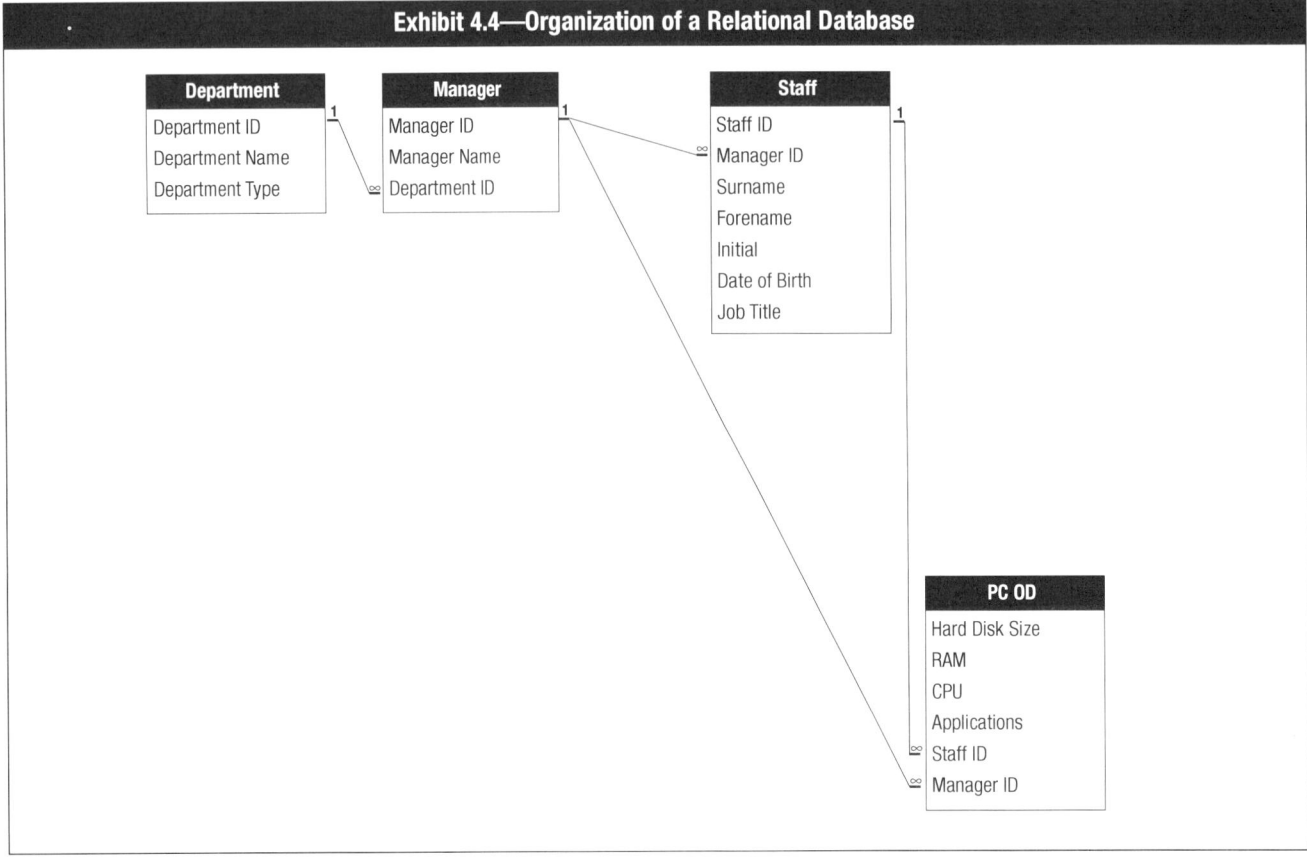

**Exhibit 4.4—Organization of a Relational Database**

Relational tables have the following properties:
• Values are atomic.
• Each row is unique.
• Column values are of the same kind.
• The sequence of columns is insignificant.
• The sequence of rows is insignificant.
• Each column has a unique name.

Certain fields may be designated as keys, so searches for specific values of that field will be quicker because of the use of indexing. If fields in two different tables take their values from the same set, a join operation can be performed to select related records in the two tables by matching values in those fields. This can be extended to joining multiple tables on multiple fields. These relationships are only specified at retrieval time, so relational databases are dynamic. The relational model is independent from the physical implementation of the data structure, and has many advantages over the hierarchical and network database models. With relational databases, it is easier:
• For users to understand and implement a physical database system
• To convert from other database structures
• To implement projection and join operations (referencing groups of related data elements not stored together)
• To create new relations for applications
• To implement access control over sensitive data
• To modify the database

A key feature of relational databases is the use of "normalization" rules to minimize the amount of information needed in tables to satisfy the users' structured and unstructured queries to the database. Generally followed, normalization rules include:
• A given instance of a data object has one and only one value for each attribute.
• Attributes represent elementary data items; they should contain no internal structure.

- Each tuple consists of a primary key that identifies some entity, together with a set of zero or more mutually independent attribute values that describes the entity in some way (fully dependent on primary key).
- Any foreign key should have a null value or should have an existing value linking to other tables; this is known as referential integrity.

Most DBMSs have internal security features that interface with the OS access control mechanism/package. A combination of the DBMS security features and security package functions is often used to cover all required security functions.

## Database Controls

It is critical that database integrity and availability be maintained. This is ensured through the following controls:
- Establish and enforce definition standards.
- Establish and implement data backup and recovery procedures to ensure database availability.
- Establish the necessary levels of access controls for data items, tables and files to prevent inadvertent or unauthorized access.
- Establish controls to ensure that only authorized personnel can update the database.
- Establish controls to handle concurrent access problems such as multiple users desiring to update the same data elements at the same time (i.e., transaction commit, locking of records/files).
- Establish controls to ensure accuracy, completeness and consistency of data elements and relationships in the database. It is important that these controls, if possible, be contained in the table/columns definitions. In this way, there is no possibility that these rules will be violated because of programming flaws or through the usage of utilities in manipulating data.
- Use database checkpoints at junctures in the job stream that minimize data loss and recovery efforts to restart processing after a system failure.
- Perform database reorganization to reduce unused disk space and verify defined data relationships.
- Follow database restructuring procedures when making logical, physical and procedural changes.
- Use database performance reporting tools to monitor and maintain database efficiency (e.g., available storage space, buffer size, CPU usage, disk storage configuration and deadlock conditions).
- Minimize the ability to use nonsystem tools, i.e., those outside security control, to access the database.

| PRACTICE QUESTION |
|---|
| **4-6**     The **PRIMARY** benefit of database normalization is the: <br><br>     A.    minimization of redundancy of information in tables required to satisfy users' needs. <br>     B.    ability to satisfy more queries. <br>     C.    maximization of database integrity by providing information in more than one table. <br>     D.    minimization of response time through faster processing of information. |
| *See answers and explanations to the practice questions at the end of the chapter. (page 357)* |

## 4.4.6 TAPE AND DISK MANAGEMENT SYSTEMS

An automated tape management system (TMS) or disk management system (DMS) is specialized system software that tracks and lists tape/disk resources needed for data center processing. The systems include the data set name and the specific tape reel or disk drive location, creation date, effective date, retention period, expiration date and contents information. A TMS/DMS minimizes computer operator time and errors caused by locating improper files or mounting the wrong data set version, and can improve space efficiency by consolidating fragmented free spaces. In addition, most of the TMS systems provide inventory control over tapes, identification of offsite rotation of backup media and security features to control tape access. A number of TMS systems work with robotic units to automatically retrieve and load tape volumes when they are requested by the computer system. Client server environments more frequently utilize storage area networks (SANs) and network attached storage (NAS). This provides greater cost effectiveness; however, it can lead to the risk of creating a single point of failure if appropriate redundancy is not built into the design.

## 4.4.7 UTILITY PROGRAMS

Utility programs are system software used to perform maintenance and routines that frequently are required during normal processing operations. Utility programs can be categorized by use, into five functional areas:
1. Understanding application systems (flowcharting software, transaction profile analyzer, executive path analyzer and data dictionary)
2. Assessing or testing data quality (data manipulation utilities, database dump utilities, data comparison utility and query facility)
3. Testing a program's ability to function correctly and maintain data integrity (test data generator, online debugging facility, output analyzer and network simulator)
4. Assisting in faster program development (visual display utility, library copy, text editor, online coding facility, report generators and code generators)
5. Improving operational efficiency (CPU and memory utilization monitors and communication line analyzers)

Smaller computer systems (i.e., PC and server operating systems) are often equipped with specific utilities to:
• Operate verification, cleaning and defragmenting of hard disk and removable memory units
• Define the file system standard (new technology file system [NTFS] or file allocation table [FAT]) to be used for each unit
• Initialize removable data volumes (floppy disk) and volumes of disk/removable memory
• Save/restore system images
• Reconstruct and restore (logically) cancelled files
• Test system units and peripherals

Many of these utility programs can perform outside the security system or can function without producing an audit trail of activity. As a result, access to and use of these sensitive and powerful utilities should be well controlled and restricted.

## 4.4.8 SOFTWARE LICENSING ISSUES

Today, more and more mission-critical applications are developed on PCs and client-server platforms. These machines can be stand-alone machines or linked to other devices across a LAN or through a WAN, and are under the purview of software copyright laws. Software copyright laws must be followed to protect against the possibility of a company paying penalties over copyright infringements and the added public embarrassment of being identified as a company that illegally uses copied software. To prevent or detect software licensing violations, the IS auditor should:
• Review the documented policies and procedures as well as any preventive controls that guard against unauthorized use or copying of software. In some cases, companies are requiring users to sign an agreement not to copy software without proper approval and without a software license agreement.
• Obtain copies of all software contracts to determine the nature of the license agreements, be it an unlimited enterprise license, per-seat license, or individual copies.
• Review the listing of all standard, used and licensed application and system software. This list should be compared with software located on various servers on the network. A random sample or an all-inclusive list of what is loaded on a specific user's PC should be reviewed as well.
• Review software currently installed on user's machines using tools such as Microsoft SMS/System Center or Peregrine.

All this will identify whether an organization has established a standard desktop environment and has implemented policies for licensed software on the network and on PCs.

Options available to prevent software license violations include:
• Centralizing control and automated distribution and installation of software (includes disabling the ability of users to install software, where possible)
• Requiring that all PCs be restricted workstations, without disk drives or working Universal Serial Bus (USB) ports, that access applications from a secure LAN
• Installing metering software on the LAN and requiring that all PCs to access applications through the metered software

- Regularly scanning user PCs, either from the LAN or directly, to ensure that unauthorized copies of software have not been loaded on the PC

Another useful control, which prevents illegal duplication of software on multiple PCs in a network, is to have site licensing agreements with vendors. With a site licensing agreement the license is based on the number of users who access the network rather than the license agreement being attached to a specific user or machine.

Alternatively, to limit license costs, a company may have a concurrent license agreement. With concurrent licensing, there are a number of users who can access the software on the network at one time. This generally requires the use of metering software. If all of the concurrent access sessions are in use, the next user will receive the message "waiting for licensing." As soon as one becomes available, the user is able to use the software. Concurrent licensing agreements also help network administrators identify the need to purchase more software licenses. For example, if a "waiting for licensing" message does occur, it will be logged so that the administrator can monitor the need to increase the number of concurrent licenses.

# 4.5 IS NETWORK INFRASTRUCTURE

IS networks were developed from the need to share information resources residing on different computer devices, which enabled organizations to improve business processes and realize substantial productivity gains.

Generally, the telecommunication links or lines for networks can be either analog or digital. They can be classified in several ways, according to the type of provider or the type of technology. Typically, they can be divided into:
- Dedicated circuit (also known as leased lines)
- Switched circuit

A dedicated circuit is a symmetric telecommunications line connecting two locations, each side of the line being permanently connected to the other. Leased lines can be used for telephone, data or Internet services. Leased lines normally range from 56/64 Kbps to 34/45 Mbps.

A switched circuit does not permanently connect two locations but can be set up on demand, based on the addressing method. There are two main types of switching mechanisms:
- Circuit switching
- Packet switching

The circuit switching mechanism is typically used over the telephone network (plain old telephone service [POTS], integrated services digital network [ISDN]). Switched circuits allow data connections that can be initiated when needed and terminated when communication is complete. This works much like a normal telephone line works for voice communication. ISDN is a good example of circuit switching. When a router has data for a remote site, the switched circuit is initiated with the circuit number of the remote network. In the case of ISDN circuits, the device actually places a call to the telephone number of the remote ISDN circuit. When the two networks are connected and authenticated, they can transfer data. When the data transmission is complete, the call can be terminated.

Packet switching is a technology in which users share common carrier resources. Because this allows the carrier to make more efficient use of its infrastructure, the cost to the customer is generally much lower than with leased lines. In a packet switching setup, networks have connections into the carrier's network, and many customers share the carrier's network. The carrier can then create virtual circuits between customers' sites by which packets of data are delivered from one to the other through the network. The section of the carrier's network that is shared is often referred to as a cloud. Some examples of packet-switching networks include Asynchronous Transfer Mode (ATM), frame relay, Switched Multimegabit Data Services (SMDS), and X.25.

Methods for transmitting signals over analog telecommunication links or lines are either baseband or broadband, as described below:

- **Baseband**—The signals are directly injected on the communication link so that one single channel is available on that link for transmitting signals. As a result, the entire capacity of the communication channel is used to transmit one data signal and communication can move in only one direction at a time (half-duplex communication). Stations are connected to the transmission medium by devices that operate alternately as transmitter and receiver, called transceivers.
- **Broadband network**—Different carrier frequencies defined within the available band, can carry analog signals, such as those generated by image processors or a data modem, as if they were placed on separate baseband channels. Interference is avoided by separating adjacent carrier frequencies with a gap that depends on the band requirements of the carried signals. The possibility of vectoring multiple independent channels on a single-carrier media enhances considerably the effectiveness of remote connections. The condition when simultaneous data or control transmission/reception takes place between two stations is called a full-duplex connection.

## 4.5.1 ENTERPRISE NETWORK ARCHITECTURES

Today's networks are part of a large, centrally managed, internetworked architecture solution of high-speed local- and wide-area computer networks serving organizations' client server-based environments. Such architectures may include clustering common types of IT functions together in network segments, each uniquely identifiable and specialized to task. For example, network segments or blocks may include web-based front-end application servers (public or private), application and database servers, and mainframe servers using terminal emulation software to allow end users to access these back-end legacy-based systems. In turn, end users can be clustered together within their own network LANs, but with rapid access capabilities to incorporate information resources. Some organizations may implement service-oriented architectures (SOA) in which web software components, using Simple Object Access Protocol (SOAP) and Extensible Markup Language (XML), interoperate in a loosely connected and distributed fashion across the network. Within this environment, information is highly accessible and available anytime and anywhere, and centrally managed for highly effective and efficient troubleshooting and performance management to achieve optimum use of network resources.

To understand the network architecture solutions offered today from a business, performance and security design standpoint, an IS auditor must understand information technologies associated with the design and development of a telecommunications infrastructure (e.g., LAN and WAN specifications). The term telecommunications comes from the Greek word *tele* (distance) and the Latin word *comunicare* (share). In modern terms, telecommunications is the electronic transmission of data, sound and images between connected end systems (two or more computers acting as sender and receiver). This process is enabled by a communications subsystem, such as a network interface card that interfaces each end user's computer to a common transmission medium, and network devices such as bridges, switches and routers, to connect computers residing on different networks.

## 4.5.2 TYPES OF NETWORKS

The types of networks common to all organizations are defined as follows:

- **Personal area networks (PANs)**—Generally, this is a microcomputer network used for communications among computer devices (including telephones, PDAs, printers, cameras, scanners, etc.) being used by an individual person. The extent of a PAN is typically within a range of 33 feet (about 10 meters). PANs can be used for communication among the personal devices themselves or to connect to a higher-level network and the Internet.

  PANs may be wired with computer buses, such as USB, Firewire and other standards. If PANs are implemented without wires, they are called wireless PANs (WPANs), which can also be made possible with network technologies such as IrDA and Bluetooth. A Bluetooth PAN is also called a piconet and is composed of up to eight active devices in a master-slave relationship. The first Bluetooth device in the piconet is the master, and all other devices are slaves that communicate with the master. A piconet typically has a range of 32.8 feet (10 meters), although ranges of up to 328 feet (100 meters) can be reached under ideal circumstances.
- **Local area networks**—LANs are computer networks that cover a limited area such as a home, office or campus. Characteristics of LANs, in contrast to WANs, are higher data transfer rates and smaller geographic range. Ethernet and Wi-Fi are the two most common technologies currently used.

• **Wide Area Networks**—-WANs are computer networks that cover a broad area such as a city, region, nation or an international link. The Internet is an example (the largest example) of a WAN. WANs are used to connect LANs and other types of networks together so that users and computers in one location can communicate with users and computers in other locations. Many WANs are built for one particular organization and are private. Others, built by Internet service providers (ISPs), provide connections from an organization's LAN to the Internet.

• **Metropolitan Area Networks (MANs)**—MANs are WANs that are limited to a city or region; usually, MANs are characterized by higher data transfer rates than WANs.

• **Storage area networks (SANs)**—SANs are a variation of LANs and are dedicated to connecting storage devices to servers and other computing devices. SANs centralize the process for the storage and administration of data.

## 4.5.3 NETWORK SERVICES

Network services are functional features made possible by appropriate OS applications. They allow orderly utilization of the resources on the network. Instead of having a single operating system that controls its own resources and shares them with the requesting programs, the network relies on standards and on a specific protocol or set of rules, enacted and operated through the basic system software of the various network devices that are capable of supporting the individual network services. Users and business applications can request network services through specific calls/interfaces. The following are network application services commonly used in organizations' networked environments:

• **Network file system**—Allows users to share files, printers and other resources in a network

• **E-mail services**—Provide the ability, via a terminal or PC connected to a communication network, to send an unstructured message to another individual or group of people

• **Print services**—Provide the ability, typically through a print server on a network, to manage and execute print request services from other devices on the network

• **Remote access services**—Provide remote access capabilities where a computing device appears, as if directly attached to the remote host

• **Directory services**—Store information about the various resources on a network and help network devices locate services, much like a conventional telephone directory. Directory services also help network administrators manage user access to network resources.

• **Network management**—Provides a set of functions to control and maintain the network. Network management provides detailed information about the status of all components in the network such as line status, active terminal, the length of message queues, error rate on a line and the traffic over a line. Network management enables computers to share information and resources within a network and provides network reliability. Network management also provides the operator with an early warning signal of network problems before they affect network reliability, allowing the operator to take timely preventive or remedial actions.

• **Dynamic Host Configuration Protocol (DHCP)**—A protocol used by networked computers (clients) to obtain IP addresses and other parameters such as the default gateway, subnet mask, and IP addresses of domain name servers (DNSs) from a DHCP server. The DHCP server ensures that all IP addresses are unique; e.g., no IP address is assigned to a second client while the first client's assignment is valid (its lease has not expired). Thus, IP address pool management is done by the server and not by a human network administrator.

• **DNS**—Translates the names of network nodes into network (IP) addresses

## 4.5.4 NETWORK STANDARDS AND PROTOCOLS

Network architecture standards facilitate the process of creating an integrated environment that applications can work within by providing a reference model that organizations can use for structuring intercomputer and network communication processes.

Besides the convenience of using compatible architectures, one major advantage of network standards is that they help organizations meet the challenge of designing and implementing an integrated, efficient, reliable, scalable and secure network of LANs and WANs with external connectivity (public Internet). This is a major challenge due to the requirements of:

• **Interoperability**—Connecting various systems to support communication among disparate technologies where different sites may use different types of media that may operate at differing speeds

- **Availability**—The organization requires that end users have continuous, reliable and secure service—24/7.
- **Flexibility**—For network scalability needed to accommodate network expansion and requirements for new applications and services
- **Maintainability**—Centralized support and troubleshooting over heterogeneous, but highly integrated systems

To accomplish these tasks, organizations need to have the ability to define specifications for the types of networks to be established (e.g., LANs/WANs) when creating an integrated environment that their applications can work within. Organizations must also provide centralized support and troubleshooting over heterogeneous, but highly integrated systems.

## 4.5.5 OSI ARCHITECTURE

The purpose of network architecture standards is to facilitate this process by providing a reference model that organizations can use for building intercomputer and network communication processes, respectively.

The benchmark standard for this process, the Open Systems Interconnection (OSI) reference model, was developed by the ISO in 1984. The OSI is a proof of a concept model composed of seven layers, each specifying particular specialized tasks or functions. Each layer is self-contained and relatively independent of the other layers in terms of its particular function. This enables solutions offered by one layer to be updated without adversely affecting the other layers.

> **Note:** The OSI reference model might be something that an IS auditor should know, however, the CISA candidate will not be tested on this specific standard in the exam.

The objective of the OSI reference model is to provide a protocol suite used to develop data-networking protocols and other standards to facilitate multivendor equipment interoperability. The OSI program was derived from a need for international networking standards and was designed to facilitate communication between hardware and software systems despite differences in underlying architectures.

It is important to note that in the OSI model each layer communicates not only with the layers above and below it in the local stack, but also with the same layer on the remote system. For example, the application layer on the local system appears to be communicating with the application layer on the remote system. All of the details of how the data are processed further down the stack are hidden from the application layer. This is true at every level of the model. Each layer appears to have a direct (virtual) connection to the same layer on the remote system.

The functions of the specific layers of the ISO/OSI model are described as follows:
- **Application layer**—The application layer provides a standard interface for applications that must communicate with devices on the network (e.g., print files on a network-connected printer, send an e-mail or store data on a file server). Thus, the application layer provides an interface to the network. In addition, the application layer may communicate the computer's available resources to the rest of the network. The application layer should not be confused with application software. Application software uses the application layer interface to access network-connected resources.
- **Presentation layer**—This layer transforms data to provide a standard interface for the application layer and provides common communication services such as encryption, text compression and reformatting (e.g., conversion of Extended Binary-coded for Decimal Interchange Code [EBCDIC] to ASCII code). The presentation layer converts the outgoing data into a format acceptable by the network standard and then passes the data to the session layer. Similarly, the presentation layer converts data received from the session layer into a format acceptable to the application layer.
- **Session layer**—The session layer controls the dialogs (sessions) between computers. It establishes, manages and terminates the connections between the local and remote application layers. All conversations, data exchanges and dialogs between the application layers are managed by the session layer.
- **Transport layer**—The transport layer provides reliable and transparent transfer of data between end points, end-to-end error recovery and flow control. The transport layer ensures that all of the data sent to it by the session layer are successfully received by the remote system's transport layer. The transport layer is responsible for acknowledging every data packet received from the remote transport layer, ensuring that an acknowledgement is received from the remote transport layer for every packet sent. If an acknowledgement is not received for a packet, then that packet will be re-sent.

- **Network layer**—The network layer creates a virtual circuit between the transport layer on the local device and the transport layer on the remote device. This is the layer of the stack that understands IP addresses and is responsible for routing, forwarding, error handling and congestion control. This layer prepares the packets for the data link layer by converting each packet into one or more datagrams. The network layer will also determine the Media Access Control (MAC) address (see the following material on the data link layer) of the next "hop" and pass this information to the data link layer. In most cases, the next "hop" will be a router or a routing switch (layer 3 switch).
- **Data link layer**—This layer provides for the reliable transfer of data across a physical link. It receives packets of data from the network layer, reformats them into frames and sends them as a bit stream to the physical layer. These frames consist of the original data as well control fields necessary to provide for synchronization, error detection and flow control. Error detection is accomplished through the use of a cyclic redundancy check (CRC) that is calculated for and then added to each frame of data. The receiving network layer calculates the CRC value for the data portion of the received frame and discards the frame if the calculated and received values do not match. A CRC calculation will detect all single-bit and most multiple-bit errors.

A bit stream received from the physical layer is similarly converted to data packets and sent to the network layer. The data link layer logically connects to another device on the same network segment using a MAC address. Each device on the network has a unique MAC hardware address that is assigned to it at the time of manufacture. The MAC address can be overridden, but this practice is not recommended. The data link layer normally only "listens" to data intended for its MAC address. An important exception to this rule is that a network interface may be configured as a "promiscuous" interface, which will listen to all data that the physical layer sends it.

- **Physical layer**—This layer provides the hardware that transmits and receives the bit stream as electrical, optical or radio signals over an appropriate medium or carrier. This layer defines the cables, connectors, cards and physical aspects of the hardware required to physically connect a device to the network. Error correction and detection is not usually implemented in the physical layer, with a few notable exceptions. Cell phones and digital microwave systems will typically implement some form of error correction code, not only detecting but actually correcting errors. The most sophisticated forms of these are used by the US National Aeronautics and Space Administration (NASA) program for communicating with their deep space probes.

ISO formulated the OSI model to establish standards for vendors developing protocols supporting open system architecture. The intent is to make different proprietary systems work seamlessly within the same network. The actual implementation of the functions defined in each layer is based on protocols developed for each layer. A protocol is an agreed-upon set of rules and procedures to follow when implementing the tasks associated with a given layer of the OSI model.

The intent of the OSI model is to provide a standard interface at each layer and to ensure that each layer does not have to be concerned with the details of how the other layers are implemented.

This approach supports system-to-system communication (peer-to-peer relationship) where each layer on the sender side provides information to its peer layer on the receiving side. The process also is characterized as a data traversal process with the following actions occurring:
- Data travels down through layers at the local end.
- Protocol-control information (headers/trailers) is used as an envelope at each layer to pick up control information.
- Data travels up through the layers at the receiving/destination end.
- Protocol-control information (headers/trailers) is removed as the information is passed up.

A traditional OSI model showing this process is depicted in **exhibit 4.5**.

## 4.5.6 APPLICATION OF THE OSI MODEL IN NETWORK ARCHITECTURES

The concepts of the OSI model are used in the design and development of organizations' network architectures. This includes LANs, WANs, MANs and use of the public TCP/IP-based global Internet. The following sections will provide a detailed technical discussion of each and will show how the OSI reference model applies to the various architectures.

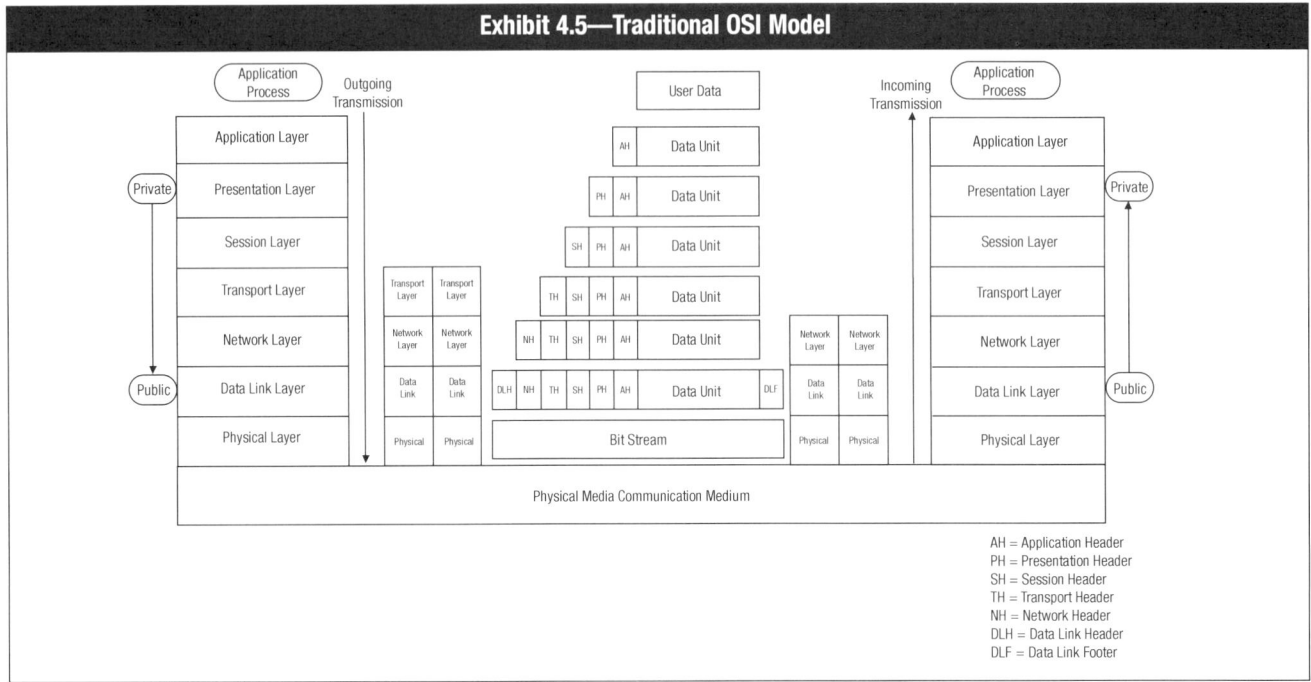

Exhibit 4.5—Traditional OSI Model

## Local Area Network

A LAN is a computer network covering a small local area. A few workstations and printers in a single room could comprise a LAN, so could three or four buildings in close proximity with thousands of devices connected together.

The great increase in reasonably priced bandwidth has reduced the design effort required to provide cost-effective LAN solutions for organizations of any size.

New LANs are almost always implemented using switched Ethernet (802.3). Twisted-pair cabling (100-Base-T or better and, increasingly, wireless LANs) connects floor switches to the workstations and printers in the immediate area. Floor switches can be connected to each other with 1000-Base-T or fiber-optic cabling. In larger organizations, the floor switches may all be connected to larger, faster switches whose only purpose is to properly route the switch to switch data.

As LANs get larger and traffic increases, the requirement to carefully plan the logical configuration of the network becomes more and more important. Network planners need to be highly skilled and very knowledgeable. Their tools include traffic monitors that allow them to monitor traffic volumes on critical links. Tracking traffic volumes, error rates and response times is every bit as important on larger LANs as it is on distributed servers and mainframes.

### LAN DESIGN FUNDAMENTALS AND SPECIFICATIONS
To set up a LAN, an organization must assess cost, speed, flexibility and reliability. The issues include:
• Assessing media for physically transmitting data
• Assessing methods for the physical network medium
• Understanding from a performance and security standpoint how data will be transmitted across the network and how the actual LAN network is organized and structured in terms of optimizing the performance of the devices connected to it

In the following sections we will discuss network physical media specifications, topology and protocols.

### NETWORK PHYSICAL MEDIA SPECIFICATIONS
Physical media used to connect various types of computing devices together in a network are:
• Twisted pairs
• Fiber optics for high-capacity and specific architectures
• Infrared and radio (wireless)

Generally, twisted-pair cabling is still the most commonly used media for LANs; however, this is increasingly being supplanted by wireless for LAN connectivity. The type and characteristics of physical media (e.g., speed, sensitivity to external disturbances, signal loss and propagation, security) not only affect the cost of implementation and support but also impact the capacity, flexibility and reliability of the network.

LANs can be implemented using various types of media including:

- **Copper (twisted-pair) circuits**—Two insulated wires are twisted around each other, with current flowing through them in opposite directions. This reduces the opportunity for cross talk between pairs in the same bundle and allows for lower sensitivity to electromagnetic disturbances (shielded twisted-pair circuits) within each individual pair. Twisted-pair circuits can also be used for some dedicated data networks. Today, the common standards for twisted-pair circuits are CAT5 and CAT6. Organizations should buy certified cables from reputable suppliers and segment problem areas with switches. Additionally, assurance should be provided that maximum cabling lengths are not exceeded since this will produce intermittent failures. A disadvantage of unshielded twisted-pair cabling is that it is not immune to the effects of electromagnetic interference (EMI) and should be run in dedicated conduits, away from sources of potential interference such as fluorescent lights. Parallel runs of cable over long distances should also be avoided since the signals on one cable can interfere with signals on adjacent cables— an EMI condition known as cross talk.
- **Fiber-optic systems**—Glass fibers are used to carry binary signals as flashes of light. Fiber-optic systems have a low transmission loss as compared to twisted-pair circuits. Optical fibers do not radiate energy nor conduct electricity. In addition, they are not affected by EMI and present a significantly lower risk of security problems such as wiretaps. Optical fiber is a more fragile medium and is more attractive for applications where changes are infrequent. Optical fiber is smaller and lighter than metallic cables of the same capacity. Fiber is the preferred choice for high-volume, longer-distance runs. One example would be using fiber to connect floor switches to enterprise data switches. In addition, fiber-optic cable is often used to connect servers to SANs.
- **Radio systems (wireless)**—Data are communicated between devices using low-powered systems that broadcast (or radiate) and receive electromagnetic signals representing data.

## IMPLEMENTATION OF WANS

Fiber-optic cables are commonly used these days for most high-capacity network connections, both between buildings and between cities. Other systems that may be used include:

- **Microwave radio systems**—Microwave radio provides line-of-site transmission of voice and data through the air. Historically, analog microwave circuits supplied the majority of long-haul low-speed data and voice transmission. This technology was used because it provided a lower-cost alternative to the low-capacity cable carrier systems of the time. Many if not most "heavy route" microwave systems have since been replaced by fiber-optic cable systems providing greatly increased capacity and greatly improved reliability at a cost per channel mile that is a tiny fraction of the cost for microwave circuits of similar capacity. All new microwave construction uses digital signals, providing greatly increased data rates and reduced error rates when compared with analog circuitry. Microwave radio circuits are still in common use on "light routes" where the economics do not favor installation of fiber. Most electrical utility companies will use microwave systems to connect their **S**upervisory **C**ontrol **A**nd **D**ata **A**cquisition (SCADA) systems together. Design of microwave circuits must take into account the physical topology of the area and the climate. Microwave antennae must be able to "see" each other. Climate conditions, such as rainfall, can adversely affect microwave links.

- **Satellite radio link systems**—These contain several receiver/amplifier/transmitter sections called transponders. Each transponder has a bandwidth of 36 megahertz (MHz), operates at a slightly different frequency, has individual transmitter sites and sends narrow beams of microwave signals to the satellite. Like microwaves, satellite signals can be affected by weather. Although satellite signals can carry large amounts of information at a time, the disadvantage is a bigger delay compared to all of the previous media, due to the "jump" from the Earth to the satellite and back (estimated at about 300 milliseconds).

**Exhibit 4.6** identifies the advantages and disadvantages of each physical layer medium available to networks. These physical specifications are applicable to WAN technologies, which will be addressed later in this chapter.

| Exhibit 4.6—Transmission Media | | | |
|---|---|---|---|
| **Type** | **Information** | **Advantages** | **Disadvantages** |
| Copper | • Used for short distances (<200 feet [60.96 meters])<br>• Supports voice and data | • Cheap<br>• Simple to install<br>• Readily available<br>• Simple to modify | • Easy to tap<br>• Easy to splice<br>• Cross talk<br>• Interference<br>• Noise |
| Coaxial cable | • Supports data and video | • Ease of installation<br>• Straightforward<br>• Readily available | • Thick<br>• Expensive<br>• Does not support many LANs<br>• Distance sensitive<br>• Difficult to modify |
| Fiber optics | • Used for long distances<br>• Supports voice data, image and video | • High bandwidth capabilities<br>• Secure<br>• Difficult to tap<br>• No cross talk<br>• Smaller and lighter than copper | • Expensive<br>• Hard to splice<br>• Difficult to modify |
| Radio systems | • Used for short distances | • Cheap | • Easy to tap<br>• Interference<br>• Noise |
| Microwave radio systems | • Line of sight carrier for voice and data signals | • Cheap<br>• Simple to install<br>• Available | • Easy to tap<br>• Interference<br>• Noise |
| Satellite radio link systems | • Uses transponders to send information | • High bandwidth and different frequencies | • Interference<br>• Noise<br>• Easy to tap |

## LAN TOPOLOGIES AND PROTOCOLS

LAN topologies define how networks are organized from a physical standpoint, whereas protocols define how information transmitted over the network is interpreted by systems.

LAN physical topology was previously tied fairly tightly to the protocols that were used to transfer information across the wire. This is no longer true. For current technology, the physical topology is driven by ease of construction, reliability and practicality. Of the physical topologies that have been commonly used—bus, ring and star—only the star is used to any great extent in new construction. **Exhibit 4.7** illustrates commonly used physical topologies.

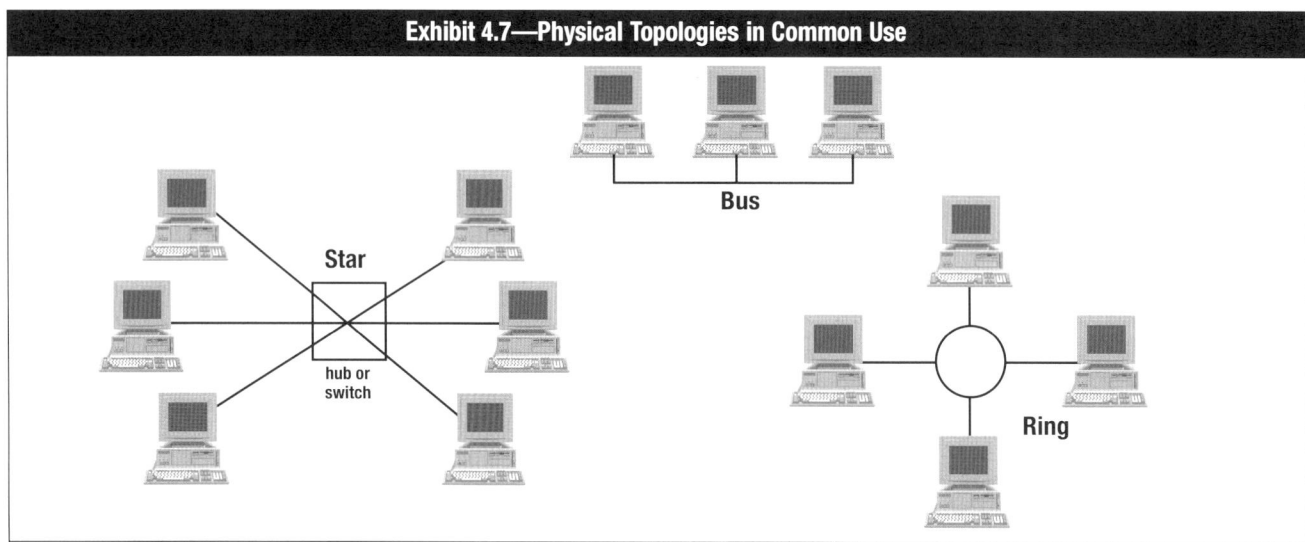

Exhibit 4.7—Physical Topologies in Common Use

# IT Service Delivery and Support

## LAN MEDIA ACCESS TECHNOLOGIES

LAN media access technologies for accessing physical transmission media are primarily either Ethernet or token passing. These technologies give devices shared access to the network, while also preventing a single device from monopolizing the network.

Ethernet has evolved from its original bus configuration, providing 10 Mbps speed with two coaxial cable versions (thin and thick), to star configurations initially using 10-Base-T (Ethernet using twisted-pair cabling) and now using today's more modern versions: Fast Ethernet (100 Mbps) and Gigabit Ethernet (1 Gbps).

A critical aspect of any communication is determining who is the recipient of a message. At this level, considering Fast Ethernet and Gigabit Ethernet, a MAC address is used to specify the recipient. Each and every network interface that is manufactured has a unique MAC address, which is only used for the last hop of any communication. See the section on the ISO model later in this chapter to see how this fits in with real-world addresses, such as 192.168.4.5. Every network interface card (NIC) connected to the network listens to every conversation on the network. Normally, a NIC device driver (software) only collects the data with its address. A NIC that has been placed in promiscuous mode will read all data passing over the network (including user IDs and passwords).

The initial bus arrangement typically provides an effective throughput of 5 Mbps among all of the systems connected to a bus segment. Bus segments could be connected together with repeaters or bridges. Repeaters would regenerate signals— allowing a longer span for the network. Bridges would connect multiple buses together—blocking any traffic that could not be delivered on a given segment. Bridges also served another critical function—that of breaking the network into multiple collision domains.

Ethernet (IEEE 802.3) is a Carrier-sense Multiple Access/Collision Detection (CSMA/CD) protocol. This is analogous to a car attempting to turn into a street. The driver's view is limited to only the street in front of him/her. If nothing is visible, the driver attempts to turn on to the street. If the driver collides with another vehicle, he/she backs up and tries again later. It should be apparent that if the street is very busy, a lot of collisions would occur. Similarly, if all of the traffic is coming from one particular house, then many cars can be handled in an efficient manner. If there are cars coming from a lot of different houses, then the overall traffic volume that can be handled is much lower. This is the way in which Ethernet behaves in a bus arrangement.

The use of coaxial cable in this example is rather problematic. The cable itself is a single point of failure. Adding a new station would not solve the problem and there exists a distinct lack of flexibility with such an implementation.

To alleviate this problem, a new physical implementation using a twisted-pair telephone cable is developed. This medium is much cheaper than coaxial cable and can be implemented using a star topology. The first implementation has all of the points of the star connected together using an unintelligent device called a hub—basically, a panel of connectors that allows all of the wires to be joined together. Circuitry within the hub electrically disconnects any branch that is not active. A problem on a single branch can still cause problems with the entire network, but the circuitry is simpler and a technician can easily isolate the problem at the hub. The traffic jam problem still exists, though.

The next big advance is replacing hubs with switches. A switch is an intelligent device that basically provides a private path for each pair of connections on the switch. If A is transferring data to B, it can do so without requiring C to transfer data to D. Further, transfers from A to D can be handled without fear of collision. This is analogous to a traffic light on a LAN. Collisions are then only an issue if more than one car is going to the same destination and a traffic light can take care of that problem.

This latest advance provides some significant breakthroughs. While the traffic volume to or from any given device is still limited to the constraints set by the used technology (e.g., 10, 100, 1000 Mbps), this volume can be maintained between many pairs of devices. Additionally, the problem of collisions is eliminated; the switch ensures that they cannot happen. A packet may be delayed—while it waits for other traffic to clear the intersection—but it never encounters a delay caused by a collision or needs to be re-sent.

From a security perspective, switches provide another significant improvement. Each device on the network can only see traffic destined for its MAC address and cannot eavesdrop on network traffic intended for other destinations.

Today, switches are so inexpensive that there is little justification for continuing to use hubs. Switches that provide individual devices with 100 Mbps service and provide 1 Gbps connection to higher-level switches are in common use. Switches are increasingly providing additional functionality that can be used to implement corporate security policy.

Another media access technology used in LANs relates to the token ring medium access method associated with ring networks. Ring networks are usually implemented as a physical ring.

Devices using this method gain access to the network on the basis of a unique frame, called a token, that is passed around the network. The purpose of the token is to attach itself to a user or device when transmitting messages/data for its intended recipient. When unattached to a user or device, a free token's header, data field and trailer components are empty and are filled by devices needing to transmit. Token ring technologies have almost disappeared in today's networks.

## LAN COMPONENTS
Components commonly associated with LANs are repeaters, hubs, bridges, switches and routers:
- **Repeaters**—Physical layer devices that extend the range of a network or connect two separate network segments together. Repeaters receive signals from one network segment and amplify (regenerate) the signal to compensate for signals (analog or digital) that are distorted due to a reduction of signal strength during transmission (i.e., attenuation).
- **Hubs**—Physical layer devices that serve as the center of a star-topology network or a network concentrator. Hubs can be active (if they repeat signals sent through them) or passive (if they merely split signals).
- **Bridges**—Data link layer devices developed in the early 1980s to connect LANs or create two separate LAN or WAN network segments from a single segment to reduce collision domains. The two segments work as different LANs below the data link level of the OSI reference model, but from that level and above, they behave as a single logical network. Bridges act as store-and-forward devices in moving frames toward their destination. This is done by analyzing the MAC header of a data packet, which represents the hardware address of a NIC. Bridges can also filter frames based on Layer 2 information. For example, they can prevent frames sent from predefined MAC addresses from entering a particular network. Bridges are software-based, and they are less efficient than other similar hardware-based devices such as switches. Therefore, bridges are not major components in today's enterprise network designs.
- **Layer 2 switches**—Layer 2 switches are data link level devices that can divide and interconnect network segments and help to reduce collision domains in Ethernet-based networks.

Furthermore, switches store and forward frames, filtering and forwarding packets among network segments, based on Layer 2 MAC source and destination addresses, as bridges and hubs do at the data link layer. Switches, however, provide more robust functionality than bridges, through use of more sophisticated data link layer protocols which are implemented via specialized hardware called application-specific integrated circuits (ASICs). The benefits of this technology are performance efficiencies gained through reduced costs, low latency or idle time, and a greater number of ports on a switch with dedicated high-speed bandwidth capabilities (e.g., many ports on a switch are available with 10/100 Ethernet and/or Gigabit Ethernet speeds).

Switches are also applicable in WAN technology specifications. See the WAN section later in this chapter for a discussion of switches in WAN environments.
- **Routers**—Similar to bridges and switches in that they link two or more physically separate network segments. The network segments linked by a router, however, remain logically separate and can function as independent networks. Routers operate at the OSI network layer by examining network addresses (i.e., routing information encoded in an IP packet). By examining the IP address, the router can make intelligent decisions to direct the packet to its destination. Routers differ from switches operating at the data link layer in that they use logically based network addresses, use different network addresses/segments off all ports, block broadcast information, block traffic to unknown addresses, and filter traffic based on network or host information.

Routers are often not as efficient as switches since they are generally software-based devices and they examine every packet coming through, which can create significant bottlenecks within a network. Therefore, careful consideration should

be taken as to where routers are placed within a network. This should include leveraging switches in network design as well as applying load balancing principles with other routers for performance efficiency considerations.

- **Layer 3 and 4 switches**—Advances in switch technology have also provided switches with operating capabilities at Layer 3 and Layer 4 of the OSI reference model. A Layer 3 switch goes beyond the Layer 2—MAC addressing, acting at the network layer of the OSI model like a router. The Layer 3 switch looks at the incoming packet's networking protocol, e.g., IP. The switch compares the destination IP address to the list of addresses in its tables, to actively calculate the best way to send a packet to its destination. This creates a "virtual circuit"; i.e., the switch has the ability to segment the LAN within itself and will create a pathway between the receiving and the transmitting device to send the data. It then forwards the packet to the recipient's address. This provides the added benefit of reducing the size of network broadcast domains. A broadcast domain is the domain segment or segments where all connected devices may be simultaneously addressed by a message using a special common network address range, referred to as a broadcast address. This is needed for specific network management functions. As the broadcast domain grows larger, this may cause performance inefficiencies and major security concerns in terms of information leakage within a network (e.g., enumerating network domains, specific computers within a domain). Broadcast domains should be limited or aligned with business functional areas/workgroups within an organization, to reduce the risk of information leakage to those without a need to know, where systems can be targeted and their vulnerabilities exploited. The major difference between a router and a Layer 3 switch is that a router performs packet switching using a microprocessor, whereas a Layer 3 switch performs the switching using application ASIC hardware.

In creating separate broadcast domains, Layer 3 switches also enable the concept of establishing a virtual LAN (VLAN). A VLAN is a group of devices on one or more logically segmented LANs. A VLAN is set up by configuring ports on a switch, so devices attached to these ports may communicate as if they were attached to the same physical network segment, although the devices are located on different LAN segments. A VLAN is based on logical rather than physical connections and, thus, allow great flexibility. This flexibility enables administrators to restrict users' access of network resources to only those specified and segment network resources for optimal performance.

In Layer 4 switching, some application information is taken into account along with Layer 3 addresses. For IP, this information includes the port numbers from protocols such as User Datagram Protocol (UDP) and TCP. These devices, unlike Layer 3 switches, are more resource intensive since they have to store application-based protocol information. Only address information is stored at the Layer 2 and Layer 3 levels.

A layer 4 (transport layer) switch allows for policy-based switching. With this functionality, Layer 4 switches can off-load a server by balancing traffic across a cluster of servers, based on individual session information and status.

- **Layer 4-7 switches**—Also known as content-switches, content services switches, web-switches or application-switches. They are typically used for load balancing among groups of servers. Load balancing can be based on HTTP, HTTPS and/or VPN, or for any application TCP/IP traffic using a specific port. Content switches can also be used to perform standard operations such as SSL encryption/decryption to reduce the load on the servers receiving the traffic, and to centralize the management of digital certificates.
- **Gateways**—Devices that are protocol converters. Typically, they connect and convert between LANs and the mainframe, or between LANs and the Internet, at the application layer of the OSI reference model. Depending on the type of gateway, the operation occurs at various OSI layers. The most common form of gateway is a systems network architecture (SNA) gateway, converting between a TCP/IP, NetBios or Inter-network Packet Exchange (IPX) session (terminal emulator) and the mainframe.

## LAN TECHNOLOGY SELECTION CRITERIA
Some of the more relevant selection criteria are:
- What are the applications?
- What are the bandwidth needs?
- What is the area to be covered and what are the physical constraints?
- What is the budget?
- What are the remote management needs?
- What are the security needs?

## Wide Area Network

A WAN is a data communications network that transmits information across geographically dispersed LANs such as among plant sites, cities and nations.

WAN characteristics include:
• They are applicable to the physical, data link and network layers of the OSI reference model.
• Data flow can be either simplex (one-way flow), half duplex (one way at a time) or full duplex (both ways at one time without turnaround delay).
• Communication lines can be either switched or dedicated.

### WAN MESSAGE TRANSMISSION TECHNIQUES

WAN message transmission techniques include:
• **Message switching**—Sends a complete message to the concentration point for storage and routing to the destination point as soon as a communications path becomes available. Transmission cost is based on message length.
• **Packet switching**—A sophisticated means of maximizing transmission capacity of networks. This is accomplished by breaking a message into transmission units, called packets, and routing them individually through the network, depending on the availability of a channel for each packet. Passwords and all types of data can be included within the packet. The transmission cost is by packet and not by message, route or distance. Sophisticated error and flow control procedures are applied to each link by the network.
• **Circuit switching**—A physical communications channel is established between communicating equipment, through a circuit-switched network. This network can be, for instance, point-to-point (e.g., leased line) multipoint, a public-switched telephone network (PSTN) or an integrated services digital network (ISDN). The connection, once established, is used exclusively by the two subscribers for the duration of the call. The network does not provide any error or flow control on the transmitted data, so this task must be performed by the user.
• **Virtual circuits** —A logical circuit between two network devices that provides for reliable data communications. Two types are available—switched virtual circuits (SVCs) or permanent virtual circuits (PVCs). SVCs dynamically establish on-demand connectivity and PVCs establish an always-on connection.
• **WAN dial-up services**—Dial-up services using asynchronous and synchronous connectivity are widely available and well suited for organizations with a large number of mobile users. Their disadvantages are low bandwidth and limited performance.

### WAN DEVICES

The following devices, typically operating at either the physical or data link layer of the OSI reference model, are specific to the WAN environment:
• **WAN switches**—Data link layer devices used for implementing various WAN technologies such as Asynchronous Transfer Mode (ATM), point-to-point frame relay and ISDN. These devices are typically associated with carrier networks providing dedicated WAN switching and router services to organizations via T-1/E-1 or T-3/E-3 connections.
• **Routers**—A device that operates at the network layer of the OSI reference model and provides an interface between different network segments on an internal network or connects the internal network to an external network
• **Modems (modulator/demodulator)**—Data communications equipment (DCE) devices that make it possible to use analog lines (generally, the public telephone network) as transmission media for digital networks. Modems convert computer digital signals into analog data signals and analog data back to digital. When a link is established, modems operating at both ends of it automatically negotiate the fastest and safest standard that the line and the modems themselves can use, establishing speed, parity, cryptographic algorithm and compression.

For transmission purposes, modems disassemble bytes into a sequence of bits that are sent sequentially to the line. At the receiving end, these bits must be reassembled into bytes.

A main task of the modems at both ends is to maintain their synchronization so the receiving device knows when each byte starts and ends. Two methods can be used for this purpose:
– **Synchronous transmission**—Bits are transmitted without interruption at a constant speed. The sending modem uses a specific character when it starts transmitting a data block to "synchronize" the receiving device. This mode allows maximum efficiency, but only if blocks are not too short. Specific technical rules must be observed to maintain synchronization.

– **Asynchronous transmission**—The transmitting device marks the beginning and end of a byte by sending a "start" and a "stop" bit before and after each data byte. The efficiency of the line is lower, but the asynchronous standard is simpler and works well for character and block mode transmissions.

Communication links can be operated both ways. Refer to the TCP/IP and Its Relation to the OSI Reference Model section (discussed later in this chapter) and to **exhibit 4.8**.

| ISO Model | Reference | TCP/IP Conceptual Layers | Protocols |
|---|---|---|---|
| Exhibit 4.8—Standards Associated With the TCP/IP Suite | | | |
| 7 | Application | Application | File Transport Protocol (FTP) |
| | | | Remote Terminal Control Protocol (TELNET) |
| | | | Simple Mail Transport Protocol (SMTP) |
| | | | Name Server Protocol (NSP) |
| | | | Simple Network Management Protocol (SNMP) |
| 6 | Presentation | | |
| 5 | Session | | |
| 4 | Transport | Transport | Transmission Control Protocol (TCP) User Datagram Protocol (UDP) |
| 3 | Network | Network interface | Internet Protocol (IP) |
| 2 | Data link | LAN or WAN interface | Ethernet, Token Ring, Point-to-point Protocol (PPP) |
| 1 | Physical | | |

- **Access servers**—These provide centralized access control for managing remote access dial-up services.
- **Channel service unit/digital service unit (CSU/DSU)**—This interfaces at the physical layer of the OSI reference model, data terminal equipment (DTE) to DCE, for switched carrier networks.
- **Multiplexors**—A physical layer device used when a physical circuit has more bandwidth capacity than required by individual signals. The multiplexor can allocate portions of its total bandwidth and use each portion as a separate signal link. It can also link several low-speed lines to one high-speed line to enhance transmission capabilities.

Generally, methods for multiplexing data include the following:
- **Time-division multiplexing (TDM)**—Information from each data channel is allocated bandwidth, based on preassigned time slots, regardless of whether there are data to transmit.
- **Asynchronous time division multiplexing (ATDM)**—Information from data channels is allocated bandwidth as needed, via dynamically assigned time slots.
- **Frequency division multiplexing (FDM)**—Information from each data channel is allocated bandwidth, based on the signal frequency of the traffic.
- **Statistical multiplexing**—Bandwidth is allocated dynamically to any data channels that have the information to transmit.

## WAN TECHNOLOGIES
Some common types of WAN technologies used to manage the communication links are provided in the following sections.

### Point-to-point Protocol
Point-to-point protocol (PPP) works in the data link layer. PPP provides a single, preestablished WAN communication path from the customer premises to a remote network, usually reached through a carrier network such as a telephone company. PPP is a widely available remote access solution that supports asynchronous and synchronous links, and operates over a wide range of media. Because PPP is more stable than the older Serial Line Internet Protocol (SLIP), PPP is the Internet standard for transmission of IP packets over serial lines. PPP makes use of two primary protocols for operation. The first,

Link Control Protocol (LCP), is used when establishing, configuring and testing the data link connection. The second, Network Control Protocol (NCP), establishes and configures different network layer protocols (e.g., Internetwork packet exchange [IPX]). PPP features include address notification, authentication, support for multiple protocols and link monitoring.

## X.25

As a packet-switched or virtual-circuit implementation, X.25 is a telecommunication standard (ITU-T) that defines how connections between data terminal equipment and data communications or circuit terminating equipment are maintained for remote terminal access and computer communications in public data networks (PDNs). Developed in 1976, X.25 operates at the lower three layers of the OSI reference model, but is no longer widely available today, primarily because it is resource-intensive in providing error control capabilities.

## Frame Relay

As a packet-switched or virtual-circuit implementation, Frame Relay is a data link layer protocol for switch devices that uses a standard encapsulation technique to handle multiple virtual circuits between connected devices. The encapsulation method is high-level data link control (HDLC) for synchronous serial links using frame characters and checksums. Frame Relay is more efficient than X.25, the protocol for which it is generally considered a replacement. Contrary to X.25, Frame Relay relies more on upper layer protocols for significant error handling processes in data transmissions. Frame Relay is a low-cost, widely available LAN technology used in WAN point-to-point connections.

## Integrated Services Digital Network

As a circuit-switched implementation, ISDN corresponds to integrated voice, data and video, and is an architecture for worldwide telecommunications. This service integrates voice, data and video communication through digital switching and transmission over digital public carrier lines. The ISDN technologies now implemented are narrowband (basic-rate and primary-rate, not aggregated) ISDN; broadband ISDN has never been widely implemented. Separate channels are used for customer information (i.e., B, bearer channels—voice, data and video) and to send signals and control information (i.e., D, data channels). ISDN uses a packet-node layered protocol, based on the CCITT's X.25 standard. Unlike Frame Relay, it is moderately available to all.

## Asynchronous Transfer Mode

As a packet-switched implementation operating at the data link layer, asynchronous transfer mode (ATM) is based on the use of a cell (a fixed-size data block) switching and multiplexing technology standard that combines the benefits of circuit switching (guaranteed capacity and constant transmission delay) with those of packet switching (flexibility and efficiency for intermittent traffic). Because ATM is asynchronous, time slots are available on demand with information identifying the source of the transmission contained in the header of each ATM cell. ATM is considered relatively expensive as a dedicated leased line option in comparison to other available WAN options.

## Multiprotocol Label Switching

Multiprotocol label switching (MPLS) provides a mechanism for engineering network traffic patterns that is independent of routing tables. MPLS assigns short labels to network packets that describe how to forward them through the network. MPLS is independent of any routing protocol and can be used for unicast packets.

In traditional Level 3 forwarding, as a packet travels from one router to the next, an independent forwarding decision is made at each hop. The IP network layer header is analyzed, and the next hop is chosen based on this analysis and on the information in the routing table. In an MPLS environment, the analysis of the packet header is performed just once, when a packet enters the MPLS cloud. The packet is then assigned to a stream, which is identified by a label—a short (20-bit), fixed-length value at the front of the packet. Labels are used as lookup indexes into the label forwarding table. This table stores forwarding information for each label. Additional information, such as class-of-service (CoS) values, which can be used to prioritize packet forwarding, can be associated with a label.

## Digital Subscriber Lines
DSL is a network provider service using modem technology over existing twisted-pair telephone lines to transport high-bandwidth data such as multimedia and video. Characteristics of digital subscriber lines (DSL) include:
• Dedicated, point-to-point, public network access on the local loop. Local loops are generally the "last mile" between a network service provider's (NSP) central office and the customer site.
• Delivers high-bandwidth data rates to dispersed customers at low cost through the existing telecommunications infrastructure
• Always-on access, which eliminates call setup and makes it ideal for Internet/intranet and remote LAN access

DSL services vary in their speed and type of modulation:
• ADSL (Asymmetric Digital Subscriber Line)
• SDSL (Symmetric Digital Subscriber Line)
• HDSL and HDSL-2 (High bit-rate Digital Subscriber Line)
• HDSL-2 (High speed digital subscriber line version 2)
• SHDSL (Single-Pair high-speed digital subscriber line)
• G.SHDSL (An international standard for symmetric DSL also known as G.001.2)
• VDSL (Very High Speed Digital Subscriber Line)

## Virtual Private Networks
A VPN extends the corporate network securely via encrypted packets sent out via virtual connections over the public Internet to distant offices, home workers, salespeople and business partners. Rather than using expensive dedicated leased lines, VPNs take advantage of the public worldwide IP infrastructure, thereby enabling remote users to make a local call (versus dialing-in at long distance rates) or use an Internet cable modem or DSL connections for inexpensive public network connectivity.

VPNs are platform independent. Any computer system that is configured to run on an IP network can be connected through a VPN with no modifications, except for the installation of remote software.

There are three types of VPNs:
1. **Remote-access VPN**—Used to connect telecommuters and mobile users to the enterprise WAN in a secure manner. It lowers the barrier to telecommuting by ensuring that information is reasonably protected on the open Internet.
2. **Intranet VPN**—Used to connect branch offices within an enterprise WAN
3. **Extranet VPN**—Used to give business partners limited access to each others' corporate network. One example of this is an automotive manufacturer with its suppliers.

The only difference between a traditional, intracompany VPN (intranet) and an intercompany VPN (extranet) is the way the VPN is managed. With an intranet VPN, all network and VPN resources are managed by a single organization. When an organization's VPN is used for an extranet, management control becomes weak. Therefore, it is recommended that in extranet VPN, each constituent company manage their own VPN and maintain control over it.

VPNs allow:
• Network managers to cost-efficiently increase the span of the corporate network
• Remote network users to securely and easily access their corporate enterprise
• Corporations to securely communicate with business partners
• Supply chain management to be efficient and effective
• Service providers to grow their businesses by providing substantial incremental bandwidth with value-added services

Determining which network resources should be linked via a VPN depends on the applications used on the various systems. Requirements often used to determine network connectivity include security policies, business models, intranet server access, application requirements, data sharing and application server access.

The process of encrypting packets, which makes VPN an effective protection scheme, uses the Internet Engineering Task Force's (IETF) IP Security (IPSec) standard. Proprietary variants exist, but IPSec is the dominant standard. This standard enables the entire packet, including its header, to be encrypted, which is known as the IPSec tunnel encryption mode. Alternatively, the transport mode encrypts the data portion of the packet only. These encryption modes are representative of the encapsulation process whereby a new packet is created.

<table>
| PRACTICE QUESTIONS |
|---|
</table>

**PRACTICE QUESTIONS**

4-7  Which of the following would allow a company to extend its enterprise's intranet across the Internet to its business partners?

    A.    Virtual private network
    B.    Client-server
    C.    Dial-up access
    D.    Network service provider

4-8  Which of the following statements relating to packet switching networks is **CORRECT**?

    A.    Packets for a given message travel the same route.
    B.    Passwords cannot be embedded within the packet.
    C.    Packet lengths are variable and each packet contains the same amount of information.
    D.    The cost charged for transmission is based on the packet, not the distance or route traveled.

*See answers and explanations to the practice questions at the end of the chapter. (pages 357-358)*

**Note:** For the security implications of VPN and for information on IPSec encryption and VPN, see the Remote Access Security section in chapter 5, Protection of Information Assets.

# Wireless Networks

Wireless technologies, in the simplest sense, enable one or more devices to communicate without physical connections, i.e., without requiring network or peripheral cabling. Wireless is a technology that enables organizations to adopt e-business solutions with tremendous growth potential. Wireless technologies use radio frequency transmissions as the means for transmitting data, whereas wired technologies use cables. Wireless technologies range from complex systems (such as wireless wide area networks [WWANs], wireless local area networks [WLANs] and cell phones) to simple devices (such as wireless headphones, microphones and other devices that do not process or store information). They also include Bluetooth devices with a miniradio frequency transceiver and infrared (IR) devices, such as remote controls, some cordless computer keyboards and mice, and wireless Hi-Fi stereo headsets, all of which require a direct line of sight between the transmitter and the receiver to close the link.

However, going wireless introduces new elements that must be addressed. For example, existing applications may need to be retrofit to make use of wireless interfaces. Network standardization must take place so that wireless devices using different operating systems can communicate with one another. Also, decisions need to be made regarding general connectivity—to facilitate the development of completely wireless mobile applications or other applications that rely on synchronization of data transfer between mobile computing systems and corporate infrastructure. Other issues include narrow bandwidth, the lack of a mature standard, and unresolved security and privacy issues.

Wireless networks serve as the transport mechanism between devices, and among devices and the traditional wired networks. Wireless networks are many and diverse but are frequently categorized into four groups based on their coverage range:
• WANs
• LANs
• Wireless personal area networks (WPANs)
• Wireless *ad hoc* networks

### WIRELESS WIDE AREA NETWORKS

Wireless wide area networking is the process of linking different networks over a large geographical area to allow wider IT resource sharing and connectivity. While computers are often connected to traditional WANs using cable networking solutions (such as telephone systems), wireless wide area networks are connected via radio, satellite and mobile phone technologies.

WWANs, using radio, satellite and mobile phone technologies, can complement and compete with more traditional systems of cable-based networking. These include wide coverage area technologies such as 2G cellular, Cellular Digital Packet Data (CDPD), Global System for Mobile Communications (GSM) and Mobitex.

For some organizations, such as those in rural areas where laying cable is too expensive, wireless technology offers the only networking solution. For others, wireless wide area networking provides greater system flexibility, as well as the opportunity to control costs where the equipment is owned.

Implementing a WWAN requires careful attention to the planning and surveying of the network. The total cost of ownership involved in switching to this rapidly evolving system of networking should also be considered.

## WIRELESS LOCAL AREA NETWORKS
WLANs allow greater flexibility and portability than traditional wired LANs. Unlike a traditional LAN, which requires a wire to connect a user's computer to the network, a WLAN connects computers and other components to the network using an access point device. An access point, or wireless networking hub, communicates with devices equipped with wireless network adaptors within a specific range of the access point; it connects to a wired Ethernet LAN via an RJ-45 port. Access point devices typically have coverage areas of up to 300 feet (approximately 100 meters). This coverage area is called a cell or range. Users move freely within the cell with their laptop or other network devices. Access point cells can be linked together to allow users to "roam" within a building or between buildings. WLAN includes 802.11, HyperLAN, HomeRF and several others.

WLAN technologies conform to a variety of standards and offer varying levels of security features. The principal advantages of standards are to encourage mass production and to allow products from multiple vendors to interoperate. The most useful standard used currently is the IEEE 802.11 standard.

> **Note:** The CISA candidate will not be tested on these IEEE standards in the exam.

802.11 refers to a family of specifications for wireless LAN technology. 802.11 specifies an over-the-air interface between a wireless client and a base station or between two wireless clients. The IEEE accepted the specification in 1997.

There are several specifications in the 802.11 family:
- **802.11 legacy**—The original version of the IEEE 802.11 standard released in 1997, called 802.11 legacy, specifies two data rates of 1 and 2 Mbps to be transmitted via IR signals or in the ISM band at 2.4 gigahertz (GHz). IR has been dropped from later revisions of the standard, because it could not succeed against the well-established Infrared Data Association (IrDA) protocol and had no known implementations. Legacy 802.11 was rapidly succeeded by 802.11b.
- **802.11b**—802.11b has a range of about 164 feet (50 meters) with the low-gain, omnidirectional antennas typically used in 802.11b devices. 802.11b has a maximum throughput of 11 Mbps; however, a significant percentage of this bandwidth is used for communications overhead. In practice, the maximum throughput is about 5.5 Mbps. Metal, water and particularly thick walls absorb 802.11b signals and decrease the range drastically. 802.11 operates in the 2.4 GHz spectrum and uses CSMA/CA as its media access method. With high-gain external antennas, this protocol can also be used in fixed point-to-point scenarios at ranges up to 5 miles (8 kilometers). It can be used to replace costly leased lines or in place of microwave communications gear. Current cards can operate at 11 Mbps, but will lose speed if signal strength is an issue. 802.11b divides the spectrum into 14 overlapping, staggered channels of 22 MHz each. Different channels or ranges are legal in different countries. Three or four channels may be used simultaneously in the same area with little or no overlap, typically channels 1, 6 and 11. Extensions have been made to the 802.11b protocol to increase speed to 22, 33 and 44 Mbps, but the extensions are proprietary and not endorsed by the IEEE. Many companies call these enhanced versions 802.11b+.
- **802.11a**—Even though the faster 802.11a standard was ratified in 1999, it did not start shipping until 2001. The 802.11a standard uses the 5 GHz band and operates at a raw speed of 54 Mbps; however, more generally, it operates at speeds in the mid-20-Mbps range. The speed is reduced to 48, 36, 34, 18, 12, 9 and then 6 Mbps, if required. 802.11a has 12 nonoverlapping channels, eight dedicated to indoor and four to point-to-point. 802.11a has not seen wide adoption because of the high adoption rate of 802.11b, and concerns about its range. At five GHz, 802.11a cannot reach as far as 802.11b (assuming the same power limitations) and may be absorbed more readily. Most manufacturers of 802.11a equipment countered the lack of market success by releasing dual-band/dual-mode or tri-mode cards that can automatically handle multiple standards (802.11a and 802.11b for dual-band/dual-mode card or 802.11a, 802.11b and 802.11g for tri-mode cards) or by releasing access points that can support all standards simultaneously.
- **802.11g**—In June 2003, a third standard for encoding was ratified, 802.11g. This standard works in the 2.4-GHz band, like 802.11b, but operates at a raw speed of 54 Mbps or, more generally, at about 24.7 Mbps net throughput, like 802.11a. The 802.11g standard is fully backwards compatible with 802.11b.

• **802.11i (WPA2)**—In June 2004, the IEEE ratified the 802.11i (WPA2) security standard for wireless. The 802.11i standard replaced the weak Wired Equivalent Privacy (WEP) protocol and introduced the Robust Secure Network (RSN) protocol for establishing secure communications. The IEEE 802.11i standard utilizes the authentication schemes of 802.1X and Extensible Authentication Protocol (EAP) in addition to a new encryption scheme Advanced Encryption Standard (AES) and dynamic key distribution scheme—Temporal Key Integrity Protocol (TKIP). An alternative authentication scheme, which utilizes a preshared key methodology, is also supported in the 802.11i specification. This is designed for small office-home office (SOHO) environments that may not have the resources to deploy an 802.1X infrastructure.

• **802.11n**—The 802.11n standard has a speed of 100 Mb (even 250 Mb at the physical layer [PHY]) and is four to five times faster than the 802.11b and 802.11g standards. The 802.11n standard offers a better operating distance in current networks that do not utilize higher speeds.

**Exhibit 4.9** summarizes the main points of several WLAN standards.

| Standard | Data Rate | Modulation Scheme | Security | Pros/Cons |
|---|---|---|---|---|
| IEEE 802.11 | Up to 2Mbps in the 2.4GHz band | Frequency-hopping spread spectrum (FHSS) or direct-sequence spread spectrum (DSSS) | Wired equivalent protection and Wi-Fi-protected Access (WPA) | • This specification has been extended into 802.11b. |
| IEEE 802.11a (Wi-Fi) | Up to 54Mbps in the 5GHz band | OFDM | Wired equivalent protection and WPA | • Products that adhere to this standard are considered Wi-Fi certified.<br>• Eight available channels and less potential for radio frequency (RF) interference than 802.11b and 802.11g<br>• Better than 802.11b at supporting multimedia voice, video and large-image applications in densely populated user environments<br>• Relatively shorter range than 802.11b<br>• Not interoperable with 802.11b |
| IEEE 802.11b (Wi-Fi) | Up to 11Mbps in the 2.4GHz band | DSSS with complementary code keying (CCK) | Wired equivalent protection and WPA | • Products that adhere to this standard are considered Wi-Fi-certified.<br>• Not interoperable with 802.11a<br>• Requires fewer access points than 802.11a for coverage of large areas<br>• Offers high-speed access to data at up to 300 feet (91.46 m) from base station<br>• 14 channels available in the 2.4GHz band with only three nonoverlapping channels |
| IEEE 802.11g (Wi-Fi) | Up to 54Mbps in the 2.4GHz band | Orthogonal frequency division multiplexing (OFDM) above 20 Mbps, DSSS with CCK below 20 Mbps | Wired equivalent protection and WPA | • Products that adhere to this standard are considered Wi-Fi certified<br>• May replace 802.11b<br>• Improved security enhancements over 802.11<br>• Compatibility available in the 2.4GHz band with only three nonoverlapping channels |
| IEEE 802.11i (Wi-Fi) | Up to 11Mbps in the 2.4GHz band | DSSS | WPA, TKIP, WPA2, AES, remote authentication dial-in user service (RADIUS) | • Adds the AES security protocol to the 802.11 standard for WLANs<br>• Security has been a primary concern to deploy wireless networks. |
| Bluetooth | Up to 2Mbps in the 2.45GHz band | FHSS | Point-to-Point Tunneling Protocol (PPTP), SSL or VPN | • No native support for IP, so it does not support TCP/IP and WLAN applications well<br>• Not originally created to support WLANs<br>• Best suited for connecting PDAs, cell phones and PCs in short intervals |

| Exhibit 4.9—Summary of WLAN Standards (*cont.*) | | | | |
|---|---|---|---|---|
| **Standard** | **Data Rate** | **Modulation Scheme** | **Security** | **Pros/Cons** |
| HomeRF | Up to 10 Mbps in the 2.4 GHZ band | FHSS | • Independent network IP addresses for each network<br>• Data are sent with a 56-bit encryption algorithm | • HomeRF is no longer being supported by any vendors or working groups<br>• Intended for use in homes, not enterprises<br>• Range is only 150 feet (45.7 meters) from base station<br>• Relatively inexpensive to set up and maintain<br>• Voice quality is always good because it continuously reserves a chunk of bandwidth for voice services<br>• Responds well to interference because of frequency-hopping modulation |
| Hiper LAN/1 (Europe) | Up to 20 Mbps in the 5 GHz band | CSMA/CA | Per-session encryption and individual authentication | • Only available in Europe<br>• HiperLAN is totally *ad hoc*, requiring no configuration and no central controller<br>• Does not provide real isochronous services<br>• Relatively expensive to operate and maintain<br>• No guarantee of bandwidth |
| HiperLAN/2 (Europe) | Up to 54 Mbps in the 5 GHz band | OFDM | Strong security features with support for individual authentication and per-session encryption keys | • Only available in Europe<br>• Designed to carry ATM cells, IP packets, Firewire packets (IEEE 1394) and digital voice (from cellular phones)<br>• Better quality of service than HiperLAN/1 and guarantees bandwidth |

## WIRED EQUIVALENT PRIVACY AND WI-FI PROTECTED ACCESS (WPA/WPA2)

IEEE 802.11's Wired Equivalent Privacy (WEP) encryption uses symmetric, private keys, which means the end user's radio-based NIC and access point must have the same key. This leads to difficulties periodically involved with distributing new keys to each NIC. As a result, keys remain unchanged on networks for extended times. With static keys, several hacking tools easily break through the relatively weak WEP encryption mechanisms.

Because of the key reuse problem and other flaws, the current standardized version of WEP does not offer strong enough security for most corporate applications. Newer security protocols, such as 802.11i (WPA2) and Wi-Fi Protected Access (WPA), however, utilize public key cryptography techniques to provide effective authentication and encryption between users and access points.

## WIRELESS PERSONAL AREA NETWORKS

WPANs are short-range wireless networks that connect wireless devices to one another. The most dominant form of WPAN technology is Bluetooth, which links wireless devices at very short distances. The oldest way to connect devices in a WPAN fashion is IR communications.

Bluetooth is an open source standard that borrows many features from existing wireless standards, such as IEEE 802.11, IrDA, Digital Enhanced Cordless Telecommunications (DECT), Motorola's Piano and TCP/IP, to connect portable devices without wires, via short-range radio frequencies (RF).

Bluetooth is a wireless protocol that connects devices within a range of up to 49 feet (15 meters) and has become a feature on some PDAs, mobile phones, PC keyboards, mice, printers, etc. It is a system that changes frequencies from moment to moment using a technique called frequency-hopping and has an effective range of about 32.8 feet (10 meters), but occasionally farther. Bluetooth is just now beginning to appear in computer systems, especially notebooks, as a replacement for physical cables and for infrared connections, which are limited to line of sight. Bluetooth devices find one another when they are in range and automatically set up a background connection.

Bluetooth allows for high data speeds (between 1 Mbps and 2 Mbps), but is designed only for peer-to-peer data transfer. A new emerging form of WPAN technology, called ZigBee, offers slower data speeds (250 Kbps) than Bluetooth, but is both cheaper than Bluetooth and requires far less energy to power.

## AD HOC NETWORKS
*Ad hoc* networks are networks designed to dynamically connect remote devices such as cell phones, laptops and PDAs. These networks are termed *ad hoc* because of their shifting network topologies. Whereas WLANs or WPANs use a fixed network infrastructure, *ad hoc* networks maintain random network configurations, relying on a system of mobile routers connected by wireless links to enable devices to communicate. Bluetooth networks can behave as *ad hoc* networks, as mobile routers control the changing network topologies of these networks. The routers also control the flow of data between devices that are capable of supporting direct links to each other. As devices move about in an unpredictable fashion, these networks must be reconfigured to handle the dynamic topology. The routing protocol employed in Bluetooth allows the routers to establish and maintain these shifting networks.

The mobile router is commonly integrated in a device such as a PDA, notebook or mobile phone. This mobile router, when configured, ensures that a remote, mobile device, such as a mobile phone, stays connected to the network. The router maintains the connection and controls the flow of communication.

## WIRELESS APPLICATION PROTOCOL
Wireless Application Protocol (WAP) is a general term used to describe the multilayered protocol and related technologies that bring Internet content to wireless mobile devices such as PDAs and cell phones. WAP protocols are largely based on Internet technologies. The motivation for developing WAP was to extend Internet technologies to wireless networks, bearers and devices.

WAP supports most wireless networks and is supported by all operating systems specifically engineered for handheld devices and some mobile phones.

These kinds of devices that use displays and access the Internet run what are called microbrowsers, which have small file sizes that can accommodate the low-memory constraints of handheld devices and the low-bandwidth constraints of a wireless handheld network. Although WAP supports Hypertext Markup Language (HTML) and XML, the Wireless Markup Language (WML) language (an XML application) is designed specifically for small screens and one-hand navigation without a keyboard.
WML is scalable from two-line text displays through graphic screens found on items such as smart phones, pocket PCs and communicators. WAP also supports WMLScript. WML is similar to JavaScript but makes minimal demands on memory and CPU power, because it is more specific and does not contain many of the unnecessary functions found in JavaScript and other scripting languages.

The following are general issues and exposures related to wireless access:
- **The interception of sensitive information**—Information is transmitted through the air, which increases the potential for unprotected information to be intercepted by unauthorized individuals.
- **The loss or theft of devices**—Wireless devices tend to be relatively small, making them much easier to steal or lose. If encryption is not strong, a hacker can easily get at the information that is password- or PIN-protected.
- **The misuse of devices**—Devices can be used to gather information or intercept information that is being passed over wireless networks for financial or personal benefit.
- **The loss of data contained in the devices**—Theft or loss can result in the loss of data that have been stored on these devices. Storage capacity can range from a few megabytes to several gigabytes of data, depending on the device.
- **Distractions caused by the devices**—The use of wireless devices distract the user. If these devices are being used in situations where an individual's full attention is required (e.g., driving a car), they could result in an increase in the number of accidents.
- **Possible health effects of device usage**—The safety or health hazards have not, as yet, been identified. However, there are currently a number of concerns with respect to electromagnetic radiation, especially for those devices that must be held beside the head.

- **Wireless user authentication**—There is a need for stronger wireless user authentication and authorization tools at the device level. The current technology is just emerging.
- **File security**—Wireless phones and PDAs do not use the type of file access security that other computer platforms can provide.
- **WEP security encryption**—WEP security depends particularly on the length of the encryption key and on the usage of static WEP (many users on a WLAN share the same key) or dynamic Wired Equivalent Privacy (WEP) (per-user, per-session, dynamic WEP key tied to the network logon). The 64-bit encryption keys that are in use in the WEP standard encryption can be easily broken by the currently available computing power. Static WEP, used in many WLANs for flexibility purposes, is a serious security risk, as a static key can easily be lost or broken, and, once this has occurred, all of the information is available for viewing and use. An attacker possessing the WEP key could also sniff packets being transmitted and decrypt them.
- **Interoperability**—Most vendors offer 128-bit encryption modes. However, they are not standardized, so there is no guarantee that they will interoperate. The use of the 128-bit encryption key has a major impact on performance with 15-20 percent degradation being experienced. Some vendors offer proprietary solutions; however, this only works if all access points and wireless cards are from the same vendor.
- **The use of wireless subnets**—To increase security, it is possible to create special subnets for wireless traffic and require authorization before packets are routed.
- **Translation point**—The location where information being transmitted via the wireless network is converted to the wired network. At this point, the information, which has been communicated via the WAP security model using Wireless Transport Layer Security, is converted to the secure socket layer, where the information is decrypted and then encrypted again for communication via TCP/IP.

## Public "Global" Internet Infrastructure

The Internet is comprised of networks distributed over the entire world and interconnected via pathways that allow the exchange of information, data and files. Being connected to the Internet means to be logically part of it. By using these pathways, a connected computer can send or receive packets of data to/from any other Internet device.

Today, the Internet is a vast, global network of networks, ranging from university networks to corporate LANs to large online services. The Internet is not run or controlled by any single person, group or organization. The only thing that is centrally controlled is the availability and assignment of Internet addresses and the attached symbolic host names. Addresses and names are used for locating the source or destination networks.

Users can access the Internet through either a LAN or by dialing into a computer connected to the Internet via an online service or a dial-in Internet service provider (ISP). Routers, which connect networks, perform most of the work of directing traffic on the Internet. Networks are connected in many different ways. There are dedicated telephone lines that can transmit data at up to 56 Kbps. There are an increasing number of T-1/E-1 leased lines that can carry 1.544/2.048 Mb of data per second. Higher-speed T-3/E-3 lines, which can carry data at 44.746/34 Mbps, are used as well. Satellites can also link networks, as can fiber-optic cables or ISDN telephone lines.

The networks in a particular geographic area are connected into a large regional network. Regional networks are connected to one another via high-speed backbones (connections that can send data at extremely high speeds). When data are sent from one regional network to another, they first travel to a network access point (NAP). NAPs then route the data to high-speed backbones, such as the high-speed backbone network services (BNS), which can transmit data at 155 Mbps. The data are then sent along the backbone to another regional network and then to a specific network and computer within that regional network.

### TCP/IP AND ITS RELATION TO THE OSI REFERENCE MODEL

The protocol suite used as the *de facto* standard for the Internet is known as the TCP/IP. The TCP/IP suite includes both network-oriented protocols and application support protocols. **Exhibit 4.10** shows some of the standards associated with the TCP/IP suite and where these fit within the ISO model. It is interesting to note that the TCP/IP set of protocols was developed before the ISO/OSI framework; therefore, there is no direct match between the TCP/IP standards and the layers of the framework.

| Exhibit 4.10—Standards Associated With the TCP/IP Suite | | | |
|---|---|---|---|
| ISO Model | Reference | TCP/IP Conceptual Layers | Protocols |
| 7 | Application | Application | File Transport Protocol (FTP)<br>Remote Terminal Control Protocol (Telnet)<br>Simple Mail Transport Protocol (SMTP)<br>Name Server Protocol (NSP)<br>Simple Network Management Protocol (SNMP) |
| 6 | Presentation | | |
| 5 | Session | | |
| 4 | Transport | Transport | Transmission Control Protocol (TCP)<br>User Datagram Protocol (UDP) |
| 3 | Network | Network interface | Internet Protocol (IP) |
| 2 | Data link | LAN or WAN | Ethernet, Token Ring, |
| 1 | Physical | interface | Point-to-Point Protocol (PPP) |

## TCP/IP INTERNET WORLD WIDE WEB SERVICES

The most common way users access resources on the Internet is through the TCP/IP Internet World Wide Web (WWW) application service. This Internet application service is the fastest growing and most innovative part of the Internet. When the user browses the web, they view multimedia pages composed of text, graphics, sound and video. The web uses hypertext links that allow connections to jump from any place on the web to any other place on the web; the language that allows the user to activate hypertext links and view web pages is called hypertext markup language (HTML). The web is implemented based on a client-server model. The client software, such as Microsoft Internet Explorer or Mozilla Firefox, is known as a web browser, and users utilize it on their local computer to access server-based resources on the web. The server software runs on a web host, which is often a Windows or UNIX system.

Other terminology associated with access and use of the WWW service include:
• **URL**—The uniform resource locator (URL) identifies the address on the WWW where a specific resource is located. To access a web site, a user enters the site's location into their browser's URL space, or they click on the hypertext link that will send them to the location. The web browser looks up the IP address of the site, and sends a request for the URL via the Hypertext Transfer Protocol (HTTP). This protocol defines how the web browser and web server communicate with one another.
URLs contain several parts. An example of a URL is *http://www.isaca.org/certification*:
• The *http://* details which Internet Protocol to use.
• The *www.isaca.org* identifies the domain and server to be contacted. It is normal to use the first part of the server name ("*www*" in this case) to identify the kind of Internet resource which is being contacted, or to address several servers in the same domain (i.e., *isaca.org*), such as: web1, web2, ftp1, ftp2. A URL may use an IP address instead of a domain name.
• The */certification* identifies a specific directory, a home page, document or other object on the server.

A URL can also be used to access other TCP/IP Internet services:
– *ftp://isaca.org*
– *telnet://isaca.org*

The URL is the location of specific resources (e.g., pages, data) or services on the Internet. In the example, the resource is a web page called "certification" and is found on the web server of ISACA. This request is sent over the Internet and the routers transfer the request to the addressed web server, which activates the HTTP protocol and processes the request. When the server finds among its resources the requested home page, document or object, it sends the request (**exhibit 4.11**) back to the web browser. In the case of an HTML page, the information sent back contains data and formatting specifications. These are in the form of a program that is executed by the client web browser and produce the screen displayed for the user. After the page is sent by the server, the HTTP connection is closed and can be reopened. **Exhibit 4.12** displays the path.

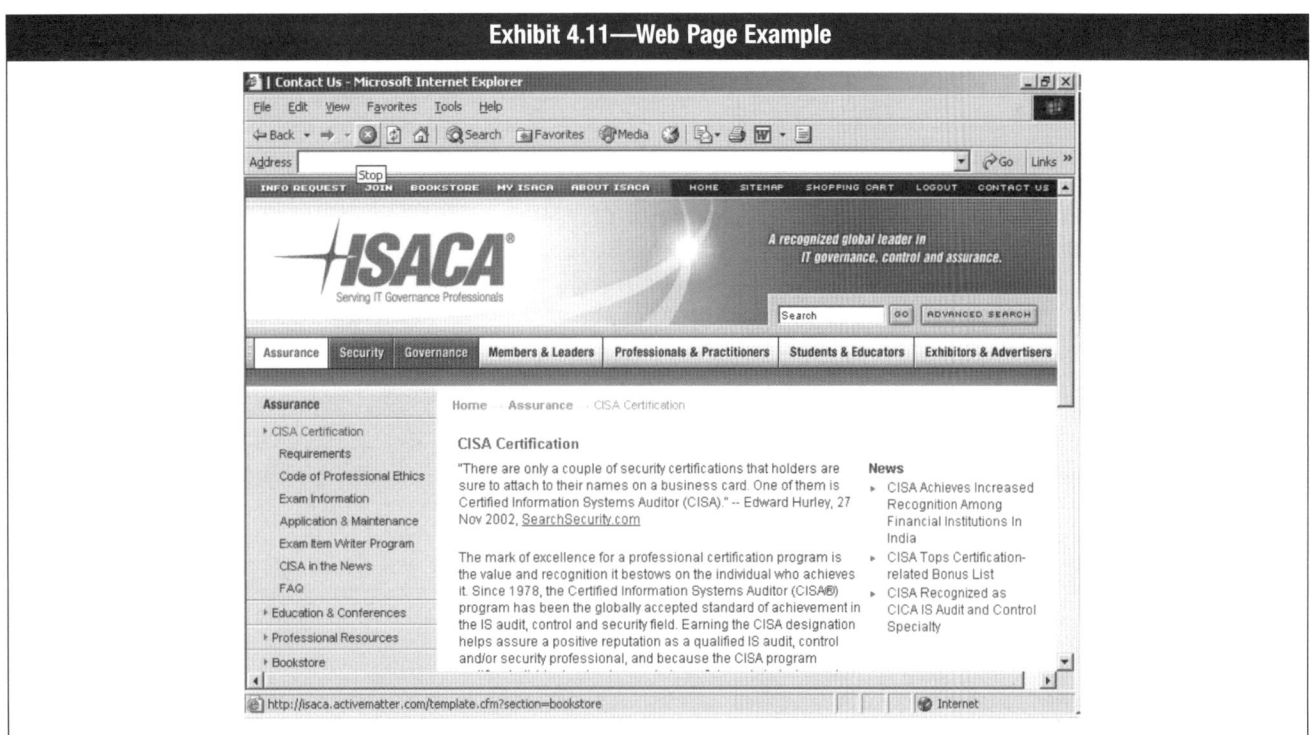

**Exhibit 4.11—Web Page Example**

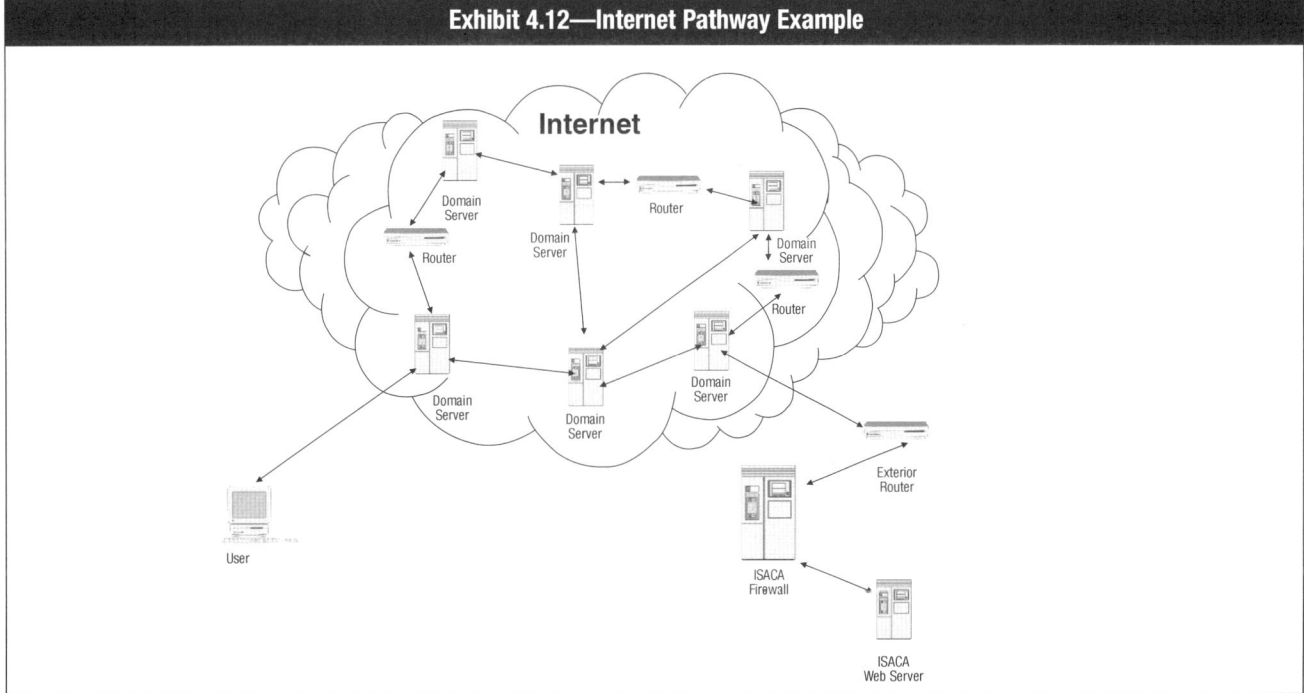

**Exhibit 4.12—Internet Pathway Example**

- **Common gateway interface scripts**—An executable, machine-independent software program run on the server that can be called and executed by a web server. CGI scripts perform a specific set of tasks, such as processing input received from a client who typed information into a form on a web page. CGI scripts are generally coded in languages such as Perl or C. Note that CGI scripts need to be closely evaluated because as they are run in the server; a bug in the scripts may allow a user to get unauthorized access to the server and, from there, eventually to the organization's network.

- **Cookie**—A message kept in the web browser for the purpose of identifying users and possibly preparing customized web pages for them. Depending on the browser, the implementation may vary, but the process is as follows. When entering (for the first time) a web site that uses cookies, the user may be asked to go through a registration process such as filling out a form that provides information, including name and interests. The web server will send back a cookie with information (text message in HTTP header), which will be kept as a text message by the browser. Afterward, whenever the user's browser requests a page from that particular server, the cookie's message is sent back to the server so that the customized view, based on that user's particular interests and preferences, can be produced. Cookies are a very important functionality because the HTTP protocol does not natively support the concept of a session. Cookies allow the web server to discern whether a known or new user is connected and to keep track of information previously sent to that user. The browser's implementation of cookies has, however, brought several security concerns, allowing breaches of security and the theft of personal information (e.g., user passwords that validate the user's identity and enable restricted web services).
- **Applets**—Programs downloaded from web servers that execute in web browsers on client machines to run any web-based application (e.g., generate web page input forms, run audio/video programs, etc.). The most common languages used for applets are Java, JavaScript and Visual Basic. Applets expose the user's machine to risks if the applets are not properly controlled by the browser. For example, the user's browser should be configured to not allow an applet to access a machine's information without prior authorization of the user.
- **Servlets**—This term typically indicates a Java applet or a small program that runs within a web server environment. A Java servlet is similar to a CGI program. But unlike a CGI program, once it is started it stays in memory and can fulfill multiple requests, thereby saving server execution time and speeding up the services.
- **Bookmark**—A marker or address that identifies a document or a specific place in a document

## GENERAL INTERNET TERMINOLOGY
The following terms are related to the use of Internet:
- **ISP**—Companies that provide the communication lines and services for connecting users (e.g., firms, universities, individuals, etc.) to the Internet
- **NAP**—These are traffic concentration spots, usually the point of convergence for Internet accesses by many Internet service providers.
- **Internet link**—The connection between the Internet users and the Internet service provider.
- **Remote Terminal Control Protocol (Telnet)**—A standard terminal emulation protocol used for remote terminal connections, enabling users to log into remote systems and use resources as if they were local. An IS auditor should note that standard Telnet traffic is not encrypted by default, and consider this risk for any production Telnet use.
- **Domain name service (DNS)**—A distributed database system that translates hostnames to IP addresses and IP addresses to hostnames, where a hostname is the name of a server on the WWW. This task is done through specialized DNS servers, which are queried via a specific protocol.
- **Direct connection**—LANs or large computers, such as mainframes, that can be directly connected to the Internet. When a LAN is connected to the Internet, all the computers on the network can have full access to the Internet.
- **Internet appliance (handheld portable device, also called Net or smart appliance)**—A low-cost Internet device, such as a PDA or a smart phone, that is built for specific functions, has a web-enabled processor, is easy to use and can be used anywhere, anytime. Such a device has a computer with limited processing power, a modem, a monitor, storage capacity and only essential features. Programs are run from an Internet server rather than from the appliance, and data are stored on the server rather than on the appliance.
- **Online services**—All of the major online services allow users to tap the full power of the Internet. No special setup is required. When users dial into the online services, they are able to use the Internet resources, including browsing the World Wide Web.
- **FTP**—A protocol that supports one of the most popular uses of the Internet, downloading files (i.e., transferring files from a computer on the Internet to the user's computer). These files can be of many types, including programs that the user can run on their computer—files with graphics, sounds and music, or text files that can be read. Most Internet files are downloaded using FTP. FTP can also be used to upload files from the computer to another computer on the Internet. To log onto an FTP site and download files, an account (or user name) and a password may need to be entered before the server or system allows the user to download or upload files. Some sites allow anyone to enter and download files. These sites are often referred to as anonymous FTP sites. As the definition suggests, anonymous FTP requires only a fictitious ID and password to transfer files. Anonymous FTP sites can be potentially dangerous if the network administrator setting

up the site does not fully understand the risks associated with anonymous FTP. If file permissions have not been specified, the anonymous FTP user could also freely upload files to the server, introducing new files or changing existing files.
• **Simple Mail Transport Protocol (SMTP)**—An Internet protocol that provides electronic mail services

## TRANSBORDER DATA FLOW
Transborder data flow refers to data transmission between two countries. Information such as e-mail, invoices, payment advice, etc. can be transmitted via suboceanic cables, telephone, television links and satellites. The selection of transmission alternatives should consider cost and possible transmission delays. The country of origin or the country of destination could have several laws applicable to transborder data flow that should be addressed. Legal compliance and protection, as well as data security and integrity, are a concern with transborder transmissions.

Privacy also is an issue since laws regarding protection and access to personal information may be different or conflicting between the source and destination countries.

Some countries also have laws concerning the encryption of data/information sent via transborder communications, thereby affecting the security and protection of data that may be exchanged between countries.

This is a particularly important issue in Internet communications, because the itinerary of the information is determined by the routers, is not fixed and, therefore, may cross a country border even while connecting two computers located on the same country.

## Network Administration and Control
Network administration ensures that the network is functioning properly from a performance and security perspective. These duties include monitoring usage and throughput, load balancing, reacting to security violations and failure conditions, saving and restoring data, and making changes for scalability as the network usage grows. Therefore, an appropriate knowledge of network structure and topology, the protocols used, and the available administration tools is required.

The software used to monitor the network and enact changes should be accessible to the network administrator only. This software is the network operating systems software associated with specific network devices, principally switches and routers.

The network operating systems provide many functions aimed at shaping the network as a unified, controlled and uniform computing environment, including:
• Supporting local and remote terminal access to hosts and servers
• Supporting sharing of common network resources, such as file and print services
• Establishing links to hosts and servers

Network operating systems have the following user-oriented features:
• Allow transparent access to the various resources of the network hosts.
• Check the user authorization to particular resources.
• Mediate and simplify the access to remote resources as easily as local resources.
• Establish uniform logon and logging procedures throughout the network.
• Make available up-to-the-minute online network documentation.
• Permit more reliable operation than possible on a single host or server, particularly when groups of equivalent hosts are used.

## NETWORK PERFORMANCE METRICS
The metrics of performance are measurements of latency and throughput (e.g., errors, retransmittals).

Latency is the delay that a message or packet will experience on its way from source to destination. Latency appears because the information needs to cross through different devices (switching and routing times) and, to a lesser extent, because signals must travel some distance (propagation delay). When a network device is busy, the packets either must wait, queued in a buffer, or have to be dropped. A very easy way to compute latency in a TCP/IP network is to use the ping command.

Chapter 4:

# IT Service Delivery and Support

Throughput is the quantity of useful work made by the system per unit of time. In telecommunications, it is the number of bytes per second that are passing through a channel.

## NETWORK MANAGEMENT ISSUES
It is much more common today to see WANs communicating with a mix of LAN and host systems network architecture (SNA) traffic, or pure LAN-oriented traffic. Almost all organizations are standardizing their telecommunications, infrastructure on TCP/IP and modern routers.

This trend to a different technical design approach is also made evident by the specific name (WAN) that designates telecommunication networks in a TCP/IP environment. A WAN needs to be monitored and managed similarly to a LAN. ISO, as part of its communications modeling effort, has defined five basic tasks related to network management:
• **Fault management**—Detects the devices that present some kind of technical fault
• **Configuration management**—Allows users to know, define and change, remotely, the configuration of any device
• **Accounting resources**—Holds the records of the resource usage in the WAN (who uses what)
• **Performance management**—Monitors usage levels and sets alarms when a threshold has been surpassed
• **Security management**—Detects suspicious traffic or users, and generates alarms accordingly

## NETWORK MANAGEMENT TOOLS
In an organization's modern inter-networking environment, all of the above tasks could be accomplished by a set of tools generically called network management tools.

The following are some of these tools:
• **Response time reports**—Identify the time necessary for a command entered by a user at a terminal to be answered by the host system. Response time is important because end users experiencing slow response time will be reluctant to utilize IS resources to their fullest extent. These reports typically identify average, worst and best response times over a given time interval for individual telecommunication lines or systems. These reports should be reviewed by IS management and system support personnel to track potential problems. If response time is slow, all possible causes, such as I/O channel bottlenecks, bandwidth utilization and CPU capacity, should be investigated; various solutions should be analyzed; and an appropriate and cost-justified corrective action should be taken.
• **Downtime reports**—Track the availability of telecommunication lines and circuits. Interruptions due to power/line failure, traffic overload, operator error or other anomalous conditions are identified on this report. If downtime is excessive, IS management should consider the following remedies:
  – Adding or replacing telecommunications lines
  – Switching to a more dependable transmission link (such as dedicated lines versus shared lines)
  – Installing backup power supplies
  – Improving access controls
  – Closely monitoring line utilization to better forecast user needs, both in the near and long term
• **Online monitors**—Check data transmission accuracy and errors. Monitoring can be done by echo checking (received data are bounced back to sender for verification) and status checking all transmissions, ensuring that messages are not lost or transmitted more than once.
• **Network monitors**—real time display of network nodes and status
• **Protocol analyzers**—Diagnostic tools attached to a network link that use network protocols' intelligence for monitoring the packets flowing along the link and produce network usage reports. Protocol analyzers are typically hardware-based and operate at the data link and/or network level. Output includes the following information:
  – Protocol(s) in use
  – The type of packets flowing along the monitored link
  – Traffic volume analysis
  – Hardware errors, noise and software problems
  – Other performance statistics (e.g., percentage of used bandwidth)
  – Problems and possible solutions
• **Simple Network Management Protocol**—This TCP/IP-based protocol monitors and controls different variables throughout the network, manages configurations, and collects statistics on performance and security. A master console polls all the network devices on a regular basis and displays the global status. Simple Network Management Protocol (SNMP) software is

capable of accepting, in real-time, specific operator requests. Based on the operator instructions, SNMP software sends specific commands to an SNMP-enabled device and retrieves the required information. To perform all of these tasks, each device (routers, switches, hubs, PCs, servers) needs to have a SNMP agent running. The actual SNMP communications occur between all the agents and the console.

- **Help desk reports**—Prepared by the help desk, which is staffed or supported by IS technicians trained to handle problems occurring during normal IS usage. If an end user encounters any problem, he/she can contact the help desk for assistance. Help desk facilities are critical to the telecommunication environment since they provide end users with an easy means of identifying and resolving problems quickly, before they have a major impact on IS performance and end-user resource utilization. Reports prepared by the help desk provide a history of the problems and their resolution.

## Applications in a Networked Environment

There are different types of applications used in a networked architecture.

### CLIENT-SERVER TECHNOLOGY

Client-server is a network architecture in which each computer or process on the network is either a server (a source of services and data) or a client (a user of these services and data that relies on servers to obtain them). In a client-server technology, the available computing power can be distributed and shared among the client workstations. Use of client-server technology is one of the most popular trends in building applications aimed at networked environments. Often, in a client-server network environment, the server provides data distribution and security functions to other computers that are independently running various applications.

The client-server architecture has a number of advantages, such as distributing the work among servers and performing as much computational work as possible on the client workstation to save bandwidth and server computing power. Important tasks, such as manipulating and changing data, may be performed locally and without the need for controlling resources on the main processing unit. In this way, the applications may run more efficiently.

To achieve these advantages, client-server application systems are divided into separate pieces or tasks. The systems are split so that processing may take place on different machines (e.g., the server and the desktop computer). Each processing component is mutually dependent on the others. This is a main difference between client-server processing and the traditional mainframe/distributed processing.

The typical client is a single PC or workstation that carries software providing a graphical interface to server computing resources. Presentation usually is provided by visually enhanced processing software, known as a graphical user interface (GUI). The server is one or more multiuser computers (these may be mainframes, minicomputers, or PCs). Server functions include any centrally supported role such as file sharing, printer sharing, database access and management, communication services, facsimile services, and application development. Multiple functions may be supported by a single server.

Client/server architecture is normally based on three levels of computing tasks (i.e., three-tier architectures).
A three-tier architecture is composed of:
- A low-power (thin) client, focused on doing GUI tasks (could be an Internet browser)
- A group (one or more) of application servers, focused on running the application logic
- A group (one or more) of database servers

This architecture does not have the limitations mentioned for two-tier applications and has other advantages such as:
- Thin clients, which are less complex and less costly to buy and maintain
- More scalability (up to several thousands of concurrent users) because the load is balanced among different servers
- Can be implemented in applications for internal usage only or in e-business applications (in this case, there could be another tier represented by the web server)
- All of the program logic is separated from the rest of the code (via application servers)

Designs that contain more than two tiers are referred to as *multi-tiered* or *n*-tiered.

Three-tier architecture applications are, however, more complex to build and more difficult to maintain than two-tier ones.

The advantages of *n*-tiered architectures is that they are far more scalable since they balance and distribute the processing load among multiple, often redundant, specialized server nodes. This, in turn, improves overall system performance and reliability since more of the processing load can be accommodated simultaneously.

Each instance of the client software can send data requests to one or more connected servers. In turn, the servers can accept these requests, process them, and return the requested information to the client. Although this concept can be applied for a variety of reasons to many different kinds of applications, the architecture remains fundamentally the same.

These days, clients are most often web browsers, although that has not always been the case. Servers typically include web servers, database servers and mail servers.

The interaction between client and server is often described using sequence diagrams. Sequence diagrams are standardized in the Unified Modeling Language.

Clients may be thick or thin. A thin client (sometimes also called a lean client) is a client computer or client software in client server architecture networks which depends primarily on the central server for processing activities, and mainly focuses on conveying input and output between the user and the remote server. In contrast, a thick or fat client does as much processing as possible and passes only data for communications and storage to the server.

Many thin client devices run only web browsers or remote desktop software, meaning that all significant processing occurs on the server. However, recent devices marketed as thin clients can run complete operating systems such as Debian GNU/Linux, qualifying them as diskless nodes or hybrid clients.

Whether the clients are thin or thick, and client server architectures are two-tiered, three-tiered or n-tiered, delivery and support are equally important to achieve the business objectives.

## Business Challenges
- The rapid growth and complexity of client/server environments challenge the ability of IT staff to effectively manage and support the environments.
- More companies are moving to client/server platforms as their mission-critical applications become more complex.
- Companies are experiencing more problems integrating application changes to existing network systems.
- High staff turnover as a result of the demand for client/server technicians adds to the cost of hiring and training, making a major impact on budgets.
- Companies that have numerous office locations and an aggressive merger/acquisition strategy do not have centrally located client/server systems. Additionally, multiple locations require support for a wide variety of operating systems, making it difficult to administer and support systems properly.

## The Solution
To address the business challenges, management has to address CobiT's 'Deliver and Support' (DS) domain. The domain is concerned with the actual delivery of required services, which includes service delivery, management of security and continuity, service support for users, and management of data and operational facilities. It typically addresses the following management questions:
- Are IT services being delivered in line with business priorities?
- Are IT costs optimized?
- Is the workforce able to use the IT systems productively and safely?
- Are adequate confidentiality, integrity and availability in place for information security?

To address the above questions, management has to address each of CobiT's Control Objectives of the following processes belonging to domain 'Delivery and Support' (DS):
DS1 Define and Manage Service Levels
DS2 Manage Third-party Services

DS3 Manage Performance and Capacity
DS4 Ensure Continuous Service
DS5 Ensure Systems Security
DS6 Identify and Allocate Costs
DS7 Educate and Train Users
DS8 Manage Service Desk and Incidents
DS9 Manage the Configuration
DS10 Manage Problems
DS11 Manage Data
DS12 Manage the Physical Environment
DS13 Manage Operations

**Note:** Implicit in two- to three-tier architectures is the presence of middleware that supports not just the communications between clients and servers, but the more advanced features such as load balancing and fail over, dynamic location of components, and establishing synchronous connections or asynchronous queue-based messages.

### MIDDLEWARE
Middleware is a client-server-specific term used to describe a unique class of software employed by client-server applications. Middleware serves as the glue between two otherwise distinct applications, and provides services such as identification, authentication, authorization, directories and security. This software resides between an application and the network, and manages the interaction between the GUI on the front end and data servers on the back end. Middleware facilitates the client-server connections over the network, and allows client applications to access and update remote databases and mainframe files.

Middleware is commonly used for:
• **Transaction processing (TP) monitors**—Programs that handle and monitor database transactions, and are used primarily for load balancing
• **Remote procedure calls (RPC)**—A protocol that enables a program on the client computer to execute another program on a remote computer (usually a server)
• **Object request broker (ORB) technology**—The use of shared, reusable business objects in a distributed computing environment. This provides the ability to support interoperability across languages and platforms, as well as enhance maintainability and adaptability of the system. Examples of such technologies are CORBA and Microsoft's COM/DCOM.
• **Messaging servers**—Programs which asynchronously prioritize, queue and/or process messages using a dedicated server

Risks and controls associated with middleware in a client-server environment are:
• **Risks**—System integrity may be adversely affected because of the very purpose of middleware, which is intended to support multiple operating environments interacting concurrently. Lack of proper software to control portability of data or programs across multiple platforms could result in a loss of data or program integrity.
• **Controls**—Management should implement compensating controls to ensure the integrity of the client-server networks. Management should ensure that systems are properly tested and approved, modifications are adequately authorized and implemented, and appropriate version control procedures are followed.

# 4.6 AUDITING INFRASTRUCTURE AND OPERATIONS

The changing technological infrastructure and the manner in which to operate it have led to evolving ways to perform audits and specific reviews of hardware, operating systems, databases, LANs, network operating controls, IS operations, lights-out, problem management reporting, hardware availability and utilization reporting, and scheduling.

## 4.6.1 HARDWARE REVIEWS

When auditing infrastructure and operations, hardware reviews should include the following:
• Review the capacity management procedures for hardware and performance evaluation procedures to determine:
  – Whether they will ensure continuous review of hardware and system software performance and capacity
  – Whether the criteria used in IS management's hardware performance monitoring plan are based on historical data and analysis obtained from the IS trouble logs, processing schedules, job accounting system reports, and preventive maintenance schedules and reports
• Review the hardware acquisition plan to determine:
  – Whether the hardware acquisition plan is compared regularly to management's business plan to ensure proper linkage and alignment with business requirements
  – Whether the environment is adequate to accommodate the currently installed hardware and new hardware to be added under the approved hardware acquisition plan
  – Whether IS management's hardware acquisition plan is in synchronization with the IS plans to identify any deficiencies in the hardware acquisition plan
  – Whether IS management's hardware acquisition plan has taken into consideration technological obsolescence of the installed equipment, as well as the new equipment in the acquisition plan
  – The adequacy of documentation for hardware and software specifications, installation requirements and the likely lead-time associated with planned acquisitions
• Review the PC acquisition criteria to determine whether:
  – IS management has issued written policy statements regarding the acquisition and use of microcomputers, and these statements have been communicated to the users
  – Criteria for the acquisition of microcomputers have been developed, and procedures and forms have been established to facilitate the acquisition approval process
  – Requests for the acquisition of microcomputers are supported by a cost-benefit analysis
  – All microcomputers are purchased through the IS purchasing department to take advantage of volume discounts or other quantity and quality benefits
• Review (hardware) change management controls to determine whether:
  – A formal change control process is in effect (documented, accepted and approved)
  – The individual responsible for scheduling was advised in a timely manner regarding changes to hardware configuration
  – IS management has developed and enforced change schedules that allow time for adequate installation and testing of new hardware
  – The operator documentation used in the IS department is appropriately revised prior to implementation of changes in hardware
  – The plans for changes are being addressed in a timely manner, by selecting a sample of hardware changes that have affected the scheduling of IS processing
  – All hardware changes have been communicated to the system programmers, application programmers and the IS staff to ensure that changes and tests are coordinated properly
  – Changes are put into place effectively so that they do not interfere with normal application production processing

## 4.6.2 OPERATING SYSTEM REVIEWS

When auditing operating software development, acquisition or maintenance, the following approach may be adopted:
• Interview technical service and other personnel regarding:
  – Review and approval process of option selection
  – Test procedures for software implementation
  – Review and approval procedures for test results
  – Implementation procedures
  – Documentation requirements
• Review system software selection procedures to determine whether they:
  – Comply with short and long range IS plans
  – Include IS processing and control requirements

- Include an overview of the capabilities of the software and control options
- Meet the IS requirements
- Ensure proper alignment with the objectives of the business
- Review the feasibility study and selection process to determine that:
  - The proposed system objectives and purposes are consistent with the request/proposal
  - The same selection criteria are applied to all proposals
- Review cost-benefit analysis of system software procedures to determine whether they have addressed:
  - Direct financial costs associated with the product
  - Cost of product maintenance
  - Hardware requirements and capacity of the product
  - Training and technical support requirements
  - Impact of the product on processing reliability
  - Impact on data security
  - Financial stability of the vendor's operations
- Review controls over the installation of changed system software to determine that:
  - All appropriate levels of software have been implemented and that predecessor updates have taken place
  - System software changes are scheduled for when they least impact IS processing
  - A written plan was established for testing changes to system software
  - Tests are being completed as planned
  - Problems encountered during testing were resolved and the changes were retested
  - Test procedures are adequate to provide reasonable assurance that changes applied to the system correct known problems and do not create new problems
  - Software will be identified before it is placed into the production environment
  - Fallback or restoration procedures are in place in case of production failure
- Review system software maintenance activities to determine that:
  - Changes made to the system software are documented
  - Current versions of the software are supported by the vendor
- Review system software change controls to determine that:
  - Access to the libraries containing the system software is limited to individual(s) needing to have such access
  - Changes to the software must be adequately documented and tested prior to implementation
  - Software are authorized properly prior to moving from the test environment to the production environment
- Review systems documentation specifically in the areas of:
  - Installation control statements
  - Parameter tables
  - Exit definitions
  - Activity logs/reports
- Review and test system software implementation to determine the adequacy of controls in:
  - Change procedures
  - Authorization procedures
  - Access security features
  - Documentation requirements
  - Documentation of system testing
    Audit trails
  - Access controls over the software in production
- Review authorization documentation to determine whether:
  - Additions, deletions, or changes to access authorization have been documented
  - The attempted violation reporting review and follow-up have been documented
- Review system software security to ensure:
  - Procedures have been established to restrict the ability to circumvent logical security access controls
  - Procedures have been established to limit access to the system interrupt capability
  - Procedures have been established to manage software patches and keep the system software up-to-date
  - Security provided by the system software is adequate
  - Existing physical and logical security provisions are adequate to restrict access to the master consoles
  - System software vendor-supplied installation passwords were changed at the time of installation

## 4.6.3 DATABASE REVIEWS

When auditing a database, an IS auditor should review the design, access, administration, interfaces, portability and database supported IS controls, as follows:
- **Design**—The existence of a database model should be verified, and all entities should have a significant name and identified primary and foreign keys. The relations should have explicit cardinality and coherent and significant names, and the business rules should be expressed in the diagram. Finally, the entity-relation model should be synchronized with the database's physical schema.

  The logical schema should be reviewed to ensure all entities in the entity-relation diagram exist as tables or views. All relations should be represented through primary or foreign keys, and all attributes should have a logical name, an indicator specifying it as a primary or foreign key, and an indicator of whether null values are allowed or not allowed. Nulls should not be allowed for primary keys, while nulls for foreign keys could be allowed as long as they are in accordance with the cardinality expressed in the entity-relation model.

  The physical schema should be reviewed for allocation of initial and extension space (storage) for tables, logs, indexes and temporary areas. Indexes by primary key and frequency of access should exist. If the database is not normalized, the justification should be reviewed.
- **Access**—The main accesses to the database as well as stored procedures and triggers should be analyzed. The use of indexes to minimize access time should be verified, and open searches, if not based on indexes, should be justified. If the DBMS allows the selection of the methods or types of indexes, the correct use should be verified.
- **Administration**—Security levels for all users and their roles should be identified within the database, and access rights for all users and/or groups of users should be justified. Backup and disaster recovery procedures should exist to assure the reliability and availability of the database. Mechanisms and procedures should be in place to assure the adequate handling of consistency and integrity during concurrent accesses.
- **Interfaces**—To ensure the integrity and confidentiality of data, information import and export procedures should be verified with other systems.
- **Portability**—Whenever possible, a Structured Query Language (SQL) should be used.
- **Database-supported IS controls**—To determine that:
  - The access to shared data is appropriate
  - The data organization is appropriate
  - Adequate change procedures are utilized to ensure the integrity of the database management software
  - The integrity of the database management system's data dictionary is maintained
  - Data redundancy is minimized by the database management system; where redundant data exists, appropriate cross-referencing is maintained within the system's data dictionary or other documentation

| PRACTICE QUESTION |
|---|
| 4-9    When conducting an audit of client-server database security, the IS auditor should be most concerned about the availability of:<br><br>     A.    system utilities.<br>     B.    application program generators.<br>     C.    systems security documentation.<br>     D.    access to stored procedures. |
| *See answers and explanations to the practice questions at the end of the chapter. (page 358)* |

## 4.6.4 NETWORK INFRASTRUCTURE AND IMPLEMENTATION REVIEWS

The IS auditor should review controls over network implementations to ensure that standards are in place for designing and selecting a network architecture, and for ensuring that the costs of procuring and operating the network does not exceed the benefits.

The unique nature of each network makes it difficult to define standard audit procedures. Modern networks are mixed with several kinds of devices and topologies (PANs, LANs, WANs, WPANs, WLANs, VLANs, etc.).

To effectively perform a review, the IS auditor should identify the following:
- Network topology and network design
- Significant network components (such as servers, routers, switches, hubs, modems, wireless devices and so forth)
- Interconnected boundary networks
- Network uses (including significant traffic types and main applications used over the network)
- Network gateway to the Internet
- Network administrator and operator
- Significant groups of network users
- In addition, the IS auditor should gain an understanding of the following:
  - Functions performed by the network administrators and operators
  - The company division or department procedures and standards relating to network design, support, naming conventions and data security
  - Network transmission media and techniques, including bridges, routers, gateways, switches and other relevant components

Understanding the above information should enable the IS auditor to make an assessment of the significant threats to the network, together with the potential impact and probability of occurrence of each threat. Having assessed the risks to the network, the IS auditor should evaluate the controls used to minimize the risks.

Physical controls should protect network components (hardware and software) and the access points by limiting access to those individuals authorized by management. Unlike most mainframes, the computers in a mixed network are usually decentralized. Company data stored on a file server are easier to damage or steal than those on a mainframe, and they should be physically protected. The IS auditor should review the following:
- Network hardware devices, particularly the file server and documentation, should be located in a secure facility and restricted to the network administrator. Other components included in the physical security review should be the wiring closet and cabling.
- Keys to the network file servers facility should be controlled to prevent or minimize the risk of unauthorized access.
- Network file servers housing should be locked or otherwise secured to prevent removal of boards, chips or the computer itself.

To test physical security, an IS auditor should perform the following:
- Inspect the network wiring closet and transmission wiring and verify they are physically secured.
- Review the electromagnetic environment protection in case wireless devices are used within the organization.
- Observe the network file server computers, and verify they are secured in a manner to reduce the risk of removal of components and the computer itself.
- Obtain a copy of the key logs for the file server room and the wiring closet, match the key logs to actual keys that have been issued, and determine that all keys held are assigned to the appropriate people, e.g., the network administrator and support staff.
- Select a sample of keys held by people without authorized access to the network file servers facility and wiring closet, and determine that these keys do not permit access to these facilities.
- Look for network operating manuals and documentation that are not properly secured.

Environmental controls for network components are similar to those considered in the computer room (see Chapter 5, Environmental Exposures and Controls). However, in the case of PAN or LAN implementations, the equipment may not require as extensive atmospheric controls as some central WAN or MAN components. The following should be considered:
- Network file server equipment should be protected from the effects of static electricity (antistatic rug) and electrical surges (surge protector).
- Air conditioning and humidity control systems should be adequate to maintain temperatures within manufacturers' specifications.
- The network components power supply should be properly controlled to ensure it remains within the manufacturer's specifications.
- The main network components should be equipped with an uninterruptible power supply (UPS) that will allow the network to operate in case of minor power fluctuations or to be brought down gracefully in case of a prolonged power outage.

- The network file server facility should be kept free of dust, smoke and other matter, particularly food.
- Backup diskettes, tapes, CD-ROMs and other offline storage devices should be protected from environmental damage.

To test environmental controls, an IS auditor should perform the following:
- Visit the server facility and verify that:
  - Temperature and humidity are adequate
  - Static electricity guards are in place
  - Electric surge protectors are in place
  - A fire suppression system is in place and tested/inspected regularly
  - Fire extinguishers are nearby and inspected regularly
  - Ensure that electromagnetic insulation is in place
  - Observe the network file servers facility, looking for food/beverage containers and tobacco products in the area, and in the garbage cans
  - Observe the storage methods and media for backup, and verify they are protected from environmental damage

Network logical security controls should be in place to restrict, identify and report authorized and unauthorized users of the network. Examples of such controls include:
- Users should be required to have unique passwords and be required to change them periodically. Passwords should be encrypted and not displayed on the computer screen when entered.
- Network user access should be based on written authorization and given on a need-to-know/need-to-do basis. This should include documenting requests for adds, changes and deletions of network logical access.
- A network workstation should be disabled automatically after a short period of inactivity.
- Remote access to the system supervisor should be prohibited.
- All logon attempts to the supervisor account should be logged on the computer system. Additionally, activities by supervisor or administrative accounts should be logged and subject to independent review.
- Up-to-date information about all communication lines connected to the outside should be maintained by the network supervisor.

To test logical security, an IS auditor should perform the following:
- Interview the person responsible for maintaining network security to ensure that the person is:
  - Aware of the risks associated with physical and logical access that must be minimized
  - Aware of the need to actively monitor logons and to account for employee changes
  - Knowledgeable in how to maintain and monitor access
- Interview users to assess their awareness of management policies regarding network security and confidentiality.
- Evaluate a sample of network users' access/security profiles to ensure access is appropriate and authorized, based on the individual's responsibilities.
- Review a sample of the security reports to:
  - Ensure only authorized access is occurring
  - Verify timely and effective review of these reports is occurring, and that there is evidence of the review
  - Look for unauthorized users. If found, determine the adequacy and timeliness of follow-up procedures.
- Attempt to gain access using a variety of unauthorized logon IDs/passwords. Verify that access is denied and logged. Log on to and briefly use the network. Then, verify that access and use are properly recorded on the automated activity report.
- If the network logon session automatically logs off after a short period of inactivity, log onto the terminal and visually verify the automatic logoff feature.
- Visually search for written passwords in the general areas of the computers that utilize the network.
- If the network is connected to an outside source through a modem, dial-up network or Internet, attempt to gain access to the network through these telecommunication media using authorized and unauthorized management.
- Review a sample of network access change requests, and determine whether they are authorized by the appropriate management and whether the standard form has been utilized.

The unique nature of each LAN makes it difficult to define standard audit procedures. To effectively perform a review, the IS auditor should identify the following:
- LAN topology and network design
- Significant LAN components (such as servers, routers, switches, hubs and modems)

- Network topology (including internal LAN configuration as well as interconnections to other LANs, WANs or public networks)
- LAN uses (including significant traffic types and main applications used over the network)
- LAN administrator
- Significant groups of LAN users

In addition, the IS auditor should gain an understanding of the following:
- Functions performed by the LAN administrator
- The company division or department procedures and standards relating to network design, support, naming conventions and data security
- LAN transmission media and techniques, including bridges, routers and gateways

Understanding the above information should enable the IS auditor to make an assessment of the significant threats to the LAN, together with the potential impact and probability of occurrence of each threat. Having assessed the risks to the LAN, the IS auditor should evaluate the controls used to minimize the risks.

| PRACTICE QUESTION |
|---|
| 4-10    When reviewing a network used for Internet communications, an IS auditor will **FIRST** examine the:<br><br>     A.    validity of password change occurrences.<br>     B.    architecture of the client-server application.<br>     C.    network architecture and design.<br>     D.    firewall protection and proxy servers. |
| *See answers and explanations to the practice questions at the end of the chapter. (page 358)* |

# 4.6.5 NETWORK OPERATING CONTROL REVIEWS

An IS auditor should review the network operations controls to determine that:
- Appropriate implementation, conversion and acceptance test plans were developed for the organization's distributed data processing network
- Implementation and testing plans for the network's hardware and communications links were established
- Operating provisions for distributed data processing networks exist to ensure consistency with the laws and regulations governing transmission of data
- Procedures to ensure that data compatibility are applied properly to all the network's datasets and that the requirements for their security have been determined
- All sensitive files/datasets in the network have been identified and that the requirements for their security have been determined
- Procedures were established to assure effective controls over the hardware and software used by the departments served by the distributed processing network
- Adequate restart and recovery mechanisms have been installed at every user location served by the distributed processing network
- The IS distributed network has been designed to assure that failure of service at any one site will have a minimal effect on the continued service to other sites served by the network
- All changes made at the user sites or by IS management to the operating systems software used by the network are controlled and can be detected promptly by the network administrator or by those responsible for the network
- Individuals have access only to authorized applications, transaction processors and datasets
- System commands affecting more than one network site are restricted to one terminal and to an authorized individual with an overall network control responsibility and security clearance
- Encryption is being used on the network to encode sensitive data
- Appropriate security policies and procedures have been implemented in one of the following environments:
  - **Highly distributed**—Is security under the control of individual user management?

– **Distributed**—Is security under the direction of user management, but adheres to the guidelines established by IS management?
– **Mixed**—Is security under the direction of individual user management, but the overall responsibility remains with IS management?
– **Centralized**—Is security under the direction of IS management, but it maintains a close relationship with user management?
– **Highly centralized**—Is security under the complete control of IS management?

## 4.6.6 IS OPERATIONS REVIEWS

Because processing environments vary among different installations, a tour of the information processing facility generally provides the IS auditor with a better understanding of operations tasks, procedures and control environment.

Audit procedures should include observation of IS personnel performing their duties to determine whether controls are in place to ensure efficiency of operations, adherence to established standards and policies, adequate supervision, IS management review, and data integrity and security.

### *Computer Operations*

Some computer operations controls relate to the day-to-day operation of hardware and software within the IS organization. Some responsibilities of the operator of the computers include mounting files located on secondary storage media, changing printer forms and discontinuing the use of devices requiring maintenance. Computer operations controls include:
• Restricting operator access capabilities such as:
  – Access to files and documentation libraries should be restricted.
  – The responsibilities to the operation of the computer and related peripheral equipment should be limited.
  – Correcting program and data problems should be restricted.
  – Access to utilities that allow system fixes to software and/or data should be restricted.
  – Access to production source code and data libraries, including run procedures, should be limited.
• Scheduling:
  – Operators should record jobs that are to be processed and their required data files.
  – Operators should schedule jobs for processing on a predetermined basis and perform them using either automated scheduling software or a manual schedule.
• Using exception-processing procedures to obtain written or electronic approval from application owners to run jobs or programs in another sequence:
  – Operators should obtain written or electronic approval from owners when scheduling on request-only jobs.
  – Operators should record all exception-processing requests.
  – Operators should review the exception-processing request log to determine the appropriateness of procedures performed.
• Executing rerun handling:
  – All re-execution of jobs should be properly authorized and logged for IS management review.
  – Procedures should be established for rerunning jobs to ensure the correct input files are used and that subsequent jobs in the sequence also are rerun if appropriate.

Computer operations audit procedures should include a review of the operator manuals to determine whether instructions are adequate to address the operation of the computer and its peripheral equipment, start-up and shutdown procedures, actions to be taken in the event of machine/program failure, records to be retained, routine job duties, and restricted activities. In addition, the IS auditor should conduct tests to determine whether procedures in the following areas are being followed in accordance with management's intent and authorization:
• Librarian access capabilities:
  – The librarian should not have computer hardware access.
  – The librarian should only have access to the tape management system.
  – Access to library facilities should be restricted to authorized staff.
  – Removal of files should be restricted by production scheduling software.
  – The librarian should handle the receipt and return of foreign media entering the library.

# IT Service Delivery and Support

– Logs of the sign-in and sign-out of data files and media should be maintained.
- Contents and location of offline storage:
  – Offline file storage media containing production system programs and data should be clearly marked as to content.
  – Offline library facilities should be located away from the computer room.
  – Audit procedures should include a review of policies and procedures for:
    · Administering the offline library
    · Checking out/in media including requirements for signature authorizations
    · Identifying, labeling, delivering and retrieving offsite backup files
    · Inventorying the system for onsite and offsite media including the specific storage locations of each tape
    · Scratching, deleting and securing disposal/destruction of media including requirements for signature authorizations

## File Handling Procedures

Procedures should be established to control the receipt and release of files and secondary storage media to/from other locations. Internal tape labels should be used to help ensure the correct media are mounted for processing. Audit procedures should include a review of these procedures to determine whether they are adequate and are in accordance with management's intent and authorization. In addition, the IS auditor should determine whether these procedures are being followed.

## Data Entry Control

Data entry controls include:
- Authorization of input documents
- Reconciliation of batch totals
- Segregation of duties between the person who keys the data and the person who reviews the keyed data for accuracy and errors

Audit procedures for data entry should include a review of the controls and the procedures to determine whether:
- Adequate controls exist
- IS personnel are adhering to the established policies
- Proper segregation of duties is being maintained
- Control reports are being produced, maintained and reviewed
- The control reports are accurate and complete
- Authorization forms are complete and contain appropriate signatures

# 4.6.7 LIGHTS-OUT OPERATIONS

Several control concerns arise from a lights-out operation. These concerns include:
- Remote access to the master console is often granted to standby operators for contingency purposes such as automated software failure. Therefore, communication access security must be sufficient to guard against unauthorized use.
- Contingency plans must allow for the proper identification of a disaster in the unattended facility. In addition, the automated operation software or manual contingency procedures must be documented and tested adequately at the recovery site.
- The application of proper program change controls and access controls, because vital IS operations are performed by software systems. Also, tests of the software should be performed on a periodic basis, especially after changes or updates are applied.
- Assurances exist that errors are not hidden by the software and that all errors result in operator notification.

# 4.6.8 PROBLEM MANAGEMENT REPORTING REVIEWS

When performing an audit of this area, the IS auditor should ensure that adequate and documented procedures have been developed to guide IS operations personnel in logging, analyzing, resolving and escalating problems in a timely manner in accordance with management's intent and authorization.

The IS auditor should perform procedures to ensure that the problem management mechanism is being maintained properly and that outstanding errors are being addressed adequately and resolved in a timely manner. These procedures include:
- Interviews with IS operations personnel
- Reviews of the procedures used by the IS department for recording, evaluating, and resolving or escalating any operating or processing problems to determine whether they are adequate for service analysis
- Reviews of the performance records to determine whether problems exist during processing
- Reviews of the reasons for delays in application program processing to determine whether they are valid
- Reviews of the procedures used by the IS department to collect statistics regarding online processing performance to determine whether the analysis is accurate and complete
- The determination that the IS department has established procedures for handling data processing problems
- The determination that all problems identified by IS operations are being recorded for verification and resolution
- The determination that significant and recurring problems have been identified, and actions are being taken to prevent their recurrence
- The determination that processing problems were resolved on a timely basis, and the resolution was complete and reasonable
- Reviews of IS management reports produced by the problem management system to ensure evidence of proper management review
- Reviews of outstanding error log entries, describing problems to be resolved to determine whether proper documentation exists and to ensure that problems are being addressed in a timely manner
- Reviews of operations documentation to ensure that procedures have been developed for the escalation of unresolved problems to a higher level of IS management
- Reviews of help desk call logs to determine if there are any reoccurring problems that are not being reported to IS management

## 4.6.9 HARDWARE AVAILABILITY AND UTILIZATION REPORTING REVIEWS

Hardware availability and utilization can be obtained from the problem log, processing schedules, job accounting system reports, preventive maintenance schedules and reports, and the hardware performance monitoring plan.

The following are some of the audit procedures that can help determine whether proper reporting of system activities occurs to ensure optimal hardware availability and utilization:
- Review the hardware performance monitoring plan, and compare it with the problem log, processing schedules, job accounting system reports, and preventive maintenance schedules and reports to determine the validity of the process.
- Review the problem log to determine whether hardware malfunctions, reruns, the use of software utilities, abnormal system terminations and operator actions have been reviewed by IS management.
- Review the preventive maintenance schedule to determine if the prescribed maintenance frequency recommended by the respective hardware vendors is being observed.
- Review the preventative maintenance schedule to verify that maintenance is not done during peak workload periods, thereby, avoiding impairment of hardware availability.
- Review the preventative maintenance schedule to determine that it is not being performed while the system is processing critical or sensitive applications.
- Review the control and management of equipment that has the ability to contact its manufacturer without manual intervention in case of equipment failure.
- Review the hardware availability and utilization reports to determine that scheduling is adequate to meet workload schedules and user requirements.
- Review the workload schedule, hardware availability and utilization reports to determine whether scheduling is sufficiently flexible to accommodate required hardware preventive maintenance.
- Determine whether IS resources are readily available for processing those application programs that require a high level of resource availability.

## 4.6.10 SCHEDULING REVIEWS

The following audit approach should be considered when reviewing workload job scheduling and personnel scheduling:
- Obtain a list of regularly scheduled applications and related information, such as input deadlines, data preparation time, estimated processing time and output deadlines, to determine whether they have been included in service agreements.
- Review the console log to determine if scheduled jobs were completed as scheduled. If processing occurs in other than the scheduled manner, analyze the reasons for validity.
- Review the schedule to determine if processing priorities have been established for each application.
- Review the individuals capable of changing job schedules or job priorities for appropriateness.
- Determine whether the scheduling of rush/rerun jobs is consistent with their assigned priority.
- Determine whether critical applications have been identified, so that they are given highest priority in the event of resource/capacity shortage.
- Determine whether scheduling procedures are used to facilitate optimal use of computer resources while meeting service requirements.
- Determine whether the number of personnel assigned to each shift is adequate to support the workload.
- Ensure the daily job schedule provides each shift of computer operators with the work to be done, the sequence in which programs are to be run and when lower-priority work can be done. At the end of a shift, each operator should pass to the work scheduler or the next shift of operators a statement of the work completed and the reasons any scheduled work was not finished. This documentation serves as the audit trail, relating to jobs scheduled for the environment.
- Review the procedures for collecting, reporting and analyzing key performance indicators, as defined in the SLAs.

# 4.7 CHAPTER 4 QUICK REFERENCE REVIEW

## Chapter 4 Quick Reference Review

Chapter 4 addresses the need for IT service delivery and support. IT service management practices are important to provide assurance to users as well as management that the expected level of service will be delivered. Service level expectations are derived from the organization's business objectives. IT service delivery includes IS operations, IT services and management of IS and the groups responsible for supporting them.

CISA candidates should have a sound understanding of the following items, not only within the context of the present chapter, but also to correctly address questions in related subject areas. It is important to keep in mind that it is not enough to know these concepts from a definitional perspective. The CISA candidate must also be able to identify which elements among those presented may represent the greatest risk and which controls are most effective at mitigating this risk.

- Key elements of IT service delivery and support include: the management of operations, architecture and software, network infrastructure and hardware. The importance of service level agreements (SLAs) established for measuring performance in each of these areas cannot be overemphasized.
- CISA candidates should understand IT delivery services which include: service-level management, IT financial management, capacity management, IT service continuity management and availability management.
- It is important to understand the process of incident handling as it relates to IT service management. Essential to this process is to prioritize items after determining their impact and urgency. Unresolved incidents should be escalated based on criteria set by management.
- CISA candidates should possess an understanding of wireless technologies, the methods for securing transmissions, and general issues and exposures related to wireless access.
- CISA candidates should be familiar with Internet services such as: URL, CGI scripts, cookies, applets, and servlets. The use of and risks associated with Telnet and FTP should also be understood. Special attention should be focused on network administration and control, including the use of network performance metrics, management issues and tools.
- Client server technology should be understood within the context of thin clients, application servers, database servers, middleware and how the interaction of these elements may result in specific risks to the organization. The CISA candidate will be expected to exercise good judgment in determining which controls would be most effective in mitigating risks inherent in a client server environment.
- Key steps to be conducted when performing reviews of operating systems, databases, network infrastructure and operations should also be understood by the CISA candidate. This may include knowing which controls are most important or most effective in ensuring a controlled environment.  ·

# 4.8 CHAPTER 4 CASE STUDIES

The following case study is included as a learning tool to reinforce the concepts introduced in this chapter. Exam candidates should note that the CISA exam does not currently use this format for testing.

## 4.8.1 CASE STUDY A

### Case Study A Scenario

The IS auditor has recently been asked to perform an external and internal network security assessment for an organization that processes health benefit claims. The organization has a complex network infrastructure with multiple local area and wireless networks, a Frame Relay network crosses international borders. Additionally, there is an Internet site that is accessed by doctors and hospitals. The Internet site has both open areas and sections containing medical claim information that requires an ID and password to access. An intranet site is also available that allows employees to check on the status of their personal medical claims and purchase prescription drugs at a discount using a credit card. The frame relay network carries unencrypted, nonsensitive statistical data that are sent to regulatory agencies but do not include any customer identifiable information. The last review of network security was performed more than five years ago. At that time, numerous exposures were noted in the areas of firewall rule management and patch management for application servers. Internet applications were also found to be susceptible to SQL injection. It should be noted that wireless access as well as the intranet portal had not been installed at the time of the last review. Since the last review, a new firewall has been installed and patch management is now controlled by a centralized mechanism for pushing patches out to all servers. Internet applications have been upgraded to take advantage of newer technologies. Additionally, an intrusion detection system has been added, and reports produced by this system are monitored on a daily basis. Traffic over the network involves a mixture of protocols, as a number of legacy systems are still in use. All sensitive network traffic traversing the Internet is first encrypted prior to being sent. Traffic on the internal local area and wireless networks is encoded in hexadecimal so that no data appear in cleartext. A number of devices also utilize Bluetooth to transmit data between PDAs and laptop computers.

| CASE STUDY A QUESTIONS |
| --- |

A1.    In performing an external network security assessment, which of the following should normally be performed **FIRST**?

        A.    Exploitation
        B.    Enumeration
        C.    Reconnaissance
        D.    Vulnerability scanning


A2.    Which of the following presents the **GREATEST** risk to the organization?

        A.    Not all traffic traversing the Internet is encrypted.
        B.    Traffic on internal networks is unencrypted.
        C.    Cross-border data flow is unencrypted.
        D.    Multiple protocols are being used.

*See answers and explanations to the practice questions at the end of the chapter. (page 359)*

## 4.8.2 CASE STUDY B

### Case Study B Scenario

The IS auditor has been asked to represent the internal audit department on a task force to define the requirements for a new branch automation project for a community bank with 17 branches. This new system would handle deposit and loan information as well as other confidential customer information. The branches are all located within the same geographic area, so the director of branch operations has suggested the use of a microwave radio system to provide connectivity due to its low cost of operation and the fact that it is a private (and not a public) network. The director has also strongly suggested that it would be preferable to provide each branch with a direct coaxial connection to the Internet (using the local cable television provider) as a backup should the microwave system develop a fault. The direct Internet connection would also be connected to a wireless access point at each branch to provide free wireless access to customers. The director also asked that each branch be provided with mail and application servers that would be administered by the administrative manager of each branch. The IS auditor was informed by the IT manager for the bank that the cable service provider will encrypt all traffic sent over the direct coaxial connection to the Internet.

---

**CASE STUDY B QUESTIONS**

B1.    In reviewing the information for this project, what would be the **MOST** important concern regarding the use of microwave radio systems based on the above scenario?

    A.    Lack of encryption
    B.    Lack of scalability
    C.    Likelihood of a service outage
    D.    Cost overruns in implementation


B2.    Which of the following would **BEST** reduce the likelihood of business systems being attacked through the wireless network?

    A.    Scanning all connected devices for malware
    B.    Placing the wireless network on a firewalled subnet
    C.    Logging all access and issuing alerts for failed logon attempts
    D.    Limiting access to regular business hours and standard protocols

*See answers and explanations to the practice questions at the end of the chapter. (page 359)*

---

# 4.9 ANSWERS TO PRACTICE QUESTIONS

4-1    **C**    Benchmarking provides a means of determining the level of performance offered by similar information processing facility environments. User satisfaction is the measure to ensure that an effective information processing operation meets user requirements. Goal accomplishment evaluates effectiveness involved in comparing performance with predefined goals. Capacity and growth planning are essential due to the importance of IT in organizations and the constant change in technology.

4-2    **C**    The first consideration in reviewing the agreement is to ensure that the business is asking for the most appropriate services to meet its business requirements. There should be evidence that the business has considered what services are required, both at present and in the future. The cost is important (choice A), since the business may be paying for levels of services that are not required or are not appropriate, but it is not of first importance. Both, audit access (choice D) and security objectives, rather than security mechanisms (choice B), are issues to be considered as part of the review, but are also not of first importance.

4-3    **A**    The FSO agreed to an inadequate SLA. To meet business needs, the FSO should renegotiate as soon as possible. It is clear that IT is meeting the required availability standard. Acquiring additional computing resources may be inefficient or cost prohibitive. Streamlining month-end closing may not be possible and/or may not affect availability.

4-4    **A**    When testing change management, the IS auditor should always start with system-generated information, containing the date and time a module was last updated, and trace from there to the documentation authorizing the change. To trace in the opposite direction would run the risk of not detecting undocumented changes. Similarly, focusing exclusively on the accuracy or completeness of the documentation examined does not ensure that all changes were, in fact, documented.

4-5    **C**    Capacity management is the planning and monitoring of computer resources to ensure that available resources are used efficiently and effectively. This does not mean that all resources must be fully utilized; full utilization (100 percent) is an indication that management should consider adding capacity. New applications will not always require new resources since existing capacity may be sufficient to accommodate them. Utilization should routinely be between 85 and 95 percent, but occasional dips are also acceptable.

4-6    **A**    The normalization means the elimination of redundant data. Therefore, the objective of normalization in relational databases is to minimize the quantum of information by eliminating redundant data in tables, quickly processing users' requests and maintaining data integrity. Maximizing the quantum of information is against the rules of normalization. If particular information is provided in different tables, the objective of data integrity may be violated because one table may be updated and not others. Normalization rules advocate storing data in only one table, therefore, minimizing the response time through faster processing of information.

4-7    **A**    VPN technology allows external partners to securely participate in the extranet using public networks as a transport or shared private network. Because of low cost, using public networks (Internet) as a transport is the principal method. VPNs rely on tunneling/encapsulation techniques, which allow the Internet Protocol (IP) to carry a variety of different protocols (e.g., SNA, IPX, NETBEUI). Client-server does not address extending the network to business partners (i.e., client-servers refers to a group of computers within an organization connected by a communications network where the client is the requesting machine and the server is the supplying machine). A network service provider may provide services to a shared private network by providing Internet services, but it does not extend an organization's intranet.

4-8    **D**    D is the correct choice since transmission charges are based on packets transmitted, not the distance or route traveled. Passwords and other data can be placed within a packet, making choice B incorrect. Choices A and C are not correct because a complete message is broken into transmission units (packets), which are routed individually through the network.

4-9    **A**    System utilities may enable unauthorized changes to be made to data on the client-server database. In an audit of database security, the controls over such utilities would be the primary concern of the IS auditor. Application program generators are an intrinsic part of client-server technology, and the IS auditor would evaluate the controls over the generators access rights to the database rather than their availability. Security documentation should be restricted to authorized security staff, but this is not a primary concern, nor is access to stored procedures.

4-10    **C**    The first step in auditing a network is to understand the network architecture and design. This would provide an overall picture of the network and its connectivity. This will be the starting point for identifying the various layers of information and the access architecture across the various layers such as proxy servers, firewalls and client-server applications. Reviewing validity of password changes would be performed as part of substantive testing.

# 4.10 ANSWERS TO CASE STUDY QUESTIONS

## ANSWERS TO CASE STUDY A QUESTIONS

A1.   **C**   Information reconnaissance should be performed first to establish the "footprint" of the target organization (e.g., Internet-facing IP address ranges) and search for any "information leakage" that would inadvertently disclose technical details about the organization's network. Such leakage can occur as the result of Internet postings in which a network administrator asks a question regarding how to correct a network problem and identifies their organization, or when a job posting requests specific experience in a certain firewall or security package. Enumeration involves mapping the network services, protocols and devices and would normally occur after the initial reconnaissance. Vulnerability scanning and exploitation would occur in the later stages of the assessment.

A2.   **B**   The internal network is used to transmit sensitive information such as patient information and credit card numbers. Since the internal network also includes wireless, these factors create a major risk when such transmissions are not encrypted. With regard to the other choices, it is not necessary that all Internet traffic be encrypted. The fact that sensitive traffic traversing the Internet is encrypted should be sufficient. Since cross-border data flow does not include any sensitive information, this does not present a significant risk. The use of multiple protocols is typical and does not present a significant risk to the organization.

## ANSWERS TO CASE STUDY B QUESTIONS

B1.   **A**   Lack of encryption is the most important concern since microwave radio systems are easy to tap. Lack of scalability and the likelihood of a service outage or cost overruns in implementation are important, but not as important as ensuring the confidentiality and integrity of customer data.

B2.   **B**   Isolating the wireless network by placing it on a firewalled subnet would best reduce the likelihood of attack. Scanning for malware would not detect the use of investigative tools designed to harvest passwords or reveal network vulnerabilities. Logging access and limiting access to normal business hours would not prevent a successful attack.

# 4.11 SUGGESTED RESOURCES FOR FURTHER STUDY

**Braag, Roberta; Mark Rhode-Ousley; Keith Strassburg;** *The Complete Reference Network Security*, McGraw Hill, **USA, 2003**

National Institute of Standards and Technology, "Security Considerations for Voice Over IP Systems," USA, 2005

Schneier, Bruce; *Secrets & Lies: Digital Security in a Networked World*, John Wiley & Sons, USA, 2004

*Note:* Publications in bold are stocked in the ISACA Bookstore. *Information Systems Control Journal* articles are available at *www.isaca.org/archives*. The articles are available online to ISACA members only during their first year of release, and then are opened to the public. All referenced *Journal* articles are available on the CISA Practice Question Database v9.

360                                                                                                    *CISA Review Manual 2009*

*Chapter 5:*

# Protection of
# Information Assets

Chapter 5:

# Protection of Information Assets

# Protection of Information Assets

# 5.1 INTRODUCTION

The chapter on Protection of Information Assets addresses the key components that ensure confidentiality, integrity and availability (CIA) of information assets. The chapter evaluates design, implementation and monitoring of logical and physical access controls to ensure CIA. The chapter also evaluates network infrastructure security, environmental controls and processes and procedures used to store, retrieve, transport and dispose of confidential information assets. The chapter describes the various methods and procedures followed by organizations and focuses on the auditor's role in evaluating these procedures.

## 5.1.1 OBJECTIVE

The objective of this area is to ensure that the CISA candidate understands and can provide assurance that the security architecture (policies, standards, procedures and controls) ensures the confidentiality, integrity and availability of information assets.

> **This area represents 31 percent of the CISA examination (approximately 62 questions).**

## 5.1.2 TASKS

There are five (5) tasks within the protection of information assets area:

T5.1    Evaluate the design, implementation and monitoring of logical access controls to ensure the confidentiality, integrity, availability and authorized use of information assets.

T5.2    Evaluate network infrastructure security to ensure confidentiality, integrity, availability and authorized use of the network and the information transmitted.

T5.3    Evaluate the design, implementation and monitoring of environmental controls to prevent or minimize loss.

T5.4    Evaluate the design, implementation and monitoring of physical access controls to ensure that information assets are adequately safeguarded.

T5.5    Evaluate the processes and procedures used to store, retrieve, transport and dispose of confidential information assets.

## 5.1.3 KNOWLEDGE STATEMENTS

There are 17 knowledge statements within the protection of information assets area:

KS5.1    Knowledge of the techniques for the design, implementation and monitoring of security (e.g., threat and risk assessment, sensitivity analysis, privacy impact assessment)

KS5.2    Knowledge of logical access controls for the identification, authentication and restriction of users to authorized functions and data (e.g., dynamic passwords, challenge/response, menus, profiles)

KS5.3    Knowledge of logical access security architectures (e.g., single sign-on, user identification strategies, identity management)

KS5.4    Knowledge of attack methods and techniques (e.g., hacking, spoofing, Trojan horses, denial of service, spamming)

KS5.5    Knowledge of processes related to monitoring and responding to security incidents (e.g., escalation procedures, emergency incident response team)

KS5.6    Knowledge of network and Internet security devices, protocols and techniques (e.g., secure sockets layer [SSL], secure electronic transaction [SET], virtual private network [VPN], network address translation [NAT])

KS5.7    Knowledge of intrusion detection systems and firewall configuration, implementation, operation and maintenance

KS5.8    Knowledge of encryption algorithm techniques (e.g., Advanced Encryption Standard [AES], RSA)

KS5.9    Knowledge of public key infrastructure (PKI) components (e.g., certification authorities, registration authorities) and digital signature techniques

KS5.10    Knowledge of virus detection tools and control techniques

KS5.11    Knowledge of security testing and assessment tools (e.g., penetration testing, vulnerability scanning)

KS5.12    Knowledge of environmental protection practices and devices (e.g., fire suppression, cooling systems, water sensors)

KS5.13    Knowledge of physical security systems and practices (e.g., biometrics, access cards, cipher locks, tokens)

KS5.14 Knowledge of data classification schemes (e.g., public, confidential, private and sensitive data)

KS5.15 Knowledge of voice communications security (e.g., voice-over IP)

KS5.16 Knowledge of the processes and procedures used to store, retrieve, transport and dispose of confidential information assets

KS5.17 Knowledge of controls and risks associated with the use of portable and wireless devices (e.g., personal digital assistants [PDAs], Universal Serial Bus [USB] devices, Bluetooth devices)

## 5.1.4 RELATIONSHIP OF TASK TO KNOWLEDGE STATEMENTS

The task statements are what the CISA candidate is expected to know how to do. The knowledge statements delineate what the CISA candidate is expected to know in order to perform the tasks. The task and knowledge statements are approximately mapped in **exhibit 5.1** insofar as it is possible to do so. Note that although there is often overlap, each task statement will generally map to several knowledge statements.

| Exhibit 5.1— Tasks and Knowledge Statements Mapping | |
|---|---|
| **Task Statements** | **Knowledge Statements** |
| T5.1 Evaluate the design, implementation and monitoring of logical access controls to ensure the confidentiality, integrity, availability and authorized use of information assets. | KS5.1 Knowledge of the techniques for the design, implementation and monitoring of security (e.g., threat and risk assessment, sensitivity analysis, privacy impact assessment) <br><br> KS5.2 Knowledge of logical access controls for the identification, authentication and restriction of users to authorized functions and data (e.g., dynamic passwords, challenge/response, menus, profiles) <br><br> KS5.3 Knowledge of logical access security architectures (e.g., single sign-on, user identification strategies, identity management) <br><br> KS5.5 Knowledge of processes related to monitoring and responding to security incidents (e.g., escalation procedures, emergency incident response team) <br><br> KS5.14 Knowledge of data classification schemes (e.g., public, confidential, private and sensitive data) <br><br> KS5.17 Knowledge of controls and risks associated with the use of portable and wireless devices (e.g., personal digital assistants [PDAs], Universal Serial Bus [USB] devices, Bluetooth devices) |
| T5.2 Evaluate network infrastructure security to ensure confidentiality, integrity, availability and authorized use of the network and the information transmitted. | KS5.1 Knowledge of the techniques for the design, implementation and monitoring of security (e.g., threat and risk assessment, sensitivity analysis, privacy impact assessment) <br><br> KS5.4 Knowledge of attack methods and techniques (e.g., hacking, spoofing, Trojan horses, denial of service, spamming) <br><br> KS5.5 Knowledge of processes related to monitoring and responding to security incidents (e.g., escalation procedures, emergency incident response team) <br><br> KS5.6 Knowledge of network and Internet security devices, protocols and techniques (e.g., secure sockets layer [SSL], secure electronic transaction [SET], virtual private network [VPN], network address translation [NAT]) |

| Exhibit 5.1— Tasks and Knowledge Statements Mapping | |
|---|---|
| **Task Statements** | **Knowledge Statements** |
| T5.2 (*cont.*) | KS5.7 Knowledge of intrusion detection systems and firewall configuration, implementation, operation and maintenance<br><br>KS5.8 Knowledge of encryption algorithm techniques (e.g., Advanced Encryption Standard [AES], RSA)<br><br>KS5.9 Knowledge of public key infrastructure (PKI) components (e.g., certification authorities, registration authorities) and digital signature techniques<br><br>KS5.10 Knowledge of virus detection tools and control techniques<br><br>KS5.11 Knowledge of security testing and assessment tools (e.g., penetration testing, vulnerability scanning)<br><br>KS5.15 Knowledge of voice communications security (e.g., voice-over IP)<br><br>KS5.17 Knowledge of controls and risks associated with the use of portable and wireless devices (e.g., personal digital assistants [PDAs], Universal Serial Bus [USB] devices, Bluetooth devices) |
| T5.3 Evaluate the design, implementation and monitoring of environmental controls to prevent or minimize loss. | KS5.1 Knowledge of the techniques for the design, implementation and monitoring of security (e.g., threat and risk assessment, sensitivity analysis, privacy impact assessment)<br><br>KS5.12 Knowledge of environmental protection practices and devices (e.g., fire suppression, cooling systems, water sensors) |
| T5.4 Evaluate the design, implementation and monitoring of physical access controls to ensure that information assets are adequately safeguarded. | KS5.1 Knowledge of the techniques for the design, implementation and monitoring of security (e.g., threat and risk assessment, sensitivity analysis, privacy impact assessment)<br><br>KS5.5 Knowledge of processes related to monitoring and responding to security incidents (e.g., escalation procedures, emergency incident response team)<br><br>KS5.13 Knowledge of physical security systems and practices (e.g., biometrics, access cards, cipher locks, tokens) |
| T5.5 Evaluate the processes and procedures used to store, retrieve, transport and dispose of confidential information assets. | KS5.1 Knowledge of the techniques for the design, implementation and monitoring of security (e.g., threat and risk assessment, sensitivity analysis, privacy impact assessment)<br><br>KS5.8 Knowledge of encryption algorithm techniques (e.g., Advanced Encryption Standard [AES], RSA)<br><br>KS5.14 Knowledge of data classification schemes (e.g., public, confidential, private and sensitive data)<br><br>KS5.16 Knowledge of the processes and procedures used to store, retrieve, transport and dispose of confidential information assets<br><br>KS5.17 Knowledge of controls and risks associated with the use of portable and wireless devices (e.g., personal digital assistants [PDAs], Universal Serial Bus [USB] devices, Bluetooth devices) |

# 5.2 IMPORTANCE OF INFORMATION SECURITY MANAGEMENT

The most critical factor in protecting information assets and privacy is laying the foundation for effective information security management. The advent of electronic trading through service providers and directly with customers; the loss of organizational barriers through use of remote access facilities; and high-profile security exposures, such as viruses, denial of service (DoS) attacks, intrusions, unauthorized access, disclosures and identity theft over the Internet, have raised the profile of information and privacy risk, and the need for effective information security management.

In light of these issues, the International Organization for Standardization (ISO) published ISO 17799 as a comprehensive set of controls comprising best practices for information security management. The ISO has developed 133 controls under 39 main categories (generally referred to as objectives), which are categorized under 11 clauses within its publication titled *Information Technology Security Techniques—Code of Practice for Information Security Management*. Similarly, the IT Governance Institute provides good practices for IT processes that are defined within four domains, which include 220 controls classified under 34 high-level objectives, through its publication CoBiT 4.0. Additionally, legislation relating to information technology is becoming more prolific, with many countries enacting laws on issues such as copyright and software piracy, privacy, digital signatures and intellectual property. These commercial, competitive and legislative pressures require the implementation of proper security policies, procedures and related system access controls, monitoring, and review.

Security objectives to meet organization's business requirements include the following:
• Ensure the continued availability of their information systems.
• Ensure the integrity of the information stored on their computer systems and while in transit.
• Preserve the confidentiality of sensitive data while stored and in transit.
• Ensure conformity to applicable laws, regulations and standards.
• Ensure adherence to trust and obligation requirements in relation to any information relating to an identified or identifiable individual (i.e., data subject) in accordance with its privacy policy or applicable privacy laws and regulations.

It is important to recognize that these objectives are necessary, but not sufficient since patterns often come into play regarding the objective of retaining competitive advantage. However, this section will not deal with ways and approaches to maintain such advantage, but rather how to protect the IT systems from security pitfalls affecting their contents or operation.

With regard to data integrity as it relates to security objectives, it should be noted that the data/information integrity concept refers generally to the accuracy, completeness, consistency (or neutrality), validity and verifiability of the data once they have been input or loaded on the system (see chapter 4, IT Service Delivery and Support, on system usage and for ways of validating information to be fed into the system). Integrity refers to reliability of the data. Information relevance is an afterthought.

## 5.2.1 KEY ELEMENTS OF INFORMATION SECURITY MANAGEMENT

Security failures can be costly to business. Losses may be suffered as a result of the failure itself or costs may be incurred when recovering from the incident, followed by more costs to secure systems and prevent further failure.

An IT system packed with security features and devices will not be protected unless it is properly implemented and managed and carefully operated, monitored and reviewed; just as the best safety features of a vehicle might prove useless if the driver is careless or if that vehicle is not properly maintained. Security objectives cannot be met by simply affecting technical and procedural protections. An educated security attitude and attention by all employees, management as well as external service providers and external trusted IT users/partners are vital to the achievement of security objectives. IS security is more than just a mechanism. IS security also includes cultural aspects that must be embraced by all individuals within an organization for IS security to be effective.

> **Note:** For a detailed overview of Information Security Governance, please refer to that section in chapter 2 on IT governance.

Related key elements include the following:
- **Senior management commitment and support**—Commitment and support from senior management are important for successful establishment and continuance of an information security management program.
- **Policies and procedures**—The policy framework should be established with a concise top management declaration of direction, addressing the value of information assets, the need for security, and the importance of defining a hierarchy of classes of sensitive and critical assets. After approval by the governing body of the organization and by related roles and responsibilities, the information security program will be substantiated with procedures, standards to develop minimum security baselines, measurement criteria and methods, specific guidelines, and practices. These elements are applied to the resources—information assets, people, systems and IT infrastructure. The policy should ensure resource conformity with laws and regulations, integrity, confidentiality, and availability of the data. Security policies and procedures must be up to date and reflect business objectives, generally accepted security standards and practices.

  The development of the IS security policy, which provides the framework for designing and developing logical and physical access controls, is the responsibility of the top level of management in an organization which delegates its implementation to the appropriate level of management with required control such as audits, monitoring of security processes and technical vulnerability assessments. The policy contributes to the protection of information assets. Its objective is to protect the information capital against all types of risks: accidental, intentional or acts of God. An existing and enforced IS security policy is of paramount importance for an organization's survival and controlled development. The policy should ensure systems conformity with laws and regulations, integrity, confidentiality, and availability of the data.
- **Organization**—Responsibilities for the protection of individual assets should be clearly defined. The information security policy should provide general guidance on the allocation of security roles and responsibilities in the organization and, where necessary, detailed guidance for specific sites, assets, services and related security processes, such as IT recovery and business continuity planning.
- **Security awareness and education**—All employees of an organization and, where relevant, third-party users should receive appropriate training and regular updates to foster security awareness and compliance with written security policies and procedures. For new employees, this training should occur before access to information or service is granted. A number of different mechanisms available for raising security awareness include:
  - Regular updates to written security policies and procedures
  - Formal information security training
  - Internal certification program for relevant personnel
  - Statements signed by employees agreeing to follow the written security policy and procedures, including nondisclosure obligations
  - Use of appropriate publication media for distribution of security-related material (e.g., company newsletter, web page, videos, etc.)
  - Visible enforcement of security rules
  - Security drills and simulated security incidents
  - Rewarding employees who report suspicious events
  - Periodic audits
- **Monitoring and compliance**—IS auditors are usually charged to assess, on a regular basis, the effectiveness of an organization's security program(s). To fulfill this task, they must have an understanding of the protection schemes, the security framework and the related issues, including compliance with applicable laws and regulations. As an example, these issues may relate to organizational due diligence for security and privacy of sensitive information, particularly as it relates to specific industries (e.g., banking and financial institutions, health care).
- **Incident handling and response**—A computer security incident is an event adversely affecting the processing of computer usage. This includes loss of confidentiality of information, compromise of integrity of information, denial of service, unauthorized access to systems, misuse of systems or information, theft and damage to systems. Other incidents include virus attacks and intrusion by humans within or outside the organization.

## 5.2.2 INFORMATION SECURITY MANAGEMENT ROLES AND RESPONSIBILITIES

All defined and documented responsibilities and accountabilities must be established and communicated to all relevant personnel and management. Responsibilities to consider by position include:
• **IS security steering committee**—Security policies, guidelines and procedures affect the entire organization and, as such, should have the support and suggestions of end users, executive management, auditors, security administration, IS personnel and legal counsel. Therefore, individuals representing various management levels should meet as a committee to discuss these issues, and establish and approve security practices. The committee should be formally established with appropriate terms of reference.
• **Executive management**—Responsible for the overall protection of information assets, and for issuing and maintaining the policy framework
• **Security advisory group**—Responsible for the review of the security plans of the organization. This group should include people involved in the business. The group should provide comments on security issues to the chief security officer and communicate to the business whether the security programs meet the business objectives.
• **Chief privacy officer (CPO)**—A senior level corporate official responsible for articulating and enforcing the policies that companies use to protect their customers' and employees' privacy rights
• **Chief information security officer (CISO)**—A senior level corporate official responsible for articulating and enforcing the policies that companies use to protect their information assets. This is a much broader roll than a chief security officer (CSO), who is normally only responsible for physical security within the organization.
• **Process owners**—Ensure appropriate security measures are consistent with organizational policy and are maintained
• **Information assets owners and data owners**—Ownership entails responsibility for the owned asset
• **Users**—Follow procedures set out in the organization's security policy and adhere to privacy and security regulations, which are often specific to sensitive application fields (e.g., health, finance, legal, etc.).
• **External parties**—Include third-party providers and trading partners that deal with the organization's information assets or have access to company premises
• **Security administrator**—Staff level position responsible for providing adequate physical and logical security for IS programs, data and equipment. Normally, the information security policies will provide the basic guidelines under which the security administrator will operate.
• **Security specialists/advisors**—Assist with the design, implementation, management and review of the organization's security policy, standards and procedures.
• **IT developers**—Implement information security within their applications
• **IS auditors**—Provide independent assurance to management on the appropriateness and effectiveness of information security objectives and the controls related to these objectives

## 5.2.3 INVENTORY AND CLASSIFICATION OF INFORMATION ASSETS

Effective control requires a detailed inventory of information assets. Such a list is the first step in classifying the assets and determining the level of protection to be provided to each asset.

The inventory record of each information asset should include:
• A clear and distinct identification of the asset
• Its relative value to the organization
• Its location
• Its security/risk classification
• Its asset group (where the asset forms part of a larger information system)
• Its owner
• Its designated custodian

Information assets have varying degrees of sensitivity and criticality in meeting business objectives. By assigning classes or levels of sensitivity and criticality to information resources and establishing specific security rules for each class, it is

possible to define the level of access controls that should be applied to each information asset. Classification of information assets reduces the risk and cost of over- or under-protecting information resources in tying security in with business objectives since it helps to build and maintain a consistent and homogeneous perspective of the security requirements for information assets throughout the organization. ISO 17799 recommends the inventorying of information assets and the classification, handling and labeling of information in accordance with preset guidelines.

Classifications should be simple such as designations by differing degrees for sensitivity and criticality. End-user managers and security administrators can then use these classifications in their risk assessment process to assist with determining who should be able to access what, and the most appropriate level of such access. Most organizations use a classification scheme with three to five levels of sensitivity. The number of classification categories should take into consideration the size and nature of the organization and the fact that complex schemes may become too impractical to use.

Data should always be treated as a core IS asset. Categorizing data is a major part of he task of classifying all information assets. Data classification as a control measure should define:
• the owner of the information asset
• who has access rights (need to know)
• the level of access to be granted
• which individual is responsible for determining the access rights and access levels
• What approvals are needed for access
• The extent and depth of security controls

For instance, application programmers or system development programmers should not have access to production data or programs.

Data classification must take into account legal/regulatory/contractual/internal requirements for maintaining privacy, confidentiality, integrity and availability of information. Data classification is also useful to identify who should have access to the production data used to run the business versus those who are permitted to access test data and programs under development.

Adopting a classification scheme and assigning the information to one sensitivity level enables uniform treatment of data, through applying level-specific policies and procedures rather than addressing each type of information. It is highly difficult to follow information security policies if documents and media are not assigned to a sensitivity level and users are not instructed how to deal with each piece of information. If documents or media are not labeled according to a classification scheme, this is an indicator of a potential misuse of information. Users might reveal confidential information because they did not know that the requirements prohibited disclosure. Social engineering relies on this kind of misunderstanding at the end user level.

## 5.2.4 SYSTEM ACCESS PERMISSION

System access permission is the prerogative to act on a computer resource. This usually refers to a technical privilege, for example, the ability to read, create, modify or delete a file or data; execute a program; or open or use an external connection.

System access to computerized information resources is established, managed and controlled at the physical and/or logical level. Physical system access controls restrict the entry and exit of personnel to an area such as an office building, suite, data center or room containing information processing equipment such as a local area network (LAN) server. There are many types of physical access controls including badges, memory cards, guard keys, true floor-to-ceiling wall construction fences, locks and biometrics. Logical system access controls restrict the logical resources of the system (transactions, data, programs, applications) and are applied when the subject resource is needed. On the basis of identification and authentication of the user that requires a given resource and by analyzing the security profiles of the user and the resource, it is possible to determine if the requested access is to be allowed, i.e., what information users can utilize, the programs or transactions they can run, and the modifications they can make. Such controls may be built into the operating system, invoked through separate access control software and incorporated into application programs, database systems, network control devices and utilities (e.g., real-time performance monitors).

Physical or logical system access to any computerized information should be on a documented need-to-know basis where there is a legitimate business requirement based on least privilege and segregation of duties. These principles should be used by IS auditors when they evaluate the appropriateness of criteria for defining permissions and granting security privileges. Organizations should establish such basic criteria for assigning technical access to specific data, programs, devices and resources, including who will have access and what level of access they will be allowed. For instance, it may be desirable for everyone in the organization to have access to specific information on the system such as the data displayed on an organization's daily calendar of nonconfidential meetings. The program that formats and displays the calendar, however, might be modifiable by only a few system administrators, while the operating system controlling that program might be directly accessible by still fewer.

The IT assets under logical security can be grouped in four layers—networks, platforms (operating systems), databases and applications. This concept of layered security for system access provides greater scope and granularity of control to information resources. For example, network and platform layers provide pervasive general systems control over users authenticating into systems, system software and application configurations, data sets, load libraries, and any production data set libraries. Database and application controls generally provide a greater degree of control over user activity within a particular business process by controlling access to records, specific data fields and transactions.

The information owner or manager who is responsible for the accurate use and reporting of information should provide written authorization for users or defined roles to gain access to information resources under their control. The manager should hand over this documentation directly to the security administrator to ensure that mishandling or alteration of the authorization does not occur.

Logical access capabilities are implemented by security administration in a set of access rules that stipulate which users (or groups of users) are authorized to access a resource at a particular level (e.g., read-, update- or execute-only) and under which conditions (e.g., time of the day or a subset of computer terminals). The security administrator invokes the appropriate system access control mechanism upon receipt of a proper authorization request from the information owner or manager to grant a specified user the rights for access to, or use of, a protected resource. The IS auditor should be aware that access is granted to the organization's information systems utilizing the principles of need-to-know, least privilege and segregation of duties.

Reviews of access authorization should be evaluated regularly to ensure that they are still valid. Personnel and departmental changes, malicious efforts, and just plain carelessness result in authorization creep and can impact the effectiveness of access controls. Many times, when personnel leave their organization, their systems access is not removed, which increases the risk of unauthorized access. For this reason, the information asset owner should review access controls periodically with a predetermined authorization matrix that defines the least-privileged access level and authority for an individual/role with reference to his/her job roles and responsibilities. Any access exceeding the access philosophy in authorized matrix or in actual access levels granted on a system should be updated and changed accordingly. One of the best practices is to integrate the review of access rights with human resource processes. When an employee transfers to a different function, i.e., promotions, lateral transfers or demotions, access rights are adjusted at the same time. Development of a security-conscious culture increases the effectiveness of access controls.

Nonemployees with access to corporate IS resources should also be held responsible for security compliance and be accountable for security breaches. Nonemployees include contract employees, vendor programmers/analysts, maintenance personnel, clients, auditors, visitors and consultants. It should be understood that nonemployees are also accountable to the organization's security requirements.

| PRACTICE QUESTION |
|---|
| **5-1**    A utility is available to update critical tables in case of data inconsistency. This utility can be executed at the operating system (OS) prompt or as one of the menu options in an application. The **BEST** control to mitigate the risk of unauthorized manipulation of data is to:<br><br>    A.   delete the utility software and install it as and when required.<br>    B.   provide access to the utility on a need-to-use basis.<br>    C.   provide access to the utility to user management.<br>    D.   define access so that the utility can be executed only in the menu option. |
| *See answers and explanations to the practice questions at the end of the chapter. (page 469)* |

## 5.2.5 MANDATORY AND DISCRETIONARY ACCESS CONTROLS

Mandatory access controls (MACs) are logical access control filters used to validate access credentials that cannot be controlled or modified by normal users or data owners; they act by default. Conversely, those controls that may be configured or modified by the users or data owners are called discretionary access controls (DACs).

MACs are a good choice to enforce a ground level of critical security without possible exception, if this is required by corporate security policies or other security rules. A MAC could be carried out by comparing the sensitivity of the information resources, such as files, data or storage devices, kept on a user-unmodifiable tag attached to the security object with the security clearance of the accessing entity such as a user or an application. With MACs, only administrators may make decisions that are derived from policy. Only an administrator may change the category of a resource, and no one may grant a right of access that is explicitly forbidden in the access control policy. MACs are prohibitive; anything that is not expressly permitted is forbidden.

DACs are a protection that may be activated or modified by the data owner, at his/her discretion. This would be the case of data owner-defined sharing of information resources, where the data owner may select who will be enabled to access his/her resource and the security level of this access. DACs cannot override MACs; DACs act as an additional filter, prohibiting still more access with the same exclusionary principle.

When information systems enforce MAC policies, the systems must distinguish between MAC and the discretionary policies that offer more flexibility. This distinction must be ensured during object creation, classification downgrading and labeling.

## 5.2.6 PRIVACY MANAGEMENT ISSUES AND THE ROLE OF IS AUDITORS

Privacy means adherence to trust and obligation in relation to any information relating to an identified or identifiable individual (data subject).

Privacy is an organizationwide matter that, by its nature, requires a consistent and homogeneous approach throughout the organization. Therefore, it should be systematically built into policies, standards and procedures from the very beginning. To best meet this challenge, management should perform privacy impact analysis or assessments that IT auditors may be called in to support or review. Such assessment should:
• Pinpoint the nature of personally identifiable information associated with business processes
• Document the collection, use, disclosure and destruction of personally identifiable information
• Ensure that accountability for privacy issues exists
• Be the foundation for informed policy, operations and system design decisions based on an understanding of privacy risk and the options available for mitigating that risk

Based on the results, it should be possible to create a consistent format and structured process for analyzing technical and legal compliance with relevant regulations and internal policies. This structured process would provide a framework to ensure that privacy is considered in all IT projects, from the conceptual and requirements analysis stage to the final design approval, funding, implementation and communication stage, so that privacy compliance is built into projects rather than retrofit.

The focus and extent of privacy impact analysis or assessment may vary depending on changes in:
• Technology:
  – New programs
  – Changes in existing programs
  – Additional system linkages
  – Data warehouse
  – New products
• Processes:
  – Change management
  – Business process reengineering
  – Enhanced accessibility rules
  – New systems
  – New operations
• People:
  – Business partners
  – Vendors
  – Service providers

The IS auditor may also be called on to give assurance to the responsible managers concerning the compliance with privacy policy, laws and other regulations. To fulfill this role, the IS auditor should:
• Identify and understand legal requirements regarding privacy from laws, regulations and contract agreements
• Check whether personal data are correctly managed in respect to these requirements
• Verify that the correct security measures are adopted
• Review management's privacy policy to ascertain that it takes into consideration the requirement of applicable privacy laws and regulations. This includes transborder data flow requirements such as Safe Harbor and Organization for Economic Cooperation and Development (OECD) guidelines governing the protection of privacy and transborder flows of personal data.

In assessing applicable privacy laws and regulations that need to be complied with by any particular organization, particularly for organizations operating in a different part of the globe, IS auditors should seek expert opinion on the requirements of any laws and regulations, and should carry out the necessary compliance and substantive tests to form an opinion and report on the compliance of such laws and regulations.

# 5.2.7 CRITICAL SUCCESS FACTORS TO INFORMATION SECURITY MANAGEMENT

Managers and employees within an organization often tend to consider information security as a secondary priority if compared with their own efficiency or effectiveness matters because these have a direct and material impact on the outcome of their work.

For this reason, a strong commitment and support by the senior management on security training is needed, over and above the aforementioned role concerning the information security policy. This commitment should be supported with a comprehensive program of formal security awareness training. This may require special management-level training since security is not necessarily a part of management expertise. The security training for different functions within the organization needs to be customized to address specific security needs. Different functions have different levels of risk. For example infrastructure staff need technical security training, whereas security management requires training that will show the link between information security management and the needs of the organization.

A second vital point is that a professional risk-based approach must be used systematically to identify sensitive and critical information resources and to ensure that there is a clear understanding of threats and risks. Thereafter, appropriate risk assessment activities should be undertaken to mitigate unacceptable risks and ensure that residual risks are at an acceptable level.

# Protection of Information Assets

## 5.2.8 INFORMATION SECURITY AND EXTERNAL PARTIES

The security of the organization's information and information processing facilities that are accessed, processed, communicated to or managed by external parties should be maintained, and should not be reduced by the introduction of external party products or services. Any access to the organization's information processing facilities and processing, and communication of information by external parties should be controlled. Where there is a business need for working with external parties that may require access to the organization's information and information processing facilities, or in obtaining or providing a product and service from or to an external party, a risk assessment should be carried out to determine security implications and control requirements. Controls should be agreed to and defined in an agreement with the external party. Organizations shall gain the right to audit the implementation and operation of the resulting security controls.

These external party arrangements can include:
• Service providers such as internet service providers (ISPs), network providers, telephone services, maintenance and support services
• Managed security services
• Customers
• Outsourcing facilities and/or operations, e.g., IT systems, data collection services, call center operations
• Management and business consultants and auditors
• Developers and suppliers, e.g., of software products and IT systems
• Cleaning, catering and other outsourced support services
• Temporary personnel, student placement and other casual short-term appointments

Such agreements can help to reduce the risks associated with external parties.

## Identification of Risks Related to External Parties

The risks to the organization's information and information processing facilities from business processes involving external parties should be identified and appropriate controls implemented before granting access. Where there is a need to allow an external party access to the information processing facilities or information of an organization, a risk assessment should be carried out to identify any requirements for specific controls. The identification of risks related to external party access should take into account the following issues:

• The information processing facilities an external party is required to access
• The type of access the external party will have to the information and information processing facilities, for example:
  – Physical access, e.g., to offices, computer rooms and filing cabinets
  – Logical access, e.g., to an organization's databases and information systems
  – Network connectivity between the organization's and the external party's network(s), e.g., permanent connection and remote access
  – Whether the access is taking place onsite or offsite
• The value and sensitivity of the information involved, and its criticality for business operations
• The controls necessary to protect information that is not intended to be accessible by external parties
• The external party personnel involved in handling the organization's information
• How the organization or personnel authorized to have access can be identified, the authorization verified, and how often this needs to be reconfirmed
• The different means and controls employed by the external party when storing, processing, communicating, sharing and exchanging information
• The impact of access not being available to the external party when required, and the external party entering or receiving inaccurate or misleading information
• Practices and procedures to deal with information security incidents and potential damages, and the terms and conditions for the continuation of external party access in the case of an information security incident
• Legal and regulatory requirements and other contractual obligations relevant to the external party that should be taken into account
• How the interests of any other stakeholders may be affected by the arrangements

Access by external parties to the organization's information should not be provided until the appropriate controls have been implemented and, where feasible, a contract has been signed defining the terms and conditions for the connection or access and the working arrangement. Generally, all security requirements resulting from work with external parties or internal controls should be reflected by the agreement with the external party. It should be ensured that the external party is aware of its obligations, and accepts the responsibilities and liabilities involved in accessing, processing, communicating or managing the organization's information and information processing facilities.

External parties might put information at risk if their security management is inadequate. Controls should be identified and applied to administer external party access to information processing facilities. For example, if there is a special need for confidentiality of the information, nondisclosure agreements might be used. Organizations may face risks associated with interorganizational processes, management and communication, if a high degree of outsourcing is applied or where there are several external parties involved.

## Addressing Security When Dealing With Customers

All identified security requirements should be addressed before giving customers access to the organization's information or assets.

The following terms should be considered to address security prior to giving customers access to any of the organization's assets (depending on the type and extent of access given, not all of them may apply):
- Asset protection, including:
  - Procedures to protect the organization's assets, including information and software, and management of known vulnerabilities
  - Procedures to determine whether any compromise of the assets, e.g., loss or modification of data, has occurred
  - Integrity
  - Restrictions on copying and disclosing information
- Description of the product or service to be provided
- The different reasons, requirements and benefits for customer access
- Access control policy, covering:
  - Permitted access methods, and the control and use of unique identifiers such as user IDs and passwords
  - An authorization process for user access and privileges
  - A statement that all access that is not explicitly authorized is forbidden
  - A process for revoking access rights or interrupting the connection between systems
- Arrangements for reporting, notification and investigation of information inaccuracies (e.g., of personal details), information security incidents and security breaches
- The target level of service and unacceptable levels of service
- The right to monitor and revoke any activity related to the organization's assets
- The respective liabilities of the organization and the customer
- Responsibilities with respect to legal matters and ensuring that the legal requirements are met, e.g., data protection legislation, especially taking into account different national legal systems if the agreement involves cooperation with customers in other countries
- Intellectual property rights (IPRs) and copyright assignment, and protection of any collaborative work

The security requirements related to customers accessing organizational assets can vary considerably depending on the information processing facilities and information being accessed. These security requirements can be addressed using customer agreements that contain all identified risks and security requirements.

Agreements with external parties may also involve other parties. Agreements granting an external party access should include an allowance for designation of other eligible parties and conditions for their access and involvement.

## Addressing Security in Third-party Agreements

Agreements with third parties involving accessing, processing, communicating or managing the organization's information or information processing facilities, or adding products or services to information processing facilities should cover all relevant security requirements. The agreement should ensure that there is no misunderstanding between the organization and

# Protection of Information Assets

the third party. The organization should ensure that the agreement includes adequate indemnification provisions to protect against potential losses caused by the actions of the third party.

The following terms should be considered for inclusion in the agreement to satisfy the identified security requirements:
- The information security policy
- Controls to ensure asset protection, including:
  - Procedures to protect organizational assets, including information, software and hardware
  - Any required physical protection controls and mechanisms
  - Controls to ensure protection against malicious software
  - Procedures to determine whether any compromise of the assets, e.g., loss or modification of information, software and hardware, has occurred
  - Controls to ensure the return or destruction of information and assets at the end of or at an agreed point in time during the agreement
  - Confidentiality, integrity, availability and any other relevant property of the assets
  - Restrictions on copying and disclosing information, and using confidentiality agreements
- User and administrator training in methods, procedures and security
- A means to ensure user awareness of information security responsibilities and issues
- Provision for the transfer of personnel, where appropriate
- Responsibilities regarding hardware and software installation, and maintenance
- A clear reporting structure and agreed reporting formats
- A clear and specified process for change management
- Access control policy, covering
  - The different reasons, requirements and benefits that make the access by the third party necessary
  - Permitted access methods, and the control and use of unique identifiers such as user IDs and passwords
  - An authorization process for user access and privileges
  - A requirement to maintain a list of individuals authorized to use the services being made available, and what their rights and privileges are with respect to such use
  - A statement that all access that is not explicitly authorized is forbidden
  - A process for revoking access rights or interrupting the connection between systems
- Arrangements for reporting, notification, and investigation of information security incidents and security breaches as well as violations of the requirements stated in the agreement
- A description of the product or service to be provided, and a description of the information to be made available along with its security classification
- The target level of service and unacceptable levels of service
- The definition of verifiable performance criteria, their monitoring and reporting
- The right to monitor and revoke any activity related to the organization's assets
- The right to audit responsibilities defined in the agreement, to have those audits carried out by a third party, and to enumerate the statutory rights of auditors (and, where appropriate, the provision of a service auditor's report)
- The establishment of an escalation process for problem resolution
- Service continuity requirements, including measures for availability and reliability, in accordance with an organization's business priorities
- The respective liabilities of the parties to the agreement
- Responsibilities with respect to legal matters and ensuring that the legal requirements are met, e.g., data protection legislation, especially taking into account different national legal systems if the agreement involves cooperation with organizations in other countries
- IPRs and copyright assignment and protection of any collaborative work
- Involvement of the third party with subcontractors, and the security controls these subcontractors need to implement
- Conditions for renegotiation/termination of agreements such as:
  - A contingency plan in case either party wishes to terminate the relationship before the end of the agreements
  - A provision for renegotiation of agreements if the security requirements of the organization change
- Current documentation of asset lists, licenses, agreements or rights relating to them

In general, it is very difficult to ensure the return or destruction of confidential information disclosed to a third party at the end of the agreement. To prevent unauthorized copies or use, printed documents should be consulted on site. Using technical controls such as digital rights management (DRM) should be considered to set up the desired constraints such as the printing of the document, copying, authorized readers, or using it after a certain date.

The agreements can vary considerably for different organizations and among the different types of third parties. Therefore, care should be taken to include all identified risks and security requirements in the agreements. Where necessary, the required controls and procedures can be expanded in a security management plan.

If information security management is outsourced, the agreements should address how the third party will guarantee that adequate security, as defined by the risk assessment, will be maintained, and how security will be adapted to identify and deal with changes to risks. Some of the differences between outsourcing and the other forms of third-party service provision include the question of liability, planning the transition period and potential disruption of operations during this period, contingency planning arrangements and due diligence reviews, and collection and management of information on security incidents. Therefore, it is important that the organization plans and manages the transition to an outsourced arrangement and has suitable processes in place to manage changes and the renegotiation/termination of agreements.

The procedures for continuing processing in the event that the third party becomes unable to supply its services need to be considered in the agreement to avoid any delay in arranging replacement services. Agreements with third parties may also involve other parties. Agreements granting third-party access should include allowance for designation of other eligible parties and conditions for their access and involvement.

Generally, agreements are primarily developed by the organization. There may be occasions in some circumstances where an agreement may be developed and imposed upon an organization by a third party. The organization needs to ensure that its own security is not unnecessarily impacted by third-party requirements stipulated in imposed agreements.

## 5.2.9 HUMAN RESOURCES SECURITY AND THIRD PARTIES

Proper information security practices should be in place to ensure that employees, contractors and third-party users understand their responsibilities, and are suitable for the roles they are considered for, and to reduce the risk of theft, fraud or misuse of facilities, specifically:
• Security responsibilities should be addressed prior to employment in adequate job descriptions, and in terms and conditions of employment.
• All candidates for employment, contractors and third-party users should be adequately screened, especially for sensitive jobs.
• Employees, contractors and third-party users of information processing facilities should sign an agreement on their security roles and responsibilities, including the need to maintain confidentiality.

Security roles and responsibilities of employees, contractors and third-party users should be defined and documented in accordance with the organization's information security policy.

## Screening
Background verification checks on all candidates for employment, contractors and third-party users should be carried out in accordance with relevant laws, regulations and ethics, and proportional to the business requirements, the classification of the information to be accessed and the perceived risks. A screening process should be carried out for contractors and third-party users. Where contractors are provided through an agency, the contract with the agency should clearly specify the agency's responsibilities for screening and the notification procedures they need to follow if screening has not been completed or if the results give cause for doubt or concern. In the same way, the agreement with the third party should clearly specify all responsibilities and notification procedures for screening.

## Terms and Conditions of Employment

As part of their contractual obligation, employees, contractors and third-party users should agree and sign the terms and conditions of their employment contract, which should state their and the organization's responsibilities for information security. The terms and conditions of employment should reflect the organization's security policy in addition to clarifying and stating:

- That all employees, contractors and third-party users who are given access to sensitive information should sign a confidentiality or nondisclosure agreement prior to being given access to information processing facilities
- The employee, contractor and any other user's legal responsibilities and rights, e.g., regarding copyright laws or data protection legislation
- Responsibilities for the classification of information and management of organizational assets associated with information systems and services handled by the employee, contractor or third-party user
- Responsibilities of the employee, contractor or third-party user for the handling of information received from other companies or external parties
- Responsibilities of the organization for the handling of personal information, including personal information created as a result of, or in the course of, employment with the organization
- Responsibilities that are extended outside the organization's premises and outside normal working hours, e.g., in the case of working at home
- Actions to be taken if the employee, contractor or third-party user disregards the organization's security requirements

The organization should ensure that employees, contractors and third-party users agree to terms and conditions concerning information security appropriate to the nature and extent of access they will have to the organization's assets associated with information systems and services. Where appropriate, responsibilities contained within the terms and conditions of employment should continue for a defined period after the end of the employment.

## During Employment

Management should require employees, contractors and third-party users to apply security in accordance with the established policies and procedures of the organization. Specific responsibilities should be documented in approved job descriptions. This will help ensure that employees, contractors and third-party users are aware of information security threats and concerns, their responsibilities and liabilities, and are equipped to support organizational security policy in the course of their normal work and to reduce the risk of human error. Management responsibilities should be defined to ensure that security is applied throughout an individual's employment within the organization. An adequate level of awareness, education and training in security procedures and the correct use of information processing facilities should be provided to all employees, contractors and third-party users to minimize possible security risks. A formal disciplinary process for handling security breaches should be established.

## Termination or Change of Employment

Responsibilities should be in place to ensure that an employee, contractor or third-party user's exit from the organization is managed and that the return of all equipment and the removal of all access rights are completed. The communication of termination responsibilities should include ongoing security requirements and legal responsibilities and, where appropriate, responsibilities contained within any confidentiality agreement, and the terms and conditions of employment continuing for a defined period after the end of the employee, contractor or third-party user's employment. Responsibilities and duties still valid after termination of employment should be contained in the employee, contractor or third-party user's contracts.

## Removal of Access Rights

The access rights of all employees, contractors and third-party users to information and information processing facilities should be removed upon termination of their employment, contract or agreement, or adjusted upon change. The access rights that should be removed or adapted include physical and logical access, keys, identification cards, information processing facilities, subscriptions, and removal from any documentation that identifies them as a current member of the organization. This should include notifying partners and relevant third parties—if a departing employee has access to the third party premises. If a departing employee, contractor or third-party user has known passwords for accounts remaining active, these should be changed upon termination or change of employment, contract or agreement. Access rights for

information assets and information processing facilities should be reduced or removed before the employment terminates or changes, depending on the evaluation of risk factors such as:
• Whether the termination or change is initiated by the employee, contractor or third-party user, or by management and the reason of termination
• The current responsibilities of the employee, contractor or any other user
• The value of the assets currently accessible

Procedures should be in place to ensure that information security management is promptly informed of all employee movements, including employees leaving the organization.

## 5.2.10 COMPUTER CRIME ISSUES AND EXPOSURES

Computer systems can be used to steal money, goods, software or corporate information. Crimes can also be committed when the computer application process or data are manipulated to accept false or unauthorized transactions. There is also the simple, nontechnical method of computer crime: stealing computer equipment.

Computer crime can be performed with absolutely nothing physically being taken or stolen, and it can be done at the ease of logging in from home or a coffee shop. Simply viewing computerized data can provide an offender with enough intelligence to steal ideas or confidential information (intellectual property). In case of the systems connected to wide area networks (WANs) or the Internet, the crime scene could be anywhere in the world, making the investigation very difficult. Cyber-criminals take advantage of existing gaps in the legislation of different countries when planning cyber attacks in order to avoid prosecution.

Committing crimes that exploit the computer and the information it contains can be damaging to the reputation, morale and the continued existence of an organization. Loss of customers or market share, embarrassment to management and legal actions against the organization can be a result. Threats to business include the following:
• **Financial loss**—These losses can be direct, through loss of electronic funds, or indirect, through the costs of correcting the exposure.
• **Legal repercussions**—There are numerous privacy and human rights laws an organization should consider when developing security policies and procedures. These laws can protect the organization but also can protect the perpetrator from prosecution. In addition, not having proper security measures could expose the organization to lawsuits from investors and insurers if a significant loss occurs from a security violation. Most companies must also comply with industry-specific regulatory agencies' requirements. The IS auditor should obtain legal assistance when reviewing the legal issues associated with computer security.
• **Loss of credibility or competitive edge**—Many organizations, especially service firms such as banks, savings and loans and investment firms, need credibility and public trust to maintain a competitive edge. A security violation can damage this credibility severely, resulting in loss of business and prestige.
• **Blackmail/industrial espionage/organized crime**—By gaining access to confidential information or the means to adversely impact computer operations, a perpetrator can extort payments or services from an organization by threatening to exploit the security breach or publicly disclose the confidential information of the organization. Also, by gaining access, the perpetrator could obtain proprietary information and sell it to a competitor.
• **Disclosure of confidential, sensitive or embarrassing information**—As noted previously, such events can damage an organization's credibility and its means of conducting business. Legal or regulatory actions against the company may also be the result of disclosure.
• **Sabotage**—Some perpetrators are not looking for financial gain. They merely want to cause damage due to a dislike of the organization or for self-gratification. The recent wave of hacktivism is one, in which the perpetrators intend to sabotage for the sake of expressing their agendas—commonly, political.

It is important that the IS auditor know and understand the differences between computer crime and computer abuse to support risk analysis methodologies and related control practices. What constitutes a crime depends on the statistics of the jurisdiction and the extent to which a breach of security is a civil or criminal offense. This brings into play requirements for what the organization needs to do should a crime be suspected, i.e., protection of evidence, reporting of a crime, etc.

Perpetrators in computer crimes are often the same people who exploit physical exposures, although the skills needed to exploit logical exposures are more technical and complex. Possible perpetrators include:

- **Hackers**—Persons with the ability to explore the details of programmable systems and the knowledge to stretch or exploit their capabilities, whether ethical or not. Hackers are typically attempting to test the limits of access restrictions to prove their ability to overcome the obstacles. Some often do not access a computer with the intent of destruction, although this is often the result. Other categories of hackers include "hacktivists" and criminal hackers. As described previously, "hacktivists" refer to individuals using their skills to forward an ideological agenda, possibly breaking the law in the process, but justifying their actions for ideological reasons (end justifies the means). Finally, there are hackers more malicious in nature whose primary purpose or intent is to commit a crime through their actions for some level of personal gain or satisfaction (also referred to as crackers). The terms hack and crack are often used interchangeably.
- **Script kiddies**—Script kiddies refer to individuals who use scripts and programs written by others to perform their intrusions and are often incapable of writing similar scripts on their own.
- **Crackers**—Persons who try to break the security of, and gain access to, someone else's system without being invited to do so
- **Employees (authorized or unauthorized)**—Affiliated with the organization and given system access based on job responsibilities, these individuals can cause significant harm to an organization. Therefore, screening prospective employees through appropriate background checks is an important means of preventing computer crimes within the organization.
- **IS personnel**—These individuals have the easiest access to computerized information, since they are the custodians of this information. In addition to logical access controls, good segregation of duties and supervision help in reducing logical access violations by these individuals.
- **End users**—Personnel who often have broad knowledge of the information within the organization and have easy access to internal resources
- **Former employees**—Be wary of former employees who have left on unfavorable terms, they still may have the access if it was not immediately removed at the time of the employee's termination.
- **Interested or educated outsiders**—These may include:
  - Competitors
  - Terrorists
  - Organized criminals
  - Hackers looking for a challenge
  - Script kiddies for the purpose of curiosity, joyriding and testing their newly acquired tools/scripts and exploits
  - Crackers
  - Phreakers
- **Part-time and temporary personnel**—Remember that office cleaners often have a great deal of physical access and could perpetrate a computer crime.
- **Third parties**—Vendors, visitors, consultants or other third parties who, through projects, gain access to the organization's resources and could perpetrate a crime
- **Accidental ignorant**—Someone who unknowingly perpetrates a violation

Other examples of criminals include small-time crooks, organized crime and state-sponsored criminal activities.

Although collaboration has been improved in solving cyber crimes committed from one country to another, political issues existing between some countries might hinder an investigation. Therefore, additional preventive measures should be taken to protect information systems vulnerable to international attacks.

To better understand how computers are used in or are the target of computer crimes see **exhibit 5.2**.

Perpetrators may use one or more methods in tandem to commit a crime.

| Exhibit 5.2—Computer Crimes | | |
|---|---|---|
| **Source of the Attack** | **Target of the Attack** | **Examples** |
| Computer is the object of the crime.<br><br>Perpetrator uses another computer to launch an attack. | Specific identified computer | • Denial of service (DoS)<br>• Hacking |
| Computer is the subject of the crime.<br><br>Perpetrator uses computer to commit crime and the target is another computer. | Target may or may not be defined. Perpetrator launches the attack with no specific target in mind. | • Distributed DoS<br>• Virus |
| Computer is the tool of the crime.<br><br>Perpetrator uses computer to commit crime but the target is not the computer. | Target is data or information stored on the computer. | • Fraud<br>• Unauthorized access<br>• Phishing<br>• Installing key loggers |
| Computer symbolizes the crime.<br><br>Perpetrator lures the user of computers to get confidential information. | Target is user of the computers. | • Social engineering methods:<br>  – Phishing<br>  – Fake websites<br>  – Scam mails<br>  – Spam Mails<br>  – Fake resumes for employment |

## 5.2.11 SECURITY INCIDENT HANDLING AND RESPONSE

To minimize damage from security incidents, to recover and to learn from such incidents, a formal incident response capability should be established, and it should include the following phases:
• Planning and preparation
• Detection
• Initiation
• Recording
• Evaluation
• Containment
• Eradication
• Escalation
• Response
• Recovery
• Closure
• Reporting
• Postincident review
• Lessons learned

The organization and management of an incident response capability should be coordinated or centralized with the establishment of key roles and responsibilities. This should include:
• A coordinator who acts as the liaison to business process owners
• A director who oversees the incident response capability
• Managers who manage individual incidents
• Security specialists who detect, investigate, contain and recover from incidents
• Nonsecurity technical specialists who provide assistance based on subject-matter expertise
• Business unit leader liaisons (legal, human resources, public relations, etc.)

In establishing this process, employees and contractors are made aware of procedures for reporting the different types of incidents (e.g., security breach, threat, weakness or malfunction) that might have an impact on the security of organizational assets. They should be required to report any observed or suspected incidents as quickly as possible to the designated point of contact. The organization should establish a formal disciplinary process for dealing with those who commit security

breaches such as employees, third parties, etc. To address incidents properly, it is necessary to collect evidence as soon as possible after the occurrence. Legal advice may be needed in the process of evidence collection and protection.

Incidents occur because vulnerabilities are not addressed properly. Incident management processes should include vulnerabilities management practices. A postincident review phase should determine which vulnerabilities were not addressed and why, and input provided for improvement to the policies and procedures implemented to address vulnerabilities. Also, analyzing the cause of incidents may reveal errors in the risk analysis, indicating that the residual risks are higher than the calculated values and inappropriate countermeasures have been taken to reduce inherent risks.

Ideally, an organizational computer security incident response team (CSIRT) or computer emergency response team (CERT) should be formulated with clear lines of reporting, and responsibilities for standby support should be established. Organizational CSIRT will act as an efficient detective and corrective control. Additionally, with its members' participation and involvement in security awareness programs, exercises and workshops, it can demonstrate a preventive control.

Organizational CSIRT should also disseminate security alerts such as recent threats, security guidelines and security updates to the users and assist them in understanding the security risk of errors and omissions. Organizational CSIRT should act as single point of contact for all incidents and issues related to information security, should also respond to abuse reports pertaining to the network of its constituency.

An IS auditor should ensure that the CSIRT is actively involved with users to assist them in the mitigation of risks arising from security failures and also to prevent security incidents. Auditors should also ensure that there is a formal, documented plan and that it contains vulnerabilities identification, reporting and incident response procedures to common, security-related threats/issues, such as:
• Virus outbreak
• Web defacement
• Abuse notification
• Unauthorized access alert from audit trails
• Security attack alerts from intrusion detection systems
• Hardware/software theft
• System root compromises
• Physical security breach
• Spyware/malware/Trojans detected on PCs
• Fake defamatory information in media, including on web sites
• Forensic investigations

Additionally, automated intrusion detection systems should be in place, for example, in networks and platforms, to notify administrators in a real-time manner of a potential incident and define a process for determining the severity of the incidents and the steps to take in high-risk situations.

# 5.3 LOGICAL ACCESS

As explained in the above paragraphs, logical access controls are the primary means used to manage and protect information assets. Logical access controls enact and substantiate management-designed policies and procedures intended to protect these assets and the controls are designed to reduce risks to a level acceptable to an organization. IS auditors need to understand this relationship. In doing so, IS auditors should be able to analyze and evaluate the effectiveness of a logical access control in accomplishing information security objectives and avoiding losses resulting from exposures. These exposures can result in minor inconveniences to a total shutdown of computer functions.

*Chapter 5:*

# Protection of Information Assets

## 5.3.1 LOGICAL ACCESS EXPOSURES

Technical exposures are one type of exposure that exist from accidental or intentional exploitation of logical access control weaknesses. Intentional exploitation of technical exposures might lead to computer crime. However, all computer crimes do not exploit technical exposures.

Technical exposures are the unauthorized (intentional or unintentional) activities interfering with normal processing, such as implementation or modification of data and software, locking or misusing user services, destroying data, compromising system usability, distracting processing resources, or spying data flow or users activities at either the network, platform (operating system), database or application level. Technical exposures include:

- **Data leakage**—Involves siphoning or leaking information out of the computer. This can involve dumping files to paper, or can be as simple as stealing computer reports and tapes. Unlike product leakage, data leakage leaves the original copy, so it may go undetected.
- **Wire tapping**—Involves eavesdropping on information being transmitted over telecommunications lines
- **Trojan horses/backdoors**—Involve hiding malicious, fraudulent code in an authorized or falsely authorized computer program. This hidden code will be executed whenever the authorized program is executed. An example is a web application with a Trojan horse that enables a perpetrator to execute arbitrary OS commands in gaining unauthorized access to a web server.
- **Viruses**—Involve the insertion of malicious program code into other executable code that can self-replicate and spread from computer to computer, via sharing of removable computer media, USB removable devices, transfer of logic over telecommunication lines or direct link with an infected machine/code. A virus can harmlessly display cute messages on computer terminals, dangerously erase or alter computer files, or simply fill computer memory with junk to a point where the computer can no longer function. An added danger is that a virus may lie dormant for some time until triggered by a certain event or occurrence, such as a date or being copied a prespecified number of times, during which time the virus has silently been spreading.
- **Worms**—Destructive programs that may destroy data or use up tremendous computer and communication resources but do not replicate like viruses. Such programs do not change other programs, but can run independently and travel from machine to machine across network connections by exploiting vulnerability and application/system weaknesses. Worms also may have portions of themselves running on many different machines.
- **Spyware**—Malware, similar to viruses, such as keystroke loggers and system analyzers, that collects potentially sensitive information, such as credit card numbers, bank details etc. from the host, and transmits the information to the originator when an online connection is detected.
- **Logic bombs**—While similar to computer viruses, they do not self-replicate. The creation of logic bombs requires some specialized knowledge, as it involves programming the destruction or modification of data at a specific time in the future. However, unlike viruses or worms, logic bombs are very difficult to detect before they become active; thus, of all the computer crime schemes, they have the greatest potential for damage. "Detonation" can be timed to cause maximum damage and to take place long after the departure of the perpetrator. The logic bomb may also be used as a tool of extortion, with a ransom being demanded in exchange for disclosure of the location of the bomb.
- **Denial-of-service (DoS) attack**—Disrupts or completely denies service to legitimate users, networks, systems or other resources. The intent of any such attack usually is malicious in nature and often takes little skill because the requisite tools are readily available.
- **Computer shutdown**—Initiated through terminals or personal computers connected directly (online) or remotely (dial-up lines) to the computer. Only individuals who know a high-level logon ID usually can initiate the shutdown process, but this security measure is effective only if proper security access controls are in place for the high-level logon ID and the telecommunications connections into the computer. Some systems have proven to be vulnerable to shutting themselves down under certain conditions of overload.
- **War driving**—Involves receiving wireless data from a laptop (ideally while driving) and cracking the encryption controls to gain access or to simply eavesdrop the information being transferred over the wireless communication link
- **Piggybacking**—The act of following an authorized person through a secured door or electronically attaching to an authorized telecommunications link to intercept and possibly alter transmissions
- **Trap doors**—Exit points out of any area of authorized operating system code, used for insertion of specific logic such as program interrupts, i.e. to inspect data during processing. Trap doors may exist because of testing or maintenance reasons. Sometimes programmers insert code that allows them to bypass an operating system's integrity for the purpose of debugging

at the development time or later during maintenance and system improvements. Trapdoors should be eliminated in the final editing of code, but sometimes they are forgotten or intentionally left for future access into the system. The exposure is due to the fact that the logic inserted could be used to obtain unauthorized access rights. Additionally, logic design flaws and programming errors in complex programs may also introduce unwanted trap doors into a system. A well-known example is an incomplete parameter checking. If the system does not inspect fully the parameters of a request, an improper parameter value might cause a system fault and this could result in granting the requestor undue privileged access.

- **Asynchronous attacks**—This term indicates a very technical exposure, resulting from how the system operates in allocating resources to the different jobs in a multiprocessing environment, where several jobs are competing for computer central processing unit (CPU) resources. For example, when resources are requested and are not available at the very moment of the requests, the operating system stores these requests and, as resources become available, performs them in the order in which resources are available or according to a job's stated priority. An intruder taking advantage of this would cause a large application program running with checkpoint/restart capabilities to initiate a checkpoint restart. This is a feature that stops the program at a specified intermediate point for later restart in an orderly manner without losing data. This also means that the OS must save the copy of the computer programs and data in their current state at the checkpoint as well as several system parameters describing mode and security level of the program at the time of the stop. Individuals with access to this information might be able to gain access to the checkpoint restart copy of system parameters and change those parameters such that upon restart, the program would function at a higher-priority security level or privileged level in the computer, resulting in unauthorized access to data, other programs or the operating system. This is a very complex and technical exposure and the IS auditor may require the assistance of a network or systems administrator and/or a system software programmer/analyst to evaluate it.
- **Rounding down**—Involves drawing off small amounts of money (the rounding fraction) from a computerized transaction or account and rerouting this amount to the perpetrator's account. Since the amounts are so small, they are rarely noticed, although it would be easy to detect irregularities by summing the rounding fractions. Proper rounding should give a total close to zero.
- **Salami technique**—Involves the slicing of small amounts of money from a computerized transaction or account. It is similar to the rounding down technique. The difference between the rounding down technique and the salami technique is that in rounding down the program rounds off by the smallest money fraction. For example, in the rounding down technique, a US $1,235,954.39 transaction may be rounded to US $1,235,954.35. On the other hand, the salami technique truncates the last few digits from the transaction amount, so US $1,235,954.39 becomes US $1,235,954.30 or $1,235,954.00, depending on the algorithm/formula built into the program. In fact, other variations of the same technique are applied to rates and percentages.

## 5.3.2 SOCIAL ENGINEERING

The human factor is considered to be the weakest link in the information security chain. Social engineering is the human side of breaking into a computer system. It relies on interpersonal relations and deception. Organizations with strong technical security countermeasures, such as authentication processes, firewalls and encryption, may still fail to protect their information systems. This may happen if an employee unknowingly gives away confidential information (e.g., passwords and IP addresses) by answering questions over the phone with someone they do not know or replying to an e-mail from an unknown person. Some examples of social engineering include impersonation through telephone call, dumpster diving and shoulder surfing. The best means of defense for social engineering is an ongoing security awareness program, wherein all employees and third parties (who have access to the organization's facilities) are educated about the risks involved in falling prey to social engineering attacks.

### *Phishing*

One particular form of attack about which users should be warned is phishing. This normally takes the form of an e-mail, though it may be a personal or telephone approach, pretending to be an authorized person or organization legitimately requesting information. It may be a bank asking for confirmation of the users access codes to their Internet banking service, warning that failure to respond will result in future access being denied. The unsuspecting users provide the information and find that their bank account has been cleared of funds.

Phishing attacks may also seek to obtain apparently innocuous business information such as the individuals Social Security number (in the US) or staff number, which might subsequently be used to obtain a replacement password giving unauthorized access to company systems.

## Pharming

An attack that aims to redirect the traffic of a web site to a bogus web site. Pharming can be conducted either by changing the host's file on a victim's computer or by exploiting a vulnerability in DNS server software. DNS servers are computers responsible for resolving Internet names into their real addresses—they are the "signposts" of the Internet. Compromised DNS servers are sometimes referred to as "poisoned." In recent years both pharming and phishing have been used to steal identity information. Pharming has become a major concern to businesses hosting e-commerce and to online banking web sites. Sophisticated measures known as anti-pharming are required to protect against this serious threat. Antivirus software and spyware removal software cannot protect against pharming.

An IS auditor should review the procedures of the help desk staff for caller identification and to identify weaknesses in that mechanism. For instance, if help desk staff authenticates based on their birth date, then an attacker can obtain this information for any authorized user.

## 5.3.3 FAMILIARIZATION WITH THE ORGANIZATION'S IT ENVIRONMENT

For IS auditors to effectively assess logical access controls within their organization, they first need to gain a technical and organizational understanding of the organization's IT environment. The purpose of this is to determine which areas from a risk standpoint warrant IS auditing attention in planning current and future work. This includes reviewing the network, OS platform, database and application security layers associated with the organization's IT information systems architecture.

## 5.3.4 PATHS OF LOGICAL ACCESS

Access or points of entry to an organization's IS infrastructure can be gained through several avenues. Each avenue is subject to appropriate levels of access security.

Sometimes these paths are direct, as is the case for a PC terminal user tying directly into a mainframe. This happens when the IS environment is under direct control of the main system and when the users are locally known individuals, with well-defined access profiles. More complex is the case of a LAN, where many specific IS resources are tied to a common linking structure. The LAN resources may have different access paths/levels, normally mediated through LAN connectivity, and the network itself is considered an important IS resource at a higher access level. In today's environment it is most common to find a combination of direct/local network/remote access paths. Complexity is increased by a number of intermediate devices that act as "security doors" among the various environments and by the need of crossing low-security or totally open IT spaces such as the Internet. A frequent example of access path through common nodes is either a back-end or front-end interconnected network of systems for internally or externally based users. Front-end systems are network-based systems connecting an organization to outside, untrusted networks, such as corporate web sites, where a customer can access the web site externally to initiate transactions that connect to a proxy server application which in turn connects, for example, to a back-end database system to update a customer database. Front-end systems can also be internally based to automate business, paperless processes that tie into back-end systems in a similar manner.

## General Points of Entry

General points of entry to either front-end or back-end systems control the accesses from an organization's networking or telecommunications infrastructure into their information resources (e.g., applications, databases, facilities, networks). The approach followed is based on a client-server model. A large organization can have thousands of interconnected network servers. Connectivity in this environment needs to be controlled through a smaller set of primary domain controllers (servers), which enable a user to obtain access to specific secondary points of entry (e.g., application servers, databases, etc.).

General modes of access into this infrastructure occur through the following:
• **Network connectivity**—Access is gained by linking a PC to a segment of an organizations' network infrastructure, either through a physical or a wireless connection. At a minimum, such access requires user identification and authentication to a domain-controlling server. More specific access to a particular application or database may also require the users to identify

and authenticate themselves to that particular server (secondary point of entry). Other modes of access into the infrastructure can also occur through network management devices, such as routers and firewalls, which should be strictly controlled.

• **Remote access**—A user dials in remotely to an organization's server, which generally requires the user to identify and authenticate him/herself to the server for access to specific functions that can be performed remotely (e.g., e-mail, File Transfer Protocol [FTP], Internet or some application-specific function). Complete access to view all network resources usually requires a virtual private network (VPN), which allows a secure authentication and connection into those resources where privileges have been granted. Remote access points of entry can be extensive and should be centrally controlled where possible.

From a security standpoint, it is incumbent upon the organization to know all of the points of entry into its information resource infrastructure which, in many organizations, will not be a trivial task (e.g., thousands of remote access users). This is significant because any point of entry not appropriately controlled can potentially compromise the security of an organization's sensitive and critical information resources. IS auditors, in performing detailed network assessments and access control reviews, should determine whether all points of entry are known and should support management's effort in obtaining the resources to identify and manage all access paths.

## 5.3.5 LOGICAL ACCESS CONTROL SOFTWARE

Information technology has made it possible for computer systems to store and contain large quantities of sensitive data, increase the capability of sharing resources from one system to another, and permit many users to access the system through Internet/intranet technologies. All of these factors have made organizations' IS resources more widely and promptly accessible and available.

To protect an organization's information resources, access control software has become even more critical in assuring the confidentiality, integrity and availability of information resources. The purpose of access control software is to prevent the unauthorized access and modification to an organization's sensitive data and the use of system critical functions.

To achieve this kind of control, it is necessary to apply access controls across all layers of an organization's IS architecture, including networks, platforms or operating system, databases, and application systems. Each of them usually features some form of identification and authentication, access authorization, checking of specific information resources, and logging and reporting of user activities.

The greatest degree of protection in applying access control software against internal and external users' unauthorized access is at the network and platform/operating system levels. These systems are also referred to as general support systems, and they make up the primary infrastructure on which applications and database systems will reside.

Operating system access control software interfaces with network access control software and resides on network layer devices (e.g., routers, firewalls) that manage and control external access to organizations' networks. Additionally, operating system access control software interfaces with database and/or application systems access controls to protect system libraries and user data sets.

General operating and/or application systems access control functions include the following:
• Create or change user profiles
• Assign user identification and authentication
• Apply user logon limitation rules
• Notification concerning proper use and access prior to initial login
• Create individual accountability and auditability by logging user activities
• Establish rules for access to specific information resources (e.g., system-level application resources and data)
• Log events
• Report capabilities

Database and/or application-level access control functions include the following:
• Create or change data files and database profiles
• Verify user authorization at the application and transaction level

- Verify user authorization within the application
- Verify user authorization at the field level for changes within a database
- Verify subsystem authorization for the user at the file level
- Log database/data communications access activities for monitoring access violations

In summary, access control software is provided at different levels within an IS architecture, where each level provides a certain degree of security. Properties of such relationships are that lower layers (applications, databases) are dependent on upper, infrastructure-type layers to protect general system resources. Lower layers provide the granularity needed at the application level in segregating duties by function.

| PRACTICE QUESTION |
|---|
| 5-2    Which of the following **BEST** provides access control to payroll data being processed on a local server? |
|     A.    Logging access to personal information<br>    B.    Using separate passwords for sensitive transactions<br>    C.    Using software that restricts access rules to authorized staff<br>    D.    Restricting system access to business hours |
| *See answers and explanations to the practice questions at the end of the chapter. (page 469)* |

# 5.3.6 IDENTIFICATION AND AUTHENTICATION

Identification and authentication (I&A) in logical access control software is the process of establishing and proving one's identity. It is the process by which the system obtains from a user his/her claimed identity and the credentials needed to authenticate this identity, and validates both pieces of information.

I&A is a critical building block of computer security since it is needed for most types of access control and is necessary for establishing user accountability. User accountability requires the linking of activities on a computer system to specific individuals and, therefore, requires the system to identify users. For most systems, I&A is the first line of defense because it prevents unauthorized people (or unauthorized processes) from entering a computer system or accessing an information asset. If users are not properly identified and authenticated, particularly in today's open-system-networked environments, organizations have a higher exposure to risk of unauthorized access.

Some of I&A's more common vulnerabilities that may be exploited to gain unauthorized system access include:
- Weak authentication methods
- The potential for users to bypass the authentication mechanism
- The lack of confidentiality and integrity for the stored authentication information
- The lack of encryption for authentication and protection of information transmitted over a network
- The user's lack of knowledge on the risks associated with sharing authentication elements (e.g., passwords, security tokens)

Authentication is typically categorized as "something you know" (e.g., password), "something you have" (e.g., token card) and "something you are (or do)" (a biometric feature). These techniques can be used independently or in combination to authenticate and identify a user. For example, a single-factor technique (something you know) involves the use of the traditional logon ID and password. Something you know, such as a pin number, combined and associated with something you have, such as a token card, is known as a two-factor authentication technique. Something you are is a biometric authentication technique, such as a palm or iris biometric scan. Each of these techniques is described in detail in the following sections.

A combination of more than one method, such as token and password (or personal identification number [PIN] or token and biometric device, is referred to as "multifactor" authentication.

Identifying and authenticating are altogether separate system matters. While some types of authentication elements might, by themselves, suffice to identify a user, identification and authentication differ because of:
• Meaning
• Methods, peripherals and techniques supporting them
• Requirements in terms of secrecy and management
• Attributes. Authentication does not have attributes in itself, while an identity may have a defined validity in time and other information attached to itself.
• The fact that identity does not normally change, while authentication tokens bound to secrecy must be regularly replaced to preserve their reliability

## Logon IDs and Passwords

Logon IDs and passwords are the components of a user identification and authentication process, where the authentication is based on something you know. The computer can maintain an internal list of valid logon IDs and a corresponding set of access rules for each logon ID. These access rules are related to the computer resources. As a minimum requirement, access rules are usually specified at the operating system level (controlling access to files) or within individual application systems (controlling access to menu functions and types of data or transactions).

The logon ID should be restricted to provide individual, but not group identification. If a group of users is to be formed for interchangeability, the system usually offers the ability to attach a logon ID to a named group, with common rights. Each user gets a unique logon ID that can be identified by the system. The format of logon IDs is typically standardized.

The password provides individual authentication.

### FEATURES OF PASSWORDS

A password should be easy for the user to remember but difficult for an intruder to guess.

Initial passwords may be allocated by the security administrator or generated by the system itself. When the user logs on for the first time, the system should force a password change to improve confidentiality. Initial password assignments should be randomly generated. The ID and password should be communicated in a controlled manner to ensure that only the appropriate user receives this information. New accounts without an initial password assignment should be suspended.

If the wrong password is entered a predefined number of times, typically three, the logon ID should be automatically locked out. Locking-out may be made permanent (only the administrator may unlock the ID) or temporary (the system automatically unlocks the ID after a prespecified time period).

Users that have forgotten their password must notify the security administrator. This is the only person with sufficient privileges to reset the password and, in case this is necessary, to unlock the logon ID. The security administrator should reactivate the logon ID only after verifying the user's identification (challenge/response system), much like a bank verifies an account holder's ID before giving information over the phone (such as mother's maiden name). To verify, the security administrator should return the user's call after verifying his/her extension or calling his/her supervisor for verification.

Passwords should be one-way encrypted internally to the computer, using a sufficiently strong algorithm. This allows checking passwords without the need of recording them explicitly. To reduce the risk of an intruder gaining access to other users' logon IDs, passwords should not be displayed in any form. Passwords are normally masked on a computer screen, and they are not shown on computer reports. Passwords should not be kept on index or card files, or written on pieces of paper taped somewhere near the terminal or inside a person's desk. These are the first places a potential intruder will look.

Passwords should be changed periodically. On a regular basis (for example, every 30 days), the user should change his/her password. The frequency of changing the password should depend upon the criticality of the information access level, the nature of the organization, the IS architecture and technologies used, etc. Passwords should be changed by the user at his/her terminal/workstation, rather than at the administrator's terminal or in any location where their new password might be observed. The best method is to force the change by notifying the user prior to the password expiration date. The risk of

allowing voluntary password changes is that, generally, users will not change their passwords unless forced to do so. Password management is stronger if a history of previously used passwords is maintained by the system and their re-use prohibited for a period, such as no re-use of the last 12 passwords.

Passwords must be unique to an individual; if a password is known to more than one person, the responsibility of the user for all activity within the account cannot be enforced.

Special treatment should be applied to supervisor or administrator accounts. These accounts allow full access to the system. Normally there are a limited number of such accounts per system/authentication level. For accountability, the administrator password should be known only by one individual. On the other hand, the organization should be able to access the system in an emergency situation when the administrator is not available. To enable this, practices such as keeping the administrator password in a sealed envelope, kept in a locked cabinet and available only to top managers, should be implemented. This is sometimes referred to as a "firecall" ID.

All of the guidelines above should be formalized in a password policy and made as a mandatory requirement. An "acceptable use" policy should also include the requirement to follow the policy.

## IDENTIFICATION AND AUTHENTICATION BEST PRACTICES
Logon ID requirements include:
• Logon ID syntax should follow an internal naming rule, however this rule should be kept as confidential as the IDs themselves.
• Default system accounts, such as Guest, Administrator and Admin, should be renamed whenever technically possible.
• Logon IDs not used after a predetermined period of time should be deactivated to prevent possible misuse. This can be done automatically by the system or manually by the security administrator.
• The system should automatically disconnect a logon session if no activity has occurred for a period of time. This reduces the risk of misuse of an active logon session left unattended, because the user went to lunch, left for home, went to a meeting or otherwise forgot to log off. This is often referred to as a session time out. Regaining access should require the reentry of the authentication method, password, token etc.

The password syntax rules include:
• Ideally, passwords should be a minimum of eight characters in length. The length of the password will, at times, depend on the capability of the system being used. A passphrase is generally accepted as a more secure password.
• Passwords should require a combination of at least three of the following characteristics: alpha, numeric, upper and lower case, and special characters.
• Passwords should not be particularly identifiable with the user (such as first name, last name, spouse name, pet's name, etc.). Some organizations prohibit the use of vowels, making word association/guessing of passwords more difficult.
• The system should enforce regular password changes every 30 days and enforce regular password changes every 30 days and not permit previous password(s) to be used for at least a year after being changed.

At a minimum, the above rules should be applied to individuals with privileged system account authority (system administrators, security administrators, etc.) vs. general users. Users with privileged authority need such access in establishing and managing appropriate system configurations. However, such privileges enable the user to bypass any access control software restrictions that may exist on the system. The general rule to apply is that, the greater the degree of sensitivity of the access rights, the stricter the access controls should be.

| PRACTICE QUESTION |
|---|
| 5-3    An IS auditor has just completed a review of an organization that has a mainframe and a client-server environment where all production data reside. Which of the following weaknesses would be considered the **MOST** serious? <br><br>    A.    The security officer also serves as the database administrator. <br>    B.    Password controls are not administered over the client-server environment. <br>    C.    There is no business continuity plan for the mainframe system's noncritical applications. <br>    D.    Most local area networks (LANs) do not back up file-server-fixed disks regularly. |
| *See answers and explanations to the practice questions at the end of the chapter. (page 469)* |

## Token Devices, One-time Passwords

In a common two-factor authentication technique, the user is assigned a microprocessor-controlled smart card or a USB key, synchronized with a specific authentication device on the system. This smart card/key is set to generate unique, time-dependent, pseudo-random strings that are called "session passwords" and are recognized by the authenticating device and program. They attest that the caller is currently in possession of his/her own smart device. Each string is valid for only one logon session. Users must either physically read out and retype the string, or insert the smart card/USB key in a reader/USB slot along with typing in their own memorized password to gain access to the system. This technique involves something you have (a device subject to theft) and something you know (a personal identification number).

## Biometrics

Biometric access controls are the best means of authenticating a user's identity based on a unique, measurable attribute or trait for verifying the identity of a human being. This control restricts computer access based on a physical (something you are) or behavioral (something you do) characteristic of the user. Traditionally, biometric systems have been used very little as an access control technique. However, due to advances in hardware efficiencies and storage, biometric systems are becoming a more viable option as an access control mechanism.

Using a biometric generally involves use of a reader device that interprets the individual's biometric features before permitting authorized access. However, this is not a flawless process because certain biometric features can change (e.g., scarred fingerprints, signature irregularities and change in voice). For this reason, biometric access control systems are not all equally effective and easy to use.

Entering a user's biometric into a system occurs through an enrollment process by storing a user's particular biometric feature. This occurs through an iterative averaging process of acquiring a physical or behavioral sample, extracting unique data from the sample (converted into a mathematical code), creating an initial template, comparing new sample(s) with what has been stored and developing a final template that can be used to authenticate the user. Subsequent samples will be used in determining whether a match or nonmatch condition exists for granting access.

Performance of biometric control devices is determined through three quantitative measures (percentage based) to rank the level of matching accuracy. One type, the false-rejection rate (FRR), or type-I error rate, is the number of times an individual granted authority to use the system is falsely rejected by the system. An aggregate measure of type-I error rates is the failure-to-enroll rate (FER), the proportion of people who fail to be enrolled successfully. The other, referred to as the false-acceptance rate (FAR), or type-II error rate, is the number of times an individual not granted authority to use a system is falsely accepted by the system. Each biometric system may be adjusted to lower FRR or FER, but as a general rule when one decreases, the other increases (and *vice versa*), and there is an adjustment point where the two errors are equal. An overall metric related to the two error types is the equal error rate (EER), which is the percent showing when false rejection and acceptance are equal. The lower the overall measure, the more effective the biometric.

Generally, the ordering of biometric devices with the best response times and lowest EERs are palm, hand, iris, retina, fingerprint and voice, respectively.

An analysis of each physically oriented biometric is shown below:
1. **Palm**—Analyzes physical characteristics associated with the palm such as ridges and valleys found on the palm. Use of this biometric entails placing the hand on a scanner where physical characteristics are captured.
2. **Hand geometry**—As one of the oldest biometric techniques, hand geometry is concerned with measuring the physical characteristics of the users' hands and fingers from a three-dimensional perspective. The user places his hand, palm-down, on a metal surface with five guidance pegs to ensure that fingers are placed properly and in the correct hand position. The template is built from measurements of physical geometric characteristics of a person's hand (usually 90 measurements) for example, length, width, thickness and surface area.

Advantages of these systems are the social acceptance that they have received as well as the very little computer storage space that is required for the template, generally 10-20 bytes. The main disadvantage compared to other biometrics methods is the lack of uniqueness of hand geometry data. Moreover, an injury to the hand may cause the measurements to change, resulting in recognition problems.

3. **Iris**—An iris, which has patterns associated with the colored portions surrounding the pupils, is unique for every individual and, therefore, a viable method for user identification. To capture this information, the user is asked to center his/her eye onto a device by seeing the reflection of their iris in the device. Upon this alignment occurring, a camera takes a picture of the user's iris and compares it with a stored image. The iris is stable over time, having over 400 characteristics, although only approximately 260 of these are used to generate the template. As is the case with fingerprint scanning, the template carries less information than a high-quality image.

The key advantage to iris identification is that contact with the device is not needed, which contrasts with other forms of identification such as fingerprint and retinal scans. Disadvantages of iris recognition are the high cost of the system, as compared to other biometric technologies, and the high amount of storage requirements needed to uniquely identify a user.

4. **Retina**—Retina scan uses optical technology to map the capillary pattern of the eye's retina. The user has to put his eye within 0.4 to 0.8 inches (1-2 centimeters) of the reader while an image of the pupil is taken. The patterns of the retina are measured at over 400 points to generate a 96-byte template. Retinal scan is extremely reliable, and it has the lowest false-acceptance rate among the current biometric methods. Disadvantages of retinal scanning include the need for fairly close physical contact with the scanning device, which impairs user acceptance, and the high cost.

5. **Fingerprint**—This access control method is commonly used; the user places his/her finger on an optical device or silicon surface to get his/her fingerprint scanned. The template generated for the fingerprint, named "minutiae," measures bifurcations, divergences, enclosures, endings and valleys in the ridge pattern. It contains only specific data about the fingerprint (the minutiae), not the whole image of the fingerprint itself. Additionally, the full fingerprint cannot be reconstructed from the template. Depending on the provider, the fingerprint template may use between 250 bytes to more than 1,000 bytes. More storage space implies lower error rates. Fingerprint characteristics are described by a set of numeric values. While the user puts the finger in place for between two and three seconds, a typical image containing between 30 and 40 finger details is obtained and an automated comparison to the user's template takes place.

Advantages of fingerprint scanning are low cost, small size of the device, ability to physically interface into existing client-server-based systems, and ease of integration into existing access control methods. Disadvantages include the need for physical contact with the device and the possibility of poor-quality images due to residues, such as dirt and body oils, on the finger. Additionally, fingerprint biometrics are not as effective as other techniques.

6. **Face**—In this method, the biometric reader processes an image captured by a video camera, which is usually within 24 inches (60 cm) of the human face, isolating it from the other objects captured within the image. The reader analyzes images captured for general facial characteristics. The template created is based on either generating two- or three-dimensional mapping arrays or by combining facial-metric measurements of the distance between specific facial features, such as the eyes, nose and mouth. Some vendors also include thermal imaging in the template.

The face is considered to be one of the most natural and most "friendly" biometrics, and it is acceptable to users because it is fast and easy to use. The main disadvantage of face recognition is the lack of uniqueness, which means that people who look alike may fool the device. Moreover, some systems cannot maintain high levels of performance as the database grows in size.

An analysis of each behavior-oriented biometric is shown below:

1. **Signature recognition**—Also referred to as signature dynamics, the information from the reader is used to analyze two different areas of an individual's signature: the specific features of the signature and the specific features of the signing process. It includes speed, pen pressure, directions, stroke length and the points in time when the pen is lifted from the paper.

Advantages of this method are that it is fast, easy to use and has a low implementation cost. Other advantages are that even though a person might be able to duplicate the visual image of someone else's signature, it is difficult if not impossible to duplicate the dynamics (e.g., time duration in signing, pen-pressure, how often pen leaves signing block, etc.).

# Protection of Information Assets

The main disadvantage is capturing the uniqueness of a signature particularly when a user does not sign his/her name in a consistent manner. For example this may occur due to illness/disease or use of initials versus a complete signature. Additionally, users' signing behavior may change when signing onto signature identification and authentication "tablets" versus writing the signature in ink onto a piece of paper.

2. **Voice recognition**—Voice recognition involves taking the acoustic signal of a person's voice, saying a "passphrase," and converting it to a unique digital code that can then be stored in a template (approximately 1,500-3,000 bytes). Voice recognition incorporates several variables or parameters to recognize one's voice/speech pattern including pitch, dynamics and waveform.

The main attraction of this method is that it can be used for telephone applications, where it can be deployed with no additional user hardware costs. It also has a high rate of acceptance among users.

Disadvantages of this method include:
• The large volume of storage requirements
• Changes to people's voices
• The possibility of misspoken phrases
• A clandestine recording of the user's voice saying the passphrase could be made and played back to gain access.
• Background noises can interfere with the system.

## MANAGEMENT OF BIOMETRICS
Management of biometrics should address effective security for the collection, distribution and processing of biometric data, encompassing:
• Data integrity, authenticity and nonrepudiation
• Management of biometric data across its life cycle—comprised of the enrollment, transmission and storage, verification, identification, and termination processes
• Usage of biometric technology, including one-to-one and one-to-many matching, for the identification and authentication of its users
• Application of biometric technology for internal and external, as well as logical and physical, access control
• Encapsulation of biometric data
• Techniques for the secure transmission and storage of biometric data
• Security of the physical hardware used throughout the biometric data life cycle
• Techniques for integrity and privacy protection of biometric data

Management should develop and approve a biometric information management and security (BIMS) policy. The auditor should use the BIMS policy to gain a better understanding of the biometric systems in use. With respect to testing, the auditor should make sure this policy has been developed and the biometric information is being secured appropriately.

As is the case with any critical information system, logical and physical controls including business continuity plans should address this area.

Life cycle controls for the development of biometric solutions should be in place to cover the enrollment request, the template creation and storage, and the verification and identification procedures. The identification and authentication procedures for individual enrollment and template creation should be specified in the BIMS policy. Management needs to have controls in place to ensure that these procedures are being followed in accordance with this policy. If the biometric device malfunctions or is inoperable, backup authentication methods should also be developed. Controls should also be in place to protect the sample data as well as the template from modification during transmission.

# Single Sign-on
Users normally require access to a number of resources during the course of their daily routine. For example, users would first log into an operating system and thereafter into various applications. For each OS application or other resource in use, users are required to provide a separate set of credentials to gain access. This results in a situation where users' ability to remember passwords is significantly reduced. This also increases the chance that users will write them down on or near their workstation or area of work, and thereby increase the risks that a security breach within the organization may occur.

To address this situation, the concept of single sign-on (SSO) was developed. SSO can generally be defined as the process for consolidating all organization platform-based administration, authentication and authorization functions into a single centralized administrative function. This function would provide the appropriate interfaces to the organization's information resources, which may include:
• Client-server and distributed systems
• Mainframe systems
• Network security including remote access mechanisms

The SSO process begins with the first instance where the user credentials are introduced into the organization's IT computing environment. The information resource or SSO server handling this function is referred to as the primary domain. Every other information resource, application or platform that uses those credentials is called a secondary domain.

The challenges in managing diverse platforms through single sign-on principally involve overcoming the heterogeneous nature of diverse networks, platforms, databases and applications often found in organizations when establishing a set of credentials acceptable to all of these information resources. To effectively integrate into the SSO process, SSO administrators need to obtain an understanding of how each system manages credentialing information, access control list (ACL) authorization rules, and audit logs and reports. Requirements developed in this regard should be based on security domain policies and procedures.

SSO advantages include:
• Multiple passwords are no longer required; therefore, a user may be more inclined and motivated to select a stronger password.
• It improves an administrator's ability to manage users' accounts and authorizations to all associated systems.
• It reduces administrative overhead in resetting forgotten passwords over multiple platforms and applications.
• It reduces the time taken by users to log into multiple applications and platforms.

SSO disadvantages include:
• Support for all major operating system environments is difficult. SSO implementations will often require a number of solutions integrated into a total solution for an enterprise's IT architecture.
• The costs associated with SSO development can be significant when considering the nature and extent of interface development and maintenance that may be necessary.
• The centralized nature of SSO presents the possibility of a single point of failure and total compromise of an organization's information assets. For this reason, "strong authentication" in the form of complex password requirements and the use of biometrics is frequently implemented.

One example of SSO is Kerberos. Created by the Massachusetts Institute of Technology (USA) through its Project Athena in the 1980s, it is an authentication service used to validate services and users in a distributed computing environment (DCE). The role of the authentication service is to allow principals to positively identify themselves and participate in a DCE. Both users and servers authenticate themselves in a DCE environment, unlike security in most other client-server systems where only users are authenticated. There are two distinct steps to authentication. At initial logon time, the Kerberos third-party protocol is used within DCE to verify the identity of a client requesting to participate in a DCE network. This process results in the client obtaining credentials initially registered with the trusted third party and cryptographically protected. These credentials form the basis for setting up secure sessions with DCE servers when the user tries to access resources.

| PRACTICE QUESTION |
|---|
| 5-4     An organization is proposing to install a single sign-on facility giving access to all systems. The organization should be aware that:<br><br>    A.    maximum unauthorized access would be possible if a password is disclosed.<br>    B.    user access rights would be restricted by the additional security parameters.<br>    C.    the security administrator's workload would increase.<br>    D.    user access rights would be increased. |
| *See answers and explanations to the practice questions at the end of the chapter. (page 469)* |

## 5.3.7 AUTHORIZATION ISSUES

The authorization process used for access control requires that the system be able to identify and differentiate among users.

Access rules (authorization) specify who can access what. For example, access control is often based on least privilege, which refers to the granting to users of only those accesses required to perform their duties. Access should be on a documented need-to-know and need-to-do basis by type of access.

Computer access can be set for various levels, i.e., files, tables, data items etc. When IS auditors review computer accessibility, they need to know what can be done with the access and what is restricted. For example, access restrictions at the file level generally include the following:
• Read, inquiry or copy only
• Write, create, update or delete only
• Execute only
• A combination of the above

The least dangerous type of access is read only, as long as the information being accessed is not sensitive or confidential. This is because the user cannot alter or use the computerized file beyond basic viewing or printing.

The following is a list of computerized files and facilities generally applicable across security layers for networks, platforms, databases and applications that should be protected by logical access controls:
• Data
• Data dictionary/directory
• Telecommunications lines and dial-up lines
• Operating systems for network routers and switches
• Domain name servers
• System software
• Libraries/directories
• Application software (test and production)
• Web applications (public or private intranet-based)
• System procedure libraries
• System utilities
• Operator system exits
• Access control software
• Passwords
• Logging files
• Temporary disk files
• Tape files
• Bypass label processing feature
• Spool queues

## *Access Control Lists*

To provide security authorizations for the files and facilities listed previously, logical access control mechanisms utilize access authorization tables, also referred to as access control lists (ACLs) or access control tables. ACLs refer to a register of:
• Users (including groups, machines, processes) who have permission to use a particular system resource
• The types of access permitted

ACLs vary considerably in their capability and flexibility. Some only allow specifications for certain preset groups (e.g., owner, group and world), while more advanced ACLs allow much more flexibility such as user-defined groups. Also, more advanced ACLs can be used to explicitly deny access to a particular individual or group. With more advanced ACLs, access can be at the discretion of the policy maker (and implemented by the security administrator) or individual user, depending upon how the controls are technically implemented. When a user changes job roles within an organization, often their old

access rights are not removed before adding their new required accesses. Without removing the old access rights, there could be a potential segregation of duties issue.

## Logical Access Security Administration

In today's client-server environment, the access identification and authentication, and the authorization process, can be administered either through a centralized or decentralized environment. The advantages of conducting security in a decentralized environment are:
- The security administration is onsite at the distributed location.
- Security issues are resolved in a timely manner.
- Security controls are monitored on a more frequent basis.

The risks associated with distributed responsibility for security administration include:
- The possibility that local standards might be implemented rather than those required by the organization
- Levels of security management might be below what can be maintained by a central administration.
- Unavailability of management checks and audits that are often provided by central administration to ensure that standards are maintained

There are many ways to control remote and distributed sites:
- Software controls over access to the computer, data files and remote access to the network should be implemented.
- The physical control environment should be as secure as possible, with additions such as lockable terminals and a locked computer room.
- Access from remote locations via modems and laptops to other microcomputers should be controlled appropriately.
- Opportunities for unauthorized people to gain knowledge of the system should be limited by implementing controls over access to system documentation and manuals.
- Controls should exist for data transmitted from remote locations such as sales in one location that update accounts receivable files at another location. The sending location should transmit control information, such as transaction control totals, to enable the receiving location to verify the update of its files. When practical, central monitoring should ensure that all remotely processed data have been received completely and updated accurately.
- When replicated files exist at multiple locations, controls should ensure that all files used are correct and current and, when data are used to produce financial information, that no duplication arises.

## Remote Access Security

Today's organizations require remote access connectivity to their information resources for different types of users, such as employees, vendors, consultants, business partners and customer representatives. In providing this capability, a variety of methods and procedures are available to satisfy an organization's business need for this level of access.

Remote access users can connect to their organization's networks with the same level of functionality that exists within their office. In doing so, the remote access design uses the same network standards and protocols applicable to the systems that they are accessing, Transmission Control Protocol/Internet Protocol (TCP/IP)-based systems and systems network architecture (SNA) systems, for the mainframe where the user uses terminal emulation software to connect to a mainframe-based legacy application. Support for these connections includes asynchronous point-to-point modem connectivity, integrated services digital network (ISDN) dial-on-demand connectivity, and dedicated lines (e.g., frame relay and digital subscriber lines [DSL]).

A viable option gaining increased use is TCP/IP Internet-based remote access. This access method is a cost-effective, "inexpensive" approach that enables organizations to take advantage of the public network infrastructures and connectivity options available, under which ISPs manage modems and dial-in servers, and DSL and cable modems reduce costs further to an organization. To effectively use this option, organizations establish a virtual private network over the Internet to securely communicate data packets over this public infrastructure. Available VPN technologies apply the Internet Engineering Task Force (IETF) IPSec security standard (see encryption section later in this chapter for more details on IPSec. Advantages are their ubiquity, ease of use, inexpensive connectivity, and read, inquiry or copy only access. Disadvantages include that they are significantly less reliable than dedicated circuits, lack a central authority, and can have difficulty troubleshooting problems.

# Protection of Information Assets

Organizations should be aware that using VPNs to allow remote access to their systems can create holes in their security infrastructure. The encrypted traffic can hide unauthorized actions or malicious software that can be transmitted through such channels. Intrusion detection systems and virus scanners able to decrypt the traffic for analysis and then encrypt and forward it to the VPN endpoint should be considered as preventive controls. A good practice will terminate all VPNs to the same end-point in a so called VPN concentrator, and will not accept VPNs directed at other parts of the network.

## COMMON CONNECTIVITY METHODS

In connecting to an organization's network, a common method is to use dial-up lines (modem asynchronous point-to-point or ISDN) in accessing an organization's network access server (NAS) that works in concert with an organization's network firewall and router configuration. The NAS handles user authentication, access control and accounting, while also maintaining connectivity. The most common protocol for doing this is the Remote Access Dial-in User Service (RADIUS) and Terminal Access Control Access (TACACS). In a typical NAS implementation, calls into the network are received, and as a good security practice, the call is terminated after recording the calling number and performing preliminary authentication procedures. The standard security practice has been for the NAS to initiate a call back to a predetermined number of the user. This control can be circumvented through effective implementation of call-forwarding mechanisms.

Dial-up connectivity, not based on centralized control and least preferred from a security and control standpoint, is an organization's server whose operating system is set up to accept remote access, which is referred to as a remote access server (RAS). The latter approach is not recommended, as it is extremely difficult to control remote access from many servers using its own RAS capability.

Advantages of dial-up connectivity are its low end-user costs (local phone calls) and that it is intuitive and easy to use (familiarity). Disadvantages are related to performance; for example, reliability in establishing connections with the NAS (phone networks' electrical interference) and time-sensitive media-rich applications or a service's failure when data-rate throughput is low.

Another common connectivity method often used for remote access is dedicated network connections. Using private, often proprietary, network circuits is the approach generally considered the safest because the only network traffic carried belongs to the same organization. It is commonly used by branch/regional offices or with business partners.

Advantages of dedicated network connections include greater performance gains in data throughput and reliability, and data on a dedicated link belonging to the subscribing organization, where an intruder would have to compromise the telecommunications provider itself to access the data link. A disadvantage is that cost is typically two- to five-times higher than connections to the Internet.

Remote access risks include:
- Denial of service where remote users may not be able to gain access to data or applications that are vital for them to carry out their day-to-day business
- Malicious third parties, who may gain access to critical applications or sensitive data by exploiting weaknesses in communications software and network protocols
- Misconfigured communications software, which may result in unauthorized access or modification of an organization's information resources
- Misconfigured devices on the corporate computing infrastructure
- Host systems not secured appropriately, which could be exploited by an intruder gaining access remotely
- Physical security issues over remote users' computers

Remote access controls include:
- Policy and standards
- Proper authorizations
- Identification and authentication mechanisms
- Encryption tools and techniques such as use of a VPN
- System and network management

## Remote Access Using PDAs

The use of PDAs is widespread and they are increasingly being used to augment desktop and laptop PCs within many organizations because of their ease of use and functionality. Connectivity options for PDAs have significantly expanded. Those now being marketed include wireless connectivity, integrated telephone and camera features. As business use increases, it is likely that use of these devices will increase and include access to business information considered sensitive or confidential. Consequently, risks to the organization will increase due to PDA's inherent lack of security (e.g., easy to steal or lose due to their size). Also, if a PDA is connected or synchronized without the appropriate controls being in place, this could potentially allow unauthorized access to the organization's infrastructure.

The level of controls implemented to protect PDAs must, therefore, be consistent with the nature of the information that is stored and/or processed. It is important that organizations set appropriate policies and procedures, and individuals are made fully aware of their responsibilities regarding use of PDAs for business purposes. Control issues to address include the following:
- **Compliance**—PDAs should comply with the security requirements as defined in corporate standards. Existing policies that define computing resources should be expanded to include PDAs in addition to standard workstations and laptops.
- **Approval**—PDA use should be appropriately authorized and approved in accordance with the organization's policies and procedures.
- **Standard PDA applications**—Configuration and use of the PDA should be baselined and controlled.
- **Due care**—Employees should exercise due care within office environments and especially during travel. Any loss or theft of a PDA must be treated as a security breach and reported immediately in accordance to security management policies and procedures.
- **Awareness training**—Employee orientation and security awareness training should include coverage of PDA policy and guidelines. The training will allow propagation of awareness that PDAs are important business tools when used properly and have risks associated with them, if not managed accordingly.
- **PDA applications**—Only applications that either meet with the corporate security architecture or are delivered as standard on the PDA should be authorized for use, and all software applications must be appropriately licensed and installed by the organization's IS support team.
- **Synchronization**—PDAs should be backed up and updated. PDA information should be synchronized only with the organization's resources contained on both laptops and desktop PCs. Remote access to the organization's infrastructure must be made only via an approved route where multiple methods of synchronization are available. Generally, the only methods that should be approved and supported are a direct connection to a workstation (via cable or infrared link) and dial-in access via an approved, secure remote access service. Additionally, when the PDA is connected to the organization's infrastructure, it must not have additional communication sessions running. For example, dial-up connections must be blocked when the PDA is synchronizing directly with the LAN. This may also include users authentication to the PDA and the workstation prior to synchronization. The synchronization process must log all actions performed during synchronization and report any irregularities.
- **Encryption**—PDAs used to store sensitive or confidential information should be encrypted in accordance with the organization's information security policies, mandating use of a strong encryption mechanism. The encryption standard should be established by the organization's IT function in establishing a standard baseline.
- **Virus detection and control**—The threat associated with computer viruses applies equally to PDAs as it does to laptops and PCs. Therefore, the same level of control should be maintained for PDAs (see the Virus and Worm Controls section later in this chapter).
- **Device registration**—PDAs authorized for business use should be registered in a database to distinguish them from those that are individually owned. Organizations can push updates or manage authorized devices and exclude personally-owned PDAs.
- **Camera use**—Some PDAs include megapixel cameras that could expose sensitive information through pictures taken. The policy should regulate the use of cameras and sensitive locations should have clearly visible signage to disallow such use.

## Access Issues With Mobile Technology

Devices used to move data from networks and desktop systems to mobile equipment such as miniature hard and flash drives which simply plug into the USB port are popular and can be purchased at a low cost. Individual employees may even use

such devices of their own to misuse enterprise computing facilities and take unauthorized copies of data or even programs. These devices should be strictly controlled both by policy and denial of use. Possible actions include:
• Banning all use of transportable drives in the security policy
• Where no authorized use of USB ports exists, disabling use with a logon script which removes them from the system directory
• If they are considered necessary for business use, encrypting all data transported or saved by these devices

## Audit Logging in Monitoring System Access
Most access control software today has security features that enable a security administrator to automatically log and report all levels of access attempts—successes and failures. For example, access control software can log computer activity initiated through a logon ID or computer terminal. This information provides management an audit trail to monitor activities of a suspicious nature, such as a hacker attempting brute-force attacks on a privileged logon ID. Also, keystroke logging can be turned on for users that have sensitive access privileges. What is logged is determined by the action of the organization. Issues include what is logged, who/what has access to the logs and how long logs are retained (record-retention item).

### ACCESS RIGHTS TO SYSTEM LOGS
Access rights to system logs for security administrators to perform the above activities should be strictly controlled.

Computer security managers and system administrators/managers should have access for review purposes; however, security and/or administration personnel who maintain logical access functions may have no need for access to audit logs.

It is particularly important to ensure the integrity of audit trail data against modification. One way to do this is to use digital signatures. Another way is to use write-once devices. The audit trail files need to be protected since, for example, intruders may try to cover their tracks by modifying audit trail records. Audit trail records should be protected by strong access controls to help prevent unauthorized access. The integrity of audit trail information may be particularly important when legal issues arise, such as the use of audit trails as legal evidence. (This may, for example, require daily printing and signing of the logs.) Questions regarding such legal issues should be directed to the appropriate legal counsel.

The confidentiality of audit trail information may also be protected, for example, if the audit trail is recording information about users that may be disclosure-sensitive, such as transaction data containing personal information (e.g., before and after records of modification to income tax data). Strong access controls and encryption can be particularly effective in preserving confidentiality.

The logging of media is used to support accountability. Logs can include control numbers (or other tracking data) such as the times and dates of transfers, names and signatures of individuals involved, and other relevant information. Periodic spot checks or audits may be conducted to determine that no controlled items have been lost and that all are in the custody of individuals named in control logs. Automated media tracking systems may be helpful for maintaining inventories of tape and disk libraries.

A periodic review of system-generated logs can detect security problems, including attempts to exceed access authority or gain system access during unusual hours. Certain reports are generated for security recorded in activity logs.

| PRACTICE QUESTION |
|---|
| 5-5    An IS auditor reviewing the log of failed logon attempts would be **MOST** concerned if which of the following accounts was targeted?<br><br>A.    Network administrator<br>B.    System administrator<br>C.    Data administrator<br>D.    Database administrator |
| *See answers and explanations to the practice questions at the end of the chapter. (page 469)* |

## TOOLS FOR AUDIT TRAIL (LOGS) ANALYSIS

Many types of tools have been developed to help reduce the amount of information contained in audit records and to delineate useful information from the raw data.

On most systems, audit trail software can create large files, which can be extremely difficult to analyze manually. The use of automated tools is likely to be the difference between unused audit trail data and a robust program. Some of the types of tools include:

• **Audit reduction tools**—They are preprocessors designed to reduce the volume of audit records to facilitate manual review. Before a security review, these tools can remove many audit records known to have little security significance. (This alone may cut in half the number of records in the audit trail.) These tools generally remove records generated by specified classes of events, for example, records generated by nightly backups might be removed.
• **Trend/variance-detection tools**—They look for anomalies in user or system behavior. It is possible to construct more sophisticated processors that monitor usage trends and detect major variations. For example, if a user typically logs in at 09.00, but appears at 04.30 one morning, this may indicate a security problem that may need to be investigated.
• **Attack-signature-detection tools**—They look for an attack signature, which is a specific sequence of events indicative of an unauthorized access attempt. A simple example would be repeated failed logon attempts.

## COST CONSIDERATIONS

Audit trails involve many costs. First, some system overhead is incurred while recording the audit trail. Additional system overhead will be incurred to store and process the records. The more detailed the records, the more overhead is required. Another cost involves human and machine time required to do the analysis. This can be minimized by using tools to perform most of the analysis. Many simple analyzers can be constructed quickly and inexpensively from system utilities, but they are limited to audit reduction and the identification of particularly sensitive events. More complex tools that identify trends or sequences of events are slowly becoming available as off-the-shelf software. (If complex tools are not available for a system, development may be prohibitively expensive. Some intrusion detection systems, for example, have taken years to develop.)

The final cost of audit trails is the cost of investigating unexpected and anomalous events. If the system is identifying too many events as suspicious, administrators may spend undue time reconstructing events and questioning personnel.

The frequency of the security administrator's review of computer access reports should be commensurate with the sensitivity of the computerized information being protected. The IS auditor should ensure that the logs cannot be tampered with, or altered, without leaving an audit trail.

When reviewing or performing security access follow-up, the IS auditor should look for:
• Patterns or trends that indicate abuse of access privileges, such as concentration on a sensitive application
• Violations (such as attempting computer file access that is not authorized) and/or use of incorrect passwords

Today, attempted violations can be detected on either a real-time basis and acted upon immediately or after the fact. For interconnected networks, often the first line of defense for attacks on organization's trusted networks and systems is an intrusion detection system (IDS).

IDS refers to the process of discovering unauthorized use of computers and networks through the use of software designed for this purpose. An IDS works in a variety of ways, either host- or network-based, to analyze and log incoming data for signs of unauthorized access. In designing these systems, organization's should identify scenarios or events that constitute a high risk, provide for real-time notifications to system administrators of these events occurring (IDSs interfacing with pagers), and establish procedures for responding to those events.

An intrusion prevention system (IPS) is an extension of an IDS. An IDS detects and reports, while an IPS takes this a step further to block unauthorized connection in real time and in an automated manner. IPS works with access control devices such as routers, firewalls and application-level proxy. Once unauthorized access is detected, the IPS will send a message to appropriate devices to block such access. The drawback of an IPS is that it may inadvertently block legitimate traffic.

Once a violation has been identified:
- The person who identified the violator should refer the problem to the security administrator for investigation.
- The security administrator and responsible management should work together to investigate and determine the severity of the violation. Generally, most violations are accidental.
- If a violation attempt is serious, executive management should be notified, not law enforcement officials. Executive management normally is responsible for notifying law enforcement officials. Involvement of external agencies may result in adverse publicity that is ultimately more damaging than the original violation; therefore, the decision to involve external agencies should be left to executive management.
- Procedures should be in place to manage public relations and the press.
- To facilitate proper handling of access violations, written guidelines should exist that identify various types and levels of violations and how they will be addressed. This effectively provides direction for judging the seriousness of a violation.
- Disciplinary action should be a formal process that is applied consistently. This may involve a reprimand, probation or immediate termination. The procedures should be legally and ethically sound to reduce the risk of legal action against the company.
- Corrective measures should include a review of the computer access rules, not only for the perpetrator but for interested parties. Excessive or inappropriate access rules should be eliminated.

## Restricting and Monitoring Access

There should be restrictions and procedures of monitoring access to computer features that bypass security. Generally, only system software programmers should have access to these features:
- **Bypass label processing (BLP)**—BLP bypasses the computer reading of the file label. Since most access control rules are based on file names (labels), this can bypass access control programs.
- **System exits**—This system software feature permits the user to perform complex system maintenance, which may be tailored to a specific environment or company. They often exist outside of the computer security system and, thus, are not restricted or reported in their use.
- **Special system logon IDs**—These logon IDs often are provided with the computer by the vendor. The names can be determined easily because they are the same for all similar computer systems. Passwords should be changed immediately, upon installation, to secure the systems.

## Naming Conventions for Logical Access Controls

Access capabilities are implemented by security administration in a set of access rules that stipulates which users (or groups of users) are authorized to access a resource (such as a dataset or file) and at what level (such as read or update). The access control mechanism applies these rules whenever a user attempts to access or use a protected resource.

Access control naming conventions are structures used to govern user access to the system and user authority to access/use computer resources such as files, programs and terminals. These general naming conventions and associated files are required in a computer environment to establish and maintain personal accountability and segregation of duties in the access of data. The owners of the data or application, with the help of the security officer, usually set up naming conventions. The need for sophisticated naming conventions over access controls depends on the importance and level of security that is needed to ensure that unauthorized access has not been granted. It is important to establish naming conventions that both promote the implementation of efficient access rules and simplify security administration.

Naming conventions for system resources (e.g., as datasets, volumes, programs and employees workstations) are an important prerequisite for efficient administration of security controls. Naming conventions can be structured so that resources beginning with the same high-level qualifier can be governed by one or more generic rule(s). This reduces the number of rules required to adequately protect resources which, in turn, facilitates security administration and maintenance efforts.

## 5.3.8 STORING, RETRIEVING, TRANSPORTING AND DISPOSING OF CONFIDENTIAL INFORMATION

Management should define and implement procedures to prevent access to, or loss of, sensitive information and software from computers, disks, and other equipment or media when they are stored, disposed of or transferred to another user.

This should be done for:
- **Backup files of databases**—Backup files onto magnetic tapes are often unencrypted, so that even secret information may be obtained by simply loading backup databases to other systems for data analysis. Security problems of data media storage and transportation technologies involve ensuring that contractors used to transport and store backup tapes have adequate policies and procedures to protect the integrity and confidentiality of the information.
- **Data banks**—Research and commercial institutions collect the result of important survey or research projects on large tape libraries (i.e., geodynamic data). These data have a high commercial value and may be subject to a requirement of availability and confidentiality that persist for many years, longer than the duration of the media containing them. Preserving the value requires precautions and possibly a planned media verification or duplication activity. In general, a solution is required to the problems of long-term computer storage of sensitive information. Optical disks are a possible media, but their durability and standardization are currently not sufficient to grant permanence.
- **Disposal of media previously used to hold confidential information**—Procedures should be implemented to identify and erase the sensitive information and software inside computers, disks and other equipment or media that have been identified for disposal so that deleted data may not be retrieved by any internal or third party. Care must be taken not only to meet the requirements of data protection, but also when a machine is transferred to another user. The original user should remove any personal data that are not confidential by nature. If previously held data were sensitive, the disk should be reformatted and then a secure wipe of the disk should be carried out with appropriate utility software.

   In some cases, when information is highly confidential, it may prove insufficient to wipe the media. Random access memory (RAM) is included because favorable circumstances and appropriate technical analysis of these media could expose the data. This may require that such equipment or media should be disposed of in a secure manner (e.g., destruction). This may include "degaussing" (demagnetizing) the magnetic media, such as tapes or PC hard drives, as well as their removal and physical destruction.
- **Management of equipment sent for offsite maintenance**—Data files and proprietary software should be backed up, so that they can be erased from the equipment prior to sending it offsite for maintenance (e.g., computers, PDAs, miniature hard/flash drives). Computers holding confidential data should not be sent out for repair, unless memories are withheld.
- **Public agencies and organizations concerned with sensitive, critical or confidential information**—These organizations may have particular obligations to develop a comprehensive records management program. Policies addressing these needs must mainly refer to institutional laws concerning availability, substance, degree of confidentiality and disposal compared to available technical solutions and organizational needs. For instance, public records may be destroyed only in accordance with precise record-retention schedules, and the record holder may not mutilate, destroy, sell, loan or otherwise dispose of any record, except under a record-retention schedule or with the written consent of the owner. Proper record retention requires the preparation of separate retention schedules depending on subject files (administrative vs. other legal requirements).
- **E-token electronic keys**—For such sensitive information, data transportation on floppy and hard disks is not safe and should not be used. Taking proper care of the media can significantly reduce the chances of data loss.
- **Storage records**—Many commercial organizations fulfill legal or institutional obligations to preserve specific types of records, which may be confidential in nature, for a given number of years, by outputting microfiche images of the related documents. In some cases, however, these obligations are fulfilled by preserving database images and the source of the documents, either online or on backup tapes. In these cases, the conditions of recreating the original document must be integrally retained as well.

## *Preserving Information During Shipment or Storage*

Manufacturers publish recommended temperature and humidity levels in which to store media. These recommendations should be consulted and adhered to before storing or shipping important media. However, some general tips can be followed to help avoid potential damage to media during shipping and storage. The following environmental issues are applicable to all types of media:
• Keep out of direct sunlight.
• Keep free of dust.
• Keep free of liquids.
• Keep media away from exposure to magnetic fields, radio equipment or any sources of vibration.
• Do not air transport in areas and at times of exposure to a strong magnetic storm.

## *Media-specific Storage Precautions*

Some precautions need to be considered regarding media-specific storage (see **exhibit 5.3**).

| Exhibit 5.3—Media-specific Storage Precautions ||
| --- | --- |
| **Media Storage** | **Precautions** |
| **Hard drives** | • Store hard drives in antistatic bags, and be sure that the person removing them from the bag is static-free.<br>• If the original box and padding for the hard drive is available, use it for shipping.<br>• Avoid styrofoam packaging products or other materials that can cause static electricity.<br>• Quick drops or spikes in temperature are a danger, since such changes can lead to hard drive crashes.<br>• If the hard drive has been in a cold environment, bring it to room temperature prior to installing and using it.<br>• Avoid sudden mechanical shocks or vibrations. |
| **Magnetic media** | • Store tapes vertically.<br>• Store tapes in acid-free containers.<br>• Write-protect tapes immediately. |
| **Floppy disks** | • When handling the floppy, pick it up by the label. The mylar surface must never be touched.<br>• Write labels using a felt tip pen only. |
| **CDs and DVDs** | • Handle by the edges or by the hole in the middle.<br>• Be careful not to bend the CD.<br>• Avoid long-term exposure to bright light.<br>• Store in a hard jewel case, not in soft sleeves. |

# 5.4 NETWORK INFRASTRUCTURE SECURITY

Communication networks (wide area or local area networks) generally include devices connected to the network as well as programs and files supporting the network operations. Control is accomplished through a network control terminal and specialized communications software.

The following are controls over the communication network:
• Network control functions should be performed by individuals possessing adequate training and experience.
• Network control functions should be separated, and the duties should be rotated on a regular basis, where possible.
• Network control software must restrict operator access from performing certain functions (e.g., the ability to amend/delete operator activity logs).
• Network control software should maintain an audit trail of all operator activities.
• Audit trails should be periodically reviewed by operations management to detect any unauthorized network operations activities.
• Network operation standards and protocols should be documented and made available to the operators, and should be reviewed periodically to ensure compliance.
• Network access by the system engineers should be monitored and reviewed closely to detect unauthorized access to the network.

• Analysis should be performed to ensure workload balance, fast response time and system efficiency.
• A terminal identification file should be maintained by the communications software to check the authentication of a terminal when it tries to send or receive messages.
• Data encryption should be used, where appropriate, to protect messages from disclosure during transmission.

To improve the control and maintenance of the infrastructure and its use, besides the direct management of the network devices, consolidate the logs of these devices with the firewall's logs and the client-server operating system's logs.

In recent years, the management of large capacity storage units is frequently based on fiber channel connections (e.g., [SysAdmin, Audit, Network, Security] SANs).

Systems security is improved when a dynamic inventory of the devices is possible. In the case of an incident, it is important to know which computer is used by whom.

Another important security improvement is the ability to identify users at every step of their activity. Some application packages use predefined names (e.g., SYSTEM). New monitoring tools have been developed to resolve this problem.

The requirements above help facilitate the practice of IT governance. There are different models employed in this process. One of them is is Information Technology Infrastructure Library (ITIL). Its main goal is not only the support of resource management, but also the overall management of information systems through the use of service level agreements.

## 5.4.1 LAN SECURITY

LANs facilitate the storage and retrieval of programs and data used by a group of people. LAN software and practices also need to provide for the security of these programs and data. Unfortunately, most LAN software provides a low level of security. The emphasis has been on providing capability and functionality rather than security. As a result, risks associated with use of LANs include:
• Loss of data and program integrity through unauthorized changes
• Lack of current data protection through inability to maintain version control
• Exposure to external activity through limited user verification and potential public network access from dial-in connections
• Virus and worm infection
• Improper disclosure of data because of general access rather than need-to-know access provisions
• Violating software licenses by using unlicensed or excessive numbers of software copies
• Illegal access by impersonating or masquerading as a legitimate LAN user
• Internal user's sniffing (obtaining seemingly unimportant information from the network that can be used to launch an attack such as network address information)
• Internal user's spoofing (reconfiguring a network address to pretend to be a different address)
• Destruction of the logging and auditing data

The LAN security provisions available depend on the software product, product version and implementation. Commonly available network security administrative capabilities include:
• Declaring ownership of programs, files and storage
• Limiting access to a read-only basis
• Implementing record and file locking to prevent simultaneous update
• Enforcing user ID/password sign-on procedures, including the rules relating to password length, format and change frequency
• Using switches to implement port security policies rather than hubs or non-manageable switches. This will prevent unauthorized hosts, with unknown MAC addresses, to connect to the LAN.
• Encrypting local traffic using IPSec (IP security) protocol

The use of these security procedures requires administrative time to implement and maintain. Network administration is often inadequate, providing global access because of the limited administrative support available when limited access is appropriate.

To gain a full understanding of the LAN, the IS auditor should identify and document the following:
- LAN topology and network design
- LAN administrator/LAN owner
- Functions performed by the LAN administrator/owner
- Distinct groups of LAN users
- Computer applications used on the LAN
- Procedures and standards relating to network design, support, naming conventions and data security

## LAN Risks and Issues

The administrative and control functions available with network software might be limited. Software vendors and network users have recognized the need to provide diagnostic capabilities to identify the cause of problems when the network goes down or functions in an unusual manner. The use of logon IDs and passwords with associated administration facilities is only becoming standard now.

Read, write and execute permission capabilities for files and programs are options available with some network operating system versions, but detailed automated logs of activity (audit trails) are seldom found on LANs. Fortunately, newer versions of network software have significantly more control and administration capabilities.

LANs can represent a form of decentralized computing. Decentralized local processing provides the potential for a more responsive computing environment; however, organizations do not always give the opportunity to efficiently develop staff to address the technical, operational and control issues that the complex LAN technology represents. As a result, local LAN administrators frequently lack the experience, expertise and time to effectively manage the computing environment.

The various alternatives of media, protocol, hardware, transmission techniques, topology and network software ensure that each LAN is unique. This mix of vendors and unique environments makes it difficult to implement standard management, operating and auditing practices. As a result, the costs of resolving problems can be substantial.

Normal LAN users recognize only one attribute of the LAN—it works. In a well-structured LAN, the unsophisticated user is not able to judge whether the technology is appropriate, the software is installed and documented properly, or the necessary control and security measures are taken. Audit trails are considered only after a problem occurs.

## Dial-up Access Controls

It is possible to break LAN security through the dial-in route. Without dial-up access controls, a caller can dial in and try passwords until they gain access. Once in, they can hide pieces of software anywhere, pass through WAN links to other systems, and generally create as much or as little havoc as they like.

To minimize the risk of unauthorized dial-in access, remote users should never store their passwords in plaintext login scripts on notebooks or desktops. Furthermore, portable PCs should be protected by physical keys and/or basic input/output system (BIOS)-based passwords to limit access to data if stolen.

A dial-back modem may be used to prevent access by guessing passwords. When a call is answered by the modem, the caller must enter a code. The modem then hangs up the connection, looks up a corresponding phone number that has been authorized for dial-in access, and calls the number back if it is authenticated. However, several things may go wrong with this method. First, a hacker can record the caller's code and use it later, if the hacker can be at the target location, turn on call forwarding or otherwise divert the call. Some dial-back modems do not listen for dial tone when they call back. If the hacker stays on the line, dialing occurs but has no effect. Then the hacker can start trying passwords until they achieve entry. There are solutions to this issue. Packet-sized authenticating and encrypting modems with random password generation can be used by authenticating the user's personal identification number (PIN). Other methods include the use of one-time password generator devices held by the user. These generate unique passwords that can be authenticated by the host. Alternatively, dial-back modems are available that are connected to two individual telephone lines: one is used for incoming calls and the second is used to call back the user on the known telephone number. These are generally referred to as twin-line dial-back modems.

# 5.4.2 CLIENT-SERVER SECURITY

A client-server system typically contains numerous access points. Security procedures for these server environments are usually not as well understood or as protected as a mainframe-based processing environment. Client-server systems utilize distributed techniques, creating increased risk of access to data and processing. To effectively secure the client-server environment, all access points should be identified. In mainframe-based applications, centralized processing techniques require the user to go through one predefined route to access all resources. In a client-server environment, several access routes exist, as application data may exist on the server or on the client. Each of these routes, therefore, must be examined individually and in relation to each other to ensure that no exposures are left unchecked. To increase the security in a client-server environment, an IS auditor may want to see that the following control techniques are in place:

- Securing access to the data or application on the client-server may be performed by disabling the floppy disk drive, much like a keyless workstation that has access to a mainframe. Diskless workstations prevent access control software from being bypassed and rendering the workstation vulnerable to unauthorized access. By securing the automatic boot or startup batch files, unauthorized users may be prevented from overriding login scripts and access.
- Network monitoring devices may be used to inspect activity from known or unknown users. These devices may identify client addresses, allowing proactive session termination as well as finding evidence of unauthorized access for later investigation. However, the method of securing the client-server environment may be only as good as the administrator who monitors it. Since this is a detective control, if the network administrator does not monitor or maintain these devices, the tool becomes useless against unauthorized intruders.
- Data encryption techniques (symmetric or asymmetric encryption) can help protect sensitive or proprietary data from unauthorized access.
- Authentication systems may provide environmentwide, logical facilities that can differentiate among users. Another method, system smart cards, uses intelligent handheld devices and encryption techniques to decipher random codes provided by client-server systems. A smart card displays a temporary password that is provided by an algorithm on the system and must be reentered by the user during the logon session for access into the client-server system.
- The use of application-level access control programs and the organization of end users into functional groups is a management control that restricts access by limiting users to only those functions needed to perform their duties.

## Client-server Risks and Issues

Since the early 1990s, client-server technology has become one of the predominant ways many organizations have processed production data, and developed and delivered mission-critical products and services. Client-server technology enables business units to develop and deliver products and services to market much quicker traditional legacy methods. The trade-off is that controls over these systems typically are substandard to those associated with traditional mainframe systems. However, there are higher risks that can severely impact a company's business, if controls are not in place to prevent or detect them.

The areas of risk and concern in a client-server environment are listed below:

- Access controls may be weak in a client-server environment, if network administrators do not set up password change controls or access rules properly.
- Change control and change management procedures, whether they are automated or manual, may be inherently weak. The primary reason for this weakness is due to the relatively high level of sophistication of client-server change control tools together with inexperienced IS staff who are reluctant to introduce such tools for fear of introducing limitations on their capability.
- The loss of network availability may have a serious impact on the business or service.
- Obsolescence of the network components including hardware, software and communications
- The use of synchronous and asynchronous modems to connect the network to other networks may be unauthorized and indiscriminate.
- The connection of the network to public switched telephone networks may be weak.
- Changes to systems or data may be inaccurate, unauthorized and unapproved.
- Access to confidential data, and data modification may be unauthorized; business may be interrupted; and data may be incomplete and inaccurate.
- Application code and data may not be located on a single machine enclosed in a secure computer room, as with mainframe computing.

# Protection of Information Assets

## 5.4.3 WIRELESS SECURITY THREATS AND RISK MITIGATION

The classification of security threats may be segmented into nine categories:
• Errors and omissions
• Fraud and theft committed by authorized or unauthorized users of the system
• Employee sabotage
• Loss of physical and infrastructure support
• Malicious hackers
• Industrial espionage
• Malicious code
• Foreign government espionage
• Threats to personal privacy

All of these represent potential threats in wireless networks as well. However, the more immediate concerns for wireless communications are device theft, denial of service, malicious hackers, malicious code, theft of service, and industrial and foreign espionage.

Theft is likely to occur with wireless devices because of their portability. Authorized and unauthorized users of the system may commit fraud and theft; however, authorized users are more likely to carry out such acts. Since users of a system may know what resources a system has and the system's security flaws, it is easier for them to commit fraud and theft.

Malicious hackers, sometimes called crackers, are individuals who break into a system without authorization, usually for personal gain or to do harm. Such hackers may gain access to the wireless network access point by eavesdropping on wireless device communications.

Malicious code involves viruses, worms, Trojan horses, logic bombs or other unwanted software that is intended to damage files or bring down a system. Theft of service occurs when an unauthorized user gains access to the network and consumes network resources. In wireless networks, the unauthorized access threat stems from the relative ease with which eavesdropping can occur on radio transmissions.

Ensuring confidentiality, integrity, authenticity and availability are the prime objectives in wireless networks.

Security requirements include the following:
• **Authenticity**—A third party must be able to verify that the content of a message has not been changed in transit.
• **Nonrepudiation**—The origin or the receipt of a specific message must be verifiable by a third party.
• **Accountability**—The actions of an entity must be uniquely traceable to that entity.
• **Network availability**—The IT resource must be available on a timely basis to meet mission requirements or to avoid substantial losses. Availability also includes ensuring that resources are used only for intended purposes.

Risks in wireless networks are equal to the sum of the risk of operating a wired network plus the new risks introduced by weaknesses in wireless protocols. To mitigate these risks, an organization must adopt security measures and practices that help bring their risks to a manageable level.

To date, the list below includes some of the more salient threats and vulnerabilities of wireless systems:
• All the vulnerabilities that exist in a conventional wired network apply to wireless technologies.
• Weaknesses in wireless protocols increase the threat of disclosure of sensitive information. Many wireless networks are either not secure or use outdated encryption algorithms.
• Malicious entities may gain unauthorized access to an agency's computer or voice (IP telephony) network through wireless connections, potentially bypassing any firewall protections.
• Sensitive information that is not encrypted (or that is encrypted with poor cryptographic techniques) and is transmitted between two wireless devices may be intercepted and disclosed.
• DoS attacks may be directed at wireless connections or devices.
• Malicious entities may steal the identity of legitimate users and masquerade as them on internal or external corporate networks.

- Sensitive data may be corrupted during improper synchronization.
- Malicious entities may be able to violate the privacy of legitimate users and track their physical movements.
- Malicious entities may deploy unauthorized equipment (e.g., client devices and access points) to surreptitiously gain access to sensitive information.
- Handheld devices are easily stolen and can reveal sensitive information.
- Data may be extracted without detection from improperly configured devices.
- Viruses or other malicious code may corrupt data on a wireless device and be subsequently introduced to a wired network connection.
- Malicious entities may, through wireless connections, connect to other agencies for the purposes of launching attacks and concealing their activity.
- Interlopers, from inside or out, may be able to gain connectivity to network management controls and thereby disable or disrupt operations.
- Malicious entities may use a third-party, untrusted wireless network service to gain access to the network resources.

Currently, there are many ways that malicious entities may gain access to wireless devices. Those related to WLANs include, but are not limited to, war driving, war walking and war chalking.

## War Driving

War driving is the practice of driving around businesses or residential neighborhoods scanning with a notebook, hacking tool software and sometimes with a global position system (GPS) for wireless network names. Someone driving around the vicinity of a wireless network might be able to see the wireless network name, but whether they will be able to do anything beyond viewing the wireless network name is determined by the use of wireless security.

With wireless security-enabled and properly configured, war drivers cannot see the network name and are unable to send data, interpret data sent on the wireless network, access the resources of the wireless or wired network (shared files, private Web sites), or use the Internet connection.

Without wireless security enabled and properly configured, war drivers can send data, interpret the data sent on the wireless network, access the shared resources of the wireless or wired network (shared files, private web sites), install viruses, modify or destroy confidential data, and use the Internet connection without the knowledge or consent of the owner. For example, a malicious user might use the Internet connection to send thousands of spam e-mails or launch attacks against other computers. The malicious traffic would be traced back to the owner's home.

## War Walking

War walking is similar to war driving; however, a vehicle is not used. The potential hacker walks around the vicinity with a handheld device or a PDA. Currently, there are several free hacking tools that fit in these minidevices.

## War Chalking

War chalking is the practice of marking a series of symbols (outward-facing crescents) on sidewalks and walls to indicate nearby wireless access points. These markings are used to identify hotspots, where other computer users can connect to the Internet wirelessly and at no cost. War chalking was inspired by the practice of unemployed migrant workers, during the Great Depression in the US, using chalk marks to indicate which homes were friendly.

On the wireless personal area network (WPAN) side, one of the important risks is the man-in-the-middle attack, where an attacker seeking (unauthorized) access to a Bluetooth device inserts himself in between two authorized devices. Communications between the two devices then pass through the man in the middle, who intercepts and manipulates data packets. The following scenarios are possible:

- The attacker actively establishes a connection to the two devices. The attacker connects to both devices and pretends to each of them to be the other device. Should the attacker's device be required to authenticate itself to one of the devices, it passes the authentication request to the other device and then sends the response back to the first device. Having authenticated him/her in this way, the attacker can then interact with the device as he/she wishes. To successfully execute this attack, both devices have to be connectable.

- The attacker interferes while the devices are establishing a connection. During connection establishment, the devices have to synchronize the hop sequence to be used. The aggressor can prevent this synchronization, so both devices use the same sequence but a different offset within the sequence.

The other problem with WPANs is the uncontrolled propagation of radio waves; for example, the radio traffic on Bluetooth connections can be passively intercepted and recorded using Bluetooth protocol sniffers such as Red Fang, Bluesniff and others. If the device addresses are known, then even if the devices are currently in nondiscoverable mode, it is possible to synchronize to the frequency hopping sequence. All the layers of the Bluetooth protocol stack can be examined and analyzed offline. If encryption is not used, then it is possible to extract and monitor the transmitted user data. Use of an antenna with a strong directional characteristic and electronics capable of amplifying Bluetooth signals can make passive listening attacks possible from a greater distance than the functional range. Transmitting power control is optional and is not supported by every Bluetooth device.

The growing prevalence of people using Bluetooth-enabled equipment is likely to follow the trend of Wi-Fi war driving, in which people try to identify inadequately secured networks by driving around with a laptop.

## 5.4.4 INTERNET THREATS AND SECURITY

The nature of the Internet makes it vulnerable to attack. The Internet is a global TCP/IP-based system that enables public and private heterogeneous networks to communicate with one another. Estimates claim that there are over 300 million computers connected via the Internet. Originally designed to allow for the freest possible exchange of information, it is widely used today for commercial purposes. This poses significant security problems for organizations when protecting their information assets. For example, hackers and virus writers try to attack the Internet and computers connected to the Internet. Some want to invade others' privacy and attempt to crack into databases of sensitive information or sniff information as it travels across Internet routes. Consequently, it becomes more important for IS auditors to understand the risks and security factors that are needed to ensure that proper controls are in place when a company connects to the Internet.

The Internet Protocol is designed solely for the addressing and routing of data packets across a network. It does not guarantee or provide evidence on the delivery of messages; there is no verification of an address; the sender will not know if the message reaches its destination at the time it is required; the receiver does not know if the message came from the address specified as the return address in the packet. Other protocols correct some of these drawbacks.

### *Network Security Threats*

One class of network attacks involves probing for network information. These passive attacks can lead to actual active attacks or intrusions/penetrations into an organization's network. By probing for network information, the intruder obtains network information that can be used to target a particular system or set of systems during an actual attack.

### *Passive Attacks*

Examples of passive attacks that gather network information include the following:
- **Network analysis**—The intruder applies a systematic and methodical approach known as footprinting to create a complete profile of an organization's network security infrastructure. During this initial reconnaissance phase, the intruder uses a combination of tools and techniques to build a repository of information about a particular company's internal network. This probably would include information about system aliases, functions, internal addresses, and potential gateways and firewalls. Next, the intruder focuses on systems within the targeted address space that responded to these network queries when targeting a system for an actual attack. Once a system has been targeted, the intruder scans the system's ports to determine what services and operating system are running on the targeted system, possibly revealing vulnerable services that could be exploited.
- **Eavesdropping**—The intruder gathers the information flowing through the network with the intent of acquiring and releasing the message contents for either personal analysis or for third parties who might have commissioned such eavesdropping. This is significant particularly when considering that sensitive information, traversing a network, can be seen by all other machines, including e-mail, passwords and, in some cases, keystrokes, in real time. These activities can enable the intruder to gain unauthorized access, to fraudulently use information such as credit card accounts, and to compromise the confidentiality of sensitive information that could jeopardize or harm an individual's or an organization's reputation.

• **Traffic analysis**—The intruder determines the nature of traffic flow between defined hosts, and through an analysis of session length, frequency and message length, he/she is able to guess the type of communication taking place. This typically is used when messages are encrypted and eavesdropping would not yield any meaningful results. For example, intruders on communication networks of law enforcement agencies have often been able to predict major action being contemplated by using traffic analysis methods.

## Active Attacks

Once enough network information has been gathered, the intruder will launch an actual attack against a targeted system to either gain complete control over that system or enough control to cause certain threats to be realized. This may include obtaining unauthorized access to modify data or programs, causing a denial of service, escalating privileges, accessing other systems, and obtaining sensitive information for personal gain. These types of penetrations or intrusions are known as active attacks. They affect the integrity, availability and authentication attributes of network security. Common forms of active attacks may include any of the following:

• **Brute-force attack**—An intruder launches an attack, using many of the password cracking tools available at little or no cost, on encrypted passwords and attemps to gain unauthorized access to an organization's network or host-based systems.

• **Masquerading**—An active attack in which the intruder presents an identity other than the original identity. In this attack, the purpose is to gain access to sensitive data or computing/network resources to which access is not allowed under the original identity. Masquerading also attacks the authentication attribute by letting a genuine session authentication take place and later enters the information flow masquerading as one of the authenticated users of the session. Impersonation both by people and by machines falls under this category; masquerading by machines that present a forged IP address is referred to as IP spoofing. This form of attack often is used as a means of breaking a firewall.

• **Packet replay**—This is a combination of passive and active modes of attacks. The intruder passively captures a stream of data packets as it moves along an unprotected or vulnerable network. These packets are then actively inserted into the network as if it were another genuine message stream. This form of attack is effective particularly where the receiving end of the communication channel is automated and will act on receipt and interpretation of information packets without human intervention.

• **Phishing**—This is a type of e-mail attack that attempts to convince a user that the originator is genuine, but with the intention of obtaining information for use in social engineering. These attacks may take the form of masquerading as a lottery organization advising the recipient of a large win or the user's bank; in either case, the intent is to obtain account and PIN details. Alternative attacks may seek to obtain apparently innocuous business information which may be used in another form of active attack.

• **Message modification**—Modification involves the capturing of a message and making unauthorized changes or deletions (of full streams or parts of the message), changing the sequence, or delaying transmission of captured messages. This could have disastrous effects if, for example, the message was an instruction to a bank to pay money.

• **Unauthorized access through the Internet or web-based services**—Many Internet software packages contain vulnerabilities that render systems subject to attack. Additionally, many of these systems are large and difficult to configure, resulting in a large percentage of unauthorized access incidents. Examples include:
  – E-mail forgery (simple mail transfer protocol)
  – Telnet passwords transmitted in the clear (via path between client and server)
  – Altering the binding between IP addresses and domain names to impersonate any type of server. As long as the domain name server (DNS) is vulnerable and is used to map universal resource locators (URLs) to sites, there can be no integrity on the web.
  – Releasing common gateway interface (CGI) scripts as shareware. CGI scripts often run with privileges that give them complete control of a server.
  – Client-side execution of scripts (via JAVA in JAVA Applets), which presents the danger of running code from an arbitrary location on a client machine

• **Denial of service (DoS)**—Denial-of-service attacks occur when a computer connected to the Internet is inundated (flooded) with data and/or requests that must be serviced. The machine becomes so tied up with dealing with these messages that it is rendered useless. Well advertised after a series of successful attacks on well-known web resources, this form of attack is more about frustrating or creating a financial loss for the owner of the network resource rather than any direct financial gain for the attacker. However, this type of attack can be particularly devastating if critical information systems are down for a prolonged period of time. At the end of a successful attack of this nature, the attacked network resource is paralyzed and unavailable for

genuine users. This disruption can be achieved through a variety of means. The most common is to overload the resource with requests that overwhelm the system so that it ceases to function. Common types are:
  – Killing user threads
  – Flooding the machine with bogus requests (SYN floods)
  – Filling up disk or memory
  – Isolating machine through domain name server (DNS) attacks
  – Denying service through DNS poisoning
- **Dial-in penetration attacks (i.e., war dialing)**—In this type of attack, an intruder determines the dial-in phone number ranges from external sources, such as the Internet. The intruder may also employ social engineering tactics to get information from a company receptionist or to obtain the information from a knowledgeable employee inside the company.
- **E-mail bombing and spamming**—E-mail bombing is characterized by abusers repeatedly sending an identical e-mail message to a particular address. E-mail spamming is a variant of bombing; it refers to sending e-mail to hundreds or thousands of users (or to lists that expand to that many users). E-mail spamming can be made worse if recipients reply to the e-mail, causing all the original addressees to receive the reply. It may also occur innocently, as a result of sending a message to mailing lists and not realizing that the list explodes to thousands of users, or as a result of an incorrectly set up responder message such as a vacation alert. E-mail bombing/spamming may be combined with e-mail spoofing, which alters the identity of the account sending the e-mail, making it more difficult to determine from whom the e-mail is coming.
- **E-mail spoofing**—E-mail spoofing may occur in different forms, but all have a similar result: a user receives an e-mail that appears to have originated from one source when it actually was sent from another source. E-mail spoofing is often an attempt to trick the user into making a damaging statement or releasing sensitive information such as passwords. Examples of spoofed e-mail that could affect the security of a site include:

  – E-mail claiming to be from a system administrator requesting users to change their passwords to a specified string and threatening to suspend their account if they do not do this
  – E-mail claiming to be from a person in authority requesting users to send them a copy of a password file or other sensitive information

## Threat Impact

It is difficult to assess the impact of the attacks described above but, in generic terms, the following types of impact could occur:
- Loss of income
- Increased cost of recovery (correcting information and reestablishing services)
- Increased cost of retrospectively securing systems
- Loss of information (critical data, proprietary information, contracts)
- Loss of trade secrets
- Damage to reputation
- Degraded performance in network systems
- Legal and regulatory noncompliance
- Failure to meet contractual commitments
- Legal action by customers for loss of confidential data

## Causal Factors for Internet Attacks

Generally, Internet attacks of both a passive and active nature occur for a number of reasons including:
- Availability of tools and techniques on the Internet or as commercially available software that an intruder can download easily. For example, to scan ports, an intruder can easily obtain network scanners such as strobe, netcat, jakal, nmap or Asmodeous (Windows). Additionally, password cracking programs such as John the Ripper and LophtCrack are available free or at a minimal cost.
- Lack of security awareness and training among an organization's employees
- Exploitation of known security vulnerabilities in network- and host-based systems. Many organizations fail to properly configure their systems and to apply security patches or fixes when vulnerabilities are discovered. Most problems can be reduced significantly by keeping network- and host-based systems properly configured and up to date.
- Inadequate security over firewalls and host-based operating systems allowing intruders to view internal addresses and use network services indiscriminately

With careful consideration when designing and developing network security controls and supporting processes, an organization can effectively prevent and detect most intrusive attacks on their networks. In this situation, it becomes important for IS auditors to understand the risks and security factors that are needed to ensure proper controls are in place when a company connects to the Internet. There are several areas of control risks that must be evaluated by the IS auditor to determine the adequacy of Internet security controls.

## Internet Security Controls

To establish effective Internet security controls, an organization must develop controls within an information systems security framework from which Internet security controls can be implemented and supported. Generally, the process for establishing such a framework entails defining, through corporate policies and procedures, the rules the organization will follow to control Internet usage. For example, one set of rules should address appropriate use of Internet resources with rules that might reserve Internet privileges for those with a business need, define what information resources should be available for outside users, and define trusted and untrusted networks within and outside the organization.

Another set of rules should address the classification of the sensitivity or criticality of corporate information resources. This will help to determine what information will be available for use on the Internet and the level of security to be used for corporate resources of a sensitive or critical nature on the Internet.

From an evaluation of these issues, an organization will be able to develop guidelines specific to their situations for defining the level of security controls related to the confidentiality, integrity and availability of information resources (i.e., business applications) on the Internet. For example, operating system security hardening guidelines can be developed which define how the operating system should be configured, detail which Internet services should be blocked from use or exploitation by external untrusted users, and define how the system will be protected by firewalls. Additionally, supporting processes over these controls should be defined including:
• Risk assessments performed periodically over the development and redesign of Internet-based web applications
• Security awareness and training for employees, tailored to their levels of responsibilities
• Firewall standards and security to develop and implement firewall architectures
• Intrusion detection standards and security to develop and implement IDS architectures
• Remote access for coordinating and centrally controlling dial-up access on the Internet via corporate resources
• Incident handling and response for detection, response, containment and recovery
• Configuration management for controlling the security baseline when changes do occur
• Encryption techniques applied to protect information assets passing over the Internet
• A common desktop environment to control, in an automated fashion, what is displayed on a user's desktop
• Monitoring Internet activities for unauthorized usage and notification to end users of security incidents via CERT bulletins or alerts

In summary, Internet usage is drastically changing the way business is done and is creating opportunities for organizations to compete in what has become a global virtual market. To compete and survive in this new marketplace, organizations have to go through a paradigm shift in the way they regard security. Security, as it relates to the Internet, will have to be considered an enabler for success and treated as an essential business tool.

## Firewall Security Systems

Every time a corporation connects its internal computer network to the Internet, it faces potential danger. Because of the Internet's openness, every corporate network connected to it is vulnerable to attack. Hackers on the Internet could theoretically break into the corporate network and do harm in a number of ways: steal or damage important data, damage individual computers or the entire network, use the corporate computer's resources, or use the corporate network and resources as a way of posing as a corporate employee. Companies should build firewalls as one means of perimeter security for their networks. Likewise, this same principle holds true for sensitive or critical systems that need to be protected from untrusted users inside the corporate network (internal hackers). Firewalls are defined as a device installed at the point where network connections enter a site; they apply rules to control the type of networking traffic flowing in and out. Most commercial firewalls are built to handle the most commonly used Internet protocols.

To be effective, firewalls should allow individuals on the corporate network to access the Internet and, at the same time, stop hackers or others on the Internet from gaining access to the corporate network to cause damage. Generally, most organizations will follow a deny-all philosophy, which means that access to a given resource will be denied unless a user can provide a specific business reason or need for access to the information resource. The converse of this access philosophy, not widely accepted, is the accept-all philosophy under which everyone is allowed access unless someone can provide a reason for denying access.

## Firewall General Features

Firewalls are hardware and software combinations that are built using routers, servers and a variety of software. They should control the most vulnerable point between a corporate network and the Internet, and they can be as simple or complex as the corporate information security policy demands. There are many different types of firewalls, but most enable organizations to:
• Block access to particular sites on the Internet
• Limit traffic on an organization's public services segment to relevant addresses and ports
• Prevent certain users from accessing certain servers or services
• Monitor communications between an internal and an external network
• Monitor and record all communications between an internal network and the outside world to investigate network penetrations or detect internal subversion
• Encrypt packets that are sent between different physical locations within an organization by creating a VPN over the Internet (i.e., IP security [IPSec], VPN tunnels)

The capabilities of some firewalls can be extended so they can also provide for protection against viruses and attacks directed to exploit known operating system vulnerabilities.

## Firewall Types

Generally, the types of firewalls available today fall into three categories:
• Router packet filtering
• Application firewall systems
• Stateful inspection

These firewalls are discussed in the following sections.

### ROUTER PACKET FILTERING FIREWALLS

The simplest and earliest kinds of firewalls (i.e., first generation of firewalls) were packet filtering-based firewalls deployed between the private network and the Internet. In packet filtering, a screening router examines the header of every packet of data traveling between the Internet and the corporate network. Packet headers have information in them, including the IP address of the sender and receiver, and the authorized port numbers (application or service) allowed to use the information transmitted. Based on that information, the router knows what kind of Internet service, such as web-based or or FTP, is being used to send the data as well as the identities of the sender and receiver of the data. Using that information, the router can prevent certain packets from being sent between the Internet and the corporate network. For example, the router could block any traffic except for e-mail or traffic to and from suspicious destinations.

The advantages of this type of firewall are its simplicity and generally stable performance since the filtering rules are performed at the network layer. Its simplicity is also a disadvantage, because it is vulnerable to attacks from improperly configured filters and attacks tunneled over permitted services. Since the direct exchange of packets is permitted between outside systems and inside systems, the potential for an attack is determined by the total number of hosts and services to which the packet filtering router permits traffic. Also, if a single packet filtering router is compromised, every system on the private network may be compromised and organizations with many routers may face difficulties in designing, coding and maintaining the rule base. This means that each host directly accessible from the Internet needs to support sophisticated user authentication and needs to be regularly examined by the network administrator for signs of attack.

Some of the more common attacks against packet filter firewalls are:
- **IP spoofing**—The attacker fakes the IP address of either an internal network host or a trusted network host so that the packet being sent will pass the rule base of the firewall. This allows for penetration of the system perimeter. If the spoofing uses an internal IP address, the firewall can be configured to drop the packet on the basis of packet flow direction analysis. However, if the attacker has access to a secure or trusted external IP address and spoofs on that address, the firewall architecture is defenseless.
- **Source routing specification**—It is possible to define the routing that an IP packet must take when it traverses from the source host to the destination host, across the Internet. In this process, it is possible to define the route so it bypasses the firewall. Only those that know of the IP address, subnet mask and default gateway settings at the firewall routing station can do this. A clear defense against this attack is to examine each packet and, if the source routing specification is enabled, drop that packet. However, if the topology permits a route, skipping the choke point, this countermeasure will not be effective.
- **Miniature fragment attack**—Using this method, an attacker fragments the IP packet into smaller ones and pushes it through the firewall in the hope that only the first of the sequence of fragmented packets would be examined and the others would pass without review. This is true if the default setting is to pass residual packets. This can be countered by configuring the firewall to drop all packets where IP fragmentation is enabled or offset to value 1.

## APPLICATION FIREWALL SYSTEMS

There are two types of application firewall systems. They are referred to as application- and circuit-level firewall systems and provide greater protection capabilities than packet filtering routers. Packet filtering routers allow the direct flow of packets between internal and external systems. Application and circuit gateway firewall systems allow information to flow between systems but do not allow the direct exchange of packets. The primary risk of allowing packet exchange between internal and external systems is that the host applications residing on the protected network's systems must be secure against any threat posed by the allowed packets.

The two types of application firewall systems sit atop hardened (tightly secured) operating systems, such as Windows NT and UNIX, and work at the application level of the Open Systems Interconnection (OSI) model. The application-level gateway firewall is a system that analyzes packets through a set of proxies—one for each service (e.g., Hypertext Transmission Protocol [HTTP] proxy for web traffic, FTP proxy). This kind of work could reduce network performance. Circuit-level firewalls are more efficient and also operate at the application level—where Transmission Control Protocol (TCP) and User Datagram Protocol (UDP) sessions are validated, typically through a single, general-purpose proxy before opening a connection. Commercially, circuit-level firewalls are quite rare.

Both application firewall systems employ the concept of bastion hosting in that they handle all incoming requests from the Internet to the corporate network, such as FTP or web requests. Bastion hosts are heavily fortified against attack. By having only a single host handling incoming requests, it is easier to maintain security and track attacks. Therefore, in the event of a break-in, only the firewall system has been compromised, not the entire network. In this way, none of the computers or hosts on the corporate network can be contacted directly for requests from the Internet, providing an effective level or layer of security.

Additionally, application-based firewall systems are set up as proxy servers to act on the behalf of someone inside an organization's private network. Rather than relying on a generic packet filtering tool to manage the flow of Internet services through the firewall, a special-purpose code called a proxy server is incorporated into the firewall system. For example, when someone inside the corporate network wants to access a server on the Internet, a request from the computer is sent to the proxy server, the proxy server contacts the server on the Internet, and the proxy server then sends the information from the Internet server to the computer inside the corporate network. By acting as a go-between, proxy servers can maintain security by examining a service's (e.g., FTP, Telnet) program code and modifying and securing it to eliminate known vulnerabilities. The proxy server can also log all traffic between the Internet and the network.

The application-level firewall implementation of proxy server functions is based on providing a separate proxy for each application service (e.g., FTP, Telnet, HTTP). This differs from circuit-level firewalls, which do not need a special proxy for each application-level service. In other words, one proxy server is used for all services.

Advantages of these types of firewalls are that they provide security for commonly used protocols and generally hide the internal network from outside untrusted networks. For example, a feature available on these types of firewall systems is the

network address translation (NAT) capability. This capability takes private internal network addresses (unusable on the Internet) and maps them to a table of public IP addresses, assigned to the organization, which can be used across the Internet.

Disadvantages are poor performance and scalability as Internet usage grows. To offset this problem, the concept of load balancing is applicable in cases where a redundant fail-over firewall system may be used.

## STATEFUL INSPECTION FIREWALLS

A stateful inspection firewall keeps track of the destination IP address of each packet that leaves the organization's internal network. Whenever the response to a packet is received, its record is referenced to ascertain and ensure that the incoming message is in response to the request that went out from the organization. This is done by mapping the source IP address of an incoming packet with the list of destination IP addresses that is maintained and updated. This approach prevents any attack initiated and originated by an outsider.

The advantages of this approach over application firewall systems is that stateful inspection firewalls control the flow of IP traffic by matching information contained in the headers of connection-oriented or connectionless IP packets at the transport layer, against a set of rules specified by the firewall administrator. This provides a greater degree of efficiency when compared to typical CPU-intensive, full-time application firewall systems' proxy servers, which may perform extensive processing on each data packet at the application level.

The disadvantages include that stateful inspection firewalls can be relatively complex to administer compared to the other two types of firewalls.

# Examples of Firewall Implementations

Firewall implementations can take advantage of the functionality available in a variety of firewall designs to provide a robust layered approach in protecting an organization's information assets. Commonly used implementations available today include:

- **Screened-host firewall**—Utilizing a packet filtering router and a bastion host, this approach implements basic network layer security (packet filtering) and application server security (proxy services). An intruder in this configuration has to penetrate two separate systems before the security of the private network can be compromised. This firewall system is configured with the bastion host connected to the private network with a packet filtering router between the Internet and the bastion host. Router filtering rules allow inbound traffic to access only the bastion host, which blocks access to internal systems. Since the inside hosts reside on the same network as the bastion host, the security policy of the organization determines whether inside systems are permitted direct access to the Internet, or whether they are required to use the proxy services on the bastion host.
- **Dual-homed firewall**—A firewall system that has two or more network interfaces, each of which is connected to a different network. In a firewall configuration, a dual-homed firewall usually acts to block or filter some or all of the traffic trying to pass between the networks. A dual-homed firewall system is a more restrictive form of a screened-host firewall system, in which a dual-homed bastion host is configured with one interface established for information servers and another for private network host computers.
- **Demilitarized zone (DMZ) or screened-subnet firewall**—Utilizing two packet filtering routers and a bastion host, this approach creates the most secure firewall system since it supports network- and application-level security while defining a separate DMZ network. The DMZ functions as a small, isolated network for an organization's public servers, bastion host information servers and modem pools. Typically, DMZs are configured to limit access from the Internet and the organization's private network. Incoming traffic access is restricted into the DMZ network by the outside router and protects the organization against certain attacks by limiting the services available for use. Consequently, external systems can access only the bastion host (and its proxying service capabilities to internal systems) and possibly information servers in the DMZ. The inside router provides a second line of defense, managing DMZ access to the private network, while accepting only traffic originating from the bastion host. For outbound traffic, the inside router manages private network access to the DMZ network. It permits internal systems to access only the bastion host and information servers in the DMZ. The filtering rules on the outside router require the use of proxy services by accepting only outbound traffic on the bastion host. The key benefits of this system are that an intruder must penetrate three separate devices, private network addresses are not disclosed to the Internet, and internal systems do not have direct access to the Internet.

## Firewall Issues

Problems faced by organizations that have implemented firewalls include:
- A false sense of security may exist where management feels that no further security checks and controls are needed on the internal network (i.e., the majority of incidents are caused by insiders, who are not controlled by firewalls).
- The circumvention of firewalls through the use of modems may connect users directly to ISPs. Management should provide assurance that the use of modems when a firewall exists is strictly controlled or prohibited altogether.
- Misconfigured firewalls may allow unknown and dangerous services to pass through freely.
- What constitutes a firewall may be misunderstood (e.g., companies claiming to have a firewall merely have a screening router).
- Monitoring activities may not occur on a regular basis (i.e., log settings not appropriately applied and reviewed).
- Firewall policies may not be maintained regularly.
- Most firewalls operate at the network layer; therefore, they do not stop any application-based or input-based attacks. Examples of such attacks include Structured Query Language (SQL) injection and buffer-overflow attacks. Newer-generation firewalls are able to inspect traffic at the application layer and stop some of these attacks.

## Firewall Platforms

Firewalls may be implemented using hardware or software platforms. When implemented in hardware, it will provide good performance with minimal system overhead. Although hardware-based firewall platforms are faster, they are not as flexible or scalable as software-based firewalls. Software-based firewalls are generally slower with significant systems overhead; however, they are flexible with additional services. They may include content and virus checking, before traffic is passed to users.

It is generally better to use appliances, rather than normal servers, for the firewall. Appliances are normally installed with hardened operating systems. When server-based firewalls are used, operating systems in servers are often vulnerable to attacks. When the attacks on operating systems succeed, the firewall would be compromised. Appliance-type firewalls are, generally, significantly faster to set up and recover.

## Intrusion Detection Systems

Another element to securing networks complementing firewall implementations is an IDS. An IDS works in conjunction with routers and firewalls by monitoring network usage anomalies. It protects a company's IS resources from external as well as internal misuse.

An IDS operates continuously on the system, running in the background and notifying administrators when it detects a perceived threat. Broad categories of IDSs include:
- **Network-based IDSs**—They identify attacks within the monitored network and issue a warning to the operator. If a network-based IDS is placed between the Internet and the firewall, it will detect all the attack attempts, whether or not they enter the firewall. If the IDS is placed between a firewall and the corporate network, it will detect those attacks that enter the firewall (it will detect intruders). The IDS is not a substitute for a firewall, but it complements the function of a firewall.
- **Host-based IDSs**—They are configured for a specific environment and will monitor various internal resources of the operating system to warn of a possible attack. They can detect the modification of executable programs, detect the deletion of files and issue a warning when an attempt is made to use a privileged command.

Components of an IDS are:
- Sensors that are responsible for collecting data. The data can be in the form of network packets, log files, system call traces, etc.
- Analyzers that receive input from sensors and determine intrusive activity
- An administration console
- A user interface

Types of IDSs include:
- **Signature-based**—These IDS systems protect against detected intrusion patterns. The intrusive patterns they can identify are stored in the form of signatures.
- **Statistical-based**—These systems need a comprehensive definition of the known and expected behavior of systems.

- **Neural networks**—An IDS with this feature monitors the general patterns of activity and traffic on the network, and creates a database. This is similar to the statistical model but with added self-learning functionality.

Signature-based IDSs will not be able to detect all types of intrusions due to the limitations of the detection rules. On the other hand, statistical-based systems may report many events outside of the defined normal activity but which are normal activities on the network. A combination of signature- and statistical-based models provides better protection.

## FEATURES
The features available in an IDS include:
- Intrusion detection
- Gathering evidence on intrusive activity
- Automated response (i.e., termination of connection, alarm messaging)
- Security policy
- Interface with system tools
- Security policy management

## LIMITATIONS
An IDS cannot help with the following weaknesses:
- Weaknesses in the policy definition
- Application-level vulnerabilities
- Backdoors into applications
- Weaknesses in identification and authentication schemes

In contrast to IDSs, which rely on signature files to identify an attack as (or after) it happens, an intrusion prevention system (IPS) predicts an attack before it can take effect. It does this by monitoring key areas of a computer system, and looks for "bad behavior" such as worms, Trojans, spyware, malware and hackers. It complements firewall, antivirus and antispyware tools to provide complete protection from emerging threats. It is able to block new (zero-day) threats that bypass traditional security measures since it is not reliant on identifying and distributing threat signatures or patches. The above is taken from ISACA Guideline G33 General Considerations on the Use of the Internet, Section 3.5.2.

| PRACTICE QUESTION |
|---|
| **5-6**    A B-to-C e-commerce web site as part of its information security program wants to monitor, detect and prevent hacking activities and alert the system administrator when suspicious activities occur. Which of the following infrastructure components could be used for this purpose?<br><br>    A.    Intrusion detection systems<br>    B.    Firewalls<br>    C.    Routers<br>    D.    Asymmetric encryption |
| *See answers and explanations to the practice questions at the end of the chapter. (page 469)* |

## *Honeypots and Honeynets*
A honeypot is a software application that pretends to be an unfortunate server on the Internet and is not set up to actively protect against break-ins. Rather, it acts as a decoy system that lures hackers and, therefore, is attractive to hackers. The more a honeypot is targeted by an intruder, the more valuable it becomes. Although honeypots are technically related to intrusion detection systems and firewalls, they have no real production value as an active sentinel of networks.

There are two basic types of honeypots:
- **High-interaction**—Essentially give hackers a real environment to attack
- **Low-interaction**—Emulate production environments and, therefore, provide more limited information

A honeynet is multiple honeypots networked together to simulate a larger network installation, as part of an architecture to let hackers break into the false network while allowing investigators to watch their every move by a combination of surveillance technologies.

An IDS triggers a virtual alarm whenever an attacker breaches security of any networked computers. A stealthy keystroke logger watches everything the intruder types. A separate firewall cuts off the machines from the Internet anytime an intruder tries to attack another system from the honeynet.

All traffic on honeypots or honeynets are assumed to be suspicious because the systems are not meant for internal use and the information collected about these attacks are used proactively to update vulnerabilities on a company's live network.

Eventually, the honeynet—the concept expanded from the honeypot—could become an important part of network security.

If a honeypot is designed to be accessible from the Internet, there is a risk that external monitoring services that create lists of untrusted sites may report the organization's system as vulnerable, without knowing that the vulnerabilities belong to the honeypot and not to the system itself. Such independent reviews made public can affect the organization's reputation. Therefore, prior to implementing a honeypot in the network, careful judgment should be exercised.

# 5.4.5 ENCRYPTION

Encryption is the process of converting a plaintext message into a secure-coded form of text, called ciphertext, which cannot be understood without converting back, via decryption (the reverse process), to plaintext. This is done via a mathematical function and a special encryption/decryption password called the key. In many countries, encryption is subject to governmental laws and regulations.

Encryption generally is used to:
• Protect data in transit over networks from unauthorized interception and manipulation
• Protect information stored on computers from unauthorized viewing and manipulation
• Deter and detect accidental or intentional alterations of data
• Verify authenticity of a transaction or document

Encryption is limited in that it cannot prevent the loss or modification of data, and it is possible to compromise encryption programs if encryption keys are not protected adequately. Therefore, encryption should be regarded as an essential, but incomplete, form of access control that should be incorporated into an organization's overall computer security program.

## *Key Elements of Encryption Systems*
Key elements of encryption systems include:
• **Encryption algorithm**—A mathematically based function or calculation that encrypts/decrypts data
• **Encryption keys**—A piece of information that is used within an encryption algorithm (calculation) to make the encryption or decryption process unique. Similar to passwords, a user needs to use the correct key to access or decipher a message. The wrong key will decipher the message into an unreadable form.
• **Key length**—A predetermined length for the key. The longer the key, the more difficult it is to compromise in a brute-force attack where all possible key combinations are tried.

Effective encryption systems depend upon algorithm strength, secrecy and difficulty of compromising a key, the nonexistence of back doors by which an encrypted file can be decrypted without knowing the key, the inability to decrypt an entire ciphertext message if the way a portion of it decrypts is known (called a known-text attack), and the properties of the plaintext known by a perpetrator.

Most encrypted transactions over the Internet use a combination of private keys/public keys, secret keys, hash functions (fixed values derived mathematically from a text message) and digital certificates to achieve confidentiality, message integrity, authentication and nonrepudiation by either sender or recipient (also known as a PKI). This encryption process

# Protection of Information Assets

allows data to be stored and transported with reduced exposure so a company's corporate data are secure as they move across the Internet or other networks. There are two types of cryptographic systems: symmetric (or private key) and asymmetric (or public key).

## Private Key Cryptographic Systems

Private key cryptographic systems are based on a symmetric encryption algorithm, which uses a secret (private) key to encrypt the plaintext to the ciphertext and the same key to decrypt the ciphertext to the corresponding plaintext. In this case, the key is symmetric because the encryption key is the same as the decryption key.

The most common private key cryptographic system is the Data Encryption Standard (DES). DES is based on a public algorithm that operates on plaintext in blocks (strings or groups) of bits. This type of algorithm is known as a block cipher. DES uses blocks of 64 bits. A key of 56 bits is used for the encryption and decryption of plaintext. An additional 8 bits are used for parity checking. Any 56-bit number can be used as a key and there are 72,057,594,037,927,936 possible keys in the key space.

DES is no longer considered a strong cryptographic solution since its entire key space can be brute-forced (every possible key tried) by large computer systems within a relatively short period of time. In this regard, private key cryptographic key spaces are susceptible to compromise. DES is being replaced with AES, a public algorithm that supports keys from 128 bits to 256 bits in size.

There are two main advantages to private key cryptosystems such as DES or AES. The first is that the user has to remember/know only one key for both encryption and decryption. The second is that private key cryptosystems are generally less complicated and, therefore, use up less processing power than asymmetric techniques. This makes private key cryptosystems ideally suited for bulk data encryption. The major disadvantage of this approach is how to get the keys into the hands of those with whom you want to exchange data, particularly in e-commerce environments where customers are unknown, untrusted entities. Also, a symmetric key cannot be used to sign electronic documents or messages due to the fact that the mechanism is based on a shared secret.

## Public Key Cryptographic Systems

Public key cryptographic systems developed for key distribution solve the problem of getting symmetric keys into the hands of two people, who do not know each other, but who want to exchange information in a secure manner. Based on an asymmetric encryption process, two keys work together as a pair. One key is used to encrypt data, the other is used to decrypt data. Either key can be used to encrypt or decrypt, but once the key has been used to encrypt data, only its partner can be used to decrypt the data (even the key that was used to encrypt the data cannot be used to decrypt it).

The keys are asymmetric in that they are inversely related to each other. Based on mathematical integer factorization, the idea is to generate a single product from two large prime numbers, where it is impracticable to factor the number and recover the two factors. This integer factorization process forms the basis for public key cryptography (i.e., function easy to compute in one direction, but very difficult or impractical in the other). The system involves modular arithmetic, exponentiation, and large prime numbers thousands of bits long. Asymmetric keys are often used for short messages such as encrypting DES symmetric keys or creating digital signatures. If asymmetric keys were used to encrypt bulk data (long messages), the process would be very slow; this is the reason they are used to encrypt short messages such as digests or signatures.

A common form of asymmetric encryption is RSA. RSA is a public key cryptosystem for encryption and authentication; it was invented in 1977 by Ron Rivest, Adi Shamir and Leonard Adleman (RSA stands for the initials of their last names). It works as follows: take two large primes randomly generated, p and q, and find their product n = pq; n is called the modulus. Choose a number, e, less than n and relatively prime to (p minus 1)(q minus 1), which means that e and (p minus 1) (q minus 1) have no common factors except 1. Find another number, d, such that (ed minus 1) is divisible by (p minus 1) (q minus 1). The values e and d are called the public and private exponents, respectively. The public key is the pair (n, e); the private key is (n, d). The factors p and q may be kept with the private key, or destroyed.

It is extremely unlikely that one could obtain the private key d from the public key (n, e). If one could factor n into p and q, however, then one could obtain the private key d. Thus, the security of RSA is related to the assumption that factoring is difficult.

With asymmetric encryption, one key—the secret or private key—is known only to one person; the other key—the public key—is known by many people. In other words, a message that has been sent encrypted by the secret (private) key of the sender can be deciphered by anyone with the corresponding public key. In this way, if the public key deciphers the message satisfactorily, one can be sure of the origin of the message because only the sender (owner of the correspondent private key) could have encrypted the message. This forms the basis of authentication and nonrepudiation (i.e., the sender cannot later claim that he/she did not generate the message). A message that has been sent encrypted using the public key of the receiver can be generated by anyone, but can only be read by the receiver. This is one basis of confidentiality. In theory, a message that has been encrypted twice, first, by the sender's secret key and, second, by the receiver's public key, achieves both authentication and confidentiality objectives, but it is not commonly used because it could generate performance issues.

## Elliptical Curve Cryptography

A variant and more efficient form of public key cryptography (how to manage more security out of minimum resources) gaining prominence is the elliptical curve cryptography (ECC). ECCs work well on networked computers requiring strong cryptography but have some limitations such as bandwidth and processing power. This is even more important with devices such as smart cards, wireless phones and other mobile devices.

As explained earlier, under public key cryptography, a person makes one key publicly available and holds a second, private key. A message is encrypted with the public key, sent and decrypted with the private key. Such systems depend mostly on long key sizes and complex mathematical problems to ensure security. But now cryptographers are looking to a mathematical system known as an elliptical curve to solve the efficiency riddle. It is believed that ECC demands less computational power and, therefore, offers more security per bit. For example, an ECC with a 160-bit key offers the same security as an RSA-based system with a 1,024-bit key.

Elliptical curve systems revolve around the discrete logarithm. Instead of straight integer algebra, elliptical curve systems use an algebraic formula to determine the relationship between public and private keys within the universe created by an elliptical curve. The security of elliptic curve cryptosystem depends on finding k, the integer given the two points G and H, on an elliptic curve such that H = kG (i.e., H is G added to itself k times). This mathematical problem is commonly referred to as the elliptic curve discrete algorithm problem.

## Quantum Cryptography

Quantum cryptography is the next generation of cryptography that will solve existing problems associated with current cryptographic systems. Proven in laboratory research as a commercially viable technology, quantum cryptography taps the natural uncertainty of the quantum world (using interaction of light pulses as a way of transmitting keys and secure information). For example, key generation is based on polarization metrics, where a receiver correctly records the direction (horizontal, vertical, left diagonal, right diagonal) of light photon pulses randomly emitted to him/her from a sender. Using a prearranged code, the sender and receiver then can translate polarization measurements into bits where the horizontal and right diagonal equal 1 and the vertical and left diagonal equal 0 (i.e., 1001). On the average, the receiver will correctly guess the correct setting 50 percent of the time, so the sender will have to send 2n photon pulses to generate n bits. The bits created can be used as a secret key for a symmetric algorithm or a one-time pad encryption scheme.

## Advanced Encryption Standard

AES has replaced the DES as the cryptographic algorithm standard. Due to its short key length, the former standard for symmetric encryption—DES—reached the end of its life cycle. Secure applications need a new standard for symmetric encryption.

In 1997, NIST announced the initiation of the AES development effort and made a formal call for algorithms. On 2 October 2000, NIST announced that it had selected Rijndael to propose the algorithm for the AES. This algorithm was developed by Dr. Joan Daemen and Dr. Vincent Rijmen. Because AES was to replace DES, it was easy to see that a lot of existing and future applications and protocols would use this algorithm for secure information exchange.

Rijndael is a symmetric block cipher with variable block and key length. For AES the block length was fixed to 128 bits and three different key sizes (128, 192 and 256 bits) were specified. Therefore, AES-128, AES-192 and AES-256 are three different versions of AES. The cipher is based on round operations. Each round has a 128-bit round key and the result of the previous round as input. The round keys can be precomputed or generated out of the input key. Due to its regular structure, it can be implemented efficiently in hardware. Decryption is computed by applying inverse functions of the round operations. The sequence of operations for the round function differs from encryption which often results in separated encryption and decryption circuits. Computational performance of software implementations often differ between encryption and decryption because the inverse operations in the round function are more complex than the respective operation for encryption.

## Digital Signatures

A digital signature is an electronic identification of a person or entity created by using a public key algorithm and intended to verify to a recipient the integrity of the data and the identity of the sender. To verify the integrity of the data, a cryptographic hashing algorithm is computed against the entire message or electronic document, which generates a small fixed string message, in general, usually about 128 bits in length. This process, also referred to as a digital signature algorithm, creates a message digest (i.e., smaller extrapolated version of the original message).

Common types of message digest algorithms are SHA1, MD2, MD4 and MD5. These algorithms are one-way functions unlike private and public key encryption algorithms. The process of creating message digests cannot be reversed. They are meant for digital signature applications where a large electronic document or string of characters, such as word processor text, a spreadsheet, a database record, the content of a hard disk or a jpg image, has to be compressed in a secure manner before being signed with the private key. All digest algorithms take a message of arbitrary length and produce a 128-bit message digest. While the structures of these algorithms are somewhat similar, the design of MD2 is quite different from that of MD4 and MD5. MD2 was optimized for 8-bit machines, whereas MD4 and MD5 were created for 32-bit machines. Descriptions and source code for the three algorithms can be found as Internet RFCs 1319-1321.

> **Note:** The IS auditor needs to be familiar with how a digital signature functions to protect data. The specific types of message digest algorithms are not tested on the CISA exam.

The next step, which verifies the identity of the sender, is to encrypt the message digest using the sender's private key, which "signs" the document with the sender's digital signature for message authenticity. To decipher, the receiver would use the sender's public key, proving that the message could only have come from the sender. This process of sender authentication is known as nonrepudiation because the sender cannot later claim that they did not generate the message.

Once decrypted, the receiver will recompute the hash using the same hashing algorithm on the electronic document and compare the results with what was sent, to ensure the integrity of the message. Therefore, digital signature is a cryptographic method that ensures:
• **Data integrity**—Any change to the plaintext message would result in the recipient failing to compute the same message hash.
• **Authentication**—The recipient can ensure that the message has been sent by the claimed sender since only the claimed sender has the secret key.
• **Nonrepudiation**—The claimed sender cannot later deny generating and sending the message.

Replay protection is not ensured by signing an electronic document using a cryptographic technique. A timestamp is used to date a document and the document hash could be used to uniquely identify or number the document. The timestamp or hashing of the document could provide replay protection or warranties.

However, digital signatures and public key encryption are vulnerable to man-in-the-middle attacks wherein the sender's digital signature private key and public key may be faked. To protect against such attacks, an independent sender

authentication authority has been designed. The public key infrastructure performs the function of independently authenticating the validity of senders' digital signatures and public keys.

## Digital Envelope

A digital envelope is used to send encrypted information, using symmetric keys, and the relevant key session along with it. It is a secure method to send electronic documents without compromising the data integrity, authentication and nonrepudiation, which were obtained with the use of asymmetric keys.

The (original plaintext) message to be sent can be encrypted by using either an asymmetric key or a symmetric key. The disadvantage in using an asymmetric key is that it takes much longer to encrypt/decrypt than a symmetric key, with the resultant performance reduction explained in the section on public key cryptographic systems. If a message is short, the time taken is not an issue. When using an asymmetric key with a longer message, memory requirements are much greater and, consequently, the message takes longer to receive, which is one of the reasons it is not generally used.

A digital envelope mechanism works as follows: the symmetric key, which is used to encrypt the message (e.g., plaintext alone or plaintext plus signature), can be referred to as the session key. If the session key is sent to the receiver in cleartext, anyone can access the session key in transit and confidentiality can easily be compromised. Therefore, it is critical that the session key be encrypted before sending it to the receiver. The session key is encrypted using the receiver's public key.

The encrypted message and the encrypted session key are sent to the receiver who, in turn, opens the session key with the receiver's private key (to ensure authentication of sender and integrity of session key). The session key is then applied to the message to get it in plaintext. The process of encrypting the bulk data using symmetric key cryptography and encrypting the symmetric key with a public key algorithm is referred to as a digital envelope.

## Public Key Infrastructure

If an individual wants to send messages or electronic documents and sign them with a digital signature using a public key cryptographic system, how does the individual distribute the public key in a secure way? If the public key is distributed electronically, then it could be intercepted and changed. To prevent this from occurring, a framework needs to be established to issue, maintain and revoke public key certificates by a trusted party. This framework is known as a PKI.

PKI allows users to interact with other users and applications, and obtain and verify identities and keys from trusted sources. The actual implementation of PKI varies according to specific requirements. Key elements of the infrastructure are as follows:
• **Digital certificates**—A digital credential is composed of a public key and identifying information about the owner of the public key. The purpose of digital certificates is to associate a public key with the individual's identity. These certificates are electronic documents, digitally signed by some trusted entity with its private key (transparent to users), that contains information about the individual and his/her public key.

The process involves proving the sender's authenticity. When a person digitally signs a document, that person attaches a digital certificate issued by a trusted entity. The receiver of the message and accompanying digital certificate relies on the public key of the trusted third-party certificate authority (that is included with the digital certificate or obtained separately) to authenticate the message. The receiver can link the message to a person, not simply to a public key, because of their trust in this third party.

The status and values of a current user's certificate should include a distinguishing username, an actual public key, the algorithm used to compute the digital signature inside the certificate and a certificate validity period.
• **Certificate authority**—A CA is an authority in a network that issues and manages security credentials and public keys for message signature verification or encryption. The CA attests, to the authenticity of the owner (entity or individual) of a public key. The process involves a CA who makes a decision to issue a certificate based on evidence or knowledge obtained in verifying the identity of the recipient. As part of a PKI, a CA checks with a registration authority (RA) to verify information provided by the requestor of a digital certificate. If the RA verifies the requestor's information, the CA can then issue a certificate. Upon verifying the identity of the recipient, the CA signs the certificate with its private key for distribution to the user. Upon receipt, the user will verify the certificate signature with the CA's public key (e.g.,

commercial CAs such as Verisign issue certificates through web browsers). The ideal CA is authoritative (someone that the user trusts) for the name or key space it represents. A certificate always includes the owner's public key, expiration date and the owner's information.

Types of CAs may include:
– Organizationally empowered, which have authoritative control over those individuals in their name space
– Liability empowered, for example, choosing commercially available options (such as Verisign, Certco., Cybertrust, etc.) in obtaining a digital certificate

The CA is responsible for managing the certificate throughout its life cycle. Key elements or subcomponents of the CA structure include the certification practice statement (CPS), RAs and certificate revocation lists (CRLs).
• **Registration authority**—An RA is an authority in a network that verifies user requests for a digital certificate and tells the CA to issue it. An optional entity separate from a CA, an RA would be used by a CA with a very large customer base. CAs use RAs to delegate some of the administrative functions associated with recording or verifying some or all of the information needed by a CA to issue certificates or CRLs and to perform other certificate management functions. However, with this arrangement, the CA still retains sole responsibility for signing either digital certificates or CRLs. RAs are part of a PKI. The digital certificate contains a public key that is used to encrypt messages and verify digital signatures. If an RA is not present in the PKI structure established, the CA is assumed to have the same set of capabilities as those defined for an RA.

The administrative functions that a particular RA implements will vary based on the needs of the CA, but must support the principle of establishing or verifying the identity of the subscriber. These functions may include the following:
– Verifying information supplied by the subject (personal authentication functions)
– Verifying the right of the subject to requested certificate attributes
– Verifying that the subject actually possesses the private key being registered and that it matches the public key requested for a certificate. This is generally referred to as proof of possession (POP).
– Reporting key compromise or termination cases where revocation is required
– Assigning names for identification purposes
– Generating shared secrets for use during the initialization and certificate pick-up phases of registration
– Initiating the registration process with the CA on behalf of the subject end entity
– Initiating the key recovery processing
– Distributing the physical tokens (such as smart cards) containing the private keys
• **Certificate revocation list**—The CRL is an instrument for checking the continued validity of the certificates for which the CA has responsibility. The CRL details digital certificates that are no longer valid because they were revoked by the CA. The time gap between two updates is critical and is also a risk in digital certificates verification.
• **Certification practice statement**—CPS is a detailed set of rules governing the certificate authority's operations. It provides an understanding of the value and trustworthiness of certificates issued by a given CA in terms of the controls that an organization observes, the method it uses to validate the authenticity of certificate applicants and the CA's expectations of how its certificates may be used.

## Applications of Cryptographic Systems

The use of cryptosystems by applications, for example, in e-mail and Internet transactions generally involves a combination of private/public key pairs, secret keys, hash functions and digital certificates. The purpose of applying these combinations is to achieve confidentiality, message integrity or nonrepudiation by either the sender or recipient. The process generally involves the sender hashing the message into a message digest or prehash code for message integrity, which is encrypted using the sender's private key for authenticity, integrity and nonrepudiation (i.e., digital signature). Using his/her secret key, the sender then will encrypt the message. Afterward, the secret key is encrypted with the recipient's public key, which has been validated through the recipient's digital certificate and provides message confidentiality.

The process on the receiving end reverses what has been done by the sender. The recipient uses his/her private key to decrypt the sender's secret key. He/she uses this secret key to decrypt the message, to expose it. If the prehash code has been encrypted with the sender's private key, the recipient verifies its authenticity using the public key contained in the sender's digital certificate and decrypts the prehash code, which provides the nonrepudiation to the recipient of the sender's message. For integrity purposes, the recipient calculates a posthash code, which should equal the prehash code.

Specific examples of this method or related variants are described below:

• **Secure Sockets Layer (SSL) and Transport Layer Security (TLS)**— These are cryptographic protocols which provide secure communications on the Internet. There are only slight differences between SSL 3.0 (the latest version) and TLS 1.0, but they are not interchangeable. For general concepts, both are called SSL. SSL is a session- or connection-layered protocol widely used on the Internet for communication between browsers and web servers, where any amount of data is securely transmitted while a session is established.

SSL provides end point authentication and communications privacy over the Internet using cryptography. In typical use, only the server is authenticated while the client remains unauthenticated. Mutual authentication requires PKI deployment to clients. The protocols allow client-server applications to communicate in a way designed to prevent eavesdropping, tampering and message forgery.

SSL involves a number of basic phases:
– Peer negotiation for algorithm support
– Public-key, encryption-based key exchange and certificate-based authentication
– Symmetric cipher-based traffic encryption

During the first phase, the client and server negotiate which cryptographic algorithms will be used. Current implementations support the following choices:
– For public-key cryptography:  RSA, Diffie-Hellman, DSA or Fortezza
– For symmetric ciphers:  RC2, RC4, IDEA, DES, Triple DES or AES
– For one-way hash functions:  MD5 or SHA-1 or SHA-256

SSL runs on layers beneath application protocols such as HTTP, SMTP and Network News Transfer Protocol (NNTP) and above the TCP transport protocol, which forms part of the TCP/IP suite. While SSL can add security to any protocol that uses TCP, it is most commonly used with HTTP to form Secured Hypertext Transmission Protocol (HTTPS). HTTPS serves to secure World Wide Web pages for applications such as electronic commerce. HTTPS uses public key certificates to verify the identity of endpoints.

SSL uses a hybrid of hashed, private and public key cryptographic processes to secure transactions over the Internet through a PKI. This provides the necessary confidentiality, integrity, authenticity and nonrepudiation features needed for e-commerce transactions over the Internet. SSL can be characterized as a two-way SSL process, for example for B-to-B e-commerce activities. It also is commonly associated as a one-way consumer process, whereby a retail customer over an Internet retail virtual store can verify, through a CA, the authenticity of an e-store's public key, which can then subsequently be used to negotiate a secure transaction.

The SSL provides for:
– Confidentiality
– Integrity
– Authentication, e.g., between client and server
– Nonrepudiation

The SSL handshake protocol is based on the application layer but provides for the security of the communication sessions, too. It negotiates the security parameters for each communication section. Multiple connections can belong to one SSL session and the parties participating in one session can take part in multiple simultaneous sessions.

• **Secure Hypertext Transfer Protocol (S/HTTP)**—As an application layer protocol, S/HTTP transmits individual messages or pages securely between a web client and server by establishing an SSL-type connection. Using the https:// designation in the URL, instead of the standard http://, directs the message to a secure port number rather than the default web port address. This protocol utilizes SSL secure features but does so as a message rather than as a session-oriented protocol.

• **IPSec**—IPSec is used for communication among two or more hosts, two or more subnets, or hosts and subnets.

This IP network layer packet security protocol establishes VPNs via transport and tunnel mode encryption methods. For the transport method, the data portion of each packet referred to as the encapsulation security payload (ESP) is encrypted, achieving confidentiality over the process. In the tunnel mode, the ESP payload and its header are encrypted. To achieve nonrepudiation, an additional authentication header (AH) is applied.

In establishing IPSec sessions in either mode, security associations (SAs) are established. SAs define which security parameters should be applied between the communicating parties as encryption algorithms, keys, initialization vectors, life span of keys, etc. Within either the ESP or AH header, respectively, an SA is established when a 32-bit security parameter index (SPI) field is defined within the sending host. The SPI is a unique identifier that enables the sending host to reference the security parameters to apply, as specified, on the receiving host.

IPSec can be made more secure by using asymmetric encryption through the use of Internet Security Association and Key Management Protocol/Oakley (ISAKMP/Oakley), which allows the key management, use of public keys, negotiation, establishment, modification and deletion of SAs and attributes. For authentication, the sender uses digital certificates. The connection is made secure by supporting the generation, authentication, distribution of the SAs and those of the cryptographic keys.

- **SSH**—A client-server program that opens a secure, encrypted command-line shell session from the Internet for remote logon. Similar to a VPN, SSH uses strong cryptography to protect data, including passwords, binary files and administrative commands, transmitted between systems on a network. SSH is typically implemented between two parties by validating each other's credentials via digital certificates. SSH is useful in securing Telnet and FTP services, and is implemented at the application layer, as opposed to operating at the network layer (IPSec implementation).
- **Secure Multipurpose Internet Mail Extensions (S/MIME)**—A standard secure e-mail protocol that authenticates the identity of the sender and receiver, verifies message integrity, and ensures the privacy of a message's contents, including attachments
- **Secure Electronic Transactions (SET)**—SET is a protocol developed jointly by VISA and Master Card to secure payment transactions among all parties involved in credit card transactions on behalf of cardholders and merchants. As an open system specification, SET is an application-oriented protocol that uses trusted third parties' encryption and digital signature processes, via a PKI infrastructure of trusted third-party institutions, to address confidentiality of information, integrity of data, cardholder authentication, merchant authentication and interoperability.

## *Encryption Risks and Password Protection*

The security of encryption methods relies mainly on the secrecy of keys. In general, the more a key is used, the more vulnerable it is to compromise. For example, password cracking tools for today's microcomputers can brute-force every possible key combination for a cryptographic hashing algorithm with a 40-bit key in a matter of a few hours.

The randomness of key generation is also a significant factor in the ability to compromise a key. When passwords are tied into key generation, the strength of the encryption algorithm is diminished, particularly when common words are used. This significantly lessens the key space combinations to search for the key (e.g., an eight-character password is comparable to a 32-bit key). Therefore, a 128-bit encryption algorithm's capabilities are diminished when encrypting keys based on passwords, and the passwords lack randomness. Therefore, it is important that effective password syntax rules are applied, and easily guessed passwords are prohibited.

<table>
<tr><td colspan="2" align="center">PRACTICE QUESTIONS</td></tr>
</table>

**5-7** Which of the following **BEST** determines whether complete encryption and authentication protocols for protecting information while being transmitted exist?

    A.    A digital signature with RSA has been implemented.

    B.    Work is being done in tunnel mode with the nested services of authentication header (AH) and encapsulating security payload (ESP).

    C.    Digital certificates with RSA are being used.

    D.    Work is being done in transport mode with the nested services of AH and ESP.

**5-8** Which of the following concerns about the security of an electronic message would be addressed by digital signatures?

    A.    Unauthorized reading

    B.    Theft

    C.    Unauthorized copying

    D.    Alteration

**5-9** Which of the following would be **MOST** appropriate to ensure the confidentiality of transactions initiated via the Internet?

    A.    Digital signature

    B.    Data Encryption Standard (DES)

    C.    Virtual private network (VPN)

    D.    Public key encryption

*See answers and explanations to the practice questions at the end of the chapter. (page 469)*

# 5.4.6 VIRUSES

The term virus is a generic term applied to a variety of malicious computer programs that send out requests to the operating system of the host system under attack to append the virus to other programs. In this way, viruses are self-propagating to other programs. They can be relatively benign (e.g., web application defacement) or malicious (e.g., deleting files, corrupting programs or causing a denial of service). Generally, viruses attack four parts of the computer:
• Executable program files
• The file-directory system, which tracks the location of all the computer's files
• Boot and system areas, which are needed to start the computer
• Data files

Another variant of a virus frequently encountered is a worm, which, unlike a virus, does not physically attach itself to another program. To propagate itself to the host systems, a worm typically exploits security weaknesses in operating systems' configurations. These problems are particularly severe in today's highly decentralized client-server environments.

Today, viruses or worms are transmitted easily from the Internet by downloading files to computers' web browsers. Viruses are also transmitted as attachments to e-mail, so that when word processing software opens the attachment, the system becomes infected if it is not using scanning software to review unopened attachments. Other methods of infection occur from files received through online services, computer bulletin board systems, LANs and even shrink-wrapped software that the user may buy from a retail store.

## *Virus and Worm Controls*

To effectively reduce the risk of computer viruses and worms infiltrating an organization, a comprehensive and dynamic antivirus program needs to be established. There are two major ways to prevent and detect viruses and worms that infect

computers and network systems. The first is by having sound policies and procedures in place (preventive controls) and the second is by technical means (detective controls), including antivirus software. Neither is effective without the other.

## Management Procedural Controls

Some of the policy and procedure controls that should be in place are:
- Build any system from original, clean master copies. Boot only from original media whose write protection has always been in place, if applicable.
- Allow no disk or other kind of media, such as miniature hard/flash drives which simply plug into the USB port, to be used until it has been scanned on a stand-alone machine that is used for no other purpose and is not connected to the network.
- Update virus software scanning definitions/signatures frequently.
- Write-protect all diskettes.
- Have vendors run demonstrations on their machines.
- Enforce a rule of not using shareware without first scanning the shareware thoroughly for a virus.
- Scan before any new software is installed since commercial software occasionally is supplied with a Trojan horse (viruses or worms).
- Insist that field technicians scan their disks on a test machine before they use any of their disks on the system.
- Ensure the network administrator uses workstation and server antivirus software.
- Ensure all servers are equipped with an activated current release of the virus-detection software.
- Consider encrypting files and then decrypting them before execution.
- Ensure bridge, router and gateway updates are authentic.
- Because backups are a vital element of an antivirus strategy, ensure a sound and effective backup plan is in place. This plan should account for scanning selected backup files for virus infection once a virus has been detected.
- Educate users so they will heed these policies and procedures. For example, many viruses and worms today are propagated in the form of e-mail attachments where the attachment, such as an executable Visual Basic script, infects the user's system upon opening the attachment. The hacker relies upon social engineering tactics in getting the user to open the attachment (e.g., "Love" viruses).
- Review antivirus policies and procedures at least once a year.
- Prepare a virus eradication procedure and identify a contact person.

## Technical Controls

Technical methods of preventing viruses can be implemented through hardware and software means. The following are hardware tactics that can reduce the risk of infection:
- Use boot virus protection (i.e., built-in, firmware-based virus protection).
- Use remote booting (e.g., diskless workstations).
- Use a hardware-based password.
- Use write-protected tabs on diskettes.
- Ensure that insecure protocols are blocked by the firewall from external segments and the Internet.

However, antivirus software is, by far, the most common antivirus tool and is considered the most effective means of protecting networks and host-based computer systems against viruses. Antivirus software is both a preventive and a detective control. Unless updated periodically, antivirus software will not be an effective tool against viruses.

Antivirus software contains a number of components that address the detection of viruses via scanning technologies from different angles. There are different types of antivirus software:
- **Scanners**—Look for sequences of bits called signatures that are typical of virus programs. The two primary types are:
  - **Virus masks or signatures**—Antivirus scanners check files, sectors and system memory for known and new (unknown to scanner) viruses, on the basis of virus masks or signatures. Virus masks or signatures are specific code strings that are recognized as belonging to a virus. For polymorphic viruses, the scanner sometimes has algorithms that check for all possible combinations of a signature that could exist in an infected file.
  - **Heuristic scanners**—Analyzes the instructions in the code being scanned and decides on the basis of statistical probability whether it could contain malicious code. Heuristic scanning results could indicate that a virus may be present, i.e., possibly infected. Heuristic scanners tend to generate a high level of false-positive errors (i.e., they indicate that a virus may be present when, in fact, no virus is present).

Scanners examine memory, disk-boot sectors, executables, data files and command files for bit patterns that match a known virus. Scanners, therefore, need to be updated periodically to remain effective.

- **Active monitors**—Interpret DOS and read-only memory (ROM) BIOS calls, looking for virus-like actions. Active monitors can be problematic because they cannot distinguish between a user request and a program or virus request. As a result, users are asked to confirm actions, including formatting a disk or deleting a file or set of files.
- **Integrity CRC checkers**—Compute a binary number on a known virus-free program that is then stored in a database file. The number is called a cyclical redundancy check (CRC). On subsequent scans, when that program is called to execute, it checks for changes to the files as compared to the database and reports possible infection if changes have occurred. A match means no infection; a mismatch means a change in the program has occurred. A change in the program could mean a virus within it. These scanners are effective in detecting infection; however, they can do so only after infection has occurred, i.e., it is often too late to save files. Also, CRC checkers can only detect subsequent changes to files, because they assume files are virus free in the first place. Therefore, they are ineffective against new files that are virus-infected and that are not recorded in the database. Integrity checkers take advantage of the fact that executable programs and boot sectors do not change often, if at all.
- **Behavior blockers**—Focus on detecting potentially abnormal behavior such as writing to the boot sector or the master boot record, or making changes to executable files. Blockers can potentially detect a virus at an early stage. Most hardware-based antivirus mechanisms are based on this concept.
- **Immunizers**—Defend against viruses by appending sections of themselves to files—somewhat in the same way that file viruses append themselves. Immunizers continuously check the file for changes and report changes as possible viral behavior. Other types of immunizers are focused to a specific virus and work by giving the virus the impression that the virus has already infected the computer. This method is not always practical since it is not possible to immunize files against all known viruses.

Once a virus has been detected by antivirus software, an eradication program can be used to wipe the virus from the hard disk. Sometimes eradication programs can kill the virus without having to delete the infected program or data file, while other times those infected files must be deleted. Still, other programs, sometimes called inoculators, do not allow a program to be run if it contains a virus.

## Antivirus Software Implementation Strategies

Organizations have to develop virus implementation strategies to effectively control and prevent the spread of viruses throughout their IS infrastructure. An important means of controlling the spread of viruses is to detect the virus at its point of entry—before it has the opportunity to cause damage. This includes everything from networks, server platforms and end-user workstations.

The user server or workstation level could include screening of software and data as they enter the machine, where antivirus programs can be set to perform:
- Scheduled virus scans (daily, weekly, etc.)
- Manual/on-demand scans, where the virus scan is requested by the user
- Continuous/on-the-fly scanning, where files are scanned as they are processed

At the corporate network level, in cases of interconnected networks, virus scanning software is used as an integral part of firewall technologies, referred to as virus walls. Virus walls scan incoming traffic with the intent of detecting and removing viruses before they enter the protected network. Virus walls normally work at the following levels:
- SMTP protection, to scan inbound and outbound SMTP traffic for viruses in coordination with the mail server
- HTTP protection, to prevent virus-infected files from being downloaded and to offer protection against malicious Java and ActiveX programs
- FTP protection, to prevent infected files from being downloaded

Virus walls most often are updated automatically with new virus signatures by their vendors on a scheduled basis or on an as-needed basis when dangerous, new virus strains emerge. Virus walls also provide facilities to log virus incidents and deal with the incident in accordance with preset rules. The presence of virus walls does not preclude the necessity for virus-detection software to be installed on computers within a network because the virus wall only addresses one channel through which viruses enter the network. Virus-detection software should be loaded on all computers within the network. Virus

signature files should be kept updated. The facility of automatic "live update" has become fairly popular and allows organizations to update the virus scanner signature files as soon as updates are available.

For virus scanners to be acceptable and viable, they should have the following features:
• Reliability and quality in the detection of viruses
• Memory resident, which is a continuous checking facility
• Efficiency, such as a reasonable working speed and usage of resources

| PRACTICE QUESTION |
| --- |
| 5-10    Which of the following is the **MOST** effective antivirus control? <br><br>     A.    Scanning e-mail attachments on the mail server <br>     B.    Restoring systems from clean copies <br>     C.    Disabling USB ports <br>     D.    An online antivirus scan with up-to-date virus definitions |
| *See answers and explanations to the practice questions at the end of the chapter. (page 470)* |

# 5.4.7 VOICE-OVER IP

IP telephony, also known as Internet telephony, is the technology that makes it possible to have a voice conversation over the Internet or over any dedicated IP network instead of dedicated voice transmission lines. The protocols used to carry the signal over the IP network are commonly referred to as Voice-over IP (VoIP). VoIP is a technology where voice traffic is carried on top of existing data infrastructure. Sounds are digitized into IP packets and transferred through the network layer before being decoded back into the original voice. VoIP has significantly reduced long-distance costs in a number of large organizations.

VoIP allows the elimination of circuit switching and the associated waste of bandwidth. Instead, packet switching is used, where IP packets with voice data are sent over the network only when data needs to be sent.

It has advantages over traditional telephony:
• Unlike traditional telephony, VoIP innovation progresses at market rates rather than at the rates of the multilateral committee process of the International Telecommunications Union (ITU)
• Lower costs per call or even free calls, especially for long-distance calls
• Lower infrastructure costs. Once IP infrastructure is installed, no or little additional telephony infrastructure is needed.

VoIP introduces security risks and opportunities. VoIP has a different architecture than traditional circuit-based telephony, and these differences result in significant security issues.

VoIP systems take a wide variety of forms, including traditional telephone handsets, conferencing units and mobile units. In addition to end-user equipment, VoIP systems include a variety of other components, including call processors/call managers, gateways, routers, firewalls and protocols. Most of these components have counterparts used in data networks, but the performance demands of VoIP mean that ordinary network software and hardware must be supplemented with special VoIP components.

When designing a VoIP system, the backup has to be considered. While telecom companies usually operate under the requirement to have 99.9999 percent uptime, data traffic normally has less reliability. For this reason, the backup has to be designed to ensure that communication will not be interrupted should undesirable events occur on the data backbone. Bandwidth capacity should be baselined to determine the current levels of data traffic and adjust the necessary additional bandwidth for voice traffic. Quality of service will need to be defined so that voice traffic will be given priority over data traffic. Other considerations are laws and regulations. Certain countries may ban the use of VoIP.

## *VoIP Security Issues*

With the introduction of VoIP, the need for security is more important because it is needed to protect two assets—the data and the voice.

Protecting the security of conversations in VoIP is vital now. In a conventional office telephone system, security is a more valid assumption. Intercepting conversations requires physical access to telephone lines or compromise of the office PBX. Only particularly security-sensitive organizations bother to encrypt voice traffic over traditional telephone lines. It cannot be said for Internet-based connections.

In VoIP, packets are sent over the network from a user's computer or VoIP phone to similar equipment on the other end. Packets may pass through several intermediate systems that are not under the control of the user's ISP. The current Internet architecture does not provide the same physical wire security as the phone lines. The key to securing VoIP is to use the security mechanisms such as those deployed in data networks (e.g., firewalls, encryption) to emulate the security level currently used by public switched telephone network (PSTN) network users.

The main concern with VoIP solutions is that while, in the case of traditional telephones, if the system is disrupted, then different sites of the organization could still be reached via telephone. With VoIP, a computer system disruption also terminates the telephone since both are supported by the same devices. In this case, only mobile phones will function. Thus, a backup communications facility should be planned for if the availability of communications is vital to the organization. This would be the case with branches of financial institutions.

Another issue might arise with the fact that IP telephones and their supporting equipment require the same care and maintenance as computer systems do.

Thus, operating system patches and virus signature updates must be promptly applied to prevent a potential system outage. To enhance the protection of the telephone system and data traffic, the VoIP infrastructure should be segregated using virtual local area networks (VLANs). Any connections between these two infrastructures should be protected using firewalls that can interpret VoIP protocols.

In many cases, session border controllers (SBCs) are utilized to provide security features for VoIP traffic similar to that provided by firewalls. SBCs can be configured to filter specific VoIP protocols, monitor for DoS attacks, and provide network address and protocol translation features.

# 5.4.8 PRIVATE BRANCH EXCHANGE

A private branch exchange (PBX) is a sophisticated computer-based switch that can be thought of as essentially a small, in-house phone company for the organization that operates it. Protection of the PBX is, thus, a high priority. Failure to secure a PBX can result in exposing the organization to toll fraud, theft of proprietary or confidential information, loss of revenue, or legal entanglements.

PBXs have been a part of organizations' communication infrastructures since the early 1920s, originally using analog technology. PBXs of today use digital technology; digital signals are converted to analog for outside calls on the local loop using Plain Old Telephone Service (POTS), which refers to the standard telephone service that most homes use.

Digital PBXs are widespread throughout industry and public organizations, having replaced their analog predecessors. The advent of software-based PBXs has provided a wealth of communications capabilities within these switches. Today, even the most basic PBX systems have a wide range of capabilities that were previously available only in large-scale switches. These new features have opened up many new opportunities for an intruder to attempt to exploit the PBX, particularly by usage of these features for a purpose that was never intended.

Attributes of today's PBXs include:
• More than two telephone trunk (multiple phone) lines that terminate at the PBX
• The use of digital phones that permit integrated voice/data workstations

- Scalable computer-based PBX systems with memory that manages the switching of the calls within the PBX
- Distributed architecture with multiple switches in a hierarchical or meshed configuration with distributed intelligence providing enhanced reliability
- Nonblocking configurations where all attached devices can be engaged in calls simultaneously
- The network of lines within the PBX
- An operator console or switchboard for a human operator

One of the principal purposes of a PBX is to save the cost of requiring a line for each user to the telephone company's central office. Also, it is easier to call someone within a PBX because only three or four digits need to be dialed.

## PBX Risks

PBX environments involve many security risks, presented by people both internal and external to the organization. If a PBX is not correctly configured, backdoors can be easily established for unauthorized purposes. The threats to PBX telephone systems are many, depending on the goals of these attackers, and include:
- **Theft of service**—Toll fraud, probably the most common of motives for attackers
- **Disclosure of information**—Data disclosed without authorization, either by deliberate action or by accident. Examples include eavesdropping on conversations and unauthorized access to routing and address data.
- **Data modification**—Data altered in some meaningful way by reordering, deleting or modifying it. For example, an intruder may change billing information or modify system tables to gain additional services.
- **Unauthorized access**—Actions that permit an unauthorized user to gain access to system resources or privileges
- **Denial of service**—Actions that prevent the system from functioning in accordance with its intended purpose. A piece of equipment or entity may be rendered inoperable or forced to operate in a degraded state; operations that depend on timeliness may be delayed.
- **Traffic analysis**—A form of passive attack in which an intruder observes information about calls (although not necessarily the contents of the messages) and makes inferences, e.g., from the source and destination numbers or frequency and length of the messages. For example, an intruder observes a high volume of calls between a company's legal department and patent office, and concludes that a patent is being filed.

PBXs are sophisticated computer systems, and many of the threats and vulnerabilities associated with operating systems are shared by PBXs. But there are two important ways in which PBX security is different from conventional OS security:
- **External access/control**—As with larger telephone switches, PBXs typically require remote maintenance by the vendor. Instead of relying on local administrators to make OS updates and patches, organizations normally have updates installed remotely by the switch manufacturer. This, of course, requires remote maintenance ports and access to the switch by a potentially large pool of outside parties.
- **Feature richness**—The wide variety of features available on PBXs, particularly administrative features and conference functions, provide the possibility of unexpected attacks. A feature may be used by an attacker in a manner that was not intended by its designers. Features may also interact in unpredictable ways, leading to system compromise, even if each component of the system conforms to its security requirements and the system is operated and administered correctly.

Some additional control weaknesses include:
- Uncontrolled definition of direct inward dial (DID) lines, which would allow an external party to request a dial tone locally, enabling that person to make unauthorized long-distance phone calls
- Lack of system access controls over long-distance phone calls (e.g., default system vendor passwords unchanged, 24/7 availability of PBX lines)
- Lack of blocking controls for long-distance phone calls to particular numbers (e.g., hot numbers, cellular numbers, etc.)
- Lack of control over the numbers destined for fax machines and modems
- Not activating the option to register calls, which enables the use of call-tracking logs

Although most features are common from PBX to PBX, the implementation of these features may vary and the degree of vulnerability, if any, will depend on how each feature is implemented. For example, many PBX vendors have proprietary designs for the Digital Signaling Protocol between the PBX and the user instruments.

Knowing the design implementation will aid in determining if an intruder or an insider have an easy way to exploit weaknesses or normal functions.

## PBX Audit

When planning a PBX audit, the type of skills, the number of auditors and the length of time required to perform the audit cannot be determined without a preliminary assessment of the PBX system, since these depend on the size and complexity of the chosen PBX. The type of perceived threat and the seriousness of any discovered vulnerabilities must be decided by the auditor. Consequently, any corrective actions must also be determined based on the cost of the loss compared with the cost of the corrective action.

A list of critical items of PBX structure, usage and setup will be given, together with specific risks and applicable controls.

## PBX System Features

Many features sometimes available to the system may be used by phreakers (security crackers) or intruders for illegal purposes, including:
• Eavesdropping on conversations, without the other parties being aware of it
• Eavesdropping on conference calls
• Illegally forwarding calls from specific instruments to remote numbers
• Forwarding a user's instrument to an unused or disabled number, thereby making it unreachable by external calls

## PBX System Attacks

PBX system features and capabilities may present significant vulnerabilities. This occurs because, with such a large number of features available, it becomes difficult for the manufacturer to consider all individual risks and the potential problems caused by the manner in which different features may interact. This may result in vulnerabilities that allow an intruder unwanted access to the PBX and its instruments. See **exhibit 5.4**, showing PBX system features and corresponding risks.

Protecting against all of the risks is not easy. Knowing that a given vulnerability, in fact, exists is already a vital indication. A conservative approach of enabling only the needed features is advisable. Following are some controls to minimize PBX system attacks:
• Where possible, configure and secure separate and dedicated administrative ports.
• Control the definition of DID lines to avoid an external party requesting a dial tone locally, disabling that person's ability to make unauthorized long-distance phone calls.
• Establish system access controls over long-distance phone calls (e.g., change default system vendor passwords, limit the 24/7 availability of PBX lines).
• Block controls for long-distance phone calls to particular numbers (e.g., hot numbers, cellular numbers).
• Establish control over the numbers destined for fax machines and modems.
• Activate the option to register calls, enabling the use of call-tracking logs.

## Hardware Wiretapping

A PBX's susceptibility to tapping depends on the methods used for communication between the PBX and its attached devices. This communication may include voice, data and signaling information. The signaling information is typically composed of commands to the devices (e.g., turn on indicators, microphones and speakers) and status from the devices (e.g., hook status and keys pressed). Communications methods use analog voice with or without separate control signals, analog voice with inclusive control signals, and digital voice with inclusive control signals.

Tapping or intrusion in the control sequences is possible using various appropriate hardware technologies, which often include device modification.

The following measures provide controls to minimize risks:
• Physical security of the PBX facilities
• Usage of appropriate antitamper devices on critical hardware components

| Exhibit 5.4—PBX System Features and Risks | | |
|---|---|---|
| **System Feature** | **Description** | **Risk** |
| Automatic call distribution | Allows a PBX to be configured so that incoming calls are distributed to the next available agent or placed on hold until one becomes available | Tapping and control of traffic |
| Call forwarding | Allows specifying an alternate number to which calls will be forwarded based on certain conditions | User tracking |
| Account codes | Used to:<br>• Track calls made by certain people or for certain projects for appropriate billing<br>• Dial-in system access (user dials from outside and gains access to the normal features of the PBX)<br>• Changing the user class of service so a user can access a different set of features (i.e., the override feature) | Fraud, user tracking, nonauthorized features |
| Access codes | Key for access to specific features from the part of users with simple instruments, i.e., traditional analog phones | Nonauthorized features |
| Silent monitoring | Silently monitors other calls | Eavesdropping |
| Conferencing | Allows for conversation among several users | Eavesdropping, by adding unwanted/unknown parties to a conference |
| Override (Intrude) | Provides for the possibility to break into a busy line to inform another user of an important message | Eavesdropping |
| Autoanswer | Allows an instrument to automatically go when called—usually gives an audible or visible warning which can easily be turned off. | Gaining information not normally available, for various purposes (i.e., eavesdropping through the automatic answering of an instrument in a conference room) |
| Tenanting | Limits system user access to only those users who belong to the same tenant group—useful when one company leases out part of its buildings to other companies and tenants share an attendant, trunk lines, etc. | Illegal usage, fraud, eavesdropping |
| Voice mail | Stores messages centrally and—by using a password—allows for retrieval from inside or outside lines | Disclosure or destruction of all messages of a user when that user's password is known or discovered by an intruder, disabling of the voice mail system and even the entire switch by lengthy messages or embedded codes, illegal access to external lines |
| Privacy release | Supports shared extensions among several devices, ensuring that only one device at a time can use an extension. Privacy release disables the security by allowing devices to connect to an extension already in use. | Eavesdropping |
| Nonbusy extensions | Allows calls to an in-use extension to be added to a conference PBX when that extension is on conference and already off-hook | Eavesdropping a conference in progress |
| Diagnostics | Allows for bypassing normal call restriction procedures. This kind of diagnostic is sometimes available from any connected device. It is a separate feature, in addition to the normal maintenance terminal or attendant diagnostics. | Fraud and illegal usage |
| Camp-on or call waiting | When activated, sends a visual or audible warning to an off-hook instrument that is receiving another call. Another option of this feature is to conference with the camped-on or call-waiting party. | Making the called individual a party to a conference without knowing it |
| Dedicated connections | Connections made through the PBX without using the normal dialing sequence. It can be used to create hot-lines between devices, i.e., one rings when the other goes off-hook. It is also used for data connections between devices and the central processing facility. | Eavesdropping on a line |

## Hardware Conferencing

When implemented in hardware, the conferencing feature may employ a circuit card known as a conference bridge or a signal processor chip. This allows multiple lines to be bridged to create a conference where all parties can both speak and listen. Some PBXs have a feature where all parties can hear, but only certain parties can speak. An intruder could try to obtain a connection to the bridge where the conference could be overheard. A hardware modification to the bridge itself may make it possible to cause the output of the bridge available to a specific port. As in device modifications, some additional steps must be taken to receive this information. This may include modifying the database to make the intruder a permanent member of the bridge so any conference on that bridge could be overheard.

The following measures provide controls to minimize risks:
• Establish a strong physical security to prevent unauthorized access to telephone closets and to the PBX facilities. Whenever possible, the PBX should be kept in a locked room with restricted access.
• Lock critical hardware with antitamper devices.

## Remote Access

Remote access is frequently an unavoidable necessity of maintenance, but it can represent a serious vulnerability. The maintenance features may be accessible via a remote terminal with a modem, a system console or other device or over an outside dial-in line. This allows for systems to be located over a large area (perhaps around the world) and have one central location from which maintenance can be performed. Often it is necessary for the switch manufacturer to have remote access to the switch, to install software upgrades or to restart a switch that has experienced service degradation. Dial-back modem usage is a basic precaution, but does not offer full protection. When possible, remote access should be left closed and should be opened only upon need or upon a verified request from the organization that performs the maintenance. It is easy for attackers to contact the switch manufacturer on the pretext of needing help with a particular type of switch, obtain the names of the manufacturer's remote maintenance personnel, and then masquerade as these personnel to obtain access to the victim's switch.

The following measures provide controls to minimize risks:
• A dial-back scheme
• Careful scrutiny and proper authentication of requests to open the remote control

## Maintenance

A common maintenance feature is maintenance out of service (MOS). This feature allows maintenance personnel to place a line out of service for maintenance. It is typically used when a problem is detected with a line or when it is desired to disable a line. However, if a line is placed into MOS while it is in operation, the PBX may terminate its signaling communication with the instrument and leave the instrument's voice channel connection active, even after the telephone device is placed on-hook. If the MOS feature were to function in this manner, the potential exists for someone to use the MOS feature to establish a live microphone connection to a user's location without the user's knowledge and, thereby, eavesdrop on the area surrounding the user's telephone.

Another common maintenance feature is the ability to connect two lines together to transmit data from one line to the other and verify whether or not the second line receives the data properly. This feature would allow someone with maintenance access to connect a user's instrument to an instrument at another location to eavesdrop on the area surrounding the user's telephone without the user's knowledge.

Also, the PBX may support some maintenance features that are not normally accessible to the owner/operator of the PBX for several reasons. These types of utilities vary greatly from one PBX to another, so a general approach to finding them cannot be detailed. Some suggested courses of action to verify the existence of such features are listed below:
• Ask the manufacturer or maintenance company if any such features exist.
• Attempt to learn about undocumented usernames/passwords.
• Attempt to search the system's programmable read-only memory (PROM) or disks for evidence of such features.
• View the system load files with a binary editor to determine whether this reveals the names of undocumented commands among a list of known maintenance commands, which can be recognized in the binaries.

• Verify the existence of alarm features.
• Enable and review usage and intervention logs.

## Special Manufacturer's Features

These types of features would most likely be accessible via undocumented username/password access to the maintenance and/or administrative tools. Some possible undocumented features and their associated risks are listed below:

• **Database upload/download utility**—Such a utility allows the manufacturer to download the database from a system that is malfunctioning and examine it at their location to try to determine the cause of the malfunction. It would also allow the manufacturer to upload a new database to a PBX in the event that the database became so corrupted that the system became inoperable. Compromise of such a utility could allow an adversary to download a system's database, insert a Trojan horse or otherwise modify it to allow special features to be made available to the adversary, and upload the modified database back into the system.

• **Database examine/modify utility**—Such a utility allows the manufacturer to remotely examine and modify a system's database to repair damage caused by incorrect configuration, design bugs or tampering. This utility could also provide an intruder with the ability to modify the database to gain access to special features.

• **Software debugger/update utility**—This type of utility gives the manufacturer the ability to remotely debug a malfunctioning system. It also allows the manufacturer to remotely update systems with bug fixes and software upgrades. Such a utility could also grant an adversary the same abilities. This is perhaps the most dangerous vulnerability because access to the software would give an adversary virtually unlimited access to the PBX and its associated instruments.

## Manufacturer's Development and Test Features

There may be features that were added to the system during its development phase that were forgotten and not removed when production versions were released. There also may be hidden features that were added by a person on the development team with the intent of creating a backdoor into the customer's systems. The test features are probably easy to access for ease of development and have few restrictions to reduce development time.

Potential forms of attack include:
• Undocumented username/passwords
• Entering out-of-range values in database fields
• Dialing undocumented access codes on instruments
• Pressing certain key sequences on instruments

The following measures provide controls to minimize risks:
• Establish strong authentication of external technicians.
• Keep maintenance terminals in a locked, restricted area.
• Turn off maintenance features when not needed, if possible.

## Software Loading and Update Tampering

When software is initially loaded onto a PBX and when any software updates/patches are loaded, the PBX is particularly vulnerable to software tampering. A software update sent to a PBX administrator could be intercepted by an adversary. The update could be modified to allow special access or special features to the adversary. The modified update would then be sent to the PBX administrator who would install the update and unknowingly give the adversary unwanted access to the PBX.

A control for software loading and updates would be strong modification—tamper detection based on cryptography used in software packages. Conventional error detection codes, such as checksums or cyclic redundancy checks (CRCs), are not sufficient to ensure tamper detection.

## Crash-restart Attacks

System crashes may indicate a DoS vulnerability. The means by which a system may be crashed vary significantly from one system to another. The following list suggests a few features and conditions that can sometimes trigger a system crash:
• Call forwarding
• Voice mail
• Physical removal of hardware or media from the PBX
• Use of administrative/maintenance terminal functions
• Direct modification of the system or the database. This may be possible if the media can be read by utilities typically found on a PC or workstation.
• Normal system shutdown procedures

These approaches should be tested as possible ways of exposing the weaknesses discussed in the remainder of this section. One possible additional weakness is that a crash may interrupt the control flow and leave microphones open, so an intruder could overhear what is said after the crash. A further danger is that, in some cases, embedded logons and passwords are restored upon rebooting the system, making it possible for a remote operator to complete the remote restart. However, this also makes it possible for an attacker to gain administrator privileges on a system by crashing the system, then applying a known embedded logon ID/password combination.

Controls for these types of attacks include:
• Crash-restart vulnerability tests and preventing or forbidding, if possible, the triggering of events
• Restart procedures that eliminate the vulnerability from loss of control. This may require doing a cold start (i.e., complete shutdown, power-off, and restart) in the event of a system crash.
• If embedded passwords are found, patching the load module to replace them. Authorized manufacturer personnel can be given the new password, if needed.
• A PBX firewall to enhance the protection of the PBX. In recent years, specialized firewalls have been developed specifically for the protection of PBX systems.

## Passwords

Most PBXs grant administrative access to the system database through a system console or a generic dumb terminal. Username/password combinations are often used to protect the system from unwanted changes to the database. If remote access to the maintenance features is available, it is usually restricted by some form of password protection. There may be a single fixed maintenance account, multiple fixed maintenance accounts or general user-defined maintenance accounts. The documentation provided with the PBX should state what type of maintenance access is available. The documentation should also indicate how passwords function. Dangers from improper definition and usage of passwords are the loss of control and confidentiality, illegal usage and tampering of the database.

Controls for passwords include:
• Passwords resistant to cracking by automated tools. A password generator that creates random passwords can go a long way in defeating password crackers.
• Monitoring of multilevel password rights
• Setting an appropriate time-out period for logins

# 5.5 AUDITING INFORMATION SECURITY MANAGEMENT FRAMEWORK

Auditing the information security framework of an organization involves the audit of logical access, the use of techniques for testing security and the use of investigation techniques.

# Protection of Information Assets

## 5.5.1 AUDITING INFORMATION SECURITY MANAGEMENT FRAMEWORK

The information security management framework should be reviewed per the basic elements in an information security framework.

### Reviewing Written Policies, Procedures and Standards

Policies and procedures provide the framework and guidelines for maintaining proper operation and control. The IS auditor should review the policies and procedures to determine if they set the tone for proper security and provide a means for assigning responsibility for maintaining a secure computer processing environment.

### Logical Access Security Policies

These policies should encourage limiting logical access on a need-to-know basis. They should reasonably assess the exposure to the identified concerns.

### Formal Security Awareness and Training

Effective security will always be dependent on people. As a result, security can only be effective if employees know what is expected of them and what their responsibilities are. They should know why various security measures, such as locked doors and use of logon IDs, are in place and the repercussions of violating security.

Promoting security awareness is a preventive control. Through this process, employees become aware of their responsibilities for maintaining good physical and logical security. This can also be a detective measure, because it encourages people to identify and report possible security violations.

Training should start with the new employee orientation or induction process. Ongoing awareness can be provided in company newsletters through visible and consistent security enforcement and short reminders during staff meetings. The security administrator should direct the program. To determine the effectiveness of the program, the IS auditor should interview a sample of employees to determine their overall awareness.

### Data Ownership

Data ownership refers to the classification of data elements and the allocation of responsibility for ensuring that they are kept confidential, complete and accurate. A key point of ownership is that, by assigning responsibility for protecting computer data to particular employees, accountability is established. The IS auditor can use this information to determine if proper ownership has been assigned and whether the data owner is aware of the assignment. The IS auditor should also review a sample of job descriptions to ensure that responsibilities and duties are consistent with the information security policy. The auditor should review the classification of data and evaluate their appropriateness, as they relate to the area under review.

### Data Owners

These people are generally managers and directors responsible for using information for running and controlling the business. Their security responsibilities include authorizing access, ensuring that access rules are updated when personnel changes occur, and regularly review access rules for the data for which they are responsible.

### Data Custodians

These people are responsible for storing and safeguarding the data, and include IS personnel such as systems analysts and computer operators.

### Security Administrator

Security administrators are responsible for providing adequate physical and logical security for IS programs, data and equipment. (The physical security may be handled by someone else, not always by the security administrator.) Normally, the information security policy will provide the basic guidelines under which the security administrator will operate.

## New IT Users

New IT users (employees or third parties) and, in general, all new users assigned PCs or other IT resources should sign a document containing the main IT security obligations that they are thereby engaged to know and observe. These are:
- Reading and agreeing to follow security policies
- Keeping logon IDs and passwords secret
- Creating quality passwords according to policy
- Locking their terminal screens when not in use
- Reporting suspected violations of security
- Maintaining good physical security by keeping doors locked, safeguarding access keys, not disclosing access door lock combinations and questioning unfamiliar people
- Conforming to applicable laws and regulations
- Use of IT resources only for authorized business purposes

## Data Users

Data users, including the internal and the external user community, are the actual users of the computerized data. Their levels of access into the computer should be authorized by the data owners, and restricted and monitored by the security administrator. Their responsibilities regarding security are to be vigilant regarding the monitoring of unauthorized people in the work areas and comply with general security guidelines and policies.

## Documented Authorizations

Data access should be identified and authorized in writing. The IS auditor can review a sample of these authorizations to determine if the proper level of written authority was provided. If the facility practices data ownership, only the data owners provide written authority.

## Terminated Employee Access

Termination of employment can occur in the following circumstances:
- On the request of the employee (voluntary resignation from service)
- Scheduled (on retirement or completion of contract)
- Involuntary (forced by management in special circumstances)

In case of involuntary termination of employment, the logical and physical access rights of employees to the IT infrastructure should either be withdrawn completely or highly restricted, as early as possible, before the employee becomes aware of the termination or its likelihood. This ensures that terminated employees cannot continue to access potentially confidential or damaging information from the IT resources or perform any action that would result in damage of any kind to the IT infrastructure, applications and data. Similar procedures should be in place to terminate access for third parties upon terminating their activities with the organization.

When it is necessary for employees to continue to have access, such access must be monitored carefully and continuously and should take place with senior management's knowledge and authorization.

In case of voluntary or scheduled termination of employment, it is management's prerogative to decide whether access is restricted or withdrawn. This depends on:
- The specific circumstances associated with each case
- The sensitivity of the employee's access to the IT infrastructure and resources
- The requirements of the organization's information security policies, standards and procedures

## Security Baselines

A baseline security plan is meant to be used as a first step to IT security. The baseline plan should be followed with a full security evaluation and plan. **Exhibit 5.5** illustrates baseline security topics and their associated recommendations. **Exhibit 5.6** depicts a checklist for a baseline security evaluation.

# Protection of Information Assets

## Access Standards

Access standards should be reviewed by the IS auditor to ensure they meet organizational objectives for separating duties, prevent fraud or error, and meet policy requirements for minimizing the risk of unauthorized access.

| Exhibit 5.5—IT Security Baseline Recommendations | | |
|---|---|---|
| **Topics** | **Objective** | **Recommendations** |
| Inventory | Establish and maintain an inventory | Users are expected to follow standards for managing computers connected to the network, and have registered network addresses. The operating system and owner should be included along with the data provided. |
| Antivirus | Install antivirus software with automatic updating | Antivirus software with an automatic DAT file should be updated at regular intervals—no less than weekly. |
| Passwords | Recognize the importance of passwords | Users must use only strong passwords. The IT department should provide password guidance. Departmental accounts are created for workgroups to prevent/avoid password sharing. |
| Patching | Make it automatic—less work for you, less chance for compromise | Each machine should be configured to patch automatically for operating system and basic software patching. A process should be set up that works for the department and minimizes disruptions at inconvenient times. Workstations should be more automated to enable system administrators the time to give servers the attention required to minimize the impact on services offered. |
| Minimizing services offered by systems | Eliminate unnecessary services—reducing security risk and saving time in the long run | To improve basic security and minimize effort to maintain systems, workstations should offer only needed services. Many operating systems are installed with services turned on. By removing services, a workstation's chances of being compromised are reduced and security risks are minimized. |
| Addressing vulnerabilities | Eliminate many vulnerabilities with good system administration | System compromises can be time-consuming and damage credibility and the business's integrity. Information from enterprisewide scans helps to identify vulnerabilities on each system and provide a baseline for comparison when system integrity is in question. |
| Backups | Allow easy recovery from user mistakes and hardware failure with backups | Backups should be made offsite for increased security. |

| Exhibit 5.6—Baseline Security Evaluation Checklist | |
|---|---|
| **Topics** | **Evaluation Questions** |
| Environment/ inventory | • What types of data are maintained by the enterprise (i.e., financial, statistical, graphical)?<br>• In what form are they maintained (i.e., spreadsheets, databases, etc.)?<br>• Is there any critical or confidential information maintained or handled? If so, how is it protected?<br>• Are there any specific requirements for handling data (legal or regulatory requirements)<br>• Have you identified machines that store or require access to confidential information?<br>• What type of operating systems exist?<br>• How many subnets exist?<br>• How many workstations/servers exist?<br>• In how many locations is there IT infrastructure?<br>• Has the wireless infrastructure been deployed? How is it secured?<br>• Is staff instructed on how to lock workstations when they step away?<br>• Are users aware that unexpected e-mail attachments should not be opened?<br>• Is staff aware that many compromises are due to social engineering and the sharing of information?<br>• Does the enterprise have a network diagram that includes IP addresses, room numbers and responsible parties?<br>• Has the enterprise limited and secured physical and remote access to network services?<br>• Is corporate hardware upgraded at regular intervals?<br>• Does the enterprise have a current documented inventory of hardware and software?<br>• Is all corporate software licensed?<br>• Is license documentation available (licenses, purchase orders) if a software audit is required? |

| Exhibit 5.6—Baseline Security Evaluation Checklist (*cont.*) | |
|---|---|
| **Topics** | **Evaluation Questions** |
| Antivirus | • Does the enterprise have an antivirus policy?<br>• Are all workstations running the latest version of antivirus software, scanning engine and virus signature file?<br>• Are DAT files downloaded automatically or manually? If manually, how often and why?<br>• Does staff know who to contact when a virus is found? |
| Passwords | • Is there a corporate policy requiring strong passwords?<br>• Is the enterprise using software that enforces strong passwords?<br>• Is password caching disabled on all workstations?<br>• Are passwords changed? If so, how often?<br>• Are employees aware that passwords and accounts are not to be shared?<br>• Does the system administrator have written authorization to check for weak passwords? |
| Patching | • Are software patches applied to all operating systems automatically when possible? If done manually, how often?<br>• Are patches applied to web browsers and applications? If yes, how frequently?<br>• Do you backup each machine before applying a patch?<br>• Do you test patches prior to applying?<br>• Does the department have a documented process for patching?<br>• Do you subscribe to sufficient newsletters and groups to be aware of patches to all relevant hardware and software? |
| Minimizing services offered by systems | • Have you identified services that each user needs to accomplish job assignments?<br>• Have you removed unnecessary services which were installed by default?<br>• Does the technical staff review security settings and policies?<br>• Have you identified what services your systems are offering?<br>• Have you taken security measures for remote access?<br>• Have you transitioned to secure services? |
| Addressing vulnerabilities/ auditing | • Have you resolved vulnerabilities discovered by enterprise-wide scans?<br>• Who is the contact for vulnerability scans?<br>• Does the IT staff complete an independent vulnerability scan for the enterprise?<br>• Has the enterprise deployed any form of firewalls or IDS (host or network based)? Any under consideration? |
| Backup and recovery/business continuity | • Are files regularly backed up?<br>• Are files kept onsite in a secure location?<br>• Are backup files sent offsite to a physically secure location?<br>• Are backup files periodically restored as a test to verify whether they are a viable alternative?<br>• Can you ensure that any forms of media containing confidential and sensitive information are sanitized before disposal?<br>• Is there redundant hardware to allow work to continue in the event of a single hardware failure?<br>• Does the enterprise have the ability to continue to function if central services is not available?<br>• Does the enterprise have the ability to continue to function in the event of a wide area network failure?<br>• Have you responded to and recovered from any abuse issues/incidents? |
| IT staff | • How many of IT staff are employed full-time/part-time?<br>• Does each IT staff member have a current job description?<br>• Do job descriptions and evaluations include IT security duties?<br>• Does the department have sufficient documentation to ease the transition of incoming/outgoing staff?<br>• Does the enterprise have a privacy policy?<br>• Are all staff aware of privacy considerations?<br>• Are management/department users aware of the types of (private/nonpublic) information available to systems administrators?<br>• Does the enterprise have a privacy policy to address this privileged information (confidentiality agreement/nondisclosure agreement)?<br>• Does the enterprise have a firewall or IDS, or other software for network diagnosis? Does the enterprise have tools requiring privileges and access to confidential information acquired via routers, switches, IDS, firewalls, etc.? |

Standards for security may be defined:
• At a generic level (e.g., all passwords must be at least five characters long)
• For specific machines (e.g., all UNIX machines can be configured to enforce password changes)
• For specific application systems (e.g., sales ledger clerks can access menus that allow entry of sales invoices, but may not access menus that allow check authorization)

# Protection of Information Assets

## 5.5.2 AUDITING LOGICAL ACCESS

When evaluating logical access controls the IS auditor should:
• Obtain a general understanding of the security risks facing information processing, through a review of relevant documentation, inquiry, observation, risk assessment and evaluation techniques
• Document and evaluate controls over potential access paths into the system to assess their adequacy, efficiency and effectiveness by reviewing appropriate hardware and software security features and identifying any deficiencies or redundancies
• Test controls over access paths to determine whether they are functioning and effective by applying appropriate audit techniques
• Evaluate the access control environment to determine if the control objectives are achieved by analyzing test results and other audit evidence
• Evaluate the security environment to assess its adequacy by reviewing written policies, observing practices and procedures, and comparing them with appropriate security standards or practices and procedures used by other organizations

### Familiarization With the IT Environment

This is the first step of the audit and involves obtaining a clear understanding of the technical, managerial and security environment of the IS processing facility. This typically includes interviews, physical walk-throughs, review of documents and risk assessments.

### Documenting the Access Paths

The access path is the logical route an end user takes to access computerized information. This starts with a terminal/workstation and typically ends with the data being accessed. Along the way, numerous hardware and software components are encountered. The IS auditor should evaluate each component for proper implementation and physical and logical access security. The typical sequence of the components is as follows:
• A PC, which is part of the LAN, is used by an end user to sign on. The PC should be physically secure and the logon ID/password used for sign-on should be subject to the restrictions identified previously in this chapter in the Features of Password and Password Syntax Rules sections.
• The telecommunications software (LAN server or terminal emulator if connecting to a mainframe) intercepts the logon to direct it to the appropriate telecommunication link. The telecommunication software can restrict PCs to specific data or application software. A key audit issue with telecommunication software is to ensure that all applications have been defined within the software and that the various optional telecommunication control and processing features used are appropriate and approved by management. This analysis typically requires the assistance of a system software analyst.
• The transaction processing software may be the next component in the access path. This software routes transactions to the appropriate application software. Key audit issues include ensuring proper identification/authentication of the user (logon ID and password) and authorization of the user to gain access to the application. This analysis is performed by reviewing internal tables that reside in the transaction processing software or in separate security software. Access to these should be restricted to the security administrator.
• The application software then is encountered and should process transactions in accordance with program logic. Audit issues include restricting access to the production software library to only the implementation coordinator.
• The DBMS directs access to the computerized information. Audit issues include ensuring that all data elements are identified in the data dictionary, that access to the data dictionary is restricted to the database administrator (DBA) and that all data elements are subject to logical access control. The application data now can be accessed.
• The access control software can wrap logical access security around all of the above components. This is done via internal security tables. Audit issues include ensuring all of the above components are defined to the access control software, providing access rules that define who can access what on a need-to-know basis and restricting security table access to the security administrator.

## Interviewing Systems Personnel

To control and maintain the various components of the access path, as well as the operating system and computer mainframe, technical experts often are required. These people can be a valuable source of information to the IS auditor when gaining an understanding of security. To determine who these people are, the IS auditor should meet with the IS manager and review organizational charts and job descriptions. Key people include the security administrator, network control manager and systems software manager.

The security administrator should be asked to identify the responsibilities and functions of the position. If the answers provided to this question do not support sound control practices or do not adhere to the written job description, the IS auditor should compensate by expanding the scope of the testing of access controls. Also, the IS auditor should determine whether the security administrator is aware of the logical accesses that must be protected, has the motivation and means to actively monitor logons to account for employee changes, and is knowledgeable in how to maintain and monitor access.

A sample of end users should be interviewed to assess their awareness of management policies regarding logical security and confidentiality.

## Reviewing Reports From Access Control Software

The reporting features of access control software provide the security administrator with the opportunity to monitor adherence to security policies. By reviewing a sample of security reports, the IS auditor can determine if enough information is provided to support an investigation and if the security administrator is performing an effective review of the report.

Unsuccessful access attempts should be reported and should identify the time, terminal, logon, and file or data element for which access was attempted.

## Reviewing Application Systems Operations Manual

An application systems manual should contain documentation on the programs that generally are used throughout a data processing installation to support the development, implementation, operations and use of application systems. This manual should include information about the platform the application can run on, DBMSs, compilers, interpreters, telecommunication monitors and other applications that can run with the application.

# 5.5.3 TECHNIQUES FOR TESTING SECURITY

Auditors can use different techniques for testing security. Some methods are described in the following subsections.

## Terminal Cards and Keys

The IS auditor can take a sample of these cards or keys and attempt to gain access beyond that which is authorized. Also, the IS auditor will want to know if the security administrator followed up on any unsuccessful attempted violations.

## Terminal Identification

The IS auditor can work with the network manager to get a listing of terminal addresses and locations. This list can then be used to inventory the terminals, looking for incorrectly logged, missing or additional terminals. The IS auditor should also select a sample of terminals to ensure that they are identified in the network diagram.

## Logon IDs and Passwords

To test confidentiality, the IS auditor could attempt to guess the password of a sample of employees' logon IDs (though this is not necessarily a test). This should be done discreetly to avoid upsetting employees. The IS auditor should tour end-user and programmer work areas looking for passwords taped to the side of terminals, the inside of desk drawers or located in card files. Another source of confidential information is the wastebasket. The IS auditor might consider going through the office wastebasket looking for confidential information and passwords. Users could be asked to give their password to the IS

auditor. However, unless specifically authorized for a particular situation and supported by the security policy, no user should ever disclose his/her password. Another way to test password strength is to analyze global configuration settings for password strength in the system application and compare this with the organization's security policy.

To test encryption, the IS auditor should work with the security administrator to attempt to view the internal password table. If viewing is possible, the contents should be unreadable. Being able to view encrypted passwords can still be dangerous. On some systems, although the passwords are impossible to decipher, if an individual can obtain the encryption program, they can encrypt common passwords and look for matches. This was a method used to break into UNIX computers prior to the development of shadow password files. Application logs should also be reviewed to ensure that passwords and logon IDs are not recorded in a clear form.

To test access authorization, the IS auditor should review a sample of access authorization documents to determine if proper authority has been provided and if the authorization was granted on a need-to-know basis. Conversely, the IS auditor should get a computer-generated report of computer access rules, take a sample to determine if the access is on a need-to-know basis, and attempt to match the sample of these rules to authorizing documents. If no written authorization is found, this indicates a breakdown in control and may warrant further review to determine the exposures and implications.

Account settings for minimizing authorized access should be available from most access control software or from the operating system. To verify that these settings actually are working, the IS auditor can perform the following manual tests:
- To test periodic change requirements, the IS auditor can draw on his/her experiences using the system and interview a sample of users to determine if they are forced to change their password after the prescribed time interval.
- To test for disabling or deleting of inactive logon IDs and passwords, the IS auditor should obtain a computer-generated list of active logon IDs. On a sample basis, the IS auditor should match this list to current employees, looking for logon IDs assigned to employees or consultants who are no longer with the company.
- To test for password syntax, the IS auditor should attempt to create passwords in a format that is invalid, such as too short, too long, repeated from the previous password, incorrect mix of alpha or numeric characters, or the use of inappropriate characters.
- To test for automatic logoff of unattended terminals, the IS auditor should log on to a number of terminals. The IS auditor then simply waits for the terminals to disconnect after the established time interval. Before beginning this test, the IS auditor should verify with the security administrator that this automatic logoff feature applies to all terminals.
- To test for automatic deactivation of terminals after unsuccessful access attempts, the IS auditor should attempt to log on, purposefully entering the wrong password a number of times. The logon ID should deactivate after the established number of invalid passwords has been entered. The IS auditor will be interested in how the security administrator reactivates the logon ID. If a simple telephone call to the security administrator with no verification of identification results in reactivation, then this function is not controlled properly.
- To test for masking of passwords on terminals, the IS auditor should log on to a terminal and observe if the password is displayed when entered.

## Controls Over Production Resources
Computer access controls should extend beyond application data and transactions. There are numerous high-level utilities, macro or job control libraries, control libraries, and system software parameters for which access control should be particularly strong. Access to these libraries would provide the ability to bypass other access controls.

The IS auditor should work with the system software analyst and operations manager to determine if access is on a need-to-know basis for all sensitive production resources. Working with the security administrator, the IS auditor should determine who can access these resources and what can be done with this access.

## Logging and Reporting of Computer Access Violations
To test the reporting of access violations, the IS auditor should attempt to access computer transactions or data for which access is not authorized. The attempts should be unsuccessful and identified on security reports. This test should be coordinated with the data owner and security administrator to avoid violation of security regulations.

## Follow-up Access Violations

To test the effectiveness and timeliness of the security administrator and data owner's responses to reported violation attempts, the IS auditor should select a sample of security reports and look for evidence of follow-up and investigation of access violations. If such evidence cannot be found, the IS auditor should conduct further interviews to determine why this situation exists.

## Bypassing Security and Compensating Controls

This is a technical area of review. As a result, the IS auditor should work with the system software analyst, network manager, operations manager and security administrator to determine ways to bypass security. This typically includes bypass label processing, special system maintenance logon IDs, operating system exits, installation utilities and input/output (I/O) devices. Working with the security administrator, the IS auditor should determine who can access these resources and what can be done with this access. The IS auditor should determine if access is on a need-to-know/have basis or if compensating detective controls exist.

Since many of these bypassing security features can be exploited by technically sophisticated intruders, the IS auditor should also ensure that:
• All uses of these features are logged, reported and investigated by the security administrator or system software manager
• Unnecessary bypass security features are deactivated
• If possible, the bypass security features are subject to additional logical access controls

## Review Access Controls and Password Administration

Access controls and password administration are reviewed to determine that:
• Procedures exist for adding individuals to the list of those authorized to have access to computer resources, changing their access capabilities and deleting them from the list
• Procedures exist to ensure that individual passwords are not inadvertently disclosed
• Passwords issued are of an adequate length, cannot be easily guessed and do not contain repeating characters
• Passwords are periodically changed
• User organizations periodically validate the access capabilities currently provided to individuals in their department
• Procedures provide for the suspension of user identification codes (logon IDs or accounts) or the disabling of terminal, microcomputer or data entry device activity—after a particular number of security procedure violations

# 5.5.4 INVESTIGATION TECHNIQUES

Investigation techniques include the investigation of computer crime and the protection of evidence and chain of custody, among others.

## Investigation of Computer Crime

Computer crimes are not reported in most cases simply because they are not detected at all. In many cases where computer crimes are detected, companies hesitate to report them since they generate a large amount of negative publicity that can affect their business. In such cases, the management of the affected company seeks to simply plug the loopholes used for the crime and move on. In addition, in many countries current laws are directed toward protecting physical property. It is very difficult to use such laws against computer crime. Even in jurisdictions where the laws have been updated, the investigation procedures are not always widely known and the necessary hardware and software tools are not always available to collect the digital evidence.

In the aftermath of a computer crime, it is very important that proper procedures are used to collect evidence from the crime scene. If data and evidence is not collected in the proper manner, it could be damaged and, even if the perpetrator is eventually identified, prosecution will not be successful in the absence of undamaged evidence. Therefore, after a computer crime, the environment and evidence must be left unaltered and specialist law enforcement officials must be called in. If the incident is to be handled in-house, the company must have a suitably qualified and experienced incident response team.

## *Protection of Evidence and Chain of Custody*

The evidence of a computer crime exists in the form of log files, file time stamps, contents of memory, etc. Rebooting the system or accessing files could result in such evidence being lost, corrupted or overwritten. Therefore, one of the first steps taken should be copying one or more images of the attacked system. Memory content should also be dumped to a file before rebooting the system. Any further analysis must be performed on an image of the system and on copies of the memory dumped—not on the original system in question.

In addition to protecting the evidence, it is also important to preserve the chain of custody. Chain of custody is a term that refers to documenting, in detail, how evidence is handled and maintained, including its ownership, transfer and modification. This is necessary to satisfy legal requirements that mandate a high level of confidence regarding the integrity of evidence.

# 5.6 AUDITING NETWORK INFRASTRUCTURE SECURITY

When performing an audit of the network infrastructure, the IS auditor should:
- Review network diagrams (campus LAN networks, WANS, metropolitan area networks [MANs]) that identify the organizations internetworking infrastructure, which would include gateways, firewalls, routers, switches, hubs, access servers, modems, etc. This information is important because the IS auditor will want to evaluate these links to determine whether proper physical and logical access controls are in effect and to inventory the various terminal connections to ensure that the diagram is accurate.
- Identify the network design implemented, including the IP strategy used, segmentation of routers and switches for campus environments, and WAN configurations and protocols
- Determine that applicable security policies, standards, procedures and guidance on network management and usage exist and have been distributed to staff, network management and administration. He/she should also ensure that staff members have been trained in their duties and responsibilities.
- Identify who is responsible for security and operation of Internet connections, and evaluate whether they have sufficient knowledge and experience to undertake this role
- Determine whether consideration has been given to the legal problems arising from use of the Internet. (Considerations should include liability arising from inaccurate web pages; legislation regarding sale or advertising of regulated products, such as financial services; implications of selling/buying in different countries; and the state of application of standard contract terms to electronic trading.)
- If the service is outsourced, review review service level agreements (SLAs) to ensure that they include provisions for security in addition to availability and quality of service
- Review network administrator procedures to ensure that hardware and software components are upgraded in response to new vulnerabilities

## 5.6.1 AUDITING REMOTE ACCESS

Remote use of information resources dramatically improves business productivity, but generates control issues and security concerns. In this regard, IS auditors should determine that all remote access capabilities used by an organization provide for effective security of the organization's information resources. Remote access security controls should be documented and implemented for authorized users operating outside of the trusted network environment.

In reviewing existing remote access architectures, IS auditors should assess remote access points of entry in addressing how many (known/unknown) exist and whether greater centralized control of remote access points is needed. IS auditors should also review access points for appropriate controls, such as in the use of VPNs, authentication mechanisms, encryption, firewalls and IDSs.

As part of this review, the IS auditor should also test dial-up access controls. To test for dial-up access authorization, the IS auditor should dial the computer from a number of authorized and unauthorized telephone lines. If controls are adequate, successful connection will occur with the authorized numbers only. The IS auditor should test the logical controls invoked

after authorized connections to the computer are achieved by using the successful dial-up connections to attempt to gain unauthorized file access. Performance of this test should be coordinated through the security administrator to avoid violating security regulations.

In reviewing future remote access initiatives, IS auditors should first determine whether design and development of remote access approaches are based on a cost-effective, risk-based solution taking into account business requirements. This includes assessing the types of remote environments applicable, the integrity and availability of telecommunication services, and required measures to take in protecting the corporate infrastructure.

## Auditing Internet Points of Presence

When auditing an organization's presence on the Internet, the IS auditor should review the use of the Internet to ensure that a business case has been demonstrated for the following possible uses:
• E-mail (i.e., communications to/from business partners, customers and the general public)
• Marketing (e.g., mechanism for communicating customer values such as online home shopping catalogue)
• Sales channel/electronic commerce (e.g., electronic ordering of goods/services, purchasing of goods from home shopping catalogue using credit cards, or electronic transmission of standard electronic data interchange (EDI)-formatted order messages by business partners)
• Channel of delivery for goods/services (such as online bookstores, and Internet banking)
• Information gathering (e.g., staff browsing the web for information)

## Network Penetration Tests

Combinations of procedures, whereby an IS auditor uses the same techniques as a hacker, are called penetration tests, intrusion tests or ethical hacking. This is an effective method of identifying the real-time risks to an information processing environment. During penetration testing, an auditor attempts to circumvent the security features of a system and exploits the vulnerabilities to gain access that would otherwise be unauthorized.

Scope can vary based on the terms and conditions of the client and requirements. However, from an audit risk perspective, the following should be mentioned clearly in the audit scope:
• Precise IP addresses/ranges to be tested
• Host restricted (i.e., hosts not to be tested)
• Acceptable testing techniques (i.e., social engineering, DoS/distributed denial of service [DDoS], SQL injections, etc.)
• Acceptance of proposed methodology from management
• Timing of attack simulation (i.e., business hours, off hours, etc)
• IP addresses of the source of attack simulation (to identify between approved simulated attack and actual attack)
• Point of contact for both the penetration tester/auditor and the targeted system owner/administrator
• Handling of information collected by the penetration tester/auditor (i.e., nondisclosure agreement [NDA] or reference to standard rules of engagement)
• Warning notification from penetration tester/auditor, before the simulation begins to avoid false alarms to law enforcement bodies

The different phases of penetration testing appear in **exhibit 5.7** and the corresponding procedures in **exhibit 5.8**.

Penetration testing is intended to mimic an experienced hacker attacking a live site. It should only be performed by experienced and qualified professionals who are aware of the risks of undertaking such work and can limit the damage resulting from a successful break-in to a live site (e.g., avoidance of DoS attacks). It is a simulation of a real attack and maybe restricted by the law, an organization's policy and federal regulations; therefore, it is imperative to obtain management's consent in writing before finalization of the test/engagement scope.

There are several types of penetration tests depending upon the scope, objective and nature of the test. Generally accepted and common types of penetration tests are:
• **External testing**—Refers to attacks and control circumvention attempts on the target's network perimeter from outside the target's system, i.e., usually the Internet

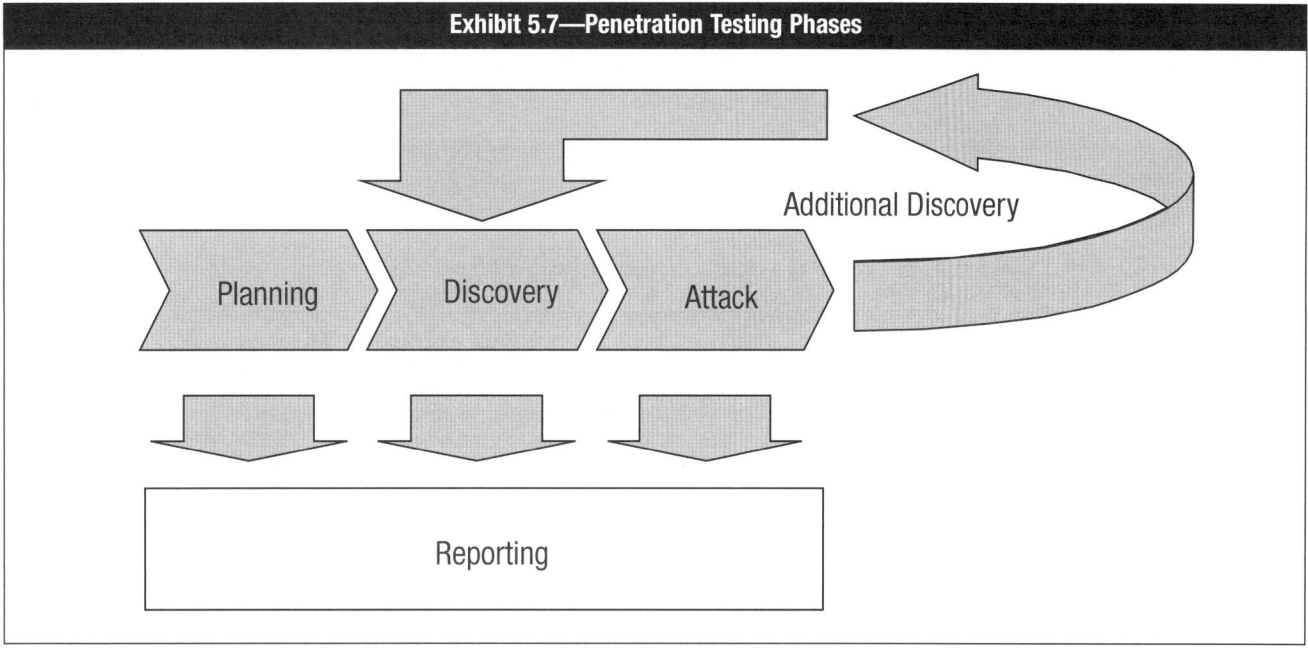

**Exhibit 5.7—Penetration Testing Phases**

Planning → Discovery → Attack → Additional Discovery

Reporting

| Exhibit 5.8—Penetration Testing Phases and Procedures ||
|---|---|
| **Phase** | **Procedures** |
| Planning | • Rules of engagement<br>• Management approval/finalization<br>• Adopted testing methodology<br>• Intrusive or nonintrusive testing<br>• Goals/objectives identified and agreed upon<br>• Timelines/deadlines agreed upon<br>• Milestones identified<br>• Assignment time tracking technique understood and communicated<br>• Deliverables agreed upon<br>• Tools collected/installed/tested in a test environment |
| Reconnaissance/<br>discovery | • Network mapping<br>• DNS interrogation<br>• WHOIS queries (from InterNIC)<br>• Searching target's web site for information<br>• Searching target's related data on search engines<br>• Searching resume/curriculum vitae of target's current and formal employees (reveals system related details)<br>• Packet capture/sniffing (during internal testing only)<br>• Host detection (Internet control message protocol [ICMP], DNS, WHOIS, PingSweep, TCP/UDP Sweep, etc.)<br>• Service detection (port scanning, stealth scanning, error/banner detection, etc.)<br>• Network topology detection (Internet control message protocol [ICMP], firewalking, etc.)<br>• OS detection (TCP stack analysis, etc.)<br>• Web site mapping<br>• Web page analysis<br>• Unused pages/scripts<br>• Broken links<br>• Hidden links/files accessible<br>• Application logic/use<br>• Points of input error page banner grabbing<br>• Vulnerability classification (based on information collected in previous steps, vulnerabilities are searched on available search engines or custom-built repositories)<br><br>Some of the attack techniques are as follows:<br>• Directory browsing<br>• Show code<br>• Error injection<br>• Type and bound checks on input |

| | Exhibit 5.8—Penetration Testing Phases and Procedures (*cont.*) |
|---|---|
| **Phase** | **Procedures** |
| Attacks | • Special character injection (metacharacters, escape characters, etc.)<br>• Cookie/session IDs analysis<br>• Authentication circumvention<br>• Long input<br>• System functions (shell escapes, etc)<br>• Logic alteration (SQL query injection, etc)<br>• Cookie/session IDs manipulation<br>• Internet service exploits (bind, mdac, unicode, apache-http, statd, sadmind, etc.)<br>• Operating system exploits<br>• Network exploits (SYN flooding, ICMP redirects, DNS poisoning etc.)<br>Furthermore, once an attack is successful, it typically follows these subprocedures of an attack phase:<br>• Privilege escalation—If only a user-level access was gained previously, then the tester will attempt to obtain super-level access (i.e., root on UNIX/Linux and administrator on NT/Windows 2000 and above)<br>• Information gathering from inside—The attacker will probe further systems on the network efficiently utilizing the compromised system as a launch pad and thereby attempt to gain access to trusted/high-risk systems<br>• Installation of further attack tools inside the system—Attacker may require installation of additional tools and penetration testing software to gain further access to the resources, trusted or high-risk systems<br><br>This phase simultaneously occurs with the rest of the three phases.<br><br>In the planning phase, rules of engagement, written consent and test plans are developed, discussed and reported.<br><br>In the discovery phase, written logs are kept and periodic reports on the status of assignment are reported to management, as appropriate.<br><br>Following the attack phase, the vulnerabilities and weaknesses discovered are reported with risk rating based on probability derived from ease of exploitation and impact derived from attack results or official advisories and resources from the vendor. In addition, the recommendations contain steps to mitigate the risks and to effectively rectify the weaknesses. |

• **Internal testing**—Refers to attacks and control circumvention attempts on the target from within the perimeter. The objective is to identify what would occur if the external network perimeter was successfully compromised and/or an authorized user from within the network wanted to compromise security of a specific resource on the network.
• **Blind testing**—Refers to the condition of testing when the penetration tester is provided with limited or no knowledge of the target's information systems. Such testing is expensive, since penetration testers have to research the target and profile it based on publicly available information only.
• **Double blind testing**—It is an extension of blind testing, since the administrator and security staff at the target are also not aware of the test. Such testing can effectively evaluate the incident handling and response capability of the target.
• **Targeted testing**—Refers to attacks and control circumvention attempts on the target, while both the target's IT team and penetration testers are aware of the testing activities. Penetration testers are provided with information related to target and network design. Additionally, they may also be provided with a limited-privilege user account to be used as a starting point to identify privilege-escalation possibilities in the system.

Although management may sponsor the activities of penetration testing, some of the risks of these activities include:
• Penetration testing does not provide assurance that all vulnerabilities are discovered and may fail to discover significant vulnerabilities.
• Miscommunication may result in the test objectives not being achieved.
• Testing activities may inadvertently trigger escalation procedures that may not have been appropriately planned.
• Sensitive information may be disclosed, heightening the target's exposure level.
• Without proper background and qualification checks of penetration testers, the penetration tester may damage the information assets or misuse the information gained for personal benefits.

Additionally, these techniques are becoming more popular for testing the reliability of firewall access controls. The IS auditor should be extremely careful if attempting to break into a live production system since, if successful, the IS auditor may cause the system to fail. Permission for the use of such techniques should always be obtained from top-level senior management. Permission from top-level senior management is also required to determine what other tests can be performed without informing the staff who are responsible for the monitoring and reporting of security violations (if any are aware that the attack will take places, they are likely to be more vigilant than normal).

# Protection of Information Assets

## Full Network Assessment Reviews

Upon completion of penetration testing, comprehensive review of all network system vulnerabilities should occur to determine whether threats to confidentiality, integrity and availability have been identified. The following reviews should occur:
• Security policy and procedures should be reviewed to determine best practices are in place.
• The network and firewall configuration should be evaluated to ensure that they have been designed to support the security of the services being provided (e.g., screening routers, dual/multihomed host, screened subnet, demilitarized zone proxy servers).
• The logical access controls should be evaluated to ensure that they support separation of duties (e.g., development vs. operation, security administration vs. audit).
• The following should be determined:
  – Intrusion detection software is in place.
  – Filtering is being performed.
  – Encryption is being used (consider VPNs/tunneling, digital signatures for e-mail, etc).
  – Strong forms of authentication are being used (consider use of smart cards, biometrics, etc., for authentication to firewalls, to internal software/hardware within the network, and to external hardware/software).
  – The firewalls have been configured properly (consider removal of all unnecessary software, addition of security and auditing software, removal of unnecessary logon IDs, disabling of unused services).
  – The application- or circuit-level gateways in use are running proxy servers for all legitimate services (e.g., teletype network [Telnet], HTTP, FTP).
  – Virus scanning is being used.
  – Periodic penetration testing is being completed.
  – Audit logging is undertaken for all key systems (e.g., firewalls, application gateways, routers, etc.) and audit logs are copied to secure file systems (consider direct writing of audit logs to write once CD-WORM).
  – The security administrators are reviewing the latest CERT, Computer Incident Advisory Capability (CIAC) and SysAdmin, Audit, Network, Security (SANS) Institute's reports for Internet security breaches.

## Development and Authorization of Network Changes

Network configuration changes to update telecommunications lines, terminals, modems and other network devices should be authorized in writing by management and implemented in a timely manner. The IS auditor can test this change control by:
• Sampling recent change requests, looking for appropriate authorization and matching the request to the actual network device
• Matching recent network changes, such as new telecommunication lines, to added terminals and authorized change requests

As an added control, the IS auditor should determine who can access the network change software. This access should be restricted to senior network administrators.

Specific development and change control procedures should be in place for network components' hardware and software. Procedures should cover:
• Firewalls
• Routers
• Switches
• Application gateways
• DNS/network topology
• Client software
• Network management software
• Web server hardware and configuration
• Application software
• Web pages

## Unauthorized Changes

One of the most important objectives of change control procedures is to prevent or detect unauthorized changes to software, configurations or parameters, and data. Unauthorized changes include any changes to software or configurations/parameters that occur without conforming to change control procedures. They include situations where changes are made to software code without authorization, in addition to legitimate changes made in accordance with change control procedures.

Controls to prevent unauthorized changes to software and software configurations include:
• Segregation of duties between software development, software administration and computer operations
• Restricting the software development team's access to the development environment only
• Restricting access to the software source codes

Controls to detect unauthorized changes to software include software code comparison utilities. Unauthorized changes to configurations/parameters can be detected through logging and monitoring system administrator activities.

Changes to data normally are controlled through the applications. Application access control mechanisms and built in application controls normally prevent unauthorized access to data. These controls can be circumvented by direct access to data. For this reason, direct access to data (specifically "write" or "change" access) should be restricted and monitored.

## Computer Forensics

By definition, computer forensics is the "process of identifying, preserving, analyzing and presenting digital evidence in a manner that is legally acceptable in any legal proceedings (i.e., a court of law)," according to D. Rodney McKemmish in his book *Computer and Intrusion Forensics*. An IS auditor may be required or asked to be involved in a forensic analysis in progress to provide expert opinion or to ensure the correct interpretation of information gathered.

Computer forensics includes activities that involve the exploration and application of methods to gather, process, interpret and use digital evidence that help to substantiate whether an incident happened such as:
• Providing validation that an attack actually occurred
• Gathering digital evidence that can later be used in judicial proceedings

Any electronic document or data can be used as digital evidence, provided there is sufficient manual or electronic proof that the contents of digital evidence are in their original state and have not been tampered with or modified during the process of collection and analysis.

It is very important to maintain evidence in any situation. Most organizations are not well equipped to deal with intrusions and electronic crimes from an operational and procedural perspective, and they respond to it only when the intrusion has occurred and the risk is realized. The evidence loses its integrity and value in legal proceedings if it has not been preserved and subject to a documented chain of custody. This happens when the incident is inappropriately managed and responded to in an *ad hoc* manner.

For evidence to be admissible in a court of law, the chain of custody needs to be maintained professionally. The chain of evidence essentially contains information regarding:
• Who had access to the evidence (chronological manner)
• The procedures followed in working with the evidence (such as disk duplication, virtual memory dump)
• Proving that the analysis is based on copies that are identical to the original evidence (could be documentation, checksums, timestamps)

It is important to use industry-specified best practices, proven tools and due diligence to provide reasonable assurance of the quality of evidence.
It is also important to demonstrate integrity and reliability of evidence for it to be acceptable to law enforcement authorities.

There are four major considerations in the chain of events in regards to evidence in computer forensics:
• **Identify**—Refers to the identification of information that is available and might form the evidence of an incident.
• **Preserve**—Refers to the practice of retrieving identified information and preserving it as evidence. The practice generally includes the imaging of original media in presence of an independent third party. The process also requires being able to document chain-of-custody so that it can be established in a court of law.
• **Analyze**—Involves extracting, processing and interpreting the evidence. Extracted data could be unintelligible binary data after it has been processed and converted into human readable format. Interpreting the data requires an in-depth knowledge of how different pieces of evidence may fit together. The analysis should be performed using an image of media and not the original.

• **Present**—Involves a presentation to the various audiences such as management, attorneys, court, etc. Acceptance of the evidence depends upon the manner of presentation (as it should be convincing), qualifications of the presenter, and credibility of the process used to preserve and analyze the evidence.

The IS auditor should give consideration to key elements of computer forensics during audit planning. These key elements are described in the following subsections.

## DATA PROTECTION
To prevent sought-after information from being altered, all measures must be in place. It is important to establish specific protocols to inform appropriate parties that electronic evidence will be sought and to not destroy it by any means.

Infrastructure and processes for incident response and handling should be in place to permit an effective response and forensic investigation if an event or incident occurs.

## DATA ACQUISITION
All information and data required should be transferred into a controlled location; this includes all types of electronic media such as as fixed disk drives and removable media. Each device must be checked to ensure that it is write-protected. This may be achieved by using a device known as a write-blocker.

It is also possible to get data and information from witnesses or related parties by recorded statements.

By volatile data, investigators can determine what is currently happening on a system. This kind of data includes open ports, open files, active processes, user logons and other data present in RAM. This information is lost when the computer is shut down.

## IMAGING
The imaging is a process that allows one to obtain a bit-for-bit copy of data to avoid damage of original data or information when multiple analyses may be performed. The imaging process is made to obtain residual data, such as deleted files, fragments of deleted files and other information present, from the disk for analysis. This is possible because imaging duplicates the disk surface, sector by sector.

With appropriate tools, it is possible to recover destroyed information (erased even by reformatting) from the disk's surface.

## EXTRACTION
This process consists of identification and selection of data from the imaged data set. This process should include standards of quality, integrity and reliability. The extraction process includes software used and media where an image was made.

The extraction process could include different sources such as system logs, firewall logs, IDS logs, audit trails and network management information.

## INTERROGATION
Interrogation is used to obtain prior indicators or relationships, including telephone numbers, IP addresses and names of individuals, from extracted data.

## INGESTION/NORMALIZATION
This process converts the information extracted to a format that can be understood by investigators. It includes conversion of hexadecimal or binary data into readable characters or a format suitable for data analysis tools.

It is possible to create relationships from data by extrapolation, using techniques such as fusion, correlation, graphing, mapping or time lining, which could be used in the construction of the investigation's hypothesis.

**REPORTING**
The information obtained from computer forensics has limited value when it is not collected and reported in the proper way.

When an IS auditor writes the report, he/she must include why the system was reviewed, how the computer data were reviewed and what conclusions were made from this analysis.

The report should achieve the following goals (taken from Mandia, Kevin; Matt Pepe; Chris Prosise; *Incident Response & Computer Forensics, 2ⁿᵈ Edition,* McGraw Hill/Osborne, USA, 2003):
• Accurately describe the details of an incident
• Be understandable to decision-makers
• Be able to withstand a barrage of legal scrutiny
• Be unambiguous and not open to misinterpretation
• Be easily referenced
• Contain all information required to explain conclusions reached
• Offer valid conclusions, opinions or recommendations when needed
• Be created in a timely manner

The report should also identify the organization, sample reports and restrictions on circulation (if any) and include any reservations or qualifications that the IS auditor has with respect to the assignment.

# 5.7 ENVIRONMENTAL EXPOSURES AND CONTROLS

## 5.7.1 ENVIRONMENTAL ISSUES AND EXPOSURES

Environmental exposures are due primarily to naturally occurring events such as lightning storms, earthquakes, volcanic eruptions, hurricanes, tornados and other types of extreme weather conditions. The result of such conditions can lead to many types of problems. One particular area of concern is power failures of computer and supporting environmental systems. Generally, power failures can be grouped into four distinct categories, based on the duration and relative severity of the failure:
• **Total failure (blackout)**—A complete loss of electrical power, which may span from a single building to an entire geographical area and is often caused by weather conditions (e.g., storm, earthquake) or the inability of an electrical utility company to meet user demands (e.g., during summer months)
• **Severely reduced voltage (brownout)**—The failure of an electrical utility company to supply power within an acceptable range (i.e., 108-125 volts AC in the US). Such failure places a strain on electronic equipment and may limit their operational life or even cause permanent damage.
• **Sags, spikes and surges**—Temporary and rapid decreases (sags) or increases (spikes and surges) in voltage levels. These anomalies can cause loss of data, data corruption, network transmission errors or physical damage to hardware devices (e.g., hard disks or memory chips).
• **Electromagnetic interference (EMI)**—Caused by electrical storms or noisy electrical equipment (e.g., motors, fluorescent lighting, radio transmitters). This interference may cause computer systems to hang or crash as well as damages similar to those caused by sags, spikes and surges.
Short-term interruptions, such as sags, spikes and surges, which last from a few millionths to a few thousandths of a second, can be prevented by using properly placed surge protectors. Intermediate-term interruptions, which last from a few seconds to 30 minutes, can be controlled by UPS devices. Finally, long-term interruptions, which last from a few hours to several days, require the use of alternate power generators. These generators may be portable devices or part of the building's infrastructure and are powered by alternative sources of energy such as diesel, gasoline or propane.

Another area of concern deals with water damage/flooding. This is a concern even with facilities located on upper floors of high-rise buildings since water damage typically occurs from broken water pipes.

Other concerns, outside of natural threats, are man-made. They include terrorist threats/attacks, vandalism, electrical shock and equipment failure.

Some questions that organizations must address related to environmental issues and exposures include the following:
- Is the power supply to the computer equipment properly controlled to ensure that power remains within the manufacturer's specifications?
- Are the air conditioning, humidity and ventilation control systems for the computer equipment adequate to maintain temperatures within manufacturers' specifications?
- Is the computer equipment protected from the effects of static electricity, using an antistatic rug or antistatic spray?
- Is the computer equipment kept free of dust, smoke and other particulate matter such as food?
- Do policies exist that prohibit the consumption of food, beverage and tobacco products near computer equipment?
- Are backup media protected from damage due to temperature extremes, the effects of magnetic fields and water damage?

## 5.7.2 CONTROLS FOR ENVIRONMENTAL EXPOSURES

Environmental exposures should be afforded the same level of protection as physical and logical exposures. Environmental controls are described in the following subsections.

### Alarm Control Panels

An alarm control panel should ideally be:
- Separated from burglar or security systems located on the premises
- Accessible to fire department personnel at all times
- Located in a weatherproof box
- In accordance with temperature requirements set by the manufacturer
- Situated in a controlled room to prevent access by unauthorized personnel
- Allocated power from a dedicated and separate circuit
- Able to control or disable separate zones within the facilities
- In adherence with local and national regulations and approved by local authorities

### Water Detectors

In the computer room, water detectors should be placed under raised floors and near drain holes, even if the computer room is on a high floor (remember water leaks). Any unattended equipment storage facilities should also have water detectors. When activated, the detectors should produce an audible alarm that can be heard by security and control personnel. The location of the water detectors should be marked on the raised computer room floor for easy identification and access. On hearing the alarm, specific individuals should be responsible for investigating the cause and initiating remedial action; other staff should be made aware by security and control personnel that there is a risk of electric shock.

### Handheld Fire Extinguishers

Fire extinguishers should be in strategic locations throughout the facility. They should be tagged for inspection and inspected at least annually.

### Manual Fire Alarms

Hand-pull fire alarms should be placed strategically throughout the facility. The resulting audible alarm should be linked to a monitored guard station.

### Smoke Detectors

Smoke detectors should be above and below the ceiling tiles throughout the facilities and below the raised computer room floor. The detectors should produce an audible alarm when activated and be linked to a monitored station (preferably by the fire department). The location of the smoke detectors above the ceiling tiles and below the raised floor should be marked on the tiling for easy identification and access. Smoke detectors should supplement, not replace, fire suppression systems.

## Fire Suppression Systems

These systems are designed to automatically activate immediately after detection of high heat, typically generated by fire. Like smoke detectors, the system should produce an audible alarm when activated and be linked to a central guard station that is regularly monitored. The system should also be inspected and tested annually. Testing intervals should comply with industry and insurance standards and guidelines. Ideally, the system should automatically trigger other mechanisms to localize the fire. This includes closing fire doors, notifying the fire department, closing off ventilation ducts and shutting down nonessential electrical equipment. In addition, the system should be segmented so a fire in one part of a large facility does not activate the entire system.

Broadly speaking, there are two methods for applying an extinguishing agent: total flooding and local application.

Systems working under a *total flooding* principle apply an extinguishing agent to a three-dimensional enclosed space in order to achieve a concentration of the agent (volume percent of the agent in air) adequate to extinguish the fire. These types of systems may be operated automatically by detection and related controls or manually by the operation of a system actuator.

Systems working under a *local application* principle apply an extinguishing agent directly onto a fire (usually a two-dimensional area), or into the three-dimensional region immediately surrounding the substance or object on fire. The main difference between local application and total flooding designs is the absence of physical barriers enclosing the fire space in the local application design.

In the context of automatic extinguishing systems, local application does normally not refer to the use of manually operated wheeled or portable fire extinguishers, although the nature of the agent delivery is similar.

The medium for fire suppression varies, but is usually one of the following:
• Water-based systems are typically referred to as sprinkler systems. These systems are effective but are also unpopular because they damage equipment and property. The system can be dry-pipe or charged (water is always in the system piping). A charged system is more reliable but has the disadvantage of exposing the facility to expensive water damage if the pipes leak or break.
• Dry-pipe sprinkling systems do not have water in the pipes until an electronic fire alarm activates the water pumps to send water into the system. This is opposed to a fully charged water pipe system. Dry-pipe systems have the advantage that any failure in the pipe will not result in water leaking into sensitive equipment from above. Since water and electricity do not mix, these systems must be combined with an automatic switch to shut down the electricity supply to the area protected.
• Halon systems release pressurized Halon gases that remove oxygen from the air, thus starving the fire. Halon was popular because it is an inert gas and does not damage equipment like water does. There should be an audible alarm and brief delay before discharge to permit personnel time to evacuate the area or to override and disconnect the system. Because Halon adversely affects the ozone layer, it was banned in the Montreal (Canada) protocol of 1987, which stopped Halon production as of 1 January 1994. The Halon substitute is FM-200, which is the most effective alternative.
• FM-200™: Also called heptafluoropropane, HFC-227 or HFC-227ea (ISO name), is a colorless odorless gaseous halocarbon. It is commonly used as a gaseous fire suppression agent. The HFC-227 fire suppression agent was the first nonozone-depleting replacement for Halon 1301. In addition, HFC-227 leaves no residue on valuable equipment after discharge. Trade names include FE-227 (DuPont), FM-200 (Chemtura) and Solkaflam 227 (Solvay Fluor). This agent suppresses fire by discharging as a gas onto the surface of combusting materials. Large amounts of heat energy are absorbed from the surface of the burning material, lowering its temperature below the ignition point. FM-200 fire suppression systems have low atmospheric lifetimes, global warming and ozone depletion potentials.
• Argonite is the brand name (a registered trademark owned by Ginge-Kerr) for a mixture of 50% argon (Ar) and 50% nitrogen (N2). It is an inert gas used in gaseous fire suppression systems for extinguishing fires where damage to equipment is to be avoided. Although argon is nontoxic, it does not satisfy the body's need for oxygen and is a simple asphyxiant. People have suffocated by breathing argon by mistake

Unlike carbon dioxide ($CO_2$) fire suppression systems, which are unable to sustain human life, FM-200 and Argonite systems are environmentally friendly. They provide an effective, safe method of fire suppression, and unlike charged sprinkler systems, they do not suffer from the disadvantage of exposing the expensive information processing facility (IPF) to water damage during the firefighting or if the pipe leaks or breaks.

- $CO_2$ systems release pressurized carbon dioxide gas into the area protected to replace the oxygen required for combustion. Unlike Halon and its later replacements, however, $CO_2$ is unable to sustain human life. Therefore, in most countries, it is illegal for such systems to be set to automatic release if any human may be in the area. Because of this, these systems are usually discharged manually, introducing an additional delay in combating the fire.

## Strategically Locating the Computer Room

To reduce the risk of flooding, the computer room should not be located in the basement or top floor. If located in a multistory building, studies show that the best location for the computer room—the location which reduces the risk of fire, smoke and water damage—is on the middle floors (e.g., third, fourth, fifth or sixth floor). Adjacent water or gas pipes should be avoided except in the case of fire suppression systems. Care should be taken to avoid locating computer rooms adjacent to areas where functions carrying a high risk are carried out, such as paper storage. The activity of neighboring organizations should be considered when establishing a computer facility. Locations adjacent to, or on the final path to, an airport or a chemical works where explosive gases may be present, for example, should be avoided.

## Regular Inspection by Fire Department

To ensure that all fire detection systems comply with building codes, the fire department should inspect the system and facilities annually. Also, the fire department should be notified of the location of the computer room, so it can be prepared with equipment appropriate for electrical fires.

## Fireproof Walls, Floors and Ceilings of the Computer Room

Walls surrounding the information processing facility should contain or block fire from spreading. The surrounding walls should have at least a two-hour fire resistance rating.

## Electrical Surge Protectors

These electrical devices reduce the risk of damage to equipment due to power spikes. Voltage regulators measure the incoming electrical current and either increase or decrease the charge to ensure a consistent current. Such protectors are typically built into the UPS system.

## Uninterruptible Power Supply/Generator

A UPS system consists of a battery or gasoline (petrol)-powered generator that interfaces with the electrical power entering the facility and the electrical power entering the computer. The system typically cleanses the power to ensure that wattage into the computer is consistent. Should a power failure occur, the UPS continues providing electrical power from the generator to the computer for a certain length of time. Depending on the sophistication of the UPS, electrical power could continue to flow for days or for just a few minutes to permit an orderly computer shutdown. A UPS system can be built into a computer or can be an external piece of equipment.

## Emergency Power-off Switch

There may be a need to immediately shut off power to the computer and peripheral devices such as during a computer room fire or emergency evacuation. Two emergency power-off switches should serve this purpose—one in the computer room, the other near, but outside, the computer room.

Switches should be clearly labeled and easily accessible, for this purpose, and yet they should still be secure from unauthorized people. The switches should be shielded to prevent accidental activation.

## Power Leads From Two Substations

Electrical power lines that feed into the facility are exposed to many environmental hazards—water, fire, lightning, cutting due to careless digging, etc. To reduce the risk of a power failure due to these events that, for the most part, are beyond the control of the organization, redundant power lines should feed into the facility. In this way, interruption of one power line does not adversely affect electrical supply.

## Wiring Placed in Electrical Panels and Conduit

Electrical fires are always a risk. To reduce the risk of such a fire occurring and spreading, wiring should be placed in fire-resistant panels and conduit. This conduit generally lies under the fire-resistant raised computer room floor.

## Inhibited Activities Within the IPF

Food, drink and tobacco use can cause fires, the buildup of contaminants or damage to sensitive equipment (especially in the case of liquids). They should be prohibited from the information processing facility (IPF). This prohibition should be overt, for example, a sign on the entryway.

## Fire-resistant Office Materials

Wastebaskets, curtains, desks, cabinets and other general office materials in the IPF should be fire-resistant. Cleaning fluids for desktops, console screens and other office furniture/fixtures should not be flammable.

## Documented and Tested Emergency Evacuation Plans

Evacuation plans should emphasize human safety, but should not leave IPFs physically unsecured. Procedures should exist for a controlled shutdown of the computer in an emergency situation, if time permits.

# 5.7.3 AUDITING ENVIRONMENTAL CONTROLS

The testing procedures noted in the following subsections should also be applied to any offsite storage and processing facilities.

## Water and Smoke Detectors

The computer room should be visited for the visual verification of the presence of water and smoke detectors. Whether the power supply to these detectors is sufficient should be determined, especially in instances of battery-operated devices. Also, the locations of the devices should be placed to give early warning of a fire, such as immediately above the computer equipment they are protecting, and should be clearly marked and visible.

## Handheld Fire Extinguishers

Handheld fire extinguishers should be in strategic highly visible locations throughout the facility and should be inspected annually.

## Fire Suppression Systems

Fire suppression systems are expensive to test and, therefore, the IS auditor's ability to determine operability is limited. IS auditors may need to limit their tests to reviewing documentation to ensure that the system has been inspected and tested within the last year. The exact testing interval should comply with industry and insurance standards and guidelines.

## Regular Inspection by Fire Department

The person responsible for fire equipment maintenance should be contacted and asked if a local fire department inspector or insurance evaluator has been recently invited to tour and inspect the facilities. If so, a copy of the report should be obtained, and how to address the noted deficiencies should be determined.

## Fireproof Walls, Floors and Ceilings of the Computer Room

With the assistance of building management, the documentation that identifies the fire rating of the walls surrounding the IPF should be located. These walls should have at least a two-hour fire resistance rating.

## Electrical Surge Protectors

The presence of electrical surge protectors on sensitive and expensive computer equipment should be visually observed.

### Power Leads From Two Substations

With the assistance of building management, documentation concerning the use and placement of redundant power lines into the IPF should be located.

### Fully Documented and Tested Business Continuity Plan

See the content area, business continuity and disaster recovery in chapter 6, for a description of testing the business continuity plans.

### Wiring Placed in Electrical Panels and Conduit

Wiring in the IPF should be placed in fire-resistant panels and conduit.

### UPS/Generator

The most recent test date should be determined and the test reports should be reviewed.

### Documented and Tested Emergency Evacuation Plans

A copy of the emergency evacuation plan should be obtained. It should be examined to determine whether it describes how to leave the IPFs in an organized manner that does not leave the facilities physically insecure. A sample of IS employees should be interviewed to determine if they are familiar with the documented plan. The emergency evacuation plans should be posted throughout the facilities.

### Humidity/Temperature Control

The IPF should be visited on regular intervals to determine if temperature and humidity are adequate.

# 5.8 PHYSICAL ACCESS EXPOSURES AND CONTROLS

Physical exposures could result in financial loss, legal repercussions, loss of credibility or loss of competitive edge. They primarily originate from natural and man-made hazards, and can expose the business to unauthorized access and unavailability of the business information.

## 5.8.1 PHYSICAL ACCESS ISSUES AND EXPOSURES

Physical access issues are a major concern in security. Exposures and possible perpetrators are described in the following subsections.

### Physical Access Exposures

Exposures that exist from accidental or intentional violation of these access paths include:
• Unauthorized entry
• Damage, vandalism or theft to equipment or documents
• Copying or viewing of sensitive or copyrighted information
• Alteration of sensitive equipment and information
• Public disclosure of sensitive information
• Abuse of data processing resources
• Blackmail
• Embezzlement

## Possible Perpetrators

Possible perpetrators include employees with authorized or unauthorized access who are:
• Disgruntled (upset by or concerned about some action by the organization or its management)
• On strike
• Threatened by disciplinary action or dismissal
• Addicted to a substance or gambling
• Experiencing financial or emotional problems
• Notified of their termination

Other possible perpetrators could include:
• Former employees
• Interested or informed outsiders such as competitors, thieves, organized crime and hackers
• An accidental ignorant, for example, someone who unknowingly perpetrates a violation (e.g., an employee or outsider)

The most likely source of exposure is from the uninformed, accidental or unknowing person, although the greatest impact may be from those with malicious or fraudulent intent.

Other questions and concerns to consider include the following:
• Are hardware facilities reasonably protected against forced entry?
• Are keys to the computer facilities adequately controlled to reduce the risk of unauthorized access?
• Are computer terminals locked or otherwise secured to prevent removal of boards, chips and the computer itself?
• Are authorized equipment passes required before computer equipment can be removed from its normal secure surroundings?

From an IS perspective, facilities to be protected include:
• Programming area
• Computer room
• Operator consoles and terminals
• Tape library, tapes, disks and all magnetic media
• Storage rooms and supplies
• Offsite backup file storage facility
• Input/output control room
• Communications closets
• Telecommunications equipment (including radios, satellites, wiring, modems and external network connections)
• Microcomputers and personal computers (PCs)
• Power sources
• Disposal sites
• Minicomputer establishments
• Dedicated telephones/telephone lines
• Control units and front-end processors
• Portable equipment (handheld scanners and coding devices, bar code readers, laptop computers and notebooks, printers, pocket LAN adapters and others)
• Onsite and remote printers
• Local area networks

Additionally, system, infrastructure or software application documentation should be protected against unauthorized access.

For these safeguards to be effective, they must extend beyond the computer facility to include any vulnerable access points within the entire organization and at organizational boundaries/interfaces with external organizations. This may include remote locations and rented, leased or shared facilities. Additionally, the IS auditor may require assurances that similar controls exist within service providers or other third parties, if they are potentially vulnerable access points to sensitive information within the organization.

## 5.8.2 PHYSICAL ACCESS CONTROLS

Physical access controls are designed to protect the organization from unauthorized access. These controls should limit access to only those individuals authorized by management. This authorization may be explicit, as in a door lock for which management has authorized you to have a key, or implicit, as in a job description that implies a need to access sensitive reports and documents. Examples of some of the more common access controls are:

- **Bolting door locks**—These locks require the traditional metal key to gain entry. The key should be stamped "do not duplicate" and should be stored and issued under strict management control.
- **Combination door locks (cipher locks)**—This system uses a numeric key pad or dial to gain entry. The combination should be changed at regular intervals or whenever an employee with access is transferred, fired or subject to disciplinary action. This reduces the risk of the combination being known by unauthorized people.
- **Electronic door locks**—This system uses a magnetic or embedded chip-based plastic card key or token entered into a sensor reader to gain access. A special code internally stored in the card or token is read by the sensor device that then activates the door locking mechanism. Electronic door locks have the following advantages over bolting and combination locks:
  - Through the special internal code, cards can be assigned to an identifiable individual.
  - Through the special internal code and sensor devices, access can be restricted based on the individual's unique access needs. Restrictions can be assigned to particular doors or to particular hours of the day.
  - They are difficult to duplicate.
  - Card entry can be easily deactivated in the event an employee is terminated or a card is lost or stolen. Silent or audible alarms can be automatically activated, if unauthorized entry is attempted. Issuing, accounting for and retrieving the card keys is an administrative process that should be carefully controlled. The card key is an important item to retrieve when an employee leaves the firm. An example of a common technique used for card entry is the swipe card. A swipe card is a physical control technique that uses a plastic card with a magnetic strip containing encoded data to provide access to restricted or secure locations. The encoded data can be read by a slotted electronic device. After a card has been swiped, the application attached to the slotted electronic device prevents unauthorized physical access to those sensitive locations, as well as logs all card users that try to gain access to the secure location.
- **Biometric door locks**—An individual's unique body features, such as voice, retina, fingerprint or signature, activate these locks. This system is used in instances when extremely sensitive facilities must be protected such as in the military.
- **Manual logging**—All visitors should be required to sign a visitor's log indicating their name, company representative, reason for visiting, person to see and date and time of entry and departure. Logging typically is at the front reception desk and entrance to the computer room. Before gaining access, visitors should also be required to provide verification of identification such as a driver's license, business card or vendor identification tag.
- **Electronic logging**—This is a feature of electronic and biometric security systems. All access can be logged, with unsuccessful attempts being highlighted.
- **Identification badges (photo IDs)**—Badges should be worn and displayed by all personnel. Visitor badges should be a different color from employee badges for easy identification. Sophisticated photo IDs can also be utilized as electronic card keys. Issuing, accounting for and retrieving the badges is an administrative process that must be carefully controlled.
- **Video (CCTV) cameras**—Cameras should be located at strategic points and monitored by security guards. Sophisticated video cameras can be activated by motion. The video surveillance recording should be retained for possible future playback.
- **Security guards**—Guards are very useful if supplemented by video cameras and locked doors. Guards supplied by an external agency should be bonded to protect the organization from loss.
- **Controlled visitor access**—All visitors should be escorted by a responsible employee. Visitors include friends, maintenance personnel, computer vendors, consultants (unless long-term, in which case special guest access may be provided) and external auditors.
- **Bonded personnel**—All service contract personnel, such as cleaning people and offsite storage services, should be bonded. This does not improve physical security but limits the financial exposure of the organization.
- **Deadman doors**—This system, which uses a pair of (two) doors, is typically found in entries to facilities such as computer rooms and document stations. For the second door to operate, the first entry door must close and lock, with only one person permitted in the holding area. This reduces the risk of piggybacking, when an unauthorized person follows an authorized person through a secured entry.

• **Not advertising the location of sensitive facilities**—Facilities such as computer rooms should not be visible or identifiable from the outside, that is, no windows or directional signs. The building or department directory should discreetly identify only the general location of the information processing facility.
• **Computer workstation locks**—These lock the device to the desk, prevent the computer from being turned on or disengage keyboard recognition, thus preventing use. Another available feature is locks that prevent turning on a PC workstation until a key lock is unlocked by a turn key or card key.
• **Controlled single entry point**—A controlled entry point monitored by a receptionist should be used by all incoming personnel. Multiple entry points increase the risk of unauthorized entry. Unnecessary or unused entry points should be eliminated or deadlocked.
• **Alarm system**—An alarm system should be linked to inactive entry points, motion detectors, and the reverse flow of enter- or exit-only doors. Security personnel should be able to hear the alarm when activated.
• **Secured report/document distribution cart**—Secured carts, such as mail carts, should be covered and locked and should not be left unattended.
• **Windows**—Ideally the computer room should not have externally facing windows: where these are present, they should be constructed of reinforced glass and, if on the ground floor of the building, further protected for example, by bars.

# 5.8.3 AUDITING PHYSICAL ACCESS

Touring the computer site is useful to gain an overall understanding and perception of the installation being reviewed. This tour provides the opportunity to begin reviewing physical access restrictions (e.g., control over employees, visitors, intruders and vendors).

The computer site (i.e., computer room, programmers' area, tape library, printer stations and management offices) and any offsite storage facilities should be included in this tour.

Much of the testing of physical safeguards can be achieved by visually observing the previously noted safeguards. Documents to assist with this effort include emergency evacuation procedures, inspection tags (recent inspection?), fire suppression system test results (successful? recently tested?) and key lock logs (all keys accounted for and not outstanding to former employees or consultants?).

Testing should extend beyond the IPF/computer room to include the following related facilities:
• Location of all operator consoles
• Printer rooms
• Computer storage rooms (this includes equipment, paper and supply rooms)
• UPS/generator
• Location of all communications equipment identified on the network diagram
• Tape library
• Offsite backup storage facility

To complete a thorough testing, the IS auditor should look above the ceiling panels and below the raised floor in the computer operations center, observing smoke and water detectors, general cleanliness, and walls that extend all the way to the real ceiling (not just the fake/suspended ceiling). For a ground-floor computer room, the auditor may also consider walking around the outside of the room, viewing the location of any windows, examining emergency exit doors for evidence that they are routinely used (such as the presence of cigarette stubs or litter) and examining the air conditioning units. The auditor should also consider whether any additional threats exist close to the room, such as storage of dangerous or flammable material.

The following paths of physical entry should be evaluated for proper security:
• All entry doors
• Emergency exit doors
• Glass windows and walls
• Movable walls and modular cubicles

• Above suspended ceilings and beneath raised floors
• Ventilation systems
• Over a curtain, fake wall

# 5.9 MOBILE COMPUTING

In today's mobile computing environment, it is not unusual to keep sensitive data on PCs and diskettes, particularly laptops, where it is more difficult to implement logical and physical access controls. Protecting laptop data confidentiality requires a layered security approach at both physical and logical access levels, as well as a laptop security policy and safe practices guidelines. In particular, the following controls will reduce the risk of disclosure of sensitive data stored on a laptop:
• Engrave or brand a serial number and company name and logo on to a laptop using an electric engraver or tamper resistant tags.
• Use a cable locking system or a locking system with a motion detector that sounds an audible alarm.
• Back up business critical or sensitive data on a regular basis, both at work (using shared folders on the company's file server) and while traveling (using floppy disks or online backup solutions).
• Encrypt data using well-known encryption software that is based on public domain algorithms and is not proprietary. The decision what to encrypt can be left to the user, by creating designated volumes in which encrypted files are stored and processed. An alternative approach encrypts the entire hard drive, leaving nothing to the discretion of the user.
• Allocate passwords to individual files to prevent them from being opened by an unauthorized person, that is, one not in possession of the password. This facility should be used with great caution since, should the password be forgotten, or the person who created it leave the organization, the data would be unavailable because the password cannot be circumvented.
• Establish a theft response team and develop procedures to follow when a laptop is stolen. These procedures should include identifying the stolen laptop, reporting the loss of the laptop to local authorities, assessing the importance or criticality of the compromised data and determining if clients or other third parties are affected by the theft.

As laptops are often used for the remote logging into the computer systems of the organization, special care should be taken to defend against malicious code. The best solution is to centralize virus signature updates and the handling of patches to the operating system.

Laptop security can be hardened by using two-factor authentication. This can be achieved using biometric readers, or devices that are possessed by the user.

# 5.10 CHAPTER 5 QUICK REFERENCE REVIEW

| Chapter 5 Quick Reference Review |
|---|
| Chapter 5 addresses the need for the protection of information assets within an organization. Protection of information assets includes the key components that ensure confidentiality, integrity and availability (CIA) of information assets. The chapter evaluates design, implementation and monitoring of logical and physical access controls to ensure CIA. The chapter also evaluates network infrastructure security, environmental controls and processes and procedures used to store, retrieve, transport and dispose of confidential information assets. The chapter describes the various methods and procedures followed by organizations and focuses on the auditor's role in evaluating these procedures. Many of these topics may, on the surface, seem very familiar to candidates; however, it is important to note that the topics addressed in this chapter require a thorough knowledge of the technologies used and the potential control weaknesses that may develop. CISA candidates should be fully aware of and conversant with the components of network infrastructure security, logical access issues and the key elements of information security management. |
| CISA candidates should have a sound understanding of the following items, not only within the context of the present chapter, but also to correctly address questions in related subject areas. It is important to keep in mind that it is not enough to know these concepts from a definitional perspective. The CISA candidate must also be able to identify which elements among those presented may represent the greatest risk and which controls are most effective at mitigating this risk. |

- Key elements of information security management include: senior management commitment and support, policies and procedures, organization, security awareness and education, monitoring and compliance, and incident handling and response.
- General points of logical entry into a system should be understood, including logical protection at the network, platform, database and application layers. CISA candidates may be asked to identify how a failure at one layer could allow an unauthorized individual to bypass certain logical security mechanisms and gain access to confidential data.
- It is important to know best practices for identification and authentication. This includes practices for handling default system accounts, normal user accounts and privileged user accounts such as system administrators.
- CISA candidates should understand the differences between various types of biometric technologies, and the advantages and disadvantages of each.
- CISA candidates should be familiar with network infrastructure security, various issues and risks associated with different technologies used in network infrastructures, and best practices for risk mitigation. Special attention should be focused on firewall implementation, the advantages and disadvantages of different types of intrusion detection/prevention systems, and encryption technologies.
- The importance of O/S patching to remediate vulnerabilities and how patching should be deployed should also be understood.

# Protection of Information Assets

# 5.11 CHAPTER 5 CASE STUDIES

The following case studies are included as a learning tool to reinforce the concepts introduced in this chapter. Exam candidates should note that the CISA exam does not currently use this format for testing.

## 5.11.1 CASE STUDY A

### *Case Study A Scenario*

Management is currently considering ways in which to enhance the physical security and protection of its data center. The IS auditor has been asked to assist in this process by evaluating the current environment and making recommendations for improvement. The data center consists of 15,000 square feet (1,395 square meters) of raised flooring on the ground floor of the corporate headquarters building. A total of 22 operations personnel require regular access. Currently, access to the data center is obtained using a proximity card, which is assigned to each authorized individual. There are three entrances to the data center, each of which utilizes a card reader and has a camera monitoring the entrance. These cameras feed their signals to a monitor at the building reception desk, which cycles through these images along with views from other cameras inside and outside the building. Two of the doors to the data center also have key locks that bypass the electronic system so that a proximity card is not required for entry. Use of proximity cards is written to an electronic log. This log is retained for 45 days. During the review, the IS auditor noted that 64 proximity cards are currently active and issued to various personnel. The data center has no exterior windows, although one wall is glass and overlooks the entry foyer and reception area for the building.

| CASE STUDY A QUESTIONS |
|---|
| A1. Which of the following risks would be mitigated by supplementing the proximity card system with a biometric scanner to provide two-factor authentication?<br><br>    A. Piggybacking or tailgating<br>    B. Sharing access cards<br>    C. Failure to log access<br>    D. Copying of keys<br><br><br>A2. Which of the following access mechanisms would present the greatest difficulty in terms of user acceptance?<br><br>    A. Hand geometry recognition<br>    B. Fingerprints<br>    C. Retina scanning<br>    D. Voice recognition |
| *See answers and explanations to the practice questions at the end of the chapter. (page 471)* |

## 5.11.2 CASE STUDY B

### Case Study B Scenario

A company needed to enable remote access to one of its servers for remote maintenance purposes. Firewall policy did not allow any external access to the internal systems. Therefore, it was decided to install a modem on that server and to activate the remote access service to permit dial-up access. As a control, a policy has been implemented to manually power on the modem only when the third party was requesting access to the server and powered off by the company's system administrator when the access is no longer needed. As more and more systems are being maintained remotely, the company is asking an IS auditor to evaluate the current risks of the existing solution and to propose the best strategy for addressing future connectivity requirements.

---

**CASE STUDY B QUESTIONS**

B1. What test is **MOST** important for the IS auditor to perform as part of the review of dial-up access controls?

  A. Dial the server from authorized and unauthorized telephone lines
  B. Determine bandwidth requirements of remote maintenance and the maximum line capacity
  C. Check if the availability of the line is guaranteed to allow remote access any time
  D. Check if call back is not used and the cost of calls is charged to the third party

B2. What is the **MOST** significant risk that the IS auditor should evaluate regarding the existing remote access practice?

  A. Modem is not powered on/off whenever is needed
  B. A nondisclosure agreement was not signed by the third-party
  C. Data exchanged over the line is not encrypted
  D. Firewall controls are bypassed

B3. Which of the following recommendations is **MOST** likely to reduce the current level of remote access risks?

  A. Maintain an access log with the date and time when the modem was powered on/off
  B. Encrypt the traffic over the telephone line
  C. Migrate the dial-up access to an Internet VPN solution
  D. Update firewall policies and implement an IDS system

B4. What control should be implemented to prevent an attack on the internal network being initiated though an Internet VPN connection?

  A. Firewall rules are periodically reviewed
  B. All VPNs terminate at a single concentrator
  C. An IDS capable to analyze encrypted traffic is implemented
  D. Antivirus software is installed on all production servers

*See answers and explanations to the practice questions at the end of the chapter. (page 471)*

---

## 5.11.3 CASE STUDY C

### *Case Study C Scenario*

"My Music" is a company dedicated to the production and distribution of video clips specializing in jazz music. Born in the Internet era, the company has actively supported the use of notebook computers by its staff so they can use them when traveling and when working from home. Through the Internet they can access the company databases and provide online information to customers. This decision has resulted in an increase in productivity and high morale among employees who are allowed to work up to two days a week from home. Based on written procedures and a training course, employees learn security procedures to avoid the risk of unauthorized access to company data. Employees' access to the company data includes using logon IDs and passwords to the application server through a VPN. Initial passwords are assigned by the security administrator. When the employee logs on for the first time, the system forces a password change to improve confidentiality.

Management is currently considering ways to improve security protection for remote access by employees. The IS auditor has been asked to assist in this process by evaluating the current environment and making recommendations for improvement.

| CASE STUDY C QUESTIONS |
|---|

C1.    Which of the following levels provides a higher degree of protection in applying access control software to avoid unauthorized access risks?

    A.    Network and operating system level
    B.    Application level
    C.    Database level
    D.    Log file level

C2.    When an employee notifies the company that he has forgotten his password, what should be done **FIRST** by the security administrator?

    A.    Allow the system to randomly generate a new password
    B.    Verify the user's identification through a challenge/response system
    C.    Provide the employee with the default password and explain that it should be changed as soon as possible
    D.    Ask the employee to move to the administrator terminal to generate a new password in order to assure confidentiality

*See answers and explanations to the practice questions at the end of the chapter. (page 472)*

## 5.11.4 CASE STUDY D

### *Case Study D Scenario*

A major financial institution has just implemented a centralized banking solution (CBS) in one of its branches. It has a secondary concern to look after marketing of the bank. Employees of a separate legal entity work on the bank premises, but they have no access to the bank's solution software. Employees of other branches get training on this solution from this branch and for training purposes temporary access credentials are also given to such employees. IS auditors observed that employees of the separate legal entity also access the CBS software through the branch employees access credentials. IS auditors also observed that there are numerous active IDs of employees who got training from the branch and have since been transferred to their original branch.

| CASE STUDY D QUESTIONS |
|---|
| D1. Which of the following should IS auditors recommend to effectively eliminate such password sharing? <br><br> A. Assimilation of security need to keep password secret <br> B. Stringent rules prohibiting sharing of password <br> C. Use of smart card along with strong password <br> D. Use of smart card along with employee's terminal ID <br><br> D2. Which of the following **BEST** addresses user ID management of trainee employees? <br><br> A. Unused user ID shall be automatically deleted periodically <br> B. To integrate access rights with human resource process <br> C. Password of unused but active user ID shall be suspended <br> D. Active user ID register shall checked frequently |
| *See answers and explanations to the practice questions at the end of the chapter. (page 472)* |

# 5.12 ANSWERS TO PRACTICE QUESTIONS

5-1    **B**    Utility software, in this case, is a data correction program for correcting any inconsistency in data. However, this utility can be used to override the incorrect update of tables directly. Therefore, access to this utility should be restricted on a need-to-use basis, and a log should be generated automatically whenever this utility is executed. Senior management should review this log periodically. Deleting the utility and installing it as and when required may not be practically feasible since there would be a time delay. Access to utilities should not be provided to user management. Defining access so that the utility can be executed in a menu option may not generate a log.

5-2    **C**    The server and system security should be defined to allow only authorized staff members access to information about the staff whose records they handle on a day-to-day basis. Choice A is a good control in that it will allow access to be analyzed if there is concern of unauthorized access. However, it will not prevent access. Choice B, restricting access to sensitive transactions, will restrict access only to some of the data. It will not prevent access to other data. Choice D, system access restricted to business hours, only restricts when unauthorized access can occur and would not prevent such access at other times. It is important to consider that the data owner is responsible for determining who is allowed access via the written software access rules.

5-3    **B**    The absence of password controls on the client-server, where production data reside, is the most critical weakness. All other findings, while they are control weaknesses, do not carry the same disastrous impact.

5-4    **A**    If a password is disclosed when single sign-on is enabled, there is a risk that unauthorized access to all systems will be possible. User access rights should remain unchanged by single sign-on, as additional security parameters are not implemented necessarily. One of the intended benefits of single sign-on is the simplification of security administration and the unlikelihood of an increased workload.

5-5    **B**    It is not possible to lock out the system administrator account after several failed logon attempts, because it would be impossible to unlock it. Therefore, it is subject to online brute force attacks. The IS auditor should be most concerned with any failed logon attempts targeted to this account. All other choices can be locked automatically by the system.

5-6    **A**    IDSs detect intrusion activity based on the intrusion rules. They can detect external and internal intrusion activity and send an automated alarm message. Firewalls and routers prevent the unwanted and well-defined communications between the internal and external networks. They do not have any automatic alarm messaging systems.

5-7    **B**    Tunnel mode provides encryption and authentication of the complete IP package. To accomplish this, the authentication header (AH) and encapsulating security payload (ESP) services can be nested. The transport mode provides primary protection for the protocols' higher layers; that is, protection extends to the data field (payload) of an IP package. The other two mechanisms provide authentication and integrity.

5-8    **D**    A digital signature includes an encrypted hash total of the size of the message as it was transmitted by its originator. This hash would no longer be accurate if the message was altered subsequently, indicating that the alteration had occurred. Digital signatures will not identify or prevent any of the other options. The signature would neither prevent nor deter unauthorized reading, copying or theft.

5-9    **D**    Encryption is the only way to ensure that Internet transactions are confidential and, of the choices available, the use of public key encryption is the best method. Digital signatures would ensure that the transaction is not changed and cannot be repudiated, but would not ensure confidentiality.

5-10    **D**    Antivirus software can be used to prevent virus attacks. By running regular scans, it can also be used to detect virus infections that have already occurred. Regular updates of the software are required to ensure it is able to update, detect and treat viruses as they emerge. Scanning e-mail attachments on the mail server is a preventive control. It will prevent infected e-mail files from being opened by the recipients, which would cause their machines to become infected. Restoring systems from clean copies is a preventive control. It will ensure that viruses are not introduced from infected copies or backups, which would reinfect machines. Disabling USB ports is a preventive control. It prevents infected files from being copied from a USB drive onto a machine, which would cause the machine to become infected.

# 5.13 ANSWERS TO CASE STUDY QUESTIONS

## *Answers to Case Study A Questions*

A1. **B** Two-factor authentication involving the use of biometrics would effectively prevent the sharing of access cards since these cards would be ineffective without the corresponding biometric. Piggybacking or tailgating would not be mitigated since traffic flow would remain unchanged. Since two entrances utilize key locks that override the electronic entry system, individuals entering using keys would not be logged by the electronic system, and keys could still potentially be copied.

A2. **C** Although the highest in terms of accuracy, many individuals feel uncomfortable with the idea of having a device scan the inside of their eye. So, even though retina scanning may be highest in terms of effectiveness from a control perspective, its lack of user acceptance may make it inappropriate for applications where customer acceptance is of prime importance. Fingerprints, hand geometry and voice recognition are all less invasive and, therefore, not as subject to adverse negative reaction by users. The objective of this area is to ensure that the CISA candidate understands and can provide assurance that the security architecture (policies, standards, procedures and controls) ensures the confidentiality, integrity and availability of information assets.

## *Answers to Case Study B Questions*

B1. **A** Dial-up access should be possible only from authorized telephone lines as a preventive control for unauthorized access when logon credentials are compromised or misused by third-party personnel. Initiating the connection by the server to an authorized phone number using the call back feature would be one implementation of this requirement. Options B, C and D address performance issues and not access control issues.

B2. **D** The company's security infrastructure relies on controls implemented on the firewall. The fact that someone from the outside can connect directly to an internal system, bypassing firewall rules, could expose the internal network to the third party, thereby facilitating unauthorized access. Choices A, B and C are risks to be considered by the IS auditor, but concern only the server being maintained remotely, and not the entire internal system.

B3. **C** Using an Internet VPN solution will eliminate the vulnerabilities of the dial-up access such as lack of encryption and bypassing firewall controls. Option A and B will address punctual issues and Option D will have no effect since security infrastructure controls are bypassed by the direct dial-up access.

B4. **C** An IDS should be able to analyze the encrypted traffic of the VPN connection to determine potential attacks. A firewall rules review and ending all VPNs in a single concentrator will prevent unauthorized connections to the internal network, but this will not prevent an attack occurring through an authorized VPN connection. Antivirus software will prevent contamination by computer viruses, but the internal system is still vulnerable to many other threats.

## Answers to Case Study C Questions

C1. **A** The greatest degree of protection in applying access control software against internal and external users' unauthorized access is at the network and platform/operating system levels. These systems are also referred to as general support systems, and they make up the primary infrastructure on which applications and database systems will reside.

C2. **B** When an employee notifies that he/she has forgotten his/her password, the security administrator should start a password process generation procedure only after verifying the user's identification using a challenge/response system or similar procedure. To verify, it is advised that the security administrator should return the user's call after verifying his/her extension or calling his/her supervisor for verification.

## Answers to Case Study D Questions

D1. **A** Assimilation of security need to keep password secret can only effectively refrain such password sharing and such assimilation is possible only through continuous and conscientious security awareness and education programs. Without assimilation of security need stringent rules prohibiting sharing of password cannot effectively stop password sharing. Use of smart card along with strong password and use of smart card along with employee's terminal ID do not deter password sharing.

D2. **B** Integration of access rights with human resource process is the best way to address user ID management. Automatic periodic deletion of unused user ID, suspension of password of unused but active user ID and frequently checking active user ID register are not the best way since vulnerability persists during the period in which user IDs remain active.

# 5.14 Suggested Resources for Further Study

**Brancik, Kenneth C.**; *Insider Computer Fraud*, Auerbach Publications, USA, 2007

**Cendrowski, Harry**; James P. Martin; Louis W. Petro; *The Handbook of Fraud Deterrence*, John Wiley & Sons Inc., USA, 2007

**Coderre, David G.**; *Fraud Detection: A Revealing Look at Fraud, 2nd Edition*, Ekaros Analytical Inc., Canada, 2004

**Davis, Chris**; Mike Schiller; Kevin Wheeler; *IT Auditing Using Controls to Protect Information Assets*, McGraw Hill, USA, 2007

**Dubin, Joel**; *The Little Black Book of Computer Security, 2nd Edition*, Penton Media Inc., USA, 2008

**Harris, Shon**; Allen Harper; Chris Eagle; Jonathan Ness; Michael Lester; *Gray Hat Hacking*, McGraw Hill, USA, 2005

**Jaquith, Andrew**; *Security Metrics*, Addison Wesley, USA, 2007

**Kairab, Sudhanshu**; *A Practical Guide to Security Assessments*, Auerbach Publications, USA, 2004

**Marcella Jr., Albert J.**; Dough Menendez; *Cyber Forensics, 2nd Edition*, Auerbach Publications, USA, 2008

**McClure, Stuart**; Joel Sambray; George Kurtz; *Hacking Exposed, 5th Edition*, McGraw Hill, USA, 2005

**Mitnick, Kevin D.**; William L. Simon; *The Art of Intrusion: The Real Stories Behind the Exploits of Hackers, Intruders & Deceivers*, Wiley Publishing, Inc., USA, 2005

**Natan, Ron Ben**; *Implementing Database Security and Auditing*, Elsevier Digital Press, USA, 2005

**Peltier, Thomas R.**; *Information Security Risk Analysis, 2nd Edition*, Auerbach Publications, USA, 2005

**Stamp, Mark**; *Information Security: Principles and Practice*, John Wiley & Sons, USA, 2005

**Stanley, Richard A.**; *Managing Risk in the Wireless Environment: Security, Audit and Control Issues*, ISACA, USA, 2005

**Tudor, Jan Killmeyer**; *Information Security Architecture: An Integrated Approach to Security in the Organization, 2nd Edition*, Auerbach Publications, USA, 2005

**Vacca, John R.**; *Biometric Technologies and Verification Systems*, Elsevier Inc., USA, 2007

**Wells, Joseph T.**; *Fraud Casebook, Lessons from the Bad Side of Business*, John Wiley & Sons Inc., USA, 2007

Page intentionally left blank

*Chapter 6:*

# Business Continuity
# and Disaster Recovery

Chapter 6:

# Business Continuity
# and Disaster Recovery

# Business Continuity and Disaster Recovery

# 6.1 INTRODUCTION

Business continuity planning (BCP) and contingency planning in support of operations are elements of a system of internal control that is established to manage availability of critical processes in the event of interruption. The most important part of such a plan deals with the cost-effective support of the information system. Availability of business data is vital to the sustainable development andor even to the survival of any organization.

The topic of business continuity is potentially wide ranging in scope. The type of business drives the types of plans that the planner should focus upon.

BCP is a continuous process rather than a project. The plans that the planner develops as part of this process will direct the response to incidents ranging from simple emergencies to full blown disasters.

The ultimate goal of the process is to be able to respond to incidents that may impact people, operations and ability to deliver goods and services to the marketplace.

This area presents an overview of business continuity and disaster recovery principles, specifically in the following areas:
• BCP and disaster recovery planning (DRP) processes
• Business impact analysis (BIA)
• Recovery strategies and alternatives
• Plan testing
• Backup and restoration
• Audit considerations

The chapter on business continuity and disaster recovery focuses on the key elements that an organization must perform in order to proactively plan for, and manage, the consequences of a disaster. The tasks required for the assurance of the business continuity and disaster recovery plans (BCPs/DRPs) are also detailed in this chapter.

## 6.1.1 OBJECTIVE

The objective of this area is to ensure that the CISA candidate understands and can provide assurance that, in the event of a disruption, the business continuity and disaster recovery processes will ensure the timely resumption of IT services while minimizing the business impact.

> **This area represents 14 percent of the CISA examination (approximately 28 questions).**

## 6.1.2 TASKS

There are three (3) tasks within the business continuity and disaster recovery area:
T6.1 Evaluate the adequacy of backup and restore provisions to ensure the availability of information required to resume processing.
T6.2 Evaluate the organization's disaster recovery plan to ensure that it enables the recovery of IT processing capabilities in the event of a disaster.
T6.3 Evaluate the organization's business continuity plan to ensure its ability to continue essential business operations during the period of an IT disruption.

## 6.1.3 KNOWLEDGE STATEMENTS

There are eight (8) knowledge statements within the business continuity and disaster recovery area:
KS6.1 Knowledge of data backup, storage, maintenance, retention and restoration processes, and practices
KS6.2 Knowledge of regulatory, legal, contractual and insurance issues related to business continuity and disaster recovery

# Business Continuity
# and Disaster Recovery

KS6.3    Knowledge of business impact analysis (BIA)

KS6.4    Knowledge of the development and maintenance of the business continuity and disaster recovery plans

KS6.5    Knowledge of business continuity and disaster recovery testing approaches and methods

KS6.6    Knowledge of human resources management practices as related to business continuity and disaster recovery (e.g., evacuation planning, response teams)

KS6.7    Knowledge of processes used to invoke the business continuity and disaster recovery plans

KS6.8    Knowledge of types of alternate processing sites and methods used to monitor the contractual agreements (e.g., hot sites, warm sites, cold sites)

## 6.1.4 RELATIONSHIP OF TASK TO KNOWLEDGE STATEMENTS

The task statements are what the CISA candidate is expected to know how to do. The knowledge statements delineate what the CISA candidate is expected to know in order to perform the tasks. The task and knowledge statements are approximately mapped in **exhibit 6.1** insofar as it is possible to do so. Note that although there is often overlap, each task statement will generally map to several knowledge statements.

| Exhibit 6.1— Tasks and Knowledge Statements Mapping ||
|---|---|
| **Task Statements** | **Knowledge Statements** |
| T6.1   Evaluate the adequacy of backup and restore provisions to ensure the availability of information required to resume processing. | KS6.1   Knowledge of data backup, storage, maintenance, retention and restoration processes and practices <br><br> KS6.5   Knowledge of business continuity and disaster recovery testing approaches and methods |
| T6.2   Evaluate the organization's disaster recovery plan to ensure that it enables the recovery of IT processing capabilities in the event of a disaster. | KS6.2   Knowledge of regulatory, legal, contractual and insurance issues related to business continuity and disaster recovery <br><br> KS6.3   Knowledge of business impact analysis (BIA) <br> KS6.4   Knowledge of the development and maintenance of the business continuity and disaster recovery plans <br> KS6.5   Knowledge of business continuity and disaster recovery testing approaches and methods <br> KS6.6   Knowledge of human resources management practices as related to business continuity and disaster recovery (e.g., evacuation planning, response team) <br> KS6.7   Knowledge of processes used to invoke the business continuity and disaster recovery plans <br> KS6.8   Knowledge of types of alternate processing sites and methods used to monitor the contractual agreements (hot sites, warm sites, cold sites) |
| T6.3   Evaluate the organization's business continuity plan to ensure its ability to continue essential business operations during the period of an IT disruption. | KS6.2   Knowledge of regulatory, legal, contractual and insurance issues related to business continuity and disaster recovery <br><br> KS6.3   Knowledge of business impact analysis (BIA) <br> KS6.4   Knowledge of the development and maintenance of the business continuity and disaster recovery plans <br> KS6.5   Knowledge of business continuity and disaster recovery testing approaches and methods <br> KS6.6   Knowledge of human resources management practices as related to business continuity and disaster recovery (e.g., evacuation planning, response team) <br> KS6.7   Knowledge of processes used to invoke the business continuity and disaster recovery plans |

# Business Continuity and Disaster Recovery

# 6.2 BUSINESS CONTINUITY/DISASTER RECOVERY PLANNING

BCP is a process designed to reduce the organization's business risk arising from an unexpected disruption of the critical functions/operations (manual or automated) necessary for the survival of the organization. This includes human/material resources supporting these critical functions/operations and assurance of the continuity of at least the minimum level of services necessary for at least the critical operations.

The first approach to any risk should be to try to remove the threat by considering business locations, building materials and duplication of key functions at remote sites. However modern business is, it cannot avoid all forms of corporate risk or potential damage. A realistic objective is to ensure the survival of an organization by establishing a culture that will identify and manage those risks that could cause it to suffer.

Examples of these corporate risks include:
• Inability to maintain critical customer services
• Damage to market share, image, reputation or brand
• Failure to protect the company assets, including intellectual properties and personnel
• Business control failure
• Failure to meet legal or regulatory requirements

The purpose of business continuity/disaster recovery is to enable a business to continue offering critical services in the event of a disruption and to survive a disastrous interruption to their activities. Rigorous planning and commitment of resources is necessary to adequately plan for such an event.

The first step of preparing a new BCP is to identify the business processes of strategic importance, which are those key processes that are responsible for both the permanent growth of the business and for the fulfillment of the business goals. Ideally, the BCP/DRP should be supported by a formal executive policy that states the organization's overall target for recovery and empowers those involved in developing, testing and maintaining the plans.

Based on the key processes, the risk management process should begin with a risk assessment. The risk is directly proportional to the impact on the organization and the probability of occurrence of the perceived threat. Thus, the result of the risk assessment should be the identification of the following:
• Those human resources, data, infrastructural elements and other resources (including those provided by third parties) that support the key processes
• A list of potential vulnerabilities—the dangers or threats to the organization
• The estimated probability of the occurrence of these threats

The management of these risks is addressed in the preparation of the BCP.

The operations part of the BCP should address all functions and assets required to continue as a viable organization. The extent of provisioning for alternate facilities that should be pursued is ultimately a business decision based on risk management.

Focus is on the availability of the key business processes to continue operations should any kind of disruption arise.

BCP is primarily the responsibility of senior management, as they are entrusted with safeguarding the assets and the viability of the organization, as defined in the BCP/DRP policy. The BCP is generally followed by the business and supporting units, alike, to provide a reduced but sufficient level of functionality in the business operations immediately after encountering an interruption, while recovery is taking place. The plan should address all functions and assets required to continue as a viable organization. This includes continuity procedures determined necessary to survive and minimize the consequences of business interruption.

# Business Continuity and Disaster Recovery

BCP takes into consideration:
• Those critical operations that are necessary to the survival of the organization
• The human/material resources supporting them

Besides the plan for the continuity of operations, the BCP includes:
• The DRP that is used to recover a facility rendered inoperable, including relocating operations into a new location
• The restoration plan that is used to return operations to normality whether in a restored or new facility

One of the most important requirements is the improvement of the security of normal operations. This involves the introduction of those countermeasures that decrease the probability of the occurrence of such events that might cause the disruption of business.

Depending on the complexity of the organization, there could be one or more plans to address the various aspects of business continuity and disaster recovery. These plans do not necessarily have to be integrated into one single plan. However, each has to be consistent with other plans to have a viable BCP strategy.

It is highly desirable to have a single integrated plan to ensure that:
• all aspects are covered.
• resources committed are used in the most effective way and there is reasonable confidence that, through its application, the organization will survive a disruption.

Even if similar processes of the same organization are handled at a different geographic location, the BCP and DRP solutions may be different for different scenarios. Solutions may be different due to contractual requirements (e.g., the same organization is processing an online transaction for one client and the back office is processing for another client). A BCP solution for the online service will be significantly different than one for the back office processing.

| PRACTICE QUESTIONS |
| --- |

**6-1** During an audit of a large bank, the IS auditor observes that no formal risk assessment exercise has been carried out for the various business applications to arrive at their relative importance and recovery time requirements. The risk to which the bank is exposed is that the:

    A.    business continuity plan may not have been calibrated to the relative risk that disruption of each application poses to the organization.

    B.    business continuity plan may not include all relevant applications and, therefore, may lack completeness in terms of its coverage.

    C.    business impact of a disaster may not have been accurately understood by the management.

    D.    business continuity plan may lack an effective ownership by the business owners of such applications.

**6-2** Which of the following is necessary to have **FIRST** in the development of a business continuity plan?

    A.    Risk-based classification of systems
    B.    Inventory of all assets
    C.    Complete documentation of all disasters
    D.    Availability of hardware and software

**6-3** An IS auditor should be involved in:

    A.    observing tests of the disaster recovery plan.
    B.    developing the disaster recovery plan.
    C.    maintaining the disaster recovery plan.
    D.    reviewing the disaster recovery requirements of supplier contracts.

*See answers and explanations to the practice questions at the end of the chapter. (page 514)*

# Business Continuity
# and Disaster Recovery

## 6.2.1 IS BUSINESS CONTINUITY/DISASTER RECOVERY PLANNING

In the case of IS business continuity planning, the approach is the same as in BCP with the exception being that the continuity of IS processing is threatened. IS processing is of strategic importance—it is a critical component since most key business processes depend on the availability of key systems and infrastructure components.

Throughout the BCP process, the overall plan of the organization should be taken into consideration: again, this should be supported by the executive policy. All IS plans must be consistent with and support the corporate BCP. This means that alternate processing facilities that support key operations must be ready and have up-to-date plans regarding their use.

Again, all possible steps must be taken to remove the likelihood of a disruption using the method described in other sections of this Manual. One example would be minimizing threats to the data center by a considering location: not on a flood plain, earthquake fault line or close to an area where explosive devices or toxic materials are regularly used. Another example is making use of resilient network topographies such as Loop or Mesh with alternative processing facilities already built into the network infrastructure.

IS business continuity/disaster recovery planning is a major component of an organization's overall business continuity and disaster recovery strategy. Therefore, there should be a ready-to-start reserved facility to support these operations in case of a disruption if the business cannot function without ongoing information processing. If it is a separate plan, the IS plan must be consistent with and support the corporate BCP.

The IS business continuity plan should also be aligned with the strategy of the organization. Thus, the classification with respect to the criticality of the various application systems deployed in the organization depends on the nature of the business as well as the value of each application to the business.

This value is directly proportional to the role of the application system in supporting the strategy of the organization. The components of the information system (including the technology infrastructure components) are then matched to the applications (e.g., the value of a computer or a network is determined by the importance of the application system that uses it.)

Once the risk assessment identifies the value of the IS components to the organization, a plan can be developed for establishing the criticality of systems and the most appropriate methods for their restoration.

> **Note:** The CISA candidate will not be tested on the actual calculation of risk analysis; however, the IS auditor should be familiar with risk analysis calculation.

An IS business continuity plan is much more than just a plan for information systems. A BCP identifies what the business will do in the event of a disaster. For example, where will employees report to work, how will orders be taken while the computer system is being restored, which vendors should be called to provide needed supplies? A subcomponent of the BCP is the IT disaster recovery plan. This typically details the process IT personnel will use to restore the computer systems. DRPs may be included in the BCP or as a separate document altogether, depending on the needs of the business.

Not all systems will require a recovery strategy. Based upon the results of the risk analysis, management may not see a tangible cost benefit for restoring certain applications in the event of a disaster. An overriding factor when determining recovery options is that the cost should never exceed the benefit.

The quality of IS elements is essential for IS disaster recovery. It is therefore recommended that the organization has an information security management system (ISMS) implemented to maintain the integrity, confidentiality and availability of IS.

# Business Continuity
# and Disaster Recovery

## 6.2.2 DISASTERS AND OTHER DISRUPTIVE EVENTS

Disasters are disruptions that cause critical information resources to be inoperative for a period of time, adversely impacting organizational operations. The disruption could be a few minutes to several months, depending on the extent of damage to the information resource. Most important, disasters require recovery efforts to restore operational status.

A disaster may be caused by natural calamities, such as earthquakes, floods, tornados, severe thunderstorms and fire, which cause extensive damage to the processing facility and the locality in general. Other disastrous events causing disruptions may occur when expected services, such as electrical power, telecommunications, natural gas supply or other delivery services are no longer supplied to the company due to a natural disaster or other cause. A disaster could also be caused by events precipitated by human beings such as terrorist attacks, hacker attacks, viruses or human error.

Not all critical disruptions in service are caused by a disaster. For example, disruption in service is sometimes caused by system malfunctions, accidental file deletions, network denial of service (DoS) attacks, intrusions and viruses. These events may require action to recover operational status in order to resume service. Such actions may necessitate restoration of hardware, software or data files. Therefore, a well-defined, risk-based classification system needs to be in place to determine whether a specific disruptive event requires initiating BCP or DRP efforts.

A good BCP will take into account all types of events impacting critical IS processing facilities and end users' normal organizational operation functions. For worst-case scenarios, short-term and long-term fallback strategies are required. For the short term, an alternate processing facility may be needed to satisfy immediate operational needs, as in the case of a major natural disaster. In the long term, a new permanent facility must be identified for disaster recovery and equipped to provide for continuation of IS processing services on a regular basis.

## *Dealing With Damage to Image, Reputation or Brand*

Damaging rumors may rise from many sources (even internal). They may or may not be associated with a serious incident or crisis. Whether they are "spontaneous" or a side effect of a business continuity or disaster recovery problem, their consequences may be devastating. One of the worst consequences of crises is the loss of trust.

Effective public relations (PR) activities in an organization may play an important role in helping to contain the damage to the image and ensure that the crisis is not made worse. Certain industries (e.g., banks, health care organizations, airlines, petroleum refineries, chemical, transportation, or nuclear power plants or other organizations with relevant social impact) should have elaborate protocols for dealing with accidents and catastrophes.

A few basic best practices should be considered and applied by an organization experiencing a major incident. Irrespective of the resultant objective consequences of an incident (delay or interruption in service, economic losses, etc.), if any, a negative public opinion or negative rumors can be costly. Reacting appropriately in public (or to the media) during a crisis is not simple. A properly trained spokesperson should be appointed and prepared beforehand. Normally, senior legal counsel or a public relations officer is the best choice. No one, irrespective of his/her rank in the organizational hierarchy, except for the spokesperson, should make any public statement.

As part of the preparation, the spokesperson should draft and keep on file a generic communiqué with blanks to be filled in with the specific circumstances. This should not be deviated from because of improvisation or time pressure. The communiqué should not state the causes of the incident but rather indicate that an investigation has been started and results will be reported. Liability should not be assumed. The system or the process should not be blamed.

> **Note:** Though dealing with damage to image, reputation or brand is a concern of many public organizations, the CISA candidate will not be tested on this topic.

# Business Continuity
# and Disaster Recovery

## 6.2.3 BUSINESS CONTINUITY PLANNING PROCESS

The BCP process can be divided into the following life cycle phases:
• Creation of a business continuity policy
• BIA
• Classification of operations and criticality analysis
• Identification of IS processes that support critical organizational functions
• Development of a BCP and IS disaster recovery procedures
• Development of resumption procedures
• Training and awareness program
• Testing and implementation of plan
• Monitoring

## 6.2.4 BUSINESS CONTINUITY POLICY

A business continuity policy should be proactive and encompass preventive, detective and corrective controls. The BCP is the most critical corrective control. It depends on other controls being effective—in particular, incident management and media backup.

The following guidelines are in line with best practices. Incidents and their impacts can, to some extent, be mitigated through preventive controls.

This requires that the incident management group be adequately staffed, supported and trained in crisis management, and that the BCP is well designed, documented, drill tested, funded and audited.

## 6.2.5 BUSINESS CONTINUITY PLANNING INCIDENT MANAGEMENT

Incidents and crises are dynamic by nature. They evolve, change with time and circumstances, and are often rapid and unforeseeable. Because of this, their management must be dynamic, proactive and well documented. An incident is any unexpected event, even if it causes no significant damage.

Depending on an estimation of the level of damage to the organization, all types of incidents should be categorized. A classification system could include the following categories: negligible, minor, major and crisis. Classification can dynamically change while the incident is resolved. These levels can be broadly described as follows:
• Negligible incidents are those causing no perceptible or significant damage, such as very brief operating system (OS) crashes with full information recovery or momentary power outages with uninterruptible power supply (UPS) backup.
• Minor events are those that, while not negligible, produce no negative material (of relative importance) or financial impact.
• Major incidents cause a negative material impact on business processes and may affect other systems, departments or even outside clients.
• Crisis is a major incident that can have serious material (of relative importance) impact on the continued functioning of the business and may also adversely impact other systems or third parties. The severity of the impact depends on the industry and circumstances, but is generally directly proportional to the time elapsed from the inception of the incident to incident resolution.

Minor, major and crisis incidents should be documented, classified, and revisited until corrected or resolved. This is a dynamic process since a major incident may decrease in extent momentarily and yet later expand to become a major crisis.

Negligible incidents can be analyzed statistically to identify any systemic or avoidable causes.

**Exhibit 6.2** provides an example of an incident classification system and reaction protocol.

| Exhibit 6.2—Incident/Crisis Levels | | | | |
|---|---|---|---|---|

**Reaction to Incident / Crisis Levels**

| | **1 LEVEL** | **MAIN CRITERION** (hours) SERVICE DOWNTIME | | **COMPLEMENTARY CRITERIA** | |
|---|---|---|---|---|---|
| | | FORECAST >= | ACTUAL >= | DATA | PLATFORMS |
| | 7 | | 24 | | |
| CRISIS | 6 | 24 | 12 | | |
| | 5 | 12 | 6 | Database loss of integrity | Hacked or Denial of Service Attack |
| | 4 | 6 | 4 | | Viruses, worms. Hardware failure. |
| MAJOR INC'T | 3 | 4 | 2 | | |
| | 2 | 2 | 1 | Lost Transactions | |
| MINOR INC'T | 1 | 1 | 0.5 | | |
| NEGLIGIBLE | 0 | | | | |

| | LEVEL | **2 ACTIONS** | | |
|---|---|---|---|---|
| | 7 | Follow Business Continuity Plan | Alert SM and evetually Reg. Agencies | |
| CRISIS | 6 | Follow Business Continuity Plan | Alert SM and evetually Reg. Agencies | |
| | 5 | Prepare for Business Continuity Plan | Alert SM | |
| | 4 | Correct / Clean / Restore / Replace | Alert SM | SM = Senior Management |
| MAJOR | 3 | Correct | If confirmed, alert SO | SO = Security Officer |
| | 2 | Correct | | |
| MINOR | 1 | Correct | | |
| NEGLIGIBLE | 0 | Log | ( Analyze logs regularly) | |

Source: Personas & Técnicas Multimedia SL © 2007. All rights reserved. Used by permission.

The security officer (SO) or other designated individual should be notified of all incidences as soon as any triggering event occurs. This person should then follow a pre-established protocol (e.g., calling in a spokesperson, alerting top management and involving regulatory agencies).

In general, the main criterion for incident severity (level) is service downtime. Other criteria may include the impact on data or platforms. Service can be defined as including commitments with clients that can be either external customers or internal departments. In most environments, severity is usually proportional to downtime. Other criteria may include the impact on data or platforms. A conservative fail-safe approach would be to assign any nonnegligible incident a starting, provisional severity level (see **exhibit 6.2**.) As the incident evolves, this level should be reevaluated regularly by the person or team in charge, often referred to as an incident response or firecall team.

## 6.2.6 BUSINESS IMPACT ANALYSIS

BIA is a critical step in developing the BCP. This phase involves identifying the various events that could impact the continuity of operations and their financial, human, legal and reputational impact on the organization.

To perform this phase successfully, one should obtain an understanding of the organization, key business processes, and IS resources used by the organization to support the key business processes. This phase requires a high level of support from senior management and extensive involvement of IT and end-user personnel. The criticality of the information resources (e.g., applications, data, networks, system software, facilities) that support an organization's business processes must be established with senior management approval. It is important to include all types of information resources and look beyond traditional information resources (i.e., mainframe operations) to be included in a BCP/DRP. For example, many end-user departments have installed sophisticated local area networks (LANs) and desktop workstations that perform critical functions on a daily basis, and many executives store vital information on laptops and personal digital assistants (PDAs). These devices are sometimes placed into operation without the involvement of IT.

# Business Continuity
# and Disaster Recovery

**Note:** The IS auditor should be able to evaluate the BIA. The task statement in the job practice states: "evaluate the organization's BCP to ensure its ability to continue essential business operations during the period of an IT disruption. The auditor needs to know what is involved in developing a BIA so that he/she can properly evaluate it. However, a CISA candidate will not be tested on how a BIA is performed or what method is used to perform a BIA.

There are different approaches for performing a BIA. One of the popular approaches is the questionnaire approach. This approach involves developing a detailed questionnaire and circulating it to key users in IT and end-user areas. The information gathered is tabulated and analyzed. In case additional information is required, the BIA team would contact the relevant users for additional information. Another popular approach is to interview groups of key users. The information gathered during these interview sessions is tabulated and analyzed for developing a detailed BIA plan and strategy. A third approach is to bring relevant IT personnel and end users together in a room to come to a conclusion regarding the potential business impact of various levels of disruptions.

Wherever possible, IS auditors should analyze past transaction volume in determining the impact to the business if the system was unavailable for an extended period of time. This would substantiate the interview process that IS auditors conduct for performing BIA.

The three main questions that should be considered during the BIA phase include the following:
1. What are the different business processes? Each process needs to be assessed to determine its relative importance. Indications of criticality include, for example:
 – The process supporting health and safety, such as hospital patient records and air traffic control systems
 – Disruption of the process causing a loss of income to the organization or exceptional unacceptable costs
 – The process meeting legal or statutory requirements
 – The number of business segments or number of users that are affected

 A process can be critical or noncritical depending on factors such as time of operation and mode of operation (e.g., business hours or ATM operations).
2. What are the critical information resources related to an organization's critical business processes? This is the first consideration because disruption to an information resource is not a disaster in itself, unless it is related to a critical business process, for example, an organization losing its revenue generating business processes due to an IS failure. Other examples of potential critical business processes may include:
 – Receiving payments
 – Production
 – Paying employees
 – Advertising
 – Dispatching of finished goods
 – Legal and regulatory compliance
3. What is the critical recovery time period for information resources in which business processing must be resumed before significant or unacceptable losses are suffered? In large part, the length of the time period for recovery depends on the nature of the business or service being disrupted. For instance, financial institutions, such as banks and brokerage firms, usually will have a much shorter critical recovery time period than manufacturing firms. Also, the time of year or day of week may affect the window of time for recovery. For example, a bank experiencing a major outage on Saturday at midnight has a longer time in which to recover than on Monday at midnight, assuming that the bank is not processing on Sunday.

To make this decision, there are two independent cost factors to consider as shown in **exhibit 6.3**. One is the downtime cost of the disaster. This component, in the short run (e.g., hours, days, weeks), grows quickly with time, where the impact of a disruption increases the longer it lasts. At a certain moment, it stops growing, reflecting the moment or point when the business can no longer function. The cost of downtime (increasing with time) has many components (depending on the industry and the specific company and circumstances), among them: cost of idle resources (e.g., in production), drop in sales (e.g., orders), financial costs (e.g., not invoicing nor collecting), delays (e.g., procurement) and indirect costs (e.g., loss of market share, image and goodwill).

# Business Continuity
# and Disaster Recovery

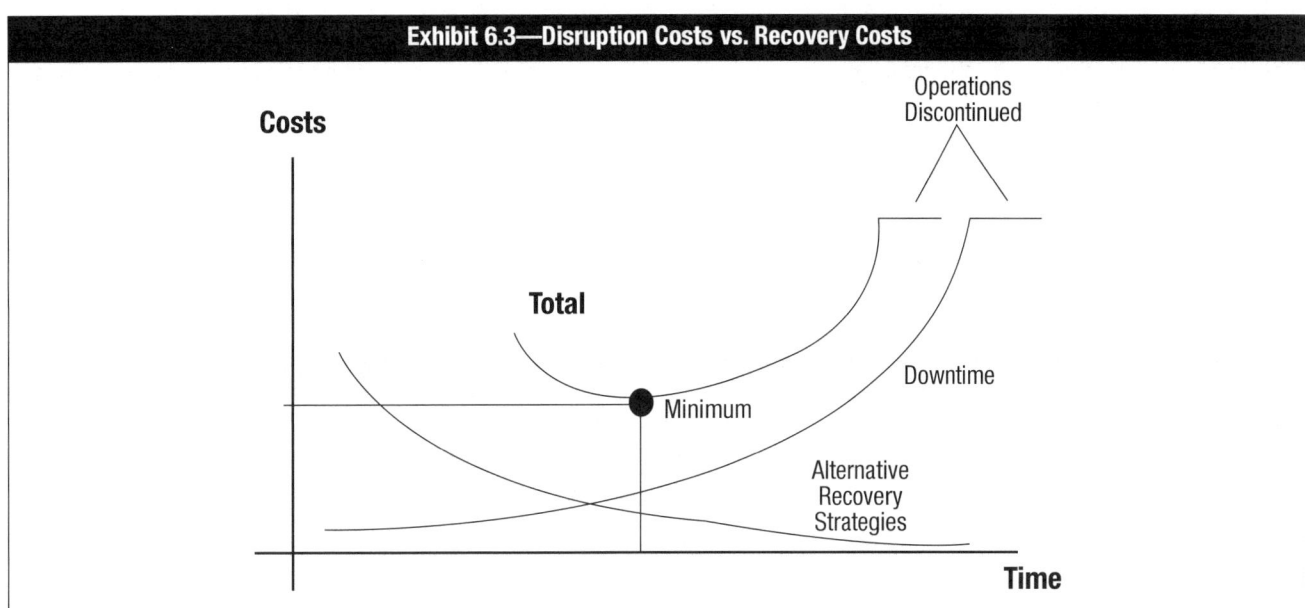

Exhibit 6.3—Disruption Costs vs. Recovery Costs

The other factor is the cost of the alternative corrective strategies (the activation of the BCP), which decreases with the target chosen for recovery time. The recovery cost also has many components (most of them rigid—inelastic). This includes the cost of preparing and periodically testing the BCP, the cost of offsite backup premises, the cost of insurance coverage, the yearly cost of alternative site arrangements, etc. Alternative recovery strategies may be represented by points using coordinates such as time frame and cost.

Upon identifying these costs, **exhibit 6.3** also shows the sum of cost curves as total costs (disruption and recovery), where an organization would want to find the point at which total cost can be minimized. This can be done by evaluating alternative development strategies where, with a few discrete strategies, the descending curve can be drawn and each point in that curve would represent a possible strategy. The curve as a whole is representative of all possible strategies. Each possible strategy has a fixed cost, i.e., does not change with time. Note that the fixed cost of each possible strategy can differ; normally, the shorter the target recovery, the higher the fixed cost. The organization pays for the strategy, even if no accident takes place. If there is an accident, variable costs will significantly increase (e.g., your "warm site" contract may provide for a flat annual fee plus a daily fee for actual occupation) due to the need for extra staff, hours, transportation and other logistics (e.g., *per diem*, new communication lines).

If the business continuity strategy aims at a longer recovery time, it will be less expensive than a more stringent requirement, but may be more susceptible to downtime costs spiraling out of control.

Terminology which the IS auditor may find associated with the development of a recovery strategy through BIA is described in section 6.2.7 below: terminology includes recovery time objective (RTO), recovery point objective (RPO), interruption window, the maximum tolerable outage (MTO) and service delivery objective (SDO).

In summary, the sum of all costs—downtime and recovery—should be minimized. The first group (downtime costs) increases with time, and the second (recovery costs) decreases with time; the sum usually is a U curve. At the bottom of the U curve, the lowest cost can be found.

> **Note:** The CISA candidate will not be tested on calculations of costs.

# Business Continuity
# and Disaster Recovery

## *Classification of Operations and Criticality Analysis*

What is the system's risk ranking? It involves a determination of risk based upon the impact derived from the critical recovery time period, as well as the likelihood that an adverse disruption will occur. Many organizations will use a risk of occurrence to determine a reasonable cost of being prepared. For example, they may determine that there is a 0.1 percent risk (or 1 in 1,000) that over the next five years the organization will suffer a serious disruption. If the assessed impact of a disruption is US $10 million, then the maximum reasonable cost of being prepared might be US $10 million x 0.1 percent = US $10,000 over five years. From this risk-based analysis process, prioritizing critical systems can take place in developing recovery strategies. The risk ranking procedure should be performed in coordination with IS processing and end-user personnel.

A typical risk ranking system may contain the classifications as found in **exhibit 6.4**.

| Exhibit 6.4—Classification of Systems | |
|---|---|
| **Classification** | **Description** |
| Critical | These functions cannot be performed unless they are replaced by identical capabilities. Critical applications cannot be replaced by manual methods. Tolerance to interruption is very low; therefore, cost of interruption is very high. |
| Vital | These functions can be performed manually, but only for a brief period of time. There is a higher tolerance to interruption than with critical systems and, therefore, somewhat lower costs of interruption, provided that functions are restored within a certain time frame (usually five days or less). |
| Sensitive | These functions can be performed manually, at a tolerable cost and for an extended period of time. While they can be performed manually, it usually is a difficult process and requires additional staff to perform. |
| Nonsensitive | These functions may be interrupted for an extended period of time, at little or no cost to the company, and require little or no catching up when restored. |

The next phase in continuity management is to identify the various recovery strategies and available alternatives for recovering from an interruption and/or disaster. The selection of an appropriate strategy based on the BIA and criticality analysis is the next step for developing BCPs and DRPs. The two metrics that help in determining the recovery strategies are the RPO and RTO.

| PRACTICE QUESTION |
|---|
| 6-4     The window of time for recovery of information processing capabilities is based on the:<br><br>     A.    criticality of the processes affected.<br>     B.    quality of the data to be processed.<br>     C.    nature of the disaster.<br>     D.    applications that are mainframe-based. |
| *See answers and explanations to the practice questions at the end of the chapter. (page 514)* |

## 6.2.7 RECOVERY POINT OBJECTIVE AND RECOVERY
##       TIME OBJECTIVE

The RPO is determined based on the acceptable data loss in case of disruption of operations. It indicates the earliest point in time in which it is acceptable to recover the data. For example, if the process can afford to lose the data up to four hours before disaster, then the latest backup available should be up to four hours before disaster or interruption, and the transactions during RPO and interruption need to be entered after recovery (known as catch-up data.)

# Business Continuity
# and Disaster Recovery

RPO effectively quantifies the permissible amount of data loss in case of interruption. It is almost impossible to recover the data completely. Even after entering incremental data, some data are still lost and are referred to as orphan data.

The RTO is determined based on the acceptable downtime in case of a disruption of operations. It indicates the earliest point in time at which the business operations must resume after disaster. **Exhibit 6.5** shows the relationship between the RTO and RPO.

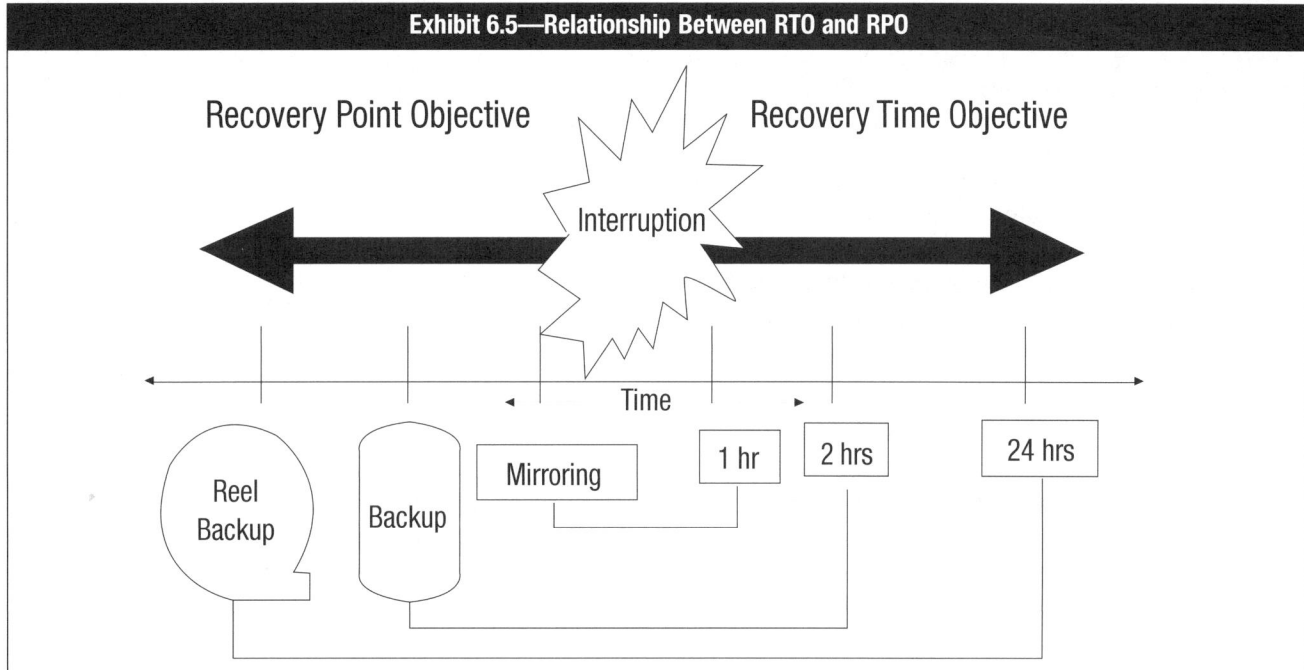

**Exhibit 6.5—Relationship Between RTO and RPO**

Both of these concepts are based on time parameters. The lower the time requirements, the higher the cost of recovery strategies, i.e., if the RPO is in minutes (lowest possible acceptable data loss), then data mirroring or duplexing should be implemented as the recovery strategy. If the RTO is lower, then the alternate site might be preferred over a hot-site contract.

Also, the lower the RTO, the lower the disaster tolerance. Disaster tolerance is the time gap within which the business can accept the nonavailability of IT critical services.

> **Note:** The CISA candidate should be familiar with which recovery strategies would be best with different RTO and RPO parameters.

In addition to RTO and RPO, there are some additional parameters that are important in defining the recovery strategies. These include:
- **Interruption window**—The time the organization can wait from the point of failure to the critical services/applications restoration. After this time, the progressive losses caused by the interruption are unaffordable.
- **Service delivery objective (SDO)**—Level of services to be reached during the alternate process mode until the normal situation is restored. This is directly related to the business needs.
- **Maximum tolerable outages**—Maximum time the organization can support processing in alternate mode. After this point, different problems may arise, especially if the alternate SDO is lower than the usual SDO, and the information pending to be updated can become unmanageable.

# Business Continuity
# and Disaster Recovery

**PRACTICE QUESTIONS**

6-5    Data mirroring should be implemented as a recovery strategy when:

    A.    recovery point objective (RPO) is low.
    B.    RPO is high.
    C.    recovery time objective (RTO) is high.
    D.    disaster tolerance is high.

6-6    When preparing a business continuity plan, which of the following must be known to establish a recovery point objective (RPO)?

    A.    The acceptable data loss in case of disruption of operations
    B.    The acceptable downtime in case of disruption of operations
    C.    Types of offsite backup facilities available
    D.    Types of IT platforms supporting critical business functions

*See answers and explanations to the practice questions at the end of the chapter. (page 514)*

# 6.2.8 RECOVERY STRATEGIES

A recovery strategy identifies the best way to recover a system in case of interruption, including disaster, and provides guidance based on which detailed recovery procedures can be developed. Different strategies should be developed, and all alternatives should be presented to senior management. Senior management should select the most appropriate strategies from the alternatives provided and accept the inherent residual risk. The selected strategies should be used to further develop the detailed BCP.

The selection of a recovery strategy would depend on:
• The criticality of the business process and the applications supporting the processes
• Cost
• Time required to recover
• Security

There are various strategies for recovering critical information resources. The appropriate strategy is the one with a cost for an acceptable recovery time that is also reasonable compared to the impact and likelihood of occurrence as determined in the BIA. The cost of recovery is the cost of preparing for possible disruptions (e.g., the fixed costs of purchasing, maintaining and regularly testing redundant computers, and maintaining alternate network routing), as well as the variable costs of putting these into use in the event of a disruption. The latter costs can often be insured against, but the former generally cannot. However, the premiums for disaster insurance usually will be lower if there is a suitable plan.

Generally, each IT platform that runs an application supporting a critical business function will need a recovery strategy. There are many alternative strategies. The most appropriate alternative, in terms of cost to recover and impact cost, should be selected based on the relative risk level identified in the business impact analysis. Recovery strategies based on the risk level identified for recovery would include developing:
• Hot sites
• Warm sites
• Cold sites
• Duplicate information processing facilities
• Mobile sites
• Reciprocal arrangements with other organizations

**Note:** The CISA candidate should know these recovery strategies and when to use them.

# Business Continuity and Disaster Recovery

## 6.2.9 RECOVERY ALTERNATIVES

Lengthier and more costly outages, particularly disasters that impair the primary physical facility, require offsite backup alternatives. The types of offsite backup hardware facilities available are:

- **Hot sites**—They are fully configured and ready to operate within several hours. The equipment, network and systems software must be compatible with the primary installation being backed up. The only additional needs are staff, programs, data files and documentation.

  Costs associated with the use of a third-party hot site are usually high, but less than creating a redundant site, and are often cost justifiable for critical applications. When properly planned, insurance coverage will usually offset the costs incurred for using this type of facility. Costs include a basic subscription cost, monthly fee, testing charges, activation costs (for the period when the site is used for an actual emergency), and hourly or daily use charges. Pricing structures vary between vendors. Some hot-site suppliers impose a high activation fee to discourage the frivolous use of the facility. Other vendors have no activation fee and encourage the use of the facility for nondisaster purposes such as overload processing. The contract must include the amount of time needed, frequency and the specified time for testing.

> **Note:** The CISA candidate will not be tested on the various cost aspects of the different disaster recovery strategies/sites.

  The hot site is intended for emergency operations of a limited time period and not for long-term extended use. Long-term use would impair the protection of other subscribers. Therefore, the hot site should be viewed as a means of accomplishing the continuation of essential operations for a period of up to several weeks following a disaster or major emergency. Further plans are still necessary to provide for subsequent operations. Several vendors offer warm- or cold-site facilities for a subscriber to migrate to after recovery of operations has been completed. This will free up the hot site for use by other subscribers.

  Components of the DRP for network connectivity to a hot site over a public-switched network should address issues such as redundancy and maintaining sufficient capacity on diverse paths to carry a rerouted path. It should also provide for late-night access routing through different central offices, so that no single point of failure can disable the entire network.

- **Warm sites**—They are partially configured, usually with network connections and selected peripheral equipment, such as disk drives and other controllers, but without the main computer. Sometimes a warm site is equipped with a less-powerful central processing unit (CPU), than the one generally used. The assumption behind the warm site concept is that the computer can usually be obtained quickly for emergency installation (provided it is a widely used model) and, since the computer is the most expensive unit, such an arrangement is less costly than a hot site. After the installation of the needed components, the site can be ready for service within hours; however, the location and installation of the CPU and other missing units could take several days or weeks.

- **Cold sites**—They have only the basic environment (i.e., electrical wiring, air conditioning, flooring, etc.) to operate an IPF reducing the cost. The cold site is ready to receive equipment, but does not offer any components at the site in advance of the need. Activation of the site may take several weeks.

- **Duplicate (redundant) information processing facilities (duplicate/redundant IPFs)**—They are dedicated, self-developed recovery sites that can back up critical applications. They can range in form from a standby hot site to a reciprocal agreement with another company installation. The assumption is that there are fewer problems in coordinating compatibility and availability in the case of duplicate IPF sites. However, larger organizations may experience problems similar to those encountered by reciprocal agreements between unrelated companies. This is particularly true whenever departmental or divisional IPFs are managed separately, or when hostile in-house political jealousies exist. Several principles must be in place to ensure the viability of this approach:
  - The site chosen should not be subject to the same natural disaster(s) as the original (primary) site.
  - There must be a coordination of hardware/software strategies. A reasonable degree of compatibility must exist to serve as a basis for backup.
  - Resource availability must be assured. The workloads of the sites must be monitored to ensure that availability for emergency backup use will not be impaired.
  - There must be agreement as to the priority of adding applications (workloads) until all the recovery resources are fully utilized.
  - Regular testing is necessary. Even though duplicate sites are under common ownership and even if the sites are under the same management, testing of the backup operation is necessary.

# Business Continuity and Disaster Recovery

- **Mobile sites**—This is a specially designed trailer that can be quickly transported to a business location or to an alternate site to provide a ready-conditioned IPF. These mobile sites can be connected to form larger work areas and can be preconfigured with servers, desktop computers, communications equipment, and even microwave and satellite data links. They are a useful alternative when there are no recovery facilities in the immediate geographic area. They are also useful in case of a widespread disaster and are a cost-effective alternative to duplicate IPFs for a multi-office organization. Needless to say, the location for the trailer, connection to the organization's networks and the installation of communications must be carefully planned; recovery will be materially delayed if a large trailer arrives at the site but there is nowhere to park it.
- **Reciprocal agreements with other organizations**—This is a less frequently used method between two or more organizations with similar equipment or applications. Under the typical agreement, participants promise to provide computer time to each other when an emergency arises.

Advantages include:
– Low cost
– May be the only option available for unique vendor equipment due to lack of availability of hot sites

Disadvantages include:
– Usually not enforceable
– Differences in equipment configuration often necessitate program changes to operate effectively
– Participants not notified of changes in workloads or equipment configurations rendering the agreement limited or useless
– Differing policies since organizations differ in procedures, personnel, and skills

Critical questions to cover in a reciprocal agreement include:
– How much time will be available at the host computer site?
– What facilities, equipment and software will be available?
– Will staff assistance be provided?
– How quickly can access be gained to the host recovery facility?
– Can data and voice communication links be established at the host site?
– How long can the emergency operation continue?
– How frequently can the system be tested for compatibility?
– How will confidentiality of data be maintained?
– What type of security will be afforded for information systems operations and data?
– How much advance notice is required for using the facility?
– Are there certain times of the year, month, etc., when the partner's facilities are not available?

## Contractual Provisions

Contractual provisions for the use of third-party sites should cover the following:
- **Configurations**—Are the vendor's hardware and software configurations adequate to meet company needs since these will vary over time?
- **Disaster**—Is the definition of disaster broad enough to meet anticipated needs?
- **Speed of availability**—How soon after a disaster will facilities be available?
- **Subscribers per site**—Does the agreement limit the number of subscribers per site?
- **Subscribers per area**—Does the agreement limit the number of subscribers in a building or area?
- **Preference**—Who gets preference if there are common or regional disasters? Is there backup for the backup facilities? Is use of the facility exclusive or does the customer have to share the available space if multiple customers simultaneously declare a disaster? Does the vendor have more than one facility available for subscriber use?
- **Insurance**—Is there adequate insurance coverage for company employees at the backup site? Will existing insurance reimburse those fees?
- **Usage period**—How long is the facility available for use? Is this period adequate? What technical support will the site operator provide? Is this adequate?
- **Communications**—Are the communications adequate? Are the communication connections to the backup site sufficient to permit unlimited communication with the alternate site if needed?

- **Warranties**—What warranties will the vendor make regarding availability of the site and the adequacy of the facilities? Are there liability limitations (there usually are) and is the company willing to live with them?
- **Audit**—Is there a right-to-audit clause permitting an audit of the site to evaluate the logical, physical and environmental security?
- **Testing**—What testing rights are included in the contract? Check with the insurance company to determine any reduction of premiums that may be forthcoming due to the backup site availability.
- **Reliability**—Can the vendor attest to the reliability of the site(s) being offered? Ideally, the vendor should have a UPS, limited subscribers, sound technical management, and guarantees of computer hardware and software compatibility.

## Procuring Alternative Hardware

There are several alternatives available for securing backup hardware and physical facilities, including:

- **A vendor or third party**—Hardware vendors are usually the best source for replacement equipment. However, this often may involve a waiting period that is not acceptable for critical operations. It is unlikely that any vendor will guarantee a specific reaction to a crisis. Vendor arrangements are utilized best when planning to move from a hot site to a warm or cold site. The arrangements should be planned in advance. Another source of equipment replacement is the used hardware market. This market can supply critical components or entire systems on relatively short notice, often at a savings. These dealer relationships should be cultivated well in advance of the actual emergency.
- **Off-the-shelf**—Such components are readily available from the inventory of suppliers on short notice and with minimum need for special arrangements. For example, modern desktop PCs generally include network cards and laptops are usually capable of connecting to a wireless network; a local area network (LAN) can, in many regions, be recreated in a relatively short time period given the necessary backup and an appropriate location. To make use of this approach, several strategies must be utilized, including:
  – Regularly updating equipment to keep current
  – Maintaining software compatibility to permit the operation of newer equipment
- **Credit agreement or emergency credit cards**—Ensuring the recovery plans include instructions on how such equipment is to be paid for. This could be by a credit agreement with suppliers or by the provision of an emergency credit card, with a sufficiently high credit limit. It should not be left to individual employees, even managers, to take responsibility for such purchases on their own account.

As data and software are required for these strategies, special arrangements need to be considered for their backup to removable media and their safe, secure storage offsite.

Additionally, part of the recovery of IT facilities will involve telecommunications, for which the strategies usually considered include:

- **Network disaster prevention**, which includes:
  – Alternative routing
  – Diverse routing
  – Long-haul network diversity
  – Protection of the local loop
  – Voice recovery
  – Availability of appropriate circuits and adequate bandwidth
- **Server disaster recovery plans**

These strategies are discussed in more detail later in this chapter.

Having developed a strategy for the recovery of sufficient IT facilities to support critical business processes, it is essential that the strategies for these functions operate until all facilities are restored. Therefore, they may include:

- Doing nothing until recovery facilities are ready
- Using manual procedures (manual fallback)
- Complying only with regulatory and legal requirements
- Focusing on the most important customers, suppliers, products, systems, etc.
- Using PC-based systems to capture data for later processing or to perform simple local processing

# Business Continuity
# and Disaster Recovery

**6-7** An IS auditor discovers that an organization's business continuity plan provides for an alternate processing site that will accommodate 50 percent of the primary processing capability. Based on this, which of the following actions should the IS auditor take?

    A.    Do nothing, because generally less than 25 percent of all processing is critical to an organization's survival; therefore, the backup capacity is adequate.

    B.    Identify applications that could be processed at the alternate site, and develop manual procedures to back up other processing.

    C.    Ensure that critical applications have been identified and that the alternate site could process all such applications.

    D.    Recommend that the IPF arrange for an alternate processing site with the capacity to handle at least 75 percent of normal processing.

*See answers and explanations to the practice questions at the end of the chapter. (page 514)*

## 6.2.10 DEVELOPMENT OF BUSINESS CONTINUITY AND DISASTER RECOVERY PLANS

Based on the inputs received from the BIA, criticality analysis and recovery strategy selected by management, a detailed BCP and DRP should be developed or reviewed. It should address all issues involved in interruption to business processes, including recovering from a disaster. The various factors that should be considered while developing/reviewing the plan are:
• Predisaster readiness covering incidence response management to address all incidences affecting business processes
• Evacuation procedures
• Procedures for declaring a disaster
• Circumstances under which a disaster should be declared. All interruptions are not disasters, but a small incident if not addressed in a timely or proper manner may lead to a disaster. For example, a virus attack not recognized and contained in time may bring down the entire IT facility.
• The clear identification of the responsibilities in the plan
• The clear identification of the persons responsible for each function in the plan
• The clear identification of contract information
• The step-by-step explanation of the recovery option
• The clear identification of the various resources required for recovery and continued operation of the organization

The plan should be documented and written in simple language, understandable to all. It is common to identify teams of personnel who are made responsible for specific tasks in case of disasters. Some important teams should be formed and their responsibilities are explained below. Copies of the plan should be maintained offsite.

## 6.2.11 ORGANIZATION AND ASSIGNMENT OF RESPONSIBILITIES

The plan should identify the teams with their assigned responsibilities in the event of an incident/disaster. To implement the strategies that have been developed for business/process recovery and key decision making, IS and end-user personnel should be identified. These individuals usually lead teams created in response to a critical function or task defined in the plan. Depending on the size of the business operation, these teams may be designated as single-person positions. The involvement of the following teams depends on the level of the disruption of service and the types of assets lost or damaged. It is a good idea to develop a matrix on the correlation between the teams needed to participate and the estimated recovery effort/level of disruption.

The recovery/continuity/response teams may include any of the following:
• **Incident response team**—A team that has been designated to receive the information about every incident that can be considered as a threat to assets/processes. This reporting can be useful for coordinating an incident in progress and or for postmortem analysis. The analysis of all incidents also provides input for updating the recovery plans.

# Business Continuity and Disaster Recovery

- **Emergency action team**—They are first responders, designated fire wardens and bucket crews, whose function is to deal with fires or other emergency response scenarios. One of their primary functions is the orderly evacuation of personnel and the securing of human life.
- **Information security team**—The main mission of this team is to develop the needed steps to maintain a similar level of information and IT resource security as was in place in at the primary site before the contingency, and implement the needed security measures in the alternative procedures environment. Additionally, this team must continually monitor the security of system and communication links, resolve any security conflicts that impede the expeditious recovery of the system, and assure the proper installation and functioning of security software. The team is also responsible for the security of the organization's assets during the disorder following a disaster.
- **Damage assessment team**—Assesses the extent of damage following the disaster. The team should be comprised of individuals who have the ability to assess damage and estimate the time required to recover operations at the affected site. This team should include staff skilled in the use of testing equipment, knowledgeable about systems and networks, and trained in applicable safety regulations and procedures. In addition, they have the responsibility to identify possible causes of the disaster and their impact on damage and predictable downtime.
- **Emergency management team**—Responsible for coordinating the activities of all other recovery/continuity/response teams and handling key decision making. They determine the activation of the BCP. Other functions entail arranging the finances of the recovery, handling legal matters evolving from the disaster, and handling public relations and media inquiries.

  This team functions as disaster overseers and is required to coordinate the following activities:
  - Retrieving critical and vital data from offsite storage
  - Installing and testing systems software and applications at the systems recovery site (hot site, cold site)
  - Identifying, purchasing, and installing hardware at the system recovery site
  - Operating from the system recovery site
  - Rerouting network communications traffic
  - Reestablishing the user/system network
  - Transporting users to the recovery facility
  - Reconstructing databases
  - Supplying necessary office goods, i.e., special forms, check stock, paper
  - Arranging and paying for employee relocation expenses at the recovery facility
  - Coordinating systems use and employee work schedules

- **Offsite storage team**—Responsible for obtaining, packaging and shipping media and records to the recovery facilities, as well as establishing and overseeing an offsite storage schedule for information created during operations at the recovery site
- **Software team**—Responsible for restoring system packs, loading and testing operating systems software, and resolving system-level problems
- **Applications team**—Travels to the system recovery site and restores user packs and application programs on the backup system. As the recovery progresses, this team may have the responsibility of monitoring application performance and database integrity.
- **Emergency operations team**—Consists of shift operators and shift supervisors who will reside at the systems recovery site and manage system operations during the entirety of the disaster and recovery projects. Another responsibility might be coordinating hardware installation, if a hot site or other equipment-ready facility has not been designated as the recovery center.
- **Network recovery team**—Responsible for rerouting wide-area voice and data communications traffic, reestablishing host network control and access at the system recovery site, providing ongoing support for data communications, and overseeing communications integrity
- **Communications team**—Travels to the recovery site where they work in conjunction with the remote network recovery team to establish a user/system network. This team also is responsible for soliciting and installing communications hardware at the recovery site and working with local exchange carriers and gateway vendors in the rerouting of local service and gateway access.
- **Transportation team**—Serves as a facilities team to locate a recovery site, if one has not been predetermined, and is responsible for coordinating the transport of company employees to a distant recovery site. It also may assist in contacting employees to inform them of new work locations, and scheduling and arranging employee lodgings.

- **User hardware team**—Locates and coordinates the delivery and installation of user terminals, printers, typewriters, photocopiers and other necessary equipment. This team also offers support to the communications team and to any hardware and facilities salvage efforts.
- **Data preparation and records team**—Working from terminals that connect to the user recovery site, the team updates the applications database. This team also oversees additional data-entry personnel and assists record salvage efforts in acquiring primary documents and other input information sources.
- **Administrative support team**—Provides clerical support to the other teams and serves as a message center for the user recovery site. This team also may control accounting and payroll functions as well as ongoing facilities management.
- **Supplies team**—Supports the efforts of the user hardware team by contacting vendors and coordinating logistics for an ongoing supply of necessary office and computer supplies
- **Salvage team**—Manages the relocation project. This team also makes a more detailed assessment of the damage to the facilities and equipment than was performed initially; provides the emergency management team with the information required to determine whether planning should be directed toward reconstruction or relocation; provides information necessary for filing insurance claims (insurance is the primary source of funding for the recovery efforts); and coordinates the efforts necessary for immediate records salvage, such as restoring paper documents and electronic media.
- **Relocation team**—Coordinates the process of moving from the hot site to a new location or to the restored original location. This involves relocating the IS processing operations, communications traffic and user operations. This team also monitors the transition to normal service levels.
- **Coordination team**—Responsible for coordinating the recovery efforts across various offices located at different geographical locations
- **Legal affairs team**—Responsible for handling the legal issues arising for various reasons due to any incident or nonavailability of services (e.g., according to new laws enacted by many countries, the organization is responsible for securing its IT assets, and will be liable for damages to innocent parties in case of incidence)
- **Recovery test team**—Responsible for testing of various plans developed and analyzing the result
- **Training team**—A team that will provide training to the users for provisions of business continuity and disaster recovery procedures

> **Note:** The IS auditor should have knowledge of these responsibilities, however the CISA candidate will not be tested on these specific assignments since they vary from organization to organization.

## 6.2.12 OTHER ISSUES IN PLAN DEVELOPMENT

The personnel who must react to the interruption/disaster scenarios are those responsible for the most critical resources. Therefore, management and user involvement is vital to the success of the BCP. User management involvement is essential to the identification of critical systems, their associated critical recovery times and the specification of needed resources. The three major divisions that require involvement in the formulation of the BCP are support services, business operations and information processing support.

Because the underlying purpose of BCP is the resumption of business operations, it is essential to consider the entire organization, not just IS processing services, when developing the plan. Where a uniform BCP does not exist for the entire organization, the plan for IS processing should be extended to include planning for all divisions and units that depend on information systems processing functions. Data processing plans must extend to the user areas to cover the sources of information, transmittal of data to the IS processing department, and delivery and deployment of processed results to the user units.

When formulating the plan, the following items should also be included:
- A list of the staff, with contact information, required to maintain critical business functions in the short, medium and long term
- The configuration of building facilities, desks, chairs, telephones, etc., required to maintain critical business functions in the short, medium and long term

## 6.2.13 COMPONENTS OF A BUSINESS CONTINUITY PLAN

Depending on the size and/or requirements of an organization, a BCP may consist of more than one plan document.

This should include:
• Continuity of operations plan
• DRP
• Business resumption plan

It may also include:
• Continuity of support plan/IT contingency plan
• Crisis communications plan
• Incident response plan
• Transportation plan
• Occupant emergency plan

The purpose and scope of the components of a BCP as suggested by NIST are shown in **exhibit 6.6**.

| Exhibit 6.6—Purpose of Scope of Buniness Continuity Plan Components | | |
|---|---|---|
| **Plan** | **Purpose** | **Scope** |
| Business continuity plan | Provide procedures for sustaining essential business operations while recovering from a significant disruption | Addresses business processes; IT addressed based only on its support for business processes |
| Business recovery (or resumption) plan (BRP) | Provide procedures for recovering business operations immediately following a disaster | Addresses business processes; not IT-focused; IT addressed based only on its support for business processes |
| Continuity of operations plan (COOP) | Provide procedures and capabilities to sustain an organization's essential, strategic functions at an alternate site for up to 30 days | Addresses the subset of an organization's missions that are deemed most critical; usually written at headquarters level; not IT-focused |
| Continuity of support plan/ IT contingency plan | Provide procedures and capabilities for recovering a major application or general support system | Same as IT contingency plan; addresses IT system disruptions; not business process focused |
| Crisis communications plan | Provides procedures for disseminating status reports to personnel and the public | Addresses communications with personnel and the public; not IT focused |
| Cyber incident response plan | Provide strategies to detect, respond to, and limit consequences of malicious cyber incident | Focuses on information security responses to incidents affecting systems and/or networks |
| Disaster recovery plan | Provide detailed procedures to facilitate recovery of capabilities at an alternate site | Often IT-focused; limited to major disruptions with long-term effects |
| Occupant emergency plan (OEP) | Provide coordinated procedures for minimizing loss of life or injury and protecting property damage in response to a physical threat | Focuses on personnel and property particular to the specific facility; not business process or IT system functionality based |

For the planning, implementation and evaluation phase of the BCP, the following should be agreed on:
• The policies that will govern all of the continuity and recovery efforts
• The goals/requirements/products for each phase
• Alternate facilities to perform tasks and operations
• Critical information resources to deploy (e.g., data and systems)

# Business Continuity and Disaster Recovery

- Persons responsible for completion
- Available resources to aid in deployment (including human)
- The scheduling of activities with priorities established

Most BCPs are created as procedures that accommodate system, user and network recovery strategies. Copies of the plan should be kept offsite—at the recovery facility, at the media storage facility and possibly at the homes of key decision-making personnel. More and more frequently, an organization places its plan on a mirrored web site. Components of this plan include key decision-making personnel, a backup of required supplies, the organization, and the assignment of responsibilities, telecommunication networks and insurance.

## *Key Decision-making Personnel*

The plan should contain a telephone list or "call tree," i.e., a notification directory, of key decision-making IS and end-user personnel required to initiate and carry out recovery efforts. This is usually a telephone directory of people who should be notified in the event of an incident/disaster or catastrophe. The point to remember when preparing the list is that in the event of a widespread disaster or a fire/explosion during normal business hours that heavily damages the organization's offices, many team leaders may not be available.

This directory should contain the following information:
- A prioritized list of contacts (i.e., who gets called first?)
- Primary and emergency telephone numbers and addresses for each critical contact person. These usually will be key team leaders responsible for contacting the members of their team.
- Phone numbers and addresses for representatives of equipment and software vendors
- Phone numbers of contacts within companies that have been designated to provide supplies and equipment or services
- Phone numbers of contact persons at recovery facilities, including hot-site representatives and predefined network communications rerouting services
- Phone numbers of contact persons at offsite media storage facilities and the contact persons within the company who are authorized to retrieve media from the offsite facility
- Phone numbers of insurance company agents
- Phone numbers of contacts at contract personnel services
- Phone numbers and contacts of legal/regulatory/governmental agencies, if required

| PRACTICE QUESTIONS |
|---|
| 6-8    In a business continuity plan, which of the following notification directories is the **MOST** important? |
|      A.    Equipment and supply vendors <br>      B.    Insurance company agents <br>      C.    Contract personnel services <br>      D.    A prioritized contact list |
| 6-9    Which of the following components of a business continuity plan is **PRIMARILY** the responsibility of an organization's IS department? |
|      A.    Developing the business continuity plan <br>      B.    Selecting and approving the strategy for the business continuity plan <br>      C.    Declaring a disaster <br>      D.    Restoring the IS systems and data after a disaster |
| *See answers and explanations to the practice questions at the end of the chapter. (pages 514-515)* |

# Business Continuity and Disaster Recovery

## *Backup of Required Supplies*

The plan should have provisions for all supplies necessary for the continuation of normal business activities in the recovery effort. This includes detailed, up-to-date hard copy procedures that can be followed easily by staff and contract personnel who are unfamiliar with the standard and recovery operations. Also, a supply of special forms, such as check stock, invoice forms and order forms, should be secured at an offsite location.

If the data entry function depends on certain hardware devices and/or software programs, these programs and equipment, including specialized electronic data interchange (EDI) equipment and programs, should be provided at the hot site. The same would apply to cryptographic equipment.

## *Telecommunication Networks Disaster Recovery Methods*

The plan should contain the organization's telecommunication networks. Today, telecommunication networks are key to business processes in large and small organizations; therefore, the procedures to ensure continuous telecommunication capabilities should be given a high priority.

Telecommunication networks are susceptible to the same natural disasters as data centers, but also are vulnerable to several disastrous events unique to telecommunications. These include central switching office disasters, cable cuts, communication software glitches and errors, security breaches connected to hacking (phone hackers are known as phreakers), and a host of other human mishaps. It is the responsibility of the organization and not the local exchange carriers to ensure constant communication capabilities. The local exchange carrier is not responsible for providing backup services, though many do back up main components within their systems. Therefore, the organization should make provisions for backing up its own telecommunication facilities.

To maintain critical business processes, the IPF's BCP should provide for adequate telecommunications capabilities. Telecommunications capabilities to consider include telephone voice circuits, wide area networks (WANs) (connections to distributed data centers), LANs (work group PC connections), and third-party EDI providers. The critical capacity requirements should be identified for the various thresholds of outage for each telecommunications capability such as two hours, eight hours or 24 hours. UPSs should be sufficient to provide backup to the telecommunication equipment as well as the computer equipment.

Methods for network protection are:
• **Redundancy**—Involves a variety of solutions, including:
  – Providing extra capacity with a plan to use the surplus capacity should the normal primary transmission capability not be available. In the case of a LAN, a second cable could be installed through an alternate route for use in the event the primary cable is damaged.
  – Providing multiple paths between routers
  – Dynamic routing protocols, such as Open Shortest Path First (OSPF) and Enhanced Interior Gateway Routing Protocol (EIGRP)
  – Providing for fail over devices to avoid single point of failures in routers, switches, firewalls, etc.
  – Saving configuration files for recovery in the event that network devices, such as those for routers and switches, fail. For example, organizations should utilize Trivial File Transport Protocol (TFTP) servers. Most network devices support TFTP for saving and retrieving configuration information.
• **Alternative routing**—The method of routing information via an alternate medium such as copper cable or fiber optics. This involves use of different networks, circuits or end points should the normal network be unavailable. Most local carriers are deploying counter-rotating, fiber-optic rings. These rings have fiber-optic cables that transmit information in two different directions and in separate cable sheaths for increased protection. Currently, these rings connect through one central switching office. However, future expansion of the rings may incorporate a second central office in the circuit. Some carriers are offering alternate routes to different points of presence or alternate central offices. Other examples include a dial-up circuit as an alternative to dedicated circuits; cellular phone and microwave communication as alternatives to land circuits; and couriers as an alternative to electronic transmissions.

# Business Continuity and Disaster Recovery

- **Diverse routing**—The method of routing traffic through split cable facilities or duplicate cable facilities. This can be accomplished with different and/or duplicate cable sheaths. If different cable sheaths are used, the cable may be in the same conduit and, therefore, subject to the same interruptions as the cable it is backing up. The communication service subscriber can duplicate the facilities by having alternate routes, although the entrance to and from the customer premises may be in the same conduit. The subscriber can obtain diverse routing and alternate routing from the local carrier, including dual entrance facilities. However, acquiring this type of access is time-consuming and costly. Most carriers provide facilities for alternate and diverse routing, although the majority of services are transmitted over terrestrial media. These cable facilities are usually located in the ground or basement. Ground-based facilities are at great risk due to the aging infrastructures of cities. In addition, cable-based facilities usually share room with mechanical and electrical systems that can impose great risks due to human error and disastrous events.
- **Long-haul network diversity**—Many vendors of recovery facilities have provided diverse long-distance network availability, utilizing T1 circuits among the major long-distance carriers. This ensures long-distance access should any single carrier experience a network failure. Several of the major carriers now have installed automatic rerouting software and redundant lines that provide instantaneous recovery should a break in their lines occur. The IS auditor should verify that the recovery facility has these vital telecommunications capabilities.
- **Last-mile circuit protection**—Many recovery facilities provide a redundant combination of local carrier T1s or E1s, microwave, and/or coaxial cable access to the local communications loop. This enables the facility to have access during a local carrier communication disaster. Alternate local carrier routing also is utilized.
- **Voice recovery**—With many service, financial and retail industries dependent on voice communication, redundant cabling and Voice-over Internet Protocol (VoIP) are common approaches to deal with it.

> **Note:** The CISA candidate should know what method is recommended for what data/information based on this criticality.

## Redundant Array of Independent (or Inexpensive) Disks

Redundant Array of Independent (or Inexpensive) Disks (RAID) provides performance improvements and fault-tolerant capabilities via hardware or software solutions, breaking up data and writing data to a series of multiple disks to simultaneously improve performance and/or save large files. These systems provide the potential for cost-effective mirroring offsite for data backup.

A variety of methods categorized into 11 levels (the most popular being 0, 3 and 5) are defined for combining several disk drives into what appears to the system as a single disk drive. RAID improves on the single-drive-only solution since it offers better performance and/or data redundancy.

### RAID LEVEL DESCRIPTIONS

The RAID level descriptions are:
- **Level 0, striped disk array without fault tolerance**—It improves performance by creating what appears to be one disk out of several separate physical disk drives. Areas that appear to be a cylinder or a track on a logical disk drive are spread across two or more physical disk drives. Benefits of this approach relate to improved performance when accessing data for maximum transfer rates and sizes. For example, when reading a block of data, the read operation can become several simultaneous separate disk reads of several physical disks. However, this approach, while improving performance, provides no redundancy or parity.
- **Level 1, mirroring**—It enables an exact copy of information on one disk area, to be copied to another. Once established, data written to the disk are also written to the free space on the other half of the mirror set. When implementing disk mirroring, issues to consider include:
  - Mirroring drives run from the same controller does not protect the data from drive-controller failure, unless mirroring to a disk is run from a separate controller.
  - For higher disk-read performance and greater fault tolerance, a separate disk controller for each half of a mirror set should be used.
  - Disk mirroring effectively cuts available disk space in half.
  - Disk mirroring has a low initial cost since only one extra drive is needed to achieve fault tolerance.

# Business Continuity and Disaster Recovery

- Disk mirroring slows down writes since the data must be written in two places every time, but will speed up reads since the I/O controller has two places from which to read information. Generally, for multiuser environments, mirroring gets the best performance of all the RAID levels.
- **Level 2, hamming code elliptical curve cryptography (ECC)**—It is the process of interweaving data across multiple drives, including parity information created using the hamming code technique. Hamming code is the hash algorithm coding technique used for recovering lost data that is duplicated on another drive. If a hash algorithm is applied to data on one disk that is also on a redundant disk and the values are equal between the two, the data are recoverable from the redundant disk. This level generally requires a large number of disks for user storage in addition to disks containing recovery error coding, where there may be, for example, one recovery disk for every four disks of user storage. This level, however, is seldom used because of the extensive amount of computer resources needed.
- **Level 3, parallel transfer with parity**—It uses byte-level parity on a dedicated-drive and stripe user data across multiple drives. It provides the level 0 feature of improved performance using disk striping and the redundancy provided with the use of parity of data on a dedicated parity drive. Parity information is calculated on data writes occurring, which may affect performance. This process provides fault tolerant capabilities through the availability of redundant drives (at least three).
- **Level 4, independent data disks with shared parity disk**—It is similar to level 3, but uses block-level parity and disk striping, rather than byte-level within a block. As with level 3, it provides fault-tolerant capabilities through the availability of redundant drives.
- **Level 5, independent data disks with distributed parity blocks**—It stripes both data and parity information across multiple drives at the block level. It differs from regular disk striping in that data from a parity stripe is recoverable, thus, providing a fault-tolerant capability to the system. When data are written to a disk, the data are written across all striped disks in a disk array, just as with regular disk striping in RAID level 0. However, parity information written to a disk occurs on a separate disk from the one where the data it corresponds to are written. Thus, if anything happens to one of the disks in the array, the data on that disk can be reconstructed from the parity information on the other disks.

  Additionally, RAID level 5 differs from level 4 in that parity information in RAID level 5 is distributed across all the disks in the array. In level 4, a specific disk is dedicated to parity information, which makes level 5 faster than level 4, because it can perform more than one write operation at a time.
- **Level 6, independent data disks with two independent distributed parity schemes**—It is similar to RAID level 5 in that it stripes both data and parity information across multiples drives. However, it differs in that it calculates two sets of parity information for each block of data. This method with parity protection provides high fault-tolerant capabilities in responding to disk or path failures.
- **Level 7, optimized asynchrony for high I/O rates as well as high data transfer rates**—It is based on an architecture characterized by asynchronous I/O transfers. These transfers are independently controlled and cached via a real-time OS-embedded array control microprocessor, where all reads and writes are centrally cached via a high-speed internal cache data transfer bus with parity generation integrated into the cache. This enables multiple attached disk drives to be designated as hot standbys. It is a highly expensive proprietary solution providing high performance levels through its caching data transfer capabilities.
- **Level 10, very high reliability combined with high performance**—It is characterized as a striped array with high I/O rates achieved by striping level 1 segments (requires a minimum of four drives to implement).
- **Level 53, high I/O rates and data transfer performance**—It is characterized as a striped level 0 array with segments as level 3 arrays. High data transfer rates are achieved due to the level 3 array segments and high I/O rates based on level 0 striping. This is an expensive solution, but with high performance capabilities.
- **Level 0+1, high data transfer performance**—It is implemented as a mirror array whose segments are level 0 arrays. Designed with the fault tolerance similar to level 5, this approach is very expensive to implement and is susceptible to a significant loss of its fault-tolerant capability if any of its drives should fail. In essence, it becomes a level 0 array upon failure of any drive where a minimum of four is need for its implementation.

In comparing the above techniques, RAID levels 0 and 1 have better performance when they are software based, whereas RAID levels 3, 5 and 6 run faster on hardware. RAID level 2 is hardware-based and resource-intensive and, therefore, is seldom used. The remaining levels are costly/high-overhead solutions with limited scalability.

> **Note:** The CISA candidate would not be tested on the specifics of RAID levels.

# Business Continuity
# and Disaster Recovery

## Insurance

The plan should contain key information about the organization's insurance. The IS processing insurance policy is usually a multiperil policy designed to provide various types of IS coverage. It should be constructed in modules so it can be adapted to the insured's particular IS environment.

> **Note:** Specifics on insurance policies are not tested on the CISA exam because they differ from country to country. The test covers what should be included in policies and third-party agreements but would not test the specific types of coverage.

Specific types of coverage available are:
- **IS equipment and facilities**—Provides coverage about physical damage to the IPF and owned equipment. (Insurance of leased equipment should be obtained when the lessee is responsible for hazard coverage.) The IS auditor is cautioned to review these policies since many policies obligate insurance vendors to replace nonrestorable equipment only with "like kind and quality," not necessarily with new equipment by the same vendor as the damaged equipment.
- **Media (software) reconstruction**—Covers damage to IS media that is the property of the insured and for which the insured may be liable. Insurance is available for on-premises, off-premises or in-transit situations and covers the actual reproduction cost of the property. Considerations in determining the amount of coverage needed are programming costs to reproduce the media damaged, backup expenses and physical replacement of media devices such as tapes, cartridges and disks.
- **Extra expense**—Designed to cover the extra costs of continuing operations following damage or destruction at the IPF. The amount of extra-expense insurance needed is based on the availability and cost of backup facilities and operations. Extra expense can also cover the loss of net profits caused by computer media damage. This provides reimbursement for monetary losses resulting from suspension of operations due to the physical loss of equipment or media. An example of a situation requiring this type of coverage is if the information processing facilities were on the sixth floor and the first five floors were burned out. In this case, operations would be interrupted even though the IPF remained unaffected.
- **Business interruption**—Covers the loss of profit due to the disruption of the activity of the company caused by any malfunction of the IS organization
- **Valuable papers and records**—Covers the actual cash value of papers and records (not defined as media) on the insured's premises against direct physical loss or damage
- **Errors and omissions**—Provides legal liability protection in the event that the professional practitioner commits an act, error or omission that results in financial loss to a client. This insurance was originally designed for service bureaus but it is now available from several insurance companies for protecting systems analysts, software designers, programmers, consultants and other IS personnel.
- **Fidelity coverage**—Usually takes the form of bankers blanket bonds, excess fidelity insurance and commercial blanket bonds, and covers loss from dishonest or fraudulent acts by employees. This type of coverage is prevalent in financial institutions, operating their own IPF.
- **Media transportation**—Provides coverage for potential loss or damage to media in transit to off-premises IPFs. Transit coverage wording in the policy usually specifies that all documents must be filmed or otherwise copied. When the policy does not state specifically that data be filmed prior to being transported and the work is not filmed, management should obtain from the insurance carrier a letter that specifically describes the carrier's position and coverage in the event data are destroyed.

Several key points are important to remember about insurance. Most insurance covers only financial losses based on the historical level of performance and not the existing level of performance. Also, insurance does not compensate for loss of image/goodwill.

## 6.2.14 PLAN TESTING

Most business continuity tests fall short of a full-scale test of all operational portions of the corporation. This should not preclude performing full or partial testing because one of the purposes of the business continuity test is to determine how well the plan works or which portions of the plan need improvement.

The test should be scheduled during a time that will minimize disruptions to normal operations. Weekends are generally a good time to conduct tests. It is important that the key recovery team members be involved in the test process and allotted the necessary time to put their full effort into it. The test should address all critical components and simulate actual primetime processing conditions, even if the test is conducted in off hours.

## Specifications

The test should strive to accomplish the following tasks:
- Verify the completeness and precision of the BCP.
- Evaluate the performance of the personnel involved in the exercise.
- Appraise the training and awareness of employees who are not members of a business continuity team.
- Evaluate the coordination among the business continuity team and external vendors and suppliers.
- Measure the ability and capacity of the backup site to perform prescribed processing.
- Assess the vital records retrieval capability.
- Evaluate the state and quantity of equipment and supplies that have been relocated to the recovery site.
- Measure the overall performance of operational and IS processing activities related to maintaining the business entity.

> **Note:** This is an important part of the IS auditor's responsibility—assessing the results and the value of the BCP and the DRP tests.

## Test Execution

To perform testing, each of the following test phases should be completed:
- **Pretest**—The set of actions necessary to set the stage for the actual test. This ranges from placing tables in the proper operations recovery area to transporting and installing backup telephone equipment. These activities are outside the realm of those that would take place in the case of a real emergency, in which there is no forewarning of the event and, therefore, no time to take preparatory actions.
- **Test**—This is the real action of the business continuity test. Actual operational activities are executed to test the specific objectives of the BCP. Data entry, telephone calls, information systems processing, handling orders, and movement of personnel, equipment and suppliers should take place. Evaluators review staff members as they perform the designated tasks. This is the actual test of preparedness to respond to an emergency.
- **Post-test**—The cleanup of group activities. This phase comprises such assignments as returning all resources to their proper place, disconnecting equipment, returning personnel, and deleting all company data from third-party systems. The post-test cleanup also includes formally evaluating the plan and implementing indicated improvements.

In addition, the following types of tests may be performed:
- **Desk-based evaluation/paper test**—A paper walk-through of the plan, involving major players in the plan's execution who reason out what might happen in a particular type of service disruption. They may walk through the entire plan or just a portion. The paper test usually precedes the preparedness test.
- **Preparedness test**—Usually a localized version of a full test, wherein actual resources are expended in the simulation of a system crash. This test is performed regularly on different aspects of the plan and can be a cost-effective way to gradually obtain evidence about how good the plan is. It also provides a means to improve the plan in increments.
- **Full operational test**—This is one step away from an actual service disruption. The organization should have tested the plan well on paper and locally before endeavoring to completely shut down operations. For purposes of the BCP testing, this is the disaster.

## Documentation of Results

During every phase of the test, detailed documentation of observations, problems and resolutions should be maintained. Each team should have a diary form, with specific steps and information to be recorded, which can be used as documentation. This documentation serves as important historical information that can facilitate actual recovery during a real disaster. Additionally, the insurance company or the local authorities may ask for it. The documentation also aids in performing detailed analysis of both the strengths and weaknesses of the plan.

# Business Continuity and Disaster Recovery

## Results Analysis

It is important to have ways to measure the success of the plan and test against the stated objectives. Therefore, results must be quantitatively gauged as opposed to an evaluation based only on observation.

Specific measurements vary depending on the test and the organization; however, these general measurements usually apply:
• **Time**—Elapsed time for completion of prescribed tasks, delivery of equipment, assembly of personnel and arrival at a predetermined site
• **Amount**—Amount of work performed at the backup site by clerical personnel and information systems processing operations
• **Count**—The number of vital records successfully carried to the backup site versus the required number, and the number of supplies and equipment requested versus actually received. Also, the number of critical systems successfully recovered can be measured with the number of transactions processed.
• **Accuracy**—Accuracy of the data entry at the recovery site versus normal accuracy (as a percentage). Also, the accuracy of actual processing cycles can be determined by comparing output results with those for the same period processed under normal conditions.

## Plan Maintenance

Plans and strategies for business continuity should be reviewed and updated on a scheduled basis to reflect continuing recognition of changing requirements or extraordinarily (unscheduled revisions) when there is an important change affecting the plans and strategies. The following factors, and others, may impact business continuity requirements and the need for the plan to be updated:
• A strategy that is appropriate at one point in time may not be adequate as the needs of the organization change (business processes, new departments, changes in key personnel.)
• New resources/applications may be developed or acquired.
• Changes in business strategy may alter the significance of critical applications or deem additional applications as critical.
• Changes in the software or hardware environment may make current provisions obsolete or inappropriate.
• New events or a change in the likelihood of events that may cause disruption

An important step in maintaining a BCP is to update and test it whenever relevant changes take place within the organization. It is also desirable to include BCP as part of the system development life cycle (SDLC) process.

The responsibility for maintaining the BCP often falls on the BCP coordinator. Specific plan maintenance responsibilities include:
• Developing a schedule for periodic review and maintenance of the plan advising all personnel of their roles and the deadline for receiving revisions and comments
• Calling for unscheduled revisions when significant changes have occurred
• Reviewing revisions and comments and updating the plan within X days (e.g., 30 days, 2 weeks) of the review date
• Arranging and coordinating scheduled and unscheduled tests of the BCP to evaluate its adequacy
• Participating in the scheduled plan tests, which should be performed at least once per year on specific dates. For scheduled and unscheduled tests, the coordinator will write evaluations and integrate changes to resolve unsuccessful test results into the BCP within X days (e.g., 30 days, 2 weeks)
• Developing a schedule for training recovery personnel in emergency and recovery procedures as set forth in the BCP. Training dates should be scheduled within 30 days of each plan revision and scheduled plan test.
• Maintaining records of BCP maintenance activities—testing, training and reviews
• Periodically updating, at least quarterly (shorter periods are recommended), the notification directory of all personnel changes including phone numbers, responsibilities or status within the company

A software tool for administering continuity and recovery plans may be useful to track and follow-up on maintenance tasks.

# Business Continuity and Disaster Recovery

| PRACTICE QUESTION |

**6-10**    In an audit of a business continuity plan, which of the following findings is of **MOST** concern?

   A.    There is no insurance for the addition of assets during the year.
   B.    The business continuity plan manual is not updated on a regular basis.
   C.    Testing of the backup of data has not been done regularly.
   D.    Records for maintenance of the access system have not been maintained.

*See answers and explanations to the practice questions at the end of the chapter. (page 515)*

## 6.2.15 BACKUP AND RESTORATION

Because it is desirable to ensure that the profit-seeking activities of a business (including the IS operations in its supportive role) are not interrupted in the event of a disaster, secondary storage media (usually removable media or mirrored disks) are used to store programs and associated data for backup purposes. These secondary storage media are stored in one or more physical facilities (referred to as offsite libraries), based on availability of use and perceived business interruption risk. It is the offsite librarian's responsibility to maintain a perpetual inventory of the contents of these libraries, to control access to library media and to rotate media between various libraries, as needed. Also, a current copy of the BCP needs to be maintained.

### Offsite Library Controls

Controls over the offsite library facilities are important to ensure the uninterrupted operation of the business in the event of disaster and to optimize IS resource utilization. Unauthorized access to this information could impact the information systems' ability to provide continuous computing services and result in lost or unauthorized changes to data.

Controls over the offsite library include:
- Securing physical access to library contents
- Ensuring that physical construction can withstand fire/heat/water (minimum of two hours)
- Locating the library away from the computer room, preferably miles/kilometers away to avoid the risk of a disaster affecting both facilities
- Ensuring that only authorized personnel have access to the library and the offline media
- Ensuring that a perpetual inventory of all storage media and files stored in the library is maintained
- Ensuring that a record of all storage media and files moved into and out of the library is maintained
- Ensuring that a record of information regarding the contents, versions and location of data files is maintained

### Security and Control of Offsite Facilities

The offsite IPF must be as secured and controlled as the originating site. This includes adequate physical access controls such as locked doors, no windows and human surveillance. The offsite facility should not be identified easily from the outside; therefore, signs identifying the vendor/company and contents of the facility should not be present. This is to prevent intentional sabotage of the offsite facility should the destruction of the originating site be from a malicious attack. The offsite facility should not be subject to the same natural disaster that affected the originating site.

The offsite facility should possess the same constant environmental monitoring and control as the originating site. This includes monitoring the humidity, temperature and surrounding air to achieve the optimum conditions for storing optical, magnetic and paper media and, if applicable, operating computing equipment and peripheral devices. The proper environmental controls include a UPS, operating on a raised floor with proper smoke and water detectors installed, and a working/tested fire extinguishing system.

### Media and Documentation Backup

A crucial element of a business continuity onsite or offsite recovery plan is the availability of adequate data. Duplication of important data and documentation, including offsite storage of such backup data and documentation, is a prerequisite for any type of recovery.

# Business Continuity and Disaster Recovery

Where information is processed and stored in a confidential environment at the primary site and backup is to be stored in a similarly secure location, care should be exercised to ensure that the means of transporting data, whether in the form of physical backup media or via electronic data communications, extend adequate protection to the information.

Copies of data taken for offsite backup must be given the same level of security as the original files. The offsite facility and transportation arrangements must, therefore, meet the security requirements for the most sensitive class of data on the backup media.

## Periodic Backup Procedures

Both data and software files should be backed up on a periodic basis in accordance with the defined RPO. The time period in which to schedule the backup may differ per application program or software system. For instance, certain application systems that run on a monthly basis—in which master or transaction files are updated—require that the backup be scheduled after the monthly production run. However, frequently updated operating systems or application software may require weekly backups. Often online/real-time systems that perform large-volume transaction processing require nightly or immediate backups or utilize mirrored master file updates at a separate processing facility.

Scheduling the periodic backups can often be easily accomplished via an automated backup/media management system and automated job scheduling software. Automating the backup procedures will prevent erroneous or missed backup cycles due to operator error.

## Frequency of Rotation

Backup for data and software must allow for the continuing occurrence of change. A copy of the file or record, as of some point in time, is retained for backup purposes. All changes or transactions that occur during the interval between the copy and the current time also are retained.

Considerations for establishing file backup schedules include the following:
- The frequency of backup cycle and depth-of-retention generations must be determined for each data file.
- The backup strategy must anticipate failure at any step of the processing cycle.
- Master files should be retained at appropriate intervals, such as at the end of an updating procedure, to provide synchronization between files and systems.
- Transaction files should be presented to coincide with master files so a prior generation of a master file can be brought completely up-to-date to recreate a current master file.
- Real-time files require special backup techniques such as duplicate logging of transactions, use of before and/or after images of master records, time stamping of transactions and communication simulation.
- Database management systems (DBMS) require specialized backup, usually provided as an integral feature of the DBMS.
- File descriptions need to be maintained to coincide with each version of a file that is retained; for DBMS systems, this may entail keeping separate versions of data dictionaries.
- It may be necessary to secure the license to use certain vendor software at an alternate site; this should be arranged in advance of the need.
- Backup for software must include object-code and source-code libraries and provisions for maintaining program patches on a current basis at all backup locations.

Likewise, any documentation required for the consistent and continual operation of the business should be preserved in an offsite backup facility. This includes source documents required for restoration of the production database. As with data files, the offsite copies should be kept up to date to ensure their usefulness. It is important to remember that adequate backup is a prerequisite to successful recovery.

## Types of Media and Documentation Rotated

Without software, the computer hardware is of little value. Software, including operating systems, programming languages, compilers, utilities and application programs, must be maintained offsite in a current status. Paper documentation also exists and copies should be stored offsite. This information, such as operational guides, user manuals, records, data files, databases and input/output documents, provides the raw materials and the finished products for the IS processing cycle.

# Business Continuity
# and Disaster Recovery

**Exhibit 6.7** describes the documentation to be backed up and stored offsite.

| Exhibit 6.7—Offsite Storage | |
|---|---|
| **Classification** | **Description** |
| Operating procedures | Application run books, job stream control instructions, operating system manuals and special procedures |
| System and program documentation | Flow charts, program source code listings, program logic descriptions, special job control language statements, error conditions and user manuals |
| Special procedures | Any procedures or instructions that are out of the ordinary such as exception processing, variations in processing and emergency processing |
| Input source documents output documents | Duplicate copies, photocopies, microfiche, microfilm reports or summaries required for auditing, historical analysis, performance of vital work, satisfaction of legal requirements or expediting insurance claims |
| Business continuity plan | A copy of the correct plan for reference |

Sensitive data that are stored offsite should be stored in a fire-resistant magnetic media container (if sensitive data are not stored on magnetic media, a fire-resistant container would be enough). When the data are shipped back to the recovery site, the data should be stored and sealed in the magnetic media container.

Every organization should have a written policy to govern what is stored and for how long. Backup schedules and rotation media to be used in an offsite location are important. This rotation of media can be performed via management software.

**METHOD OF ROTATION**
Although there are various approaches for the rotation of media, one of the more accepted techniques is referred to as the Grandfather-Father-Son method. In this method, daily backups (son) are made over the course of a week. The final backup taken during the week becomes the backup for that week (father). The earlier daily backup media are then rotated for reuse as backup media for the second week. At the end of the month, the final weekly backup is retained as the backup for that month (grandfather). Earlier weekly backup media are then rotated for reuse in subsequent months. At the end of the year, the final monthly backup becomes the yearly backup. Normally, monthly and annual tapes/other media are retained and not subject to the rotation cycle.

A key element to this approach is that backups rotated offsite should not be returned for reuse until their replacement has been sent offsite. As an example, the backup media for week 1 should not be returned from offsite storage until the month-end backup is safely stored offsite. Variations of this method can be used depending on whether quarterly backups are required and on the amount of redundancy an organization may wish to have.

## Record Keeping for Offsite Storage
An inventory of contents at the offsite storage location should be maintained. This inventory should contain information such as:
• Data set name, volume serial number, date created, accounting period and offsite storage bin number for all backup media
• Document name, location, pertinent system and date of last update for all critical documentation

Automated media management systems usually have options that help in recording and maintaining this information.

## Business Continuity Management Best Practices
The need to continually and periodically revisit and improve on the business continuity process is critical to the development of successful and robust recovery strategies for an organization, irrespective of whether the organization is at the initial stage of developing a BCP. In an effort to enhance BCM capabilities (and to comply with regulatory guidelines), some organizations

# Business Continuity
# and Disaster Recovery

have started adopting best practices from industry-independent and industry-specific entities and regulatory agencies. Some of these entities or practices/regulations/standards are:
• Business Continuity Institute (BCI)—Provides good practices for business continuity management
• US National Fire Protection Agency (NFPA)
• US Federal Emergency Management Association (FEMA)—Provides business and industry guide for emergency management
• COBIT
• US National Institute of Standards and Technology
• US Federal Financial Institutions Examination Council (FFIEC)
• US Federal Reserve Board (FRB)
• HIPAA
• US Federal Energy Regulatory Commission (FERC)
• Disaster Recovery Institute International (DRII)—Provides professional practices for business continuity professionals

> **Note:** The CISA candidate would not be tested on specific practices/regulations/standards.

## 6.2.16 SUMMARY OF BUSINESS CONTINUITY AND DISASTER RECOVERY

To ensure continuous service, a BCP should be written to minimize the impact of disruptions. This plan should be based on the long-range IT plan and should support and be aligned with the overall business continuity strategy. Therefore, the process of developing and maintaining an appropriate DRP/BCP would be to:
• Prepare business impact analysis of the effect of the loss of critical business processes
• Identify and prioritize the systems and other resources required to support critical business processes in the event of a disruption
• Choose appropriate strategies for recovering at least sufficient IS facilities to support the critical business processes until full facilities are available
• Develop the detailed plan for recovering IS facilities (DRP)
• Develop a detailed plan for the critical business functions to continue to operate at an acceptable level (BCP)
• Test the plans
• Maintain the plans as the business changes and systems develop

# 6.3 AUDITING BUSINESS CONTINUITY

The IS auditor's tasks include:
• Understanding and evaluating business continuity strategy and its connection to business objectives
• Reviewing the BIA to ensure that it reflects current business practices and known threats
• Evaluating the BCPs to determine their adequacy and currency, by reviewing the plans and comparing them to appropriate standards and/or government regulations including the RTO, RPO etc., defined by the BIA
• Verifying that the BCPs are effective, by reviewing the results from previous tests performed by IS and end-user personnel
• Evaluating offsite storage to ensure its adequacy, by inspecting the facility and reviewing its contents and security and environmental controls
• Verifying the arrangements for transporting backup media to ensure that they meet the appropriate security requirements
• Evaluating the ability of IS and user personnel to respond effectively in emergency situations, by reviewing emergency procedures, employee training and results of their tests and drills
• Ensuring that the process of maintaining plans is in place and effective and covers both periodic and unscheduled revisions
• Evaluating whether the business continuity manuals and procedures are written in a simple and easy to understand manner. This can be achieved through interviews and determining whether all the stakeholders understand their roles and responsibilities with respect to business continuity strategies.

# Business Continuity and Disaster Recovery

## 6.3.1 REVIEWING THE BUSINESS CONTINUITY PLAN

When reviewing the developed plan, IS auditors should verify that basic elements of a well-developed plan are evident. Audit procedures to address such basic elements include:
- Obtain a copy of the current business continuity policy and strategy
- Obtain a current copy of the BCP or manual.
- Obtain a copy of the most recent BIA and identify the RTO, RPO and other key strategic directives
- Sample the distributed copies of the manual and verify that they are current.
- Verify whether the BCP supports the overall business continuity strategy.
- Evaluate the effectiveness of the documented procedures for the initiation of the business continuity effort.
- Review the identification, priorities and planned support of critical applications, including PC-based or end user-developed systems.
- Determine if all applications have been reviewed for their level of tolerance in the event of a disaster.
- Determine if all critical applications (including PC applications) have been identified.
- Determine if the hot site has the correct versions of all system software. Verify that all of the software is compatible; otherwise, the system will not be able to process production data during recovery.
- Obtain a member list for each recovery/continuity/response team.
- Obtain a copy of agreements relating to use of backup facilities
- Review the list of business continuity personnel, emergency hot-site contacts, emergency vendor contacts, etc., for appropriateness and completeness.
- Call a sample of the people indicated and verify that their phone numbers and addresses are correct, as indicated, and that they possess a current copy of the business continuity manual.
- Interview them for an understanding of their assigned responsibilities in case of interruption/disaster situation.
- Evaluate the procedures for documenting the tests.
- Evaluate the procedure for updating the manual. Are updates applied and distributed in a timely manner? Are specific responsibilities documented for maintenance of the manual?
- Review the backup procedures followed for each area covered by the DRP.
- Determine if the backup and recovery procedures are being followed.

In addition to the above steps:
- Evaluate whether all written emergency procedures are complete, appropriate, accurate, current and easy to understand.
- Identify whether the transactions reentered in the system through recovery process need to be separately identified from the normal transactions.
- Determine if all recovery/continuity/response teams have written procedures to follow in the event of a disaster.
- Determine if a suitable procedure exists for updating the written emergency procedures.
- Determine if user recovery procedures are documented.
- Determine if the plan adequately addresses movement to the recovery site.
- Determine if the plan adequately addresses recovering from the recovery site.
- Determine if items necessary for the reconstruction of the information processing facility are stored offsite, such as blueprints, hardware inventory and wiring diagrams.

Questions to consider include:
- Who is responsible for administration or coordination of the plan?
- Is the plan administrator/coordinator responsible for keeping the plan up-to-date?
- Where is the DRP stored?
- What critical systems are covered by the plan?
- What systems are not covered by the plan? Why not?
- What equipment is not covered by the plan? Why not?
- Does the plan operate under any assumptions? What are they?
- Does the plan identify rendezvous points for the disaster management committee or emergency management team to meet and decide if business continuity should be initiated?
- Are the documented procedures adequate for successful recovery?

• Does the plan address disasters of varying degrees?
• Are telecommunication's backups (both data and voice line backups) addressed in the plan?
• Where is the backup facility site?
• Does the plan address relocation to a new information processing facility in the event that the original center cannot be restored?
• Does the plan include procedures for merging master file data, automated tape management system data, etc., into predisaster files?
• Does the plan address loading data processed manually into an automated system?
• Are there formal procedures that specify backup procedures and responsibilities?
• What training has been given to personnel in using backup equipment and established procedures?
• Are the restoration procedures documented?
• Are regular and systematic backups of required sensitive and/or crucial applications and data files, being taken?
• Who determines the methods and frequency of data for critical information stored?
• What type of media is being used for backups?
• Is offsite storage used to maintain backups of critical information required for processing either onsite or offsite operations?
• Is there adequate documentation to perform a recovery in case of disaster or loss of data?
• Is there a schedule for testing and training on the plan?

## 6.3.2 EVALUATION OF PRIOR TEST RESULTS

The BCP coordinator should maintain historical documentation of the results of prior business continuity tests. The IS auditor should review these results and determine whether actions requiring correction have been incorporated into the plan. Also, the IS auditor should evaluate BCP/DRP prior tests for thoroughness and accuracy in accomplishing their objectives. Test results should be reviewed to determine whether the appropriate results were achieved and to determine problem trends and appropriate resolutions of problems.

## 6.3.3 EVALUATION OF OFFSITE STORAGE

The offsite storage facility should be evaluated to ensure the presence, synchronization and currency of critical media and documentation. This includes data files, applications software, applications documentation, systems software, systems documentation, operations documentation, necessary supplies, special forms and a copy of the BCP. To verify the conditions mentioned above, the IS auditor should perform a detailed inventory review. This inventory includes testing for correct dataset names, volume serial numbers, accounting periods and bin locations of media. The IS auditor should also review the documentation, compare it for currency with production documentation, evaluate the availability of the facility and ensure it conforms with management's requirements.

## 6.3.4 INTERVIEWING KEY PERSONNEL

The IS auditor should interview key personnel required for the successful recovery of business operations. All key personnel should have an understanding of their assigned responsibilities as well as up-to-date detailed documentation describing their tasks.

## 6.3.5 EVALUATION OF SECURITY AT OFFSITE FACILITY

The security of the offsite facility should be evaluated to ensure that it has the proper physical and environmental access controls. These controls include the ability to limit access to only authorized users of the facility, raised flooring, humidity controls, temperature controls, specialized circuitry, uninterruptible power supply, water detection devices, smoke detectors and an appropriate fire extinguishing system. The IS auditor should examine the equipment for current inspection and calibration tags. This review should also consider the security requirements of media transportation.

## 6.3.6 REVIEWING ALTERNATIVE PROCESSING CONTRACT

The IS auditor should obtain a copy of the contract with the vendor of the alternative processing facility. The vendor's references should be checked to ensure reliability, and all vendor promises should be verified in writing. The contract should be reviewed against the following guidelines:

• Ensure that the contract is written clearly and is understandable.
• Reexamine and confirm the organization's agreement with the rules that apply to sites shared with other subscribers.
• Ensure that insurance coverage ties in with and covers all (or most) expenses of the disaster.
• Ensure that tests can be performed at the hot site at regular intervals.
• Review and evaluate communications requirements for the backup site.
• Ensure that enforceable source code escrow is reviewed by a lawyer specializing in such contracts.
• Determine the limitation recourse tolerance in the event of a breached agreement.

## 6.3.7 REVIEWING INSURANCE COVERAGE

It is essential that insurance coverage reflect the actual cost of recovery. Taking into consideration the insurance premium (cost), the coverage for media damage, business interruption, equipment replacement and business continuity processing should be reviewed for adequacy.

**Note:** The CISA candidate should know what critical provisions need to be included within insurance policies to safeguard the organization.

# Business Continuity and Disaster Recovery

# 6.4 CHAPTER 6 QUICK REFERENCE REVIEW

## Chapter 6 Quick Reference Review

Chapter 6 addresses the need for business continuity and disaster recovery within an organization. Most organizations have some degree of disaster recovery plans (DRPs) in place for the recovery of IT infrastructure, critical systems and associated data. However, many organizations have not taken the next step and developed plans for how key business units will function during a period of IT disruption. CISA candidates should be aware of the components of disaster recovery and business continuity plans (BCPs), the importance of aligning one with the other, and aligning DRPs and BCPs with the organization's goals and risk tolerance. Also of importance is data backup, storage and retention, and restoration.

- CISA candidates should have a sound understanding of the following items, not only within the context of the present chapter, but also to correctly address questions in related subject areas.
- DRPs for the recovery of IT must be aligned with BCPs which address the recovery of key business processes and business units. Both must properly align with the goals and risk tolerance of the organization.
- For most organizations, it is not financially feasible to immediately recover all application systems and business processes. A business impact analysis (BIA) must be performed to understand the cost of interruption and identify which applications and processes are most critical to the continued functioning of the organization. The results of the BIA can then be used to decide the order in which applications and business processes should be restored and the timing of restoration, and what techniques or mechanisms may be needed to achieve the agreed-upon recovery timetable.
- It is important to understand the difference between the recovery time objective (RTO) and the recovery point objective (RPO). The RTO is determined based on the acceptable downtime in case of a disruption of operations. It indicates the earliest point in time at which the business operations must resume after disaster. The RPO is determined based on the acceptable data loss in case of disruption of operations. It indicates the date and time or synchronization point that systems and data will be restored to based on the availability of backup media.
- CISA candidates should understand the differences between hot sites, warm sites and cold sites, and which ones are appropriate given the needs of an organization. Obviously, an organization needing the ability to recover rapidly would opt for a hot site, or in cases requiring very high availability, a redundant site with mirrored disks.
- CISA candidates should be familiar with the different teams that are utilized in the recovery process and the components of a BCP as described earlier in this chapter.
- One of the more accepted techniques for the rotation of backup media is referred to as the grandfather-father-son method. In this method, daily backups (son) are made over the course of a week. The final backup taken during the week becomes the backup for that week (father). The earlier daily backup media are then rotated for reuse as backup media for the second week. At the end of the month, the final weekly backup is retained as the backup for that month (grandfather). Earlier weekly backup media are then rotated for reuse in subsequent months. At the end of the year, the final monthly backup becomes the yearly backup.

# Business Continuity and Disaster Recovery

# 6.5 CHAPTER 6 CASE STUDY

The following case studies are included as a learning tool to reinforce the concepts introduced in this chapter. Exam candidates should note that the CISA exam does not currently use this format for testing.

## 6.5.1 CASE STUDY A

### *Case Study A Scenario*

An organization is developing revised business continuity (BCPs) and disaster recovery plans (DRPs) for its headquarters facility and network of 16 branch offices. The current plans have not been updated in more than eight years, during which time the organization has grown by over 300 percent. At the headquarters facility, there are approximately 750 employees. These individuals connect over a local area network to an array of more than 60 application, database and file print servers located in the corporate data center and over a frame relay network to the branch offices. Traveling users access corporate systems remotely by connecting over the Internet using virtual private networking. Users at both headquarters and the branch offices access the Internet through a firewall and proxy server located in the data center. Critical applications have a recovery time objective (RTO) of between three and five days. Branch offices are located between 30 and 50 miles from one another, with none closer to the headquarters' facility than 25 miles. Each branch office has between 20 and 35 employees plus a mail server and a file/print server. Backup media for the data center are stored at a third-party facility 35 miles away. Backups for servers located at the branch offices are stored at nearby branch offices using reciprocal agreements between offices. Current contracts with a third party hot site provider include 25 servers, work area space equipped with desktop computers to accommodate 100 individuals, and a separate agreement to ship up to two servers and 10 desktop computers to any branch office declaring an emergency. The contract term is for three years, with equipment upgrades occurring at renewal time. The hot site provider has multiple facilities throughout the country in case the primary facility is in use by another customer or rendered unavailable by the disaster. Senior management desires that any enhancements be as cost effective as possible.

| CASE STUDY A QUESTIONS |
|---|
| A1.  On the basis of the above information, which of the following should the IS auditor recommend concerning the hot site? <br><br>    A.  Desktops at the hot site should be increased to 750. <br>    B.  An additional 35 servers should be added to the hot site contract. <br>    C.  All backup media should be stored at the hot site to shorten the RTO. <br>    D.  Desktop and server equipment requirements should be reviewed quarterly. <br><br>A2.  On the basis of the above information, which of the following should the IS auditor recommend concerning branch office recovery? <br><br>    A.  Add each of the branches to the existing hot site contract. <br>    B.  Ensure branches have sufficient capacity to back each other up. <br>    C.  Relocate all branch mail and file/print servers to the Data Center. <br>    D.  Add additional capacity to the hot site contract equal to the largest branch. <br><br>*See answers and explanations to the practice questions at the end of the chapter. (page 516)* |

# 6.6 ANSWERS TO PRACTICE QUESTIONS

6-1    **A**    The first and key step in developing a business continuity plan is a risk assessment exercise that analyzes the various risks that an organization faces and the impact of nonavailability of individual applications. Section 14.1.2 of ISO 27002 Standard on Information Security Management states that "Depending on the results of the risk assessment, a business continuity strategy should be developed to determine the overall approach to business continuity."

6-2    **A**    A well-defined, risk-based classification system for all assets and processes of the organization is one of the most important components for initializing the business continuity planning efforts. A well-defined, risk-based classification system would assist in identifying the criticality of each of the key processes and assets used by the organization. This would assist in the easy identification of key assets and processes to be secured, and plans to be made to recover these processes and assets at the earliest moment after a disaster. An inventory of critical assets, not all assets, is required for initiating a business continuity plan. Complete documentation of all disasters is not a prerequisite for initiating a business continuity plan; rather, various disasters are considered while developing the plan and only the one having an impact on the organization is addressed in the plan. The availability of hardware and software is not required for initiating the development of a plan; however, it is considered when developing the detailed plan in accordance with the strategy adopted.

6-3    **A**    The IS auditor should always be present when disaster recovery plans are tested to ensure that the test meets the required targets for restoration, ensure that recovery procedures are effective and efficient, and report on the results, as appropriate. IS auditors may be involved in overseeing plan development, but they are unlikely to be involved in the actual development process. Similarly, an audit of plan maintenance may be conducted, but the IS auditor normally would not have any responsibility for the actual maintenance. An IS auditor may be asked to comment upon various elements of a supplier contract, but, again, this is not always the case.

6-4    **A**    The criticality of the processes affected by the disaster is the basis for computing the window of time for recovery. The quality of the data to be processed and the nature of the disaster are not the bases for determining the window of time. Being a mainframe application does not, itself, provide a window-of-time basis.

6-5    **A**    RPO is the earliest point in time to which it is acceptable to recover the data. If RPO is very low, say in minutes, it means that the process cannot afford to lose the data in such a short time. In such cases, data mirroring should be used as a recovery strategy. If RPO is high, say in hours, then other backup procedures, such as reel backup, could be used. A high RTO will mean that so much additional time would be available for the recovery strategy. Disaster tolerance is the time gap within which the business can accept nonavailability of IT facilities. If this time gap is high, recovery strategies that take a longer time can be used.

6-6    **A**    RPO indicates the earliest point in time to which it is acceptable to recover data. The acceptable downtime in case of disruption of operations is information required to establish an RTO. Types of offsite backup facilities available and types of IT platforms supporting critical business functions are information that is not used to establish an RPO or an RTO.

6-7    **C**    A business continuity plan should provide for the recovery of critical systems, not necessarily all systems. Perhaps only 50 percent of the company's systems are critical; therefore, careful assessment of critical systems and capacity requirements should be part of the IS auditor's test of the plan.

6-8    **D**    A prioritized list of contacts is most important since it will direct the process of communication and contact to various entities in order of priority. Choices A, B and C are musts, but not as important as choice D.

6-9    **D**    The correct choice is restoring the IT systems and data after a disaster. The IT department of an organization is primarily responsible for restoring the IT systems and data after a disaster at the earliest possible time. Members of the organization's senior management are primarily responsible for developing the business continuity plan for an organization. Management is also responsible for selecting and approving the strategy for developing and implementing a detailed business continuity plan. The organization should identify a person in management as responsible for declaring a disaster. Although IT is involved in the three other choices, it is not primarily responsible for them.

6-10    **C**    The most vital assets for a company are data. In a business continuity plan, it is critical to ensure that data are available. Therefore, regular testing of the backup of data must be done. If testing is not done, the organization may not be able to retrieve data when required during a disaster; hence, the company may lose its most valuable asset and may not be able to recover from the disaster. A loss on account of lack of insurance is limited to the value of assets. If the business continuity plan manual is not updated, the company may find the manual not fully relevant for recovery during a disaster. However, recovery could be still possible. Nonmaintenance of records in an access system will not directly impact the relevance of the business continuity plan.

# 6.7 ANSWERS TO CASE STUDY QUESTIONS

## ANSWERS TO CASE STUDY A QUESTIONS

A1.    **D**    As equipment needs in a rapidly growing business are subject to frequent change, quarterly reviews are necessary to ensure that the recovery capability keeps pace with the organization. Since not all employee job functions are critical during a disaster, it is not necessary to contact the same number of desktops at a recovery facility as the number of employees. Similarly, not every server is critical to the continued operation of the business. In both cases, only a subset will be required. Since there is no assurance that the hot site will not already be occupied, it would not be advisable to store backup media at the facility. These facilities are generally not designed to provide extensive media storage, and frequent testing by other customers could compromise the security of the media.

A2.    **B**    The most cost-effective solution is to recommend that branches have sufficient capacity to accommodate critical personnel from another branch. Since critical job functions would represent only perhaps 20 percent of the staff from the affected branch, accommodations for only four to seven critical staff members would be needed. Adding each of the branches to the hot site contract would be far more expensive, while adding capacity to the hot site contract would not provide coverage as hot site contracts base their pricing on each location covered. Finally, relocating branch servers to the data center could result in performance issues, and would not address the question of where to locate displaced employees.

# Business Continuity and Disaster Recovery

# 6.8 SUGGESTED RESOURCES FOR FURTHER STUDY

**Burtles, Jim;** *Principles and Practice of Business Continuity, Tools and Techniques,* **Rothstein Associates Inc., USA, 2007**

Castella, Stephen; "Foundations for Successful BCP in Your IT Department," Contingency Planning & Management web site, 2003, *www.contingencyplanning.com/Tools/BCPHandbook/BCP102.asp*

**Graham, Julia; David Kaye;** *A Risk Management Approach to Business Continuity,* **Rothstein Associates Inc., USA, 2006**

**Hiles, Andrew;** *The Definitive Handbook of Business Continuity Management, 2nd Edition*; **John Wiley & Sons Inc., USA, 2007**

**ISACA;** *Cybercrime: Incident Response & Digital Forensics,* **USA, 2005**

**Raval, Vasant; Ashok Fichdia;** *Risks, Controls, and Security: Concepts and Applications,* **John Wiley & Sons, USA, 2007, Chapter 6: System Availability and Business Continuity**

**Snedaker, Susan;** *Business Continuity & Disaster Recovery,* **Syngress Publishing Inc., USA, 2007**

Wells, April; Charlyne Walker; Timothy Walker; *Disaster Recovery: Principles and Practices,* USA, Pearson-Prentice Hall, 2007.

Page intentionally left blank

# A

**Abend**—An abnormal end to a computer job; termination of a task prior to its completion because of an error condition that cannot be resolved by recovery facilities while the task is executing

**Access control**—The process that limits and controls access to resources of a computer system; a logical or physical control designed to protect against unauthorized entry or use

**Access control list (ACL)**—Also referred to as access control tables, this is an internal computerized table of access rules regarding the levels of computer access permitted to logon IDs and computer terminals.

**Access control table**—An internal computerized table of access rules regarding the levels of computer access permitted to logon IDs and computer terminals

**Access method**—The technique used for selecting records in a file, one at a time, for processing, retrieval or storage. The access method is related to, but distinct from, the file organization, which determines how the records are stored.

**Access path**—The logical route an end user takes to access computerized information. Typically, it includes a route through the operating system, telecommunications software, selected application software and the access control system.

**Access rights**—Also called permissions or privileges, these are the rights granted to users by the administrator or supervisor. Access rights determine the actions users can perform (e.g., read, write, execute, create and delete) on files in shared volumes or file shares on the server.

**Access servers**—Provides centralized access control for managing remote access dial-up services

**Address**—The code used to designate the location of a specific piece of data within computer storage

**Address space**—The number of distinct locations that may be referred to with the machine address. For most binary machines, it is equal to 2n, where n is the number of bits in the machine address.

**Addressing**—The method used to identify the location of a participant in a network. Ideally, addressing specifies where the participant is located rather than who they are (name) or how to get there (routing).

**Administrative controls**—The actions dealing with operational effectiveness, efficiency and adherence to regulations and management policies

**Adware**—Any software package that automatically plays, displays or downloads advertising material to a computer after the software is installed on it or while the application is being used. In most cases, this is done without any notification to the user or the user's consent. The term adware may also refer to software that displays advertisements, whether or not it does so with the user's consent; such programs display advertisements as an alternative to shareware registration fees. These are classified as "adware" in the sense of advertising-supported software, but not as spyware. Adware in this form does not operate surreptitiously or mislead the user, and provides the user with a specific service.

**Alpha**—The use of alphabetic characters or an alphabetic character string

**Alternative routing**—A service that allows the option of having an alternate route to complete a call when the marked destination is not available. In signaling, alternate routing is the process of allocating substitute routes for a given signaling traffic stream in case of failure(s) affecting the normal signaling links or routes of that traffic stream.

**American Standard Code for Information Interchange**—See ASCII.

**Analog**—A transmission signal that varies continuously in amplitude and time, and is generated in wave formation. Analog signals are used in telecommunications.

**Anonymous File Transfer Protocol (FTP)**—A method for downloading public files using the File Transfer Protocol. Anonymous FTP is called anonymous because users do not need to identify themselves before accessing files from a particular server. In general, users enter the word "anonymous" when the host prompts for a username; anything can be entered for the password such as the user's e-mail address or simply the word "guest." In many cases, an anonymous FTP site will not even prompt users for a name and password.

**Antivirus software**—Applications that detect, prevent and possibly remove all known viruses from files located in a microcomputer hard drive

**Applet**—A program written in a portable, platform independent computer language such as Java, JavaScript or Visual Basic. It is usually embedded in an HTML page downloaded from web servers and then executed by a browser on client machines to run any web-based application (e.g., generate web page input forms, run audio/video programs, etc.). Applets can only perform a restricted set of operations, thus preventing, or at least minimizing, the possible security compromise of the host computers. However, applets expose the user's machine to risks if not properly controlled by the browser, which should not allow an applet to access a machine's information without prior authorization of the user.

**Application**—A computer program or set of programs that perform the processing of records for a specific function

**Application controls**—Refer to the transactions and data relating to each computer-based application system and are, therefore, specific to each such application. The objectives of application controls, which may be manual or programmed, are to ensure the completeness and accuracy of the records and the validity of the entries made therein, resulting from both manual and programmed processing. Examples of application controls include data input validation, agreement of batch totals and encryption of data transmitted.

**Application layer**—A layer within the International Organization for Standardization (ISO)/Open Systems Interconnection (OSI) model. It is used in information transfers between users through application programs and other devices. In this layer, various protocols are needed. Some of them are specific to certain applications, and others are more general for network services.

**Application tracing and mapping**—Specialized tools that can be used to analyze the flow of data, through the processing logic of the application software, and document the logic, paths, control conditions and processing sequences.

**Application program**—A program that processes business data through activities such as data entry, update or query. It contrasts with systems programs, such as an operating system or network control program, and with utility programs such as copy or sort.

**Application programming**—The act or function of developing and maintaining applications programs in production

**Application programming interface (API)**—A set of routines, protocols and tools referred to as "building blocks" used in business application software development. A good API makes it easier to develop a program by providing all the building blocks related to functional characteristics of an operating system that applications need to specify, for example, when interfacing with the operating system (e.g., provided by MS Windows, different versions of UNIX). A programmer would utilize these APIs in developing applications that can operate effectively and efficiently on the platform chosen.

**Arithmetic logic unit (ALU)**—The area of the central processing unit that performs mathematical and analytical operations

**Artificial intelligence**—Advanced computer systems that can simulate human capabilities, such as analysis, based on a predetermined set of rules

**ASCII (American Standard Code for Information Interchange)**—Representing 128 characters, the ASCII code normally uses 7 bits. However, some variations of the ASCII code set allow 8 bits. This 8-bit ASCII code allows 256 characters to be represented.

**Assembler**—A program that takes as input a program written in assembly language and translates it into machine code or machine language

**Asymmetric key (public key)**—A cipher technique in which different cryptographic keys are used to encrypt and decrypt a message (See public key encryption)

**Asynchronous Transfer Mode (ATM)**—ATM is a high-bandwidth, low-delay switching and multiplexing technology. It is a data link layer protocol. This means that it is a protocol-independent transport mechanism. ATM allows integration of real-time voice and video, as well as data. ATM allows very high-speed data transfer rates at up to 155 Mbits. The acronym ATM should not be confused with the alternate usage for ATM which refers to an automated teller machine.

**Asynchronous transmission**—Character-at-a-time transmission

**Attribute sampling**—An audit technique used to select items from a population for audit testing purposes based on selecting all those items that have certain attributes or characteristics (such as all items over a certain size)

**Audit evidence**—The information used by an IS auditor to meet audit objectives

**Audit objective**—The specific goal(s) of an audit. These often center on substantiating the existence of internal controls to minimize business risk.

**Audit program**—A step-by-step set of audit procedures and instructions that should be performed to complete an audit

**Audit risk**—The risk that information or financial reports may contain material errors or that the IS auditor may not detect an error that has occurred; also used to describe the level of risk that an auditor is prepared to accept during an audit engagement

**Audit trail**—A visible trail of evidence enabling one to trace information contained in statements or reports back to the original input source

**Authentication**—The act of verifying the identity of a user and the user's eligibility to access computerized information. Authentication is designed to protect against fraudulent logon activity. It can also refer to the verification of the correctness of a piece of data.

# B

**Backbone**—The main communications channel of a digital network. The part of a network that handles the major traffic. It employs the highest-speed transmission paths in the network and may also run the longest distances. Smaller networks are attached to the backbone, and networks that directly connect to the end user or customer are called "access networks." A backbone can span a geographic area of any size from a single building to an office complex to an entire country. Or, it can be as small as a backplane in a single cabinet.

**Backup**—Files, equipment, data and procedures available for use in the event of a failure or loss, if the originals are destroyed or out of service

**Badge**—A card or other device that is presented or displayed to obtain access to an otherwise restricted facility, as a symbol of authority (e.g., police) or as a simple means of identification. Badges are also used in advertising and publicity.

**Bandwidth**—The range between the highest and lowest transmittable frequencies. It equates to the transmission capacity of an electronic line and is expressed in bytes per second or Hertz (cycles per second).

**Bar code**—A printed machine-readable code that consists of parallel bars of varied width and spacing

**Base case**—A standardized body of data created for testing purposes. Users normally establish the data. Base cases validate production application systems and test the ongoing accurate operation of the system.

**Baseband**—A form of modulation in which data signals are pulsed directly on the transmission medium without frequency division and usually utilize a transceiver. In baseband, the entire bandwidth of the transmission medium (e.g., coaxial cable) is utilized for a single channel.

**Batch control**—Correctness checks built into data processing systems and applied to batches of input data, particularly in the data preparation stage. There are two main forms of batch controls: sequence control, which involves consecutively numbering the records in a batch so that the presence of each record can be confirmed, and control total, which is a total of the values in selected fields within the transactions.

**Batch processing**—The processing of a group of transactions at the same time. Transactions are collected and processed against the master files at a specified time.

**Bayesian filter**—A method often employed by antispam software to filter spam based on probabilities. The message header and every word or number are each considered a token and given a probability score. Then the entire message is given a spam probability score. A message with a high score will be flagged as spam and discarded, returned to its sender or put in a spam directory for further review by the intended recipient.

**Benchmarking**—A systematic approach to comparing an organization's performance against peers and competitors in an effort to learn the best ways of conducting business (e.g., benchmarking of quality, logistical efficiency and various other metrics).

**Binary code**—A code whose representation is limited to 0 and 1

**Biometrics**—A security technique that verifies an individual's identity by analyzing a unique physical attribute such as a handprint

**Black box testing**—A testing approach that focuses on the functionality of the application or product and does not require knowledge of the code intervals

**Bridge**—A device that connects two similar networks together

**Broadband**—In broadband, multiple channels are formed by dividing the transmission medium into discrete frequency segments. It generally requires the use of a modem.

**Brouters**—Devices that perform the functions of both bridges and routers are called brouters. Naturally, they operate at both the data link and the network layers. A brouter connects the same data link type LAN segments as well as different data link ones, which is a significant advantage. Like a bridge, it forwards packets based on the data link layer address to a different network of the same type. Also, whenever required, it processes and forwards messages to a different data link type network based on the network protocol address. When brouters connect same data link type networks, they are as fast as bridges.

**Buffer**—Memory reserved to temporarily hold data. Buffers are used to offset differences between the operating speeds of different devices such as a printer and a computer. In a program, buffers are reserved areas of RAM that hold data while they are being processed.

**Bus**—Common path or channel between hardware devices. It can be between components internal to a computer or between external computers in a communications network.

**Bus configuration**—All devices (nodes) are linked along one communication line where transmissions are received by all attached nodes. This architecture is reliable in very small networks, as well as easy to use and understand. This configuration requires the least amount of cable to connect the computers together and, therefore, is less expensive than other cabling arrangements. It is also easy to extend, and two cables can be easily joined with a connector to make a longer cable for more computers to join the network. A repeater can also be used to extend a bus configuration.

**Business case**—A document that provides management with sufficient information needed to enable management to decide whether to support a proposed project before significant resources are committed to development of the project. A business case includes analysis of current business process performance; associated assumptions, needs or problems; and proposed solutions and potential constraints based upon a risk-adjusted, cost-benefit analysis.

**Business continuity plan (BCP)**—Process of developing advanced arrangements and procedures that enable an organization to respond to an event in such a manner that critical business functions continue with planned levels of interruption or essential change.

**Business impact analysis (BIA)**—A process to determine the impact of losing the support of any resource. The business impact analysis assessment study will establish the escalation of that loss over time. It is predicated on the fact that senior management, when provided reliable data to document the potential impact of a lost resource, can make the appropriate decision.

**Business process reengineering (BPR)**—Modern expression for organizational development stemming from IS/IT impacts. The ultimate goal of BPR is to yield a better performing structure, more responsive to the customer base and market conditions, while yielding material cost savings. To reengineer means redesigning a structure and procedures with intelligence and skills, while being well informed about all of the attendant factors of a given situation so as to obtain the maximum benefits from mechanization as basic rationale.

**Business risk**—Potential for harm or loss in achieving business objectives

**Bypass label processing (BLP)**—A technique of reading a computer file while bypassing the internal file/data set label. This process could result in bypassing of the security access control system.

# C

**Capability Maturity Model (CMM)**—The Capability Maturity Model (CMM) for Software, from the Software Engineering Institute (SEI), is a model used by many organizations to identify best practices useful in assessing and increasing the maturity of the software development process.

**Central processing unit (CPU)**—Computer hardware that houses the electronic circuits that control/direct all operations of the computer system

**Certificate (certification) authority (CA)**—In cryptography, a certificate authority or certification authority (CA) is an entity which issues digital certificates for use by other parties. It is an example of a trusted third party. A certificate authority attests, as the trusted provider of the public/private key pairs, to the authenticity of the owner (entity or individual) to whom a public/private key pair has been given. The process involves a CA who makes a decision to issue a certificate based on evidence or knowledge obtained in verifying the identity of the recipient. Upon verifying the identity of the recipient, the CA signs the certificate with its private key for distribution to the user, where, upon receipt, the user will decrypt the certificate with the CA's public key (e.g., commercial CAs such as Verisign provide public keys on web browsers). The ideal CA is authoritative (someone that the user trusts) for the name or key space it represents. CA's are characteristic of many public key infrastructure (PKI) schemes. There are many commercial CAs that charge for their services. Institutions and governments may have their own CAs, and there are free CAs.

**Certificate revocation list (CRL)**—An instrument for checking the continued validity of the certificates for which the certification authority (CA) has responsibility. CRL details digital certificates that are no longer valid. The time gap between two updates is very critical and is also a risk in digital certificates verification.

**Certification practice statement (CPS)**—A CPS is a detailed set of rules governing the certificate authority's operations. It provides an understanding of the value and trustworthiness of certificates issued by a given CA in terms of the controls that an organization observes, the method it uses to validate the authenticity of certificate applicants and the CA's expectations of how its certificates may be used.

**Channel Service Unit/Digital Service Unit (CSU/DSU)**—Interfaces at the physical layer of the OSI reference model, data terminal equipment (DTE) to data circuit terminating equipment (DCE), for switched carrier networks

**Check digit**—A numeric value which has been calculated mathematically that is added to data to ensure that original data have not been altered or that an incorrect, but valid match has occurred. This control is effective in detecting transposition and transcription errors.

**Checklist**—A list of items that is used to verify the completeness of a task or goal. A checklist is used in quality assurance (and, in general, in information systems audit) to check process compliance, code standardization and error prevention, and other items for which consistency processes or standards have been defined.

**Checksum**—A cryptographic checksum is a mathematical value that is assigned to a file and used to "test" the file at a later date to verify that the data contained in the file have not been maliciously changed. A cryptographic checksum is created by performing a complicated series of mathematical operations (known as a cryptographic algorithm) that translates the data in the file into a fixed string of digits called a hash value, which is then used as the checksum. Without knowing which cryptographic algorithm was used to create the hash value, it is highly unlikely that an unauthorized person would be able to change data without inadvertently changing the corresponding checksum. Cryptographic checksums are used in data transmission and data storage. Cryptographic checksums are also known as message authentication codes, integrity check-values, modification detection codes or message integrity codes.

**Ciphertext**—Information generated by an encryption algorithm to protect the cleartext. The ciphertext is unintelligible to the unauthorized reader.

**Client-server**—A group of computers connected by a communications network, where the client is the requesting machine and the server is the supplying machine. Software is specialized at both ends. Processing may take place on either the client or the server, but it is transparent to the user.

**Coaxial cable**—It is composed of an insulated wire that runs through the middle of each cable, a second wire that surrounds the insulation of the inner wire like a sheath, and the outer insulation which wraps the second wire. Coaxial cable has a greater transmission capacity than standard twisted-pair cables but has a limited range of effective distance.

**Cohesion**—The extent to which a system unit—subroutine, program, module, component, subsystem—performs a single dedicated function. Generally, the more cohesive are units, the easier it is to maintain and enhance a system since it is easier to determine where and how to apply a change.

**Cold site**—An IS backup facility that has the necessary electrical and physical components of a computer facility, but does not have the computer equipment in place. The site is ready to receive the necessary replacement computer equipment in the event the users have to move from their main computing location to the alternative computer facility.

**Communication processor**—A computer embedded in a communications system that generally performs basic tasks of classifying network traffic and enforcing network policy functions. An example is the message data processor of a Digital Divide Network (DDN) switching center. More advanced communications processors may perform additional functions.

**Comparison program**—A program for the examination of data using logical or conditional tests to determine or to identify similarities or differences

**Compensating control**—An internal control that reduces the risk of an existing or potential control weakness resulting in errors and omissions

**Compiler**—A program that translates programming language (source code) into machine executable instructions (object code)

**Completely connected (mesh) configuration**—A network topology in which devices are connected with many redundant interconnections between network nodes (primarily used for backbone networks)

**Completeness check**—A procedure designed to ensure that no fields are missing from a record

**Compliance testing**—Audit tests that determine if internal controls are being applied in a manner described in the documentation and in accordance with management's intents. These are tests that are used to determine whether internal controls actually exist and are working effectively.

**Components (as in component-based development)**—Cooperating packages of executable software that make their services available through defined interfaces. Components used in developing systems may be commercial off-the-shelf software (COTS) or may be purposely built. However, the goal of component-based development is to ultimately use as many predeveloped, pretested components as possible.

**Comprehensive audit**—An audit designed to determine the accuracy of financial records as well as evaluate the internal controls of a function or department

**Computer emergency response team (CERT)**—A group of people integrated at the organization with clear lines of reporting and responsibilities for standby support in case of an information systems emergency. This group will act as an efficient corrective control, and should also act as a single point of contact for all incidents and issues related to information systems.

**Computer forensics**—The application of the scientific method to digital media to establish factual information for judicial review. This process often involves investigating computer systems to determine whether they are or have been used for illegal or unauthorized activities. As a discipline, it combines elements of law and computer science to collect and analyze data from information systems (e.g., personal computers, networks, wireless communications and digital storage devices) in a way that is admissible as evidence in a court of law.

**Computer-aided software engineering (CASE)**—The use of software packages that aid in the development of all phases of an information system. System analysis, design programming and documentation are provided. Changes introduced in one CASE chart will update all other related charts automatically. CASE can be installed on a microcomputer for easy access.

**Computer-assisted audit technique (CAAT)**—Any automated audit technique such as generalized audit software, test data generators, computerized audit programs and specialized audit utilities

**Concurrency control**—Refers to a class of controls used in database management systems (DBMS) to ensure that transactions are processed in an atomic, consistent, isolated and durable manner (ACID). This implies that only serial and recoverable schedules are permitted, and that committed transactions are not discarded when undoing aborted transactions.

**Console log**—An automated detail report of computer system activity

**Continuity**—Preventing, mitigating and recovering from disruption. The terms business resumption planning, disaster recovery planning and contingency planning also may be used in this context; they all concentrate on the recovery aspects of continuity.

**Continuous improvement**—The goals of continuous improvement (Kaizen) include the elimination of waste, defined as "activities that add cost but do not add value;" just-in-time delivery; production load leveling of amounts and types; standardized work; paced moving lines; right-sized equipment, and so on. A closer definition of the Japanese usage of Kaizen is "to take it apart and put back together in a better way." What is taken apart is usually a process, system, product or service. Kaizen is a daily activity whose purpose goes beyond improvement. It is also a process that, when done correctly, humanizes the workplace, eliminates hard work (both mental and physical), and teaches people how to do rapid experiments using the scientific method and how to learn to see and eliminate waste in business processes.

**Control group**—Members of the operations area that are responsible for the collection, logging and submission of input for the various user groups

**Control risk**—The risk that a material error exists that would not be prevented or detected on a timely basis by the system of internal controls

**Control section**—The area of the central processing unit (CPU) that executes software, allocates internal memory and transfers operations between the arithmetic-logic, internal storage and output sections of the computer

**Cookie**—A message kept in the web browser for the purpose of identifying users and possibly preparing customized web pages for them. For the first time, a user may be required to go through a registration process. Subsequent to this, whenever the cookie's message is sent to the server, a customized view based on that user's preferences can be produced. The browser's implementation of cookies has, however, brought up several security concerns such as allowing breaches of security and the theft of personal information (e.g., user passwords that validate the user's identity and enable restricted web services).

**Corporate governance**—The system by which organizations are directed and controlled. Boards of directors are responsible for the governance of their organizations. It consists of the leadership and organizational structures and processes that ensure the organization sustains and extends strategies and objectives.

**Corrective controls**—These controls are designed to correct errors, omissions and unauthorized uses and intrusions once they are detected.

**Countermeasures**—An action, process, device or system that can prevent or mitigate the effects of threats to a computer, server or network. In this context, a threat is a potential or actual adverse event that may be malicious or incidental, and that can compromise the assets of an enterprise or the integrity of a computer or network. Internal controls are countermeasures, as they mitigate the risks presented by the threats. Countermeasures can take the form of software, hardware and modes of behavior.

**Coupling**—Measure of interconnectivity among software program modules' structure. Coupling depends on the interface complexity between modules. This can be defined as the point at which entry or reference is made to a module, and what data pass across the interface. In application software design, it is preferable to strive for the lowest possible coupling between modules. Simple connectivity among modules results in software that is easier to understand and maintain, and less prone to a ripple or domino effect, caused when errors occur at one location and propagate through a system.

**Customer relationship management (CRM)**—Customer relationship management is a way to identify, acquire and retain customers. CRM is also an industry term for software solutions that help an organization manage customer relationships in an organized manner.

# D

**Data communications**—The transfer of data between separate computer processing sites/devices using telephone lines, microwave and/or satellite links

**Data custodian**—Individuals and departments responsible for the storage and safeguarding of computerized information. This typically is within the IS organization.

**Data dictionary**—A data dictionary is a database that contains the name, type, range of values, source, and authorization for access for each data element in a database. It also indicates which application programs use that data so that when a data structure is contemplated, a list of the affected programs can be generated. The data dictionary may be a stand-alone information system used for management or documentation purposes, or it may control the operation of a database.

**Data Encryption Standard (DES)**—A private key cryptosystem published by the National Bureau of Standards (NBS), the predecessor of the US National Institute of Standards and Technology (NIST). DES has been used commonly for data encryption in the forms of software and hardware implementation. (See private key cryptosystem.)

**Data leakage**—Siphoning out or leaking information by dumping computer files or stealing computer reports and tapes

**Data owner**—Individuals, normally managers or directors, who have responsibility for the integrity, accurate reporting and use of computerized data

**Data security**—Those controls that seek to maintain confidentiality, integrity and availability of information

**Data structure**—The relationships among files in a database and among data items within each file

**Database**—A stored collection of related data needed by organizations and individuals to meet their information processing and retrieval requirements

**Database administrator (DBA)**—An individual or department responsible for the security and information classification of the shared data stored on a database system. This responsibility includes the design, definition and maintenance of the database.

**Database management system (DBMS)**—A complex set of software programs that control the organization, storage and retrieval of data in a database. It also controls the security and integrity of the database.

**Database specifications**—These are the requirements for establishing a database application. They include field definitions, field requirements, and reporting requirements for the individual information in the database.

**Decentralization**—The process of distributing computer processing to different locations within an organization

**Decision support system (DSS)**—An interactive system that provides the user with easy access to decision models and data from a wide range of sources, to support semistructured decision-making tasks typically for business purposes

**Decryption**—A technique used to recover the original plaintext from the ciphertext such that it is intelligible to the reader. The decryption is a reverse process of the encryption.

**Decryption key**—A piece of information, in a digitized form, used to recover the plaintext from the corresponding ciphertext by decryption

**Detection risk**—The risk that material errors or misstatements that have occurred will not be detected by the IS auditor

**Detective control**—These controls exist to detect and report when errors, omissions and unauthorized use or entry occur.

**Dial-back**—Used as a control over dial-up telecommunications lines. The telecommunications link established through dial-up into the computer from a remote location is interrupted so the computer can dial back to the caller. The link is permitted only if the caller is from a valid phone number or telecommunications channel.

**Digital signature**—A piece of information, a digitized form of a signature, that provides sender authenticity, message integrity and nonrepudiation. A digital signature is generated using the sender's private key or applying a one-way hash function.

**Disaster tolerance**—Disaster tolerance is the time gap during which the business can accept the non-availability of IT facilities.

**Disaster recovery planning (DRP)**—A set of human, physical, technical and procedural resources to recover, within a defined time and cost, an activity interrupted by an emergency or disaster.

**Discovery sampling**—A form of attribute sampling that is used to determine a specified probability of finding at least one example of an occurrence (attribute) in a population

**Discretionary access control (DAC)**—A protection that may be activated or modified by the data owner at his/her discretion. This would be the case of data-owner-defined sharing of information resources, where the data owner may select who can access his/her resource and the security level of the access. Discretionary access controls cannot override mandatory access controls; they act as an additional filter, prohibiting still more access with the same exclusionary principle.

**Diskless workstations**—A workstation or PC on a network that does not have its own disk. Instead, it stores files on a network file server.

**Distributed data processing network**—A system of computers connected together by a communications network. Each computer processes its data, and the network supports the system as a whole. Such a network enhances communication among the linked computers and allows access to shared files.

**Diverse routing**—The method of routing traffic through split cable facilities or duplicate cable facilities. This can be accomplished with different and/or duplicate cable sheaths. If different cable sheaths are used, the cable may be in the same conduit and, therefore, subject to the same interruptions as the cable it is backing up. The communication service subscriber can duplicate the facilities by having alternate routes, although the entrance to and from the customer premises may be in the same conduit. The subscriber can obtain diverse routing and alternate routing from the local carrier, including dual entrance facilities. However, acquiring this type of access is time-consuming and costly. Most carriers provide facilities for alternate and diverse routing, although the majority of services are transmitted over terrestrial media. These cable facilities are usually located in the ground or basement. Ground-based facilities are at great risk due to the aging infrastructures of cities. In addition, cable-based facilities usually share room with mechanical and electrical systems that can impose great risks due to human error and disastrous events.

**Domain Name Server (DNS)**—A server that translates the names of network nodes into network (IP) addresses

**DNS poisoning**—Domain name system poisoning (also called DNS cache poisoning or cache poisoning) corrupts the table of an Internet server's DNS, replacing an Internet address with the address of another vagrant or scoundrel address. If a web user looks for the page with that address, the request is redirected by the scoundrel entry in the table to a different address. Cache poisoning differs from another form of DNS poisoning in which the attacker spoofs valid e-mail accounts and floods the inboxes of administrative and technical contacts. Cache poisoning is related to URL poisoning or location poisoning, where an Internet user behavior is tracked by adding an identification number to the location line of the browser that can be recorded as the user visits successive pages on the site.

**Downloading**—The act of transferring computerized information from one computer to another computer

**Downtime report**—A report that identifies the elapsed time when a computer is not operating correctly because of machine failure

**Dumb terminal**—A display terminal without processing capability. Dumb terminals depend on the main computer for processing. All entered data are accepted without further editing or validation.

**Dynamic Host Configuration Protocol (DHCP)**—A protocol used by networked computers (clients) to obtain IP addresses and other parameters such as the default gateway, subnet mask and IP addresses of DNS servers from a DHCP server. The DHCP server ensures that all IP addresses are unique, e.g., no IP address is assigned to a second client while the first client's assignment is valid (its lease has not expired). Thus IP address pool management is done by the server and not by a human network administrator.

# E

**E-mail/interpersonal messaging**—An individual using a terminal, PC or an application can access a network to send an unstructured message to another individual or group of people.

**EBCDIC (Extended Binary-coded Decimal Interchange Code)**—An 8-bit code representing 256 characters; used in most large computer systems.

**Edit controls**—Detects errors in the input portion of information that is sent to the computer for processing. The controls may be manual or automated, and allow the user to edit data errors before processing.

**Editing**—Editing ensures that data conform to predetermined criteria and enable early identification of potential errors.

**Electronic data interchange (EDI)**—The electronic transmission of transactions (information) between two organizations. EDI promotes a more efficient paperless environment. EDI transmissions can replace the use of standard documents, including invoices or purchase orders.

**Electronic funds transfer (EFT)**—The exchange of money via telecommunications. EFT refers to any financial transaction that originates at a terminal and transfers a sum of money from one account to another.

**Embedded audit module (EAM)**—A screening process that is incorporated into the regular production programs. The module selects items during the regular production runs that fulfill certain criteria established by the IS auditor and usually outputs or copies these items to a file or report.

**Encapsulation (objects)**—Encapsulation is the technique used by layered protocols in which a lower-layer protocol accepts a message from a higher-layer protocol and places it in the data portion of a frame in the lower layer.

**Encryption**—A technique used to protect the plaintext by coding the data so they are unintelligible to the reader

**Encryption key**—A piece of information, in a digitized form, used by an encryption algorithm to convert the plaintext to the ciphertext

**End-user computing**—The ability of end users to design and implement their own information system, utilizing computer software products

**Enterprise resource planning (ERP)**—An enterprise resource planning system is an integrated system containing multiple business subsystems. Examples include SAP, Oracle Financials and J.D. Edwards.

**Escrow agent**—A person, agency or organization that is authorized to act on behalf of another to create a legal relationship with a third party in regards to an escrow agreement. In other words, an escrow agent is the custodian of an asset according to an escrow agreement. As it relates to a cryptographic key, it is the agency or organization charged with the responsibility for safeguarding the key components of the unique key.

**Escrow agreement**—A legal arrangement whereby an asset (often money, but sometimes other property such as art, a deed of title, web site, software source code or a cryptographic key) is delivered to a third party (called an escrow agent) to be held in trust or otherwise pending a contingency or the fulfillment of a condition or conditions in a contract. Upon that event occurring, the escrow agent will deliver the asset to the proper recipient; otherwise, the escrow agent is bound by his/her fiduciary duty to maintain the escrow account. Source code escrow means deposit of the source code for the software into an account held by an escrow agent. Escrow is typically requested by a party licensing software (e.g., licensee or buyer) to ensure maintenance of the software. The software source code is released by the escrow agent to the licensee if the licensor (e.g., seller or contractor) files for bankruptcy or otherwise fails to maintain and update the software as promised in the software license agreement.

**Ethernet**—A popular network protocol and cabling scheme that uses a bus topology and CSMA/CD (carrier sense multiple access/collision detection) to prevent network failures or collisions when two devices try to access the network at the same time

**Evidence**—The information an auditor gathers in the course of performing an IS audit. Evidence is relevant if it pertains to the audit objectives and has a logical relationship to the findings and conclusions it is used to support.

**Exception reports**—An exception report is generated by a program that identifies transactions or data that appear to be incorrect. These items may be outside a predetermined range or may not conform to specified criteria.

**Exclusive-OR (XOR)**—The XOR operation is a Boolean operation that produces a 0 if its two Boolean inputs are the same (0 and 0 or 1 and 1) and it produces a 1 if its two inputs are different (1 and 0). In other words, the exclusive-OR operator returns a value of TRUE only if just one of its operands is TRUE. In contrast, an inclusive-OR operator returns a value of TRUE if either or both of its operands are TRUE.

**Executable code**—The machine language code that is generally referred to as the object or load module

**Expert systems**—Expert systems are the most prevalent type of computer systems that arise from the research of artificial intelligence. An expert system has a built in hierarchy of rules which are acquired from human experts in the appropriate field. Once input is provided, the system should be able to define the nature of the problem and provide recommendations to solve the problem.

**Exposure**—A potentially adverse result or consequence to be considered in the evaluation of internal controls. Strengthening internal controls can reduce exposure but seldom eliminates it.

**Extended Binary-coded Decimal Interchange Code (EBCDIC)**—An 8-bit code representing 256 characters; used in most large computer systems

**Extensible Markup Language (XML)**—Promulgated through the World Wide Web Consortium, XML is a web-based application development technique that allows designers to create their own customized tags, thus enabling the definition, transmission, validation and interpretation of data between applications and organizations

**Extranet**—A private network that resides on the Internet and allows a company to securely share business information with customers, suppliers, or other businesses as well as to execute electronic transactions. It is different from an intranet in that it is located beyond the company's firewall. Therefore, an extranet relies on the use of securely issued digital certificates (or alternative methods of user authentication) and encryption of messages. A virtual private network (VPN) and tunneling are often used to implement extranets, to ensure security and privacy.

# F

**Fallback procedures**—A plan of action or set of procedures to be performed if a system implementation, upgrade or modification does not work as intended. These may involve restoring the system to its state prior to the implementation or change. Fallback procedures are needed to ensure that normal business processes continue in the event of failure and should always be considered in system migration or implementation.

**False authorization**—Also called false acceptance; it occurs when an unauthorized person is identified as an authorized person by the biometric system.

**False enrollment**—Occurs when an unauthorized person manages to enroll into the biometric system (enrollment is the initial process of acquiring a biometric feature and saving it as a personal reference on a smart card, a PC or in a central database).

**Feasibility study**—A phase of a system development life cycle (SDLC) methodology that researches the feasibility and adequacy of resources for the development or acquisition of a system solution to a user need

**Fiber-optic cable**—Glass fibers that transmit binary signals over a telecommunications network. Fiber-optic systems have low transmission losses as compared to twisted-pair cables; they do not radiate energy or conduct electricity; they are free from corruption and lightning-induced interference and they reduce the risk of wiretaps.

**Field**—An individual data element in a computer record. Examples include employee name, customer address, account number, product unit price and product quantity in stock.

**File**—A named collection of related records

**File allocation table (FAT)**—A table used by the operating system to keep track of where every file is located on the disk. Since a file is often fragmented, and thus subdivided into many sectors within the disk, the information stored in the FAT is used when loading or updating the contents of the file.

**File layout**—Specifies the length of the file's record, and the sequence and size of its fields. A file layout also will specify the type of data contained within each field. For example, alphanumeric, zoned decimal, packed and binary are types of data.

**File server**—A high-capacity disk storage device or a computer that stores data centrally for network users and manages access to that data. File servers can be dedicated so that no process other than network management can be executed while the network is available; file servers can be nondedicated so that standard user applications can run while the network is available.

**File Transfer Protocol (FTP)**—See Anonymous File Transfer Protocol

**Financial audit**—An audit designed to determine the accuracy of financial records and information

**Firewall**—A device that enforces security policies for traffic traversing to and from different network segments. A firewall no longer only protects a company from the Internet, but also protects sensitive segments within organizations.

**Firmware**—Memory chips with embedded program code that hold their content when power is turned off

**Foreign key**—A foreign key is a value that represents a reference to a tuple (a row in a table) containing the matching candidate key value (in the relational theory it would be a candidate key, but in real DBMS implementations it is always the primary key). The problem of ensuring that the database does not include any invalid foreign key values is therefore known as the referential integrity problem. The constraint that values of a given foreign key must match values of the corresponding candidate key is known as a referential constraint. The relation (table) that contains the foreign key is referred as the referencing relation and the relations that contain the corresponding candidate key as the referenced relation or target relation.

**Fourth-generation language (4GL)**—English-like, user friendly, nonprocedural computer languages used to program and/or read and process computer files

**Frame relay**—A packet-switched wide area network (WAN) technology that provides faster performance than older packet-switched WAN technologies, such as X.25 networks, because it was designed for today's reliable circuits and performs less rigorous error detection. Frame relay is best suited for data and image transfers. Because of its variable-length packet architecture, it is not the most efficient technology for real-time voice and video. In a frame-relay network, end nodes establish a connection via a permanent virtual circuit (PVC).

# G

**Gateway**—A hardware/software package that is used to connect networks with different protocols. The gateway has its own processor and memory, and can perform protocol and bandwidth conversions.

**Generalized audit software (GAS)**—Multipurpose audit software that can be used for general processes such as record selection, matching, recalculation and reporting

**Geographical information system (GIS)**—A tool used to integrate, convert, handle, analyze and produce information regarding the surface of the earth. These data exist as maps, tridimensional virtual models, lists and tables.

**Governance**—Corporate governance.

# H

**Hardware**—Relates to the technical and physical features of the computer

**Help desk**—A service offered via phone/Internet by an organization to its clients or employees, which provides information, assistance, and troubleshooting advice regarding software, hardware, or networks. A help desk is staffed by people that can either resolve the problem on their own or escalate the problem to specialized personnel. A help desk is often equipped with dedicated customer relationship management (CRM) software that logs the problems and tracks them until they are solved.

**Heuristic filter**—A method often employed by antispam software to filter spam using criteria established in a centralized rule database. Every e-mail message is given a rank, based upon its header and contents, which is then matched against preset thresholds. A message that surpasses the threshold will be flagged as spam and discarded, returned to its sender or put in a spam directory for further review by the intended recipient.

**Hierarchical database**—A database structured in a tree/root or parent/child relationship. Each parent can have many children, but each child may have only one parent.

**Honeypot**—A trap set to detect, deflect or, in some manner, counteract attempts at unauthorized use of information systems. Generally, it consists of a computer, data or a network site that appears to be part of a network but which is actually isolated and protected, and which seems to contain information or a resource that would be of value to attackers. Honeypots can carry risks to a network, and must be handled with care. If they are not properly walled off, an attacker can use them to actually break into a system. A honeypot that masquerades as an open proxy is known as a sugarcane. A honeypot is valuable as a surveillance and early-warning tool. While often a computer, a honeypot can take on other forms such as files or data records, or even unused IP address space. Honeypots should have no production value and, therefore, should not see any legitimate traffic or activity. Whatever they capture can then be surmised as malicious or unauthorized. One very practical implication of this is that honeypots designed to thwart spam by masquerading as systems of the types abused by spammers to send spam can accurately categorize the material they trap 100 percent of the time: it is all illicit. A honeypot needs no spam-recognition capability, no filter to separate ordinary e-mail from spam. Ordinary e-mail never comes to a honeypot.

**Hot site**—A fully operational offsite data processing facility equipped with both hardware and system software to be used in the event of a disaster

**Hypertext markup language (HTML)**—A language designed for the creation of web pages with hypertext and other information to be displayed in a web browser. HTML is used to structure information—denoting certain text as headings, paragraphs, lists and so on—and can be used to describe, to some degree, the appearance and semantics of a document.

# I

**Image processing**—The process of electronically inputting source documents by taking an image of the document, thereby eliminating the need for key entry

**Impact assessment**—A study of the potential future effects of a development project on current projects and resources. The resulting document should list the pros and cons of pursuing a specific course of action.

**Impersonation**—Impersonation, as a security concept related to Windows NT, allows a server application to temporarily "be" the client in terms of access to secure objects. Impersonation has three possible levels: identification, letting the server inspect the client's identity; impersonation, letting the server act on behalf of the client; and delegation, the same as impersonation but extended to remote systems to which the server connects (through the preservation of credentials). Impersonation by imitating or copying the identification, behavior or actions of another may also be used in social engineering to obtain otherwise unauthorized physical access.

**Incident**—Any event that is not part of the standard operation of a service and that causes, or may cause, an interruption to, or a reduction in, the quality of that service.

**Independence**—An IS auditor's self-governance and freedom from conflict of interest and undue influence. The IS auditor should be free to make his/her own decisions, not influenced by the organization being audited and its people (managers and employees).

**Indexed sequential access method (ISAM)**—A disk access method that stores data sequentially while also maintaining an index of key fields to all the records in the file for direct access capability.

**Information processing facility (IPF)**—The computer room and support areas

**Information security governance**—The leadership organizational structures and processes that safeguard information

**Inherent risk**—The risk that a material error could occur, assuming that there are no related internal controls to prevent or detect the error (see control risk)

**Input controls**—Techniques and procedures used to verify, validate and edit data, to ensure that only correct data are entered into the computer

**Instant messaging**—An online mechanism or a form of real-time communication among two or more people based on typed text and multimedia data. The text is conveyed via computers or another electronic device (e.g., cell phone or personal digital assistant [PDA]) connected over a network such as the Internet.

**Integrated services digital network (ISDN)**—A public, end-to-end, digital telecommunications network with signaling, switching, and transport capabilities supporting a wide range of service accessed by standardized interfaces with integrated customer control. The standard allows transmission of digital voice, video and data over 64 Kpbs lines.

**Integrated test facilities (ITF)**—A testing methodology where test data are processed in production systems. The data usually represent a set of fictitious entities such as departments, customers and products. Output reports are verified to confirm the correctness of the processing.

**Internet**—1) Two or more networks connected by a router; 2) the world's largest network using TCP/IP protocols to link government, university and commercial institutions.

**Internet Engineering Task Force (IETF)**—The Internet standards setting organization with international affiliates from network industry representatives. This includes all network industry developers and researchers concerned with the evolution and planned growth of the Internet.

**Internet packet (IP) spoofing**—An attack using packets with the spoofed source Internet packet (IP) addresses. This technique exploits applications that use authentication based on IP addresses. This technique also may enable an unauthorized user to gain root access on the target system.

**Irregularities**—Intentional violations of established policy or willful misstatements or omissions of information

# Glossary

**IT incident**—Any event that is not part of the ordinary operation of a service that causes, or may cause, an interruption to or a reduction in the quality of that service.

**IT governance framework**—A model that integrates a set of guidelines, policies and methods that represent the organizational approach to the IT governance. Per COBIT 4.0, IT governance is the responsibility of the board of directors and executive management. It is an integral part of institutional governance, and consists of the leadership and organizational structures and processes that ensure that the organization's IT sustains and extends the organization's strategy and objectives.

**IT infrastructure**—The set of hardware, software and facilities that integrates an organization's IT assets. Specifically, the equipment (including servers, routers, switches, and cabling), software, services and products used in storing, processing, transmitting and displaying all forms of information for the organization's users.

# K

**Kaizen**—See continuous improvement.

**Key performance indicator (KPI)**—Defined measures that determine how well the process is performing in enabling the goal to be reached. They are lead indicators of whether a goal will likely be reached or not, and are good indicators of capabilities, practices and skills. They measure the activity goals, which are the actions the process owner must take to achieve effective process performance.

# L

**Librarian**—The individual responsible for the safeguard and maintenance of all program and data files

**Licensing agreement**—A contract that establishes the terms and conditions under which a piece of software is being licensed (i.e., made legally available for use) from the software developer (owner) to the user

**Limit check**—Tests of specified amount fields against stipulated high or low limits of acceptability. When both high and low values are used, the test may be called a range check.

**Literals**—Any notation for representing a value within programming language source code, e.g., a string literal; a chunk of input data that is represented "as is" in compressed data

**Local area network (LAN)**—Communications networks that serve several users within a specified geographical area. Personal computer LANs function as distributed processing systems in which each computer in the network does its own processing and manages some of its data. Shared data are stored in a file server that acts as a remote disk drive for all users in the network.

**Log**—Records details of the information or events in an organized record-keeping system, usually sequenced in the order in which they occurred

**Logon**—The process of connecting to the computer. It typically requires entry of a user ID and password into a computer terminal.

# M

**Malware**—Short for "malicious software," malware is software designed to infiltrate, damage or obtain information from a computer system without the owner's consent. Malware is commonly taken to include computer viruses, worms, Trojan horses, spyware and adware. Spyware is generally used for marketing purposes and, as such, is not really malicious although it is generally unwanted. However, spyware can also be used to gather information for identity theft or other clearly illicit purposes.

**Management information system (MIS)**—An organized assembly of resources and procedures required to collect, process and distribute data for use in decision making.

**Mandatory access controls (MAC)**—Logical access control filters used to validate access credentials that cannot be controlled or modified by normal users or data owners; they act by default. Conversely, those controls that may be configured or modified by the users or data owners are called discretionary access controls (DACs).

**Mapping**—Diagramming data that is to be exchanged electronically, including how it is to be used and what business management systems need it. It is a preliminary step for developing an applications link. (See application tracing and mapping.)

**Materiality**—An auditing concept regarding the importance of an item of information with regard to its impact or effect on the functioning of the entity being audited. An expression of the relative significance or importance of a particular matter in the context of the organization as a whole.

**Maturity model**—A collection of instructions an organization can follow to gain better control over its software development process. The Capability Maturity Model (CMM) for Software, from the Software Engineering Institute (SEI), is a model used by many organizations to identify best practices useful in assessing and increasing the maturity of the software development processes. The CMM ranks software development organizations according to a hierarchy of five process maturity levels. Each level ranks the development environment according to its capability of producing quality software. A set of standards is associated with each of the five levels. The standards for level one describe the most immature or chaotic processes and the standards for level five describe the most mature or quality processes.

**Media Access Control (MAC)**—A unique, 48-bit, hard-coded address of a physical layer device, such as an Ethernet LAN or a wireless network card. The MAC is applied to the hardware at the factory and cannot be modified.

**Media oxidation**—The deterioration of the media (e.g., tapes) upon which data is digitally stored due to exposure to oxygen and moisture, for example, tapes deteriorating in a warm, humid environment. Proper environmental controls should prevent or significantly slow this process.

**Message switching**—A telecommunications traffic controlling methodology in which a complete message is sent to a concentration point and stored until the communications path is established.

**Middleware**—Another term for an application programmer interface (API). It refers to the interfaces that allow programmers to access lower- or higher-level services by providing an intermediary layer that includes function calls to the services.

**Milestone**—A terminal element that marks the completion of a work package or phase, typically marked by a high-level event such as project completion; receipt, endorsement or signing of a previously-defined deliverable; or a high-level review meeting at which the appropriate level of project completion is determined and agreed to. Typically, a milestone is associated with some sort of decision that outlines the future of a project and, for an outsourced project, may have a payment to the contractor associated with it.

**Mirroring**—A technique where the system writes the data simultaneously to separate hard drives or drive arrays. The advantages of mirroring are minimal downtime, simple data recovery, and increased performance in reading from the disk. If one hard drive or disk array fails, the system can operate from the working hard drive or disk array, or the system can use one disk to process a read request and another disk for a different processing request. The disadvantage of mirroring is that both drives or disk arrays are processing in the writing-to-disks function, which can hinder system performance. Mirroring has a high fault tolerance and can be implemented through a hardware RAID controller or through the operating system.

**Mission-critical application**—An application that is vital to the operation of the organization. The term is very popular for describing the applications required to run the day-to-day business.

**Mobile site**—This is a specially designed trailer that can be quickly transported to a business location or to an alternate site to provide a ready-conditioned information processing facility. These mobile sites can be connected to form larger work areas and can be preconfigured with servers, desktop computers, communications equipment, and even microwave and satellite data links.

**Modulation**—The process of converting a digital computer signal into an analog telecommunications signal

**Monetary unit sampling**—A sampling technique that estimates the amount of overstatement in an account balance

# N

**Network**—A system of interconnected computers and the communications equipment used to connect them

**Network administrator**—Responsible for planning, implementing and maintaining the telecommunications infrastructure, and also may be responsible for voice networks. For smaller organizations, this may entail maintaining a LAN and assisting end users.

**Network attached storage (NAS)**—This utilizes dedicated storage devices that centralize storage of data. Such devices generally do not provide traditional file/print or application services.

**Network interface card (NIC)**—A communications card that, when inserted into a computer, allows it to communicate with other computers on a network. Most network interface cards are designed for a particular type of network or protocol.

**Noise**—Disturbances, such as static, in data transmissions that cause messages to be misinterpreted by the receiver

**Nondisclosure agreement (NDA)**—Also called a confidential disclosure agreement (CDA), confidentiality agreement or secrecy agreement, it is a legal contract between at least two parties that outlines confidential materials the parties wish to share with one another for certain purposes, but wish to restrict from generalized use. In other words, it is a contract through which the parties agree not to disclose information covered by the agreement. An NDA creates a confidential relationship between the parties to protect any type of trade secret. As such, an NDA can protect non-public business information. (Note: In the case of certain governmental entities, the confidentiality of information other than trade secrets may be subject to applicable statutory requirements and, in some cases, may be required to be revealed to an outside party requesting the information. Generally, the governmental entity will include a provision in the contract to allow the seller to review a request for information the seller identifies as confidential and the seller may appeal such a decision requiring disclosure.) NDAs are commonly signed when two companies or individuals are considering doing business together and need to understand the processes used in one another's businesses solely for the purpose of evaluating the potential business relationship. NDAs can be "mutual," meaning both parties are restricted in their use of the materials provided, or they can only restrict a single party. It is also possible for an employee to sign an NDA or NDA-like agreement with a company at the time of hiring; in fact, some employment agreements will include a clause restricting "confidential information" in general.

**Normalization**—The elimination of redundant data

# O

**Objectivity**—The ability of the IS auditor to exercise judgment, express opinions and present recommendations with impartiality

**Offsite storage**—A storage facility located away from the building housing the primary information processing facility (IPF), used for storage of computer media such as offline backup data and storage files

**Open Shortest Path First (OSPF)**—A routing protocol, developed for IP networks, that is based on the shortest-path-first or link-state algorithm.

**Operating system**—A master control program that runs the computer and acts as a scheduler and traffic controller. It is the first program copied into the computer's memory after the computer is turned on and must reside in memory at all times. It sets the standards for the application programs that run in it.

**Operational control**—These controls deal with the everyday operation of a company or organization to ensure that all objectives are achieved.

**Operator console**—A special terminal used by computer operations personnel to control computer and systems operations functions. These terminals typically provide a high level of computer access and should be properly secured.

# P

**Packet**—A block of data for data transmission. A packet contains both routing information and data.

**Packet switching**—The process of transmitting messages in convenient pieces that can be reassembled at the destination

**Paper test**—A walk-through of the steps of a regular test, but without actually performing the steps. It is usually used in disaster recovery and contingency testing, where team members review and become familiar with the plans and their specific roles and responsibilities.

**Parallel testing**—The process of feeding test data into two systems, the modified system and an alternative system (possibly the original system), and comparing results

**Parity check**—A general hardware control that helps to detect data errors when data are read from memory or communicated from one computer to another. A one-bit digit (either 0 or 1) is added to a data item to indicate whether the sum of that data item's bit is odd or even. When the parity bit disagrees with the sum of the other bits, the computer reports an error. The probability of a parity check detecting an error is 50 percent.

**Password**—A protected, generally computer-encrypted string of characters that authenticate a computer user to the computer system

**Patch management**—An area of systems management that involves acquiring, testing, and installing multiple patches (code changes) to an administered computer system, to maintain up-to-date software and often to address security risks. Patch management tasks include the following: maintaining current knowledge of available patches; deciding what patches are appropriate for particular systems; ensuring that patches are installed properly; testing systems after installation; and documenting all associated procedures such as specific configurations required. A number of products are available to automate patch management tasks. Patches are sometimes ineffective and can sometimes cause more problems than they fix. Patch management experts suggest that system administrators take simple steps to avoid problems such as performing backups and testing patches on non-critical systems prior to installations. Patch management can be viewed as part of change management.

**Payroll system**—An electronic system for processing payroll information and the related electronic (e.g., electronic timekeeping and/or human resources system), human (e.g., payroll clerk), and external party (e.g., bank) interfaces. In a more limited sense, it is the electronic system that performs the processing for generating payroll checks and/or bank direct deposits to employees.

**Performance testing**—Comparing the system's performance to other equivalent systems, using well-defined benchmarks

**Personal digital assistant (PDA)**—Also called palmtop and pocket computer; these are handheld devices that provide computing, Internet, networking and telephone characteristics.

**Personal identification number (PIN)**—A type of password (i.e., a secret number assigned to an individual) that, in conjunction with some means of identifying the individual, serves to verify the authenticity of the individual. PINs have been adopted by financial institutions as the primary means of verifying customers in an electronic funds transfer system (EFTS).

**Phishing**—This is a type of e-mail attack that attempts to convince a user that the originator is genuine, but with the intention of obtaining information for use in social engineering. These attacks may take the form of masquerading as a lottery organization advising the recipient of a large win or the user's bank; in either case, the intent is to obtain account and PIN details. Alternative attacks may seek to obtain apparently innocuous business information which may be used in another form of active attack.

**Phreakers**—Those who crack security, most frequently phone and other communication networks

**Point-of-sale (POS) systems**—Enable the capture of data at the time and place of transaction. POS terminals may include use of optical scanners for use with bar codes or magnetic card readers for use with credit cards. POS systems may be online to a central computer or may use stand-alone terminals or microcomputers that hold the transactions until the end of a specified period when they are sent to the main computer for batch processing.

**Point-to-point protocol (PPP)**—Commonly used to establish a direct connection between two nodes, it can connect computers using serial cable, phone line, trunk line, cellular telephone, specialized radio links or fiber optic links. Its main features include enhanced error detection, automatic self-configuration and looped link detection. Most Internet service providers use PPP for customers' dial-up access to the Internet. PPP is commonly used to act as a "layer 2" (the data link layer of the open systems interconnection [OSI] model) protocol for connection over synchronous and asynchronous circuits, where it has largely superseded an older nonstandard protocol (known as serial line internet protocol [SLIP]) and telephone company mandated standards (such as X.25). PPP was designed to work with numerous "layer 3" network layer protocols including IP, Novell's IPX, and AppleTalk.

**Port**—An interface point between the CPU and a peripheral device

**Privacy**—Privacy involves providing proper protection for personally identifiable information relating to an identified or identifiable individual (data subject). Management should ensure that proper controls are in place and functioning to be in compliance with its privacy policy or applicable privacy laws and regulations.

**Private branch exchange (PBX)**—A telephone exchange that is owned by a private business, as opposed to one owned by a common carrier or by a telephone company.

**Private key cryptosystem**—Used in data encryption, it uses a secret key to encrypt the plaintext to the ciphertext. It also uses the same key to decrypt the ciphertext to the corresponding plaintext. In this case, the key is symmetric such that the encryption key is equivalent to the decryption key.

**Problem escalation procedure**—The process of escalating a problem up from junior to senior support staff, and ultimately to higher levels of management. It is often used in help desk management where an unresolved problem is escalated up the chain of command until it is solved.

**Program Evaluation and Review Technique (PERT)**—A project management technique used in the planning and control of system projects

**Project portfolio**—The set of projects owned by a company; it usually includes the main guidelines relative to each project including objectives, costs, timelines and other information specific to the project.

**Protocol**—The rules by which a network operates and controls the flow and priority of transmissions

**Prototyping**—A system development technique that enables users and developers to reach agreement on system requirements. Prototyping uses programmed simulation techniques to represent a model of the final system to the user for advisement and critique. The emphasis is on end-user screens and reports. Internal controls are not a priority item since this is only a model.

**Public key cryptosystem**—Used in data encryption, it uses an encryption key, as a public key, to encrypt the plaintext to the ciphertext. It uses a different decryption key, as a secret key, to decrypt the ciphertext to the corresponding plaintext. In contrast to a private key cryptosystem, the decryption key should be secret; however, the encryption key can be known to everyone. In a public key cryptosystem, the two keys are asymmetric, such that the encryption key is not equivalent to the decryption key.

**Public key encryption**—A cryptographic system that uses two keys. One is a public key, which is known to everyone, and the second is a private or secret key, which is only known to the recipient of the message.

**Public key infrastructure (PKI)**—A system that authentically distributes user's public keys using certificates. It verifies and authenticates the validity of each party involved in an Internet transaction through digital certificates, certificate authorities and other registration authorities.

# Q

**Quality assurance**—A technique used to design, develop and implement a product or service reducing costs and preserving the quality.

**Queue**—A group of items that are waiting to be serviced or processed

# R

**Radio wave interference**—The superposition of two or more radio waves resulting in a different radio wave pattern that is more difficult to intercept and decode properly

**Random access memory (RAM)**—The computer's primary working memory. Each byte of memory can be accessed randomly regardless of adjacent bytes.

**Record**—A collection of related information treated as a unit. Separate fields within the record are used for processing the information.

**Recovery point objective (RPO)**—The recovery point objective is determined based on the acceptable data loss in case of disruption of operations. It indicates the earliest point in time to which it is acceptable to recover the data. RPO effectively quantifies the permissible amount of data loss in case of interruption.

**Recovery testing**—A test to check the system's ability to recover after a software or hardware failure

**Recovery time objective (RTO)**—The recovery time objective is determined based on the acceptable down time in case of disruption of operations. It indicates the earliest point in time at which the business operations must resume after disaster.

**Redundant Array of Inexpensive Disks (RAID)**—Provides performance improvements and fault-tolerant capabilities via hardware or software solutions by writing to a series of multiple disks to improve performance and/or save large files simultaneously

**Reengineering**—A process involving the extraction of components from existing systems and restructuring these components to develop new systems or to enhance the efficiency of existing systems. Existing software systems thus can be modernized to prolong their functionality. An example of this is a software code translator that can take an existing hierarchical database system and transpose it to a relational database system. CASE includes a source code reengineering feature.

**Registration authority (RA)**—An optional entity separate from a CA that would be used by a CA with a very large customer base. CAs use RAs to delegate some of the administrative functions associated with recording or verifying some or all of the information needed by a CA to issue certificates or CRLs and to perform other certificate management functions. However, with this arrangement, the CA still retains sole responsibility for signing either digital certificates or CRLs. If an RA is not present in the established PKI structure, the CA is assumed to have the same set of capabilities as those defined for an RA.

**Regression testing**—A testing technique used to retest earlier program abends or logical errors that occurred during the initial testing phase

**Remote access service (RAS)**—Refers to any combination of hardware and software to enable the remote access to tools or information that typically reside on a network of IT devices. Originally coined by Microsoft when referring to their built-in NT remote access tools, RAS was a service provided by Windows NT that allows most of the services that would be available on a network to be accessed over a modem link. Over the years, many vendors have provided both hardware and software solutions to gain remote access to various types of networked information. In fact, most modern routers include a basic RAS capability that can be enabled for any dial-up interface.

**Remote Procedure Call (RPC)**—The traditional Internet service protocol widely used for many years on UNIX-based operating systems and supported by the Internet Engineering Task Force (IETF) that allows a program on one computer to execute a program on another (e.g., server). The primary benefit derived from its use is that a system developer need not develop specific procedures for the targeted computer system. For example, in a client-server arrangement, the client program sends a message to the server with appropriate arguments, and the server returns a message containing the results of the program executed. Common Object Request Broker Architecture (CORBA) and Distributed Component Object Model (DCOM) are two newer object-oriented methods for related RPC functionality.

**Repeaters**—A physical layer device that regenerates and propagates electrical signals between two network segments. Repeaters receive signals from one network segment and amplify (regenerate) the signal to compensate for signals (analog or digital) distorted by transmission loss due to reduction of signal strength during transmission (i.e., attenuation).

**Replication**—In its broad computing sense, involves the use of redundant software or hardware elements to provide availability and fault-tolerant capabilities. In a database context, replication involves the sharing of data between databases to reduce workload among database servers, thereby improving client performance while maintaining consistency among all systems.

**Repository**—The central database that stores and organizes data

**Request for proposal (RFP)**—A document distributed to software vendors, requesting their submission of a proposal to develop or provide a software product

**Requirements definition**—A phase of a SDLC methodology where the affected user groups define the requirements of the system for meeting the defined needs

**Resilience**—The ability of a system or network to recover automatically from any disruption, usually with minimal recognizable effect

**Return on investment (ROI)**—A measure of operating performance and efficiency, computed in its simplest form by dividing net income by average total assets

**Reverse engineering**—A software engineering technique whereby existing application system code can be redesigned and coded using computer-aided software engineering (CASE) technology

**Ring configuration**—Used in either token ring or fiber-distributed database interface (FDDI) networks, all stations (nodes) are connected to a multistation access unit (MSAU), which physically resembles a star-type topology. A ring configuration is created when these MSAUs are linked together in forming a network. Messages in this network are sent in a deterministic fashion from sender and receiver via a small frame, referred to as a token ring. To send a message, a sender obtains the token with the right priority as the token travels around the ring, with receiving nodes reading those messages addressed to it.

**Risk**—The potential that a given threat will exploit vulnerabilities of an asset or group of assets to cause loss of/or damage to the assets. It usually is measured by a combination of impact and probability of occurrence.

**Risk analysis**—The initial steps of risk management: analyzing the value of assets to the business, identifying threats to those assets and evaluating how vulnerable each asset is to those threats. ISO/IEC 27001 information security standard defines risk analysis as the systematic use of information to identify sources and to estimate the risk. [ISO/IEC Guide 73:2002]

**Risk assessment**—The overall process of risk analysis and risk evaluation [ISO/IEC Guide 73:2002]

**Risk evaluation**—The process of comparing the estimated risk against given risk criteria to determine the significance of the risk. [ISO/IEC Guide 73:2002]

**Risk management**—The coordinated activities to direct and control an organization with regard to risk (in this International Standard the term 'control' is used as a synonym for 'measure'). [ISO/IEC Guide 73:2002]

**Risk treatment**—The process of selection and implementation of measures to modify risk [ISO/IEC Guide 73:2002]

**Rounding down**—A method of computer fraud involving a computer code that instructs the computer to remove small amounts of money from an authorized computer transaction by rounding down to the nearest whole value denomination and rerouting the rounded off amount to the perpetrator's account

**Router**—A networking device that can send (route) packets to the connected LAN segment, based on addressing at the network layer (Layer 3) in the OSI model. Networks connected by routers can use different or similar networking protocols. Routers usually are capable of filtering packets based on parameters such as source address, destination address, protocol and network application (ports).

**RSA**—A public key cryptosystem developed by R. Rivest, A. Shamir and L. Adleman. RSA has two different keys; the public encryption key and the secret decryption key. The strength of RSA depends on the difficulty of the prime number factorization. For applications with high-level security, the number of the decryption key bits should be greater than 512 bits. RSA is used for both encryption and digital signatures.

**Run-to-run totals**—Provide verification that all transmitted data are read and processed

# S

**Scheduling**—A method used in the information processing facility (IPF) to determine and establish the sequence of computer job processing

**Scope creep**—Also called requirement creep; this refers to uncontrolled changes in a project's scope. This phenomenon can occur when the scope of a project is not properly defined, documented and controlled. Typically, the scope increase consists of either new products or new features of already approved products. Therefore, the project team drifts away from its original purpose. Because of one's tendency to focus on only one dimension of a project, scope creep can also result in a project team overrunning its original budget and schedule. For example, scope creep can be a result of poor change control, lack of proper identification of what products and features are required to bring about the achievement of project objectives in the first place, or a weak project manager or executive sponsor.

**Secure Sockets Layer (SSL)**—A protocol that is used to transmit private documents through the Internet. This protocol uses a private key to encrypt the data that is to be transferred through the SSL connection.

**Security incident**—A series of unexpected events that involves an attack or series of attacks (compromise and/or breach of security) at one or more sites. A security incident normally includes an estimation of its level of impact. A limited number of impact levels are defined and, for each, the specific actions required and the people who need to be notified are identified.

**Security testing**—Making sure the modified/new system includes appropriate access controls and does not introduce any security holes that might compromise other systems

**Service set identifier (SSID)**—In Wi-Fi Wireless LAN computer networking, this is a code attached to all packets on a wireless network to identify each packet as part of that network. The code consists of a maximum of 32 alphanumeric characters. All wireless devices attempting to communicate with each other must share the same SSID. Apart from identifying each packet, SSID also serves to uniquely identify a group of wireless network devices used in a given service set. There are two major variants of the SSID. *Ad hoc* wireless networks that consist of client machines without an access point use the IBSSID (Independent Basic Service Set Identifier); whereas on an infrastructure network that includes an access point, the basic service set identifier (BSS ID) or extended service set identifier (ESS ID) is used.

**Servlet**—Typically indicates a Java applet or a small program that runs within a web server environment. A Java servlet is similar to a common gateway interface (CGI) program, but unlike a CGI program, once started, it stays in memory and can fulfill multiple requests, thereby saving server execution time and speeding up the services.

**Session border controller (SBC)**—Provide security features for VoIP traffic similar to that provided by firewalls. SBCs can be configured to filter specific VoIP protocols, monitor for denial-of-service (DOS) attacks, and provide network address and protocol translation features.

**Sign-on procedure**—The procedure performed by a user to gain access to an application or operating system. If the user is properly identified and authenticated by the system's security, the user will be able to access the software.

**Simple Object Access Protocol (SOAP)**—A platform-independent, XML-based formatted protocol enabling applications to communicate with each other over the Internet. Use of this protocol may provide a significant security risk to web application operations since use of SOAP piggybacks onto a web-based document object model and is transmitted via HTTP (port 80) to penetrate server firewalls, which are usually configured to accept port 80 and port 21 (FTP) requests. Web-based document models define how objects on a web page are associated with each other and how they can be manipulated while being sent from a server to a client browser. SOAP typically relies on XML for presentation formatting and also adds appropriate HTTP-based headers to send it. SOAP forms the foundation layer of the web services stack, providing a basic messaging framework on which more abstract layers can build. There are several different types of messaging patterns in SOAP but, by far, the most common is the Remote Procedure Call (RPC) pattern, in which one network node (the client) sends a request message to another node (the server), and the server immediately sends a response message to the client.

**Slack time (float)**—Time in the project schedule, the use of which does not affect the project's critical path (the minimum time to complete the project based upon the estimated time for each project segment and their relationships). Slack time is commonly referred to as "float" and generally is not "owned" by either party to the transaction.

**SMART (specific, measurable, achievable, relevant, time-bound)**—A development methodology for value management

**Software**—Programs and supporting documentation that enable and facilitate use of the computer. Software controls the operation of the hardware.

**Source code**—Source code is the language in which a program is written. Source code is translated into object code by assemblers and compilers. In some cases, source code may be converted automatically into another language by a conversion program. Source code is not executable by the computer directly. It must first be converted into machine language.

**Source documents**—The forms used to record data that have been captured. A source document may be a piece of paper, a turnaround document or an image displayed for online data input.

**Source lines of code (SLOC)**—Source lines of code are often used in deriving single-point software size estimations.

**Spool (simultaneous peripheral operations online)**—An automated function that can be operating system or application based in which electronic data being transmitted between storage areas are spooled or stored until the receiving device or storage area is prepared and able to receive the information. This operation allows more efficient electronic data transfers from one device to another by permitting higher speed sending functions, such as internal memory, to continue on with other operations instead of waiting on the slower speed receiving device such as a printer.

**Spyware**—Software whose purpose is to monitor a computer user's actions (e.g., web sites they visit) and report these actions to a third party, without the informed consent of that machine's owner or legitimate user. A particularly malicious form of spyware is software that monitors keystrokes (e.g., to obtain passwords) or otherwise gathers sensitive information such as credit card numbers, which it then transmits to a malicious third party. The term has also come to refer more broadly to software that subverts the computer's operation for the benefit of a third party.

**Standing data**—Permanent reference data used in transaction processing. These data are changed infrequently such as a product price file or a name and address file.

**Statistical sampling**—A method of selecting a portion of a population, by means of mathematical calculations and probabilities, for the purpose of making scientifically and mathematically sound inferences regarding the characteristics of the entire population

**Storage area networks (SANs)**—A variation of a LAN that is dedicated for the purpose of connecting storage devices to servers and other computing devices. SANs centralize the process for the storage and administration of data.

**Structured Query Language (SQL)**—The primary language used by both application programmers and end users in accessing relational databases

**Substantive testing**—Determines the integrity of actual processing, which provides evidence of the validity of the final outcome. This is done outside of a review of processes and related internal controls. For example, balances in the financial statement and the transactions to support those balances is a substantive test. General types of testing involve recalculation, confirmations, verification of outcomes from other information sources and observations. Substantive testing will be limited when there is a low risk of control failure. Conversely, if the testing of controls reveals weaknesses in control, the level of substantive testing would be increased.

**Supply chain management (SCM)**—A concept that allows an organization to more effectively and efficiently manage the activities of design, manufacturing, distribution, service and recycling of products and services to its customers

**Suspense file**—A computer file used to maintain information (i.e., on transactions, payments, or other events) until the proper disposition of that information can be determined. Once the proper disposition of the item is determined, it should be removed from the suspense file and processed in accordance with the proper procedures for that particular transaction. Two examples of items that may be included in a suspense file are receipt of a payment from a source that is not readily identified or data that do not yet have an identified match during migration to a new application.

**Switches**—Typically associated as a data link layer device, switches enable LAN network segments to be created and interconnected, which also has the added benefit of reducing collision domains in Ethernet-based networks.

**Synchronous transmission**—Block-at-a-time data transmission

**System software**—A collection of computer programs used in the design, processing and control of all applications. The programs and processing routines that control the computer hardware, including the operating system and utility programs.

**System testing**—A series of tests designed to ensure that the modified program interacts correctly with other system components. These test procedures are typically performed by the system maintenance staff in their development library.

**Systems development life cycle (SDLC)**—The phases deployed in the development or acquisition of a software system. Typical phases include the feasibility study, requirements study, requirements definition, detailed design, programming, testing, installation and postimplementation review.

# T

**Tape management system (TMS)**—A system software tool that logs, monitors and directs computer tape usage.

**Telecommunications**—Electronic communications by special devices over distances or around devices that preclude direct interpersonal exchange

**Terminal**—A device for sending and receiving computerized data over transmission lines

**Test data**—Data that are used to test a computer program. Depending on the purpose of the test, the data may be production data (files) or data created by either information systems (IS) or the customer (user).

**Throughput**—The quantity of useful work made by the system per unit of time. Throughput can be measured in instructions per second or some other unit of performance. When referring to a data transfer operation, throughput measures the useful data transfer rate and is expressed in kbps, Mbps and Gbps.

**Transaction**—Business events or information grouped together because they have a single or similar purpose. Typically, a transaction is applied to a calculation or event that then results in the updating of a holding or master file.

**Transaction log**—A manual or automated log of all updates to data files and databases

**Transmission Control Protocol/Internet Protocol (TCP/IP)**—A set of communications protocols that encompasses media access, packet transport, session communications, file transfer, electronic mail, terminal emulation, remote file access and network management. TCP/IP provides the basis for the Internet.

**Trojan horse**—Purposefully hidden malicious or damaging code within an authorized computer program

**Tunneling**—A method by which one network protocol encapsulates another protocol within itself. It is commonly used to bridge between incompatible hosts/routers or to provide encryption. When protocol A encapsulates protocol B, then a protocol A header and optional tunneling headers are appended to the original protocol B packet. Protocol A then becomes the data link layer of protocol B. Examples of tunneling protocols include IPSec, Point-to-point Protocol Over Ethernet (PPPoE), and Layer 2 Tunneling Protocol (L2TP).

**Tuple**—A tuple is a row in a database table.

**Twisted pairs**—A pair of small, insulated wires that are twisted around each other to minimize interference from other wires in the cable. This is a low-capacity transmission medium.

# U

**Unicode**—A standard for representing characters as integers. It uses 16 bits, which means that it can represent more than 65,000 unique characters as is necessary for languages such as Chinese and Japanese.

**Uninterruptible power supply (UPS)**—Provides short-term backup power from batteries for a computer system when the electrical power fails or drops to an unacceptable voltage level

**Unit testing**—A testing technique that is used to test program logic within a particular program or module. The purpose of the test is to ensure that the internal operation of the program performs according to specification. It uses a set of test cases that focus on the control structure of the procedural design.

**Universal Serial BUS (USB)**—An external bus standard that provides capabilities to transfer data at a rate of 12 Mbps. A USB port can connect up to 127 peripheral devices.

**User awareness**—The training process in security-specific issues to reduce security problems since users are often the weakest link in the security chain

**Utility programs**—Specialized system software used to perform particular computerized functions and routines that are frequently required during normal processing. Examples include sorting, backing up and erasing data.

**Utility script**—A sequence of commands input into a single file to automate a repetitive and specific task. The utility script is then executed, either automatically or manually, to perform the task. In UNIX, these are known as shell scripts.

# V

**Variable sampling**—A sampling technique used to estimate the average or total value of a population based on a sample; a statistical model used to project a quantitative characteristic such as a monetary amount

**Verification**—Checks that data are entered correctly

**Virus**—Malicious programs designed to spread and replicate from computer to computer through telecommunications links or through sharing of computer diskettes and files

**Voice mail**—A system of storing messages in a private recording medium where the called party can later retrieve the messages

**Voice-over Internet protocol (VoIP)**—Also called IP Telephony, Internet telephony and Broadband Phone, this is a technology that makes it possible to have a voice conversation over the Internet or over any dedicated Internet Protocol (IP) network instead of dedicated voice transmission lines.

# W

**WAN switch**—A data link layer device used for implementing various WAN technologies such as asynchronous transfer mode, point-to-point frame relay solutions, and integrated services digital network (ISDN). These devices are typically associated with carrier networks providing dedicated WAN switching and router services to organizations via T-1 or T-3 connections.

**Wide area network (WAN)**—A computer network connecting different remote locations that may range from short distances, such as a floor or building, to extremely long transmissions that encompass a large region or several countries

**Wi-Fi Protected Access (WPA)**—A class of systems used to secure wireless (Wi-Fi) computer networks. It was created in response to several serious weaknesses researchers found in the previous system, Wired Equivalent Privacy (WEP). WPA implements the majority of the IEEE 802.11i standard, and was intended as an intermediate measure to take the place of WEP while 802.11i was prepared. WPA is designed to work with all wireless network interface cards, but not necessarily with first generation wireless access points. WPA2 implements the full standard, but will not work with some older network cards. Both provide good security with two significant issues. First, either WPA or WPA2 must be enabled and chosen in preference to WEP; WEP is usually presented as the first security choice in most installation instructions. Second, in the "personal" mode, the most likely choice for homes and small offices, a pass phrase is required that, for full security, must be longer than the typical 6- to 8-character passwords users are taught to employ.

**Wired Equivalent Privacy (WEP)**—A scheme that is part of the IEEE 802.11 wireless networking standard to secure IEEE 802.11 wireless networks (also known as Wi-Fi networks). Because a wireless network broadcasts messages using radio, it is particularly susceptible to eavesdropping. WEP was intended to provide comparable confidentiality to a traditional wired network (in particular, it does not protect users of the network from each other), hence the name. Several serious weaknesses were identified by cryptanalysts, and WEP was superseded by Wi-Fi Protected Access (WPA) in 2003, and then by the full IEEE 802.11i standard (also known as WPA2) in 2004. Despite the weaknesses, WEP provides a level of security that can deter casual snooping.

**Wiretapping**—The practice of eavesdropping on information being transmitted over telecommunications links

**Note:** The CISA candidate may also want to be familiar with ISACA's Glossary which can be viewed at *www.isaca.org/glossary*. Also available is a list of CISA exam terminology in different languages that can be viewed at *www.isaca.org/examterm*.

The CISA candidate should be familiar with the following list of acronyms published in the *Candidate's Guide to the CISA Examination*. These acronyms are the only stand-alone abbreviations used in examination questions.

| | | | |
|---|---|---|---|
| ASCII | American Standard Code for Information Interchange | IT | Information technology |
| | | LAN | Local area network |
| Bit | Binary digit | PBX | Private branch (business) exchange |
| CASE | Computer-aided system engineering | PC | Personal computer/microcomputer |
| CCTV | Closed-circuit television | PCR | Program change request |
| CPU | Central processing unit | PDA | Personal digital assistant |
| DBA | Database administrator | PERT | Program Evaluation Review Technique |
| DBMS | Database management system | PIN | Personal identification number |
| EDI | Electronic data interchange | PKI | Public key infrastructure |
| FTP | File Transfer Protocol | RAID | Redundant Array of Inexpensive Disks |
| HTTP | Hypertext Transmission Protocol | RFID | Radio frequency identification |
| HTTPS | Secured Hypertext Transmission Protocol | SDLC | System development life cycle |
| ID | Identification | SSL | Secure Sockets Layer |
| IDS | Intrusion detection system | TCP | Transmission Control Protocol |
| IP | Internet protocol | UPS | Uninterruptible power supply |
| IS | Information systems | VoIP | Voice-over Internet Protocol |
| ISO | International Organization for Standardization | WAN | Wide area network |

In addition to the aforementioned acronyms, candidates may also wish to become familiar with the following additional acronyms. Should any of these abbreviations be used in exam questions, their meanings would be included when the acronym appears.

| | | | |
|---|---|---|---|
| 4GL | Fourth-generation language | BCI | Business Continuity Institute |
| ACID | Atomicity, consistency, isolation and durability | BCM | Business continuity management |
| | | BCP | Business continuity plan |
| ACL | Access control list | BCP | Business continuity planning |
| AES | Advanced Encryption Standard | BDA | Business dependency assessment |
| AH | Authentication header | BI | Business intelligence |
| AI | Artificial intelligence | BIA | Business impact analysis |
| AICPA | American Institute of Certified Public Accountants | BIMS | Biometric Information Management and Security |
| ALE | Annual loss expectancy | BIOS | Basic Input/Output System |
| ALU | Arithmetic-logic unit | BIS | Bank for International Settlements |
| ANSI | American National Standards Institute | BLP | Bypass label process |
| API | Application programming interface | BNS | Backbone network services |
| ARP | Address Resolution Protocol | BOM | Bill of materials |
| ASIC | Application-specific integrated circuit | BOMP | Bill of materials processor |
| ATDM | Asynchronous time division multiplexing | BPR | Business process reengineering |
| ATM | Asynchronous Transfer Mode | BRP | Business recovery (or resumption) plan |
| ATM | Automated teller machine | BSC | Balanced scorecard |
| B-to-B | Business-to-business | C-to-G | Consumer-to-government |
| B-to-C | Business-to-consumer | CA | Certificate authority |
| B-to-E | Business-to-employee | CAAT | Computer-assisted audit technique |
| B-to-G | Business-to-government | CAD | Computer-assisted design |

# Acronyms

| | | | |
|---|---|---|---|
| CAE | Computer-assisted engineering | CSMA/CD | Carrier-sense Multiple Access/ Collision Detection |
| CAM | Computer-aided manufacturing | | |
| CASE | Computer-aided software engineering | CSO | Chief security officer |
| CCK | Complementary Code Keying | CSU-DSU | Channel service unit/digital service unit |
| CCM | Constructive Cost Model | DAC | Discretionary access control |
| CD | Compact disk | DASD | Direct access storage device |
| CD-R | Compact disk-recordable | DAT | Digital audio tape |
| CD-RW | Compact disk-rewritable | DCE | Data communications equipment |
| CDDF | Call Data Distribution Function | DCE | Distributed computing environment |
| CDPD | Cellular Digital Packet Data | DCOM | Distributed Component Object Model (Microsoft) |
| CEO | Chief executive officer | | |
| CERT | Computer emergency response team | DCT | Discrete Cosine Transform |
| CGI | Common gateway interface | DD/DS | Data dictionary/directory system |
| CIA | Confidentiality, integrity and availability | DDL | Data Definition Language |
| CIAC | Computer Incident Advisory Capability | DDN | Digital Divide Network |
| CICA | Canadian Institute of Chartered Accountants | DDoS | Distributed denial of service |
| CIM | Computer-integrated manufacturing | DECT | Digital Enhanced Cordless Telecommunications |
| CIO | Chief information officer | | |
| CIS | Continuous and intermittent simulation | DES | Data Encryption Standard |
| CISO | Chief information security officer | DFD | Data flow diagram |
| CMDB | Configuration management database | DHCP | Dynamic Host Configuration Protocol |
| CMM | Capability Maturity Model | DID | Direct inward dial |
| CMMI | Capability Maturity Model Integration | DIP | Document image processing |
| CNC | Computerized Numeric Control | DLL | Dynamic link library |
| CobiT | *Control Objectives for Information and related Technology* | DMS | Disk management system |
| | | DMZ | Demilitarized zone |
| COCOMO2 | Constructive Cost Model | DNS | Domain name server |
| CODASYL | Conference on Data Systems Language | DoS | Denial of service |
| COM | Component Object Model | DOSD | Data-oriented system development |
| COM/DCOM | Component Object Model/Distributed Component Object Model | DRII | Disaster Recovery Institute International |
| | | DRP | Disaster recovery plan |
| COOP | Continuity of operations plan | DRP | Disaster recovery planning |
| CORBA | Common Object Request Broker Architecture | DSL | Digital subscriber lines |
| CoS | Class of service | DSS | Decision support systems |
| COSO | Committee of Sponsoring Organizations of the Treadway Commission | DSSS | Direct Sequence Spread Spectrum |
| | | DTE | Data terminal equipment |
| CPM | Critical Path Methodology | DTR | Data terminal ready |
| CPO | Chief privacy officer | DVD | Digital video disc |
| CPS | Certification practice statement | DVD-HD | Digital video disc-high definition/high density |
| CRC | Cyclic redundancy check | DW | Data warehouse |
| CRL | Certificate revocation list | EA | Enterprise architecture |
| CRM | Customer relationship management | EAC | Estimates at completion |
| CSA | Control self-assessment | EAI | Enterprise application integration |
| CSF | Critical success factor | EAM | Embedded audit module |
| CSIRT | Computer security incident response team | EAP | Extensible Authentication Protocol |
| CSMA/CA | Carrier-sense Multiple Access/Collision Avoidance | EBCDIC | Extended Binary-coded for Decimal Interchange Code |

| | | | | |
|---|---|---|---|---|
| EC | Electronic commerce | | HDLC | High-level data link control |
| ECC | Elliptical Curve Cryptography | | HIPAA | Health Insurance Portability and Accountability Act (USA) |
| EDFA | Enterprise data flow architecture | | | |
| EDI | Electronic data interchange | | HIPO | Hierarchy input-process-output |
| EER | Equal-error rate | | HMI | Human machine interfacing |
| EFT | Electronic funds transfer | | HTML | Hypertext Markup Language |
| EIGRP | Enhanced Interior Gateway Routing Protocol | | HW/SW | Hardware/software |
| EJB | Enterprise java beans | | I/O | Input/output |
| EMI | Electromagnetic interference | | I&A | Identification and authentication |
| EMRT | Emergency response time | | ICMP | Internet Control Message Protocol |
| ERD | Entity relationship diagram | | ICT | Information and communication technologies |
| ERP | Enterprise resource planning | | | |
| ESP | Encapsulating security payload | | IDE | Integrated development environment |
| EVA | Earned value analysis | | IDEF1X | Integration Definition for Information Modeling |
| FAR | False-acceptance rate | | | |
| FAT | File allocation table | | IETF | Internet Engineering Task Force |
| FC | Fibre channels | | IMS | Integrated manufacturing systems |
| FDDI | Fiber Distributed Data Interface | | IPF | Information processing facility |
| FDM | Frequency division multiplexing | | IPL | Initial program load |
| FEA | Federal enterprise architecture | | IPMA | International Project Management Association |
| FEMA | Federal Emergency Management Association (USA) | | IPRs | Intellectual property rights |
| | | | IPS | Intrusion prevention system |
| FER | Failure-to-enroll rate | | IPSec | IP Security |
| FERC | Federal Energy Regulatory Commission (USA) | | IPX | Internetwork Packet Exchange |
| | | | IR | Incident response |
| FFIEC | Federal Financial Institutions Examination Council (USA) | | IR | Infrared |
| | | | IRC | Internet relay chat |
| FFT | Fast Fourier Transform | | IrDA | Infrared Data Association |
| FHSS | Frequency-hopping spread spectrum | | IRM | Incident response management |
| FIPS | Federal Information Processing Standards | | IS/DRP | IS disaster recovery planning |
| FP | Function point | | ISAKMP/ Oakley | Internet Security Association and Key Management Protocol/Oakley |
| FPA | Function point analysis | | | |
| FRAD | Frame relay assembler/disassembler | | ISAM | Indexed Sequential Access Method |
| FRB | Federal Reserve Board (USA) | | ISDN | Integrated services digital network |
| FRR | False-rejection rate | | ISP | Internet service provider |
| GAS | Generalized audit software | | ITF | Integrated test facility |
| Gb | Gigabit | | ITGI | IT Governance Institute |
| GB | Gigabyte | | ITIL | Information Technology Infrastructure Library |
| GID | Group ID | | ITSM | IT service management |
| GIS | Geographic information systems | | ITT | Invitation to tender |
| GPS | Global position system | | ITU | International Telecommunications Union |
| GSM | Global system for mobile communications | | IVR | Interactive voice response |
| GUI | Graphical user interface | | JIT | Just in time |
| HA | High availability | | Kb | Kilobit |
| HD-DVD | High definition/high density-digital video disc | | KB | Kilobyte |
| | | | KB | Knowledge base |

# Acronyms

| | | | |
|---|---|---|---|
| KDSI | Thousand delivered source instructions | OOSD | Object-oriented system development |
| KGI | Key goal indicator | ORB | Object request broker (ORB) |
| KLOC | Kilo lines of code | OS | Operating system |
| KPI | Key performance indicator | OSI | Open Systems Interconnection |
| L2TP | Layer 2 Tunneling Protocol | OSPF | Open Shortest Path First |
| LCP | Link Control Protocol | PAD | Packet assembler/disassembler |
| M&A | Mergers and acquisition | PAN | Personal area network |
| MAC | Mandatory access control | PBX | Private branch exchange |
| MAC | Message Authentication Code | PDA | Personal digital assistant |
| MAC Address | Media Access Control Address | PDCA | Plan-do-check-act |
| MAN | Metropolitan area network | PDN | Public data network |
| MAP | Manufacturing accounting and production | PER | Package-enabled reengineering |
| MIS | Management information system | PHY | Physical layer |
| MODEM | Modulator/demodulator | PICS | Platform for Internet Content Selection |
| MOS | Maintenance out of service | PID | Process ID |
| MPLS | Multiprotocol label switching | PID | Project initiation document |
| MRP | Manufacturing resources planning | PLC | Programmable logic controllers |
| MSAUs | Multistation access units | PMBOK | Project Management Body of Knowledge |
| MTBF | Mean time between failures | PMI | Project Management Institute |
| MTS | Microsoft's Transaction Server | POC | Proof of concept |
| MTTR | Mean time to repair | POP | Proof of possession |
| NAP | Network access point | POS | Point of sale or (Point-of-sale systems) |
| NAS | Network access server | POTS | Plain old telephone service |
| NAS | Network attached storage | PPP | Point-to-point Protocol |
| NAT | Network address translation | PPPoE | Point-to-point Protocol Over Ethernet |
| NCP | Network Control Protocol | PPTP | Point-to-Point Tunneling Protocol |
| NDA | Nondisclosure agreement | PR | Public relations |
| NFPA | National Fire Protection Agency (USA) | PRD | Project request document |
| NFS | Network File System | PRINCE2 | Projects in Controlled Environments 2 |
| NIC | Network interface card | PROM | Programmable Read-only Memory |
| NIST | National Institute of Standards and Technology (USA) | PSTN | Public switched telephone network |
| | | PVC | Permanent virtual circuit |
| NNTP | Network News Transfer Protocol | QA | Quality assurance |
| NSP | Name Server Protocol | QAT | Quality assurance testing |
| NSP | Network service provider | RA | Registration authority |
| NT | New technology | RAD | Rapid application development |
| NTFS | NT file system | RADIUS | Remote Access Dial-in User Service |
| NTP | Network Time Protocol | RAID | Redundant Array of Inexpensive Disks |
| OBS | Object breakdown structure | RAM | Random access memory |
| OCSP | Online Certificate Status Protocol | RAS | Remote access service |
| OECD | Organization for Economic Cooperation and Development | RBAC | Role-based access control |
| | | RDBMS | Relational database management system |
| OEP | Occupant emergency plan | RF | Radio frequencies |
| OFDM | Orthogonal Frequency Division Multiplexing | RFI | Request for information |
| OLAP | Online analytical processing | RFP | Request for proposal |
| OO | Object-oriented | RIP | Routing Information Protocol |

| | | | |
|---|---|---|---|
| RMI | Remote method invocation | SPICE | Software Process Improvement and Capability dEetermination |
| ROI | Return on investment | SPOC | Single point of contact |
| ROLAP | Relational online analytical processing | SPOOL | Simultaneous peripheral operations online |
| ROM | Read-only memory | SQL | Structured Query Language |
| RPC | Remote procedure call | SSH | Secure Shell |
| RPO | Recovery point objective | SSID | Service Set IDentifier |
| RSN | Robust secure network | SSO | Single sign-on |
| RST | Reset | SVC | Switched virtual circuits |
| RTO | Recovery time objective | SYSGEN | System generation |
| RTU | Remote terminal unit | TACACS | Terminal Access Control Access Control System |
| RW | Rewritable | TCO | Total cost of ownership |
| S/HTTP | Secure Hypertext Transfer Protocol | TCP/IP | Transmission Control Protocol/Internet Protocol |
| S/MIME | Secure Multipurpose Internet Mail Extensions | TCP/UDP | Transmission Control Protocol/User Datagram Protocol |
| SA | Security Association | TDM | Time-division multiplexing |
| SAN | The SANS Institute | TELNET | Teletype network |
| SANS | SysAdmin, Audit, Network, Security | TES | Terminal emulation software |
| SAS | Statement on Auditing Standards | TFTP | Trivial File Transport Protocol |
| SBC | Session border controller | TKIP | Temporal Key Integrity Protocol |
| SCADA | Supervisory Control and Data Acquisition | TLS | Transport layer security |
| SCARF | Systems Control Audit Review File | TMS | Tape management system |
| SCARF/EAM | Systems Control Audit Review File and Embedded Audit Modules | TP monitors | Transaction processing monitors |
| SCM | Supply chain management | TQM | Total quality management |
| SCOR | Supply chain operations reference | TR | Technical report |
| SD/MMC | Secure digital multimedia card | UAT | User acceptance testing |
| SDLC | System development life cycle | UBE | Unsolicited bulk e-mail |
| SDO | Service delivery objective | UDDI | Description, discovery and integration |
| SEC | Securities and Exchange Commission (USA) | UDP | User Datagram Protocol |
| SET | Secure electronic transactions | UID | User ID |
| SLA | Service level agreement | UML | Unified Modeling Language |
| SLIP | Serial Line Internet Protocol | URI | Uniform resource identifier |
| SLM | Service level management | URL | Uniform resource locator |
| SLOC | Source lines of code | URN | Uniform resource name |
| SMARRT | Specific, measurable, actionable, realistic, results-oriented and timely | USB | Universal Serial Bus |
| SME | Subject matter expert | VLAN | Virtual local area network |
| SMF | System management facility | VoIP | Voice-over IP |
| SMTP | Simple Mail Transport Protocol | VPN | Virtual private network |
| SNA | Systems network architecture | WAP | Wireless Application Protocol |
| SNMP | Simple Network Management Protocol | WBS | Work breakdown structure |
| SO | Security officer | WEP | Wired Equivalent Privacy |
| SOA | Service-oriented architectures | WLAN | Wireless local area network |
| SOAP | Simple Object Access Protocol | WML | Wireless Markup Language |
| SOHO | Small office-home office | WORM | Write Once and Read Many |
| SOW | Statement of work | | |
| SPI | Security parameter index | | |

# Acronyms

| | |
|---|---|
| WP | Work package |
| WPA | Wi-Fi Protected Access |
| WPAN | Wireless personal area network |
| WSDL | Web Services Description Language |
| WWAN | Wireless wide area network |
| WWW | World Wide Web |
| X-to-X | Exchange-to-exchange |
| XBRL | Extensible Business Reporting Language |
| XML | Extensible Markup Language |
| Xquery | XML query |
| XSL | Extensible Stylesheet Language |

# REFERENCES

When studying for the CISA exam, we recommend that a candidate use the references at the end of each chapter for further study.

**This References section at the end of the manual** includes all of the references that appear at the end of each chapter. In addition, we have provided references from ISACA's *Information Systems Control Journal* and references to materials used during the creation of this manual. The additional references are optional, should you need further material when studying for the exam, and may be useful to you in your work experience.

**AESRM, "Imperatives Driving Security Convergence,"** *Information Systems Control Journal,* vol. 4, 2006, p. 33-36

**AlAboodi, Saad Saleh; "A New Approach for Assessing the Maturity of Information Security,"** *Information Systems Control Journal,* vol. 3, 2006, p. 36-42

Allen, Julia; "*Governing for Enterprise Security,*" Carnegie Mellon University, USA, 2005

**Anand, Sanjay; "The Why and How of Leveraging Synergies Across Sarbanes-Oxley and Other Regulations,"** *Information Systems Control Journal,* vol. 3, 2007, p. 49

**Anderson, Kent; "A Business Model for Information Security,"** *Information Systems Control Journal,* vol. 3, 2008, p. 51-52

**Anderson, Kent; "Dysfunctional Operations in IT,"** *Information Systems Control Journal,* vol. 1, 2008, p. 5-6

APM Group Limited (APMG); PRINCE2, *www.prince2.org.uk*

**Ataya, Georges; John Thorp; "Portfolio Management—Unlocking the Value of IT Investments,"** *Information Systems Control Journal,* vol. 4, 2007, p. 20

**Ayaz, Mustafa; "Lessons From a Fraud Case in Turkey,"** *Information Systems Control Journal,* vol. 2, 2008, *www.isacaorg/jonline*

**Axelrod, C. Warren; "Achieving Privacy Through Security Measures,"** *Information Systems Control Journal,* vol. 2, 2007, p. 56

**Axelrod, C. Warren; "Cybersecurity and the Critical Infrastructure: Looking Beyond the Perimeter,"** *Information Systems Control Journal,* vol. 3, 2006, p. 24-28

**Axelrod, C. Warren; "The Dynamics of Privacy Risk,"** *Information Systems Control Journal,* vol. 1, 2007, p. 51

**Bakalov, Rudy and Feisal Nanji; "Offshore Application Development Done Right,"** *Information Systems Control Journal,* vol. 5, 2005, p. 52-56

**Bakalov, Rudy; Feisal Nanji; "Offshore Application Development Done Right,"** *Information Systems Control Journal,* vol. 5, 2005, p. 52-56

**Bakshi, Sunil; "Internal Cyberforensics,"** *Information Systems Control Journal,* vol. 6, 2005, p. 50-53

*Note:* Publications in bold are stocked in the ISACA Bookstore. *Information Systems Control Journal* articles are available at *www.isaca.org/archives*. The articles are available online to ISACA members only during their first year of release, and then are opened to the public. All referenced *Journal* articles are available on the CISA Practice Question Database v9.

Barrett, Neil; *Traces of Guilt*; Transworld Publishing, UK, 2004

**Basham, Robin; "Procedure Guidelines and Controls Documentation: SDLC Controls in CoBiT® 4.0,"** *Information Systems Control Journal***, vol. 6, 2006, p. 43**

**Bassett, Richard A.; Rita Mack; Jason Foster; Andrew Swiatlon; "Security and Ownership of Personal Electronic Devices,"** *Information Systems Control Journal***, vol. 6, 2005, p. 44-48**

**Bayuk, Jennifer L.;** *Stepping Through the IS Audit, 2nd Edition***, ISACA, USA, 2004**

**Bellone, Jason; Dino Cataldo Dell'Accio; Juan Rodriguez; "A Cosmological Approach to IT Governance: Val IT and CoBiT Lessons Learned at the United Nations,"** *Information Systems Control Journal***, vol. 1, 2007, p. 37**

**Benvenuto, Nicholas A.; David Brand; "Outsourcing—A Risk Management Perspective,"** *Information Systems Control Journal***, vol. 5, 2005, p. 35-40**

**Bhatia, Mohan; "IT Merger Due Diligence: A Blueprint,"** *Information Systems Control Journal***, vol. 1, 2007, p. 46**

**Birdi, Tarlok; Kees Jansen; "Network Intrusion Detection: Know What You Do (Not) Need,"** *Information Systems Control Journal***, vol. 1, 2006,** *www.isaca.org/jonline*

**"The Birth of Val IT,"** *Information Systems Control Journal***, vol. 6, 2006, p. 25**

**Blecher, Max; "Outsourcing IT Governance to Deliver Business Value,"** *Information Systems Control Journal***, vol. 4, 2007, p. 13**

**Bloem, Jaap; Menno van Doorn; Piyush Mittal;** *Making IT Governance Work in a Sarbanes-Oxley World***, John Wiley & Sons Inc., USA, 2005**

**Bonham, Stephen S.;** *IT Project Portfolio Management***, Artech House Inc., USA, 2005**

**Bostick, John; "The Soul Searching That Comes With Sole-Sourcing,",** *Information Systems Control Journal***, vol. 5, 2007, p. 47-48**

**Braag, Roberta; Mark Rhode-Ousley; Keith Strassburg;** *The Complete Reference Network Security***, McGraw Hill, USA, 2003**

Bragg, Tom; "Designing and Building Software Projects: Lessons From the Building Trades," Agile Project Management Advisory Service, vol. 3, no. 10, 2002, *www.cutter.com/content/project/fulltext/reports/2002/10/index.html*

**Brancik, Kenneth C.;** *Insider Computer Fraud, Auerbach Publications***, USA, 2007**

**Brennan, Gerard (Rod); "Continuous Auditing Comes of Age,"** *Information Systems Control Journal***, vol. 1, 2008, p. 50-51**

Bromba, Manfred; "Biometrics," Bioidentification, http://home.t-online.de/home/manfred.bromba/biofaqe.htm

**Brooks, Dean; Rich Lanza; "Why Companies Are Not Implementing Audit, Antifraud and Assurance,"** *Information Systems Control Journal***, vol. 1, 2006, p. 30-31**

---

*Note:* Publications in bold are stocked in the ISACA Bookstore. *Information Systems Control Journal* articles are available at *www.isaca.org/archives*. The articles are available online to ISACA members only during their first year of release, and then are opened to the public. All referenced *Journal* articles are available on the CISA Practice Question Database v9.

Brotby, Krag; "Information Security Governance: Who Needs It?," *Information Systems Control Journal*, vol. 2, 2007, p. 13

Brotby, W. Krag; IT Governance Institute; *Information Security Governance: Guidance for Boards of Directors and Executive Management, 2nd Edition*, IT Governance Institute, USA, 2006

Burtles, Jim; *Principles and Practice of Business Continuity*, Tools and Techniques, Rothstein Associates Inc., USA, 2007

Busta, Bruce (Harv); Kris Portz; Joel Strong; Roger Lewis; "Expert Consensus on the Top IT Controls for a Small Business," *Information Systems Control Journal*, vol. 6, 2006, p. 22

Canal, Vicente Aceituno; "Usefulness of an Information Security Management Maturity Model," *Information Systems Control Journal*, vol. 2, 2008, p. 48-51

Carbonel, Jean-Christophe; "Assessing IT Security Governance Through a Maturity Model and the Definition of a Governance Profile," *Information Systems Control Journal*, vol. 2, 2008, p. 29-32

Cascarino, Richard; *Auditor's Guide to Information Systems Auditing*, John Wiley & Sons Inc., USA, 2007

Castella, Stephen; "Foundations for Successful BCP in Your IT Department," Contingency Planning & Management web site, 2003, *www.contingencyplanning.com/Tools/BCPHandbook/BCP102.asp*

Cendrowski, Harry; James P. Martin; Louis W. Petro; *The Handbook of Fraud Deterrence*, John Wiley & Sons Inc., USA, 2007

Cerullo, Michael J; M. Virginia Cerullo; "How the New Standards and Regulations Affect an Auditor's Assessment of Compliance With Internal Controls," *Information Systems Control Journal*, vol. 4, 2005

Champlain, Jack J.; *Practical IT Auditing*, Warren Gorham & Lamont, USA, 2002 Edition with 2007 Supplement

Chatterji, Sushil; "Bridging Business and IT Strategies With Enterprise Architecture: Realising the Real Value of Business-IT Alignment," *Information Systems Control Journal*, vol. 3, 2007, p. 17

Chavan, Umesh; "An Approach to Vulnerability Management," *Information Systems Control Journal*, vol. 5, 2005, *www.isaca.org/jonline*

Chen, T. Y.; Pak-Lok Poon; Sau-Fun Tang; "Applying Testing to Requirements Inspection for Software Quality Assurance," *Information Systems Control Journal*, vol. 6, 2006, p. 50

Cilli, Claudio; "Identity Theft: A New Frontier for Hackers and Cybercrime," *Information Systems Control Journal*, vol. 6, 2005, p. 39-42

Cilli, Claudio; "Privacy: An Opportunity for IS Auditors," *Information Systems Control Journal*, vol. 4, 2005

Cleland, D.I.; Project Management—Strategic Design and Implementation, McGraw Hill, USA, 2006

Cline, Bryan S.; "The Information Security Assessment and Evaluation Methodologies: A DoD Framework for Control Self-assessment," *Information Systems Control Journal*, vol. 2, 2007, *www.isaca.org/jonline*

# References

Cobb, Andrew T.; Jian Guan; Alan S. Levitan; "Control Considerations in Object-oriented Systems," *Information Systems Control Journal*, vol. 3, 2007, p. 28

Coe, Martin; "SAS 70 for Sarbanes-Oxley Compliance," *Information Systems Control Journal*, vol. 5, 2006, p. 48

Council, Claude L.; "Implementing COBIT in Higher Education: Practices That Work Best," *Information Systems Control Journal*, vol. 5, 2006, *www.isaca.org/jonline*

Council, Claude L.; "Implications for the Future of COBIT Systems in Higher Education: Putting Critical Research and Theory Into Practice," *Information Systems Control Journal*, vol. 1, 2007, p. 33

Cuong, Nguyen Huu; "The Need for Legislation Like Sarbanes-Oxley for IT Governance: An Australian Perspective," *Information Systems Control Journal*, vol. 3, 2007, *www.isaca.org/jonline*

Danielyan, Edgar; "IEEE802.11," *The Internet Protocol Journal*, vol. 5, no. 1, March 2002, Cisco, *www.cisco.com/warp/public/759/ipj_5-1.pdf*

Dartmouth AI Conference, "Internet History," *http://livinginternet.com/i/ii_ai.htm*

Davis, Chris; Mike Schiller; Kevin Wheeler; *IT Auditing Using Controls to Protect Information Assets*, McGraw Hill, USA, 2007

Davis, Robert E.; *IT Auditing: The Process*, Pleier Corporation, USA, 2005

De Haes, Steven; David Gilmore; Wim Van Grembergen; Gary Hardy; Alan Simmons; Paul A. Williams; Lighthouse Global; ITGI; *IT Governance Domain Practices and Competencies Series*, IT Governance Institute, USA, 2005

De Haes, Steven; Wim Van Grembergen; "Practices in IT Governance and Business/IT Alignment," *Information Systems Control Journal*, vol. 2, 2008, p. 23-27

Deshmukh, Meera; Raj, Shankar; "Environment Interaction Evaluation: Building Proactive Compliance Into Technology Solutions," *Information Systems Control Journal*, vol. 4, 2007, p. 43

Deshmukh, Meera; Raj Shankar; "Influencer Analysis: A Perspective on Reducing Audit Risk," *Information Systems Control Journal*, vol. 3, 2007, p. 53

Dimitriadis, Christos K.; "Analyzing the Security of Internet Banking Authentication Mechanisms," *Information Systems Control Journal*, vol. 3, 2007, p. 34

Dimitriadis, Christos K.; "Improving Security Management for Mobile Operators," *Information Systems Control Journal*, vol. 4, 2006, p. 28-32

Dodds, Ruppert; "How Does Information Security Fit Into a Governance Framework," *Information Systems Control Journal*, vol. 4, 2005, *www.isaca.org/jonline*

Du, Hui; Chen Zhang; "Risks and Risk Control of Wi-Fi Network Systems," *Information Systems Control Journal*, vol. 4, 2006, p. 38-44

Dubin, Joel; *The Little Black Book of Computer Security, 2nd Edition*, Penton Media Inc. USA, 2008

*Note:* Publications in bold are stocked in the ISACA Bookstore. *Information Systems Control Journal* articles are available at *www.isaca.org/archives*. The articles are available online to ISACA members only during their first year of release, and then are opened to the public. All referenced *Journal* articles are available on the CISA Practice Question Database v9.

Earl, John; "Auditing IBM AS/400 and System I," *Information Systems Control Journal*, vol. 1, 2008, *www.isacaorg/jonline*

Emmett, Mark D.; "Preparation Is the Key Ingredient to a Successful SIM," *Information Systems Control Journal*, vol. 6, 2007, p. 36-40

ERP Knowledgebase, *Information Technology Tool Box*, USA, 2004, *http://erp.ittoolbox.com*

Etges Rafael; Karen McNeil; "Understanding Data Classification Based on Business and Security Requirements," *Information Systems Control Journal*, vol. 5, 2006, *www.isaca.org/jonline*

Fabian, Robert; "Interdependence of COBIT and ITIL," *Information Systems Control Journal*, vol. 1, 2007, p. 32

Federal Reserve Bank of Chicago, *An Internet Banking Primer*, USA, 2005

Finne, Thomas; "Audit and Assurance of Information Systems and Business Processes: A Foundation for Providing Sound Governance Decision Making," *Information Systems Control Journal*, vol. 1, 2006, p. 40-42

Ford, Merilee; H. Kim Lew; Steve Spanier; Tim Stevenson; *Internetworking Technologies Handbook, 3rd Edition*, Cisco Press, USA, 2004

Fulmer, Kenneth L.; *Business Continuity Planning: A Step-by-Step Guide with Planning Forms on CD-ROM*, Rothstein Associates Inc., USA, 2005

Gallegos, Frederick; "Computer Forensics: An Overview," *Information Systems Control Journal*, vol. 6, 2005, p.13-16

Gallegos, Frederick; "Educating the Masses: Audit, Control and Security of Information Systems Today and Tomorrow," *Information Systems Control Journal*, vol. 6, 2004, p. 13-15

Gallegos, Frederick; "Red Teams: An Audit Tool, Technique and Methodology for Information Assurance," *Information Systems Control Journal*, vol. 2, 2006, p. 51-56

Goggins, Kelley; "Contingency Planning 101," Contingency Planning & Management, 2003, *www.contingencyplanning.com/Tools*

Gonzalez, Silka Maria; "New Rules Regarding E-discovery," *Information Systems Control Journal*, vol. 3, 2007, p. 51

Gonzalez, Silka Maria; "SAS 70 Reports—What Do They Really Tell You?," *Information Systems Control Journal*, vol. 2, 2008, *www.isacaorg/jonline*

Graham, Julia; David Kaye; *A Risk Management Approach to Business Continuity*, Rothstein Associates Inc., USA, 2006

Grembergen, Wim Van; Steven De Haes; "COBIT's Management Guidelines Revisited: The KGIs/KPIs Cascade," *Information Systems Control Journal*, vol. 6, 2005, p. 54-56

Grembergen, Wim Van; Steven De Haes; Hilde Van Brempt; "How Does the Business Drive IT? Identifying, Prioritising and Linking Business and IT Goals," *Information Systems Control Journal*, vol. 6, 2007, p. 45-48

Grenough, Jerry; "Tools, Techniques and Tips for IT Auditors: Strategies for Complying With Section 404," *Information Systems Control Journal*, vol. 5, 2006, p. 43

*Note:* Publications in bold are stocked in the ISACA Bookstore. *Information Systems Control Journal* articles are available at *www.isaca.org/archives*. The articles are available online to ISACA members only during their first year of release, and then are opened to the public. All referenced *Journal* articles are available on the CISA Practice Question Database v9.

# References

Guldentops, Erik; " The Rule of Four of IT Governance," *Information Systems Control Journal*, vol. 6, 2007, p. 19

Hamaker, Stacey; "Enterprise Governance and the Role of IT," *Information Systems Control Journal*, vol. 6, 2005, p. 27-30

Handscombe, Kevin; "Continuous Auditing From a Practical Perspective," *Information Systems Control Journal*, vol. 2, 2007, p. 51

Hardy, Gary; "Guidance on Aligning CoBiT, ITIL and ISO 17799," *Information Systems Control Journal*, vol. 1, 2006, p. 32-33

Hardy, Gary; *Lighthouse Global; Information Risks: Whose Business Are They?*, IT Governance Institute, USA, 2005

Hardy, Gary; Erik Guldentops; "CoBiT 4.0: The New Face of CoBiT," *Information Systems Control Journal*, vol. 6, 2005, p. 35-38

Harris, Shon; Allen Harper; Chris Eagle; Jonathan Ness; Michael Lester; *Gray Hat Hacking, 2nd Edition*, McGraw Hill, USA, 2007

Hendrawirawan, David; Huseyin Tanriverdi; Carl Zetterlund; Hunaid Hakam; Hyun Ho Kim; Hyewon Paik; Yeohoon Yoon; "ERP Security and Segregation of Duties Audit: A Framework for Building an Automated Solution," *Information Systems Control Journal*, vol. 2, 2007, p. 46

Hettigei, Nandasena T.; "The Auditor's Role in IT Development Projects," *Information Systems Control Journal*, vol. 4, 2005

Hiles, Andrew; *The Definitive Handbook of Business Continuity Management, 2nd Edition*; John Wiley & Sons Inc., USA, 2007

Huang, Song; "Survivability Strategies for the Next Generation Network," *Information Systems Control Journal*, vol. 6, 2006, p. 19

Humphrey Jr., Robert B.; "Mac OS X: A Secure Platform and a Valuable Security Tool in a Distributed Network Environment," *Information Systems Control Journal*, vol. 2, 2006, p. 43-46

Information Assurance Directorate; "US Government Biometric Verification Mode Protection Profile (PP) for Medium Robustness Environments," 15 November 2003, *www.commoncriteriaportal.org/public/files/ppfiles/pp_vid1022-pp.pdf*

Information Headquarters, "IEEE 802.11 (Wi-Fi)," 2004, *www.informationheadquarters.com/Apple_Macintosh/IEEE_80211b.shtml*

Information Processing Limited, "Software Testing and Software Development Lifecycles," *www.ipl.com/pdf/p0821.pdf*

ISACA, AESRM™, ASIS; "Imperatives Driving Security Convergence," *Information Systems Control Journal*, vol. 4, 2006, p. 33

ISACA, *Cybercrime: Incident Response & Digital Forensics*, USA, 2005

ISACA, *IS Standards, Guidelines and Procedures for Auditing and Control Professionals*, USA, 2007, *www.isaca.org/standards.*

*Note:* Publications in bold are stocked in the ISACA Bookstore. *Information Systems Control Journal* articles are available at *www.isaca.org/archives*. The articles are available online to ISACA members only during their first year of release, and then are opened to the public. All referenced *Journal* articles are available on the CISA Practice Question Database v9.

ISACA, *ITAF™: A Professional Practices Framework for IT Assurance*, USA, 2008

ISACA, Security, *Audit and Control Features SAP® R/3®—A Technical and Risk Management, 2nd Edition*, USA, 2006

IT Governance Institute, *Board Briefing on IT Governance, 2nd Edition*, USA, 2003

IT Governance Institute, CoBiT 4.1, USA, 2007

IT Governance Institute, *Enterprise Value: Governance of IT Investments*, Complete Set, USA, 2008

IT Governance Institute, *Enterprise Value: Governance of IT Investments*, The Business Case, USA, 2006

IT Governance Institute, *Enterprise Value: Governance of IT Investments*, The Val IT Framework 2.0, 2008

IT Governance Institute, *IT Control Objectives for Sarbanes-Oxley, 2nd Edition*, USA, 2008, *www.isaca.org/sox*

IT Governance Institute, *IT Governance Global Status Report 2006*, USA, 2008

IT Governance Institute, *Measuring and Demonstrating the Value of IT*, USA, 2005

Jamal, Nazam; "Containing Corporate Governance Costs: The Role of Technology," *Information Systems Control Journal*, vol. 2, 2006, p. 21-24

Janssens, Laurent; "Auditing CMMI® Maturity and Sarbanes-Oxley Compliance," *Information Systems Control Journal*, vol. 3, 2007, *www.isaca.org/jonline*

Jaquith, Andrew; *Security Metrics*, Addison Wesley, USA, 2007

Jogani, Anil; "Governance of Mobile Technology in Enterprises," *Information Systems Control Journal*, vol. 4, 2006, p. 25-27

Johnson, Everett C.; "Shedding Light on Executives' View of IT Governance," *Information Systems Control Journal*, vol. 4, 2006, p. 50

Johnstone, Dale; Ellis Chung Yee Wong; "Practicing Information Technology Auditing for Fraud," *Information Systems Control Journal*, vol. 1, 2008, p. 29-32

**Kairab, Sudhanshu; A Practical Guide to Security Assessments, Auerbach Publications, USA, 2004**

Kanter, Howard A.; Thomas J. Muscarello; Christopher Ralston; "Measuring the Readability of Software Requirement Specifications: An Empirical Study," *Information Systems Control Journal*, vol. 1, 2008, p. 40-45

Kaur, Jasber; Yap May Lin; Ainon Zarina Mohamed Nadzri; "IS Auditing Standards in Malaysia," *Information Systems Control Journal*, vol. 1, 2008, *www.isaca.org/jonline*

Kennedy, Susan; "Common Web Application Vulnerabilities," *Information Systems Control Journal*, vol. 4, 2005, p. 29-31

Kothari, Priyank; "Is Your Business Continuity Plan a Paper Tiger?," *Information Systems Control Journal*, vol. 3, 2007, p. 25

*Note:* Publications in bold are stocked in the ISACA Bookstore. *Information Systems Control Journal* articles are available at *www.isaca.org/archives*. The articles are available online to ISACA members only during their first year of release, and then are opened to the public. All referenced *Journal* articles are available on the CISA Practice Question Database v9.

Kumar, Parimal; "The Technological Auditor: How Automation Is Changing Auditors' Roles," *Information Systems Control Journal*, vol. 5, 2006, p. 41

**Lahti, Christian B.; Roderick Peterson;** *Sarbanes-Oxley IT Compliance Using Open Source Tools, 2nd Edition*, Syngress Publishing Inc., Canada, 2007

Lambeth, John; "Using CobiT as a Tool to Lead Enterprise IT Organizations," *Information Systems Control Journal*, vol. 1, 2007, p. 28

Lawton, Robert; "Transitioning IT From a Compliance to a Value-driven Enterprise Using CobiT," *Information Systems Control Journal*, vol. 6, 2007, p. 43-44

Leo, Martin; "End Point Security," *Information Systems Control Journal*, vol. 2, 2008, p. 33-35

Livingston, John; "Tips and Strategies to Protect Laptops and the Sensitive Data They Contain," *Information Systems Control Journal*, vol. 5, 2007, p. 42-44

Lonsdale, Derek; Wendy Clark; Bina Udvadia; "ITIL in a Complex World: Focusing on Success in a Multisourced Environment," *Information Systems Control Journal*, vol. 1, 2006, p. 38-39

MacLeod, Calum; "One of Today's Most Overlooked Security Threats—Six Ways Auditors Can Fight It," *Information Systems Control Journal*, vol. 5, 2007, *www.isaca.org/jonline*

**Maizlish, Bryan; Robert Handler;** *IT Portfolio Management Step-by-Step: Unlocking the Business Value of Technology*, John Wiley & Sons, USA, 2005

Malik, William J., "Information Security Governance," *Information Systems Control Journal*, vol. 3, 2006, p. 23

**Marcella Jr., Albert J.; Dough Menendez;** *Cyber Forensics, 2nd Edition*, Auerbach Publications, USA, 2008

**McClure, Stuart; Joel Sambray; George Kurtz;** *Hacking Exposed, 5th Edition*, McGraw Hill, USA, 2005

McKinney, Charles; "Capability Maturity Models and Outsourcing: A Case for Sourcing Risk Management," *Information Systems Control Journal*, vol. 5, 2005, p. 28-34

Melanon, Dwayne; "Security Controls That Work," *Information Systems Control Journal*, vol. 4, 2007, p. 29

Micallef, Mario; "The New World of Risk-based Regulation (Part 1)," *Information Systems Control Journal*, vol. 6, 2007, p. 30-35

Micallef, Mario; "The New World of Risk-based Regulation (Part 2)," *Information Systems Control Journal*, vol. 1, 2008, p. 52-56

Milligan, Patricia Mayer; Donna Hutcheson; "Business Risks and Security Assessment for Mobile Devices,", *Information Systems Control Journal*, vol. 1, 2008, p. 24-28

**Mitnick, Kevin D.; William L. Simon;** *The Art of Intrusion: The Real Stories Behind the Exploits of Hackers, Intruders & Deceivers*, Wiley Publishing Inc., USA, 2005

*Note:* Publications in bold are stocked in the ISACA Bookstore. *Information Systems Control Journal* articles are available at *www.isaca.org/archives*. The articles are available online to ISACA members only during their first year of release, and then are opened to the public. All referenced *Journal* articles are available on the CISA Practice Question Database v9.

**Moeller, Robert; "Improving Regulatory Compliance: How to Make Content Protection Controls Effective,"** *Information Systems Control Journal,* **vol. 2, 2007, p. 24**

**Mohanty, Bapi; "Embedding Policy Governance Within Business Processes,"** *Information Systems Control Journal,* **vol. 5, 2006, p. 53**

Mohay, George; Alison Anderson; Byron Collie; Olivier de Vel; D. Rodney McKemmish; *Computer and Intrusion Forensics,* UK, 2004

**Moseley, Marty; "Keys to Data Governance Success: Teamwork and an Iterative Approach,"** *Information Systems Control Journal,* **vol. 3, 2008, p. 37-39**

**Musaji, Yusuf; "A Holistic Definition of IT Security—Part 1,"** *Information Systems Control Journal,* **vol. 3, 2006, p. 43-46**

**Musaji, Yusuf; "A Holistic Definition of IT Security—Part 2,"** *Information Systems Control Journal,* **vol. 4, 2006, p. 51-56**

**Musaji, Yusuf; "Sarbanes-Oxley and Business Process Outsourcing Risk,"** *Information Systems Control Journal,* **vol. 5, 2005, p. 47-49**

**Musaji, Yusuf; "The SAP Landscape That Warrants Audit,"** *Information Systems Control Journal,* **vol. 1, 2006,** *www.isaca.org/jonline*

**Muthukrishnan, Ravi; "The Auditor's Role in Reviewing Business Continuity Planning,"** *Information Systems Control Journal,* **vol. 4, 2005, p. 52-56**

**Natan, Ron Ben;** *Implementing Database Security and Auditing,* **Elsevier Digital Press, USA, 2005**

National Emergency Management Association; "National Incident Management System," USA, 2004, *http://nemaweb.org/docs/NIMS%20-%20Final%20Draft.pdf*

National Institute of Justice; "Electronic Crime Scene Investigation: A Guide for First Responders," USA, p. 98, *www.ncjrs.org/pdffiles1/nij/187736.pdf*

National Institute of Standards and Technology, "Integration Definition for Information Modeling," USA, 2005, *www.nist.gov*

National Institute of Standards and Technology, "Security Considerations for Voice Over IP Systems," USA, 2005

**Nolan, John; "Best Practices for Establishing an Effective Workplace Policy for Acceptable Computer Usage,"** *Information Systems Control Journal,* **vol. 6, 2005, p. 32-34**

**O'Bryan, Sharon K.; "Critical Elements of Information Security Program Success,"** *Information Systems Control Journal,* **vol. 3, 2006, p. 29-31**

Office of Internal Audit Best Practices, "Internal Controls," Wayne State University, USA, 2005, *http://internalaudit.wayne.edu/Internal/Auditing%20Best%20Practices.doc*

*Note:* **Publications in bold are stocked in the ISACA Bookstore.** *Information Systems Control Journal* **articles are available at** *www.isaca.org/archives.* **The articles are available online to ISACA members only during their first year of release, and then are opened to the public. All referenced** *Journal* **articles are available on the CISA Practice Question Database v9.**

Olatilu Olu; "Identity Theft and Corporations' Due Diligence," *Information Systems Control Journal*, vol. 6, 2006, p. 27

Oliver, Derek; "MP3 Players: Today's Business Threat," *Information Systems Control Journal*, vol. 4, 2007, p. 36

Oseni, Ezekiel; "Change Management in Process Change," *Information Systems Control Journal*, vol. 1, 2007, *www.isaca.org/jonline*

Ott, John; Andrew MacLeod; Kevin Mar Fan ; "Computer-assisted Audit Techniques: Value of Data Mining for Corporate Auditors," *Information Systems Control Journal*, vol. 3, 2008, p. 45-48

Panko, Raymond R.; "Controlling Spreadsheets," *Information Systems Control Journal*, vol. 1, 2007, *www.isaca.org/jonline*

Pareek, Mukul; "Automating Controls," *Information Systems Control Journal*, vol. 3, 2007, p. 45

Pareek, Mukul; "IT Governance and Post-merger Systems Integration," *Information Systems Control Journal*, vol. 2, 2005, p. 30-33

Pareek, Mukul; "Living With Risk," *Information Systems Control Journal*, vol. 6, 2006, p. 35

Pareek, Mukul; "Optimizing Controls to Test as Part of a Risk-based Audit Strategy," *Information Systems Control Journal*, vol. 2, 2006, p. 39-41

Parkes, Hugh; "Shifting Governance Roles and Responsibilities: Improving Management Reporting as Part of Corporate and IT Governance," *Information Systems Control Journal*, vol. 5, 2006, p. 25

Pauls, Nicole; "Security Information Management: Not Just the Next Big Thing," *Information Systems Control Journal*, vol. 5, 2005, *www.isaca.org/jonline*

Peltier, Thomas R.; *Information Security Risk Analysis, Second Edition*, Auerbach Publications, USA, 2005

Pfister, John; "Case Study: Reducing the Cost of Sarbanes-Oxley Compliance Through Common Controls," *Information Systems Control Journal*, vol. 5, 2006, p. 34

Pironti, John P.; "Developing Metrics for Effective Information Security Governance," *Information Systems Control Journal*, vol. 2, 2007, p. 33

Pironti, John P.; "Information Security Governance: Motivations, Benefits and Outcomes," *Information Systems Control Journal*, vol. 4, 2006, p. 45-48

Pironti, John P.; "Key Elements of a Threat and Vulnerability Management Program," *Information Systems Control Journal*, vol. 3, 2006, p. 52-56

Pironti, John P.; "Key Elements of an Information Risk Management Program: Transforming Information Security Into Information Risk Management," *Information Systems Control Journal*, vol. 2, 2008, p. 42-47

Poole, Vernon; "Why Information Security Governance Is Critical to Wider Corporate Governance Demands—A European Perspective," *Information Systems Control Journal*, vol. 1, 2006, p. 23-25

Project Management Forum (PMFORUM), *www.pmforum.org*

*Note:* Publications in bold are stocked in the ISACA Bookstore. *Information Systems Control Journal* articles are available at *www.isaca.org/archives*. The articles are available online to ISACA members only during their first year of release, and then are opened to the public. All referenced *Journal* articles are available on the CISA Practice Question Database v9.

Project Management Institute (PMI), *A Guide to the Project Management Body of Knowledge (PMBOK Guide)*, USA, 2004

Project Management Institute (PMI), *www.pmi.org*

**Protzman, Kristine M.; Vasant Raval; "Concept Mapping—A Learning Tool for the Information Systems Audit Profession,"** *Information Systems Control Journal*, vol. 3, 2006, *www.isaca.org/jonline*

Public Company Accounting Oversight Board, Auditing Standard No. 2, "An Audit of Internal Control Over Financial Reporting Conducted in Conjunction With an Audit of Financial Statements," USA, 2004

Purdue University, "Computer Integrated Manufacturing Technology," USA, 2004, *www.tech.purdue.edu/cimt/*

**Rafeq, A.; "Using COBIT for IT Control Health Check-up,"** *Information Systems Control Journal*, vol. 5, 2005, p. 18-19

**Ramanathan, Ramanan R.; "E-business: Trust Inhibitors, "** *Information Systems Control Journal*, vol. 2, 2008, *www.isacaorg/jonline*

**Ramanathan, S.; "IT Governance—Challenges in Implementation From an Asian Perspective,"** *Information Systems Control Journal*, vol. 5, 2007, p. 26-27

**Ramirez, David; "Security Within VoIP Networks,"** *Information Systems Control Journal*, vol. 6, 2007, p. 41-42

**Ramirez, David; "Streamline ISO 27001 Implementation: Reducing the Time and Effort Required for Compliance",** *Information Systems Control Journal*, vol.5, 2006, p. 51

**Ramos, Michael J.;** *How to Comply with Sarbanes-Oxley Section 404, 3rd Edition*, John Wiley & Sons Inc., USA, 2008

**Rao, Sree Krishna; "Clause 49: An Opportunity for Indian-listed Corporations to Achieve IT Governance,"** *Information Systems Control Journal*, vol. 1, 2007, *www.isaca.org/jonline*

**Raval, Vasant; Ashok Fichdia;** *Risks, Controls, and Security: Concepts and Applications*, John Wiley & Sons, USA, 2007, Chapter 6: System Availability and Business Continuity

**Rittinghouse, John; James E. Ransome;** *Wireless Operational Security*, Elsevier Digital Press, USA, 2004

**Robinson, Nick; "The Many Faces of IT Governance: Crafting an IT Governance Architecture,"** *Information Systems Control Journal*, vol. 1, 2007, p. 14

**Rollins, Steven; Richard Lanza;** *Essential Project Investment Governance and Reporting Preventing Project Fraud and Ensuring Sarbanes-Oxley Compliance*, J. Ross Publishing, USA, 2005

**Rosenberg, Beth; "Telephone Line Scanning Still Matters,"** *Information Systems Control Journal*, vol. 6, 2006, *www.isaca.org/jonline*

**Ross, Steven J.; "Compliance and Beyond,"** *Information Systems Control Journal*, vol. 4, 2007, p. 5

**Ross, Steven J.; "Contents and Context,"** *Information Systems Control Journal*, vol. 1, 2006, p. 8-9

**Ross, Steven J; "Converging Need, Diverging Response,"** *Information Systems Control Journal*, vol. 2, 2006, p. 8-9

# References

Ross, Steven J.; "Downtime and Data Loss," *Information Systems Control Journal*, vol. 5, 2006, p. 11

Ross, Steven J.; "Falling Off the Truck," *Information Systems Control Journal*, vol. 3, 2006, p. 9-10

Ross, Steven J.; "Give 'em the New Razzle-Dazzle," *Information Systems Control Journal*, vol. 5, 2005, p. 9-10

Ross, Steven; "I'm Not the Sheriff," *Information Systems Control Journal*, vol. 6, 2007, p. 9-10

Ross, Steven J.; "Information Systems in the Time of Flu," *Information Systems Control Journal*, vol. 4, 2006, p. 11-12

Ross, Steven J.; "IS Security Matters: The Right Question," *Information Systems Control Journal*, vol. 4, 2005

Ross, Steven J.; "Resilience Transformation," *Information Systems Control Journal*, vol. 3, 2008, p.,9-10

Ross, Steven J.; "The Resilient Toothbrush," *Information Systems Control Journal*, vol. 2, 2008, p.,9-10

Ross, Steven J.; "Standard Deviation," *Information Systems Control Journal*, vol. 6, 2005, p. 9-10

SABSA® (Sherwood Applied Business Security Architecture), Enterprise Security Architecture, UK, 2008, *www.sabsa-institute.org*

Sandrino-Arndt, Bop; "People, Portfolios and Processes: The 3P Model of IT Governance," *Information Systems Control Journal*, vol. 2, 2008, p. 36-40

Sardanopoli, Vito; "CISAs and CISMs Working in Sync: How Their Individual Contributions Together Can Achieve Effective IT Risk Management," *Information Systems Control Journal*, vol. 2, 2008, p. 52-53

Sarva, Srinivas; "Continuous Auditing Through Leveraging Technology," *Information Systems Control Journal*, vol. 2, 2006, p. 47-50

Sayana, S. Anantha; "Auditing IT Service Delivery," *Information Systems Control Journal*, vol. 5, 2005, p. 13-14

Schaafsma, Siep; Paul Williams; "Empirical Research Into Val IT Supports the Use of COBIT," *Information Systems Control Journal*, vol. 5, 2007, p. 31-32

Schneier, Bruce; *Secrets & Lies: Digital Security in a Networked World*, John Wiley & Sons, USA, 2004

Senft, Sandra; Daniel P. Manson; Carol Gonzales; Frederick Gallegos; *Information Technology Control and Audit*, 3rd *Edition*, Auerbach, USA, 2008

Sethuraman, Sekar; "Framework for Measuring and Reporting Performance of Information Security Programs in Offshore Outsourcing," *Information Systems Control Journal*, vol. 6, 2006, *www.isaca.org/jonline*

Sethuraman, Sekar; "Turning a Security Compliance Program Into a Competitive Business Advantage," *Information Systems Control Journal*, vol. 5, 2007, *www.isaca.org/jonline*

Shaikh, Junaid M.; Shaharudin Jakpar; "Dispelling and Construction of Social Accounting in View of Social Audit," *Information Systems Control Journal*, vol. 2, 2007, p. 39

*Note:* Publications in bold are stocked in the ISACA Bookstore. *Information Systems Control Journal* articles are available at *www.isaca.org/archives*. The articles are available online to ISACA members only during their first year of release, and then are opened to the public. All referenced *Journal* articles are available on the CISA Practice Question Database v9.

Sharkasi, Omar Y.; "Who Audits the IT Auditor: Monitoring IT Audit Tasks," *Information Systems Control Journal*, vol. 5, 2007, *www.isaca.org/jonline*

Shue, Lily; "Guidance on Tax Compliance for Business and Accounting Software and SAF-T," *Information Systems Control Journal*, vol. 2, 2006, p. 31-32

Simmonds, Alan; David Gilmore; Lighthouse Global; It Governance Institute; *Governance of Outsourcing*, IT Governance Institute, USA, 2005

Singh, Jaspreet; "Pay Today or Pay Later—Calculating ROI to Justify Information Security and Compliance Budgets," *Information Systems Control Journal*, vol. 3, 2008, p. 49-50

Singh-Latulipe, Rob; "Val IT: From the Vantage Point of the COBIT 4.0 Pentagon Model for IT Governance," *Information Systems Control Journal*, vol. 1, 2007, p. 20

Singleton, Tommie W.; "COBIT—A Key to Success as an IT Auditor," *Information Systems Control Journal*, vol. 1, 2006, p. 11-13

Singleton, Tommie W.; "Emerging Technical Standards on Financial Audits: How IT Auditors Gather Evidence to Evaluate Internal Controls," *Information Systems Control Journal*, vol. 4, 2007, p. 9

Singleton, Tommie W.; "IT Audit Education and Professional Development," *Information Systems Control Journal*, vol. 3, 2007, p. 13

Singleton, Tommie W.; "Systems Development Life Cycle and IT Audits," *Information Systems Control Journal*, vol. 1, 2007, p. 24

Singleton, Tommie; "The COSO Model: How IT Auditors Can Use It to Evaluate the Effectiveness of Internal Controls," *Information Systems Control Journal*, vol. 6, 2007, p. 13-15

Singleton, Tommie; "The COSO Model: How IT Audtiors Can Use IT to Measure the Effectiveness of Internal Controls (Part 2)," *Information Systems Control Journal*, vol. 1, 2008, p. 9-10

Singleton, Tommie W.; "What Every IT Auditor Should Know About Cybercrimes," *Information Systems Control Journal*, vol. 2, 2008, p. 13-14

Singleton, Tommie W.; "What Every IT Auditor Should Know About Identity Theft," *Information Systems Control Journal*, vol. 6, 2006, p. 12

Singleton, Tommie W.; "What Every IT Auditor Should Know About Project Risk Management," *Information Systems Control Journal*, vol. 5, 2006, p. 17

Singleton, Tommie W.; "What Every IT Auditor Should Know About the New Risk Suite Standards," *Information Systems Control Journal*, vol. 5, 2007, p. 13-15

Singleton, Tommie W.; "What Every IT Auditor Should Know About Wireless Telecommunication," *Information Systems Control Journal*, vol. 4, 2006, p. 19-21

Snedaker, Susan; *Business Continuity & Disaster Recovery*, Syngress Publishing Inc., USA, 2007

*Note:* Publications in bold are stocked in the ISACA Bookstore. *Information Systems Control Journal* articles are available at *www.isaca.org/archives*. The articles are available online to ISACA members only during their first year of release, and then are opened to the public. All referenced *Journal* articles are available on the CISA Practice Question Database v9.

Srinivasan, S.; "Information Security Policies and Controls for a Trusted Environment," *Information Systems Control Journal*, vol. 2, 2008, p. 54-56

Staley, A. Blair; Scott Inch; Mike Shapeero; "From CSI to the Classroom: Developing a Computer Forensics Degree Program," *Information Systems Control Journal*, vol. 6, 2006, p. 39

Stamp, Mark; *Information Security: Principles and Practice*, John Wiley & Sons, USA, 2005

Stanley, Richard A.; *Managing Risk in the Wireless Environment: Security*, Audit and Control Issues, ISACA, USA, 2005

Strasser, Mathias; Edgar Weippl; "Sarbanes-Oxley Act Compliance: Strategies for Implementing an Audit Committee Complaints Procedure," *Information Systems Control Journal*, vol. 5, 2006, *www.isaca.org/jonline*

Sundberg, Barrett; Zheng-Da Tan; Trey Baublits; Hae-Joon Lee; Grant Stanis; Huseyin Tanriverdi; "A Framework for Conducting IT Due Diligence in Mergers and Acquisitions," *Information Systems Control Journal*, vol. 6, 2006, *www.isaca.org/jonline*

Swaroop, Shankar; "Managing Multiple Medium- and Small-scale Projects in Large IT Organizations," *Information Systems Control Journal*, vol. 4, 2007, p. 22

Sweren, Scott H.; "ISO 17799: Then, Now and in the Future," *Information Systems Control Journal*, vol. 1, 2006, p. 34-37

Talpade, Rajesh; Jatinder, Singh; "Compliance Assessment of IP Networks: A Necessity Today," *Information Systems Control Journal*, 4, 2007, p. 51

Tarantino, Anthony; *Manager's Guide to Compliance*, John Wiley & Sons Inc., USA, 2006

Taylor, Patrick; "Insider Threat—The Fraud That Puts Companies at Risk," *Information Systems Control Journal*, vol. 1, 2008, p. 46-47

Tharp, Tom; "The Unique Benefits and Risks of USB Mass Storage Devices," *Information Systems Control Journal*, vol. 2, 2007, p. 28

Thomas, Brian; "Data Warehouse Audits Promote and Sustain Reporting System Value," *Information Systems Control Journal*, vol. 1, 2008, *www.isaca.org/jonline*

Thorp, John; "Program Management: Seeing Both the Forest and the Trees," *Information Systems Control Journal*, vol. 3, 2007, p. 20

Thorp, John; "Value Management—Responding to the Challenge of Value," *Information Systems Control Journal*, vol. 5, 2006, p. 21

Trivedi, Tejus; "10 Things to Consider When Offshoring Operations," *Information Systems Control Journal*, vol. 5, 2007, p. 49-50

Trustweaver, "Legal Aspects of PKI: Fact Sheet," 31 August 2004, *www.tekki.se/docs/lbwp.pdf*

Tryfonas, Theodore; Paula Thomas; Paul Owen; "ID Theft: Fraudster Techniques for Personal Data Collection, the Related Digital Evidence and Investigation Issues," *Information Systems Control Journal*, vol. 1, 2006

*Note:* Publications in bold are stocked in the ISACA Bookstore. *Information Systems Control Journal* articles are available at *www.isaca.org/archives*. The articles are available online to ISACA members only during their first year of release, and then are opened to the public. All referenced *Journal* articles are available on the CISA Practice Question Database v9.

Tryfonas, Theodore; Paula Thomas; Paul Owen; "ID Theft: Fraudster Techniques for Personal Data Collection, the Related Digital Evidence and Investigation Issues," *Information Systems Control Journal*, vol. 1, 2006, *www.isaca.org/jonline*

Tschanz, David W.; "Stop Spam Now," *Redmondmag.com*, March 2005, *www.cmsconnect.com/News/CMSInPrint/Redmond-050302.htm*

**Tudor, Jan Killmeyer;** *Information Security Architecture: An Integrated Approach to Security in the Organization, 2nd Edition*, **Auerbach Publications, USA, 2005**

**Ungerman, Mark; "Creating and Enforcing an Effective Information Security Policy,"** *Information Systems Control Journal*, **vol. 6, 2005,** *www.isaca.org/jonline*

**Unwala, Huzeifa; "Information Assurance—Online Lottery Systems,"** *Information Systems Control Journal*, **vol. 3, 2006,** *www.isaca.org/jonline*

**Unwala, Huzeifa; Milind Dharmadhikari; "The Unconquered Monitoring Frontier: Time for a Real-time Command Center,"** *Information Systems Control Journal*, **vol. 1, 2008, p. 35-37**

**Vacca, John R.;** *Biometric Technologies and Verification Systems*, **Elsevier Inc., USA, 2007**

**Van Grembergen, Wim; Steven De Haes; Jan Moons; "IT Governance: Linking Business Goals to IT Goals and COBIT Processes,"** *Information Systems Control Journal*, **vol. 4, 2005**

**Velichety, Swapna; Jeongil Park ; Sungkun Jung ; Sooyoun Lee; Hüseyin Tanriverdi; " Company Perspectives on Business Value of IT Investments in Sarbanes-Oxley Compliance,"** *Information Systems Control Journal*, **vol. 1, 2007, p. 42**

**Ven, Kris; Steven De Haes; Wim Van Grembergen; Jan Verelst; "Using COBIT 4.1 to Guide the Adoption and Implementation of Open Source Software,"** *Information Systems Control Journal*, **vol. 3, 2008, p. 31-35**

**Wandrei, Philip L.; "Maximizing Backup and Recovery of Data and Systems,"** *Information Systems Control Journal*, **vol. 3, 2007, p. 42**

Wells, April; Charlyne Walker; Timothy Walker; *Disaster Recovery: Principles and Practices*, USA, Pearson-Prentice Hall, 2007.

**Wells, Joseph T.;** *Fraud Casebook, Lessons from the Bad Side of Business*, **John Wiley & Sons Inc., USA, 2007**

Wikimedia Foundation, "ACID (Atomicity, Consistency, Isolation and Durability)," Wikipedia, *www.wikipedia.org/wiki/ACID*

Wikimedia Foundation, *Wikipedia, www.wikipedia.com*

Wikipedia, "Artificial Intelligence," *http://en.wikipedia.org/wiki/Artificial_intelligence*

**Williams, Paul; "Optimising Returns From IT-related Bussiness Investments,"** *Information Systems Control Journal*, **vol. 5, 2005, p. 41-45**

*Note:* Publications in bold are stocked in the ISACA Bookstore. *Information Systems Control Journal* articles are available at *www.isaca.org/archives*. The articles are available online to ISACA members only during their first year of release, and then are opened to the public. All referenced *Journal* articles are available on the CISA Practice Question Database v9.

# References

Williams, Paul; John Spangenberg; Sonja Kovaleva; "IT and Shareholder Return: Creating Value in the Insurance Industry," *Information Systems Control Journal*, vol. 4, 2007, p. 39

Williams, Paul; "Executive and Board Roles in Information Security," *Information Systems Control Journal*, vol. 6, 2007, *www.isaca.org/jonline*

Woda, Alex; "Achieving Compliance With the PCI Data Security Standard," *Information Systems Control Journal*, vol. 4, 2007, p. 46

Wulgaert, Tim; *Security Awareness: Best Practices to Secure Your Enterprise*, ISACA, USA, 2005

Wysocki, Robert K.; *Effective Project Management*, Wiley Publishing Inc., USA, 2007

*Note:* Publications in bold are stocked in the ISACA Bookstore. *Information Systems Control Journal* articles are available at *www.isaca.org/archives*. The articles are available online to ISACA members only during their first year of release, and then are opened to the public. All referenced *Journal* articles are available on the CISA Practice Question Database v9.

# THE CISA EXAM AND COBIT

COBIT 4.1 is an initiative conducted by the IT Governance Institute. COBIT has been developed as a generally applicable and accepted framework for good IT security and control practices that provide a reference for management, users, and IS audit, control and security practitioners. COBIT is based on ITGI's control objectives, enhanced with existing and emerging international technical, professional, regulatory and industry-specific standards. The resulting control objectives have been developed for application to organizationwide information systems.

COBIT also supports a generic IT assurance/audit process which can be summarized as:
• Obtaining an understanding of business requirements, related risks and relevant control measures
• Evaluating the appropriateness of stated controls
• Assessing compliance by testing whether the stated controls are working: as prescribed, consistently and continuously
• Substantiating the risk of control objectives not being met by using analytical techniques and/or consulting alternative sources

Although knowledge of COBIT is not specifically tested on the CISA exam, the COBIT control objectives or processes reflect the tasks identified in the CISA job practice. As such, a thorough review of COBIT is recommended for candidate preparation for the CISA exam. To focus a candidate's attention on the specific COBIT processes that relate to CISA practice analysis tasks, the following table has been provided to aid in a candidate's exam preparation.

Note: The COBIT framework is available at no charge from ISACA/ITGI and can be downloaded at *www.isaca.org/cobit*.

To focus a candidate's attention on the specific COBIT processes that relate to CISA practice analysis tasks, the following table has been provided to aid in a candidate's exam preparation.

## Chapter 1:  The IS Audit Process

| CISA Review Manual | | COBIT 3rd Edition | COBIT 4.0 and 4.1 |
|---|---|---|---|
| **Tasks** | | **COBIT Processes** | |
| 1.1 | Develop and implement a risk-based IS audit strategy for the organization in compliance with IS audit standards, guidelines and best practices. | PO9 *Assess risk* <br> M3 *Obtain independent assurance* <br> M4 *Provide for independent audit* | PO9 *Assess and manage IT risks* <br> ME2 *Monitor and evaluate internal control* |
| 1.2 | Plan specific audits to ensure that IT and business systems are protected and controlled. | M3 *Obtain independent assurance* <br> M4 *Provide for independent audit* | PO9 *Assess and manage IT risks* <br> ME2 *Monitor and evaluate internal control* |
| 1.3 | Conduct audits in accordance with IS audit standards, guidelines and best practices to meet planned audit objectives. | M4 *Provide for independent audit* | ME2 *Monitor and evaluate internal control* |
| 1.4 | Communicate emerging issues, potential risks and audit results to key stakeholders. | M3 *Obtain independent assurance* <br> M4 *Provide for independent audit* | PO8 *Manage quality* <br> PO9 *Assess and manage IT risks* |
| 1.5 | Advise on the implementation of risk management and control practices within the organization while maintaining independence. | PO9 *Assess risk* <br> PO11 *Manage quality* <br> M3 *Obtain independent assurance* <br> M4 *Provide for independent audit* | |

## Chapter 2: IT Governance

| CISA Review Manual | CobiT 3rd Edition | CobiT 4.0 and 4.1 |
|---|---|---|
| **Tasks** | **CobiT Processes** | |
| 2.1 Evaluate the effectiveness of the IT governance structure to ensure adequate board control over the decisions, directions and performance of IT so that it supports the organization's strategies and objectives. | PO1 Define a strategic plan<br>PO4 Define the IT organization and relationship<br>PO5 Manage the IT investment<br>PO6 Communicate management aims and directions<br>M2 Assess internal control adequacy<br>M3 Obtain independent assurance<br>M4 Provide for independent audit | PO1 Define a strategic plan<br>PO4 Define the IT processes, organization and relationships<br>PO5 Manage the IT investment<br>PO6 Communicate management aims and directions<br>ME4 Provide IT governance |
| 2.2 Evaluate IT organizational structure and human resources (personnel) management to ensure that they support the organization's strategies and objectives. | PO4 Define the IT organization and relationships<br>PO7 Manage human resources<br>DS1 Define and manage service levels | PO4 Define the IT processes, organization and relationships<br>PO7 Manage IT human resources<br>DS1 Define and manage service levels |
| 2.3 Evaluate the IT strategy and the process for its development, approval, implementation and maintenance to ensure that it supports the organizations strategies and objectives. | PO1 Define a strategic IT plan<br>PO5 Manage the IT investment | PO1 Define a strategic IT plan<br>PO5 Manage the IT investment |
| 2.4 Evaluate the organization's IT policies, standards, procedures and processes for their development, approval, implementation, and maintenance to ensure that they support the IT strategy and comply with regulatory and legal | PO8 Ensure compliance with external requirements<br>AI6 Manage changes<br>M1 Monitor the processes | AI6 Manage changes<br>ME1 Monitor and evaluate performance<br>ME3 Ensure regulatory compliance (4.0)<br>ME3 Ensure compliance with external requirements (4.1) |
| 2.5 Evaluate management practices to ensure compliance with the organization's IT strategy, policies, standards and procedures. | PO6 Communicate management aims and direction<br>PO7 Manage human resources<br>PO10 Manage project<br>PO11 Manage quality<br>DS6 Identify and allocate costs | PO6 Communicate management aims and direction<br>PO7 Manage IT human resources<br>PO8 Manage quality<br>PO10 Manage projects<br>DS6 Identify and allocate costs |
| 2.6 Evaluate IT resource investment, use and allocation practices to ensure alignment with the organization's strategies and objectives. | PO5 Manage the IT investment<br>PO10 Manage projects | PO5 Manage the IT investment<br>PO10 Manage projects |
| 2.7 Evaluate IT contracting strategies and policies and contract management practices to ensure that they support the organization's strategies and objectives. | PO7 Manage human resources<br>PO8 Ensure compliance with external requirements<br>AI1 Identify automated solutions<br>DS2 Manage third-party services<br>DS9 Manage the configuration | PO7 Manage IT human resources<br>AI1 Identify automated solutions<br>DS2 Manage third-party services<br>DS9 Manage the configuration<br>ME3 Ensure regulatory compliance (4.0)<br>ME3 Ensure compliance with external requirements (4.1) |

## Chapter 2: IT Governance (cont.)

| CISA Review Manual | | CoBiT 3rd Edition | | CoBiT 4.0 and 4.1 | |
|---|---|---|---|---|---|
| **Tasks** | | **CoBiT Processes** | | | |
| 2.8 | Evaluate risk management practices to ensure that the organization's IT related risks are properly managed. | PO1 | Define a strategic IT plan | PO1 | Define a strategic IT plan |
| | | PO6 | Communicate management aims and directions | PO6 | Communicate management aims and directions |
| | | PO9 | Assess risk | PO9 | Assess and manage IT risks |
| | | PO10 | Manage projects | | |
| | | M1 | Monitor the process | PO10 | Manage projects |
| | | M4 | Provide for independent audit | ME4 | Provide IT governance |
| 2.9 | Evaluate monitoring and assurance practices to ensure that the board and executive management receive sufficient and timely information about IT performance. | PO8 | Ensure compliance with external requirements | PO8 | Manage quality |
| | | PO10 | Manage projects | PO10 | Manage projects |
| | | PO11 | Manage quality | ME2 | Monitor and evaluate internal control |
| | | M2 | Assess internal control adequacy | ME3 | Ensure regulatory compliance (4.0) |
| | | M3 | Obtain independent assurance | ME3 | Ensure compliance with external requirements (4.1) |

## Chapter 3: Systems and Infrastructure Life Cycle Management

| CISA Review Manual | | CoBiT 3rd Edition | | CoBiT 4.0 and 4.1 | |
|---|---|---|---|---|---|
| **Tasks** | | **CoBiT Processes** | | | |
| 3.1 | Evaluate the business case for the proposed system development/acquisition to ensure that it meets the organization's business goals. | PO3 | Determine technological direction | PO3 | Determine technological direction |
| | | PO11 | Manage quality | PO8 | Manage quality |
| | | AI1 | Identify automated solutions | AI1 | Identify automated solutions |
| | | AI2 | Acquire and maintain application software | AI2 | Acquire and maintain application software |
| | | AI3 | Acquire and maintain technology infrastructure | AI3 | Acquire and maintain technology infrastructure |
| | | DS9 | Manage the configuration | DS9 | Manage the configuration |
| 3.2 | Evaluate the project management framework and project governance practices to ensure that business objectives are achieved in a cost-effective manner while managing risks to the organization. | PO9 | Assess risks | PO8 | Manage quality |
| | | PO10 | Manage projects | PO9 | Assess and manage IT risk |
| | | PO11 | Manage quality | | |
| | | AI1 | Identify automated solutions | PO10 | Manage projects |
| | | | | AI1 | Identify automated solutions |
| | | AI2 | Acquire and maintain application software | AI2 | Acquire and maintain application software |
| 3.3 | Perform reviews to ensure that a project is progressing in accordance with project plans, it is adequately supported by documentation and the status reporting is accurate. | PO10 | Manage project | PO10 | Manage projects |
| | | AI1 | Identify automated solutions | AI1 | Identify automated solutions |
| | | AI2 | Acquire and maintain application software | AI2 | Acquire and maintain application software |
| | | M3 | Obtain independent assurance | ME2 | Monitor and evaluate internal control |
| | | M4 | Provide for independent audit | | |

## Chapter 3:  Systems and Infrastructure Life Cycle Management (cont.)

| CISA Review Manual | CoBiT 3rd Edition | CoBiT 4.0 and 4.1 |
|---|---|---|
| **Tasks** | **CoBiT Processes** | |
| 3.4 Evaluate proposed control mechanisms for systems and/or infrastructure during specification, development/acquisition, and testing to ensure that they will provide safeguards and comply with the organization's policies and other requirements. | PO10 *Manage projects*<br>PO11 *Manage quality*<br>AI1 *Identify automated solutions*<br>AI2 *Acquire and maintain application software*<br>AI5 *Install and accredit systems* | PO8 *Manage quality*<br>PO10 *Manage projects*<br>AI1 *Identify automated solutions*<br>AI2 *Acquire and maintain application software*<br>AI7 *Install and accredit solutions and changes* . |
| 3.5 Evaluate the processes by which systems and/or infrastructure are developed/acquired and tested to ensure that the deliverables meet the organization's objectives. | PO10 *Manage projects*<br>PO11 *Manage quality*<br>AI1 *Identify automated solutions* | PO8 *Manage quality*<br>PO10 *Manage projects*<br>AI1 *Identify automated solutions*<br>AI2 *Acquire and maintain application software*<br>AI7 *Install and accredit solutions and changes* |
| 3.6 Evaluate the readiness of the system and/or infrastructure for implementation and migration into production. | AI2 *Acquire and maintain application software*<br>AI5 *Install and accredit systems* | PO3 *Determine technological direction*<br>AI3 *Acquire and maintain technology infrastructure*<br>AI7 *Install and accredit solutions and changes* |
| 3.7 Perform postimplementation review of systems and/or infrastructure to ensure that they meet the organization's objectives and are subject to effective internal control. | PO3 *Determine technological direction*<br>AI3 *Acquire and maintain technology infrastructure*<br>AI5 *Install and accredit systems* | PO8 *Manage quality*<br>PO10 *Manage projects*<br>AI7 *Install and accredit solutions and changes* |
| 3.8 Perform periodic reviews of systems and/or infrastructure to ensure that they continue to meet the organization's objectives and are subject to effective internal control. | PO10 *Manage projects*<br>PO11 *Manage quality*<br>AI5 *Install and accredit systems*<br>PO6 *Communicate management aims and direction*<br>PO10 *Manage projects*<br>PO11 *Manage quality*<br>AI5 *Install and accredit systems*<br>DS1 *Define and manage service levels*<br>DS3 *Manage performance and capacity*<br>M2 *Assess internal control adequacy*<br>M3 *Obtain independent assurance*<br>M4 *Provide for independent audit* | PO6 *Communicate management aims and direction*<br>PO8 *Manage quality*<br>PO10 *Manage projects*<br>AI7 *Install and accredit solutions and changes*<br>DS1 *Define and manage service levels*<br>DS3 *Manage performance and capacity*<br>ME2 *Monitor and evaluate internal control* |

## Chapter 3: Systems and Infrastructure Life Cycle Management (cont.)

| CISA Review Manual | | CoBiT 3rd Edition | | CoBiT 4.0 and 4.1 | |
|---|---|---|---|---|---|
| **Tasks** | | **CoBiT Processes** | | | |
| 3.9 | Evaluate the process by which systems and/or infrastructure are maintained to ensure the continued support of the organization's objectives and are subject to effective internal control. | PO3<br>PO11<br>AI3<br>AI6 | Determine technological direction<br>Manage quality<br>Acquire and maintain technology infrastructure<br>Manage changes | PO3<br>PO8<br>AI3<br>AI6 | Determine technological direction<br>Manage quality<br>Acquire and maintain technology infrastructure<br>Manage changes |
| 3.10 | Evaluate the process by which systems and/or infrastructure are disposed of to ensure that they comply with the organization's policies and procedures. | PO6<br><br>AI1 | Communicate management aims and direction<br>Identify automated solutions | PO6<br><br>AI1 | Communicate management aims and direction<br>Identify automated solutions |

## Chapter 4: IT Service Delivery and Support

| CISA Review Manual | | CoBiT 3rd Edition | | CoBiT 4.0 and 4.1 | |
|---|---|---|---|---|---|
| **Tasks** | | **CoBiT Processes** | | | |
| 4.1 | Evaluate service level management practices to ensure that the level of service from internal and external service providers is defined and managed. | AI4<br>DS1<br>DS2<br>DS6<br>DS8<br>M1 | Develop and maintain procedures<br>Define and manage service levels<br>Manage third-party services<br>Identify and allocate costs<br>Assist and advise customers<br>Monitor the process | AI4<br>DS1<br>DS2<br>DS6<br>DS8<br>DS10<br>ME1 | Enable operation and use<br>Define and manage service levels<br>Manage third-party services<br>Identify and allocate costs<br>Manage service desk and incidents<br>Manage problems<br>Monitor and evaluate IT performance |
| 4.2 | Evaluate operations management to ensure that IT support functions effectively meet business needs. | PO9<br>AI4<br>AI5<br>DS13<br>M2 | Assess risks<br>Develop and maintain procedures<br>Install and accredit systems<br>Manage operations<br>Assess internal control adequacy | PO9<br>AI4<br>AI7<br>DS13<br>ME1 | Assess and manage IT risks<br>Enable operation and use<br>Install and accredit solutions and changes<br>Manage operations<br>Monitor and evaluate IT performance |
| 4.3 | Evaluate data administration practices to ensure the integrity and optimization of databases. | PO2<br>PO4<br>AI1<br>AI2<br>AI5<br>DS5<br>M1 | Define the information architecture<br>Define the IT organisation and relationships<br>Identify automated solutions<br>Acquire and maintain application software<br>Intall and accredit systems<br>Ensure systems security<br>Monitor the process | PO2<br>PO4<br>AI1<br>AI2<br>AI7<br>DS5<br>ME1 | Define the information architecture<br>Define the IT processes, organization and relationships<br>Identify automated solutions<br>Acquire and maintain application software<br>Install and accredit solutions and changes<br>Ensure systems security<br>Monitor and evaluate IT performance |

## Chapter 4: IT Service Delivery and Support (cont.)

| CISA Review Manual | COBIT 3rd Edition | COBIT 4.0 and 4.1 |
|---|---|---|
| **Tasks** | **COBIT Processes** | |
| 4.4 Evaluate the use of capacity and performance monitoring tools and techniques to ensure that IT services meet the organization's objectives. | PO11 *Manage quality*<br>AI1 *Identify automated solutions*<br>AI5 *Install and accredit systems*<br>DS1 *Define and manage service levels*<br>DS3 *Manage performance and capacity*<br>M1 *Monitor the process* | AI1 *Identify automated solutions*<br>AI7 *Install and accredit solutions and changes*<br>DS1 *Define and manage service levels*<br>DS3 *Manage performance and capacity*<br>ME1 *Monitor and evaluate IT performance* |
| 4.5 Evaluate change, configuration and release management practices to ensure that changes made to the organization's production environment are adequately controlled and documented. | AI2 *Acquire and maintain application software*<br>AI3 *Acquire and maintain technology infrastructure*<br>AI5 *Install and accredit systems*<br>AI6 *Manage changes*<br>DS9 *Manage the configuration* | PO8 *Manage quality*<br>AI2 *Acquire and maintain application software*<br>AI3 *Acquire and maintain technology infrastructure*<br>AI6 *Manage changes*<br>AI7 *Install and accredit solutions and change*<br>DS9 *Manage the configuration* |
| 4.6 Evaluate problem and incident management practices to ensure that incidents, problems or errors are recorded, analyzed and resolved in a timely manner. | DS8 *Assist and advise customers*<br>DS10 *Manage problems and incidents*<br>DS11 *Manage data adequacy* | DS8 *Manage service desk and incidents*<br>DS10 *Manage Problems*<br>DS11 *Manage data*<br>ME1 *Monitor and evaluate IT performance* |
| 4.7 Evaluate the functionality of the IT infrastructure (e.g., network components, hardware, system software) to ensure that it supports the organization's objectives. | PO3 *Determine technological direction*<br>PO11 *Manage quality*<br>AI3 *Acquire and maintain technology infrastructure*<br>AI6 *Manage changes* | PO3 *Determine technological direction*<br>PO8 *Manage quality*<br>AI3 *Acquire and maintain technology infrastructure*<br>AI6 *Manage changes* |

## Chapter 5: Protection of Information Assets

| CISA Review Manual | COBIT 3rd Edition | COBIT 4.0 and 4.1 |
|---|---|---|
| **Tasks** | **COBIT Processes** | |
| 5.1 Evaluate the design, implementation and monitoring of logical access controls to ensure the confidentiality, integrity, availability and authorized use of information assets. | AI6 *Manage changes*<br>DS4 *Ensure continous service*<br>DS5 *Ensure systems security*<br>DS10 *Manage problems and incidents*<br>M1 *Monitor the process* | AI6 *Manage changes*<br>DS4 *Ensure continous service*<br>DS5 *Ensure systems security*<br>DS10 *Manage problems*<br>ME1 *Monitor and evaluate IT performance* |
| 5.2 Evaluate network infrastructure security to ensure confidentiality, integrity, availability and authorized use of the network and the information transmitted. | DS4 *Ensure continous service*<br>DS5 *Ensure systems security*<br>DS11 *Manage data*<br>DS13 *Manage operations*<br>M1 *Monitor the process* | DS4 *Ensure continous service*<br>DS5 *Ensure systems security*<br>DS11 *Manage data*<br>DS13 *Manage operations*<br>ME1 *Monitor and evaluate IT performance* |

## Chapter 5: Protection of Information Assets (cont.)

| CISA Review Manual | | CoBiT 3rd Edition | CoBiT 4.0 and 4.1 |
|---|---|---|---|
| **Tasks** | | **CoBiT Processes** | |
| 5.3 | Evaluate the design, implementation and monitoring of environmental controls to prevent or minimize loss. | PO9 *Assess risks*<br>DS4 *Ensure continous service*<br>DS12 *Manage facilities*<br>M1 *Monitor the process* | PO9 *Assess and manage IT risk*<br>DS4 *Ensure continous service*<br>DS12 *Manage the physical environment*<br>ME1 *Monitor and evaluate IT performance*<br>ME3 *Ensure regulatory compliance (4.0)*<br>ME3 *Ensure compliance with external requirements (4.1)* |
| 5.4 | Evaluate the design, implementation and monitoring of physical access controls to ensure that information assets are adequately safeguarded. | PO4 *Define the IT organization and relationships*<br>PO8 *Ensure compliance with external requirements*<br>DS5 *Ensure systems security*<br>DS12 *Manage facilities*<br>M1 *Monitor the process* | PO4 *Define the IT processes, organization and relationships*<br>DS5 *Ensure systems security*<br>DS12 *Manage the physical environment*<br>ME1 *Monitor and Evaluate IT performance*<br>ME3 *Ensure regulatory compliance (4.0)*<br>ME3 *Ensure compliance with external requirements (4.1)* |
| 5.5 | Evaluate the processes and procedures used to store, retrieve, transport and dispose of confidential information assets. | AI *Acquire and maintain technology infrastructure*<br>DS4 *Ensure continous service*<br>DS5 *Ensure systems security*<br>DS11 *Manage data*<br>M1 *Monitor the process* | AI3 *Acquire and maintain technology infrastructure*<br>DS4 *Ensure continous service*<br>DS5 *Ensure systems security*<br>DS11 *Manage data*<br>ME1 *Monitor and evaluate IT performance*<br>ME3 *Ensure regulatory compliance (4.0)*<br>ME3 *Ensure compliance with external requirements (4.1)* |

## Chapter 6: Business Continuity and Disaster Recovery

| CISA Review Manual | | CobiT 3rd Edition | CobiT 4.0 and 4.1 |
|---|---|---|---|
| **Tasks** | | **CobiT Processes** | |
| 6.1 | Evaluate the adequacy of backup and restore provisions to ensure the availability of information required to resume processing. | PO2 *Define the information architecture*<br>DS4 *Ensure continuous service*<br>DS11 *Manage data* | PO2 *Define the information architecture*<br>DS4 *Ensure continuous service*<br>DS11 *Manage data* |
| 6.2 | Evaluate the organization's disaster recovery plan to ensure that it enables the recovery of IT processing capabilities in the event of a disaster. | DS4 *Ensure continuous service*<br>DS11 *Manage data*<br>DS12 *Manage facilites*<br>DS13 *Manage operations* | DS4 *Ensure continuous service*<br>DS11 *Manage data*<br>DS12 *Manage the physical environment*<br>ME3 *Ensure regulatory compliance (4.0)*<br>ME3 *Ensure compliance with external requirements (4.1)* |
| 6.3 | Evaluate the organization's business continuity plan to ensure its ability to continue essential business operations during the period of an IT disruption. | DS4 *Ensure continuous service* | DS4 *Ensure continuous service*<br>DS13 *Manage operations* |

## COBIT® 3RD EDITION®

The following information provides the COBIT domains and the 34 IT processes which can be identified for each CISA job practice task listed in the previous tables.

| Planning and Organisation | High Level Control Objective |
|---|---|
| **Planning and Organisation** | PO1    Define a strategic IT plan<br>PO2    Define the information architecture<br>PO3    Determine technological direction<br>PO4    Define the IT organisation and relationships<br>PO5    Manage the IT investment<br>PO6    Communicate management aims and direction<br>PO7    Manage human resources<br>PO8    Ensure compliance with external requirements<br>PO9    Assess risks<br>PO10   Manage projects<br>PO11   Manage quality |
| **Acquisition and Implementation** | AI1    Identify automated solutions<br>AI2    Acquire and maintain application software<br>AI3    Acquire and maintain technology infrastructure<br>AI4    Develop and maintain procedures<br>AI5    Install and accredit systems<br>AI6    Manage changes |
| **Delivery and Support** | DS1    Define and manage service levels<br>DS2    Manage third-party services<br>DS3    Manage performance and capacity<br>DS4    Ensure continuous service<br>DS5    Ensure systems security<br>DS6    Identify and allocate costs<br>DS7    Educate and train users<br>DS8    Assist and advise customers<br>DS9    Manage the configuration<br>DS10  Manage problems and incidents<br>DS11  Manage data<br>DS12  Manage facilities<br>DS13  Manage operations |
| **Monitoring** | M1    Monitor the process<br>M2    Assess internal control adequacy<br>M3    Obtain independent assurance<br>M4    Provide for independent audit |

## COBIT® 4.0

The following information provides the COBIT domains and the 34 IT processes which can be identified for each CISA job practice task listed in the previous tables.

| Domain | Process | |
|---|---|---|
| **Plan and Organise** | PO1 | Define a strategic IT plan. |
| | PO2 | Define the information architecture. |
| | PO3 | Determine technological direction. |
| | PO4 | Define the IT processes, organisation and relationships. |
| | PO5 | Manage the IT investment. |
| | PO6 | Communicate management aims and direction. |
| | PO7 | Manage IT human resources. |
| | PO8 | Manage quality. |
| | PO9 | Assess and manage IT risks. |
| | PO10 | Manage projects. |
| **Acquire and Implement** | AI1 | Identify automated solutions. |
| | AI2 | Acquire and maintain application software. |
| | AI3 | Acquire and maintain technology infrastructure. |
| | AI4 | Ensure operation and use. |
| | AI5 | Procure IT resources. |
| | AI6 | Manage changes. |
| | AI7 | Install and accredit solutions and changes. |
| **Deliver and Support** | DS1 | Define and manage service levels. |
| | DS2 | Manage third-party services. |
| | DS3 | Manage performance and capacity. |
| | DS4 | Ensure continuous service. |
| | DS5 | Ensure systems security. |
| | DS6 | Identify and allocate costs. |
| | DS7 | Educate and train users. |
| | DS8 | Manage service desk and Incidents. |
| | DS9 | Manage the configuration. |
| | DS10 | Manage problems. |
| | DS11 | Manage data. |
| | DS12 | Manage the physical environment. |
| | DS13 | Manage operations. |
| **Monitor and Evaluate** | ME1 | Monitor and evaluate IT performance. |
| | ME2 | Monitor and evaluate internal control. |
| | ME3 | Ensure regulatory compliance. |
| | ME4 | Provide IT governance. |

# COBIT® 4.1

The following information provides the COBIT domains and the 34 IT processes which can be identified for each CISA job practice task listed in the previous tables.

| Domain | Process | |
|---|---|---|
| **Plan and Organise** | PO1 | Define a strategic IT plan. |
| | PO2 | Define the information architecture. |
| | PO3 | Determine technological direction. |
| | PO4 | Define the IT processes, organisation and relationships. |
| | PO5 | Manage the IT investment. |
| | PO6 | Communicate management aims and direction. |
| | PO7 | Manage IT human resources. |
| | PO8 | Manage quality. |
| | PO9 | Assess and manage IT risks. |
| | PO10 | Manage projects. |
| **Acquire and Implement** | AI1 | Identify automated solutions. |
| | AI2 | Acquire and maintain application software. |
| | AI3 | Acquire and maintain technology infrastructure. |
| | AI4 | Ensure operation and use. |
| | AI5 | Procure IT resources. |
| | AI6 | Manage changes. |
| | AI7 | Install and accredit solutions and changes. |
| **Deliver and Support** | DS1 | Define and manage service levels. |
| | DS2 | Manage third-party services. |
| | DS3 | Manage performance and capacity. |
| | DS4 | Ensure continuous service. |
| | DS5 | Ensure systems security. |
| | DS6 | Identify and allocate costs. |
| | DS7 | Educate and train users. |
| | DS8 | Manage service desk and Incidents. |
| | DS9 | Manage the configuration. |
| | DS10 | Manage problems. |
| | DS11 | Manage data. |
| | DS12 | Manage the physical environment. |
| | DS13 | Manage operations. |
| **Monitor and Evaluate** | ME1 | Monitor and evaluate IT performance. |
| | ME2 | Monitor and evaluate internal control. |
| | ME3 | Ensure compliance with external requirements. |
| | ME4 | Provide IT governance. |

# IS AUDITING STANDARDS, GUIDELINES AND PROCEDURES

The specialized nature of IS auditing and the skills necessary to perform such audits require standards that apply specifically to IS auditing. One of the goals of ISACA is to advance globally applicable standards to meet its vision. The development and dissemination of the IS Auditing Standards are a cornerstone of the ISACA professional contribution to the audit community. The framework for the IS Auditing Standards provides multiple levels of guidance.

Standards define mandatory requirements for IS auditing and reporting. They inform:
- IS auditors of the minimum level of acceptable performance required to meet the professional responsibilities set out in the ISACA Code of Professional Ethics for IS auditors
- Management and other interested parties of the profession's expectations concerning the work of practitioners
- Holders of the CISA designation of requirements. Failure to comply with these standards may result in an investigation into the CISA holder's conduct by the ISACA Board of Directors or appropriate ISACA committee and, ultimately, in disciplinary action.

Guidelines provide guidance in applying IS Auditing Standards. The IS auditor should consider them in determining how to achieve implementation of the standards, use professional judgment in their application and be prepared to justify any departure. The objective of the IS Auditing Guidelines is to provide further information on how to comply with the IS Auditing Standards.

Procedures provide examples of procedures an IS auditor might follow in an audit engagement. The procedure documents provide information on how to meet the standards when performing IS auditing work, but do not set requirements. The objective of the IS Auditing Procedures is to provide further information on how to comply with the IS Auditing Standards.

COBIT resources should be used as a source of best practice guidance. The COBIT framework states, "It is management's responsibility to safeguard all the assets of the enterprise. To discharge this responsibility as well as to achieve its expectations, management must establish an adequate system of internal control." COBIT provides a detailed set of controls and control techniques for the information systems management environment. Selection of the most relevant material in COBIT applicable to the scope of the particular audit is based on the choice of specific COBIT IT processes and consideration of COBIT information criteria.

As defined in the COBIT framework, each of the following is organized by IT management process. COBIT is intended for use by business and IT management, as well as IS auditors; therefore, its usage enables the understanding of business objectives, communication of best practices and recommendations to be made around a commonly understood and well-respected standard reference. COBIT includes:
- *Board Briefing on IT Governance, 2nd Edition*—Helps executives understand why IT governance is important, what its issues are and what their responsibility is for managing it
- Management guidelines/maturity models—Help assign responsibility, measure performance, and benchmark and address gaps in capability
- Frameworks—Organize IT governance objectives and good practices by IT domains and processes, and link them to business requirements
- Control objectives— Provide a complete set of high-level requirements to be considered by management for effective control of each IT process
- *IT Governance Implementation Guide: Using COBIT® and Val IT™, 2nd Edition*—Provides a generic road map for implementing IT governance using the COBIT and Val IT™ resources
- COBIT and Val IT™ resources
- *COBIT® Control Practices: Guidance to Achieve Control Objectives for Successful IT Governance, 2nd Edition*—Provides guidance on why controls are worth implementing and how to implement them
- *IT Assurance Guide: Using COBIT®*—Provides guidance on how COBIT can be used to support a variety of assurance activities together with suggested testing steps for all the IT processes and control objectives

# RELATIONSHIP OF STANDARDS TO GUIDELINES AND PROCEDURES

IS Auditing Standards are brief mandatory reports on requirements regarding the audit and its findings for certification holders. IS Auditing Guidelines and Procedures are detailed guidance on how to follow those standards. The procedure examples show the steps performed by an IS auditor and are more informative than IS Auditing Guidelines. The examples are constructed to follow the IS Auditing Standards and the IS Auditing Guidelines and provide information on following the IS Auditing Standards. To some extent, they also establish best practices for procedures to be followed.

Codification:
• Standards are numbered consecutively as they are issued, beginning with S1.
• Guidelines are numbered consecutively as they are issued, beginning with G1.
• Procedures are numbered consecutively as they are issued, beginning with P1.

Please refer to the index of IS Auditing Standards, Guidelines and Procedures for a complete listing of those documents.

## USE

It is suggested that during the annual audit program, as well as individual reviews throughout the year, the IS auditor should review the standards to ensure compliance with them. The IS auditor is encouraged to refer to the ISACA standards in the report, stating that the review was conducted in compliance with the laws of the country, applicable audit regulations and ISACA standards.

## INDEX OF IS AUDITING STANDARDS, GUIDELINES AND PROCEDURES

Note: These documents are available for download on the ISACA web site, *www.isaca.org/standards*

### Index of IS Auditing Standards

| | Effective Date | |
|---|---|---|
| S1 Audit Charter | 1 January | 2005 |
| S2 Independence | 1 January | 2005 |
| S3 Professional Ethics and Standards | 1 January | 2005 |
| S4 Competence | 1 January | 2005 |
| S5 Planning | 1 January | 2005 |
| S6 Performance of Audit Work | 1 January | 2005 |
| S7 Reporting | 1 January | 2005 |
| S8 Follow-up Activities | 1 January | 2005 |
| S9 Irregularities and Illegal Acts | 1 September | 2005 |
| S10 IT Governance | 1 September | 2005 |
| S11 Use of Risk Assessment in Audit Planning | 1 November | 2005 |
| S12 Audit Materiality | 1 July | 2006 |
| S13 Using the Work of Other Experts | 1 July | 2006 |
| S14 Audit Evidence | 1 July | 2006 |
| S15 IT Controls | 1 February | 2008 |
| S16 E-commerce | 1 February | 2008 |

# Appendix B

## Index of IS Auditing Guidelines

| | | **Effective Date** | |
|---|---|---|---|
| G1 Using the Work of Other Auditors | 1 March | 2008 |
| G2 Audit Evidence Requirement | 1 December | 1998 |
| G3 Use of Computer-Assisted Audit Techniques (CAATs) | 1 March | 2008 |
| G4 Outsourcing of IS Activities to Other Organizations | 1 September | 1999 |
| G5 Audit Charter | 1 February | 2008 |
| G6 Materiality Concepts for Auditing Information Systems | 1 September | 1999 |
| G7 Due Professional Care | 1 March | 2008 |
| G8 Audit Documentation | 1 March | 2008 |
| G9 Audit Considerations for Irregularities | 1 March | 2000 |
| G10 Audit Sampling | 1 March | 2000 |
| G11 Effect of Pervasive IS Controls | 1 March | 2000 |
| G12 Organizational Relationship and Independence | 1 September | 2000 |
| G13 Use of Risk Assessment in Audit Planning | 1 September | 2000 |
| G14 Application Systems Review | 1 November | 2001 |
| G15 Planning Revised | 1 March | 2002 |
| G16 Effect of Third Parties on an Organization's IT Controls | 1 March | 2002 |
| G17 Effect of Nonaudit Role on the IS Auditor's Independence | 1 July | 2002 |
| G18 IT Governance | 1 July | 2002 |
| G19 Irregularities and Illegal Acts | 1 July | 2002 |
| G20 Reporting | 1 January | 2003 |
| G21 Enterprise Resource Planning (ERP) Systems Review | 1 August | 2003 |
| G22 Business-to-consumer (B2C) E-commerce Review | 1 August | 2003 |
| G23 System Development Life Cycle (SDLC) Review Reviews | 1 August | 2003 |
| G24 Internet Banking | 1 August | 2003 |
| G25 Review of Virtual Private Networks | 1 July | 2004 |
| G26 Business Process Reengineering (BPR) Project Reviews | 1 July | 2004 |
| G27 Mobile Computing | 1 September | 2004 |
| G28 Computer Forensics | 1 September | 2004 |
| G29 Postimplementation Review | 1 January | 2005 |
| G30 Competence | 1 June | 2005 |
| G31 Privacy | 1 June | 2005 |
| G32 Business Continuity Plan Review From IT Perspective | 1 September | 2005 |
| G33 General Considerations on the Use of the Internet | 1 March | 2006 |
| G34 Responsibility, Authority and Accountability | 1 March | 2006 |
| G35 Follow-up Activities | 1 March | 2006 |
| G36 Biometric Controls | 1 March | 2007 |
| G37 Configuration Management | 1 November | 2007 |
| G38 Access Control | 1 February | 2008 |
| G39 IT Organizations | 1 May | 2008 |

## Index of IS Auditing Procedures

| | | **Effective Date** | |
|---|---|---|---|
| P1 IS Risk Assessment | 1 July | 2002 |
| P2 Digital Signatures | 1 July | 2002 |
| P3 Intrusion Detection | 1 August | 2003 |
| P4 Viruses and other Malicious Code | 1 August | 2003 |
| P5 Control Risk Self-assessment | 1 August | 2003 |
| P6 Firewalls | 1 August | 2003 |
| P7 Irregularities and Illegal Acts | 1 November | 2003 |
| P8 Security Assessment—Penetration Testing and Vulnerability Analysis | 1 September | 2004 |
| P9 Evaluation of Management Controls Over Encryption Methodologies | 1 January | 2005 |
| P10 Business Application Change Control | 1 October | 2006 |
| P11 Electronic Fund Transfer (EFT) | 1 May | 2007 |

# 2009 CISA EXAM GENERAL INFORMATION

ISACA is a professional membership association composed of individuals interested in IS audit, assurance, control, security and governance. The CISA Certification Board is chartered by ISACA and is responsible for establishing policies for the CISA certification program and developing the exam.

## REQUIREMENTS FOR CERTIFICATION

The CISA designation is awarded to those individuals who have met and continue to meet the following requirements: (1) a passing score on the CISA exam, (2) submitting verified evidence of IS auditing, control or security experience, (3) abiding by the *Code of Professional Ethics*, (4) abiding by the professional continuing education policy, and (5) abiding by the IS Auditing Standards as adopted by ISACA.

## SUCCESSFUL COMPLETION OF THE CISA EXAM

The exam is open to all individuals who wish to take it. Successful exam candidates are not certified until they apply for certification and demonstrate that they have met all requirements.

## EXPERIENCE IN IS AUDITING, CONTROL AND SECURITY

A minimum of five (5) years professional IS auditing, control and security work experience is required for certification. Substitutions and waivers of such experience may be obtained as follows:
• A maximum of one (1) year of IS audit, control or security experience may be substituted for:
  – One full year of non-IS audit experience, or
  – A Bachelor's of Master's degree from a university that enforces the ISACA-sponsored Model Curriculum.
  – One full year of information systems experience, or
  – An Associate's degree (60 semester college credits or its equivalent).
• Two years IS audit, control or security experience may be substituted for a Bachelor's degree (120 semester college credits or its equivalent) or Master's degree.
• Two years as a full-time university instructor in a related field (e.g., computer science, accounting or information systems auditing) can be substituted for every one year of IS audit, control, assurance or security experience.

Experience must have been gained within the 10-year period preceding the application date for certification or within five (5) years from the date of initially passing the exam. An application for certification must be submitted within five (5) years from the passing date of the CISA exam. All experience must be independently verified with employers.

## DESCRIPTION OF THE EXAM

The CISA Certification Board oversees the development of the exam and ensures the currency of its content. Questions for the CISA exam are developed through a multitiered process designed to enhance the ultimate quality of the exam. Once the CISA Certification Board approves the questions, they go into the item pool from which all CISA exam questions are selected.

The purpose of the exam is to evaluate a candidate's knowledge and experience in conducting IS audits and reviews. The exam consists of 200 multiple-choice questions, administered during a four-hour session. Candidates may take the exam in several languages. A proctor speaking the primary language used at each test site is available. If a candidate desires to take the exam in a language other than the primary language of the test site, the proctor may not be conversant in the language chosen. However, written instructions will be available in the language of the exam.

# REGISTRATION FOR THE CISA EXAM

The CISA exam will be administered twice in 2009. The first 2009 CISA exam will be administered on Saturday, 13 June 2009 and the second 2009 CISA exam will be administered on Saturday, 12 December 2009. Please refer to the CISA 2009 Bulletin of Information for specific registration deadlines at: *www.isaca.org/cisaexam* The registration form can be obtained online at *www.isaca.org/cisaboi* or from ISACA at the following address:

> ISACA
> 3701 Algonquin Road, Suite 1010
> Rolling Meadows, Illinois 60008, USA
> Attn.:  CISA Exam Registrar
> Telephone:  +1.847. 253.1545
> Fax: +1.847. 253.1443
> E-mail: *certification@isaca.org*

The 2009 CISA exam fee must accompany the registration form. The *Candidate's Guide to the CISA Exam and Certification* will be sent upon receipt and recording of your registration and payment.

# CISA PROGRAM ACCREDITATION RENEWED UNDER ISO/IEC 17024:2003

The American National Standards Institute (ANSI) has voted to continue the accreditation for the CISA and CISM certifications under ISO/IEC 17024:2003, General Requirements for Bodies Operating Certification Systems of Persons. ANSI, a private, nonprofit organization, accredits other organizations to serve as third-party product, system and personnel certifiers.

ISO/IEC 17024 specifies the requirements to be followed by organizations certifying individual against specific requirements. ANSI describes ISO/IEC 17024 as "expected to play a prominent role in facilitating global standardization of the certification community, increasing mobility among countries, enhancing public safety, and protecting consumers."

ANSI's accreditation:
• Promotes the unique qualifications and expertise ISACA's certifications provide
• Protects the integrity of the certifications and provides legal defensibility
• Enhances consumer and public confidence in the certifications and the people who hold them
• Facilitates mobility across borders or industries

**ANSI Accredited Program**
**PERSONNEL CERTIFICATION**

Accreditation by ANSI signifies that ISACA's procedures meet ANSI's essential requirements for openness, balance, consensus and due process. With this accreditation, ISACA anticipates that significant opportunities for CISAs and CISMs will continue to open in the US and around the world.

# PREPARING FOR THE CISA EXAMINATION

The CISA exam evaluates a candidate's practical knowledge of the job practice areas listed in this manual and in the *Candidate's Guide to the CISA Exam.* That is, the exam is designed to test a candidate's knowledge, experience and judgment of the proper or preferred application of IS audit, security and control principles, methods and practices. Since the exam covers a broad spectrum of information systems audit, control and security issues, candidates are cautioned not to assume that reading CISA study guides and reference publications will fully prepare them for the exam. CISA candidates are encouraged to refer to their own experiences when studying for the exam and refer to CISA study guides and reference publications for further explanation of concepts or practices with which the candidate is not familiar.

No representation or warranties are made by ISACA in regard to CISA exam study guides, other ISACA publications, references or courses assuring candidates' passage of the exam.

# TYPES OF EXAM QUESTIONS

CISA exam questions are developed with the intent of measuring and testing practical knowledge and the application of general concepts and standards. All questions are multiple choice and are designed for one best answer.

Every CISA question has a stem (question) and four options (answer choices). The candidate is asked to choose the correct or best answer from the options. The stem may be in the form of a question or incomplete statement. In some instances, a scenario or description problem may also be included. These questions normally include a description of a situation and require the candidate to answer two or more questions based on the information provided. Many times a CISA exam question will require the candidate to choose the most likely or best answer. In every case the candidate is required to read the question carefully, eliminate known incorrect answers and then make the best choice possible. Knowing the format in which questions are asked and how to study to gain knowledge of what will be tested will go a long way toward answering them correctly.

# ADMINISTRATION OF THE EXAM

ISACA has contracted with an internationally recognized testing agency. This not-for-profit corporation engages in the development and administration of credentialing exams for certification and licensing purposes. It assists ISACA in the construction, administration and scoring of the CISA exam.

# SITTING FOR THE EXAM

Be prompt. Registration will begin at each center at the time indicated on your admission ticket. All candidates must be registered and in the test center when the chief examiner begins reading the oral instructions. **NO CANDIDATE WILL BE ADMITTED TO THE TEST CENTER ONCE THE CHIEF EXAMINER BEGINS READING THE ORAL INSTRUCTIONS, APPROXIMATELY 30 MINUTES BEFORE THE EXAM BEGINS.** Any candidate who arrives after the oral instructions have begun will not be allowed to sit for the exam and will forfeit the registration fee. Candidates can use their admission tickets only at the designated test center on the admission ticket.

Candidates will be admitted to the test center only if they have a valid admission ticket and an acceptable form of identification. Examples of acceptable identification include those with a photograph (such as a passport or photo driver's license). Any candidate who does not provide an original form of identification will not be allowed to sit at the exam and will forfeit their registration fee.

Observe the following conventions when completing the exam:
• Candidates are not allowed to bring study materials into the exam site. The complete Personal Belongings Policy is available at *www.isaca.org/cisabelongings*
• Bring several No. 2 lead pencils. Do not assume someone will provide pencils for answering the exam.
• The chief examiner or designate at each test center will read aloud the instructions for entering information on the answer sheet. It is imperative that candidates include their exam identification number as it appears on their admission ticket and any other requested information. Failure to do so may result in a delay or errors.
• Identify key words or phrases in the question (**MOST, BEST, FIRST**) before selecting and recording answers.
• Read the provided instructions carefully before attempting to answer questions. Skipping over these directions or reading them too quickly could result in missing important information and possibly losing credit points.
• It is imperative that candidates mark the appropriate area when indicating their response on the answer sheet. When correcting a previously answered question, fully erase a wrong answer before writing in the new one.
• Remember to answer all questions since there is no penalty for wrong answers. Grading is based solely on the number of questions answered correctly.

# BUDGETING YOUR TIME

- Candidates must arrive at the exam testing site by the time indicated on their admission ticket. This will give candidates time to be seated and get acclimated.
- The exam is administered over a four-hour period. This allows for a little over one minute per question. Therefore, it is advisable that candidates pace themselves to complete the entire exam. Candidates must complete an average of 50 questions per hour.
- Candidates are urged to record their answers on their answer sheet. No additional time will be allowed after the exam time has elapsed to transfer or record answers should candidates mark their answers in the question booklet.

# RULES AND PROCEDURES

- Candidates will be asked to sign the answer sheet to protect the security of the exam and maintain the validity of the scores.
- Upon the discretion of the CISA Certification Board, any candidate can be disqualified who is discovered engaging in any kind of misconduct, such as giving or receiving help; using notes, papers, or other aids; attempting to take the examination for someone else; or removing test materials or notes from the testing room. The testing agency will provide the CISA Certification Board with records regarding such irregularities. The board will review reported incidents, and all board decisions are final.
- Candidates may **not** take the exam question booklet after completion of the exam.

# GRADING THE EXAM

The CISA exam consists of 200 items. Candidate scores are reported as a scaled scored. A scaled score is a conversion of a candidate's raw score on an exam to a common scale. ISACA uses and reports scores on a common scale from 200 to 800. A candidate must receive a score of 450 or higher to pass the exam. A score of 450 represents a minimum consistent standard of knowledge as established by ISACA's CISA Certification Board. A candidate receiving a passing score may then apply for certification if all other requirements are met.

**Passing the exam does not grant the CISA designation. To become a CISA, each candidate must complete all requirements, including submitting an application for certification.**

The CISA examination contains some questions which are included for research and analysis purposes only. These questions are not separately identified and the candidate's final score will be based only on the common scored questions. There are various versions of each exam but only the common questions are scored for your results.

A candidate receiving a score less than 450 is not successful and can retake the exam during any future exam administration. To assist with future study, the result letter each candidate receives will include a score analysis by content area. There are no limits to the number of times a candidate can take the exam.

**Approximately eight weeks after the test date, the official exam results will be mailed to candidates.** Additionally, with the candidate's consent on the registration form, an e-mail containing the candidates pass/fail status and score will be sent to paid candidates. This e-mail notification will only be sent to the address listed in the candidate's profile at the time of the initial release of the results. To ensure the confidentiality of scores, exam results will not be reported by telephone or fax. To prevent the e-mail notification from being sent to the candidate's spam folder, the candidate should add *certification@isaca.org* to his/her address book, whitelist or safe senders list.

Successful candidates will receive an application for certification. For those candidates not passing the exam, the score report will contain a subscore for each job domain. The subscores can be useful in identifying those areas in which the candidate may need further study before retaking the exam. Unsuccessful candidates should note that taking either a simple or weighted average of the subscores does not derive the total scaled score.

Candidates receiving a failing score on the exam may request a rescoring of their answer sheet. This procedure ensures that no stray marks, multiple responses or other conditions interfered with computer scoring. Requests for hand scoring must be made in writing to the certification department within 90 days following the release of the exam results. All requests must include a candidate's name, exam identification number and mailing address. A fee of US $50 must accompany this request.

# Index

CSMA/CA, See Carrier-sense Multiple Access/Collision Avoidance
CSMA/CD, See Carrier-sense Multiple Access/Collision Detection
CSO, See Chief security officer
CSU/DSU, See Channel service unit/digital service unit
Cube, 263
Customer relationship management (CRM), 144, 169, 194, 256, 268, 269, 526, 532, 548
Cyber attacks, 382
Cybercrime, 35, 556
Cybertrust 425

# D

DACs, See Discretionary Access Controls
Daemen, 423
Damage assessment team, 495
Damage/flooding, 454
Data analysis, 62, 265, 404, 453
Data center, 51, 295, 307, 313, 373, 465
Data classification, 160, 368, 369, 373
Data communication, 265, 308, 309
Data communications equipment (DCE), 326, 327, 396, 524, 548
Data communications software, 281, 305, 308
Data control group, 124
Data conversion, 141, 146, 147, 148, 149, 150, 151, 152, 177, 186, 188, 189, 197, 226, 240
Data custodian, 526
Data dictionary (DD), 281, 309, 310, 314, 346, 397, 443, 526, 548
Data Dictionary/Directory System (DD/DS), 310
Data editing, 227, 268
Data encryption, 123, 252, 257, 406, 408, 421, 428, 526, 538, 539, 548
Data Encryption Standard (DES), 252, 428
Data entry, 75, 105, 118, 226, 282, 351, 446, 499, 503, 504, 520
Data file control procedures, 143, 228
Data flow diagram (DFD), 261, 548
Data input, 224, 225, 227, 520, 542
Data integrity, 40, 114, 143, 176, 188, 190, 224, 232, 233, 248, 254, 307, 310, 314, 350, 370, 395, 423, 424
Data leakage, 386, 527
Data management, 15, 117, 242, 267, 281, 306, 309
Data mining, 62, 254, 263
Data owner, 122, 375, 439, 445, 446, 469, 527, 528
Data redundancy, 309, 310, 346, 500
Data security, 114, 117, 291, 310, 339, 345, 347, 349, 407, 527
Data structure, 49, 188, 263, 265, 312, 526, 527
Data terminal equipment (DTE), 327, 524
Data transmission, 339, 340, 524, 537, 543
Data validation, 39, 143, 227, 232
Data warehouse, 263, 376, 548
Data-oriented system development, 141, 197, 548
Database administration, 33, 75, 119, 122, 283, 284, 285, 286
Database administrator (DBA), 119, 122, 443
Database controls, 281, 313
Database management system (DBMS) 281, 306, 308, 309, 310, 313, 346, 443, 506, 525, 527, 531, 547, 550
Database specifications, 46, 169, 177, 193, 198, 527

Datagram, 325, 327, 336, 416, 551
DBA, See Database administrator
DBMS, See Database management system
DCE, See Data communications equipment or Distributed computing environment
DCT, See Discrete cosine transform
DD, See Data dictionary
DD/DS, See Data Dictionary/Directory System
Deadlock, 313
Debugging, 50, 141, 178, 179, 183, 314, 386
Decentralization, 93, 527
Decision focus, 144, 266
Decision support systems (DSS), 144, 166, 169, 266, 267, 268, 527, 548
Decision trees, 261
Decompiler, 201
Decryption, 420, 421, 423, 527, 538, 539, 541
Decryption key, 421, 527, 538, 539, 541
DECT, See Digital Enhanced Cordless Telecommunications
Delineation, 106
Delivery and Support (DS), 310, 342, 548
Delta, 288
Demilitarized zone (DMZ), 417
Denial of service, 98, 367, 370, 371, 384, 399, 409, 412, 428, 433, 448, 483, 548
Deployment, 86, 97, 105, 146, 147, 148, 150, 151, 152, 167, 176, 177, 182, 187, 197, 204, 252, 426, 496, 498
DES, See Data Encryption Standard
Design and development, 105, 141, 143, 144, 166, 170, 192, 193, 239, 267, 316, 319, 448
Detailed design, 143, 166, 168, 176, 177, 178, 194, 216, 239, 544
DFD, See Data flow diagram
Dial-back, 407, 436, 527
Dial-in penetration attacks, 413
Dial-up access controls, 363, 407, 447, 466
Dictionaries, 193, 213, 309, 506
DID, See Direct inward dial
Diffie-Hellman, 426
Digital Enhanced Cordless Telecommunications (DECT), 333, 548
Digital signature, 244, 252, 367, 369, 423, 424, 425, 427, 428, 469, 527
Digital subscriber lines (DSL), 329
Digital video disc (DVD), 207, 301, 303, 305, 548, 549
Direct inward dial (DID), 214, 215, 227, 228, 433
Direct-sequence spread spectrum (DSSS), 548
Disaster recovery, 2, 4, 31, 33, 77, 79, 80, 106, 107, 114, 213, 258, 259, 290, 346, 459, 475, 476, 477, 478, 479, 480, 481, 482, 483, 484, 485, 486, 487, 488, 489, 490, 491, 492, 493, 494, 495, 496, 497, 498, 499, 500, 501, 502, 503, 504, 505, 506, 507, 508, 509, 510, 511, 514, 515, 516, 525, 527, 537, 548, 549
Disaster Recovery Institute International (DRII), 508, 548
Disaster recovery planning, 33, 476, 478, 482, 525, 548, 549
Disaster recovery procedures, 259, 346, 484, 496
Disasters and other disruptive events, 476, 483
Discovery sampling, 47, 528
Discrete cosine transform (DCT), 259, 548
Discretionary Access Controls (DACs), 123, 362, 375, 528, 535, 548
Disk management, 281, 305, 313, 548

## E

## I

IR, See Infrared

IrDA, See Infrared Data Association

Iris, 390, 393, 394

Irregularities, 18, 19, 20, 21, 40, 47, 103, 121, 124, 180, 235, 387, 393, 400, 533

IS audit function, 8, 13, 17

IS budgets, 75, 111, 171

ISACA Code of Professional Ethics, 8, 16, 17, 584

ISACA IS Auditing Guidelines, 8, 20

ISACA IS Auditing Standards, 8, 11, 12, 13, 14, 16, 17, 20, 55

ISACA IS Auditing Standards and Guidelines, 8, 16, 55

ISAM, See Indexed sequential access method

ISDN, See Integrated Services Digital Network

ISM, 331

ISO 9001, 112, 113, 218, 560

ISO/OSI, See International Organization for Standardization/Open Systems Interconnection

Isochronous, 333

ISP, See Internet service provider

Issuer, 255

IT governance, 4, 19, 20, 31, 32, 59, 74, 75, 77, 78, 81, 82, 83, 84, 85, 89, 91, 92, 103, 115, 124, 370, 406, 534, 549

IT performance, 77, 78, 80, 114, 575, 577, 578, 579, 582, 583

ITF, See Integrated test facilities

ITIL, See Information Technology Infrastructure Library

ITT, See Invitation to tender

ITU, See International Telecommunications Union

## J

J2EE, 91

Java, 179, 198, 199, 242, 338, 412, 430, 520, 542, 549

JAVAScript, 179, 334, 338, 520

JIT, See Just-in-time

Job description, 116, 442, 444, 461

Job rotation, 103

Joins, 263

Journal, 72, 124, 137, 277, 360, 473, 517

Judging the materiality of findings, 9, 52

Just-in-time (JIT), 256

## K

KDSI, See Thousand delivered source instructions

Kerberos, 396

Kernel, 307

Key decision-making personnel, 476, 498

Key goal indicator (KGI), 103, 550

Key performance indicator (KPI), 78,103, 262, 534, 550

Key personnel, 125, 477, 510

Key verification, 227

Keyboard, 299, 300, 334, 462

KGI, See Key goal indicator

Kick-off meeting, 157

Kilo lines of code (KLOC), 162, 550

KLOC, See Kilo lines of code

KPI, See Key performance indicator

## L

L2TP, See Layer 2 Tunneling Protocol

Laptop, 300, 331, 355, 386, 400, 411, 460, 463

Laserdisc, 303

Last-mile circuit protection, 500

Latency, 324, 339

Layer 2 Tunneling Protocol (L2TP), 544, 550

LCP, See Link Control Protocol

Leased lines, 329, 331, 335

Legacy systems, 194, 242, 244, 355

Librarian, 298, 350, 505, 534

Library control software, 280, 296

Licensed software, 314

Lights-out operations, 280, 282, 290, 291, 351

Limit check, 227, 534

Link Control Protocol (LCP), 328, 550

Linux, 242, 307, 342, 450

LISP, 179, 260

Load balancing, 283, 284, 285, 286, 325, 339, 343, 417

Local area network, 281, 300, 320, 373, 534, 547, 551

Lock, 436, 438, 440, 441, 461, 462, 469

Log, 123, 124, 188, 226, 227, 229, 230, 240, 241, 247, 249, 250, 251, 287, 289, 293, 308, 338, 348, 350, 352, 353, 389, 390, 392, 395, 396, 400, 401, 402, 416, 418, 430, 445, 447, 461, 465, 466, 467, 469, 525, 534, 544

Logging, 62, 124, 208, 226, 227, 230, 232, 250, 280, 306, 307, 308, 339, 351, 362, 364, 382, 389, 390, 397, 401, 406, 445, 451, 452, 461, 463, 506, 526

Logic bomb, 386

Logical access controls, 114, 363, 367, 368, 383, 385, 388, 397, 403, 443, 446, 447, 451

Logical security, 33, 34, 40, 112, 345, 348, 372, 374, 439, 444

Logoff, 289, 348, 445

Logon, 104, 214, 240, 289, 335, 339, 348, 362, 364, 386, 389, 390, 391, 392, 393, 396, 401, 402, 403, 407, 408, 427, 438, 439, 440, 443, 444, 445, 446, 451, 467, 469, 471, 519, 521, 534

Logon ID, 386, 390, 391, 392, 401, 438, 443, 445

Long-haul network diversity, 493, 500

Loopholes, 446

Lophtcrack, 413

## M

MAC, See Mandatory access controls

Machine language, 179, 521, 530, 542

Macro, 445

Magnetic card readers, 252, 538

Mainframe, 118, 162, 178, 216, 217, 296, 299, 303, 306, 307, 316, 325, 341, 343, 347, 388, 392, 396, 398, 408, 443, 444, 485, 488

Malware, 385, 419, 534

Management information system (MIS), 169, 535, 550

Management principles, 114, 221

Mandatory access controls (MAC), 319, 323, 324, 325, 375, 406, 528, 535, 550

Manual controls, 218, 230

Manufacturing resources planning (MRP), 256, 550

## Q

## R

## S

## T

## U

## V

VAN, See Value-added network
Variable sampling, 47, 48, 545
VB, See Visual Basic
Video camera, 394
Virtual private network (VPN), 428, 530
Virus, 363, 367, 369, 371, 384, 385, 386, 399, 400, 406, 411, 418, 428, 429, 430, 431, 432, 442, 451, 463, 470, 494, 545
Visual Basic (VB), 179, 338, 429, 520
Voice mail, 435, 438, 545
Voice recovery, 493, 500
Voice response ordering systems, 144, 258
Voice-over IP (VoIP), 364, 368, 369, 431, 432, 500, 542, 545, 547, , 551
VoIP, See Voice-over IP
VPN, See Virtual private network
Vulnerability, 13, 21, 50, 87, 298, 355, 367, 369, 371, 386, 433, 434, 436, 437, 438, 442, 449, 472, 473

## W

WAP, See Wireless Application Protocol
War chalking, 363, 410
Warm sites, 479, 490, 491
Web servers, 301, 338, 342, 426, 520
Web Services Description Language (WSDL), 200
Web site, ii, 91, 241, 242, 248, 336, 338, 388, 419, 449, 498, 529
Web-based, 291, 299, 316, 338, 412, 415, 520, 530, 542
Web-based EDI, 143, 247, 248
WebCams, 301, 305
WEP, See Wired Equivalent Privacy

White box testing, 182
WHOIS, 449
Wi-Fi Protected Access (WPA), 546
Wired Equivalent Privacy (WEP), 332, 333, 335
Wireless, 48, 120, 242, 254, 281, 300, 316, 320, 321, 330, 331, 332, 333, 334, 335, 347, 355, 363, 368, 369, 386, 388, 400, 409, 410, 422, 438, 441, 473, 525, 535, 542, 545, 546, 551
Wireless Application Protocol (WAP), 334
Wireless local area networks (WLAN), 281, 330, 331, 332, 333, 335, 551
Wireless Markup Language (WML), 334
Wireless personal area networks (WPAN), 281, 330, 333, 334, 410
Wireless wide area network (WWAN), 331
WLAN, See Wireless local area networks
WML, See Wireless Markup Language
Workpapers, 36, 59, 192
Workstation, 178, 225, 300, 341, 348, 391, 395, 400, 408, 429, 430, 438, 441, 443, 462, 528
Worms, 386, 409, 419, 428, 429, 534
WPA, See Wi-Fi Protected Access
WPAN, See Wireless personal area networks
WSDL, See Web Services Description Language
WWAN, See Wireless wide area network

## X

X.25, 281, 328, 531, 538
XML, 316, 334, 530, 542
XML, See Extensible Markup Language

# EVALUATION

ISACA continuously monitors the swift and profound professional, technological and environmental advances affecting the IS audit, assurance, control and security professions. Recognizing these rapid advances, the *CISA Review Manual* is updated annually.

To assist ISACA with keeping abreast of these advances, the ISACA Board of Directors would appreciate you taking a moment to comment on the *CISA Review Manual 2009*. Such feedback is invaluable to our efforts to fully serve the profession and future CISA examination candidates.

Please complete the questionnaire below and return to:

ISACA
3701 Algonquin Road, Suite 1010
Rolling Meadows, IL 60008
USA
Attention: Manager—Certification Study Program and Educational Development

1.  The *CISA Review Manual 2009* was (check one);
    ___ very helpful          ___ helpful              ___ not very helpful
    in preparing              in preparing             in preparing
    me for the exam.          me for the exam.         me for the exam.

2.  The format of the *CISA Review Manual 2009* made it (check one):
    ___ very easy to read.        ___ readable.            ___ hard to read.

3.  The content of the manual was (check one);
    ___ too detailed          ___ detailed enough      ___ not detailed enough
    in preparing              in preparing             in preparing
    me for the exam.          me for the exam.         me for the exam.

    If not detailed, or not detailed enough, please indicate where additional detail should be provided.
    _____
    _____

4.  The case studies at the end of each chapter were (check one):
    ___ very helpful          ___ helpful              ___ not very helpful
    in preparing              in preparing             in preparing
    me for the exam.          me for the exam.         me for the exam.

    If not helpful, please indicate where additional detail should be provided.
    _____
    _____

5.  The practice questions at the end of each chapter were (check one):
    ___ very helpful          ___ helpful              ___ not very helpful
    in preparing              in preparing             in preparing
    me for the exam.          me for the exam.         me for the exam.

6.  What other improvements would you recommend be made to the *CISA Review Manual* to make it more useful (be as specific as possible):
    _____
    _____

If you would like to complete this evaluation online please go to *www.isaca.org/studyaidsevaluation*.

Please also note on the back of this page (or a separate page) any specific comments and/or suggestions you may have concerning errors and omissions, enhancements, references and format. If you wish, please include your name, address and phone number so we may follow-up with you. Thank you for your support and assistance.

# OTHER COMMENTS/SUGGESTIONS

# NOTES

# *Prepare for the* **2009** *CISA Exams*

**ORDER NOW**—2009 CISA® Review Materials for Exam Preparation and Professional Development

To pass the Certified Information Systems Auditor™ (CISA) exam, a candidate should have an organized plan of study. To assist individuals with the development of a successful study plan, ISACA® offers several study aids and review courses (*www.isaca.org/cisareview*) to exam candidates.

## CISA Review Manual 2009
*ISACA*

The *CISA® Review Manual 2009* has been completely revised and updated with new content to reflect changing industry principles and practices, and is organized according to the current CISA job practice areas. The manual features detailed descriptions of the tasks performed by information systems (IS) auditors and the knowledge required to plan, manage and perform IS audits. The new edition also features new content based on the *IT Assurance Framework™ (ITAF™)*, recently published by ISACA. ITAF is a comprehensive assurance model that incorporates standards and good practices, providing guidance on the design, conduct and reporting of IT audit and assurance assignments; defines terms and concepts specific to IT assurance; and establishes standards that address IT audit and assurance professional roles and responsibilities, knowledge and skills, diligence, conduct, and reporting requirements. The *CISA Review Manual 2009* also includes brief chapter summaries focused on the main topics and new case studies to assist a candidate in understanding current practices. Also included are definitions of terms most commonly found on the exam, practice questions similar in content to what has previously appeared on the exam and references to additional study materials. This manual can be used as a stand-alone document for individual study or as a guide or reference for study groups and chapters conducting local review courses.

The 2009 edition has been developed and is organized to help prepare the CISA candidate in understanding the essential concepts and studying the following job practice areas:
• IS audit process
• IT governance
• Systems and infrastructure life cycle management
• IT service delivery and support
• Protection of information assets
• Business continuity and disaster recovery

**CRM-9**  English Edition
**CRM-9F** French Edition
**CRM-9I** Italian Edition
**CRM-9J** Japanese Edition
**CRM-9S** Spanish Edition

## CISA Review Questions, Answers & Explanations Manual 2008
*ISACA*

The *CISA® Review Questions, Answers & Explanations Manual 2008* consists of 600 multiple-choice study questions that have previously appeared in the *CISA® Review Questions, Answers & Explanations Manual 2006* and the *2007 Supplement*. Many questions have been revised or completely rewritten to recognize a change in job practice, be more representative of the current CISA exam question format, and/or to provide further clarity or explanation of the correct answer. These questions are not actual exam items, but are intended to provide the CISA candidate with an understanding of the type and structure of questions and content that have previously appeared on the exam. This publication is ideal to use in conjunction with the CISA Review Manual 2009.

To assist the user in maximizing study efforts, questions are presented in the following two ways:

• Sorted by job practice area
• Scrambled as a sample 200-question exam

**QAE-8**  English Edition
**QAE-8I** Italian Edition
**QAE-8J** Japanese Edition
**QAE-8S** Spanish Edition

## CISA Review Questions, Answers & Explanations Manual 2008 and 2009 Supplements
*ISACA*

Developed each year, the *CISA® Review Questions, Answers & Explanations Manual 2009 Supplement* and *2008 Supplement* are recommended for use when preparing for the 2009 CISA exam. Each edition consists of 100 sample questions, answers and explanations based on the current CISA job practice areas, using a process for item development similar to the process for developing actual exam items. The questions are intended to provide the CISA candidate with an understanding of the type and structure of questions that have typically appeared on past exams, and were prepared specifically for use in studying for the CISA exam.

| **2009 Editions** | | **2008 Editions** | |
|---|---|---|---|
| **QAE-9ES** | English Edition | **QAE-8ES** | English Edition |
| **QAE-9FS** | French Edition | **QAE-8FS** | French Edition |
| **QAE-9IS** | Italian Edition | **QAE-8IS** | Italian Edition |
| **QAE-9JS** | Japanese Edition | **QAE-8JS** | Japanese Edition |
| **QAE-9SS** | Spanish Edition | **QAE-8SS** | Spanish Edition |

## CISA Practice Question Database v9
*ISACA*

The CISA® Practice Question Database v9 combines the *CISA Review Questions, Answers & Explanations Manual 2008* with the *CISA Review Questions, Answers & Explanations Manual 2008 Supplement* and *2009 Supplement* into one comprehensive 800-question study guide. Sample exams with randomly selected questions can be taken and the results viewed by job practice, allowing for concentrated study one area at a time. Additionally, questions generated during a study session are sorted based upon the user's previous scoring history, allowing CISA candidates to easily and quickly identify their strengths and weaknesses, and focus their study efforts accordingly. Other features allow the user to select sample exams by specific job practice areas, view questions that were previously answered incorrectly and vary the length of their study sessions. Also included are *Information Systems Control Journal®* articles referenced in the *CISA Review Manual 2009*. Available in CD-ROM format or as a web site download.

PLEASE NOTE the following system requirements:
• Intel Pentium 3 or higher (Pentium 4 recommended)
• Windows 98SE or higher
• 256 MB RAM (512 MB recommended)
• Hard drive with 80 MB of available space
• CD-ROM drive
• Display with recommended resolution of 1024 x 768

| | |
|---|---|
| **CDB-9** | English Edition—CD-ROM |
| **CDB-9W** | English Edition—Web site download |
| **CDB-9S** | Spanish Edition—CD-ROM |
| **CDB-9SW** | Spanish Edition—Web site download |

## CISA Online Review Course
*ISACA*

A complete web-based exam review course is available at *www.isaca.org/elearningcampus*.

**To order CISA review materials for the June/December 2009 exams, visit our web site at**
**www.isaca.org/cisabooks.**